Engineering Mechanics: Dynamics

Authors
J. L. Meriam • L. G. Kraige

ISBN 9781119375548

List of Titles

Engineering Mechanics Volume 2: Dynamics, 8th edition
by J. L. Meriam and L. G. Kraige
 ISBN: 978-1-118-88584-0

Table of Contents

ENGINEERING MECHANICS

VOLUME 2

DYNAMICS

EIGHTH EDITION

ENGINEERING MECHANICS

VOLUME 2

DYNAMICS

EIGHTH EDITION

J.L. MERIAM

L.G. KRAIGE

Virginia Polytechnic Institute and State University

J.N. BOLTON

Bluefield State College

WILEY

On the cover: Liftoff of a Falcon 9 rocket from Vandenberg Air Force Base, California. This SpaceX rocket is a two-stage launch vehicle which generates 1.3 million pounds of thrust at sea level.

Vice President & Executive Publisher	Don Fowley
Executive Marketing Manager	Dan Sayre
Executive Editor	Linda Ratts
Editorial Assistants	Emily Meussner/Francesca Baratta
Content Manager	Karoline Luciano
Production Editor	Ken Santor, Production Management Services provided by Camelot Editorial Services, LLC
Marketing Manager	Christopher Ruel
Senior Designer	Maureen Eide
Cover Design	Wendy Lai
Cover Photo	SPACEX
Electronic Illustrations	Precision Graphics
Senior Photo Editor	Billy Ray
Product Designer	Jennifer Welter
Content Editor	Wendy Ashenberg

This book was set in 9.5/12 New Century Schoolbook Lt Std. by Aptara, Inc., and printed and bound by Quad Graphics Versailles. The cover was printed by Quad Graphics.

This book is printed on acid-free paper. ∞

Founded in 1807, John Wiley & Sons, Inc. has been a valued source of knowledge and understanding for more than 200 years, helping people around the world meet their needs and fulfill their aspirations. Our company is built on a foundation of principles that include responsibility to the communities we serve and where we live and work. In 2008, we launched a Corporate Citizenship Initiative, a global effort to address the environmental, social, economic, and ethical challenges we face in our business. Among the issues we are addressing are carbon impact, paper specifications and procurement, ethical conduct within our business and among our vendors, and community and charitable support. For more information, please visit our website: www.wiley.com/go/citizenship.

Library of Congress Cataloging-in-Publication Data

Meriam, J. L. (James L.)
Dynamics / J. L. Meriam, L. G. Kraige, Virginia Polytechnic Institute and State University, J. N. Bolton, Bluefield State College.—Eighth edition.
pages ; cm—(Engineering mechanics)
Includes bibliographical references and index.
ISBN 978-1-118-88584-0 (cloth)
1. Machinery, Dynamics of. I. Kraige, L. G. (L. Glenn) II. Bolton, J. N. (Jeff N.) III. Title.
TA352.M45 2015
620.1—dc23
2015016668

ISBN: 978-1-118-88584-0
ISBN: 978-1-119-02253-4 (BRV)

Printed in the United States of America

10 9 8 7 6 5 4 3 2 1

Conversion Factors

U.S. Customary Units to SI Units

To convert from	To	Multiply by
(Acceleration)		
foot/second^2 (ft/sec^2)	meter/second^2 (m/s^2)	3.048×10^{-1}*
inch/second^2 (in./sec^2)	meter/second^2 (m/s^2)	2.54×10^{-2}*
(Area)		
foot^2 (ft^2)	meter^2 (m^2)	9.2903×10^{-2}
inch^2 (in.^{2})	meter^2 (m^2)	6.4516×10^{-4}*
(Density)		
pound mass/inch^3 (lbm/in.^{3})	kilogram/meter^3 (kg/m^3)	2.7680×10^4
pound mass/foot^3 (lbm/ft^3)	kilogram/meter^3 (kg/m^3)	1.6018×10
(Force)		
kip (1000 lb)	newton (N)	4.4482×10^3
pound force (lb)	newton (N)	4.4482
(Length)		
foot (ft)	meter (m)	3.048×10^{-1}*
inch (in.)	meter (m)	2.54×10^{-2}*
mile (mi), (U.S. statute)	meter (m)	1.6093×10^3
mile (mi), (international nautical)	meter (m)	1.852×10^3*
(Mass)		
pound mass (lbm)	kilogram (kg)	4.5359×10^{-1}
slug (lb-sec^2/ft)	kilogram (kg)	1.4594×10
ton (2000 lbm)	kilogram (kg)	9.0718×10^2
(Moment of force)		
pound-foot (lb-ft)	newton-meter (N·m)	1.3558
pound-inch (lb-in.)	newton-meter (N·m)	0.1129 8
(Moment of inertia, area)		
inch^4	meter^4 (m^4)	41.623×10^{-8}
(Moment of inertia, mass)		
pound-foot-second^2 (lb-ft-sec^2)	kilogram-meter^2 (kg·m^2)	1.3558
(Momentum, linear)		
pound-second (lb-sec)	kilogram-meter/second (kg·m/s)	4.4482
(Momentum, angular)		
pound-foot-second (lb-ft-sec)	newton-meter-second (kg·m^2/s)	1.3558
(Power)		
foot-pound/minute (ft-lb/min)	watt (W)	2.2597×10^{-2}
horsepower (550 ft-lb/sec)	watt (W)	7.4570×10^2
(Pressure, stress)		
atmosphere (std)(14.7 lb/in.^{2})	newton/meter^2 (N/m^2 or Pa)	1.0133×10^5
pound/foot^2 (lb/ft^2)	newton/meter^2 (N/m^2 or Pa)	4.7880×10
pound/inch^2 (lb/in.^{2} or psi)	newton/meter^2 (N/m^2 or Pa)	6.8948×10^3
(Spring constant)		
pound/inch (lb/in.)	newton/meter (N/m)	1.7513×10^2
(Velocity)		
foot/second (ft/sec)	meter/second (m/s)	3.048×10^{-1}*
knot (nautical mi/hr)	meter/second (m/s)	5.1444×10^{-1}
mile/hour (mi/hr)	meter/second (m/s)	4.4704×10^{-1}*
mile/hour (mi/hr)	kilometer/hour (km/h)	1.6093
(Volume)		
foot^3 (ft^3)	meter^3 (m^3)	2.8317×10^{-2}
inch^3 (in.^{3})	meter^3 (m^3)	1.6387×10^{-5}
(Work, Energy)		
British thermal unit (BTU)	joule (J)	1.0551×10^3
foot-pound force (ft-lb)	joule (J)	1.3558
kilowatt-hour (kw-h)	joule (J)	3.60×10^6*

*Exact value

SI Units Used in Mechanics

Quantity	Unit	SI Symbol
(Base Units)		
Length	meter*	m
Mass	kilogram	kg
Time	second	s
(Derived Units)		
Acceleration, linear	meter/second^2	m/s^2
Acceleration, angular	radian/second^2	rad/s^2
Area	meter^2	m^2
Density	kilogram/meter^3	kg/m^3
Force	newton	N (= kg·m/s^2)
Frequency	hertz	Hz (= 1/s)
Impulse, linear	newton-second	N·s
Impulse, angular	newton-meter-second	N·m·s
Moment of force	newton-meter	N·m
Moment of inertia, area	meter^4	m^4
Moment of inertia, mass	kilogram-meter^2	kg·m^2
Momentum, linear	kilogram-meter/second	kg·m/s (= N·s)
Momentum, angular	kilogram-meter^2/second	kg·m^2/s (= N·m·s)
Power	watt	W (= J/s = N·m/s)
Pressure, stress	pascal	Pa (= N/m^2)
Product of inertia, area	meter^4	m^4
Product of inertia, mass	kilogram-meter^2	kg·m^2
Spring constant	newton/meter	N/m
Velocity, linear	meter/second	m/s
Velocity, angular	radian/second	rad/s
Volume	meter^3	m^3
Work, energy	joule	J (= N·m)
(Supplementary and Other Acceptable Units)		
Distance (navigation)	nautical mile	(= 1.852 km)
Mass	ton (metric)	t (= 1000 kg)
Plane angle	degrees (decimal)	°
Plane angle	radian	—
Speed	knot	(1.852 km/h)
Time	day	d
Time	hour	h
Time	minute	min

*Also spelled *metre.*

SI Unit Prefixes

Multiplication Factor	Prefix	Symbol
1 000 000 000 000 = 10^{12}	tera	T
1 000 000 000 = 10^9	giga	G
1 000 000 = 10^6	mega	M
1 000 = 10^3	kilo	k
100 = 10^2	hecto	h
10 = 10	deka	da
0.1 = 10^{-1}	deci	d
0.01 = 10^{-2}	centi	c
0.001 = 10^{-3}	milli	m
0.000 001 = 10^{-6}	micro	μ
0.000 000 001 = 10^{-9}	nano	n
0.000 000 000 001 = 10^{-12}	pico	p

Selected Rules for Writing Metric Quantities

1. (a) Use prefixes to keep numerical values generally between 0.1 and 1000.
   (b) Use of the prefixes hecto, deka, deci, and centi should generally be avoided except for certain areas or volumes where the numbers would be awkward otherwise.
   (c) Use prefixes only in the numerator of unit combinations. The one exception is the base unit kilogram. (*Example:* write kN/m not N/mm; J/kg not mJ/g)
   (d) Avoid double prefixes. (*Example:* write GN not kMN)
2. Unit designations
   (a) Use a dot for multiplication of units. (*Example:* write N·m not Nm)
   (b) Avoid ambiguous double solidus. (*Example:* write N/m^2 not N/m/m)
   (c) Exponents refer to entire unit. (*Example:* mm^2 means (mm)^2)
3. Number grouping
   Use a space rather than a comma to separate numbers in groups of three, counting from the decimal point in both directions. (*Example:* 4 607 321.048 72) Space may be omitted for numbers of four digits. (*Example:* 4296 or 0.0476)

PART I

DYNAMICS OF PARTICLES

PROBLEMS

Introductory Problems

Problems 2/1 through 2/8 treat the motion of a particle which moves along the s-axis shown in the figure.

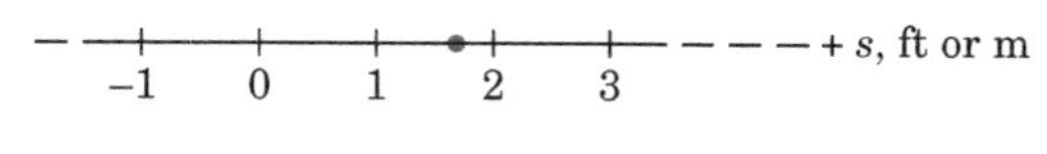

Problems 2/1–2/8

2/1 The velocity of a particle is given by $v = 25t^2 - 80t - 200$, where v is in feet per second and t is in seconds. Plot the velocity v and acceleration a versus time for the first 6 seconds of motion and evaluate the velocity when a is zero.

2/2 The position of a particle is given by $s = 0.27t^3 - 0.65t^2 - 2.35t + 4.4$, where s is in feet and the time t is in seconds. Plot the displacement, velocity, and acceleration as functions of time for the first 5 seconds of motion. Determine the positive time when the particle changes its direction.

2/3 The velocity of a particle which moves along the s-axis is given by $v = 2 - 4t + 5t^{3/2}$, where t is in seconds and v is in meters per second. Evaluate the position s, velocity v, and acceleration a when $t = 3$ s. The particle is at the position $s_0 = 3$ m when $t = 0$.

2/4 The displacement of a particle which moves along the s-axis is given by $s = (-2 + 3t)e^{-0.5t}$, where s is in meters and t is in seconds. Plot the displacement, velocity, and acceleration versus time for the first 20 seconds of motion. Determine the time at which the acceleration is zero.

2/5 The acceleration of a particle is given by $a = 2t - 10$, where a is in meters per second squared and t is in seconds. Determine the velocity and displacement as functions of time. The initial displacement at $t = 0$ is $s_0 = -4$ m, and the initial velocity is $v_0 = 3$ m/s.

2/6 The acceleration of a particle is given by $a = -kt^2$, where a is in meters per second squared and the time t is in seconds. If the initial velocity of the particle at $t = 0$ is $v_0 = 12$ m/s and the particle takes 6 seconds to reverse direction, determine the magnitude and units of the constant k. What is the net displacement of the particle over the same 6-second interval of motion?

2/7 The acceleration of a particle is given by $a = -ks^2$, where a is in meters per second squared, k is a constant, and s is in meters. Determine the velocity of the particle as a function of its position s. Evaluate your expression for $s = 5$ m if $k = 0.1\ \text{m}^{-1}\text{s}^{-2}$ and the initial conditions at time $t = 0$ are $s_0 = 3$ m and $v_0 = 10$ m/s.

2/8 The acceleration of a particle is given by $a = c_1 + c_2 v$, where a is in millimeters per second squared, the velocity v is in millimeters per second, and c_1 and c_2 are constants. If the particle position and velocity at $t = 0$ are s_0 and v_0, respectively, determine expressions for the position s of the particle in terms of the velocity v and time t.

2/9 Calculate the constant acceleration a in g's which the catapult of an aircraft carrier must provide to produce a launch velocity of 180 mi/hr in a distance of 300 ft. Assume that the carrier is at anchor.

2/10 A particle in an experimental apparatus has a velocity given by $v = k\sqrt{s}$, where v is in millimeters per second, the position s is millimeters, and the constant $k = 0.2\ \text{mm}^{1/2}\text{s}^{-1}$. If the particle has a velocity $v_0 = 3$ mm/s at $t = 0$, determine the particle position, velocity, and acceleration as functions of time, and compute the time, position, and acceleration of the particle when the velocity reaches 15 mm/s.

2/11 Ball 1 is launched with an initial vertical velocity $v_1 = 160$ ft/sec. Three seconds later, ball 2 is launched with an initial vertical velocity v_2. Determine v_2 if the balls are to collide at an altitude of 300 ft. At the instant of collision, is ball 1 ascending or descending?

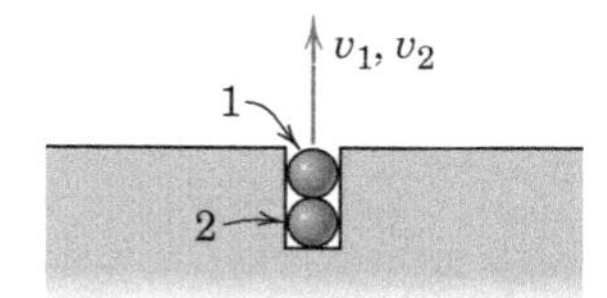

Problem 2/11

2/12 Experimental data for the motion of a particle along a straight line yield measured values of the velocity v for various position coordinates s. A smooth curve is drawn through the points as shown in the graph. Determine the acceleration of the particle when $s = 20$ ft.

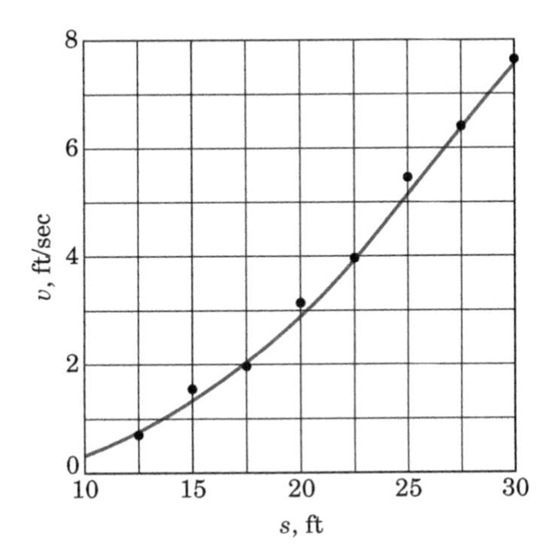

Problem 2/12

2/13 In the pinewood-derby event shown, the car is released from rest at the starting position A and then rolls down the incline and on to the finish line C. If the constant acceleration down the incline is 2.75 m/s^2 and the speed from B to C is essentially constant, determine the time duration t_{AC} for the race. The effects of the small transition area at B can be neglected.

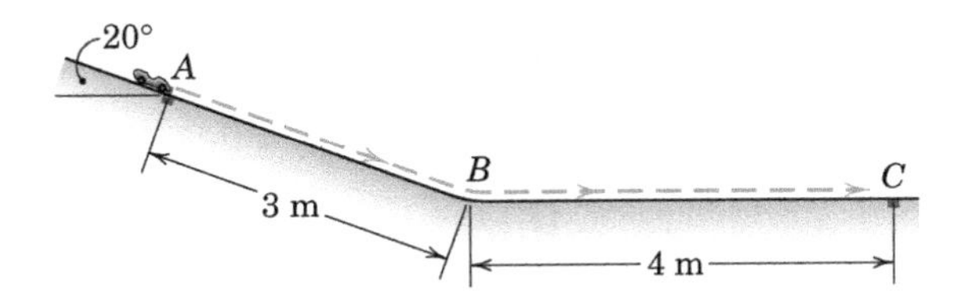

Problem 2/13

2/14 A ball is thrown vertically up with a velocity of 30 m/s at the edge of a 60-m cliff. Calculate the height h to which the ball rises and the total time t after release for the ball to reach the bottom of the cliff. Neglect air resistance and take the downward acceleration to be 9.81 m/s^2.

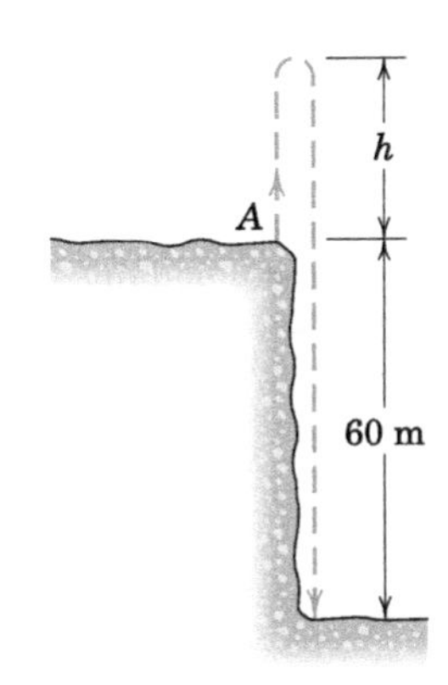

Problem 2/14

2/15 A car comes to a complete stop from an initial speed of 50 mi/hr in a distance of 100 ft. With the same constant acceleration, what would be the stopping distance s from an initial speed of 70 mi/hr?

2/16 The pilot of a jet transport brings the engines to full takeoff power before releasing the brakes as the aircraft is standing on the runway. The jet thrust remains constant, and the aircraft has a near-constant acceleration of $0.4g$. If the takeoff speed is 200 km/h, calculate the distance s and time t from rest to takeoff.

2/17 A game requires that two children each throw a ball upward as high as possible from point O and then run horizontally in opposite directions away from O. The child who travels the greater distance before their thrown ball impacts the ground wins. If child A throws a ball upward with a speed of $v_1 = 70$ ft/sec and immediately runs leftward at a constant speed of $v_A = 16$ ft/sec while child B throws the ball upward with a speed of $v_2 = 64$ ft/sec and immediately runs rightward with a constant speed of $v_B = 18$ ft/sec, which child will win the game?

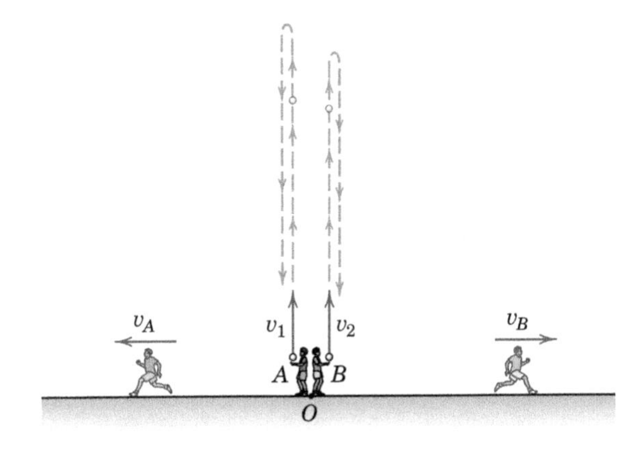

Problem 2/17

Representative Problems

2/18 During an 8-second interval, the velocity of a particle moving in a straight line varies with time as shown. Within reasonable limits of accuracy, determine the amount Δa by which the acceleration at $t = 4$ s exceeds the average acceleration during the interval. What is the displacement Δs during the interval?

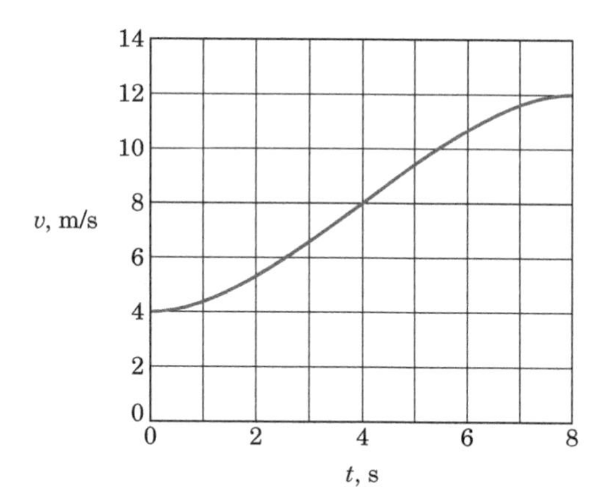

Problem 2/18

2/19 In the final stages of a moon landing, the lunar module descends under retrothrust of its descent engine to within $h = 5$ m of the lunar surface where it has a downward velocity of 2 m/s. If the descent engine is cut off abruptly at this point, compute the impact velocity of the landing gear with the moon. Lunar gravity is $\frac{1}{6}$ of the earth's gravity.

2/20 A girl rolls a ball up an incline and allows it to return to her. For the angle θ and ball involved, the acceleration of the ball along the incline is constant at $0.25g$, directed down the incline. If the ball is released with a speed of 4 m/s, determine the distance s it moves up the incline before reversing its direction and the total time t required for the ball to return to the child's hand.

Problem 2/20

2/21 At a football tryout, a player runs a 40-yard dash in 4.25 seconds. If he reaches his maximum speed at the 16-yard mark with a constant acceleration and then maintains that speed for the remainder of the run, determine his acceleration over the first 16 yards, his maximum speed, and the time duration of the acceleration.

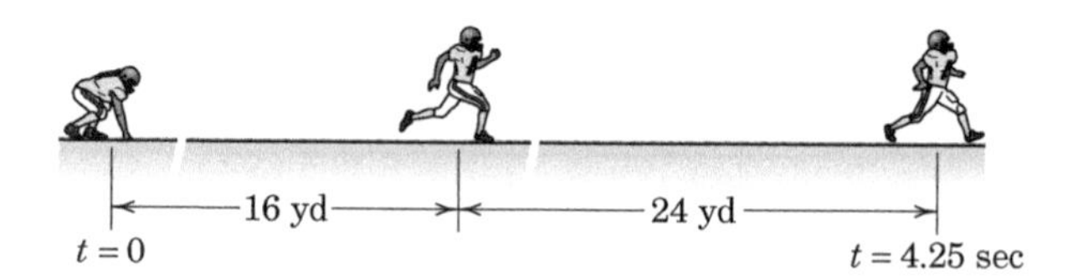

Problem 2/21

2/22 The main elevator A of the CN Tower in Toronto rises about 350 m and for most of its run has a constant speed of 22 km/h. Assume that both the acceleration and deceleration have a constant magnitude of $\frac{1}{4}g$ and determine the time duration t of the elevator run.

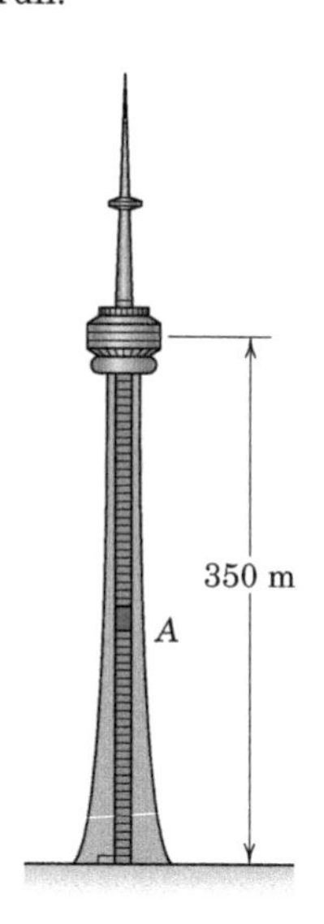

Problem 2/22

2/23 A Scotch-yoke mechanism is used to convert rotary motion into reciprocating motion. As the disk rotates at the constant angular rate ω, a pin A slides in a vertical slot causing the slotted member to displace horizontally according to $x = r\sin(\omega t)$ relative to the fixed disk center O. Determine the expressions for the velocity and acceleration of a point P on the output shaft of the mechanism as functions of time, and determine the maximum velocity and acceleration of point P during one cycle. Use the values $r = 75$ mm and $\omega = \pi$ rad/s.

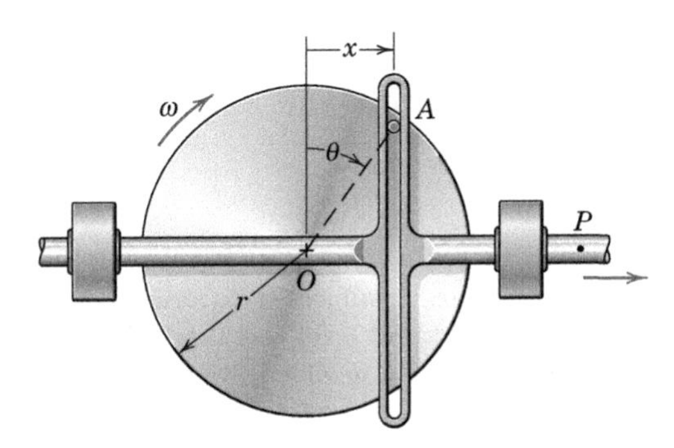

Problem 2/23

2/24 A train which is traveling at 80 mi/hr applies its brakes as it reaches point A and slows down with a constant deceleration. Its decreased velocity is observed to be 60 mi/hr as it passes a point 1/2 mi beyond A. A car moving at 50 mi/hr passes point B at the same instant that the train reaches point A. In an unwise effort to beat the train to the crossing, the driver "steps on the gas." Calculate the constant acceleration a that the car must have in order to beat the train to the crossing by 4 seconds and find the velocity v of the car as it reaches the crossing.

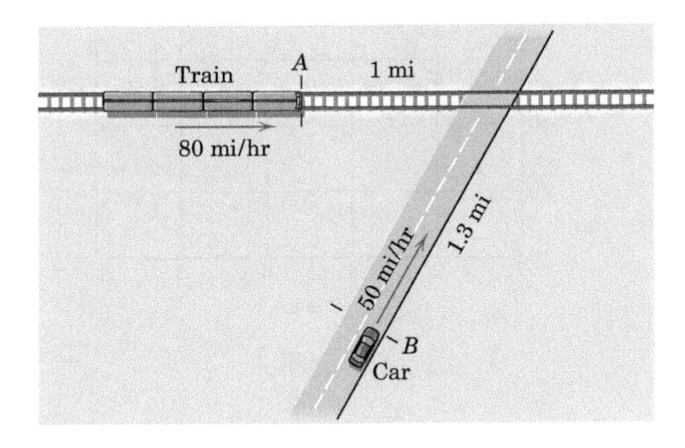

Problem 2/24

2/25 Small steel balls fall from rest through the opening at A at the steady rate of two per second. Find the vertical separation h of two consecutive balls when the lower one has dropped 3 meters. Neglect air resistance.

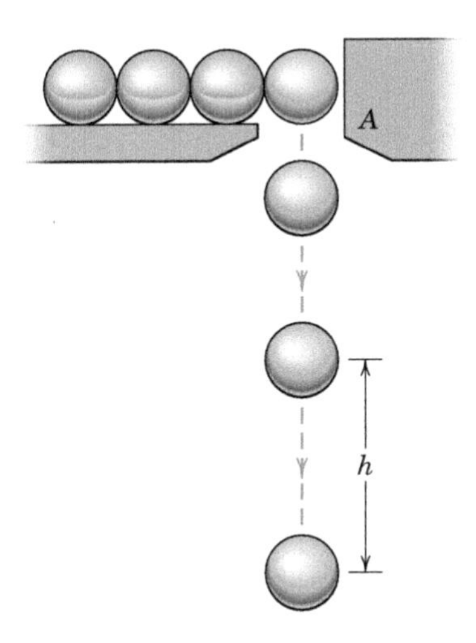

Problem 2/25

2/26 Car A is traveling at a constant speed $v_A = 130$ km/h at a location where the speed limit is 100 km/h. The police officer in car P observes this speed via radar. At the moment when A passes P, the police car begins to accelerate at the constant rate of 6 m/s^2 until a speed of 160 km/h is achieved, and that speed is then maintained. Determine the distance required for the police officer to overtake car A. Neglect any nonrectilinear motion of P.

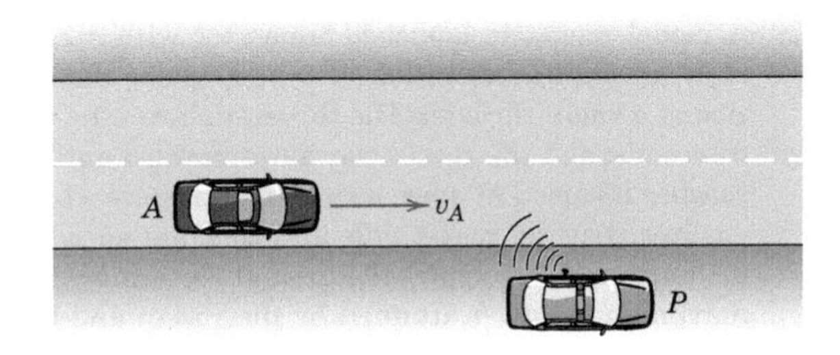

Problem 2/26

2/27 A toy helicopter is flying in a straight line at a constant speed of 4.5 m/s. If a projectile is launched vertically with an initial speed of $v_0 = 28$ m/s, what horizontal distance d should the helicopter be from the launch site S if the projectile is to be traveling downward when it strikes the helicopter? Assume that the projectile travels only in the vertical direction.

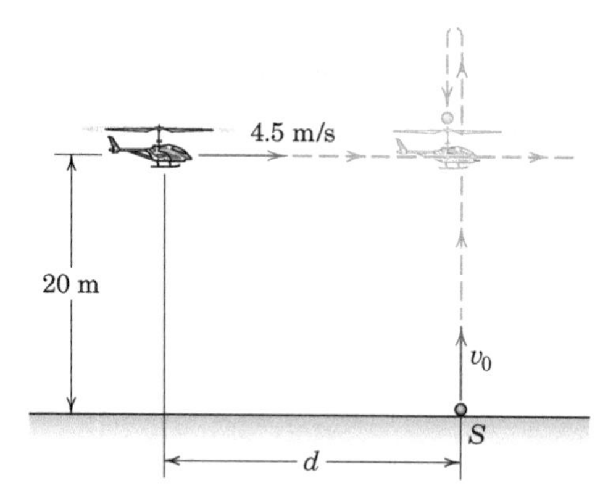

Problem 2/27

2/28 A particle moving along a straight line has an acceleration which varies according to position as shown. If the velocity of the particle at the position $x = -5$ ft is $v = -2$ ft/sec, determine the velocity when $x = 9$ ft.

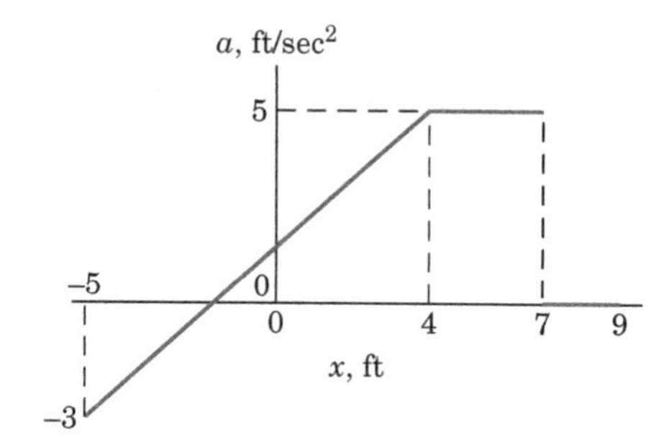

Problem 2/28

2/29 A model rocket is launched from rest with a constant upward acceleration of 3 m/s^2 under the action of a small thruster. The thruster shuts off after 8 seconds, and the rocket continues upward until it reaches its apex. At apex, a small chute opens which ensures that the rocket falls at a constant speed of 0.85 m/s until it impacts the ground. Determine the maximum height h attained by the rocket and the total flight time. Neglect aerodynamic drag during ascent, and assume that the mass of the rocket and the acceleration of gravity are both constant.

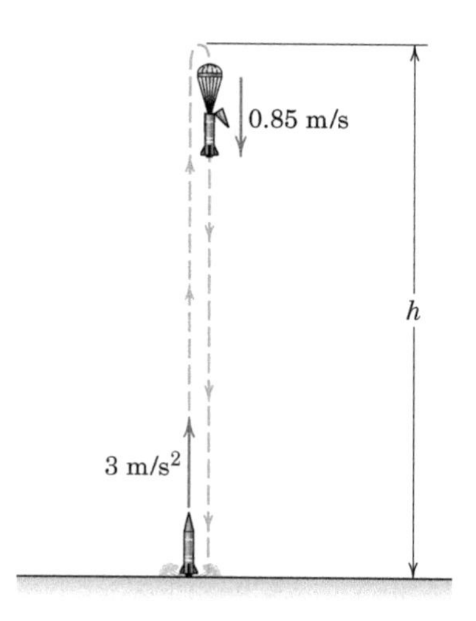

Problem 2/29

2/30 An electric car is subjected to acceleration tests along a straight and level test track. The resulting v-t data are closely modeled over the first 10 seconds by the function $v = 24t - t^2 + 5\sqrt{t}$, where t is the time in seconds and v is the velocity in feet per second. Determine the displacement s as a function of time over the interval $0 \leq t \leq 10$ sec and specify its value at time $t = 10$ sec.

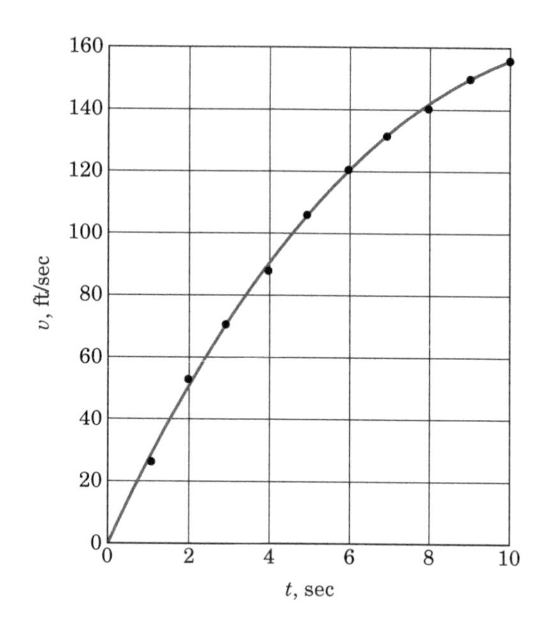

Problem 2/30

2/31 A vacuum-propelled capsule for a high-speed tube transportation system of the future is being designed for operation between two stations A and B, which are 10 km apart. If the acceleration and deceleration are to have a limiting magnitude of $0.6g$ and if velocities are to be limited to 400 km/h, determine the minimum time t for the capsule to make the 10-km trip.

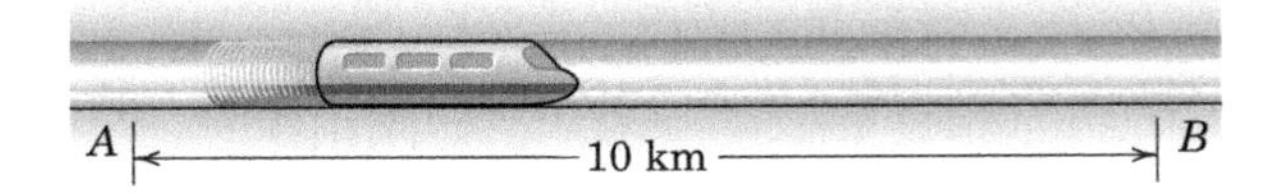

Problem 2/31

2/32 If the velocity v of a particle moving along a straight line decreases linearly with its displacement s from 20 m/s to a value approaching zero at $s = 30$ m, determine the acceleration a of the particle when $s = 15$ m and show that the particle never reaches the 30-m displacement.

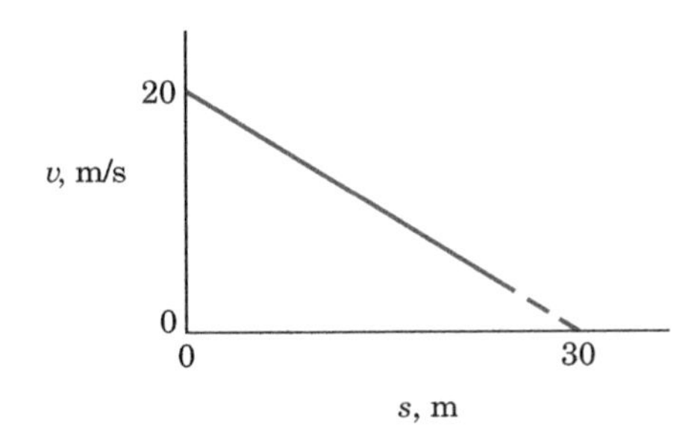

Problem 2/32

2/33 A particle moves along the x-axis with the velocity history shown. If the particle is at the position $x = -4$ in. at time $t = 0$, plot the corresponding displacement history for the time interval $0 \le t \le 10$ sec. Additionally, find the net displacement and total distance traveled by the particle for this interval.

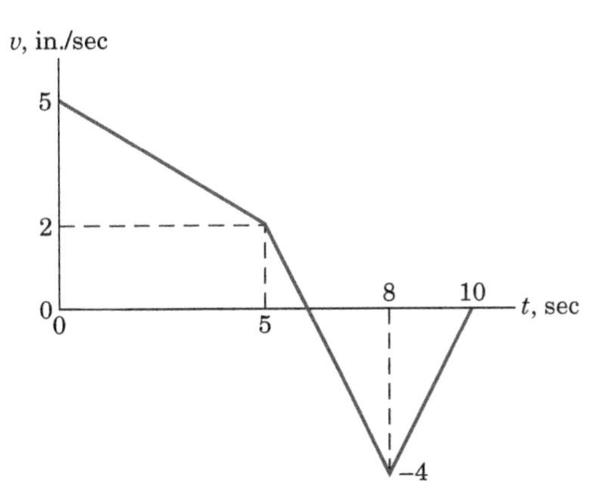

Problem 2/33

2/34 The 230,000-lb space-shuttle orbiter touches down at about 220 mi/hr. At 200 mi/hr its drag parachute deploys. At 35 mi/hr, the chute is jettisoned from the orbiter. If the deceleration in feet per second squared during the time that the chute is deployed is $-0.0003v^2$ (speed v in feet per second), determine the corresponding distance traveled by the orbiter. Assume no braking from its wheel brakes.

Problem 2/34

2/35 Reconsider the rollout of the space-shuttle orbiter of the previous problem. The drag chute is deployed at 200 mi/hr, the wheel brakes are applied at 100 mi/hr until wheelstop, and the drag chute is jettisoned at 35 mi/hr. If the drag chute results in a deceleration of $-0.0003v^2$ (in feet per second squared when the speed v is in feet per second) and the wheel brakes cause a constant deceleration of 5 ft/sec^2, determine the distance traveled from 200 mi/hr to wheelstop.

2/36 The cart impacts the safety barrier with speed $v_0 = 3.25$ m/s and is brought to a stop by the nest of nonlinear springs which provide a deceleration $a = -k_1x - k_2x^3$, where x is the amount of spring deflection from the undeformed position and k_1 and k_2 are positive constants. If the maximum spring deflection is 475 mm and the velocity at half-maximum deflection is 2.85 m/s, determine the values and corresponding units for the constants k_1 and k_2.

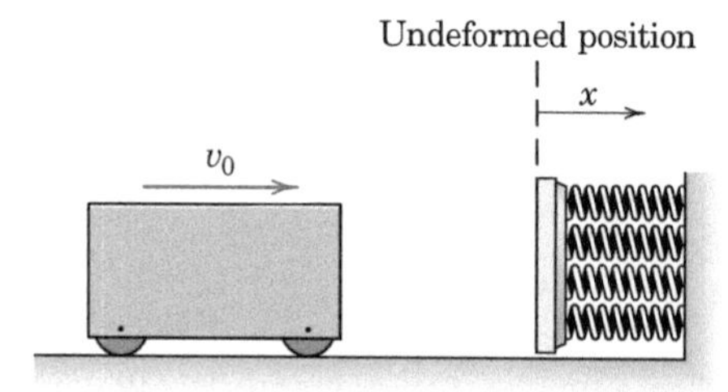

Problem 2/36

2/37 The graph shows the rectilinear acceleration of a particle as a function of time over a 12-second interval. If the particle is at rest at the position $s_0 = 0$ at time $t = 0$, determine the velocity of the particle when (*a*) $t = 4$ s, (*b*) $t = 8$ s, and (*c*) $t = 12$ s.

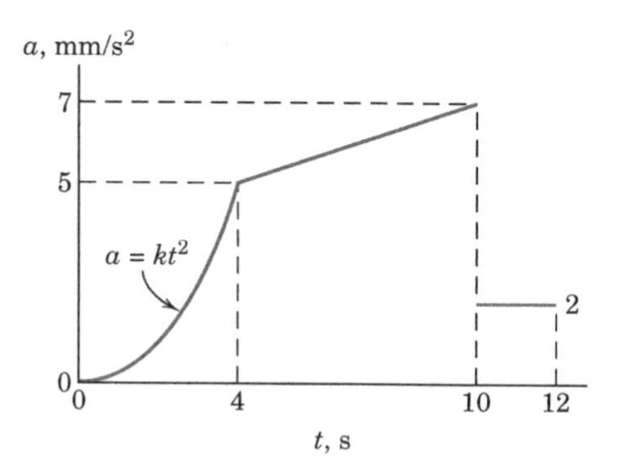

Problem 2/37

2/38 Compute the impact speed of a body released from rest at an altitude $h = 650$ miles above the surface of Mars. (*a*) First assume a constant gravitational acceleration $g_{m_0} = 12.3\ \text{ft/sec}^2$ (equal to that at the surface) and (*b*) then account for the variation of g with altitude (refer to Art. 1/5). Neglect any effects of atmospheric drag.

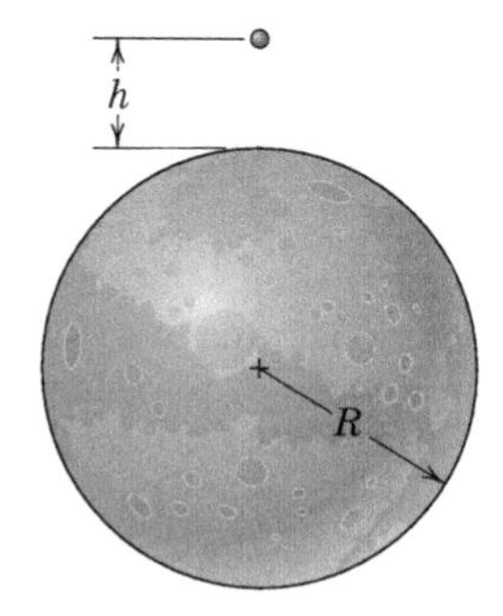

Problem 2/38

2/39 The steel ball A of diameter D slides freely on the horizontal rod which leads to the pole face of the electromagnet. The force of attraction obeys an inverse-square law, and the resulting acceleration of the ball is $a = K/(L - x)^2$, where K is a measure of the strength of the magnetic field. If the ball is released from rest at $x = 0$, determine the velocity v with which it strikes the pole face.

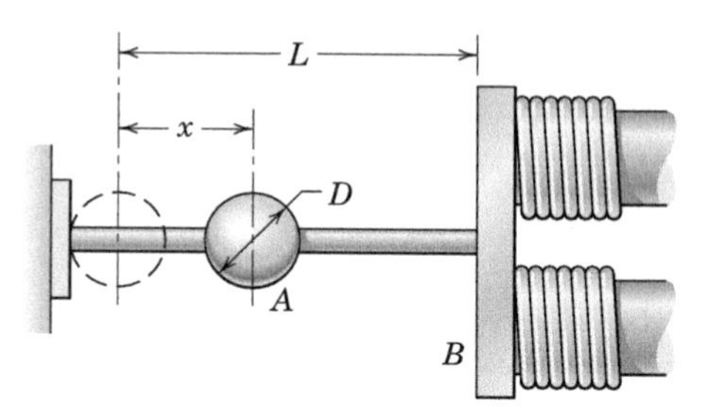

Problem 2/39

***2/40** The falling object has a speed v_0 when it strikes and subsequently deforms the foam arresting material until it comes to rest. The resistance of the foam material to deformation is a function of penetration depth y and object speed v so that the acceleration of the object is $a = g - k_1 v - k_2 y$, where v is the particle speed in inches per second, y is the penetration depth in inches, and k_1 and k_2 are positive constants. Plot the penetration depth y and velocity v of the object as functions of time over the first five seconds for $k_1 = 12\ \text{sec}^{-1}$, $k_2 = 24\ \text{sec}^{-2}$, and $v_0 = 25$ in./sec. Determine the time when the penetration depth reaches 95% of its final value.

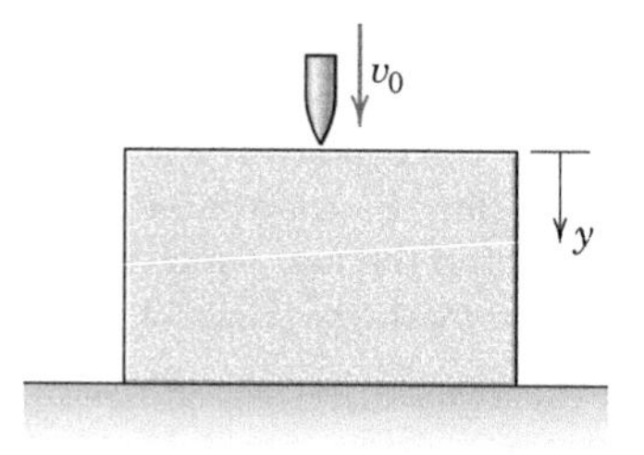

Problem 2/40

2/41 The electronic throttle control of a model train is programmed so that the train speed varies with position as shown in the plot. Determine the time t required for the train to complete one lap.

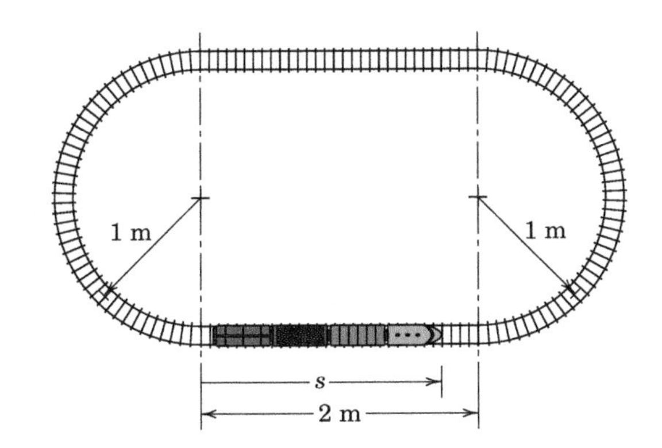

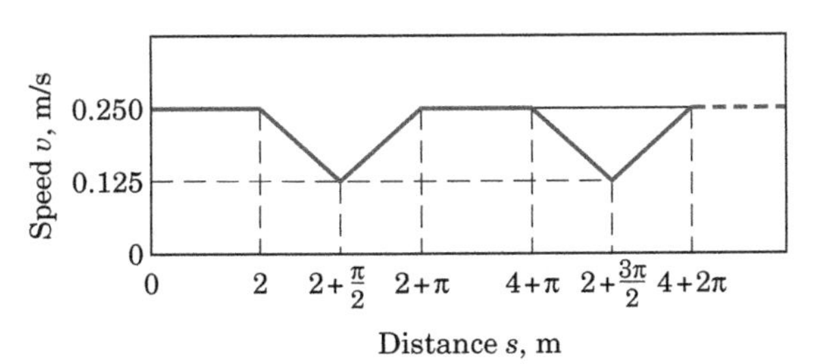

Problem 2/41

2/42 A projectile is fired downward with initial speed v_0 in an experimental fluid and experiences an acceleration $a = \sigma - \eta v^2$, where σ and η are positive constants and v is the projectile speed. Determine the distance traveled by the projectile when its speed has been reduced to one-half of the initial speed v_0. Also, determine the terminal velocity of the projectile. Evaluate for $\sigma = 0.7$ m/s^2, $\eta = 0.2$ m^{-1}, and $v_0 = 4$ m/s.

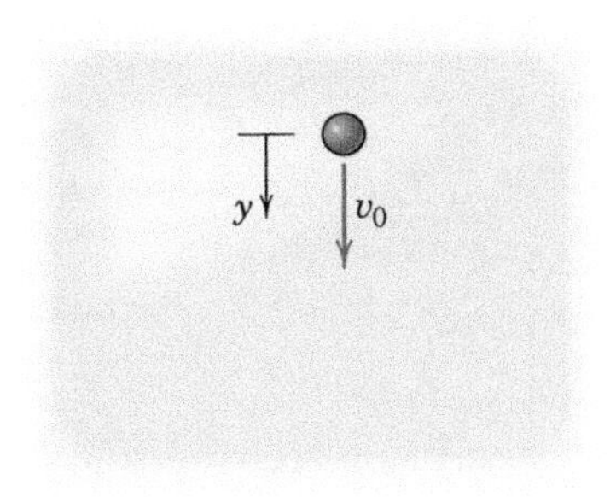

Problem 2/42

2/43 The aerodynamic resistance to motion of a car is nearly proportional to the square of its velocity. Additional frictional resistance is constant, so that the acceleration of the car when coasting may be written $a = -C_1 - C_2v^2$, where C_1 and C_2 are constants which depend on the mechanical configuration of the car. If the car has an initial velocity v_0 when the engine is disengaged, derive an expression for the distance D required for the car to coast to a stop.

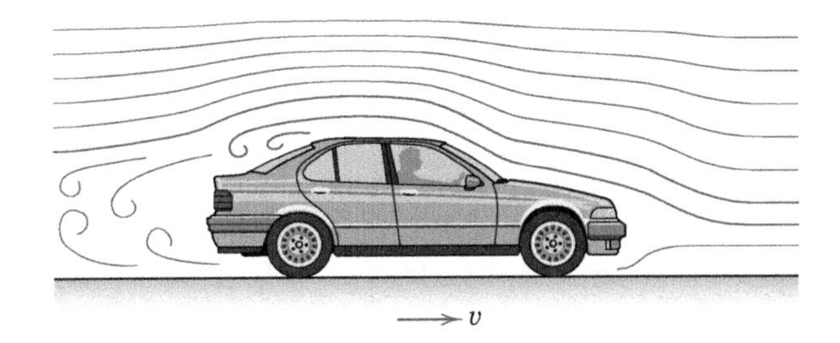

Problem 2/43

2/44 The driver of a car, which is initially at rest at the top A of the grade, releases the brakes and coasts down the grade with an acceleration in feet per second squared given by $a = 3.22 - 0.004v^2$, where v is the velocity in feet per second. Determine the velocity v_B at the bottom B of the grade.

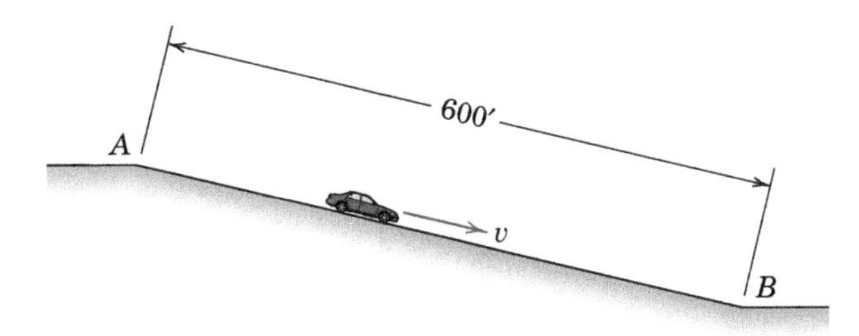

Problem 2/44

2/45 When the effect of aerodynamic drag is included, the y-acceleration of a baseball moving vertically upward is $a_u = -g - kv^2$, while the acceleration when the ball is moving downward is $a_d = -g + kv^2$, where k is a positive constant and v is the speed in meters per second. If the ball is thrown upward at 30 m/s from essentially ground level, compute its maximum height h and its speed v_f upon impact with the ground. Take k to be 0.006 m^{-1} and assume that g is constant.

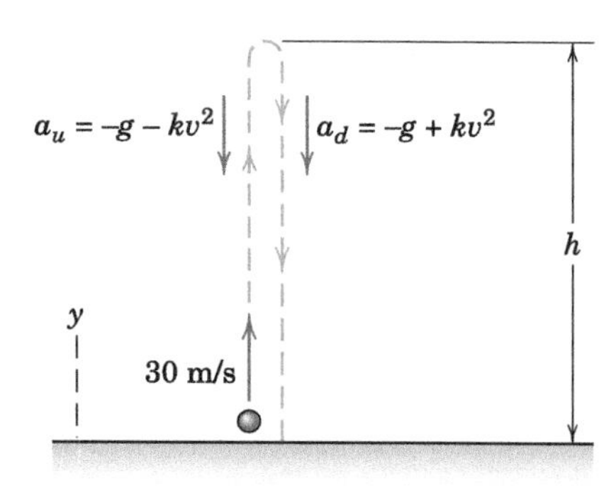

Problem 2/45

2/46 For the baseball of Prob. 2/45 thrown upward with an initial speed of 30 m/s, determine the time t_u from ground to apex and the time t_d from apex to ground.

2/47 The stories of a tall building are uniformly 10 feet in height. A ball A is dropped from the rooftop position shown. Determine the times required for it to pass the 10 feet of the first, tenth, and one-hundredth stories (counted from the top). Neglect aerodynamic drag.

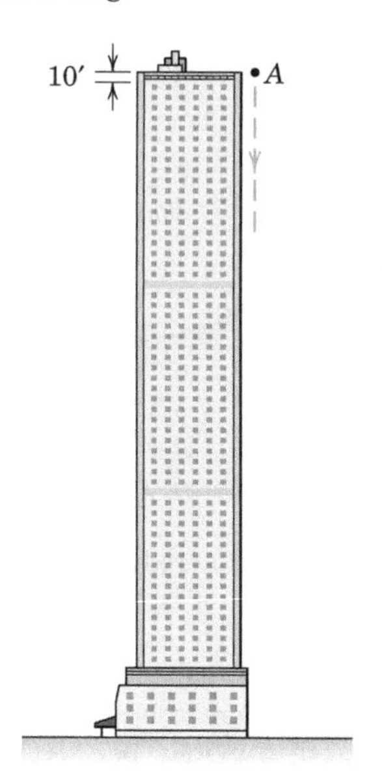

Problem 2/47

2/48 Repeat Prob. 2/47, except now include the effects of aerodynamic drag. The drag force causes an acceleration component in ft/sec^2 of $0.005v^2$ in the direction opposite the velocity vector, where v is in ft/sec.

2/49 On its takeoff roll, the airplane starts from rest and accelerates according to $a = a_0 - kv^2$, where a_0 is the constant acceleration resulting from the engine thrust and $-kv^2$ is the acceleration due to aerodynamic drag. If $a_0 = 2$ m/s^2, $k = 0.00004$ m^{-1}, and v is in meters per second, determine the design length of runway required for the airplane to reach the takeoff speed of 250 km/h if the drag term is (a) excluded and (b) included.

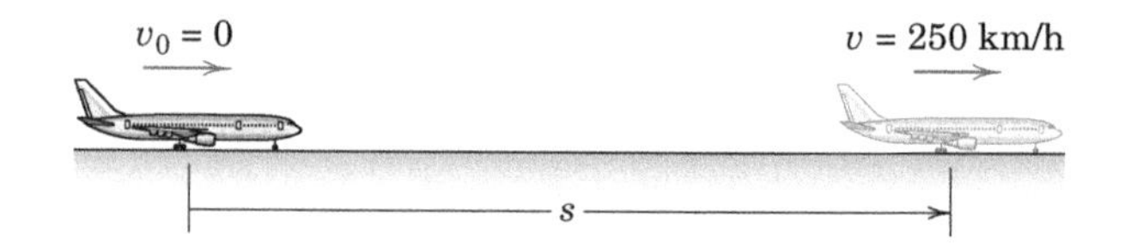

Problem 2/49

2/50 A test projectile is fired horizontally into a viscous liquid with a velocity v_0. The retarding force is proportional to the square of the velocity, so that the acceleration becomes $a = -kv^2$. Derive expressions for the distance D traveled in the liquid and the corresponding time t required to reduce the velocity to $v_0/2$. Neglect any vertical motion.

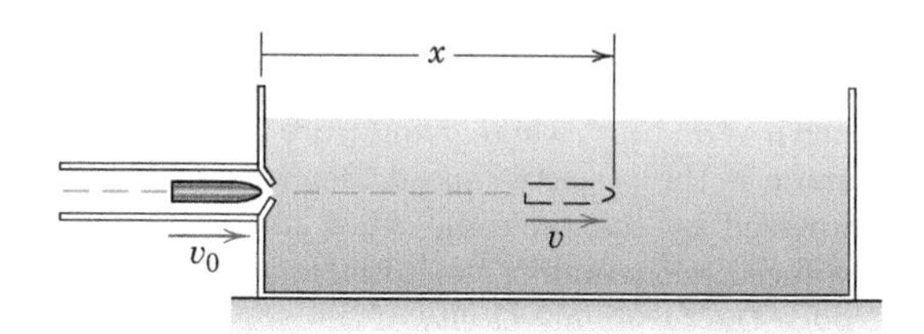

Problem 2/50

2/51 A bumper, consisting of a nest of three springs, is used to arrest the horizontal motion of a large mass which is traveling at 40 ft/sec as it contacts the bumper. The two outer springs cause a deceleration proportional to the spring deformation. The center spring increases the deceleration rate when the compression exceeds 6 in. as shown on the graph. Determine the maximum compression x of the outer springs.

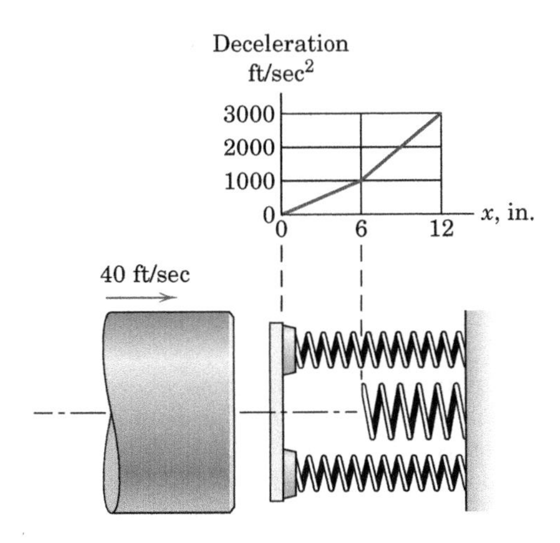

Problem 2/51

2/52 Car A travels at a constant speed of 65 mi/hr. When in the position shown at time $t = 0$, car B has a speed of 25 mi/hr and accelerates at a constant rate of $0.1g$ along its path until it reaches a speed of 65 mi/hr, after which it travels at that constant speed. What is the steady-state position of car A with respect to car B?

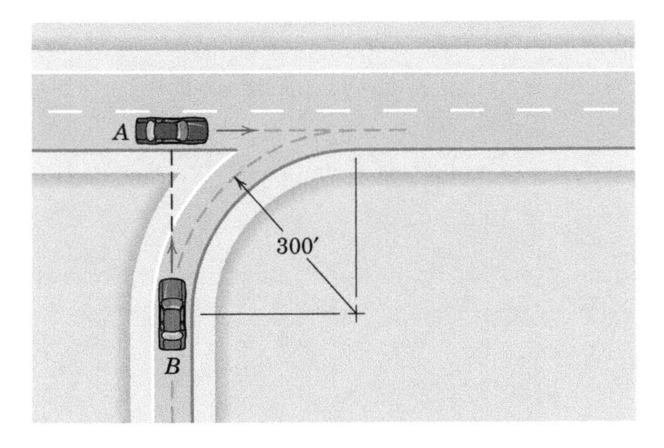

Problem 2/52

2/53 A block of mass m rests on a rough horizontal surface and is attached to a spring of stiffness k. The coefficients of both static and kinetic friction are μ. The block is displaced a distance x_0 to the right of the unstretched position of the spring and released from rest. If the value of x_0 is large enough, the spring force will overcome the maximum available static friction force and the block will slide toward the unstretched position of the spring with an acceleration $a = \mu g - \frac{k}{m}x$, where x represents the amount of stretch (or compression) in the spring at any given location in the motion. Use the values $m = 5$ kg, $k = 150$ N/m, $\mu = 0.40$, and $x_0 = 200$ mm and determine the final spring stretch (or compression) x_f when the block comes to a complete stop.

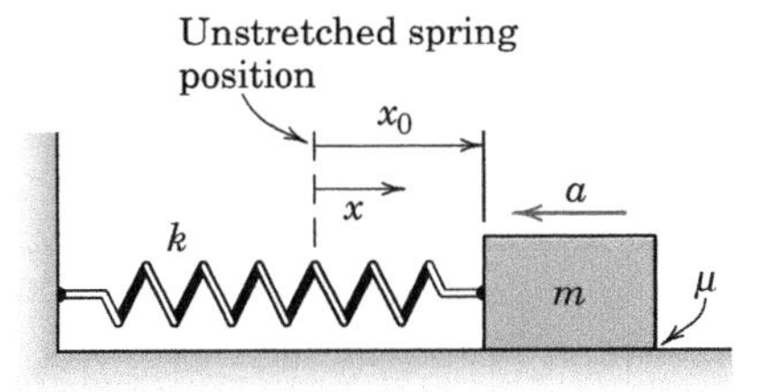

Problem 2/53

▶2/54 The situation of Prob. 2/53 is repeated here. This time, use the values $m = 5$ kg, $k = 150$ N/m, $\mu = 0.40$, and $x_0 = 500$ mm and determine the final spring stretch (or compression) x_f when the block comes to a complete stop. (*Note:* The sign on the μg term is dictated by the direction of motion for the block and always acts in the direction opposite velocity.)

▶2/55 The vertical acceleration of a certain solid-fuel rocket is given by $a = ke^{-bt} - cv - g$, where k, b, and c are constants, v is the vertical velocity acquired, and g is the gravitational acceleration, essentially constant for atmospheric flight. The exponential term represents the effect of a decaying thrust as fuel is burned, and the term $-cv$ approximates the retardation due to atmospheric resistance. Determine the expression for the vertical velocity of the rocket t seconds after firing.

▶**2/56** The preliminary design for a rapid-transit system calls for the train velocity to vary with time as shown in the plot as the train runs the 3.2 km between stations A and B. The slopes of the cubic transition curves (which are of form $a + bt + ct^2 + dt^3$) are zero at the end points. Determine the total run time t between the stations and the maximum acceleration.

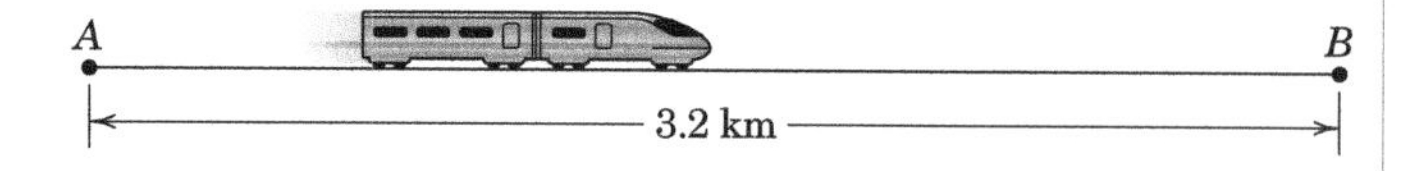

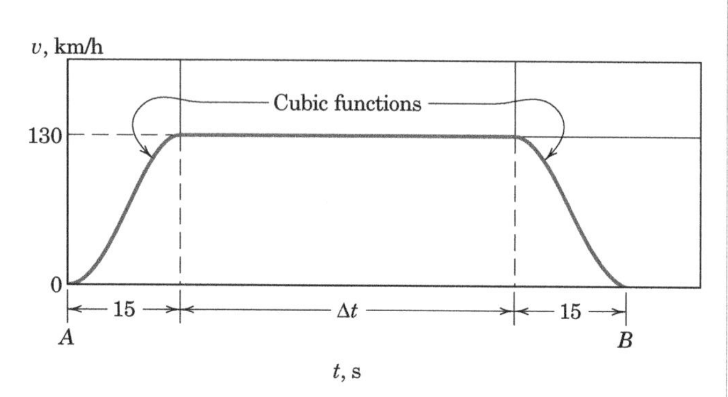

Problem 2/56

2/57 A projectile is fired vertically from point A with an initial speed of 255 ft/sec. Relative to an observer located at B, at what times will the line of sight to the projectile make an angle of 30° with the horizontal? Compute the magnitude of the speed of the projectile at each time, and ignore the effect of aerodynamic drag on the projectile.

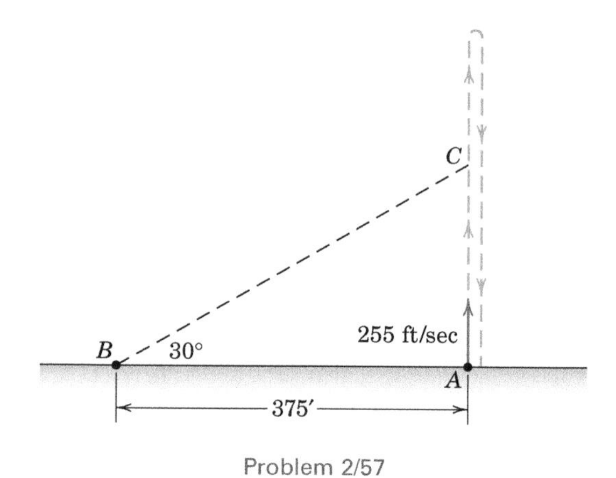

Problem 2/57

▶**2/58** Repeat Prob. 2/57 for the case where aerodynamic drag is included. The magnitude of the drag deceleration is kv^2, where $k = 3.5(10^{-3})$ ft^{-1} and v is the speed in feet per second. The direction of the drag is opposite the motion of the projectile throughout the flight (when the projectile is moving upward, the drag is directed downward, and when the projectile is moving downward, the drag is directed upward).

PROBLEMS

(In the following problems where motion as a projectile in air is involved, neglect air resistance unless otherwise stated and use $g = 9.81$ m/s^2 or $g = 32.2$ ft/sec^2.)

Introductory Problems

2/59 At time $t = 10$ s, the velocity of a particle moving in the x-y plane is $\mathbf{v} = 0.1\mathbf{i} + 2\mathbf{j}$ m/s. By time $t = 10.1$ s, its velocity has become $-0.1\mathbf{i} + 1.8\mathbf{j}$ m/s. Determine the magnitude a_{av} of its average acceleration during this interval and the angle θ made by the average acceleration with the positive x-axis.

2/60 At time $t = 0$, the position vector of a particle moving in the x-y plane is $\mathbf{r} = 5\mathbf{i}$ m. By time $t = 0.02$ s, its position vector has become $5.1\mathbf{i} + 0.4\mathbf{j}$ m. Determine the magnitude v_{av} of its average velocity during this interval and the angle θ made by the average velocity with the positive x-axis.

2/61 At time $t = 0$, a particle is at rest in the x-y plane at the coordinates $(x_0, y_0) = (6, 0)$ in. If the particle is then subjected to the acceleration components $a_x = 0.5 - 0.35t$ in./sec^2 and $a_y = 0.15t - 0.02t^2$ in./sec^2, determine the coordinates of the particle position when $t = 6$ sec. Plot the path of the particle during this time period.

2/62 The rectangular coordinates of a particle which moves with curvilinear motion are given by $x = 10.25t + 1.75t^2 - 0.45t^3$ and $y = 6.32 + 14.65t - 2.48t^2$, where x and y are in millimeters and the time t is in seconds, beginning from $t = 0$. Determine the velocity $\mathbf{v}$ and acceleration $\mathbf{a}$ of the particle when $t = 5$ s. Also, determine the time when the velocity of the particle makes an angle of 45° with the x-axis.

2/63 For a certain interval of motion the pin A is forced to move in the fixed parabolic slot by the horizontal slotted arm which is elevated in the y-direction at the constant rate of 3 in./sec. All measurements are in inches and seconds. Calculate the velocity v and acceleration a of pin A when $x = 6$ in.

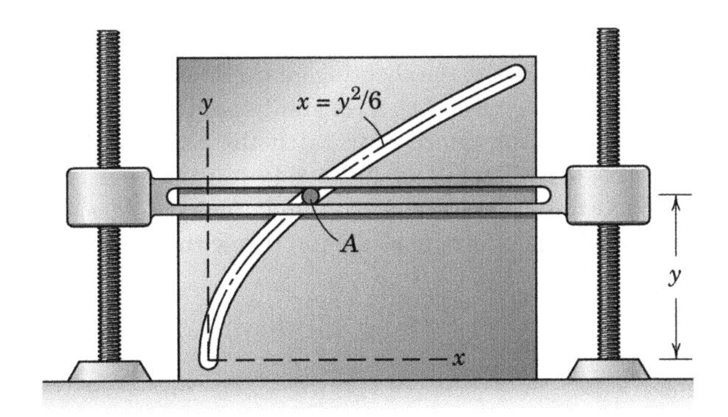

Problem 2/63

2/64 With what minimum horizontal velocity u can a boy throw a rock at A and have it just clear the obstruction at B?

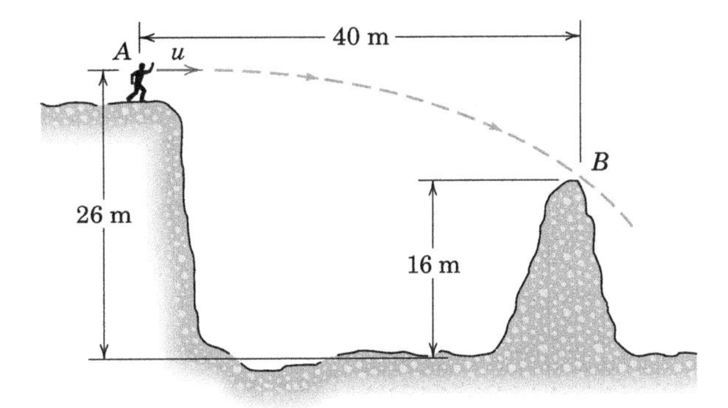

Problem 2/64

2/65 Prove the well-known result that, for a given launch speed v_0, the launch angle $\theta = 45°$ yields the maximum horizontal range R. Determine the maximum range. (Note that this result does not hold when aerodynamic drag is included in the analysis.)

2/66 A placekicker is attempting to make a 64-yard field goal. If the launch angle of the football is 40°, what is the minimum initial speed u which will allow the kicker to succeed?

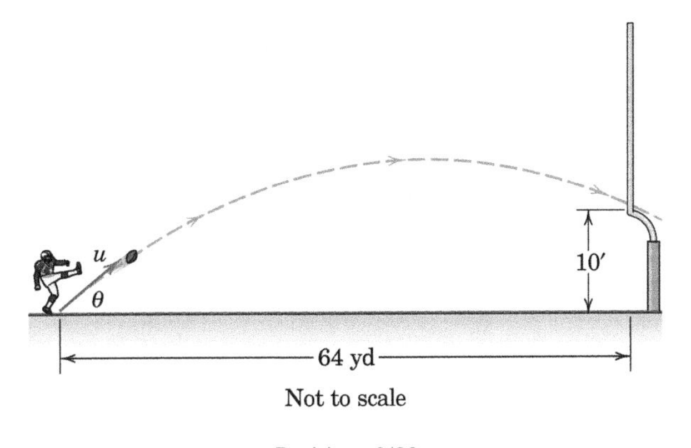

Problem 2/66

2/67 In a basketball game, the point guard A intends to throw a pass to the shooting guard B, who is breaking toward the basket at a constant speed of 12 ft/sec. If the shooting guard is to catch the ball at a height of 7 ft at C while in full stride to execute a layup, determine the speed v_0 and launch angle θ with which the point guard should throw the ball.

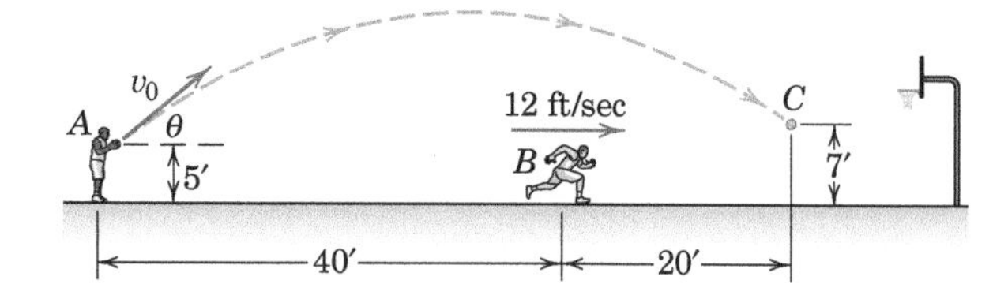

Problem 2/67

2/68 A fireworks show is choreographed to have two shells cross paths at a height of 160 feet and explode at an apex of 200 feet under normal weather conditions. If the shells have a launch angle $\theta = 60°$ above the horizontal, determine the common launch speed v_0 for the shells, the separation distance d between the launch points A and B, and the time from launch at which the shells explode.

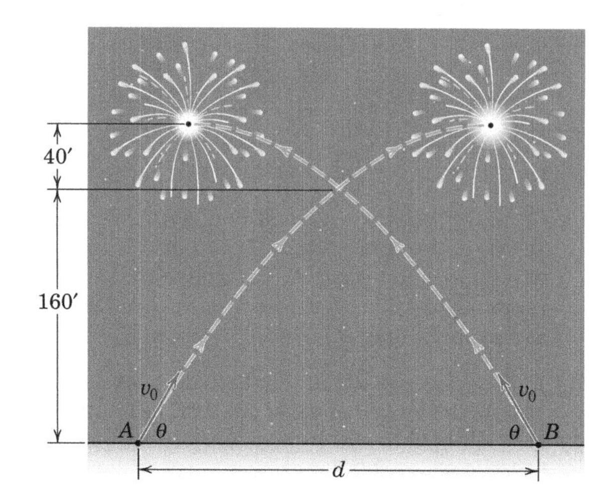

Problem 2/68

2/69 If a strong wind induces a constant rightward acceleration of 16 ft/sec^2 for the fireworks shells of Prob. 2/68, determine the horizontal shift of the crossing point of the shells. Refer to the printed answers for Prob. 2/68 as needed.

2/70 The center of mass G of a high jumper follows the trajectory shown. Determine the component v_0, measured in the vertical plane of the figure, of his takeoff velocity and angle θ if the apex of the trajectory just clears the bar at A. (In general, must the mass center G of the jumper clear the bar during a successful jump?)

Problem 2/70

Representative Problems

2/71 Electrons are emitted at A with a velocity u at the angle θ into the space between two charged plates. The electric field between the plates is in the direction E and repels the electrons approaching the upper plate. The field produces an acceleration of the electrons in the E-direction of eE/m, where e is the electron charge and m is its mass. Determine the field strength E which will permit the electrons to cross one-half of the gap between the plates. Also find the distance s.

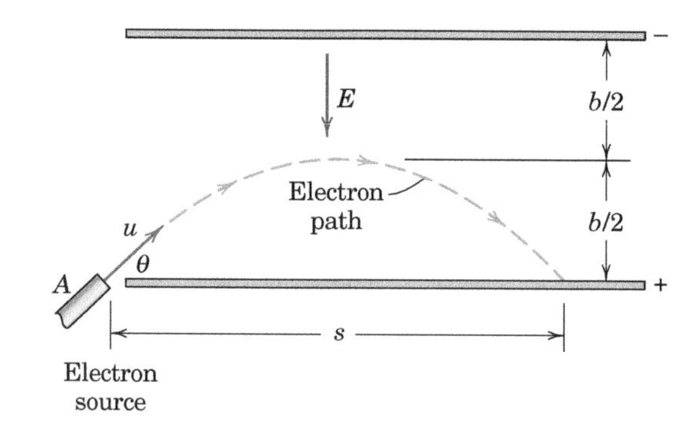

Problem 2/71

2/72 A boy tosses a ball onto the roof of a house. For the launch conditions shown, determine the slant distance s to the point of impact. Also, determine the angle θ which the velocity of the ball makes with the roof at the moment of impact.

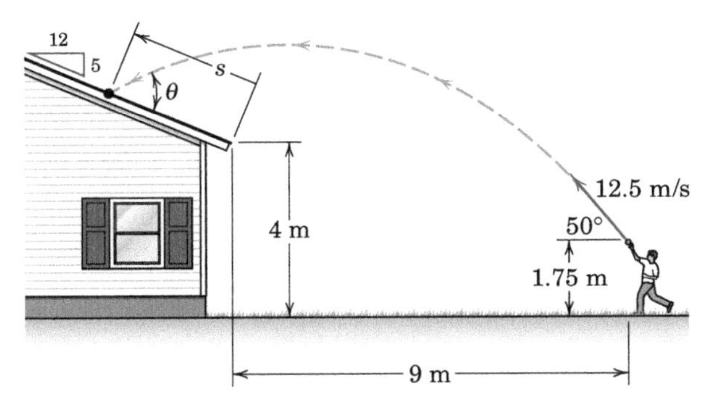

Problem 2/72

2/73 A small airplane flying horizontally with a speed of 180 mi/hr at an altitude of 400 ft above a remote valley drops an emergency medical package at A. The package has a parachute which deploys at B and allows the package to descend vertically at the constant rate of 6 ft/sec. If the drop is designed so that the package is to reach the ground 37 seconds after release at A, determine the horizontal lead L so that the package hits the target. Neglect atmospheric resistance from A to B.

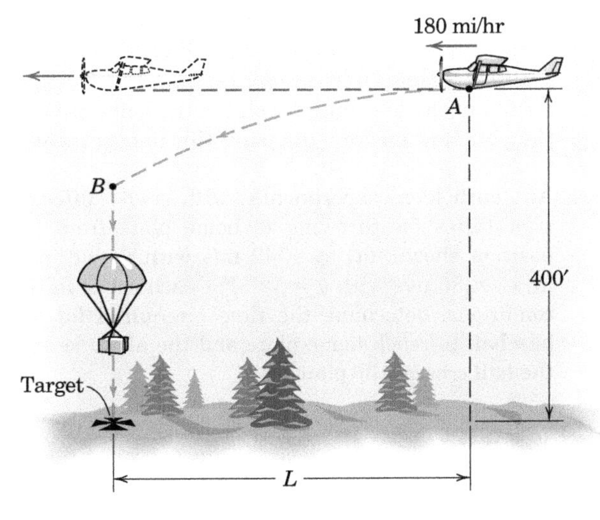

Problem 2/73

2/74 As part of a circus performance, a man is attempting to throw a dart into an apple which is dropped from an overhead platform. Upon release of the apple, the man has a reflex delay of 215 milliseconds before throwing the dart. If the dart is released with a speed $v_0 = 14$ m/s, at what distance d below the platform should the man aim if the dart is to strike the apple before it hits the ground?

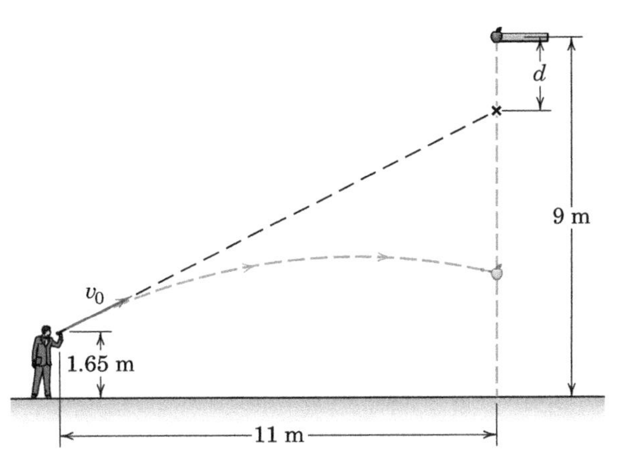

Problem 2/74

2/75 A marksman fires a practice round from A toward a target B. If the target diameter is 160 mm and the target center is at the same altitude as the end of the rifle barrel, determine the range of "shallow" launch angles θ for which the round will strike the target. Neglect aerodynamic drag and assume that the round is directed along the vertical centerline of the target. (*Note:* The word "shallow" indicates a low-flying trajectory for the round.)

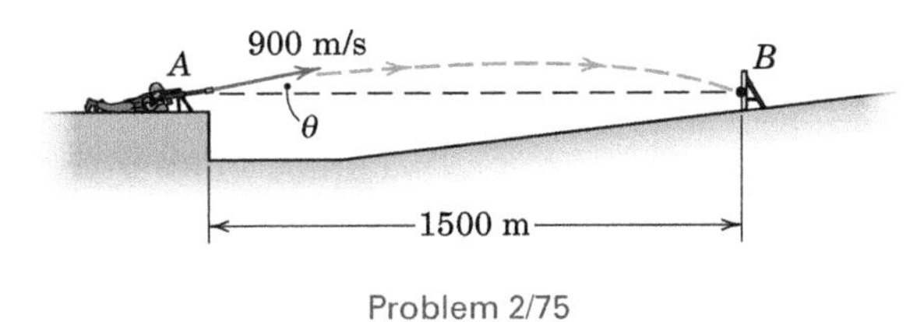

Problem 2/75

2/76 The pilot of an airplane carrying a package of mail to a remote outpost wishes to release the package at the right moment to hit the recovery location A. What angle θ with the horizontal should the pilot's line of sight to the target make at the instant of release? The airplane is flying horizontally at an altitude of 100 m with a velocity of 200 km/h.

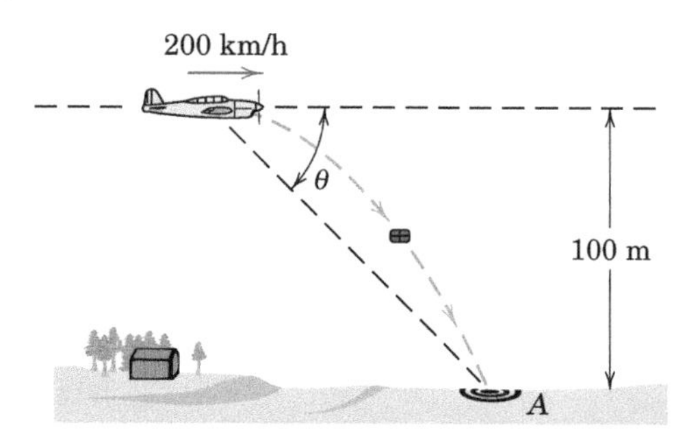

Problem 2/76

2/77 During a baseball practice session, the cutoff man A executes a throw to the third baseman B. If the initial speed of the baseball is $v_0 = 130$ ft/sec, what launch angle θ is best if the ball is to arrive at third base at essentially ground level?

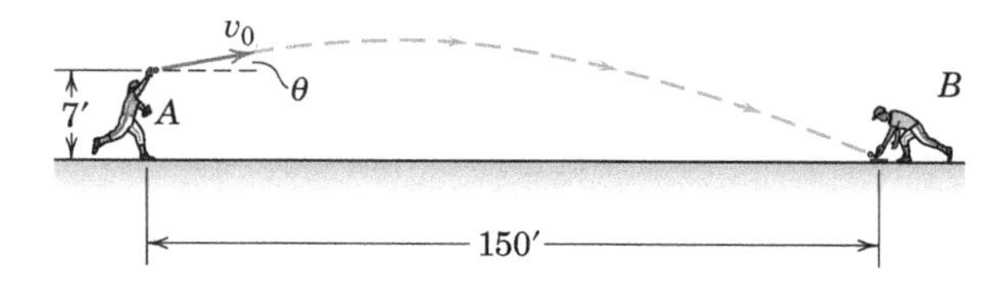

Problem 2/77

2/78 A particle is launched from point A with a horizontal speed u and subsequently passes through a vertical opening of height b as shown. Determine the distance d which will allow the landing zone for the particle to also have a width b. Additionally, determine the range of u which will allow the projectile to pass through the vertical opening for this value of d.

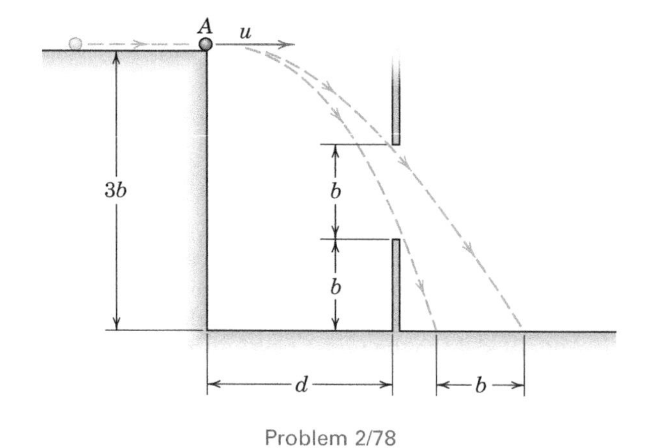

Problem 2/78

2/79 If the tennis player serves the ball horizontally ($\theta = 0$), calculate its velocity v if the center of the ball clears the 0.9-m net by 150 mm. Also find the distance s from the net to the point where the ball hits the court surface. Neglect air resistance and the effect of ball spin.

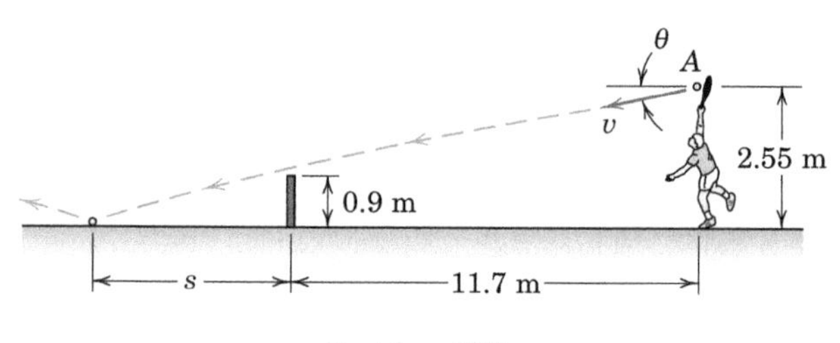

Problem 2/79

2/80 A golfer is attempting to reach the elevated green by hitting his ball under a low-hanging branch in one tree A, but over the top of a second tree B. For $v_0 = 115$ mi/hr and $\theta = 18°$, where does the golf ball land first?

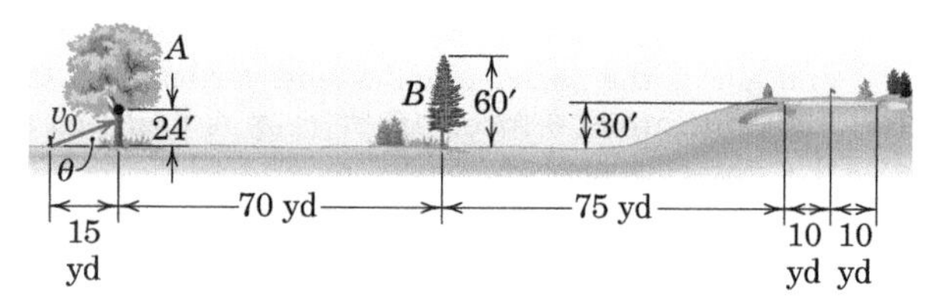

Problem 2/80

2/81 If the launch speed of the golf ball of the previous problem remains $v_0 = 115$ mi/hr, what launch angle θ will put the first impact point of the ball closest to the pin? How far from the pin is this impact point?

2/82 An outfielder experiments with two different trajectories for throwing to home plate from the position shown: (a) $v_0 = 42$ m/s with $\theta = 8°$ and (b) $v_0 = 36$ m/s with $\theta = 12°$. For each set of initial conditions, determine the time t required for the baseball to reach home plate and the altitude h as the ball crosses the plate.

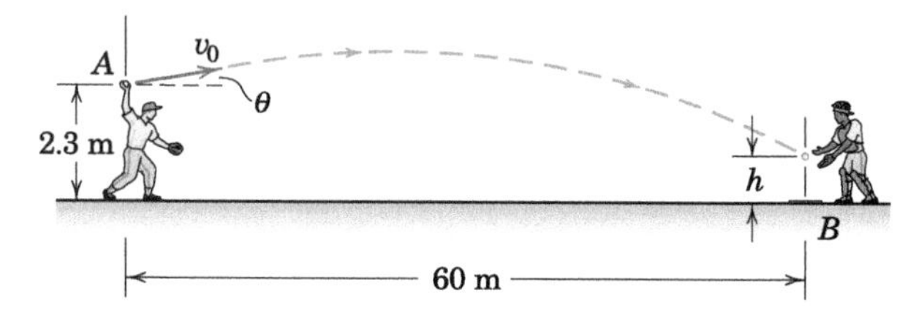

Problem 2/82

2/83 A ski jumper has the takeoff conditions shown. Determine the inclined distance d from the takeoff point A to the location where the skier first touches down in the landing zone, and the total time t_f during which the skier is in the air. For simplicity, assume that the landing zone BC is straight.

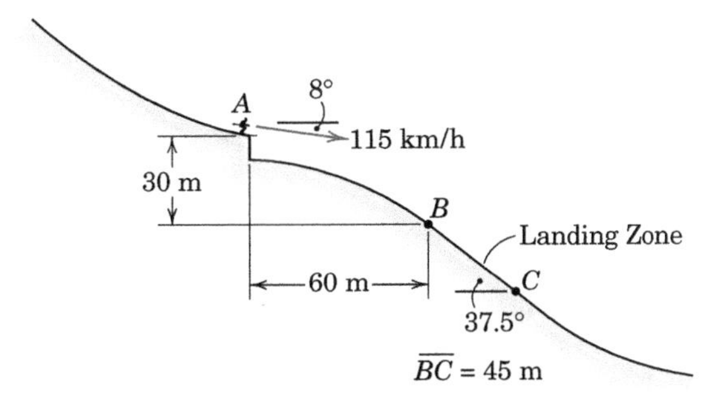

Problem 2/83

2/84 A projectile is launched with a speed $v_0 = 25$ m/s from the floor of a 5-m-high tunnel as shown. Determine the maximum horizontal range R of the projectile and the corresponding launch angle θ.

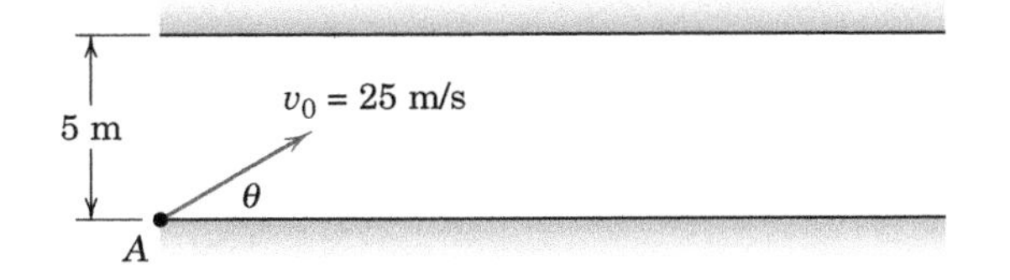

Problem 2/84

2/85 A boy throws a ball upward with a speed $v_0 = 12$ m/s. The wind imparts a horizontal acceleration of 0.4 m/s^2 to the left. At what angle θ must the ball be thrown so that it returns to the point of release? Assume that the wind does not affect the vertical motion.

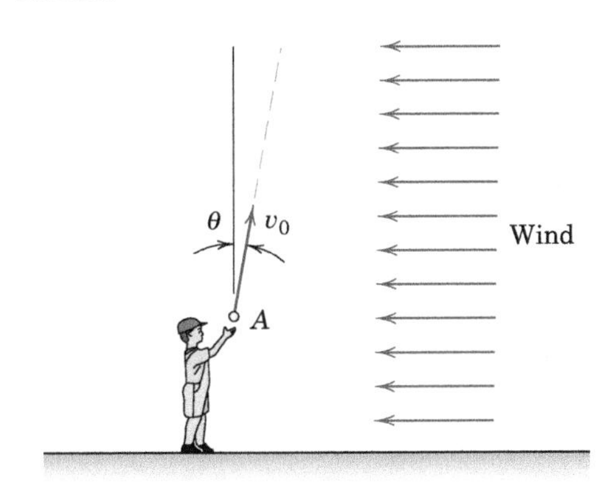

Problem 2/85

2/86 A projectile is launched from point O with the initial conditions shown. Determine the impact coordinates for the projectile if (*a*) $v_0 = 60$ ft/sec and $\theta = 40°$ and (*b*) $v_0 = 85$ ft/sec and $\theta = 15°$.

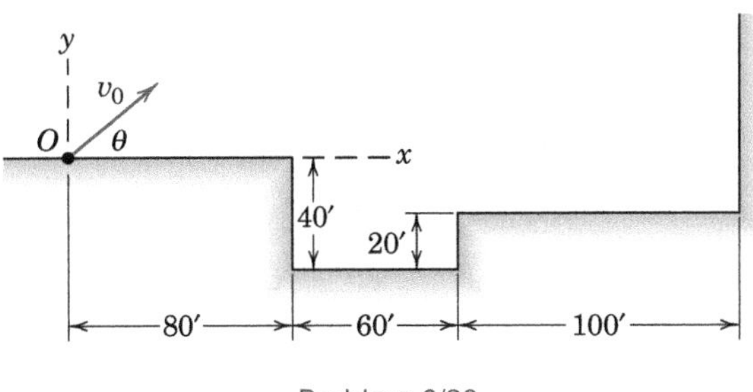

Problem 2/86

2/87 A projectile is launched from point A with the initial conditions shown in the figure. Determine the slant distance s which locates the point B of impact. Calculate the time of flight t.

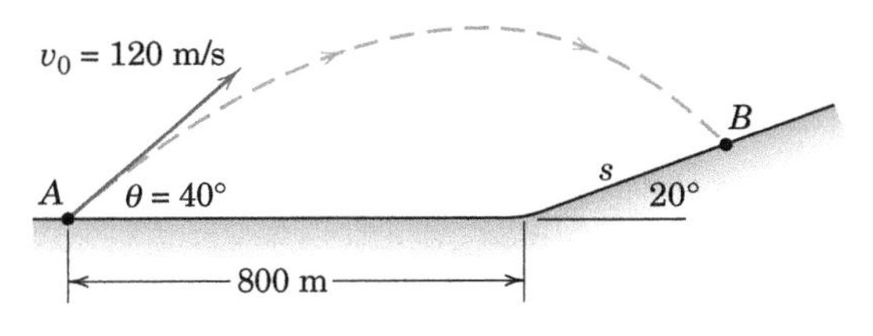

Problem 2/87

2/88 A team of engineering students is designing a catapult to launch a small ball at A so that it lands in the box. If it is known that the initial velocity vector makes a 30° angle with the horizontal, determine the range of launch speeds v_0 for which the ball will land inside the box.

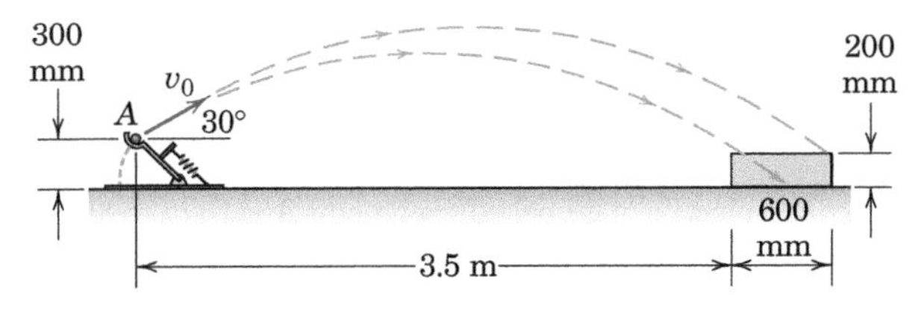

Problem 2/88

2/89 A snow blower travels forward at a constant speed $v_s = 1.4$ ft/sec along the straight and level path shown. The snow is ejected with a speed $v_r = 30$ ft/sec relative to the machine at the 40° angle indicated. Determine the distance d which locates the snow-blower position from which an ejected snow particle ultimately strikes the slender pole P. At what height h above the ground does the snow strike the pole? Points O and Q are at ground level.

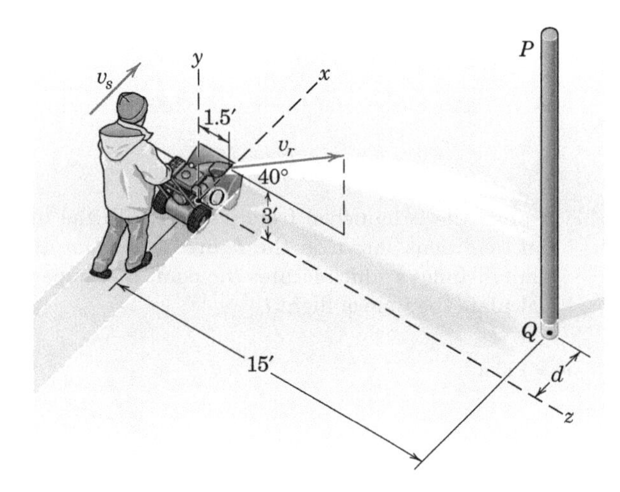

Problem 2/89

2/90 Determine the location h of the spot toward which the pitcher must throw if the ball is to hit the catcher's mitt. The ball is released with a speed of 40 m/s.

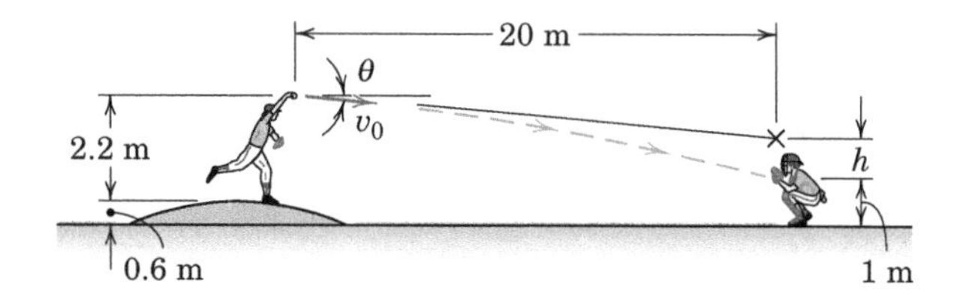

Problem 2/90

2/91 A projectile is launched from point A with $v_0 = 30$ m/s and $\theta = 35°$. Determine the x- and y-coordinates of the point of impact.

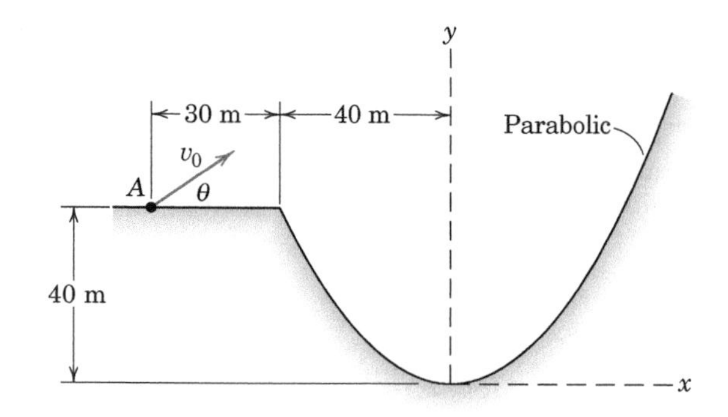

Problem 2/91

2/92 A projectile is fired with a velocity u at right angles to the slope, which is inclined at an angle θ with the horizontal. Derive an expression for the distance R to the point of impact.

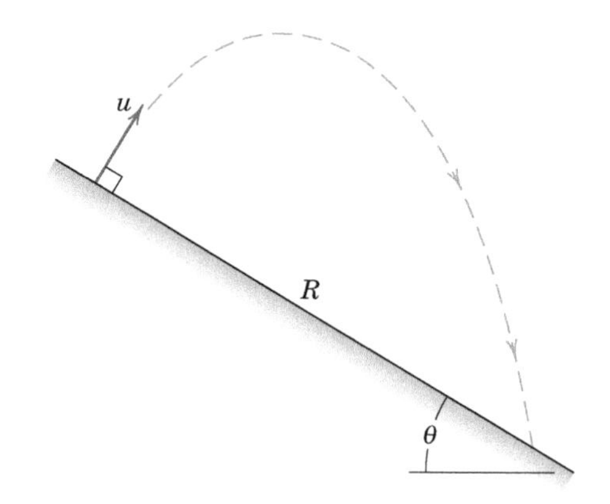

Problem 2/92

2/93 A projectile is launched from point A with an initial speed $v_0 = 100$ ft/sec. Determine the minimum value of the launch angle α for which the projectile will land at point B.

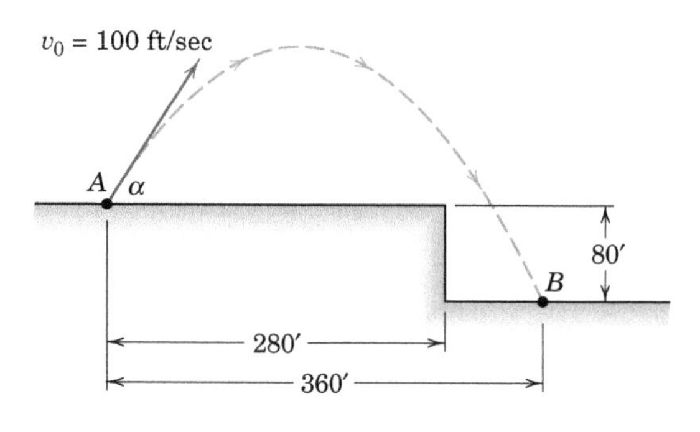

Problem 2/93

2/94 A projectile is launched from point A and lands on the same level at D. Its maximum altitude is h. Determine and plot the fraction f_2 of the total flight time that the projectile is above the level f_1h, where f_1 is a fraction which can vary from zero to 1. State the value of f_2 for $f_1 = \frac{3}{4}$.

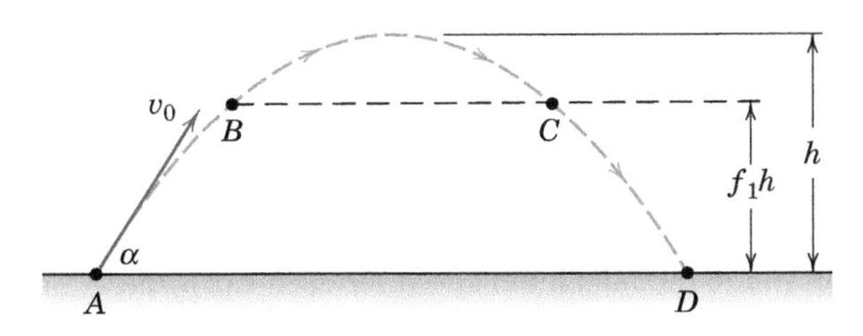

Problem 2/94

▶2/95 A projectile is launched with speed v_0 from point A. Determine the launch angle θ which results in the maximum range R up the incline of angle α (where $0 \leq \alpha \leq 90°$). Evaluate your results for $\alpha = 0, 30°$, and $45°$.

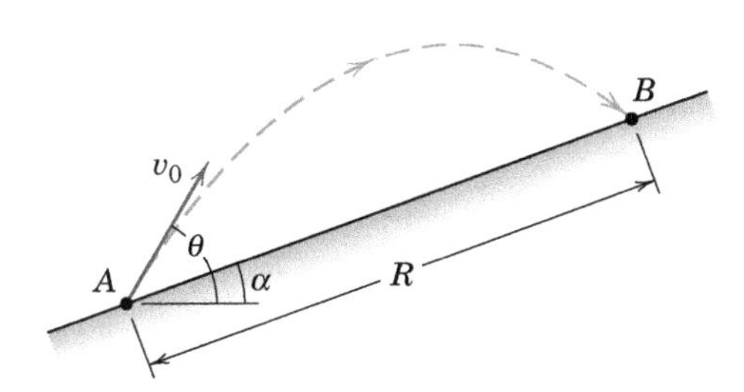

Problem 2/95

▶2/96 A projectile is ejected into an experimental fluid at time $t = 0$. The initial speed is v_0 and the angle to the horizontal is θ. The drag on the projectile results in an acceleration term $\mathbf{a}_D = -k\mathbf{v}$, where k is a constant and $\mathbf{v}$ is the velocity of the projectile. Determine the x- and y-components of both the velocity and displacement as functions of time. What is the terminal velocity? Include the effects of gravitational acceleration.

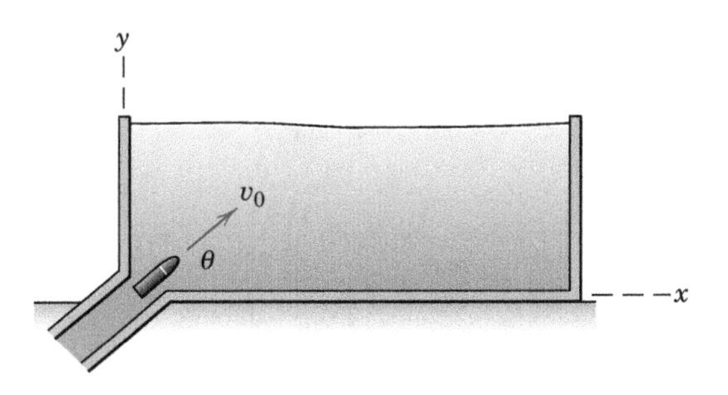

Problem 2/96

PROBLEMS

Introductory Problems

2/97 A test car starts from rest on a horizontal circular track of 80-m radius and increases its speed at a uniform rate to reach 100 km/h in 10 seconds. Determine the magnitude a of the total acceleration of the car 8 seconds after the start.

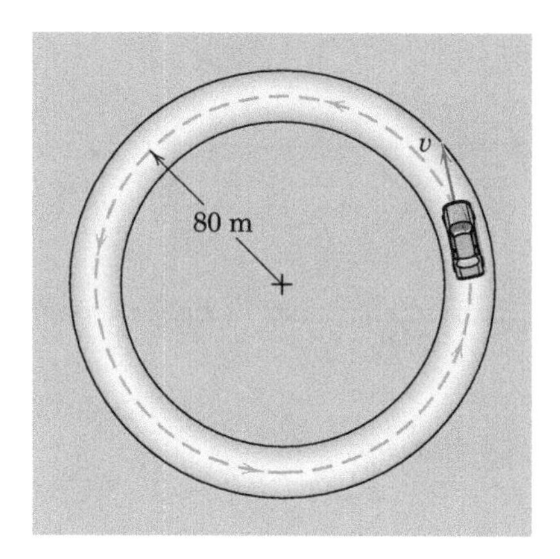

Problem 2/97

2/98 If the compact disc is spinning at a constant angular rate $\dot{\theta} = 360$ rev/min, determine the magnitudes of the accelerations of points A and B at the instant shown.

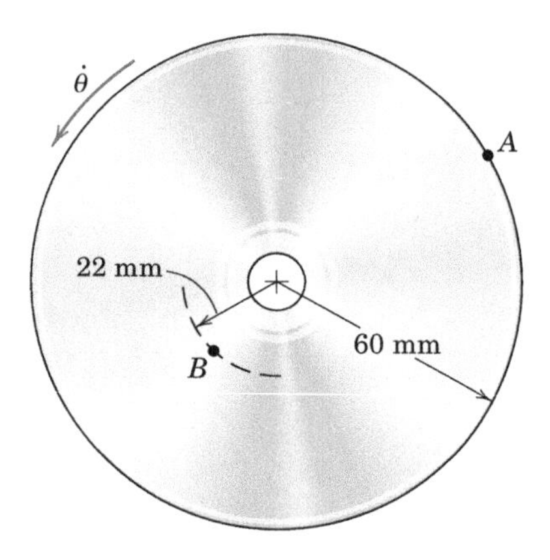

Problem 2/98

2/99 The car moves on a horizontal surface without any slippage of its tires. For each of the eight horizontal acceleration vectors, describe in words the instantaneous motion of the car. The car velocity is directed to the left as shown for all cases.

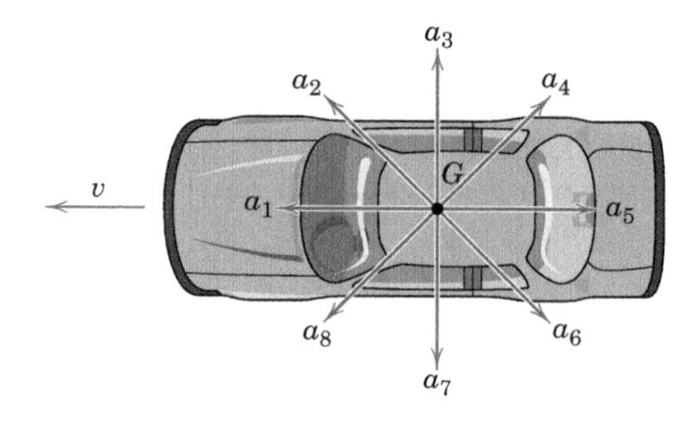

Problem 2/99

2/100 Determine the maximum speed for each car if the normal acceleration is limited to $0.88g$. The roadway is unbanked and level.

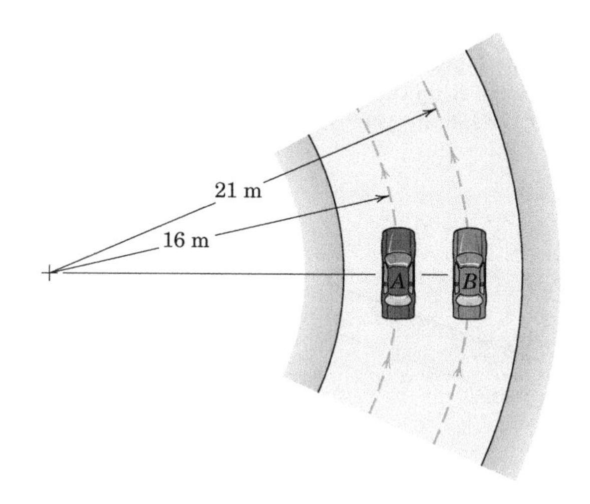

Problem 2/100

2/101 An accelerometer C is mounted to the side of the roller-coaster car and records a total acceleration of $3.5g$ as the empty car passes the bottommost position of the track as shown. If the speed of the car at this position is 215 km/h and is decreasing at the rate of 18 km/h every second, determine the radius of curvature ρ of the track at the position shown.

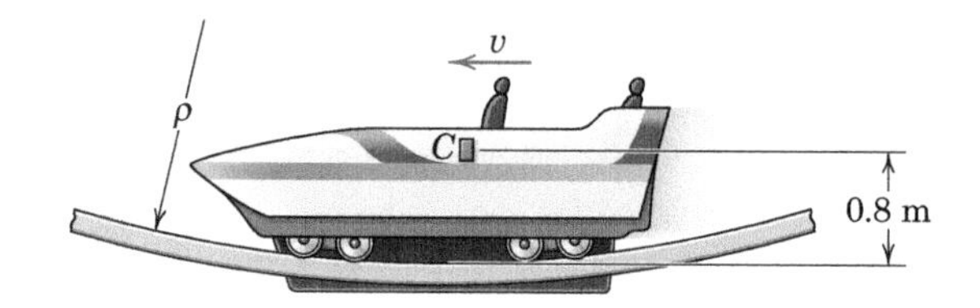

Problem 2/101

2/102 The driver of the truck has an acceleration of 0.4g as the truck passes over the top A of the hump in the road at constant speed. The radius of curvature of the road at the top of the hump is 98 m, and the center of mass G of the driver (considered a particle) is 2 m above the road. Calculate the speed v of the truck.

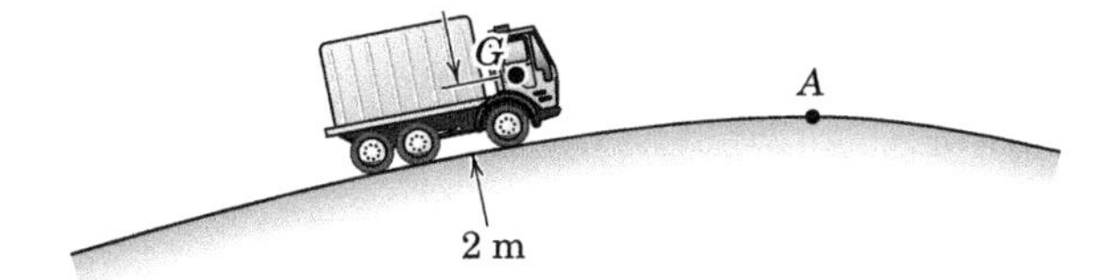

Problem 2/102

2/103 A particle moves along the curved path shown. The particle has a speed $v_A = 12$ ft/sec at time t_A and a speed $v_B = 14$ ft/sec at time t_B. Determine the average values of the normal and tangential accelerations of the particle between points A and B.

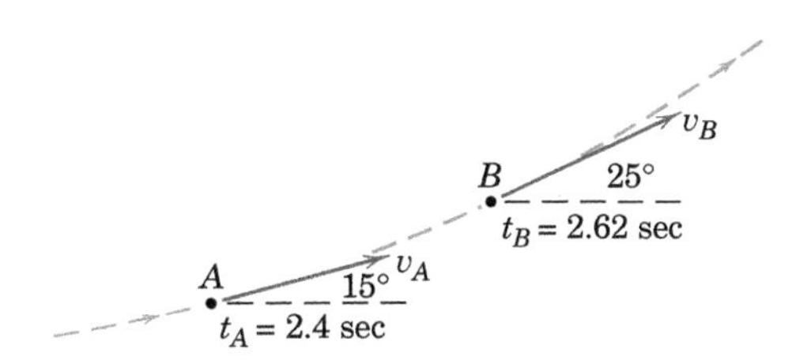

Problem 2/103

2/104 A ship which moves at a steady 20-knot speed (1 knot = 1.852 km/h) executes a turn to port by changing its compass heading at a constant counterclockwise rate. If it requires 60 seconds to alter course 90°, calculate the magnitude of the acceleration **a** of the ship during the turn.

2/105 A sprinter practicing for the 200-m dash accelerates uniformly from rest at A and reaches a top speed of 40 km/h at the 60-m mark. He then maintains this speed for the next 70 meters before uniformly slowing to a final speed of 35 km/h at the finish line. Determine the maximum horizontal acceleration which the sprinter experiences during the run. Where does this maximum acceleration value occur?

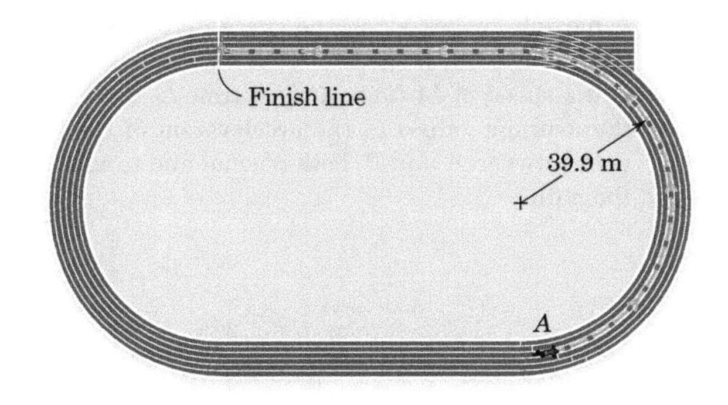

Problem 2/105

2/106 A train enters a curved horizontal section of track at a speed of 100 km/h and slows down with constant deceleration to 50 km/h in 12 seconds. An accelerometer mounted inside the train records a horizontal acceleration of 2 m/s^2 when the train is 6 seconds into the curve. Calculate the radius of curvature ρ of the track for this instant.

2/107 A particle moves on a circular path of radius r = 0.8 m with a constant speed of 2 m/s. The velocity undergoes a vector change $\Delta\mathbf{v}$ from A to B. Express the magnitude of $\Delta\mathbf{v}$ in terms of v and $\Delta\theta$ and divide it by the time interval Δt between A and B to obtain the magnitude of the average acceleration of the particle for (*a*) $\Delta\theta = 30°$, (*b*) $\Delta\theta = 15°$, and (*c*) $\Delta\theta = 5°$. In each case, determine the percentage difference from the instantaneous value of acceleration.

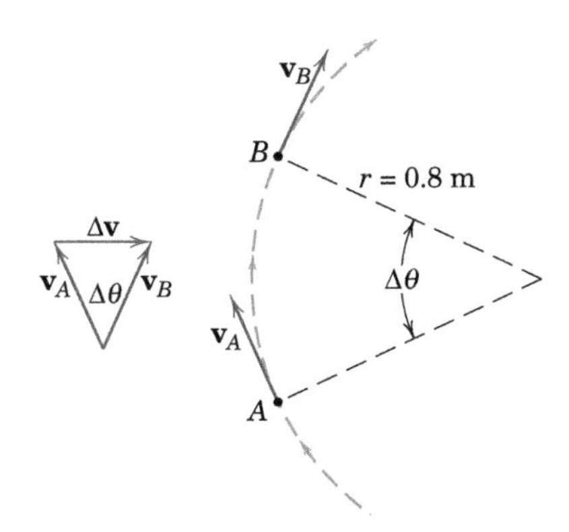

Problem 2/107

2/108 A particle moves along the curved path shown. If the particle has a speed of 40 ft/sec at A at time t_A and a speed of 44 ft/sec at B at time t_B, determine the average values of the acceleration of the particle between A and B, both normal and tangent to the path.

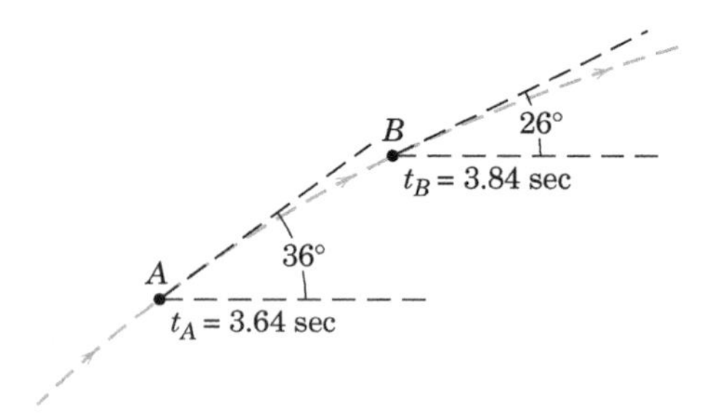

Problem 2/108

Representative Problems

2/109 An overhead view of part of a pinball game is shown. If the plunger imparts a speed of 3 m/s to the ball which travels in the smooth horizontal slot, determine the acceleration **a** of the ball (*a*) just before it exits the curve at C and (*b*) when it is halfway between points D and E. Use the values $r = 150$ mm and $\theta = 60°$.

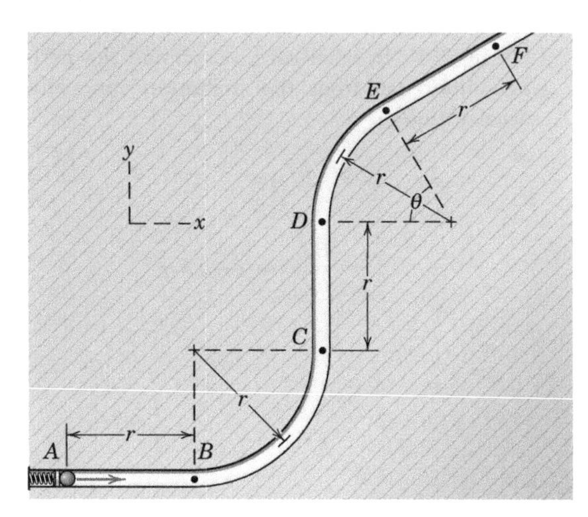

Problem 2/109

2/110 For the pinball game of Prob. 2/109, if the plunger imparts an initial speed of 3 m/s to the ball at time $t = 0$, determine the acceleration **a** of the ball (*a*) at time $t = 0.08$ s and (*b*) at time $t = 0.20$ s. At point F, the speed of the pinball has decreased by 10% from the initial value, and this decrease may be assumed to occur uniformly over the total distance traveled by the pinball. Use the values $r = 150$ mm and $\theta = 60°$.

2/111 The speed of a car increases uniformly with time from 50 km/h at A to 100 km/h at B during 10 seconds. The radius of curvature of the hump at A is 40 m. If the magnitude of the total acceleration of the mass center of the car is the same at B as at A, compute the radius of curvature ρ_B of the dip in the road at B. The mass center of the car is 0.6 m from the road.

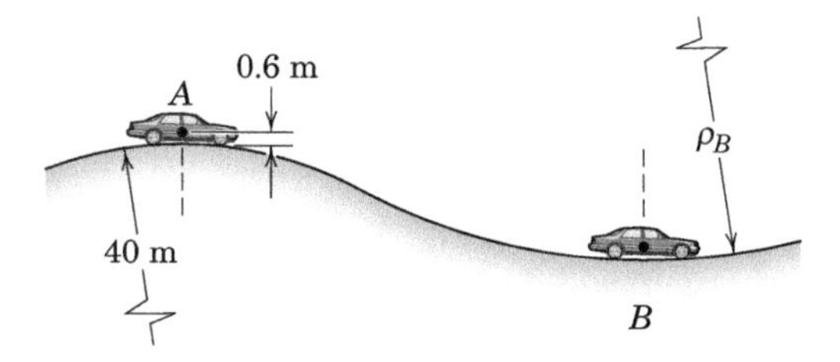

Problem 2/111

2/112 A minivan starts from rest on the road whose constant radius of curvature is 40 m and whose bank angle is 10°. The motion occurs in a horizontal plane. If the constant forward acceleration of the minivan is 1.8 m/s^2, determine the magnitude a of its total acceleration 5 seconds after starting.

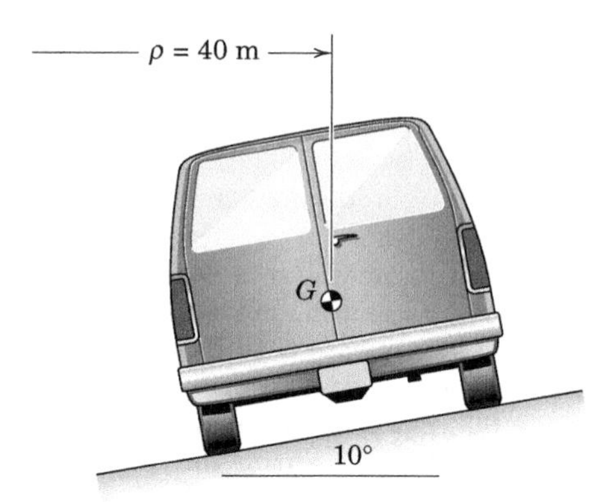

Problem 2/112

2/113 Consider the polar axis of the earth to be fixed in space and compute the magnitudes of the velocity and acceleration of a point P on the earth's surface at latitude 40° north. The mean diameter of the earth is 12 742 km and its angular velocity is $0.7292(10^{-4})$ rad/s.

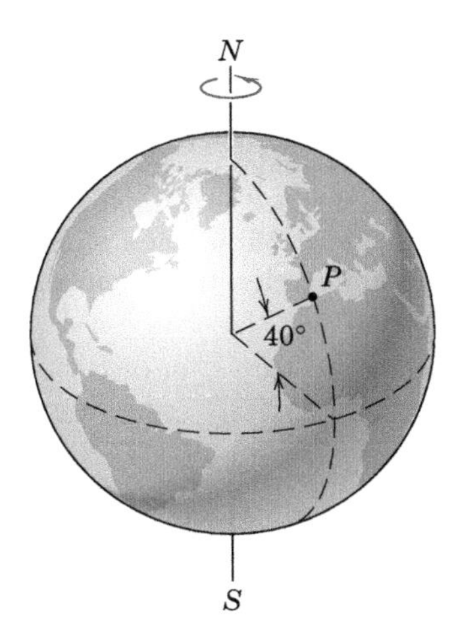

Problem 2/113

2/114 The car C increases its speed at the constant rate of 1.5 m/s^2 as it rounds the curve shown. If the magnitude of the total acceleration of the car is 2.5 m/s^2 at point A where the radius of curvature is 200 m, compute the speed v of the car at this point.

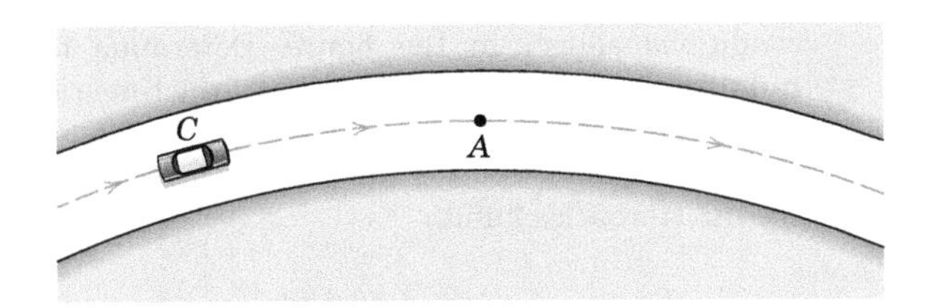

Problem 2/114

2/115 At the bottom A of the vertical inside loop, the magnitude of the total acceleration of the airplane is $3g$. If the airspeed is 800 km/h and is increasing at the rate of 20 km/h per second, calculate the radius of curvature ρ of the path at A.

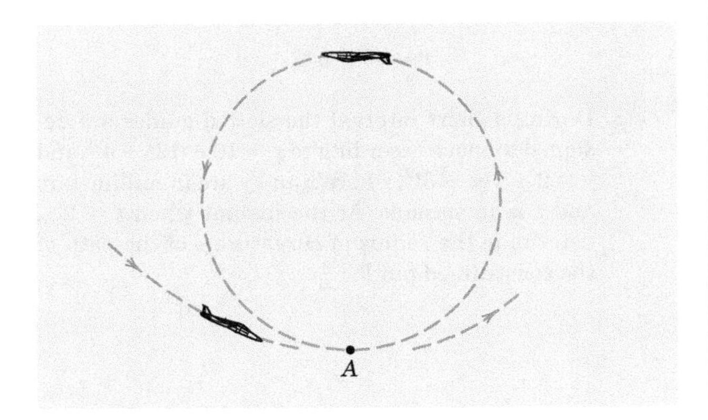

Problem 2/115

2/116 A golf ball is launched with the initial conditions shown in the figure. Determine the radius of curvature of the trajectory and the time rate of change of the speed of the ball (a) just after launch and (b) at apex. Neglect aerodynamic drag.

Problem 2/116

***2/117** If the golf ball of Prob. 2/116 is launched at time $t = 0$, determine the two times when the radius of curvature of the trajectory has a value of 1800 ft.

2/118 The preliminary design for a "small" space station to orbit the earth in a circular path consists of a ring (torus) with a circular cross section as shown. The living space within the torus is shown in section A, where the "ground level" is 20 ft from the center of the section. Calculate the angular speed N in revolutions per minute required to simulate standard gravity at the surface of the earth (32.17 ft/sec^2). Recall that you would be unaware of a gravitational field if you were in a nonrotating spacecraft in a circular orbit around the earth.

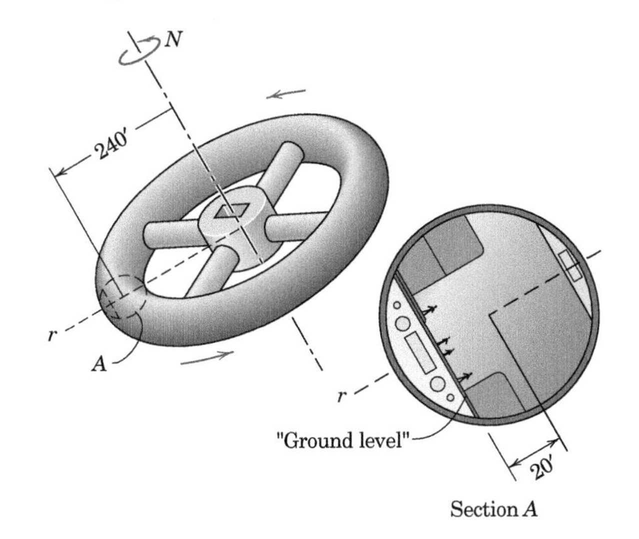

Problem 2/118

2/119 A spacecraft S is orbiting Jupiter in a circular path 1000 km above the surface with a constant speed. Using the gravitational law, calculate the magnitude v of its orbital velocity with respect to Jupiter. Use Table D/2 of Appendix D as needed.

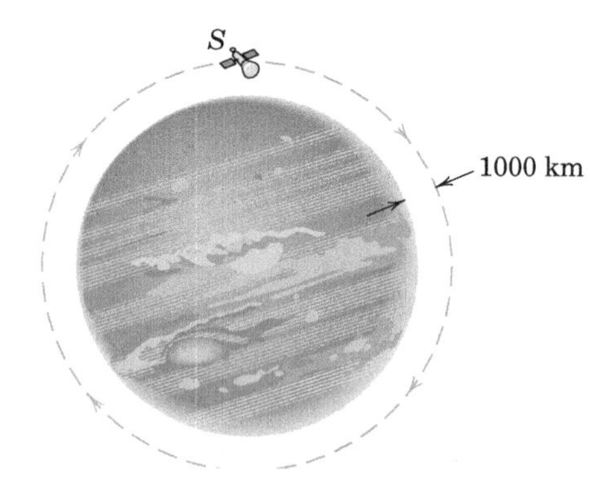

Problem 2/119

2/120 Two cars travel at constant speeds through a curved portion of highway. If the front ends of both cars cross line CC at the same instant, and each driver minimizes his or her time in the curve, determine the distance δ which the second car has yet to go along its own path to reach line DD at the instant the first car reaches there. The maximum horizontal acceleration for car A is $0.60g$ and that for car B is $0.76g$. Which car crosses line DD first?

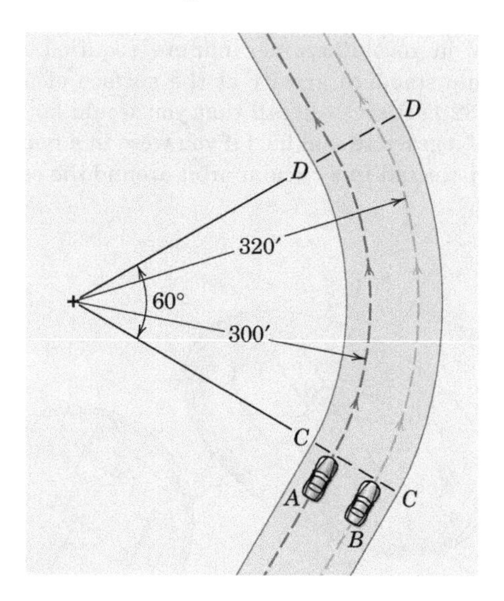

Problem 2/120

2/121 The figure shows a portion of a plate cam used in the design of a control mechanism. The motion of pin P in the fixed slot of the plate cam is controlled by the vertical guide A, which travels horizontally at a constant speed of 6 in./sec over the central sinusoidal portion of the slot. Determine the normal component of acceleration when the pin is at the position $x = 2$ in.

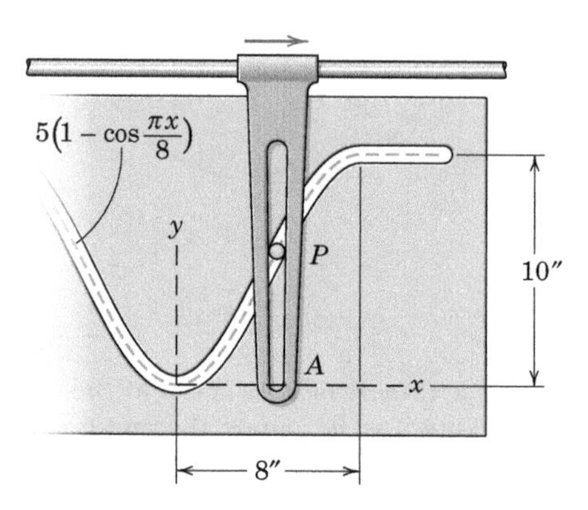

Problem 2/121

2/122 A football player releases a ball with the initial conditions shown in the figure. Determine the radius of curvature ρ of the path and the time rate of change $\dot{v}$ of the speed at times $t = 1$ sec and $t = 2$ sec, where $t = 0$ is the time of release from the quarterback's hand.

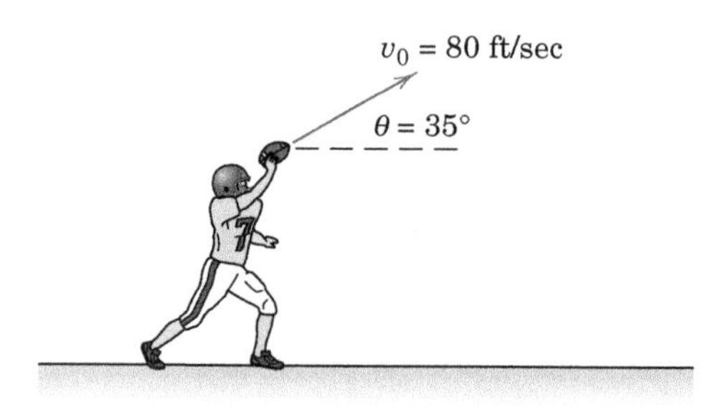

Problem 2/122

2/123 During a short interval the slotted guides are designed to move according to $x = 16 - 12t + 4t^2$ and $y = 2 + 15t - 3t^2$, where x and y are in millimeters and t is in seconds. At the instant when $t = 2$ s, determine the radius of curvature ρ of the path of the constrained pin P.

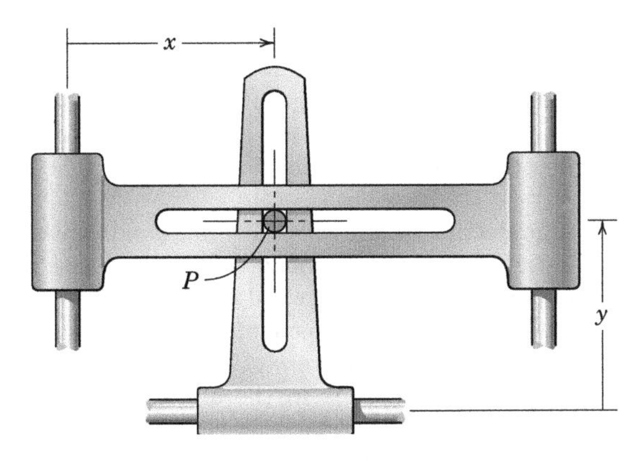

Problem 2/123

2/124 The particle P starts from rest at point A at time $t = 0$ and changes its speed thereafter at a constant rate of $2g$ as it follows the horizontal path shown. Determine the magnitude and direction of its total acceleration (a) just before it passes point B, (b) just after it passes point B, and (c) as it passes point C. State your directions relative to the x-axis shown (CCW positive).

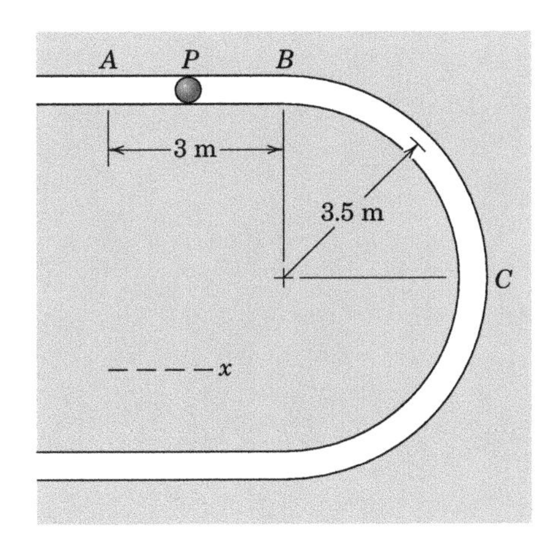

Problem 2/124

2/125 In the design of a timing mechanism, the motion of pin P in the fixed circular slot is controlled by the guide A, which is being elevated by its lead screw. Guide A starts from rest with pin P at the lowest point in the circular slot, and accelerates upward at a constant rate until it reaches a speed of 175 mm/s at the halfway point of its vertical displacement. The guide then decelerates at a constant rate and comes to a stop with pin P at the uppermost point in the circular slot. Determine the n- and t-components of acceleration of pin P once the pin has traveled 30° around the slot from the starting position.

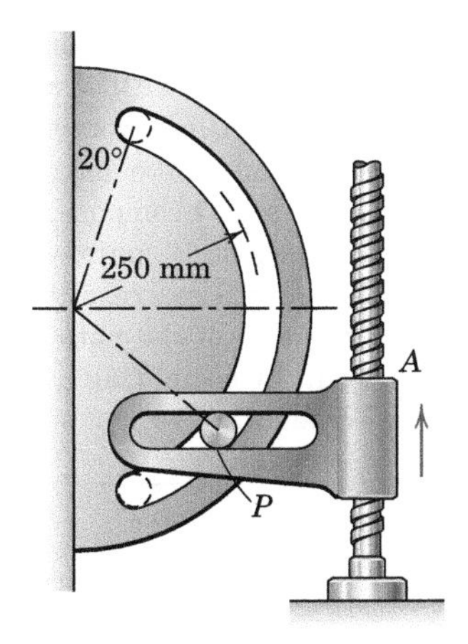

Problem 2/125

2/126 An earth satellite which moves in the elliptical equatorial orbit shown has a velocity v in space of 17 970 km/h when it passes the end of the semi-minor axis at A. The earth has an absolute surface value of g of 9.821 m/s^2 and has a radius of 6371 km. Determine the radius of curvature ρ of the orbit at A.

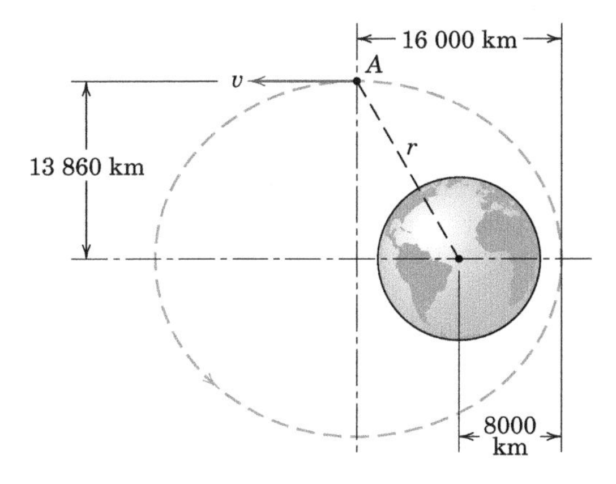

Problem 2/126

2/127 In the design of a control mechanism, the vertical slotted guide is moving with a constant velocity $\dot{x} = 15$ in./sec during the interval of motion from $x = -8$ in. to $x = +8$ in. For the instant when $x = 6$ in., calculate the n- and t-components of acceleration of the pin P, which is confined to move in the parabolic slot. From these results, determine the radius of curvature ρ of the path at this position. Verify your result by computing ρ from the expression cited in Appendix C/10.

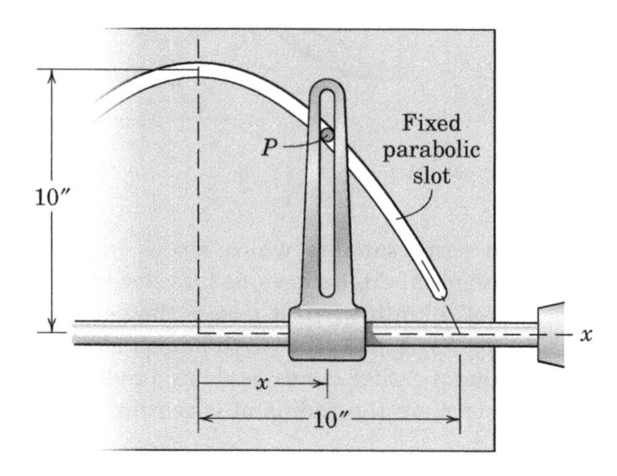

Problem 2/127

▶2/128 In a handling test, a car is driven through the slalom course shown. It is assumed that the car path is sinusoidal and that the maximum lateral acceleration is 0.7g. If the testers wish to design a slalom through which the maximum speed is 80 km/h, what cone spacing L should be used?

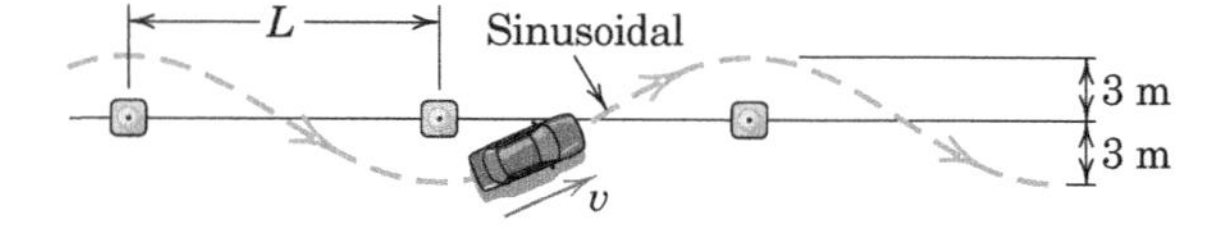

Problem 2/128

▶2/129 A particle which moves with curvilinear motion has coordinates in meters which vary with time t in seconds according to $x = 2t^2 + 3t - 1$ and $y = 5t - 2$. Determine the coordinates of the center of curvature C at time $t = 1$ s.

*2/130 A projectile is launched at time $t = 0$ with the initial conditions shown in the figure. If the wind imparts a constant leftward acceleration of 5 m/s^2, plot the n- and t-components of acceleration and the radius of curvature ρ of the trajectory for the time the projectile is in the air. State the maximum magnitude of each acceleration component along with the time at which it occurs. Additionally, determine the minimum radius of curvature for the trajectory and its corresponding time.

Problem 2/130

PROBLEMS

Introductory Problems

2/131 A car P travels along a straight road with a constant speed $v = 65$ mi/hr. At the instant when the angle $\theta = 60°$, determine the values of $\dot{r}$ in ft/sec and $\dot{\theta}$ in deg/sec.

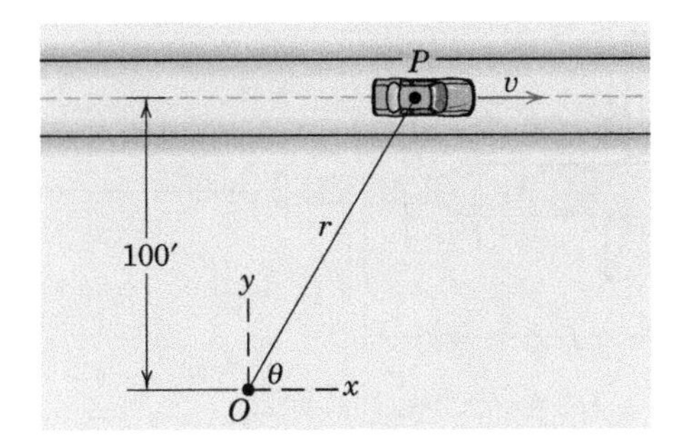

Problem 2/131

2/132 The sprinter begins from rest at position A and accelerates along the track. If the stationary tracking camera at O is rotating counterclockwise at the rate of 12.5 deg/s when the sprinter passes the 60-m mark, determine the speed v of the sprinter and the value of $\dot{r}$.

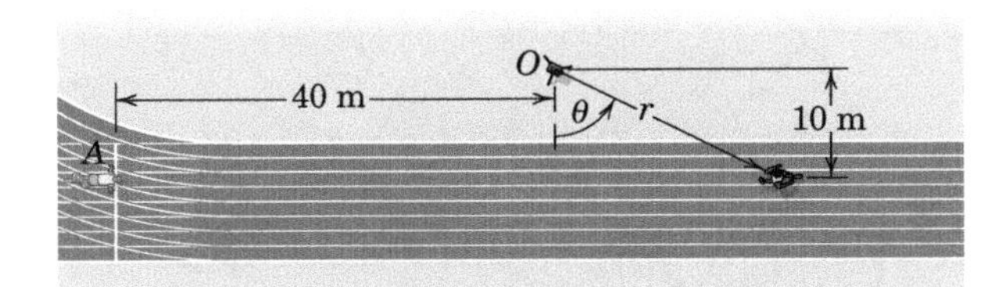

Problem 2/132

2/133 A drone flies over an observer O with constant speed in a straight line as shown. Determine the signs (plus, minus, or zero) for r, $\dot{r}$, $\ddot{r}$, θ, $\dot{\theta}$, and $\ddot{\theta}$ for each of the positions A, B, and C.

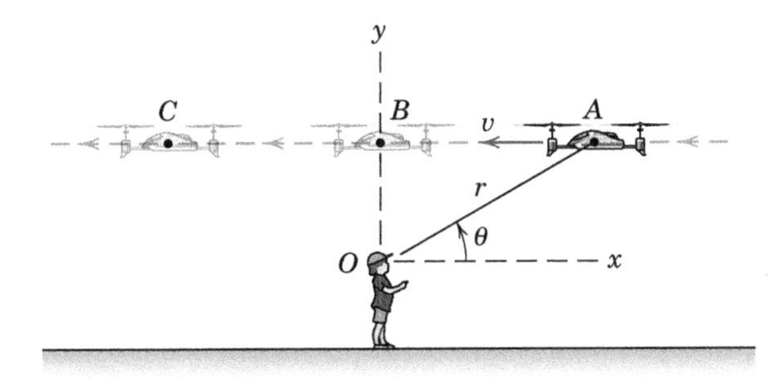

Problem 2/133

2/134 Motion of the sliding block P in the rotating radial slot is controlled by the power screw as shown. For the instant represented, $\dot{\theta} = 0.1$ rad/s, $\ddot{\theta} = -0.04$ rad/s^2, and $r = 300$ mm. Also, the screw turns at a constant speed giving $\dot{r} = 40$ mm/s. For this instant, determine the magnitudes of the velocity $\mathbf{v}$ and acceleration $\mathbf{a}$ of P. Sketch $\mathbf{v}$ and $\mathbf{a}$ if $\theta = 120°$.

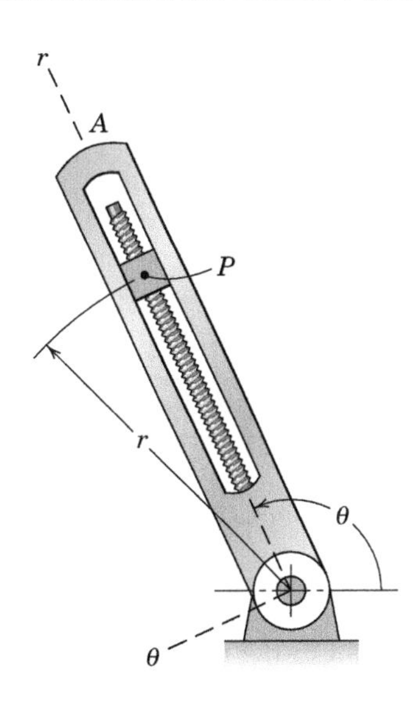

Problem 2/134

2/135 Rotation of bar OA is controlled by the lead screw which imparts a horizontal velocity v to collar C and causes pin P to travel along the smooth slot. Determine the values of $\dot{r}$ and $\dot{\theta}$, where $r = \overline{OP}$, if $h = 160$ mm, $x = 120$ mm, and $v = 25$ mm/s at the instant represented.

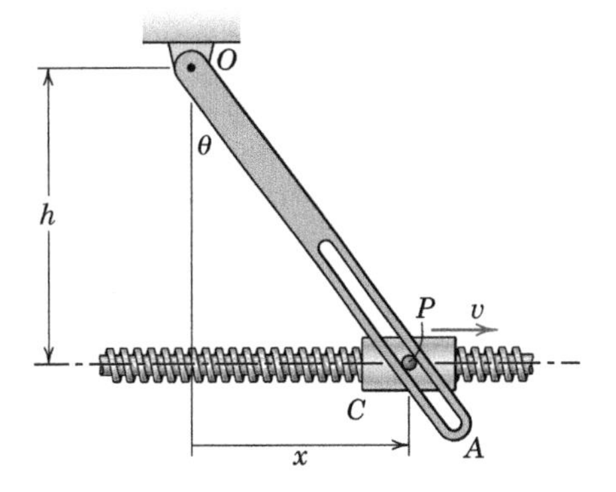

Problem 2/135

2/136 For the bar of Prob. 2/135, determine the values of $\ddot{r}$ and $\ddot{\theta}$ if the velocity of collar C is decreasing at a rate of 5 mm/s^2 at the instant in question. Refer to the printed answers for Prob. 2/135 as needed.

2/137 The boom OAB pivots about point O, while section AB simultaneously extends from within section OA. Determine the velocity and acceleration of the center B of the pulley for the following conditions: $\theta = 20°$, $\dot{\theta} = 5$ deg/sec, $\ddot{\theta} = 2$ deg/sec^2, $l = 7$ ft, $\dot{l} = 1.5$ ft/sec, $\ddot{l} = -4$ ft/sec^2. The quantities $\dot{l}$ and $\ddot{l}$ are the first and second time derivatives, respectively, of the length l of section AB.

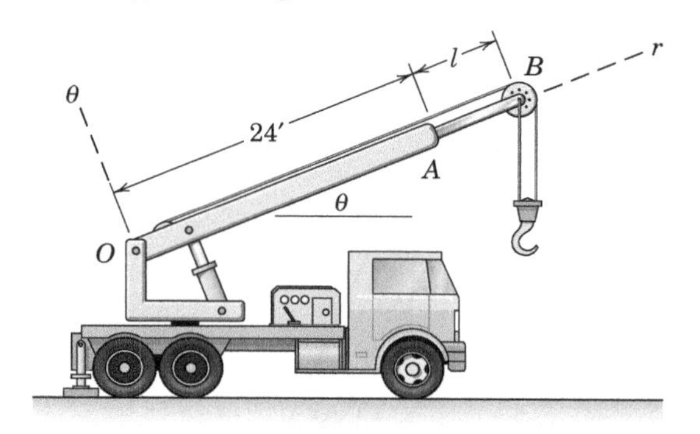

Problem 2/137

2/138 A particle moving along a plane curve has a position vector $\mathbf{r}$, a velocity $\mathbf{v}$, and an acceleration $\mathbf{a}$. Unit vectors in the r- and θ-directions are $\mathbf{e}_r$ and $\mathbf{e}_\theta$, respectively, and both r and θ are changing with time. Explain why each of the following statements is correctly marked as an inequality.

$\dot{\mathbf{r}} \neq v$ $\quad \ddot{\mathbf{r}} \neq a \quad \dot{\mathbf{r}} \neq \dot{r}\mathbf{e}_r$

$\dot{r} \neq v$ $\quad \ddot{r} \neq a \quad \ddot{\mathbf{r}} \neq \ddot{r}\mathbf{e}_r$

$\dot{r} \neq \mathbf{v}$ $\quad \ddot{r} \neq \mathbf{a} \quad \dot{\mathbf{r}} \neq r\dot{\theta}\mathbf{e}_\theta$

2/139 Consider the portion of an excavator shown. At the instant under consideration, the hydraulic cylinder is extending at a rate of 6 in./sec, which is decreasing at the rate of 2 in./sec every second. Simultaneously, the cylinder is rotating about a horizontal axis through O at a constant rate of 10 deg/sec. Determine the velocity $\mathbf{v}$ and acceleration $\mathbf{a}$ of the clevis attachment at B.

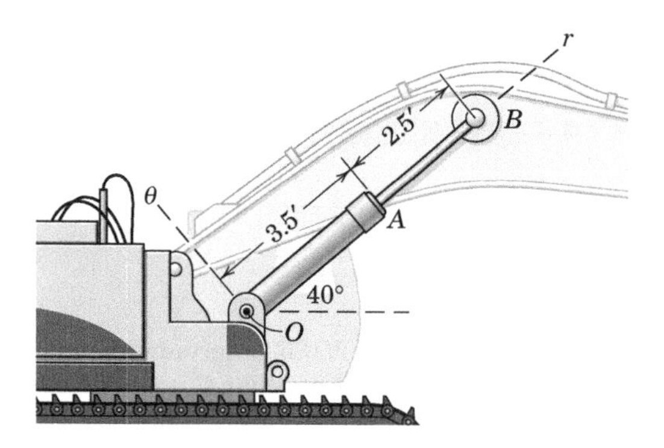

Problem 2/139

2/140 The nozzle shown rotates with constant angular speed Ω about a fixed horizontal axis through point O. Because of the change in diameter by a factor of 2, the water speed relative to the nozzle at A is v, while that at B is $4v$. The water speeds at both A and B are constant. Determine the velocity and acceleration of a water particle as it passes (*a*) point A and (*b*) point B.

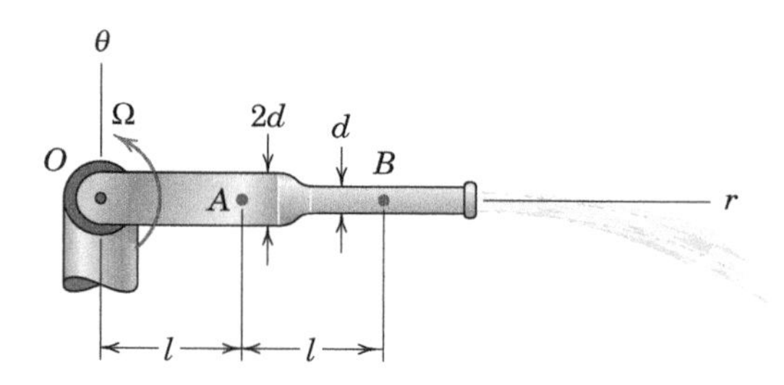

Problem 2/140

2/141 The radial position of a fluid particle P in a certain centrifugal pump with radial vanes is approximated by $r = r_0 \cosh Kt$, where t is time and $K = \dot{\theta}$ is the constant angular rate at which the impeller turns. Determine the expression for the magnitude of the total acceleration of the particle just prior to leaving the vane in terms of r_0, R, and K.

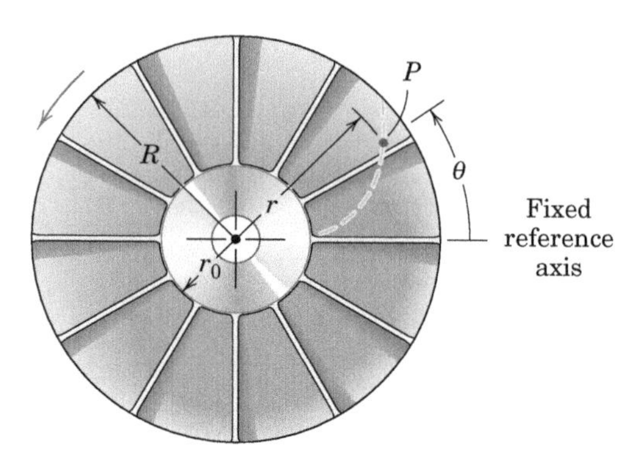

Problem 2/141

2/142 A helicopter starts from rest at point A and travels along the straight-line path with a constant acceleration a. If the speed $v = 28$ m/s when the altitude of the helicopter is $h = 40$ m, determine the values of $\dot{r}$, $\ddot{r}$, $\dot{\theta}$, and $\ddot{\theta}$ as measured by the tracking device at O. At this instant, $\theta = 40°$, and the distance $d = 160$ m. Neglect the small height of the tracking device above the ground.

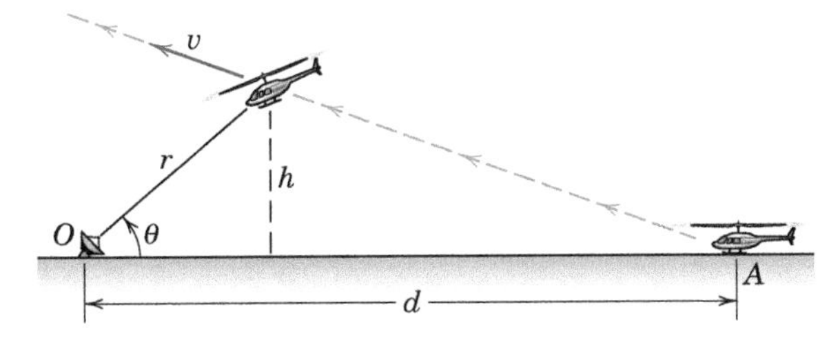

Problem 2/142

2/143 The slider P can be moved inward by means of the string S, while the slotted arm rotates about point O. The angular position of the arm is given by $\theta = 0.8t - \dfrac{t^2}{20}$, where θ is in radians and t is in seconds. The slider is at $r = 1.6$ m when $t = 0$ and thereafter is drawn inward at the constant rate of 0.2 m/s. Determine the magnitude and direction (expressed by the angle α relative to the x-axis) of the velocity and acceleration of the slider when $t = 4$ s.

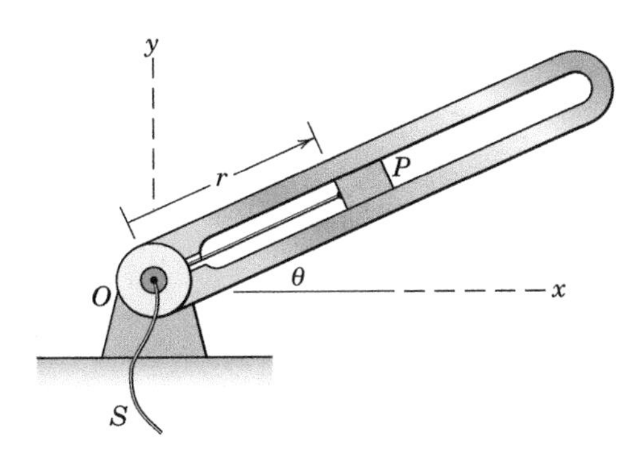

Problem 2/143

Representative Problems

2/144 Cars A and B are both moving with constant speed v on the straight and level highway. They are side-by-side in adjacent lanes as shown. If the radar unit attached to the stationary police car P measures "line-of-sight" velocity, what speed v' will be observed for each car? Use the values $v = 70$ mi/hr, $L = 200$ ft, and $D = 22$ ft.

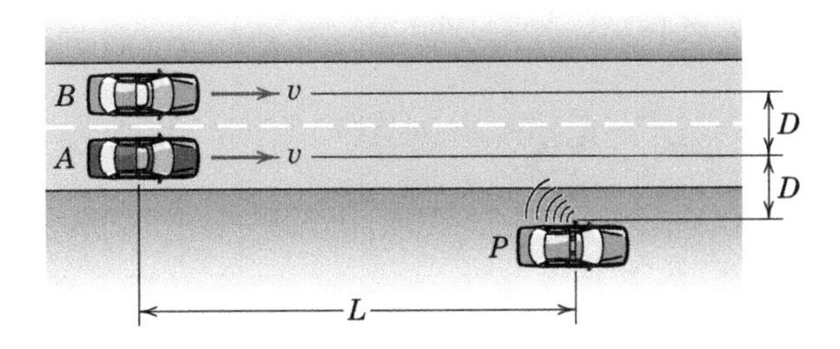

Problem 2/144

2/145 A fireworks shell P is launched upward from point A and explodes at its apex at an altitude of 275 ft. Relative to an observer at O, determine the values of $\dot{r}$ and $\dot{\theta}$ when the shell reaches an altitude $y = 175$ ft. Neglect aerodynamic drag.

Problem 2/145

2/146 For the fireworks shell of Prob. 2/145, determine the values of $\ddot{r}$ and $\ddot{\theta}$ when the shell reaches an altitude $y = 175$ ft. Refer to the printed answers for Prob. 2/145 as needed.

2/147 The rocket is fired vertically and tracked by the radar station shown. When θ reaches 60°, other corresponding measurements give the values $r = 9$ km, $\ddot{r} = 21$ m/s^2, and $\dot{\theta} = 0.02$ rad/s. Calculate the magnitudes of the velocity and acceleration of the rocket at this position.

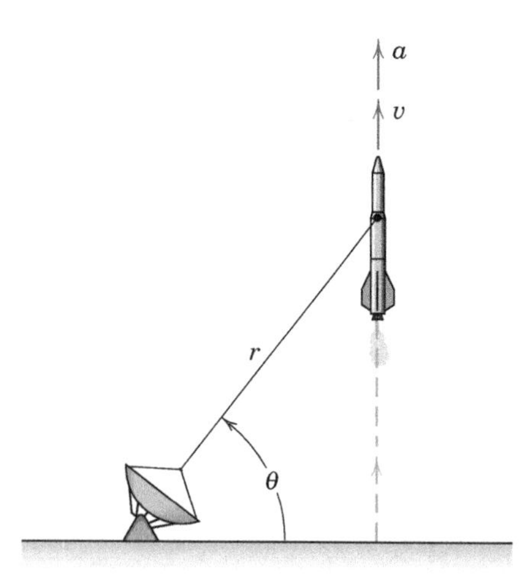

Problem 2/147

2/148 As it passes the position shown, the particle P has a constant speed $v = 100$ m/s along the straight line shown. Determine the corresponding values of $\dot{r}$, $\dot{\theta}$, $\ddot{r}$, and $\ddot{\theta}$.

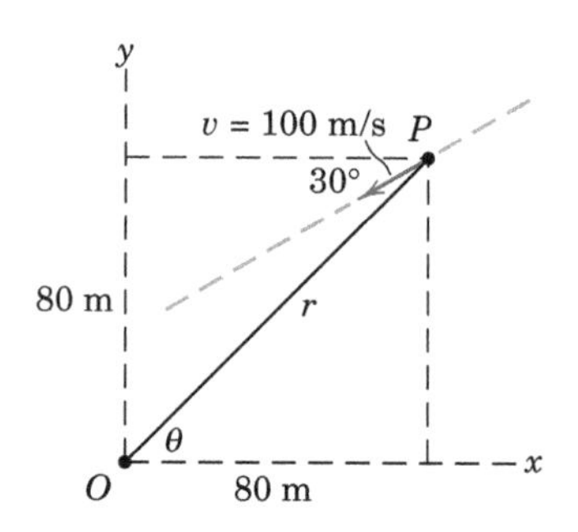

Problem 2/148

2/149 Repeat Prob. 2/148, but now the speed of the particle P is decreasing at the rate of 20 m/s^2 as it moves along the indicated straight path.

***2/150** The diver leaves the platform with an initial upward speed of 2.5 m/s. A stationary camera on the ground is programmed to track the diver throughout the dive by rotating the lens to keep the diver centered in the captured image. Plot $\dot{\theta}$ and $\ddot{\theta}$ as functions of time for the camera over the entire dive and state the values of $\dot{\theta}$ and $\ddot{\theta}$ at the instant the diver enters the water. Treat the diver as a particle which has only vertical motion. Additionally, state the maximum magnitudes of $\dot{\theta}$ and $\ddot{\theta}$ during the dive and the times at which they occur.

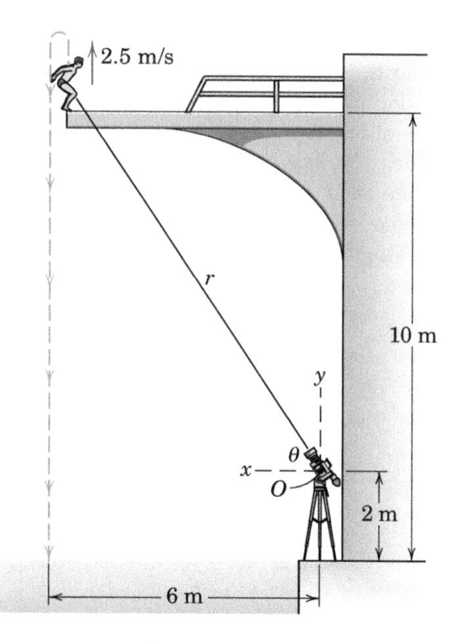

Problem 2/150

2/151 Instruments located at O are part of the ground-traffic control system for a major airport. At a certain instant during the takeoff roll of the aircraft P, the sensors indicate the angle $\theta = 50°$ and the range rate $\dot{r} = 140$ ft/sec. Determine the corresponding speed v of the aircraft and the value of $\dot{\theta}$.

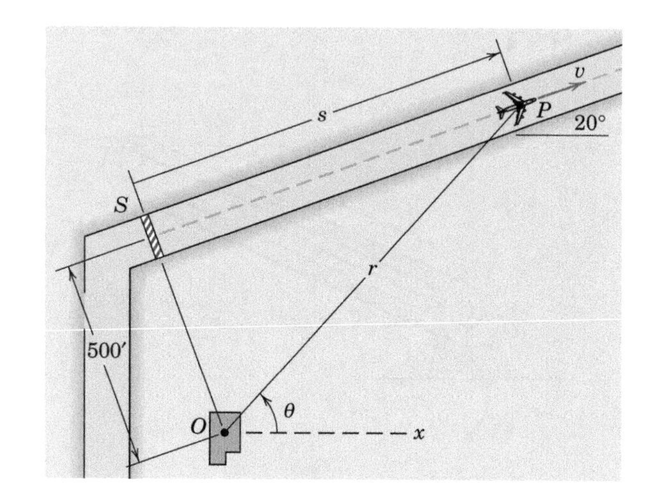

Problem 2/151

2/152 In addition to the information supplied in the previous problem, the sensors at O indicate that $\ddot{r} = 14$ ft/sec^2. Determine the corresponding acceleration a of the aircraft and the value of $\ddot{\theta}$.

2/153 At the bottom of a loop in the vertical (r-θ) plane at an altitude of 400 m, the airplane P has a horizontal velocity of 600 km/h and no horizontal acceleration. The radius of curvature of the loop is 1200 m. For the radar tracking at O, determine the recorded values of $\ddot{r}$ and $\ddot{\theta}$ for this instant.

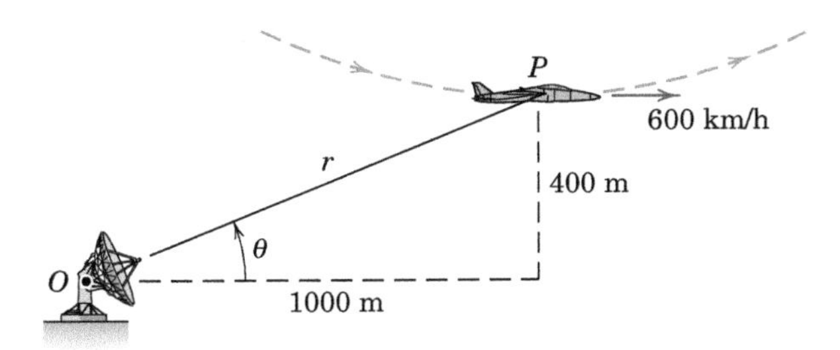

Problem 2/153

2/154 The member OA of the industrial robot telescopes and pivots about the fixed axis at point O. At the instant shown, $\theta = 60°$, $\dot{\theta} = 1.2$ rad/s, $\ddot{\theta} = 0.8$ rad/s^2, $\overline{OA} = 0.9$ m, $\dot{\overline{OA}} = 0.5$ m/s, and $\ddot{\overline{OA}} = -6$ m/s^2. Determine the magnitudes of the velocity and acceleration of joint A of the robot. Also, sketch the velocity and acceleration of A and determine the angles which these vectors make with the positive x-axis. The base of the robot does not revolve about a vertical axis.

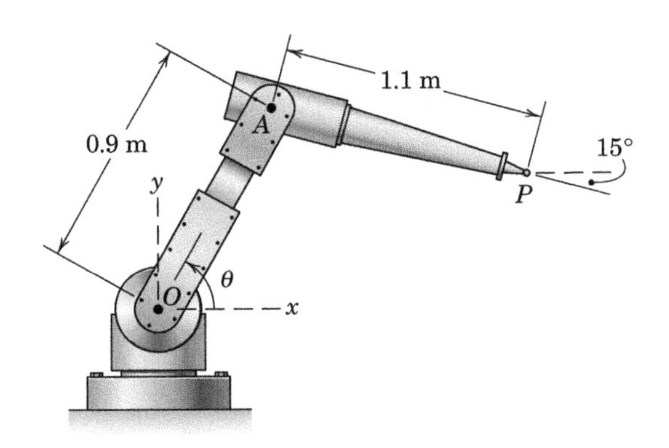

Problem 2/154

2/155 At the instant depicted in the figure, the radar station at O measures the range rate of the space shuttle P to be $\dot{r} = -12{,}272$ ft/sec, with O considered fixed. If it is known that the shuttle is in a circular orbit at an altitude $h = 150$ mi, determine the orbital speed of the shuttle from this information.

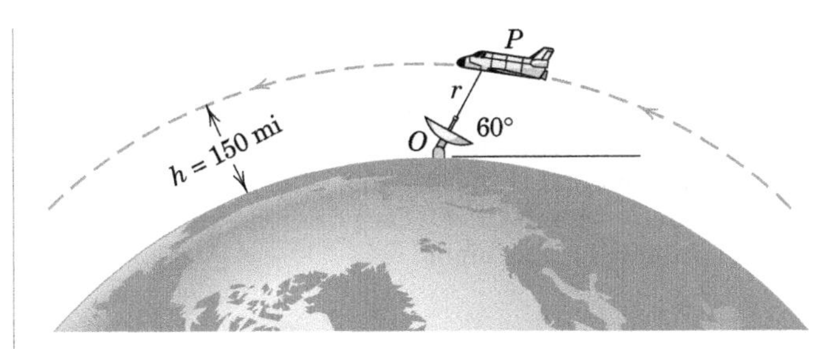

Problem 2/155

2/156 A locomotive is traveling on the straight and level track with a speed $v = 90$ km/h and a deceleration $a = 0.5$ m/s^2 as shown. Relative to the fixed observer at O, determine the quantities $\dot{r}$, $\ddot{r}$, $\dot{\theta}$, and $\ddot{\theta}$ at the instant when $\theta = 60°$ and $r = 400$ m.

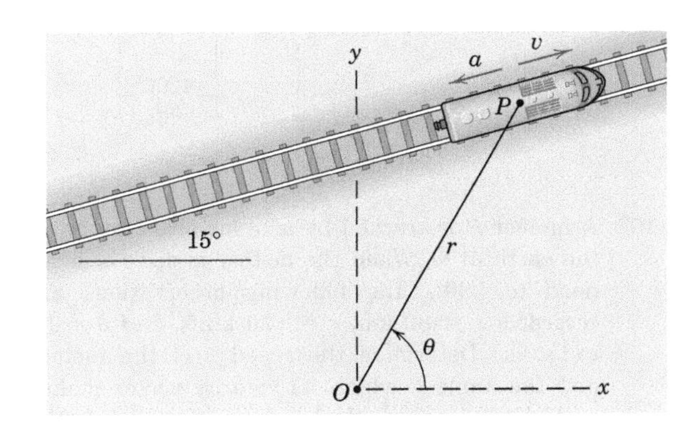

Problem 2/156

2/157 The small block P starts from rest at time $t = 0$ at point A and moves up the incline with constant acceleration a. Determine $\dot{r}$ as a function of time.

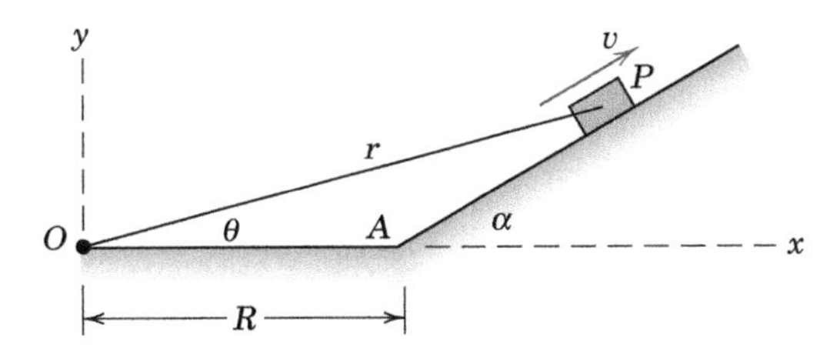

Problem 2/157

2/158 For the conditions of Prob. 2/157, determine $\dot{\theta}$ as a function of time.

2/159 An earth satellite traveling in the elliptical orbit shown has a velocity $v = 12{,}149$ mi/hr as it passes the end of the semiminor axis at A. The acceleration of the satellite at A is due to gravitational attraction and is $32.23[3959/8400]^2 = 7.159$ ft/sec^2 directed from A to O. For position A calculate the values of $\dot{r}$, $\ddot{r}$, $\dot{\theta}$, and $\ddot{\theta}$.

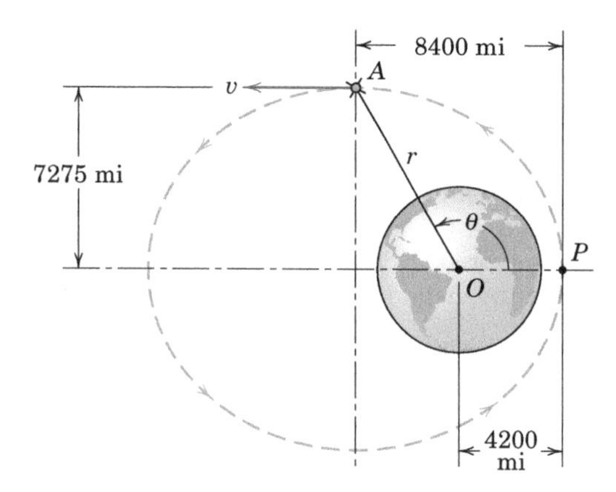

Problem 2/159

2/160 A meteor P is tracked by a radar observatory on the earth at O. When the meteor is directly overhead ($\theta = 90°$), the following observations are recorded: $r = 80$ km, $\dot{r} = -20$ km/s, and $\dot{\theta} = 0.4$ rad/s. (a) Determine the speed v of the meteor and the angle β which its velocity vector makes with the horizontal. Neglect any effects due to the earth's rotation. (b) Repeat with all given quantities remaining the same, except that $\theta = 75°$.

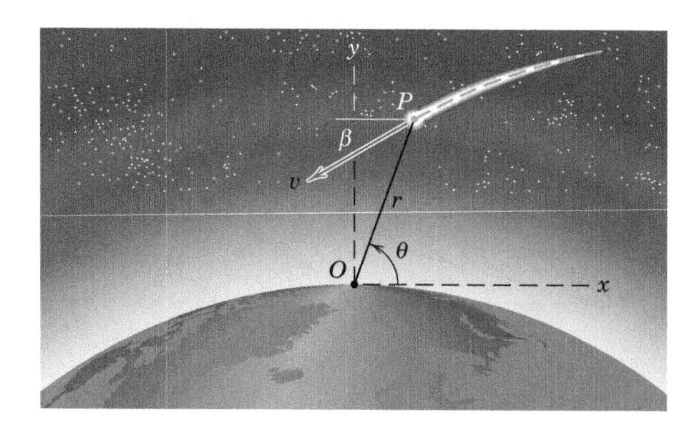

Problem 2/160

2/161 The low-flying aircraft P is traveling at a constant speed of 360 km/h in the holding circle of radius 3 km. For the instant shown, determine the quantities r, $\dot{r}$, $\ddot{r}$, θ, $\dot{\theta}$, and $\ddot{\theta}$ relative to the fixed x-y coordinate system, which has its origin on a mountaintop at O. Treat the system as two-dimensional.

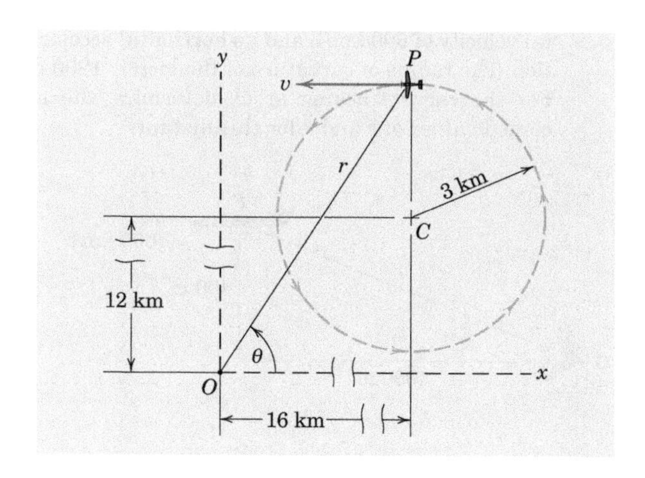

Problem 2/161

▶2/162 At time $t = 0$, the baseball player releases a ball with the initial conditions shown in the figure. Determine the quantities r, $\dot{r}$, $\ddot{r}$, θ, $\dot{\theta}$, and $\ddot{\theta}$, all relative to the x-y coordinate system shown, at time $t = 0.5$ sec.

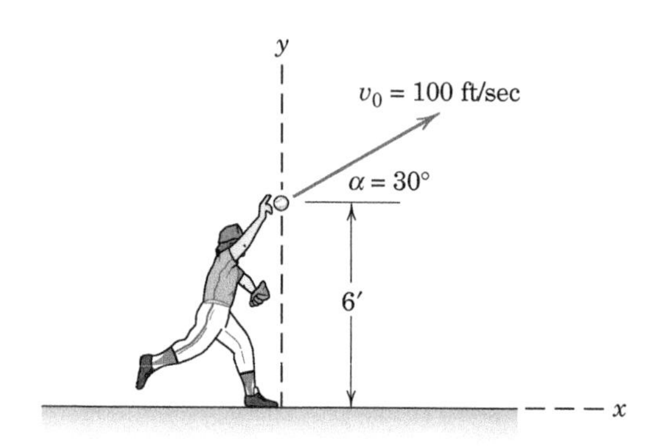

Problem 2/162

►**2/163** The racing airplane is beginning an inside loop in the vertical plane. The tracking station at O records the following data for a particular instant: $r = 90$ m, $\dot{r} = 15.5$ m/s, $\ddot{r} = 74.5$ m/s^2, $\theta = 30°$, $\dot{\theta} = 0.53$ rad/s, and $\ddot{\theta} = -0.29$ rad/s^2. Determine the values of v, $\dot{v}$, ρ, and β at this instant.

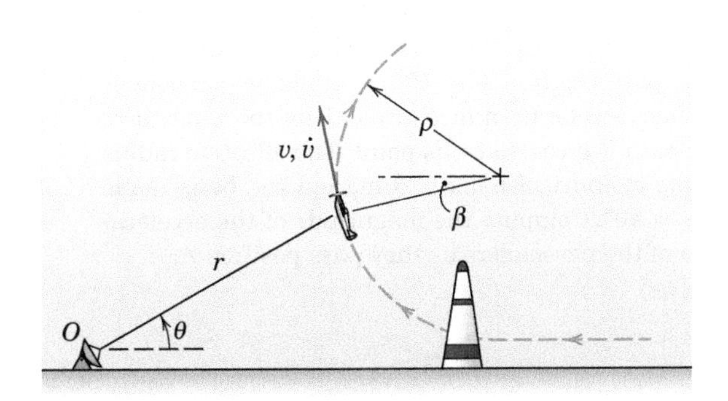

Problem 2/163

►**2/164** A golf ball is driven with the initial conditions shown in the figure. If the wind imparts a constant horizontal deceleration of 4 ft/sec^2, determine the values of r, $\dot{r}$, $\ddot{r}$, θ, $\dot{\theta}$, and $\ddot{\theta}$ when $t = 1.05$ sec. Take the r-coordinate to be measured from the origin.

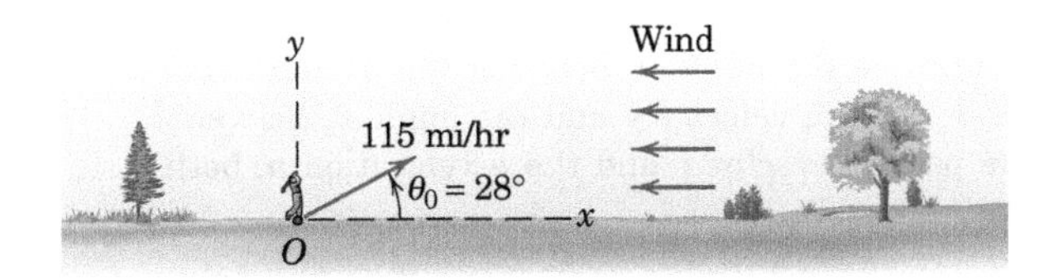

Problem 2/164

PROBLEMS

Introductory Problems

2/165 The rectangular coordinates of a particle are given in millimeters as functions of time t in seconds by $x = 30 \cos 2t$, $y = 40 \sin 2t$, and $z = 20t + 3t^2$. Determine the angle θ_1 between the position vector $\mathbf{r}$ and the velocity $\mathbf{v}$ and the angle θ_2 between the position vector $\mathbf{r}$ and the acceleration $\mathbf{a}$, both at time $t = 2$ s.

2/166 A projectile is launched from point O with an initial velocity of magnitude $v_0 = 600$ ft/sec, directed as shown in the figure. Compute the x-, y-, and z-components of position, velocity, and acceleration 20 seconds after launch. Neglect aerodynamic drag.

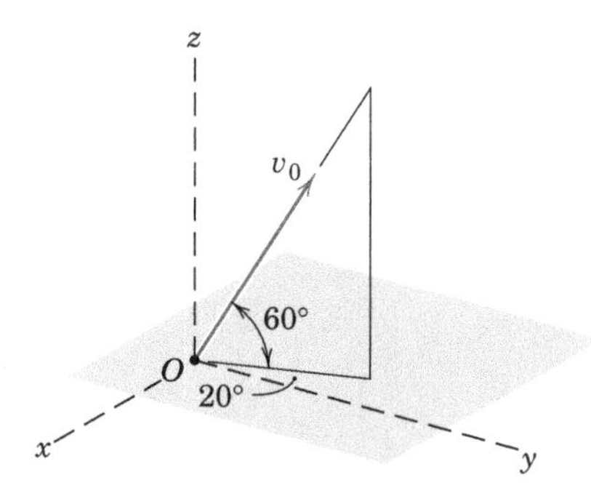

Problem 2/166

2/167 A projectile is launched from point O at a speed $v_0 = 80$ m/s with the goal of hitting the target A. At the launch instant, a strong horizontal wind begins blowing and imparts a constant acceleration of 1.25 m/s^2 to the projectile in the same direction as the wind. If the launch conditions are chosen so that the particle would impact the target in the absence of wind effects, determine the impact coordinates of the projectile. (*Note:* There is more than one solution to this problem, so choose only answers which feature positive angles less than 90°. Select the impact location which is closest to the target.)

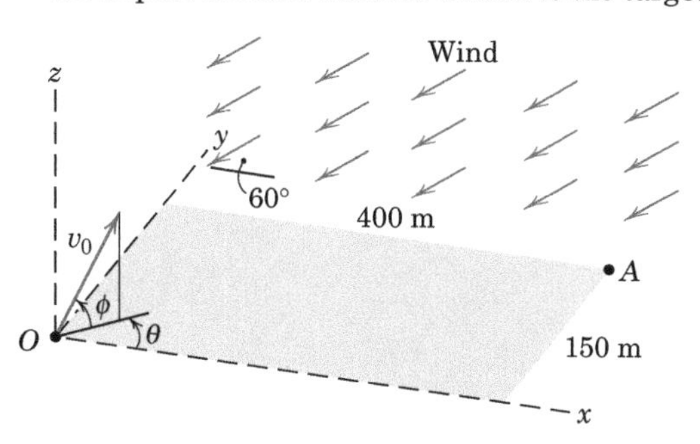

Problem 2/167

2/168 If the launch speed of the projectile in Prob. 2/167 remains unchanged, what values of θ and ϕ (positive and less than 90°) will ensure that the projectile impacts the target at A if the wind conditions are considered. Please list both possible combinations of the angles.

2/169 An amusement ride called the "corkscrew" takes the passengers through the upside-down curve of a horizontal cylindrical helix. The velocity of the cars as they pass position A is 15 m/s, and the component of their acceleration measured along the tangent to the path is $g \cos \gamma$ at this point. The effective radius of the cylindrical helix is 5 m, and the helix angle is $\gamma = 40°$. Compute the magnitude of the acceleration of the passengers as they pass position A.

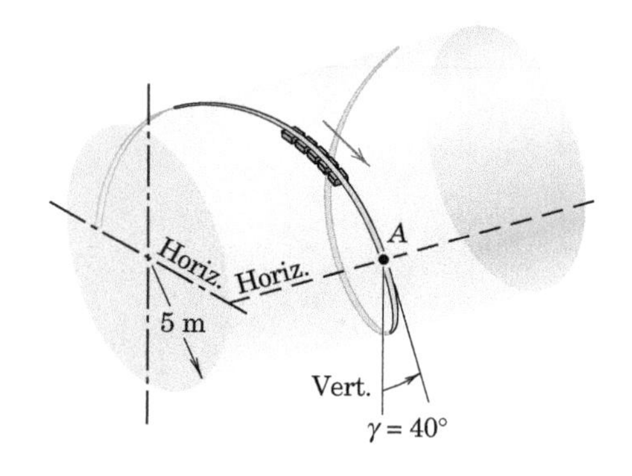

Problem 2/169

2/170 The radar antenna at P tracks the jet aircraft A, which is flying horizontally at a speed u and an altitude h above the level of P. Determine the expressions for the components of the velocity in the spherical coordinates of the antenna motion.

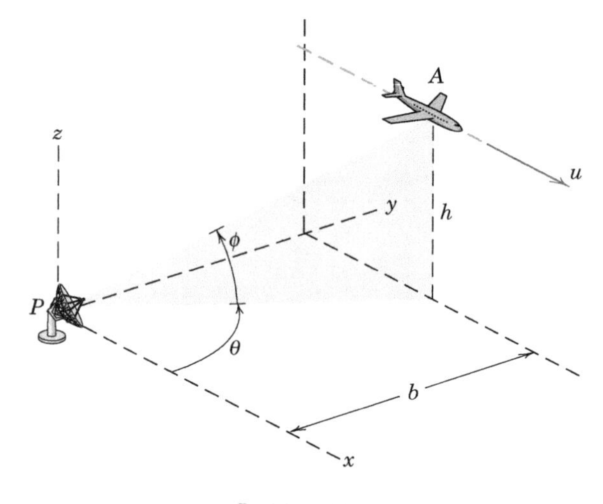

Problem 2/170

2/171 The rotating element in a mixing chamber is given a periodic axial movement $z = z_0 \sin 2\pi nt$ while it is rotating at the constant angular velocity $\dot{\theta} = \omega$. Determine the expression for the maximum magnitude of the acceleration of a point A on the rim of radius r. The frequency n of vertical oscillation is constant.

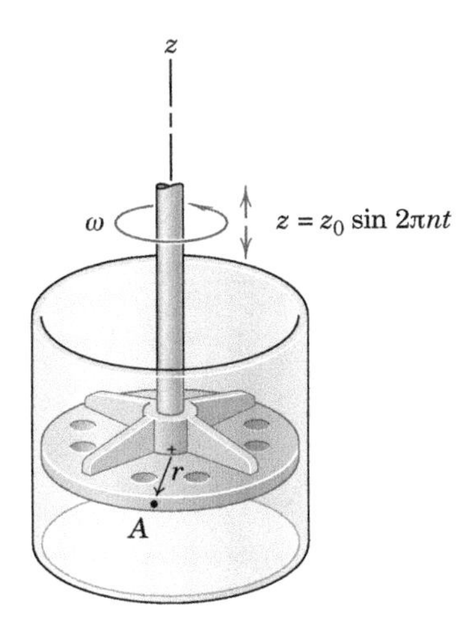

Problem 2/171

Representative Problems

2/172 A helicopter starts from rest at point A and travels along the indicated path with a constant acceleration a. If the helicopter has a speed of 60 m/s when it reaches B, determine the values of $\dot{R}$, $\dot{\theta}$, and $\dot{\phi}$ as measured by the radar tracking device at O at the instant when $h = 100$ m.

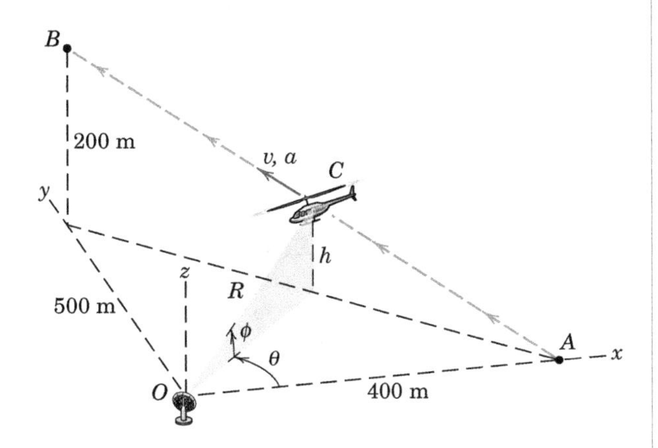

Problem 2/172

2/173 For the helicopter of Prob. 2/172, find the values of $\ddot{R}$, $\ddot{\theta}$, and $\ddot{\phi}$ for the radar tracking device at O at the instant when $h = 100$ m. Refer to the printed answers for Prob. 2/172 as needed.

2/174 The vertical shaft of the industrial robot rotates at the constant rate ω. The length h of the vertical shaft has a known time history, and this is true of its time derivatives $\dot{h}$ and $\ddot{h}$ as well. Likewise, the values of l, $\dot{l}$, and $\ddot{l}$ are known. Determine the magnitudes of the velocity and acceleration of point P. The lengths h_0 and l_0 are fixed.

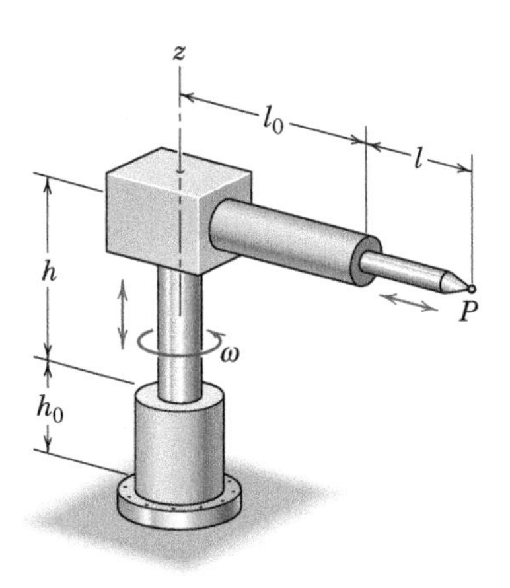

Problem 2/174

2/175 An industrial robot is being used to position a small part P. Calculate the magnitude of the acceleration **a** of P for the instant when $\beta = 30°$ if $\dot{\beta} = 10$ deg/s and $\ddot{\beta} = 20$ deg/s^2 at this same instant. The base of the robot is revolving at the constant rate $\omega = 40$ deg/s. During the motion arms AO and AP remain perpendicular.

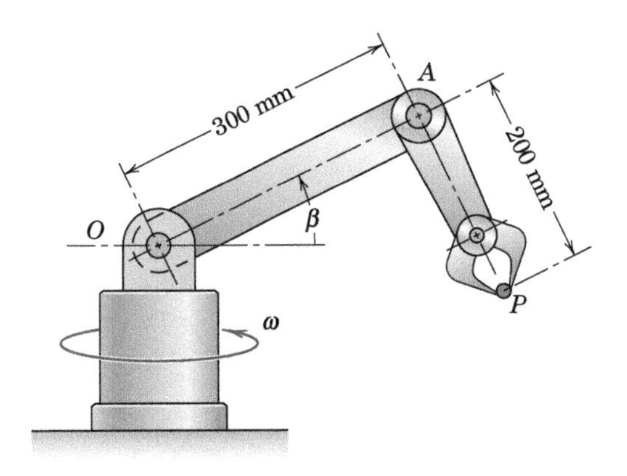

Problem 2/175

2/176 The car A is ascending a parking-garage ramp in the form of a cylindrical helix of 24-ft radius rising 10 ft for each half turn. At the position shown the car has a speed of 15 mi/hr, which is decreasing at the rate of 2 mi/hr per second. Determine the r-, θ-, and z-components of the acceleration of the car.

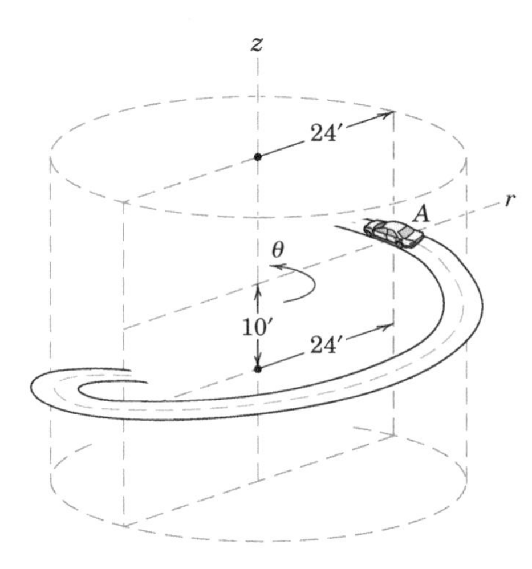

Problem 2/176

2/177 The base structure of the firetruck ladder rotates about a vertical axis through O with a constant angular velocity $\Omega = 10$ deg/s. At the same time, the ladder unit OB elevates at a constant rate $\dot{\phi} = 7$ deg/s, and section AB of the ladder extends from within section OA at the constant rate of 0.5 m/s. At the instant under consideration, $\phi = 30°$, $\overline{OA} = 9$ m, and $\overline{AB} = 6$ m. Determine the magnitudes of the velocity and acceleration of the end B of the ladder.

Problem 2/177

2/178 The rod OA is held at the constant angle $\beta = 30°$ while it rotates about the vertical with a constant angular rate $\dot{\theta} = 120$ rev/min. Simultaneously, the sliding ball P oscillates along the rod with its distance in millimeters from the fixed pivot O given by $R = 200 + 50 \sin 2\pi nt$, where the frequency n of oscillation along the rod is a constant 2 cycles per second and where t is the time in seconds. Calculate the magnitude of the acceleration of P for an instant when its velocity along the rod from O toward A is a maximum.

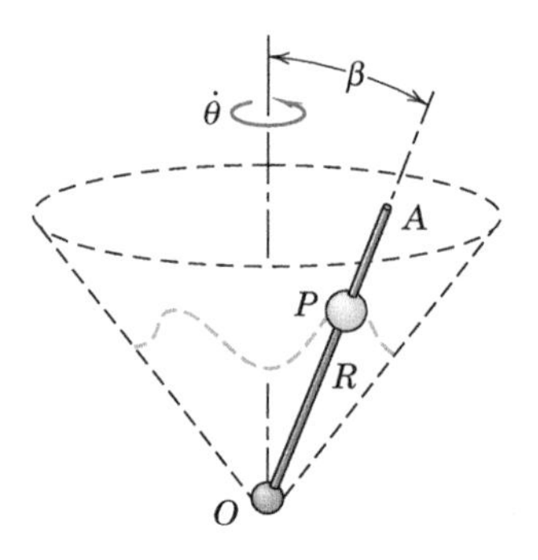

Problem 2/178

▶**2/179** Beginning with Eq. 2/18, the expression for particle velocity in spherical coordinates, derive the acceleration components in Eq. 2/19. (*Note:* Start by writing the unit vectors for the R-, θ-, and ϕ-coordinates in terms of the fixed unit vectors **i**, **j**, and **k**.)

▶**2/180** In the design of an amusement-park ride, the cars are attached to arms of length R which are hinged to a central rotating collar which drives the assembly about the vertical axis with a constant angular rate $\omega = \dot{\theta}$. The cars rise and fall with the track according to the relation $z = (h/2)(1 - \cos 2\theta)$. Find the R-, θ-, and ϕ-components of the velocity $\mathbf{v}$ of each car as it passes the position $\theta = \pi/4$ rad.

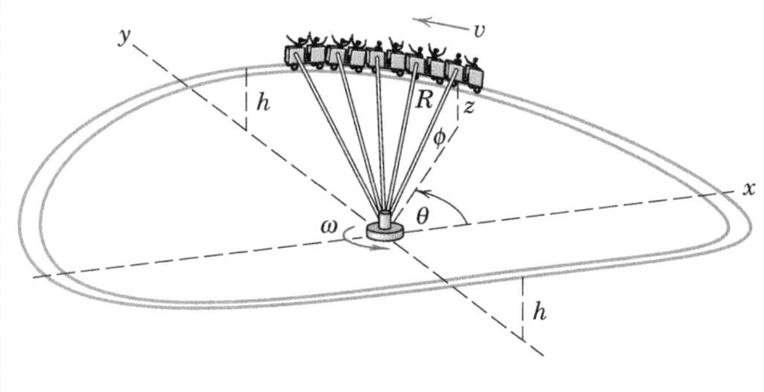

Problem 2/180

▶**2/181** The particle P moves down the spiral path which is wrapped around the surface of a right circular cone of base radius b and altitude h. The angle γ between the tangent to the curve at any point and a horizontal tangent to the cone at this point is constant. Also the motion of the particle is controlled so that $\dot{\theta}$ is constant. Determine the expression for the radial acceleration a_r of the particle for any value of θ.

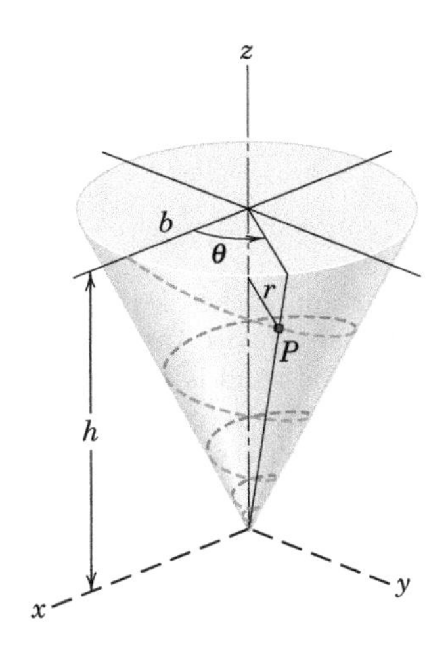

Problem 2/181

▶**2/182** The disk A rotates about the vertical z-axis with a constant speed $\omega = \dot{\theta} = \pi/3$ rad/s. Simultaneously, the hinged arm OB is elevated at the constant rate $\dot{\phi} = 2\pi/3$ rad/s. At time $t = 0$, both $\theta = 0$ and $\phi = 0$. The angle θ is measured from the fixed reference x-axis. The small sphere P slides out along the rod according to $R = 50 + 200t^2$, where R is in millimeters and t is in seconds. Determine the magnitude of the total acceleration $\mathbf{a}$ of P when $t = \frac{1}{2}$ s.

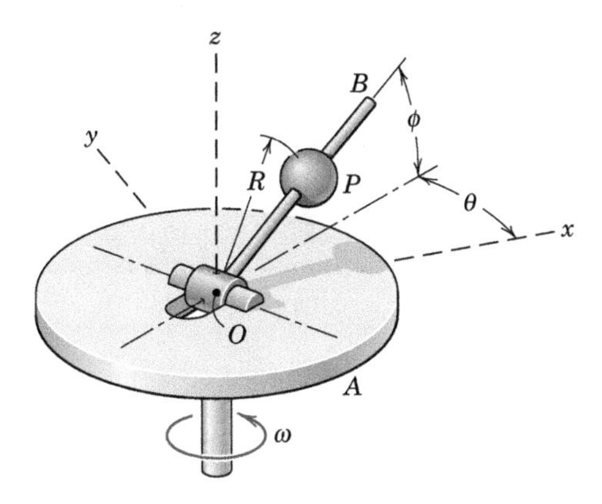

Problem 2/182

PROBLEMS

Introductory Problems

2/183 Rapid-transit trains A and B travel on parallel tracks. Train A has a speed of 80 km/h and is slowing at the rate of 2 m/s^2, while train B has a constant speed of 40 km/h. Determine the velocity and acceleration of train B relative to train A.

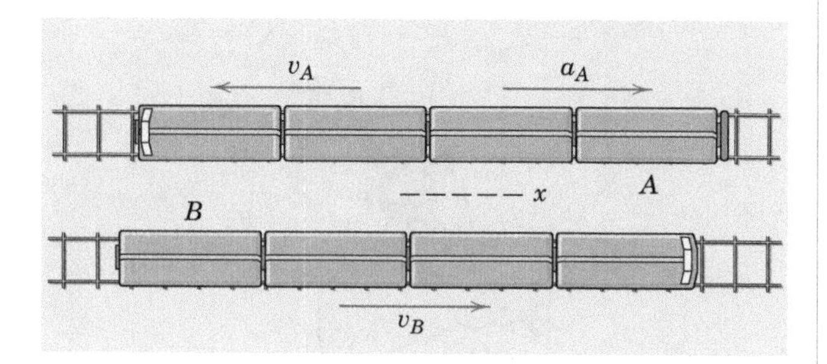

Problem 2/183

2/184 Train A is traveling at a constant speed v_A = 35 mi/hr while car B travels in a straight line along the road as shown at a constant speed v_B. A conductor C in the train begins to walk to the rear of the train car at a constant speed of 4 ft/sec relative to the train. If the conductor perceives car B to move directly westward at 16 ft/sec, how fast is the car traveling?

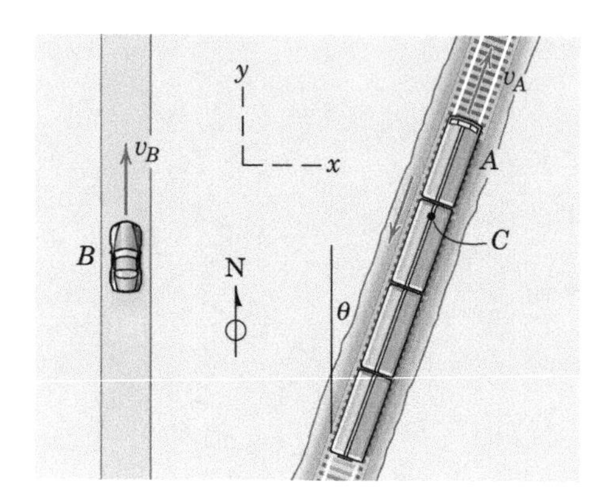

Problem 2/184

2/185 The jet transport B is flying north with a velocity v_B = 600 km/h when a smaller aircraft A passes underneath the transport headed in the 60° direction shown. To passengers in B, however, A appears to be flying sideways and moving east. Determine the actual velocity of A and the velocity which A appears to have relative to B.

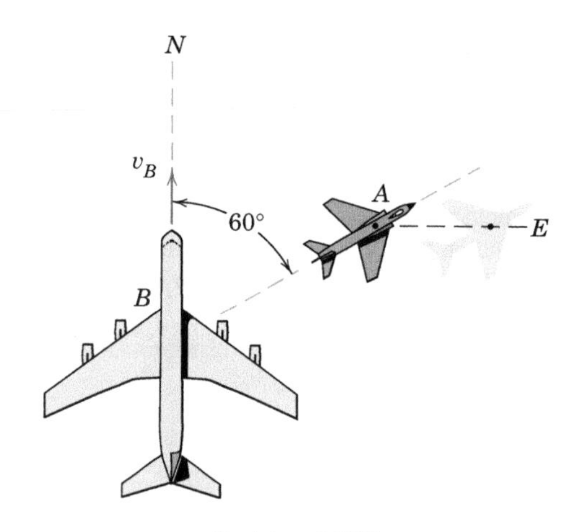

Problem 2/185

2/186 A helicopter approaches a rescue scene. A victim P is drifting along with the river current of speed v_C = 2 m/s. The wind is blowing at a speed v_W = 3 m/s as indicated. Determine the velocity relative to the wind which the helicopter must acquire so that it maintains a steady overhead position relative to the victim.

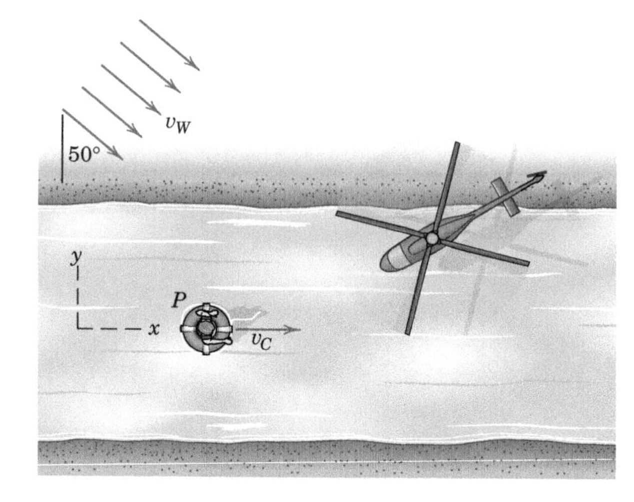

Problem 2/186

2/187 A ship capable of making a speed of 16 knots through still water is to maintain a true course due west while encountering a 3-knot current running from north to south. What should be the heading of the ship (measured clockwise from the north to the nearest degree)? How long does it take the ship to proceed 24 nautical miles due west?

2/188 Train A travels with a constant speed $v_A = 120$ km/h along the straight and level track. The driver of car B, anticipating the railway grade crossing C, decreases the car speed of 90 km/h at the rate of 3 m/s^2. Determine the velocity and acceleration of the train relative to the car.

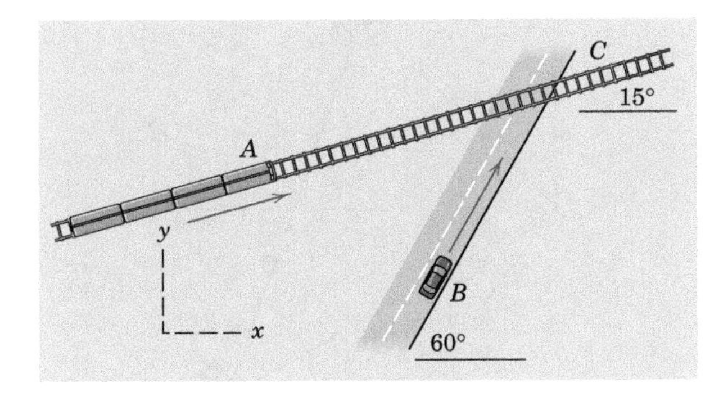

Problem 2/188

2/189 The car A has a forward speed of 18 km/h and is accelerating at 3 m/s^2. Determine the velocity and acceleration of the car relative to observer B, who rides in a nonrotating chair on the Ferris wheel. The angular rate $\Omega = 3$ rev/min of the Ferris wheel is constant.

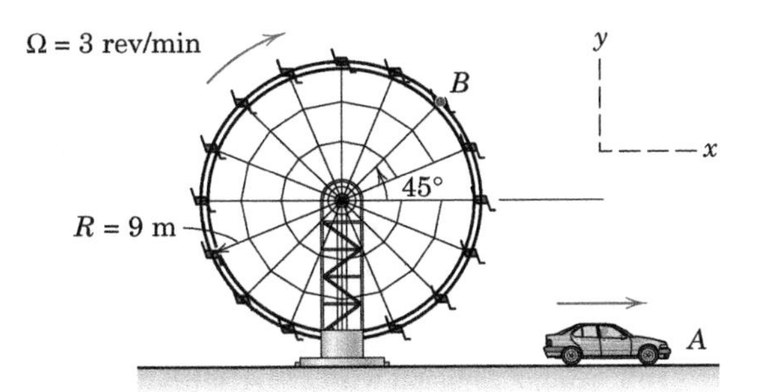

Problem 2/189

2/190 For the instant represented, car A has an acceleration in the direction of its motion, and car B has a speed of 45 mi/hr which is increasing. If the acceleration of B as observed from A is zero for this instant, determine the acceleration of A and the rate at which the speed of B is changing.

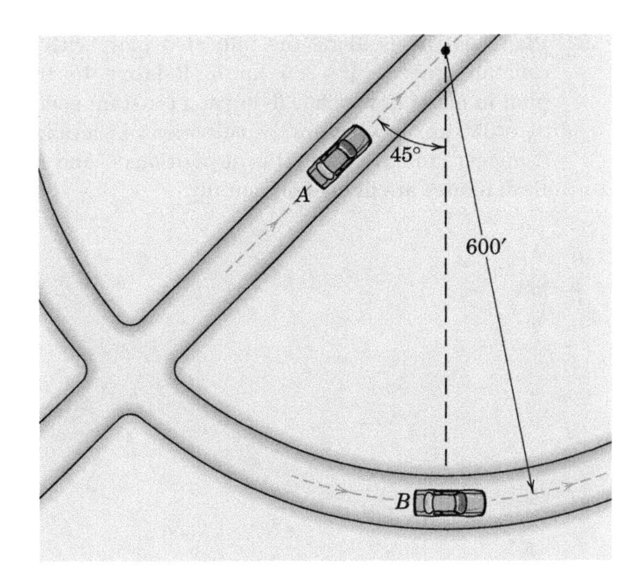

Problem 2/190

Representative Problems

2/191 A drop of water falls with no initial speed from point A of a highway overpass. After dropping 6 m, it strikes the windshield at point B of a car which is traveling at a speed of 100 km/h on the horizontal road. If the windshield is inclined 50° from the vertical as shown, determine the angle θ relative to the normal n to the windshield at which the water drop strikes.

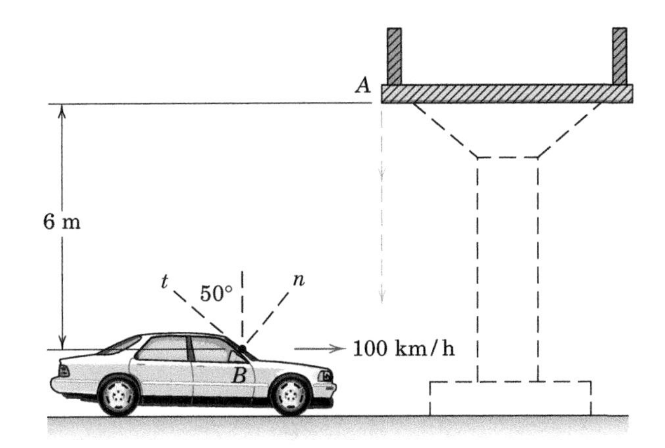

Problem 2/191

2/192 Plane A travels along the indicated path with a constant speed $v_A = 285$ km/h. Relative to the pilot in plane B, which is flying at a constant speed $v_B = 350$ km/h, what are the velocities which plane A appears to have when it is at positions C and E? Both planes are flying horizontally.

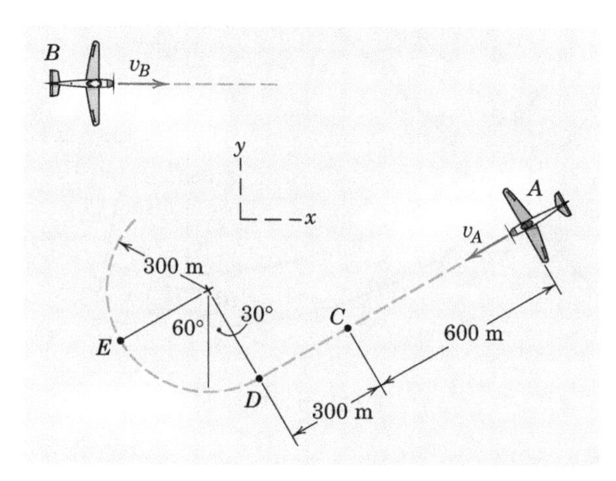

Problem 2/192

2/193 For the planes of Prob. 2/192, beginning at the position shown, plane A increases its speed at a constant rate and acquires a speed of 415 km/h by the time it reaches position E, while plane B experiences a steady deceleration of 1.5 m/s^2. Relative to the pilot in plane B, what are the velocities and accelerations which plane A appears to have when it is at positions C and E?

2/194 A sailboat moving in the direction shown is tacking to windward against a north wind. The log registers a hull speed of 6.5 knots. A "telltale" (light string tied to the rigging) indicates that the direction of the apparent wind is 35° from the centerline of the boat. What is the true wind velocity v_w?

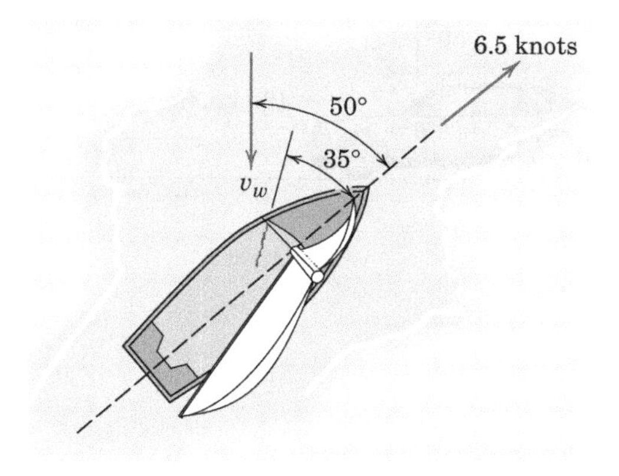

Problem 2/194

2/195 At the instant illustrated, car B has a speed of 30 km/h and car A has a speed of 40 km/h. Determine the values of $\dot{r}$ and $\dot{\theta}$ for this instant where r and θ are measured relative to a longitudinal axis fixed to car B as indicated in the figure.

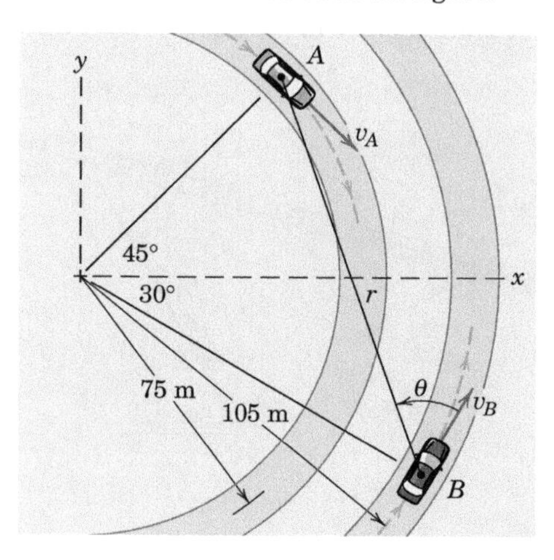

Problem 2/195

2/196 For the cars of Prob. 2/195, determine the instantaneous values of $\ddot{r}$ and $\ddot{\theta}$ if car A is slowing down at a rate of 1.25 m/s^2 and car B is speeding up at a rate of 2.5 m/s^2. Refer to the printed answers for Prob. 2/195 as needed.

2/197 Car A is traveling at 25 mi/hr and applies the brakes at the position shown so as to arrive at the intersection C at a complete stop with a constant deceleration. Car B has a speed of 40 mi/hr at the instant represented and is capable of a maximum deceleration of 18 ft/sec^2. If the driver of car B is distracted, and does not apply his brakes until 1.30 seconds after car A begins to brake, the result being a collision with car A, with what relative speed will car B strike car A? Treat both cars as particles.

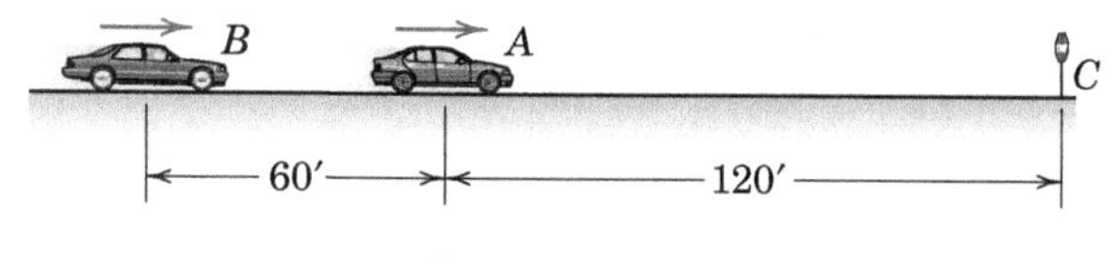

Problem 2/197

2/198 As part of an unmanned-autonomous-vehicle (UAV) demonstration, an unmanned vehicle B launches a projectile A from the position shown while traveling at a constant speed of 30 km/h. The projectile is launched with a speed of 70 m/s relative to the vehicle. At what launch angle α should the projectile be fired to ensure that it strikes a target at C? Compare your answer with that for the case where the vehicle is stationary.

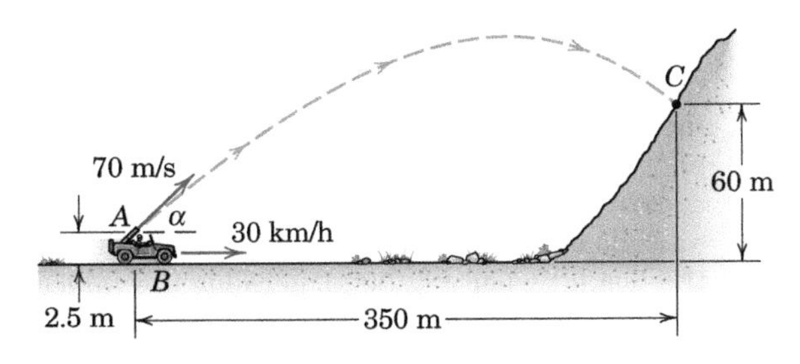

Problem 2/198

2/199 The shuttle orbiter A is in a circular orbit of altitude 200 mi, while spacecraft B is in a geosynchronous circular orbit of altitude 22,300 mi. Determine the acceleration of B relative to a nonrotating observer in shuttle A. Use $g_0 = 32.23$ ft/sec^2 for the surface-level gravitational acceleration and $R = 3959$ mi for the radius of the earth.

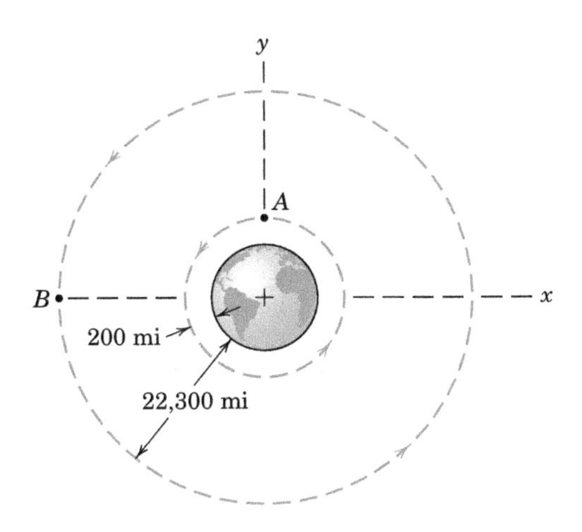

Problem 2/199

2/200 After starting from the position marked with the "x", a football receiver B runs the slant-in pattern shown, making a cut at P and thereafter running with a constant speed $v_B = 7$ yd/sec in the direction shown. The quarterback releases the ball with a horizontal velocity of 100 ft/sec at the instant the receiver passes point P. Determine the angle α at which the quarterback must throw the ball, and the velocity of the ball relative to the receiver when the ball is caught. Neglect any vertical motion of the ball.

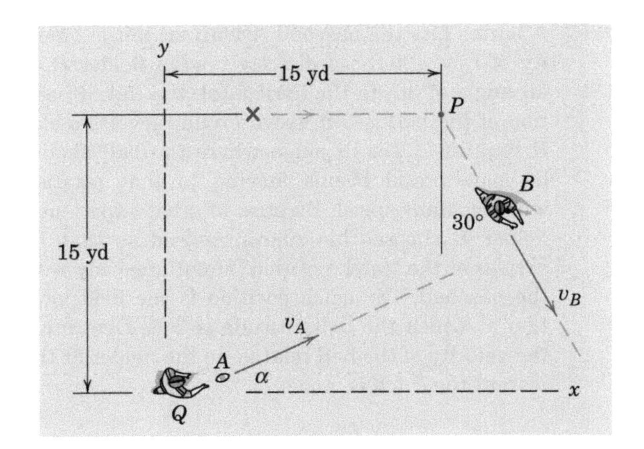

Problem 2/200

2/201 Car A is traveling at the constant speed of 60 km/h as it rounds the circular curve of 300-m radius and at the instant represented is at the position $\theta = 45°$. Car B is traveling at the constant speed of 80 km/h and passes the center of the circle at this same instant. Car A is located with respect to car B by polar coordinates r and θ with the pole moving with B. For this instant determine $v_{A/B}$ and the values of $\dot{r}$ and $\dot{\theta}$ as measured by an observer in car B.

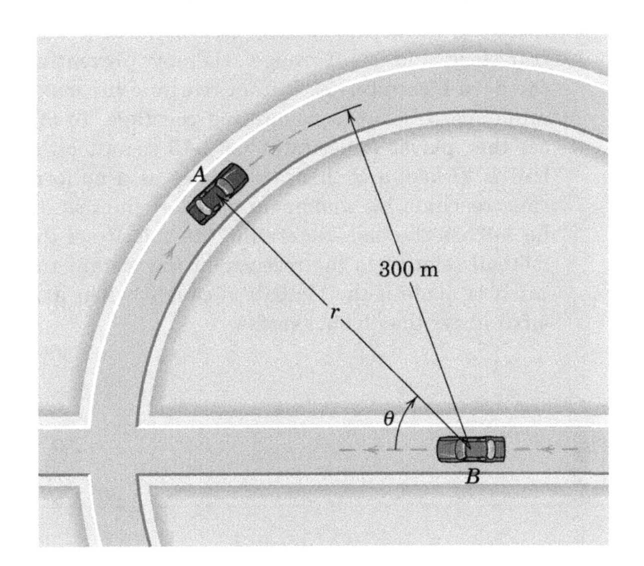

Problem 2/201

2/202 For the conditions of Prob. 2/201, determine the values of $\ddot{r}$ and $\ddot{\theta}$ as measured by an observer in car B at the instant represented. Use the results for $\dot{r}$ and $\dot{\theta}$ cited in the answers for that problem.

2/203 A batter hits the baseball A with an initial velocity of $v_0 = 100$ ft/sec directly toward fielder B at an angle of 30° to the horizontal; the initial position of the ball is 3 ft above ground level. Fielder B requires $\frac{1}{4}$ sec to judge where the ball should be caught and begins moving to that position with constant speed. Because of great experience, fielder B chooses his running speed so that he arrives at the "catch position" simultaneously with the baseball. The catch position is the field location at which the ball altitude is 7 ft. Determine the velocity of the ball relative to the fielder at the instant the catch is made.

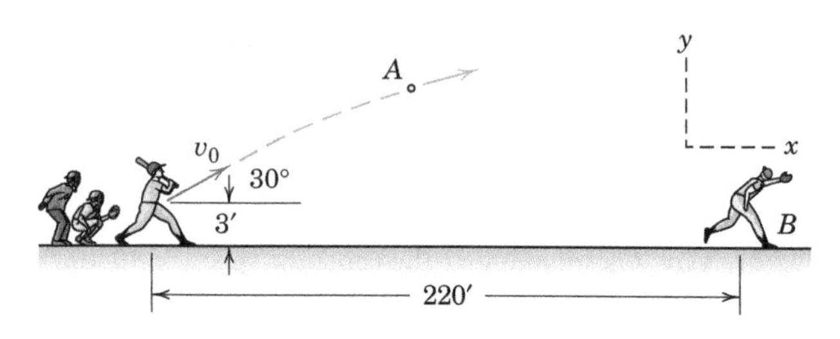

Problem 2/203

2/204 A place kicker A executes a "pooch" kick, which is designed to eliminate a potential return by the receiving team. The "pooch" kick features a high trajectory and short range, thereby preventing the deep kick returner B from reaching his maximum speed before encountering coverage. To offset this, player B hesitates for 1.15 sec after the ball is kicked, and then accelerates at a uniform rate, reaching his maximum speed at the instant he catches the ball. Determine the velocity of the football relative to the receiver at the instant the catch is made if the football is caught when it is 4.5 ft above the playing surface.

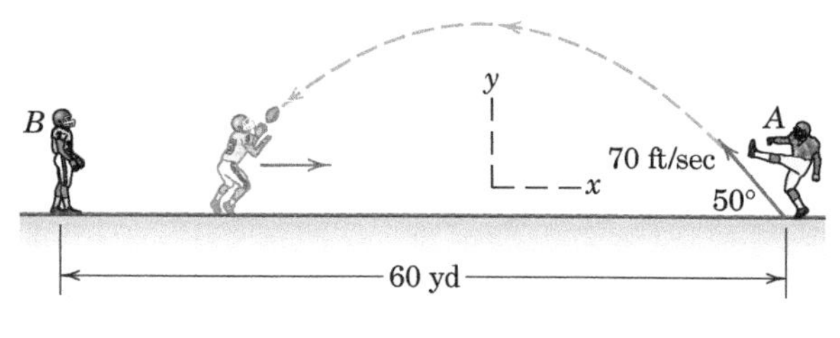

Problem 2/204

▶2/205 The aircraft A with radar detection equipment is flying horizontally at an altitude of 12 km and is increasing its speed at the rate of 1.2 m/s each second. Its radar locks onto an aircraft B flying in the same direction and in the same vertical plane at an altitude of 18 km. If A has a speed of 1000 km/h at the instant when $\theta = 30°$, determine the values of $\ddot{r}$ and $\ddot{\theta}$ at this same instant if B has a constant speed of 1500 km/h.

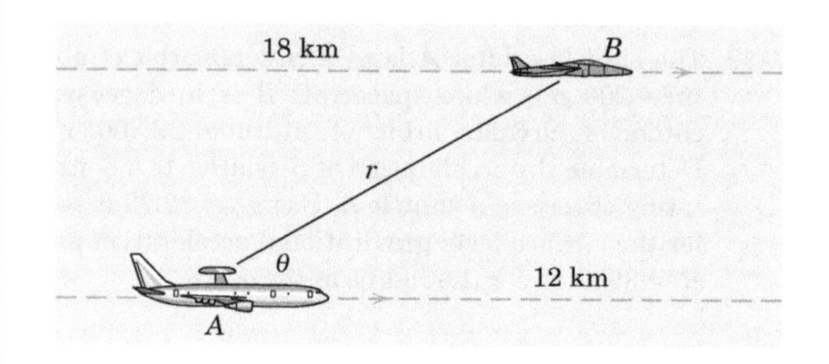

Problem 2/205

▶2/206 At a certain instant after jumping from the airplane A, a skydiver B is in the position shown and has reached a terminal (constant) speed $v_B = 50$ m/s. The airplane has the same constant speed $v_A = 50$ m/s, and after a period of level flight is just beginning to follow the circular path shown of radius $\rho_A = 2000$ m. (*a*) Determine the velocity and acceleration of the airplane relative to the skydiver. (*b*) Determine the time rate of change of the speed v_r of the airplane and the radius of curvature ρ_r of its path, both as observed by the nonrotating skydiver.

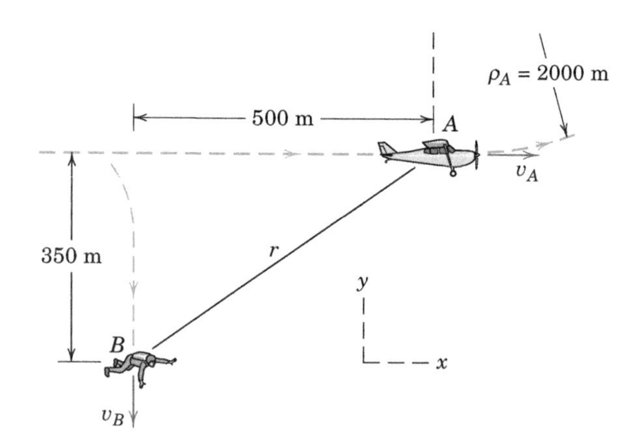

Problem 2/206

PROBLEMS

Introductory Problems

2/207 If the velocity $\dot{x}$ of block A up the incline is increasing at the rate of 0.044 m/s each second, determine the acceleration of B.

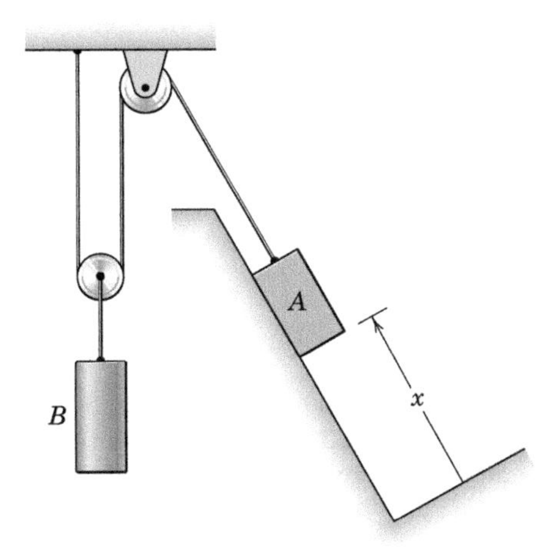

Problem 2/207

2/208 At the instant represented, $\mathbf{v}_{B/A} = 3.5\mathbf{j}$ m/s. Determine the velocity of each body at this instant. Assume that the upper surface of A remains horizontal.

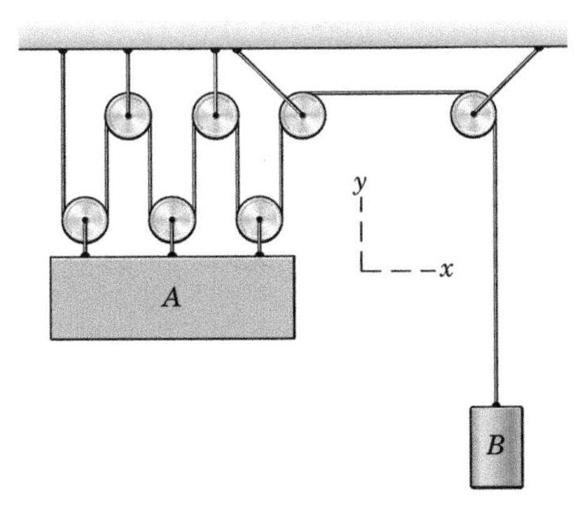

Problem 2/208

2/209 At a certain instant, the velocity of cylinder B is 1.2 m/s down and its acceleration is 2 m/s^2 up. Determine the corresponding velocity and acceleration of block A.

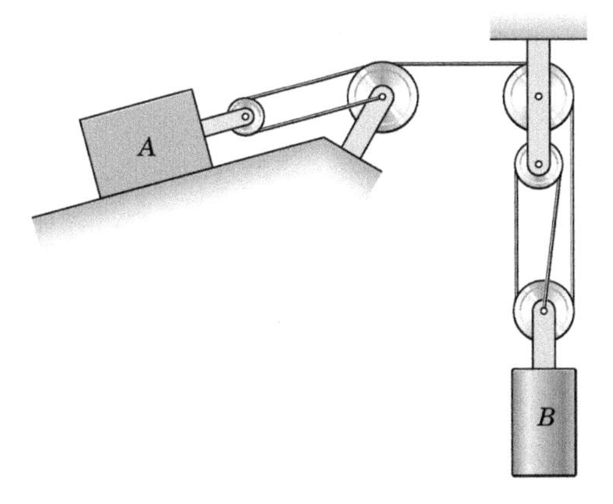

Problem 2/209

2/210 Determine the velocity of cart A if cylinder B has a downward velocity of 2 ft/sec at the instant illustrated. The two pulleys at C are pivoted independently.

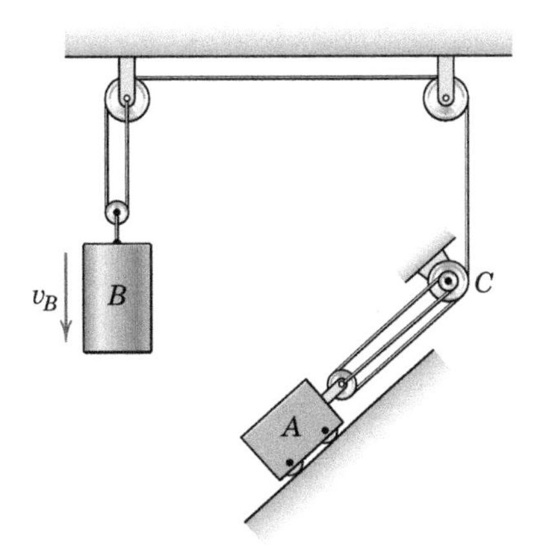

Problem 2/210

2/211 An electric motor M is used to reel in cable and hoist a bicycle into the ceiling space of a garage. Pulleys are fastened to the bicycle frame with hooks at locations A and B, and the motor can reel in cable at a steady rate of 12 in./sec. At this rate, how long will it take to hoist the bicycle 5 feet into the air? Assume that the bicycle remains level.

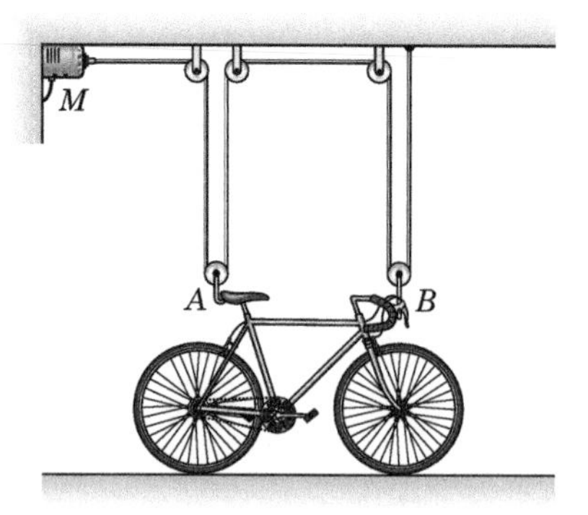

Problem 2/211

2/212 Determine the relation which governs the accelerations of A, B, and C, all measured positive down. Identify the number of degrees of freedom.

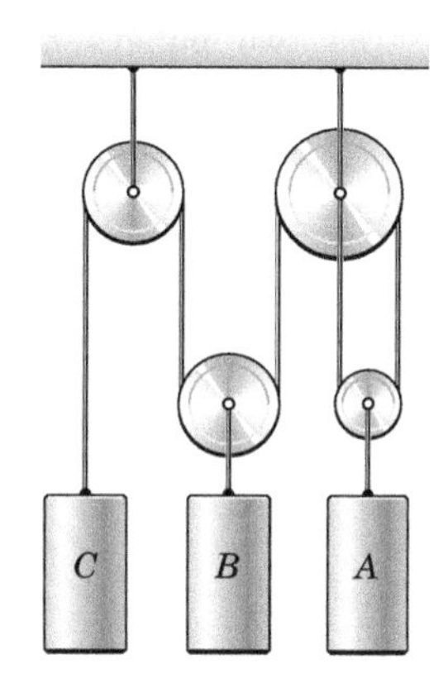

Problem 2/212

Representative Problems

2/213 Determine an expression for the velocity v_A of the cart A down the incline in terms of the upward velocity v_B of cylinder B.

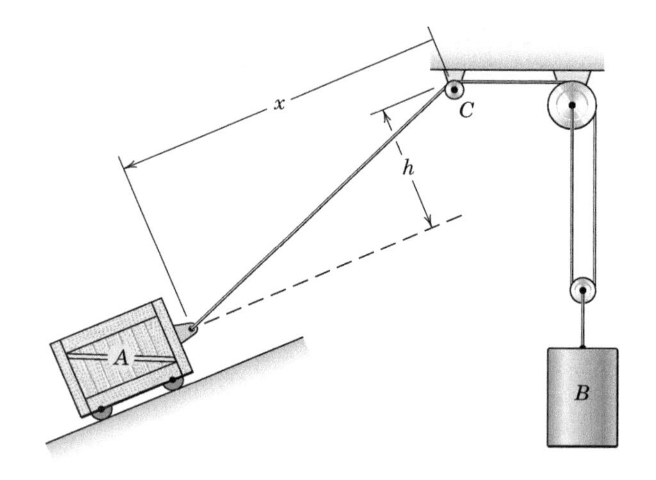

Problem 2/213

2/214 Neglect the diameters of the small pulleys and establish the relationship between the velocity of A and the velocity of B for a given value of y.

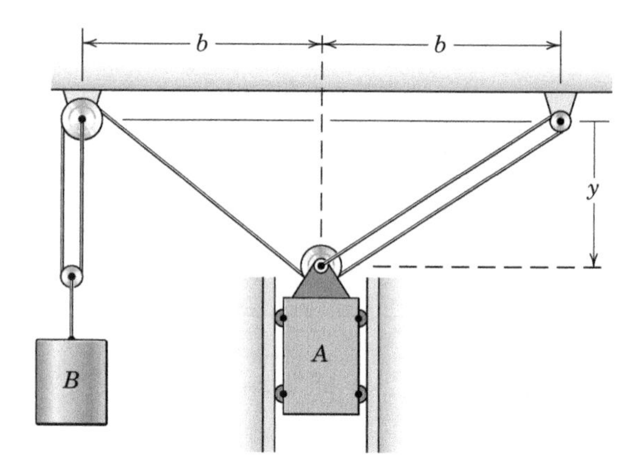

Problem 2/214

2/215 Under the action of force P, the constant acceleration of block B is 6 ft/sec^2 up the incline. For the instant when the velocity of B is 3 ft/sec up the incline, determine the velocity of B relative to A, the acceleration of B relative to A, and the absolute velocity of point C of the cable.

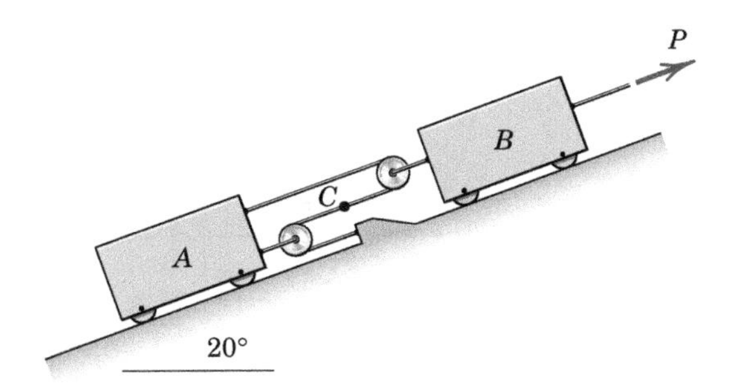

Problem 2/215

2/216 Determine the relationship which governs the velocities of the four cylinders. Express all velocities as positive down. How many degrees of freedom are there?

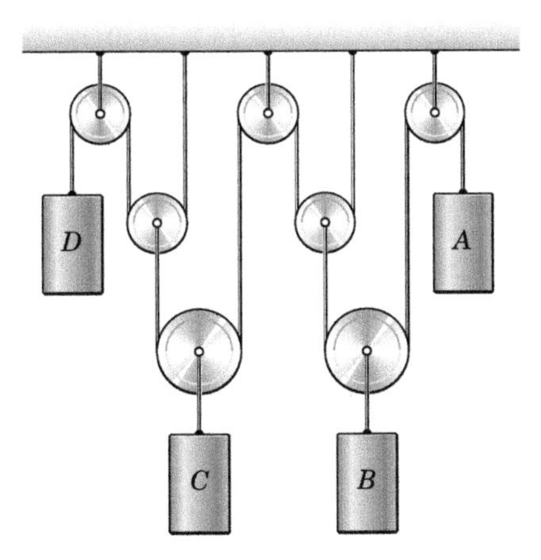

Problem 2/216

2/217 Collars A and B slide along the fixed right-angle rods and are connected by a cord of length L. Determine the acceleration a_x of collar B as a function of y if collar A is given a constant upward velocity v_A.

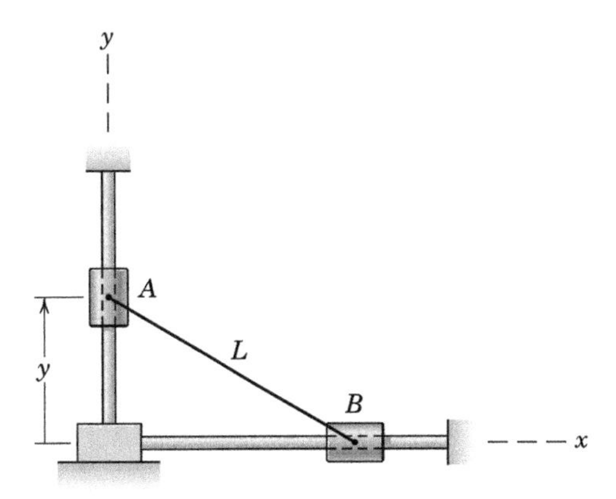

Problem 2/217

2/218 The small sliders A and B are connected by the rigid slender rod. If the velocity of slider B is 2 m/s to the right and is constant over a certain interval of time, determine the speed of slider A when the system is in the position shown.

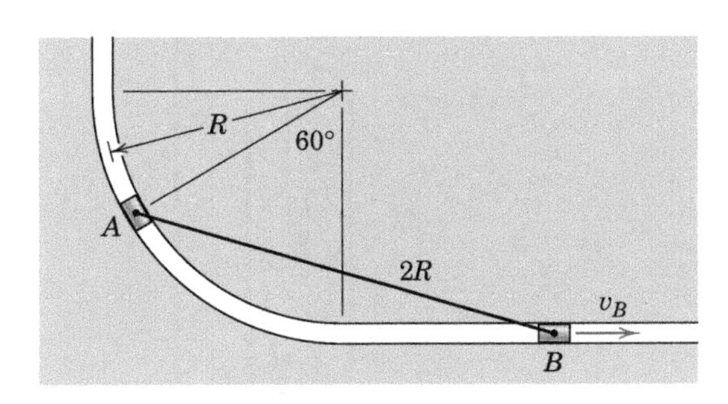

Problem 2/218

2/219 For a given value of y, determine the upward velocity of A in terms of the downward velocity of B. Neglect the diameters of the pulleys.

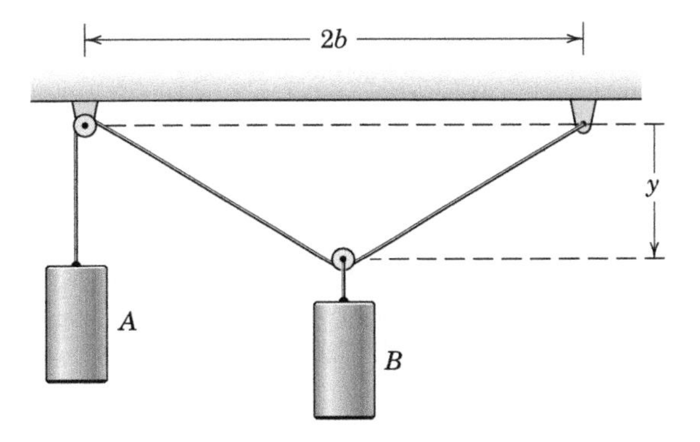

Problem 2/219

2/220 Cart A has a leftward velocity v_A and acceleration a_A at the instant represented. Determine the expressions for the velocity and acceleration of cart B in terms of the position x_A of cart A. Neglect the diameters of the pulleys and assume that there is no mechanical interference. The two pulleys at C are pivoted independently.

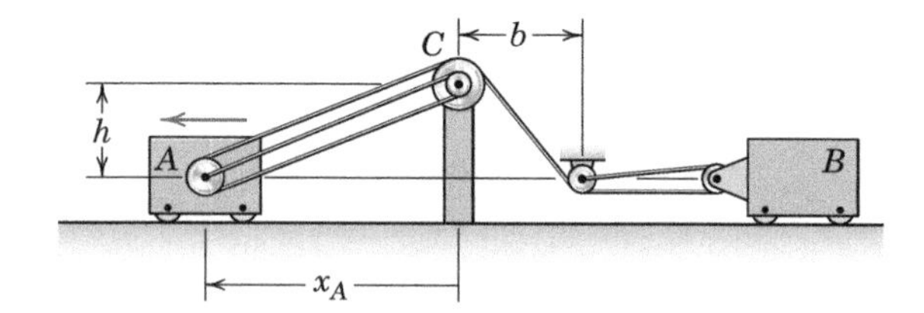

Problem 2/220

2/221 Determine the vertical rise h of the load W during 10 seconds if the hoisting drum draws in cable at the constant rate of 180 mm/s.

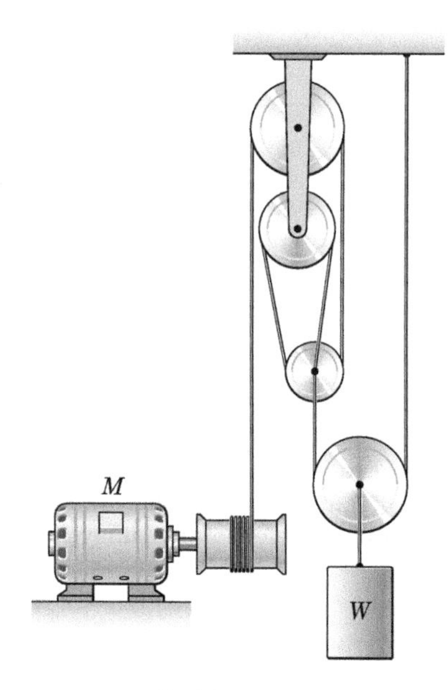

Problem 2/221

2/222 The hoisting system shown is used to easily raise kayaks for overhead storage. Determine expressions for the upward velocity and acceleration of the kayak at any height y if the winch M reels in cable at a constant rate $\dot{l}$. Assume that the kayak remains level.

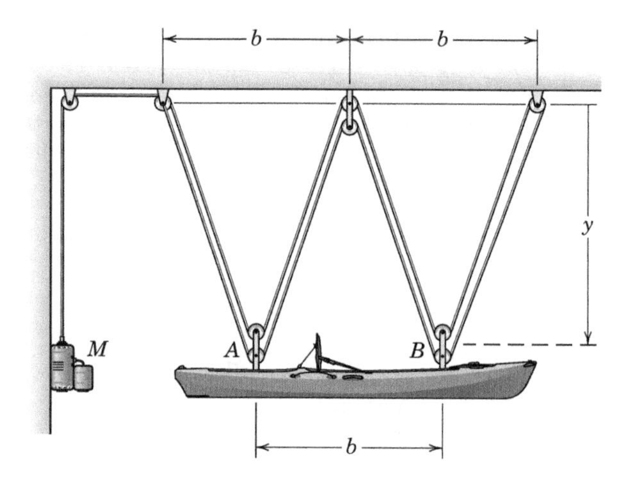

Problem 2/222

2/223 Develop an expression for the upward velocity of cylinder B in terms of the downward velocity of cylinder A. The cylinders are connected by a series of n cables and pulleys in a repeating fashion as shown.

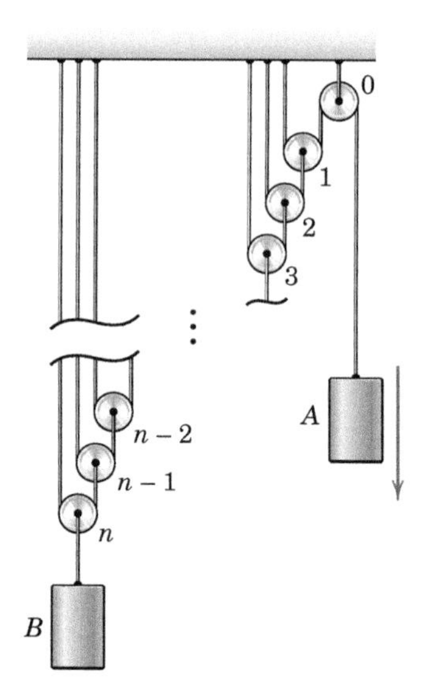

Problem 2/223

2/224 If load B has a downward velocity v_B, determine the upward component $(v_A)_y$ of the velocity of A in terms of b, the boom length l, and the angle θ. Assume that the cable supporting A remains vertical.

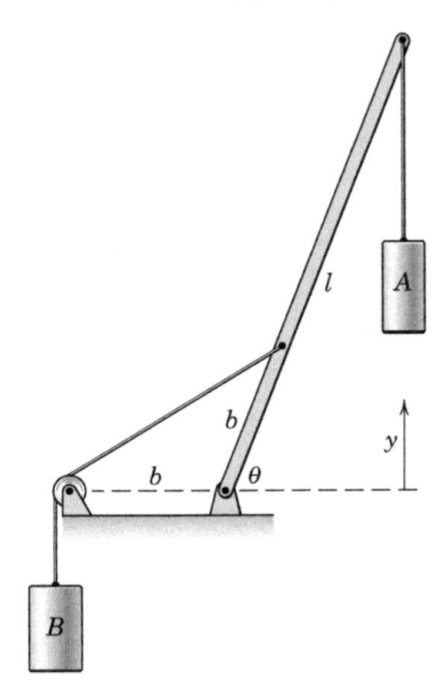

Problem 2/224

2/225 The rod of the fixed hydraulic cylinder is moving to the left with a constant speed $v_A = 25$ mm/s. Determine the corresponding velocity of slider B when $s_A = 425$ mm. The length of the cord is 1050 mm, and the effects of the radius of the small pulley A may be neglected.

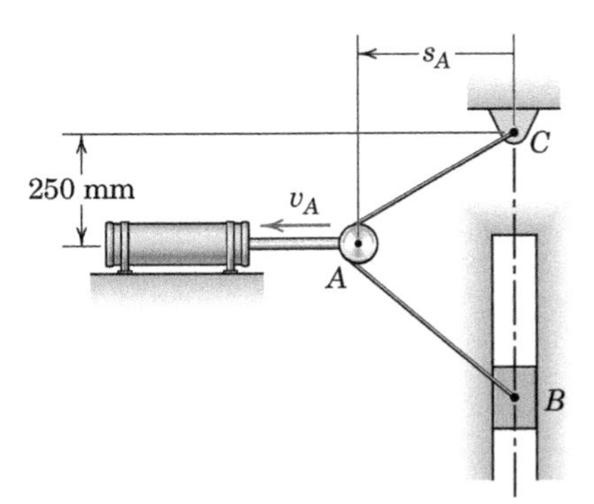

Problem 2/225

▶2/226 With all conditions of Prob. 2/225 remaining the same, determine the acceleration of slider B at the instant when $s_A = 425$ mm.

2/227 The two sliders are connected by the light rigid bar and move in the smooth vertical-plane guide. At the instant illustrated, the speed of slider A is 25 mm/s, $\theta = 45°$, and $\phi = 15°$. Determine the speed of slider B for this instant if $r = 175$ mm.

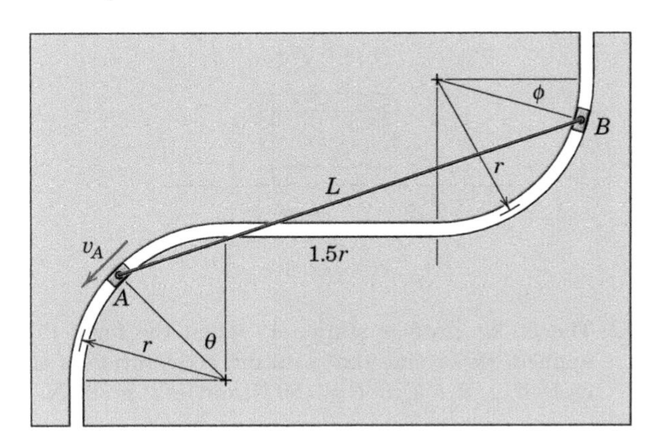

Problem 2/227

▶2/228 For the two sliders of Prob. 2/227, determine the time rate of change of speed for slider B at the location shown if the speed of slider A is constant over a short interval which includes the position shown.

PROBLEMS

Introductory Problems

3/1 The 50-kg crate is projected along the floor with an initial speed of 8 m/s at $x = 0$. The coefficient of kinetic friction is 0.40. Calculate the time required for the crate to come to rest and the corresponding distance x traveled.

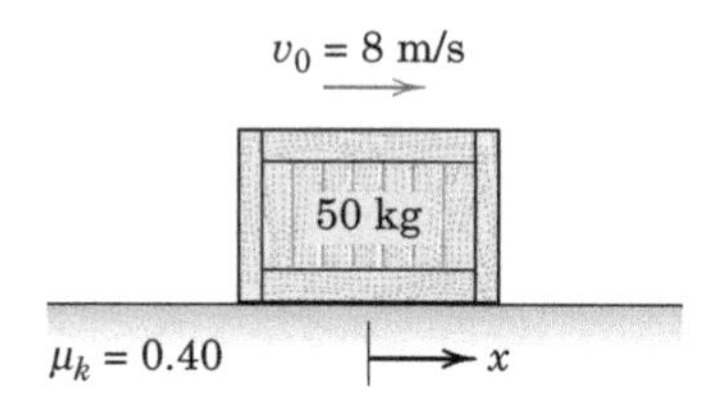

Problem 3/1

3/2 The 50-kg crate is stationary when the force P is applied. Determine the resulting acceleration of the crate if (*a*) $P = 0$, (*b*) $P = 150$ N, and (*c*) $P = 300$ N.

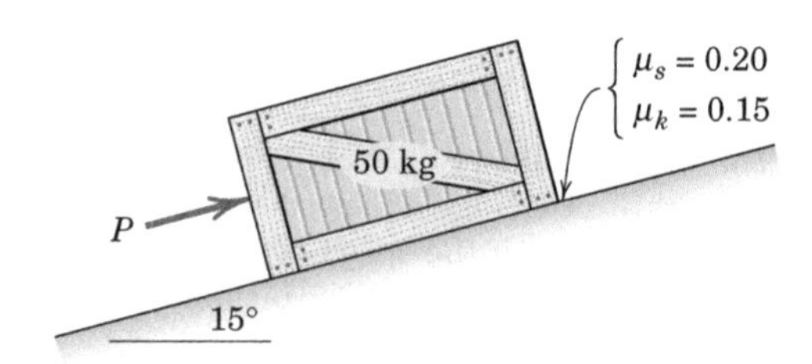

Problem 3/2

3/3 At a certain instant, the 80-lb crate has a velocity of 30 ft/sec up the 20° incline. Calculate the time t required for the crate to come to rest and the corresponding distance d traveled. Also, determine the distance d' traveled when the crate speed has been reduced to 15 ft/sec.

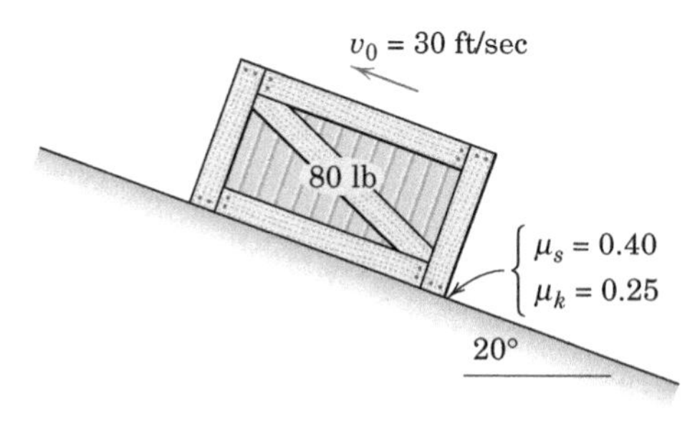

Problem 3/3

3/4 A man pulls himself up the 15° incline by the method shown. If the combined mass of the man and cart is 100 kg, determine the acceleration of the cart if the man exerts a pull of 175 N on the rope. Neglect all friction and the mass of the rope, pulleys, and wheels.

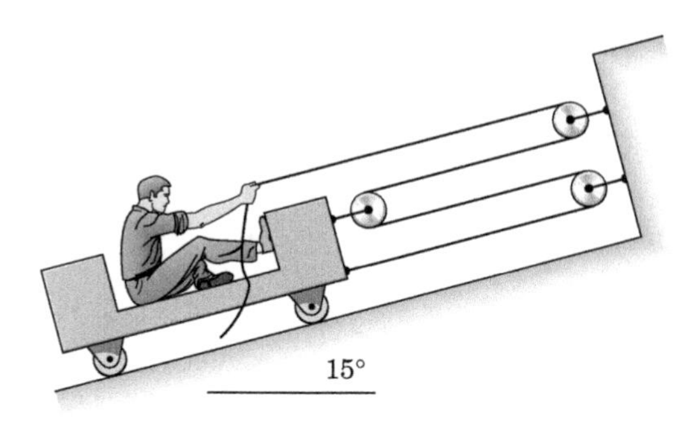

Problem 3/4

3/5 The 10-Mg truck hauls the 20-Mg trailer. If the unit starts from rest on a level road with a tractive force of 20 kN between the driving wheels of the truck and the road, compute the tension T in the horizontal drawbar and the acceleration a of the rig.

Problem 3/5

3/6 A 60-kg woman holds a 9-kg package as she stands within an elevator which briefly accelerates upward at a rate of $g/4$. Determine the force R which the elevator floor exerts on her feet and the lifting force L which she exerts on the package during the acceleration interval. If the elevator support cables suddenly and completely fail, what values would R and L acquire?

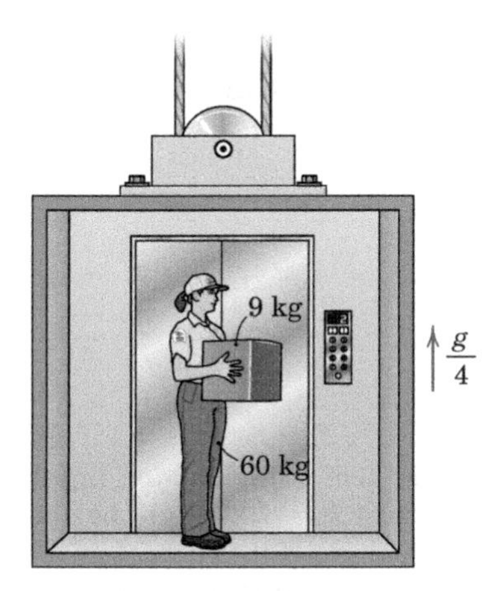

Problem 3/6

3/7 During a brake test, the rear-engine car is stopped from an initial speed of 100 km/h in a distance of 50 m. If it is known that all four wheels contribute equally to the braking force, determine the braking force F at each wheel. Assume a constant deceleration for the 1500-kg car.

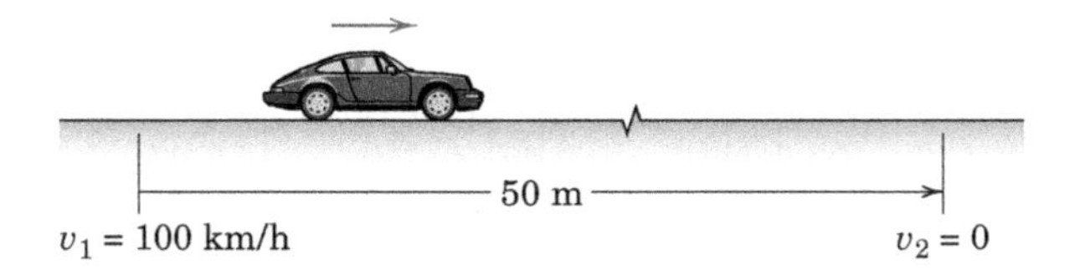

Problem 3/7

3/8 A skier starts from rest on the 40° slope at time $t = 0$ and is clocked at $t = 2.58$ s as he passes a speed checkpoint 20 m down the slope. Determine the coefficient of kinetic friction between the snow and the skis. Neglect wind resistance.

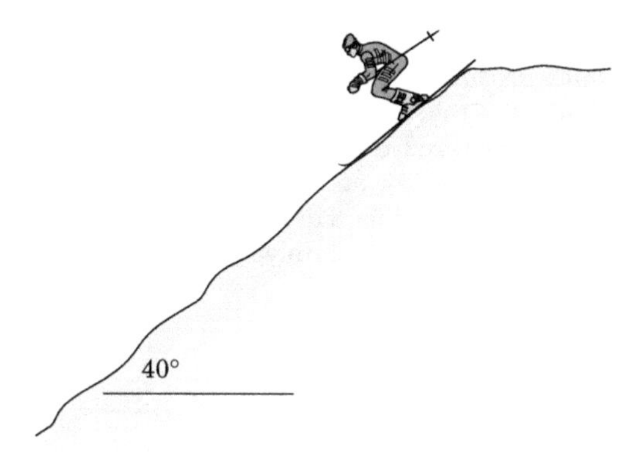

Problem 3/8

3/9 The inexperienced driver of an all-wheel-drive car applies too much throttle as he attempts to accelerate from rest up the slippery 10-percent incline. The result is wheel spin at all four tires, each of which has the same gripping ability. Determine the vehicle acceleration for the conditions of (*a*) light snow, $\mu_k = 0.12$ and (*b*) ice, $\mu_k = 0.05$.

Problem 3/9

3/10 Determine the steady-state angle α if the constant force P is applied to the cart of mass M. The pendulum bob has mass m and the rigid bar of length L has negligible mass. Ignore all friction. Evaluate your expression for $P = 0$.

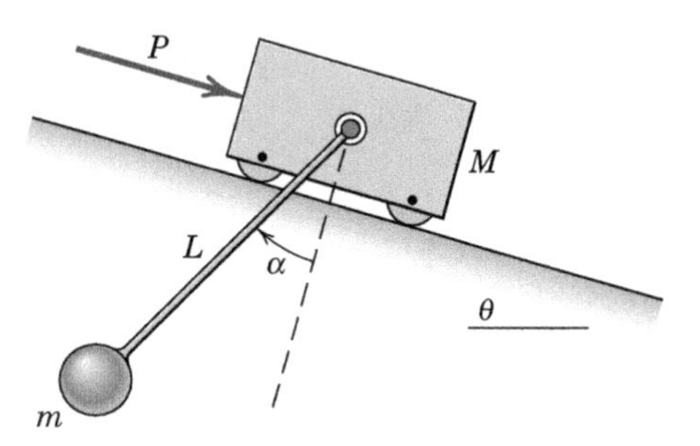

Problem 3/10

3/11 The 300-Mg jet airliner has three engines, each of which produces a nearly constant thrust of 240 kN during the takeoff roll. Determine the length s of runway required if the takeoff speed is 220 km/h. Compute s first for an uphill takeoff direction from A to B and second for a downhill takeoff from B to A on the slightly inclined runway. Neglect air and rolling resistance.

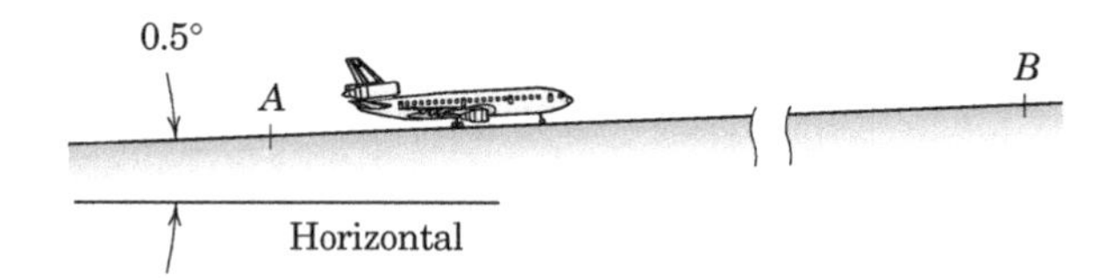

Problem 3/11

3/12 For a given horizontal force P, determine the normal reaction forces at A and B. The mass of the cylinder is m and that of the cart is M. Neglect all friction.

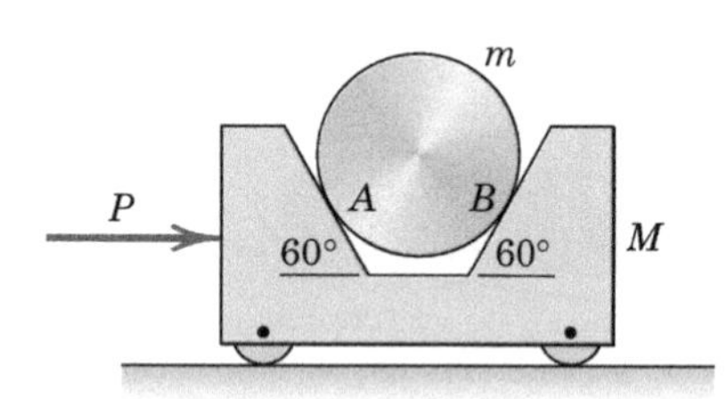

Problem 3/12

3/13 The system of the previous problem is now placed on the 15° incline. What force P will cause the normal reaction force at B to be zero?

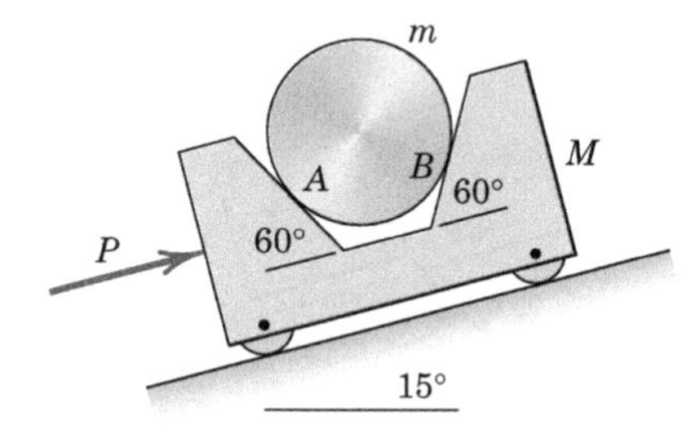

Problem 3/13

3/14 The 750,000-lb jetliner A has four engines, each of which produces a nearly constant thrust of 40,000 lb during the takeoff roll. A small commuter aircraft B taxis toward the end of the runway at a constant speed v_B = 15 mi/hr. Determine the velocity and acceleration which A appears to have relative to an observer in B 10 seconds after A begins its takeoff roll. Neglect air and rolling resistance.

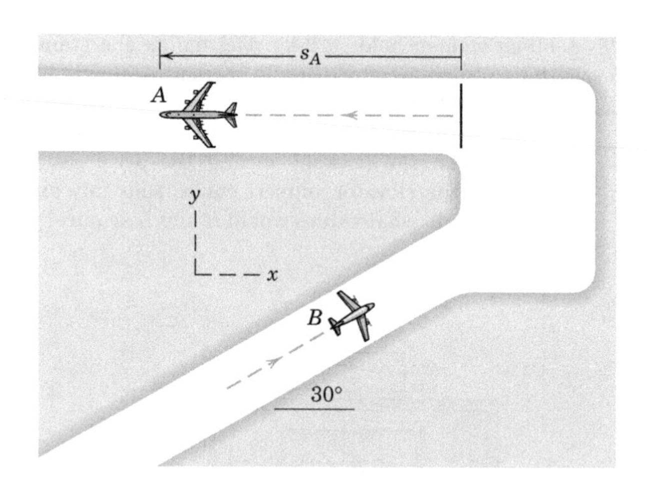

Problem 3/14

Representative Problems

3/15 The drive system of the 350-ton tugboat causes an external thrust P = 7000 lb to be applied as indicated in the figure. If the tugboat pushes an 800-ton coal barge starting from rest, what is the acceleration of the combined unit? Also, determine the force R of interaction between tugboat and barge. Neglect water resistance.

Problem 3/15

3/16 Determine the tension P in the cable which will give the 100-lb block a steady acceleration of 5 ft/sec^2 up the incline.

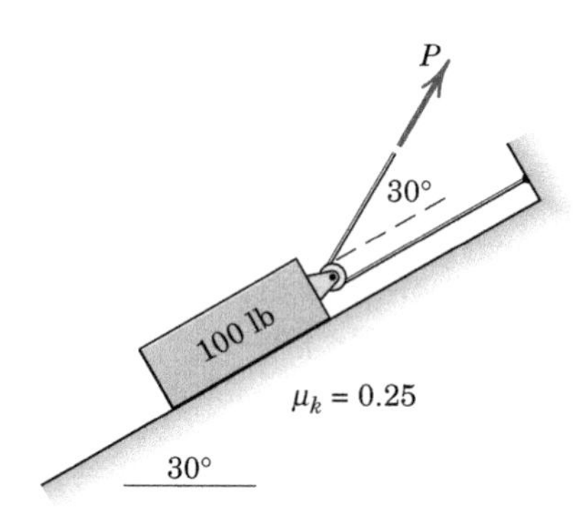

Problem 3/16

3/17 A cesium-ion engine for deep-space propulsion is designed to produce a constant thrust of 2.5 N for long periods of time. If the engine is to propel a 70-Mg spacecraft on an interplanetary mission, compute the time t required for a speed increase from 40 000 km/h to 65 000 km/h. Also find the distance s traveled during this interval. Assume that the spacecraft is moving in a remote region of space where the thrust from its ion engine is the only force acting on the spacecraft in the direction of its motion.

3/18 A toy train has magnetic couplers whose maximum attractive force is 0.2 lb between adjacent cars. What is the maximum force P with which a child can pull the locomotive and not break the train apart at a coupler? If P is slightly exceeded, which coupler fails? Neglect the mass and friction associated with all wheels.

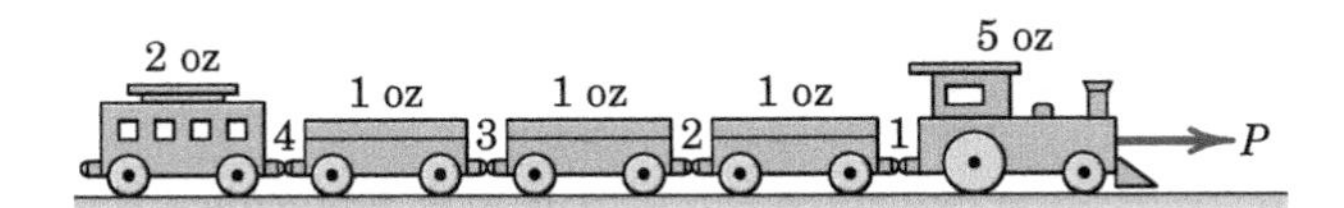

Problem 3/18

3/19 A worker develops a tension T in the cable as he attempts to move the 50-kg cart up the 20° incline. Determine the resulting acceleration of the cart if (*a*) $T = 150$ N and (*b*) $T = 200$ N. Neglect all friction, except that at the worker's feet.

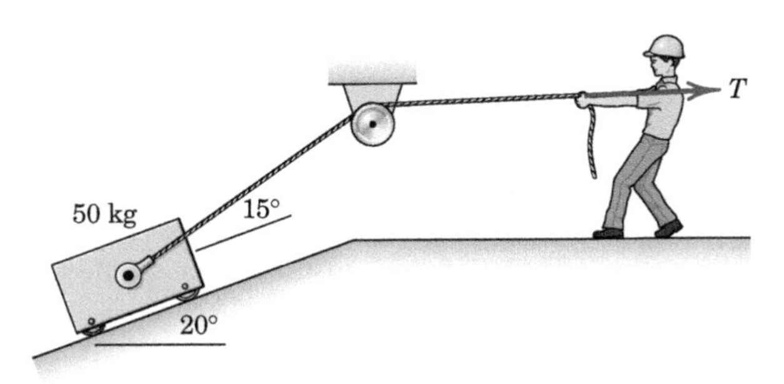

Problem 3/19

3/20 The wheeled cart of Prob. 3/19 is now replaced with a 50-kg sliding wooden crate. The coefficients of static and kinetic friction are given in the figure. Determine the acceleration of the crate if (*a*) $T = 300$ N and (*b*) $T = 400$ N. Neglect friction at the pulley.

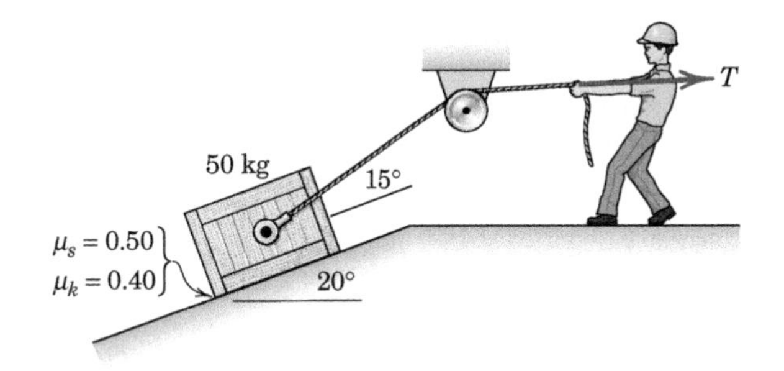

Problem 3/20

3/21 A small package is deposited by the conveyor belt onto the 30° ramp at A with a velocity of 0.8 m/s. Calculate the distance s on the level surface BC at which the package comes to rest. The coefficient of kinetic friction for the package and supporting surface from A to C is 0.30.

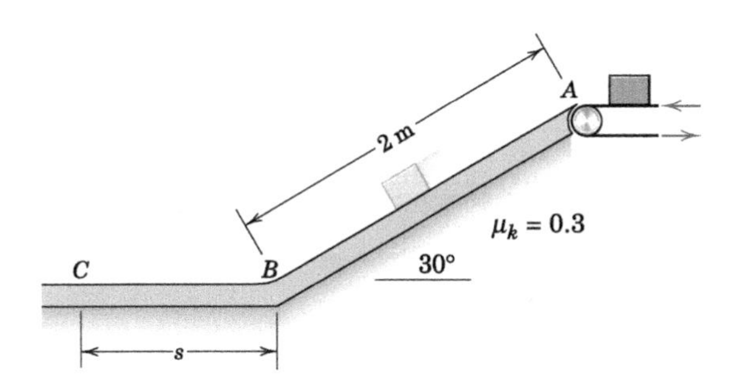

Problem 3/21

3/22 A bicyclist finds that she descends the slope $\theta_1 = 3°$ at a certain constant speed with no braking or pedaling required. The slope changes fairly abruptly to θ_2 at point A. If the bicyclist takes no action but continues to coast, determine the acceleration a of the bike just after it passes point A for the conditions (*a*) $\theta_2 = 5°$ and (*b*) $\theta_2 = 0$.

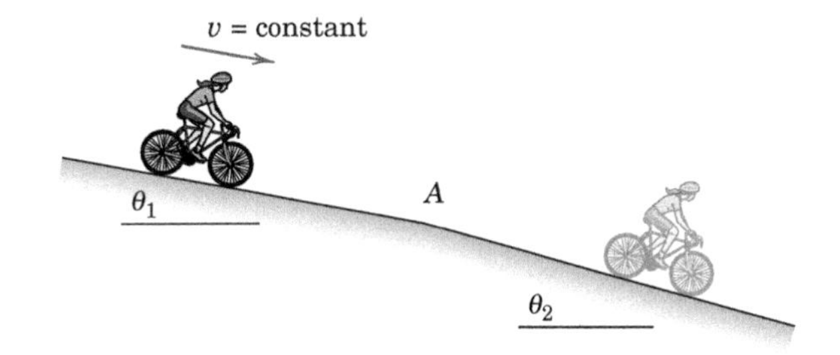

Problem 3/22

3/23 The 5-oz pinewood-derby car is released from rest at the starting line A and crosses the finish line C 2.75 sec later. The transition at B is small and smooth. Assume that the net retarding force is constant throughout the run and find this force.

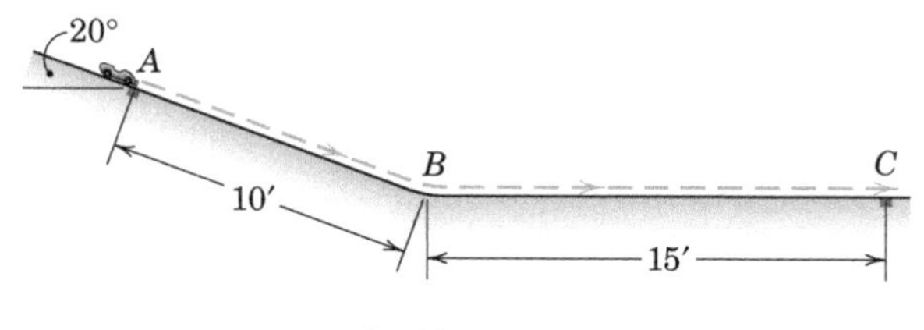

Problem 3/23

3/24 The coefficient of static friction between the flat bed of the truck and the crate it carries is 0.30. Determine the minimum stopping distance s which the truck can have from a speed of 70 km/h with constant deceleration if the crate is not to slip forward.

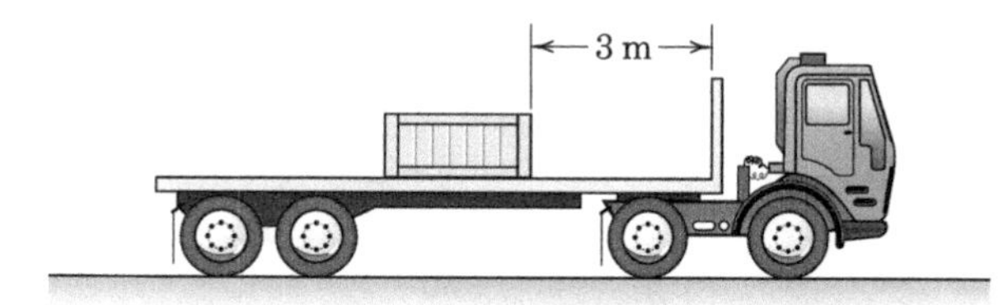

Problem 3/24

3/25 If the truck of Prob. 3/24 comes to a stop from an initial forward speed of 70 km/h in a distance of 50 m with uniform deceleration, determine whether or not the crate strikes the wall at the forward end of the flat bed. If the crate does strike the wall, calculate its speed relative to the truck as the impact occurs. Use the friction coefficients $\mu_s = 0.30$ and $\mu_k = 0.25$.

3/26 The winch takes in cable at the rate of 200 mm/s, and this rate is momentarily increasing at 500 mm/s each second. Determine the tensions in the three cables. Neglect the weights of the pulleys.

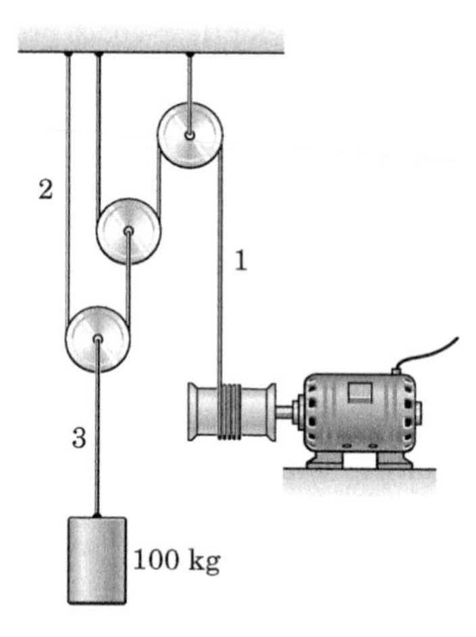

Problem 3/26

3/27 Determine the vertical acceleration of the 60-lb cylinder for each of the two cases. Neglect friction and the mass of the pulleys.

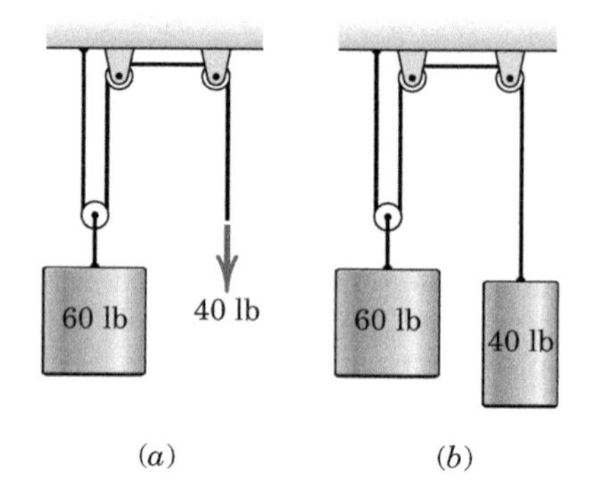

Problem 3/27

3/28 Determine the weight of cylinder B which would cause block A to accelerate (*a*) 5 ft/sec^2 down the incline and (*b*) 5 ft/sec^2 up the incline. Neglect all friction.

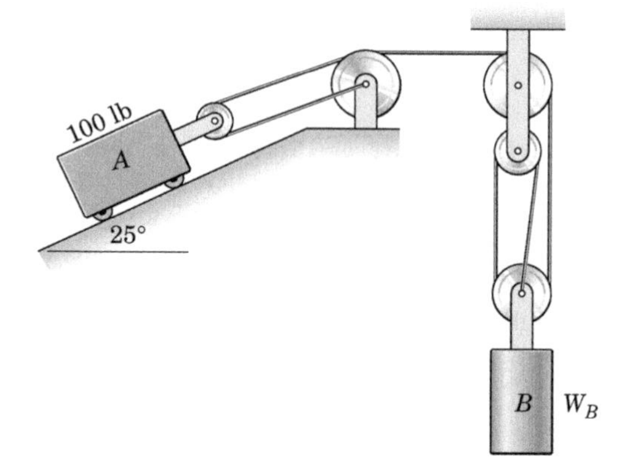

Problem 3/28

3/29 A player pitches a baseball horizontally toward a speed-sensing radar gun. The baseball weighs $5\frac{1}{8}$ oz and has a circumference of $9\frac{1}{8}$ in. If the speed at $x = 0$ is $v_0 = 90$ mi/hr, estimate the speed as a function of x. Assume that the horizontal aerodynamic drag on the baseball is given by $D = C_D(\frac{1}{2}\rho v^2)S$, where C_D is the drag coefficient, ρ is the air density, v is the speed, and S is the cross-sectional area of the baseball. Use a value of 0.3 for C_D. Neglect the vertical component of the motion but comment on the validity of this assumption. Evaluate your answer for $x = 60$ ft, which is the approximate distance between a pitcher's hand and home plate.

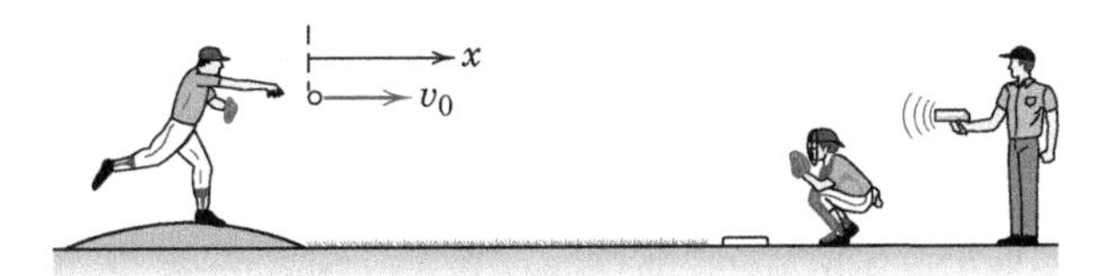

Problem 3/29

3/30 If the rider presses on the pedal with a force $P = 160$ N as shown, determine the resulting forward acceleration of the bicycle. Neglect the effects of the mass of rotating parts, and assume no slippage at the rear wheel. The radii of sprockets A and B are 45 mm and 90 mm, respectively. The mass of the bicycle is 13 kg and that of the rider is 65 kg. Treat the rider as a particle moving with the bicycle frame, and neglect drivetrain friction.

Problem 3/30

3/31 The rack has a mass $m = 50$ kg. What moment M must be exerted on the gear wheel by the motor in order to accelerate the rack up the 60° incline at a rate $a = g/4$? The fixed motor which drives the gear wheel via the shaft at O is not shown. Neglect the effects of the mass of the gear wheel.

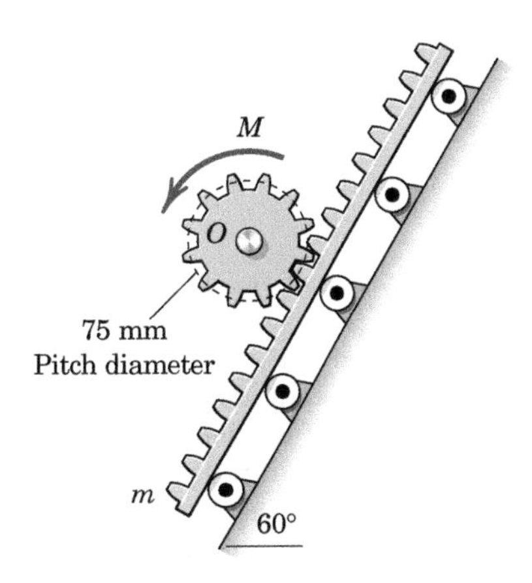

Problem 3/31

3/32 A jet airplane with a mass of 5 Mg has a touchdown speed of 300 km/h, at which instant the braking parachute is deployed and the power shut off. If the total drag on the aircraft varies with velocity as shown in the accompanying graph, calculate the distance x along the runway required to reduce the speed to 150 km/h. Approximate the variation of the drag by an equation of the form $D = kv^2$, where k is a constant.

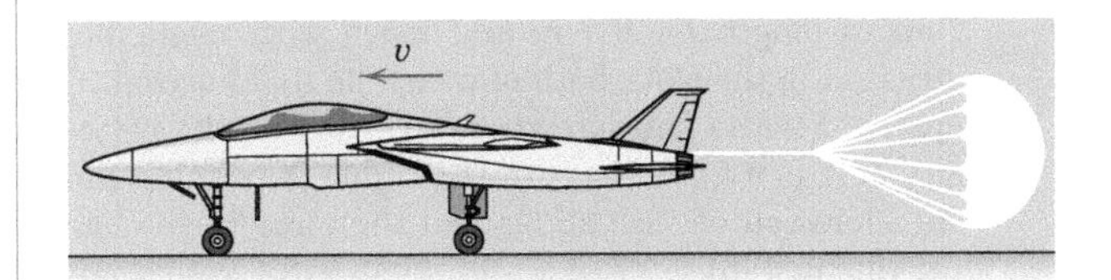

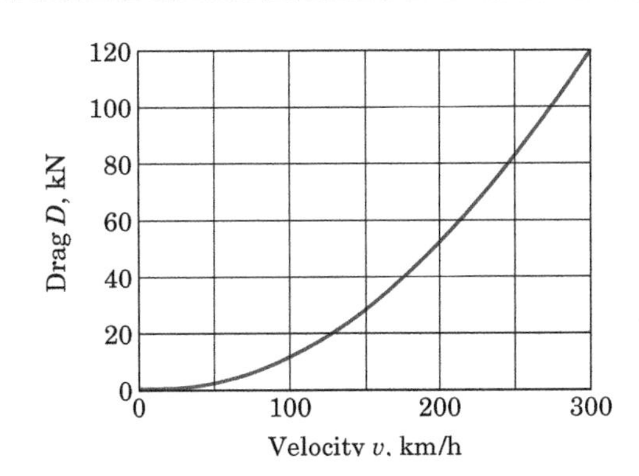

Problem 3/32

3/33 During its final approach to the runway, the aircraft speed is reduced from 300 km/h at A to 200 km/h at B. Determine the net external aerodynamic force R which acts on the 200-Mg aircraft during this interval, and find the components of this force which are parallel to and normal to the flight path.

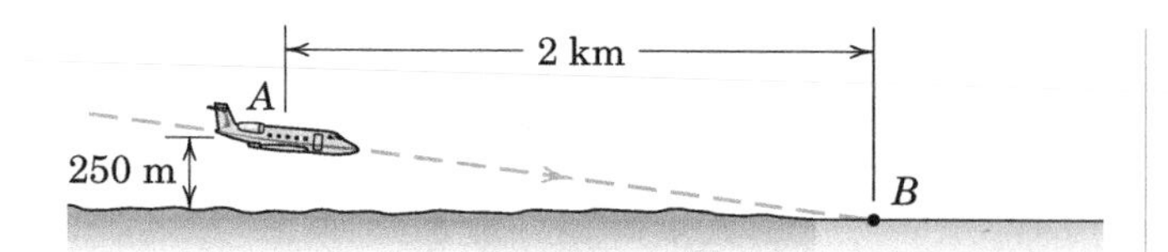

Problem 3/33

3/34 A heavy chain with a mass ρ per unit length is pulled by the constant force P along a horizontal surface consisting of a smooth section and a rough section. The chain is initially at rest on the rough surface with $x = 0$. If the coefficient of kinetic friction between the chain and the rough surface is μ_k, determine the velocity v of the chain when $x = L$. The force P is greater than $\mu_k \rho g L$ in order to initiate motion.

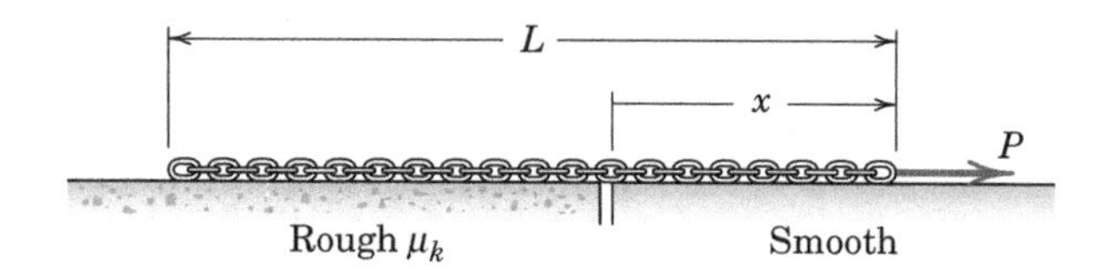

Problem 3/34

3/35 The sliders A and B are connected by a light rigid bar of length $l = 0.5$ m and move with negligible friction in the slots, both of which lie in a horizontal plane. For the position where $x_A = 0.4$ m, the velocity of A is $v_A = 0.9$ m/s to the right. Determine the acceleration of each slider and the force in the bar at this instant.

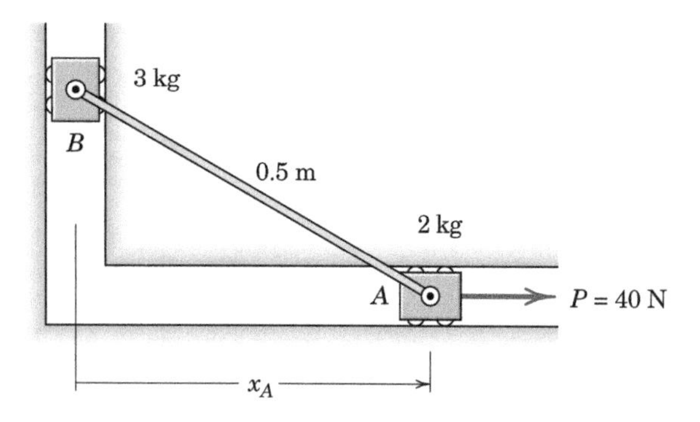

Problem 3/35

3/36 The spring of constant $k = 200$ N/m is attached to both the support and the 2-kg cylinder, which slides freely on the horizontal guide. If a constant 10-N force is applied to the cylinder at time $t = 0$ when the spring is undeformed and the system is at rest, determine the velocity of the cylinder when $x = 40$ mm. Also determine the maximum displacement of the cylinder.

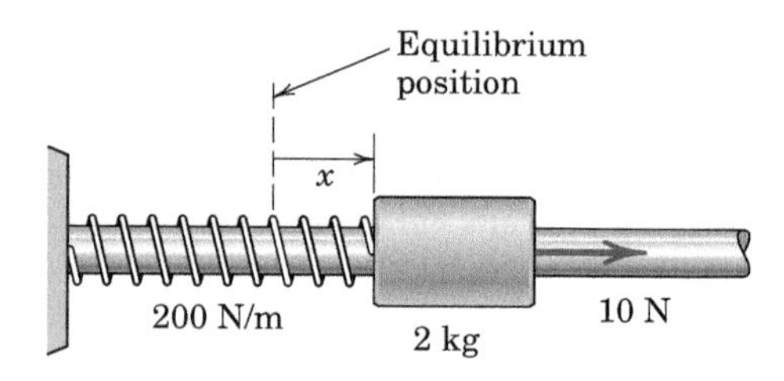

Problem 3/36

3/37 The 4-lb collar is released from rest against the light elastic spring, which has a stiffness of 10 lb/in. and has been compressed a distance of 6 in. Determine the acceleration a of the collar as a function of the vertical displacement x of the collar measured in feet from the point of release. Find the velocity v of the collar when $x = 0.5$ ft. Friction is negligible.

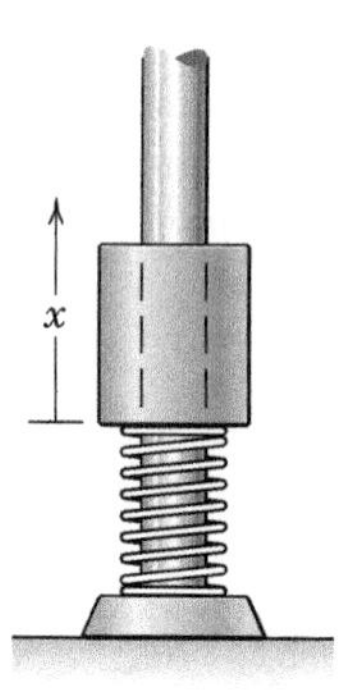

Problem 3/37

3/38 Two configurations for raising an elevator are shown. Elevator A with attached hoisting motor and drum has a total mass of 900 kg. Elevator B without motor and drum also has a mass of 900 kg. If the motor supplies a constant torque of 600 N·m to its 250-mm-diameter drum for 2 s in each case, select the configuration which results in the greater upward acceleration and determine the corresponding velocity v of the elevator 1.2 s after it starts from rest. The mass of the motorized drum is small, thus permitting it to be analyzed as though it were in equilibrium. Neglect the mass of cables and pulleys and all friction.

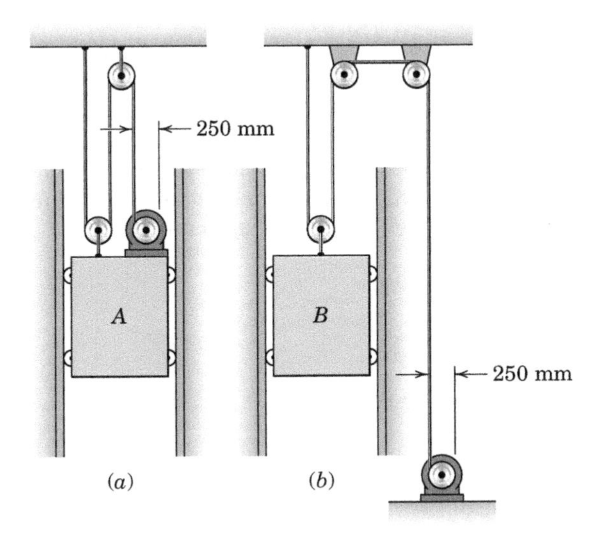

Problem 3/38

3/39 Determine the range of applied force P over which the block of mass m_2 will not slip on the wedge-shaped block of mass m_1. Neglect friction associated with the wheels of the tapered block.

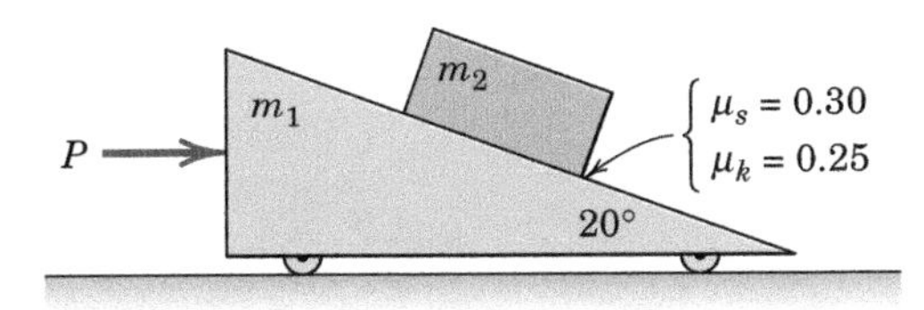

Problem 3/39

3/40 A spring-loaded device imparts an initial vertical velocity of 50 m/s to a 0.15-kg ball. The drag force on the ball is $F_D = 0.002v^2$, where F_D is in newtons when the speed v is in meters per second. Determine the maximum altitude h attained by the ball (*a*) with drag considered and (*b*) with drag neglected.

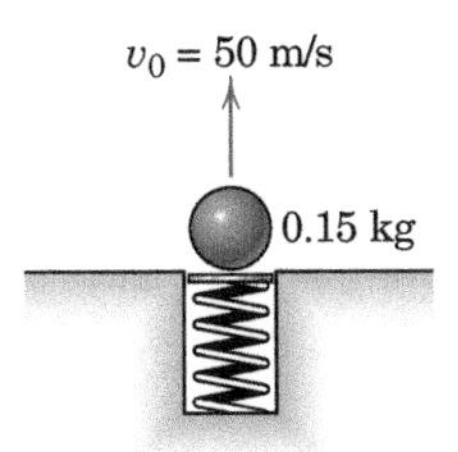

Problem 3/40

3/41 The sliders A and B are connected by a light rigid bar and move with negligible friction in the slots, both of which lie in a horizontal plane. For the position shown, the velocity of A is 0.4 m/s to the right. Determine the acceleration of each slider and the force in the bar at this instant.

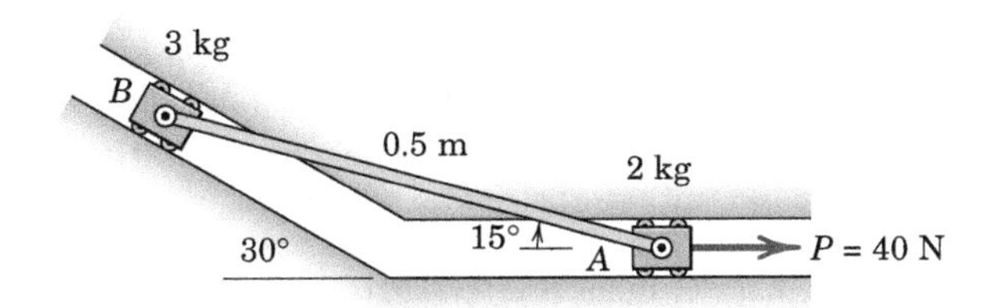

Problem 3/41

3/42 The design of a lunar mission calls for a 1200-kg spacecraft to lift off from the surface of the moon and travel in a straight line from point A and pass point B. If the spacecraft motor has a constant thrust of 2500 N, determine the speed of the spacecraft as it passes point B. Use Table D/2 and the gravitational law from Chapter 1 as needed.

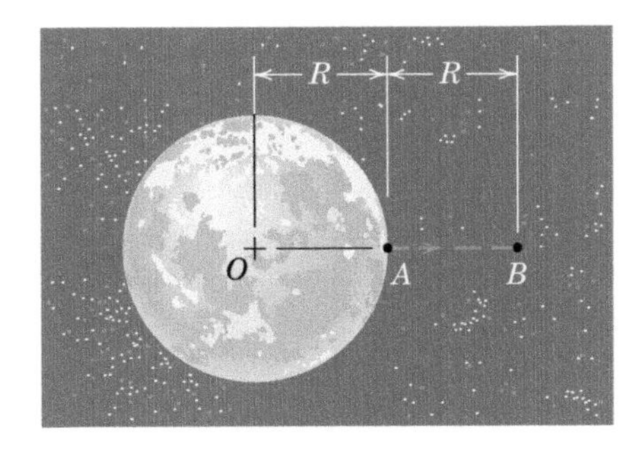

Problem 3/42

3/43 The system is released from rest in the configuration shown at time $t = 0$. Determine the time t when the block of mass m_1 contacts the lower stop of the body of mass m_2. Also, determine the corresponding distance s_2 traveled by m_2. Use the values $m_1 = 0.5$ kg, $m_2 = 2$ kg, $\mu_s = 0.25$, $\mu_k = 0.20$, and $d = 0.4$ m.

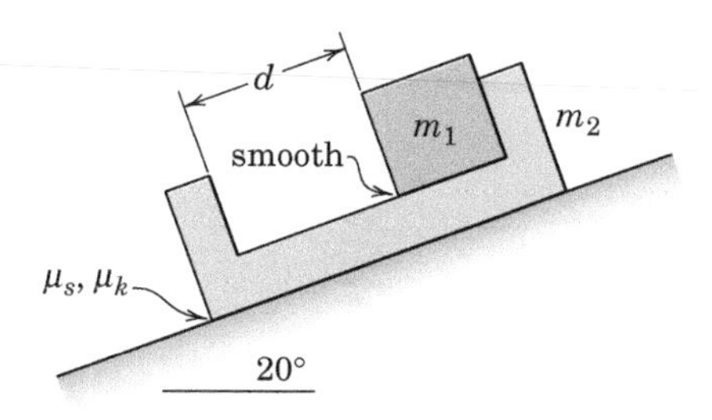

Problem 3/43

3/44 The system of Prob. 3/43 is reconsidered here, only now the interface between the two bodies is not smooth. Use the values $\mu_s = 0.10$ and $\mu_k = 0.08$ between the two bodies. Determine the time t when m_1 contacts the lower stop on m_2 and the corresponding distance s_2 traveled by m_2.

▶3/45 The rod of the fixed hydraulic cylinder is moving to the left with a speed of 100 mm/s, and this speed is momentarily increasing at a rate of 400 mm/s each second at the instant when $s_A = 425$ mm. Determine the tension in the cord at that instant. The mass of slider B is 0.5 kg, the length of the cord is 1050 mm, and the effects of the radius and friction of the small pulley at A are negligible. Find results for cases (a) negligible friction at slider B and (b) $\mu_k = 0.40$ at slider B. The action is in a vertical plane.

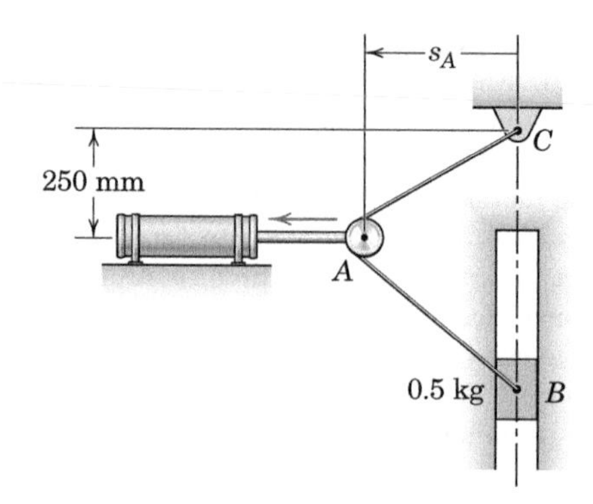

Problem 3/45

▶3/46 Two iron spheres, each of which is 100 mm in diameter, are released from rest with a center-to-center separation of 1 m. Assume an environment in space with no forces other than the force of mutual gravitational attraction and calculate the time t required for the spheres to contact each other and the absolute speed v of each sphere upon contact.

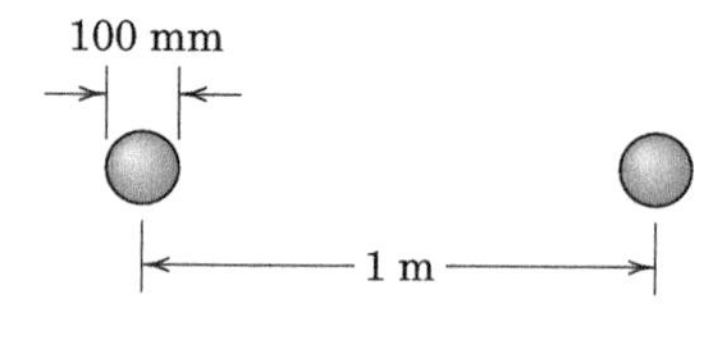

Problem 3/46

PROBLEMS

Introductory Problems

3/47 The small 2-kg block A slides down the curved path and passes the lowest point B with a speed of 4 m/s. If the radius of curvature of the path at B is 1.5 m, determine the normal force N exerted on the block by the path at this point. Is knowledge of the friction properties necessary?

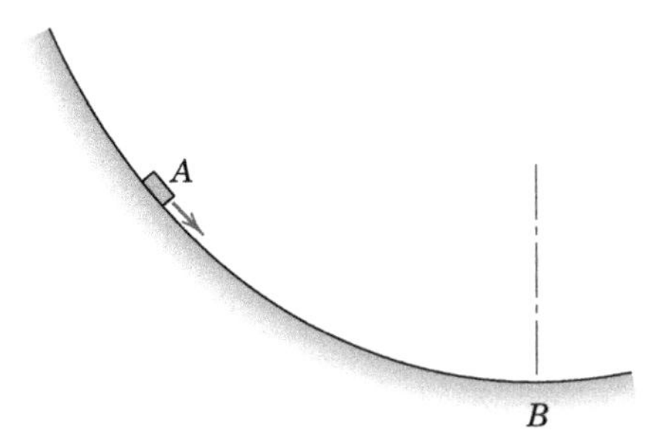

Problem 3/47

3/48 If the 2-kg block passes over the top B of the circular portion of the path with a speed of 3.5 m/s, calculate the magnitude N_B of the normal force exerted by the path on the block. Determine the maximum speed v which the block can have at A without losing contact with the path.

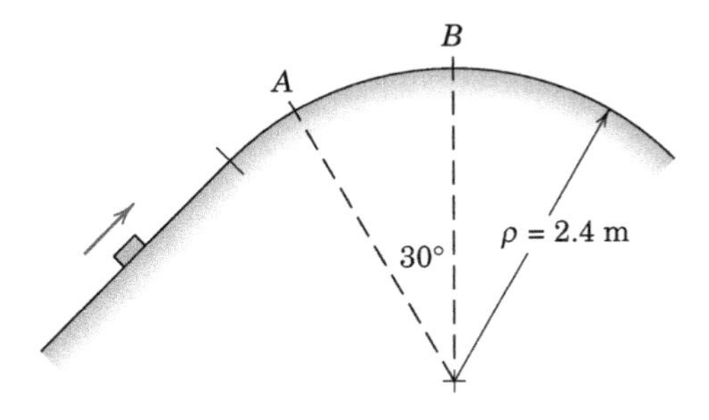

Problem 3/48

3/49 The particle of mass m is attached to the light rigid rod, and the assembly rotates about a horizontal axis through O with a constant angular velocity $\dot{\theta} = \omega$. Determine the tension T in the rod as a function of θ.

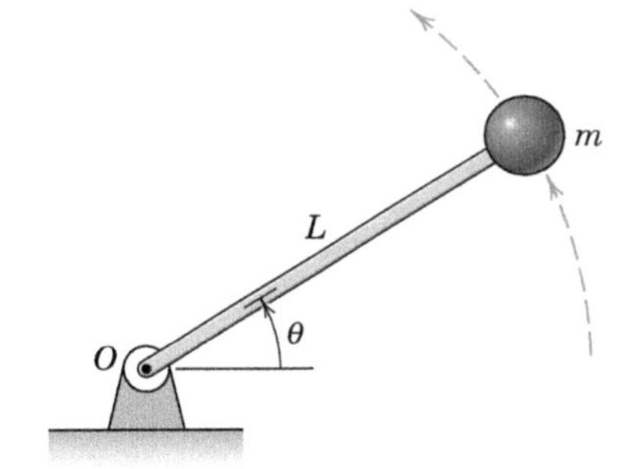

Problem 3/49

3/50 If the 180-lb ski-jumper attains a speed of 80 ft/sec as he approaches the takeoff position, calculate the magnitude N of the normal force exerted by the snow on his skis just before he reaches A.

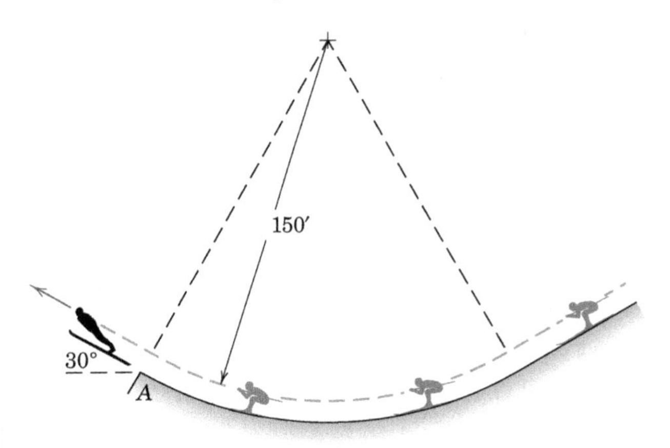

Problem 3/50

3/51 The 4-oz slider has a speed $v = 3$ ft/sec as it passes point A of the smooth guide, which lies in a horizontal plane. Determine the magnitude R of the force which the guide exerts on the slider (a) just before it passes point A of the guide and (b) as it passes point B.

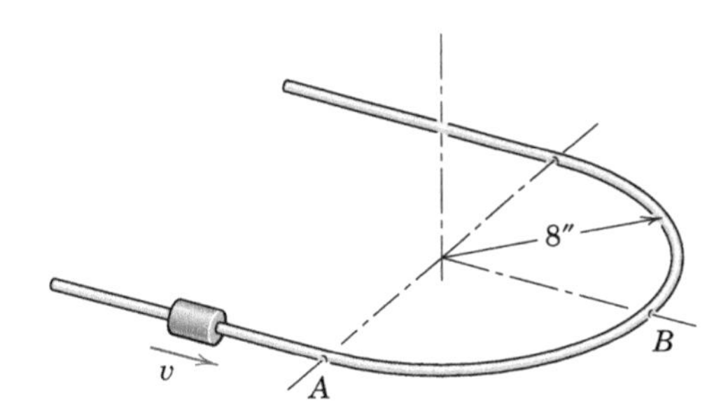

Problem 3/51

3/52 A jet transport plane flies in the trajectory shown in order to allow astronauts to experience the "weightless" condition similar to that aboard orbiting spacecraft. If the speed at the highest point is 600 mi/hr, what is the radius of curvature ρ necessary to exactly simulate the orbital "free-fall" environment?

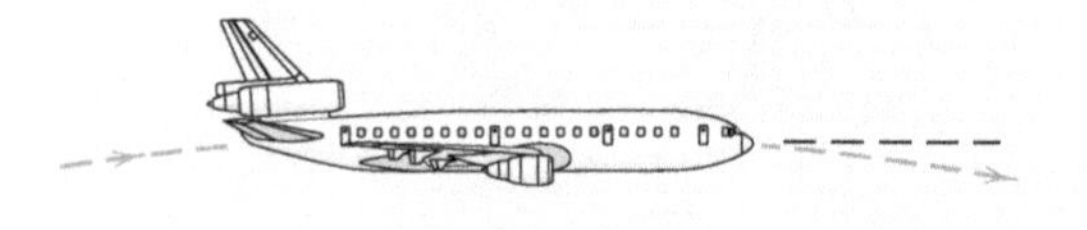

Problem 3/52

3/53 In the design of a space station to operate outside the earth's gravitational field, it is desired to give the structure a rotational speed N which will simulate the effect of the earth's gravity for members of the crew. If the centers of the crew's quarters are to be located 12 m from the axis of rotation, calculate the necessary rotational speed N of the space station in revolutions per minute.

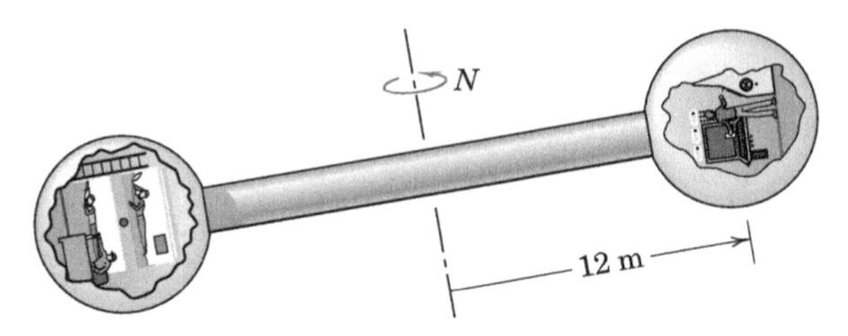

Problem 3/53

3/54 Determine the speed which the 630-kg four-man bobsled must have in order to negotiate the turn without reliance on friction. Also find the net normal force exerted on the bobsled by the track.

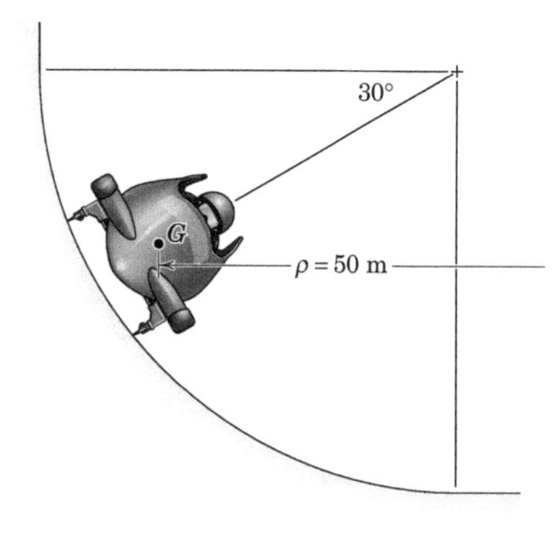

Problem 3/54

3/55 The hollow tube is pivoted about a horizontal axis through point O and is made to rotate in the vertical plane with a constant counterclockwise angular velocity $\dot{\theta} = 3$ rad/sec. If a 0.2-lb particle is sliding in the tube toward O with a velocity of 6 ft/sec relative to the tube when the position $\theta = 30°$ is passed, calculate the magnitude N of the normal force exerted by the wall of the tube on the particle at this instant.

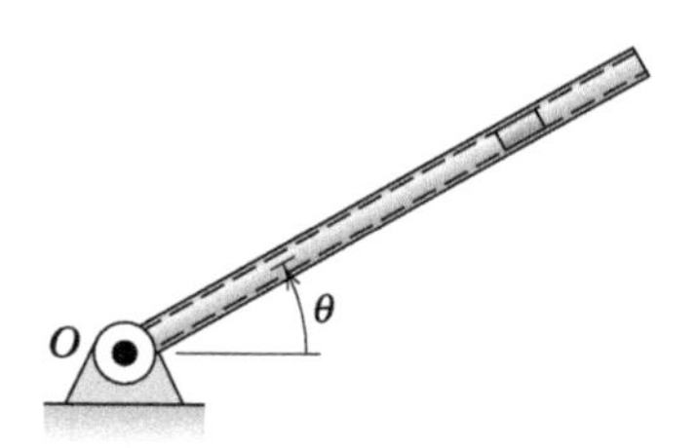

Problem 3/55

3/56 A Formula-1 car encounters a hump which has a circular shape with smooth transitions at both ends. (*a*) What speed v_B will cause the car to lose contact with the road at the topmost point B? (*b*) For a speed $v_A = 190$ km/h, what is the normal force exerted by the road on the 640-kg car as it passes point A?

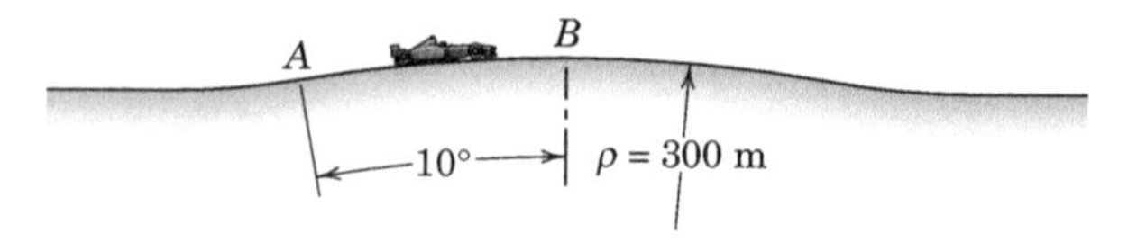

Problem 3/56

3/57 The small spheres are free to move on the inner surface of the rotating spherical chambers shown in section with radius $R = 200$ mm. If the spheres reach a steady-state angular position $\beta = 45°$, determine the angular velocity Ω of the device.

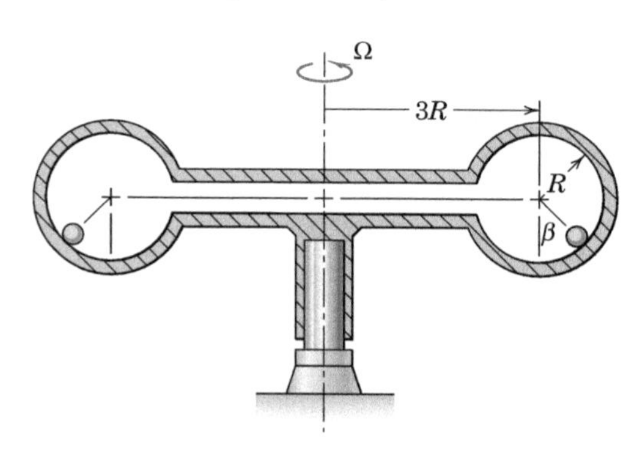

Problem 3/57

3/58 A 180-lb snowboarder has speed $v = 15$ ft/sec when in the position shown on the halfpipe. Determine the normal force on his snowboard and the magnitude of his total acceleration at the instant depicted. Use a value $\mu_k = 0.10$ for the coefficient of kinetic friction between the snowboard and the surface. Neglect the weight of the snowboard and assume that the mass center G of the snowboarder is 3 feet from the surface of the snow.

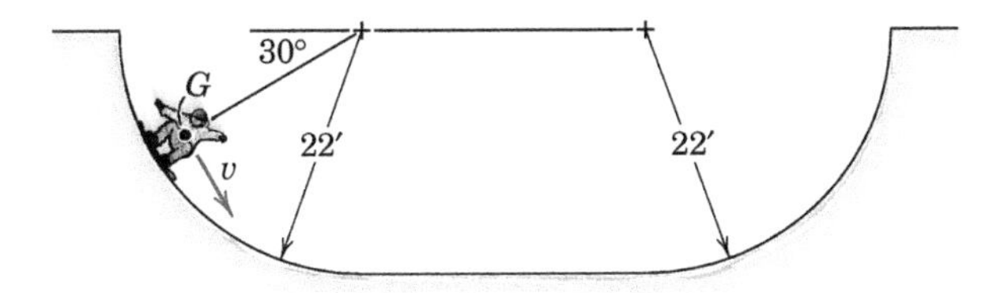

Problem 3/58

3/59 A child twirls a small 50-g ball attached to the end of a 1-m string so that the ball traces a circle in a vertical plane as shown. What is the minimum speed v which the ball must have when in position 1? If this speed is maintained throughout the circle, calculate the tension T in the string when the ball is in position 2. Neglect any small motion of the child's hand.

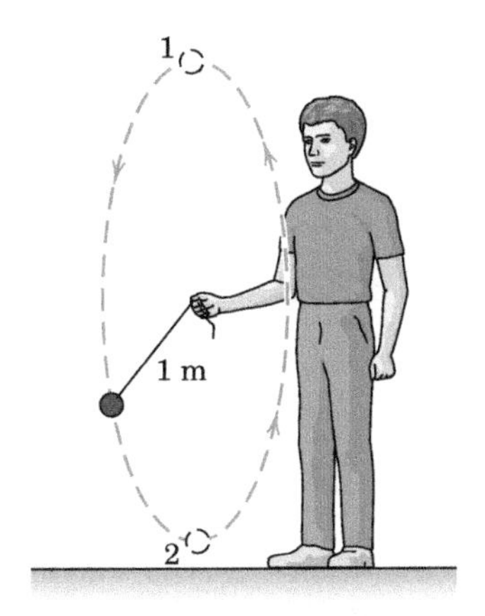

Problem 3/59

3/60 A small object A is held against the vertical side of the rotating cylindrical container of radius r by centrifugal action. If the coefficient of static friction between the object and the container is μ_s, determine the expression for the minimum rotational rate $\dot{\theta} = \omega$ of the container which will keep the object from slipping down the vertical side.

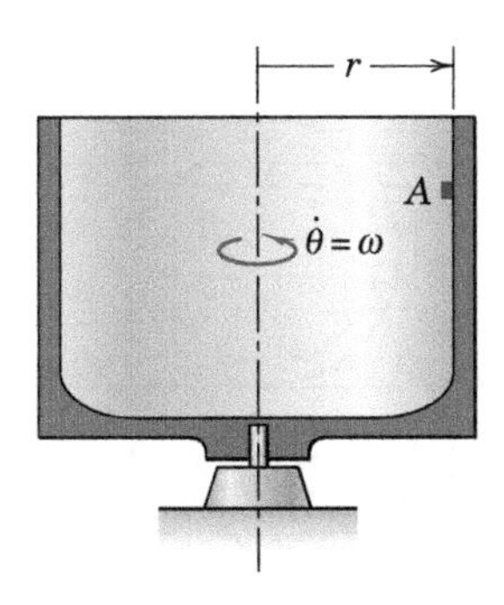

Problem 3/60

3/61 The standard test to determine the maximum lateral acceleration of a car is to drive it around a 200-ft-diameter circle painted on a level asphalt surface. The driver slowly increases the vehicle speed until he is no longer able to keep both wheel pairs straddling the line. If this maximum speed is 35 mi/hr for a 3000-lb car, determine its lateral acceleration capability a_n in g's and compute the magnitude F of the total friction force exerted by the pavement on the car tires.

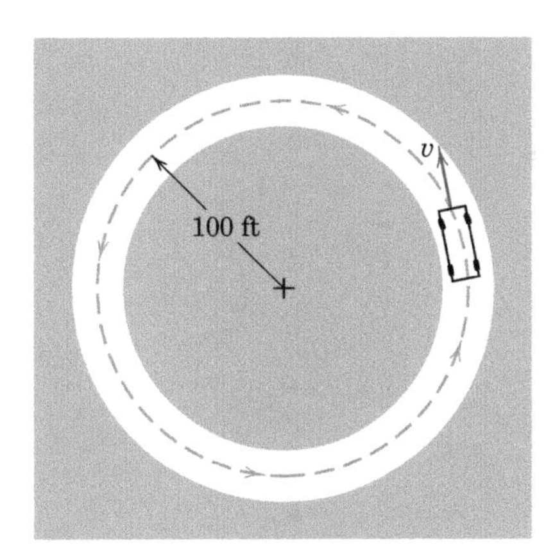

Problem 3/61

3/62 The car of Prob. 3/61 is traveling at 25 mi/hr when the driver applies the brakes, and the car continues to move along the circular path. What is the maximum deceleration possible if the tires are limited to a total horizontal friction force of 2400 lb?

Representative Problems

3/63 The flatbed truck carries a large section of circular pipe secured only by the two fixed blocks A and B of height h. The truck is in a left turn of radius ρ. Determine the maximum speed for which the pipe will be restrained. Use the values $\rho = 60$ m, $h = 0.1$ m, and $R = 0.8$ m.

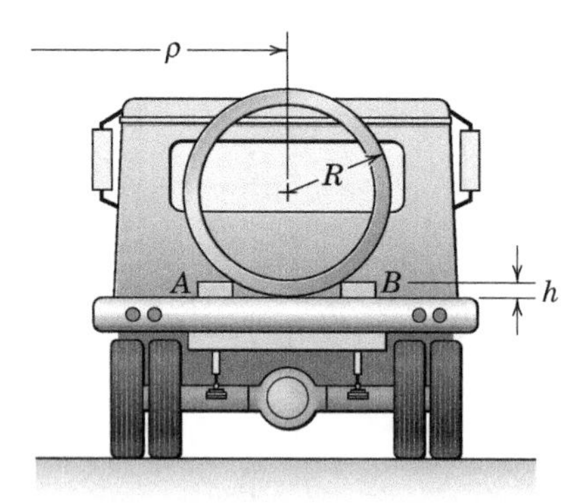

Problem 3/63

3/64 The particle of mass $m = 0.2$ kg travels with constant speed v in a circular path around the conical body. Determine the tension T in the cord. Neglect all friction, and use the values $h = 0.8$ m and $v = 0.6$ m/s. For what value of v does the normal force go to zero?

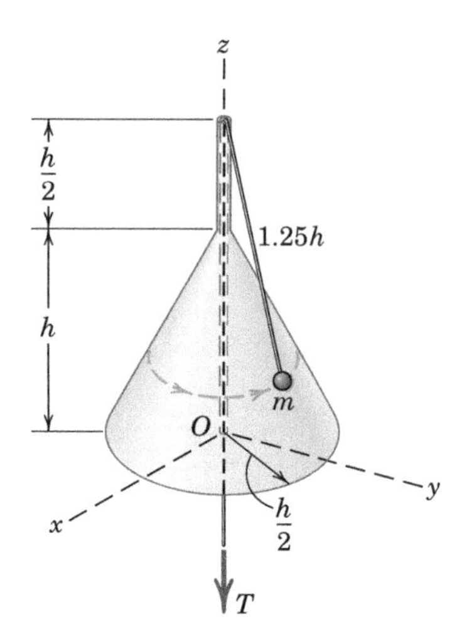

Problem 3/64

3/65 Calculate the necessary rotational speed N for the aerial ride in an amusement park in order that the arms of the gondolas will assume an angle $\theta = 60°$ with the vertical. Neglect the mass of the arms to which the gondolas are attached and treat each gondola as a particle.

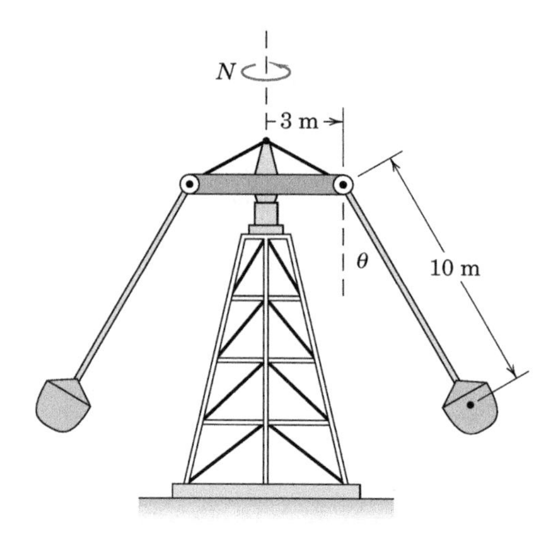

Problem 3/65

3/66 A 0.2-kg particle P is constrained to move along the vertical-plane circular slot of radius $r = 0.5$ m and is confined to the slot of arm OA, which rotates about a horizontal axis through O with a constant angular rate $\Omega = 3$ rad/s. For the instant when $\beta = 20°$, determine the force N exerted on the particle by the circular constraint and the force R exerted on it by the slotted arm.

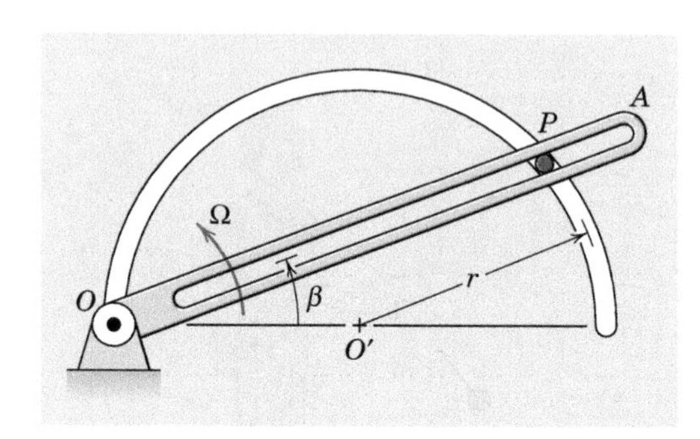

Problem 3/66

3/67 Repeat the previous problem, only now the slotted arm is rotating with angular velocity $\Omega = 3$ rad/s, and this rate is increasing at 5 rad/s^2.

3/68 At the instant under consideration, the cable attached to the cart of mass m_1 is tangent to the circular path of the cart. If the upward speed of the cylinder of mass m_2 is $v_2 = 1.2$ m/s, determine the acceleration of m_1 and the tension T in the cable. What would be the maximum speed of m_2 for which m_1 remains in contact with the surface? Use the values $R = 1.75$ m, $m_1 = 0.4$ kg, $m_2 = 0.6$ kg, and $\beta = 30°$.

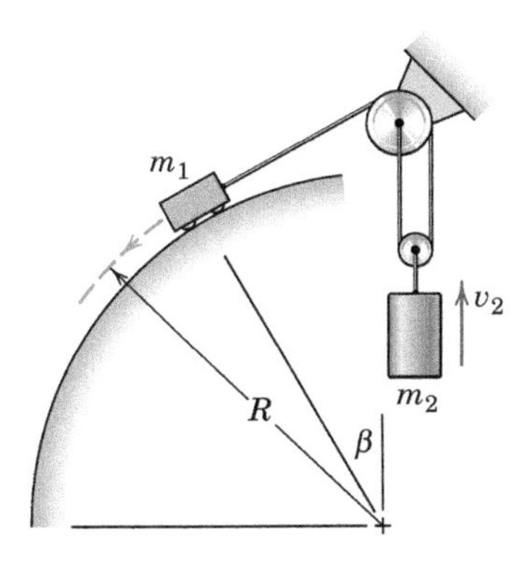

Problem 3/68

3/69 The hollow tube assembly rotates about a vertical axis with angular velocity $\omega = \dot{\theta} = 4$ rad/s and $\dot{\omega} = \ddot{\theta} = -2$ rad/s^2. A small 0.2-kg slider P moves inside the horizontal tube portion under the control of the string which passes out the bottom of the assembly. If $r = 0.8$ m, $\dot{r} = -2$ m/s, and $\ddot{r} = 4$ m/s^2, determine the tension T in the string and the horizontal force F_θ exerted on the slider by the tube.

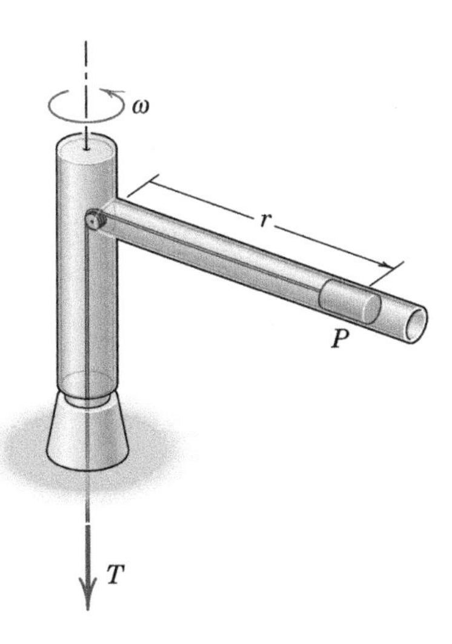

Problem 3/69

3/70 The slotted arm OA rotates about a fixed axis through O. At the instant under consideration, $\theta = 30°$, $\dot{\theta} = 45$ deg/s, and $\ddot{\theta} = 20$ deg/s^2. Determine the forces applied by both arm OA and the sides of the slot to the 0.2-kg slider B. Neglect all friction, and let $L = 0.6$ m. The motion occurs in a vertical plane.

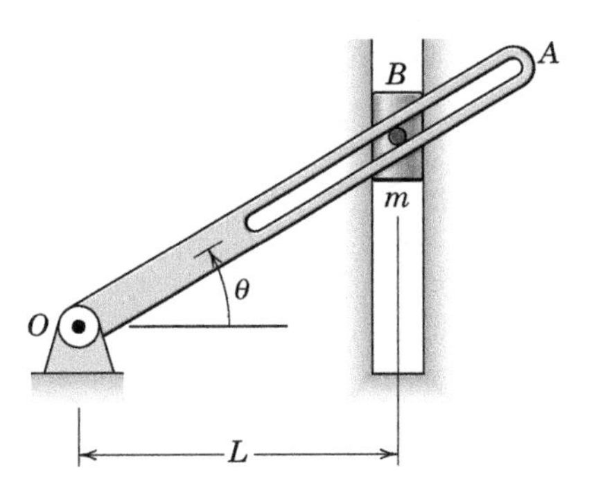

Problem 3/70

3/71 The configuration of Prob. 3/70 is now modified as shown in the figure. Use all the data of Prob. 3/70 and determine the forces applied to the slider B by both arm OA and the sides of the slot. Neglect all friction.

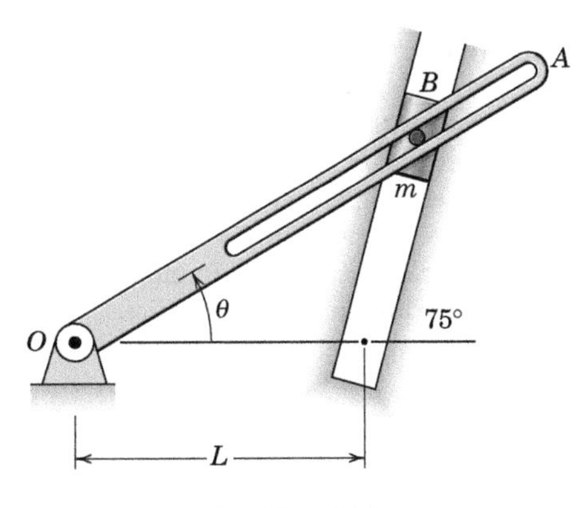

Problem 3/71

3/72 Determine the altitude h (in kilometers) above the surface of the earth at which a satellite in a circular orbit has the same period, 23.9344 h, as the earth's absolute rotation. If such an orbit lies in the equatorial plane of the earth, it is said to be geosynchronous, because the satellite does not appear to move relative to an earth-fixed observer.

3/73 The quarter-circular slotted arm OA is rotating about a horizontal axis through point O with a constant counterclockwise angular velocity Ω = 7 rad/sec. The 0.1-lb particle P is epoxied to the arm at the position $\beta = 60°$. Determine the tangential force F parallel to the slot which the epoxy must support so that the particle does not move along the slot. The value of R = 1.4 ft.

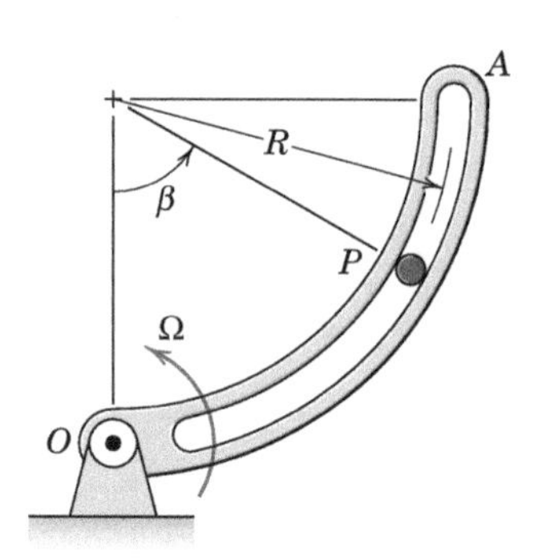

Problem 3/73

3/74 A 2-kg sphere S is being moved in a vertical plane by a robotic arm. When the angle θ is 30°, the angular velocity of the arm about a horizontal axis through O is 50 deg/s clockwise and its angular acceleration is 200 deg/s^2 counterclockwise. In addition, the hydraulic element is being shortened at the constant rate of 500 mm/s. Determine the necessary minimum gripping force P if the coefficient of static friction between the sphere and the gripping surfaces is 0.50. Compare P with the minimum gripping force P_s required to hold the sphere in static equilibrium in the 30° position.

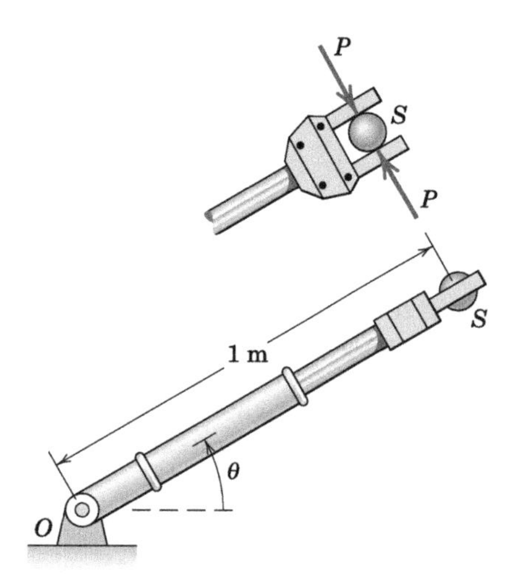

Problem 3/74

3/75 The cars of an amusement park ride have a speed v_A = 22 m/s at A and a speed v_B = 12 m/s at B. If a 75-kg rider sits on a spring scale (which registers the normal force exerted on it), determine the scale readings as the car passes points A and B. Assume that the person's arms and legs do not support appreciable force.

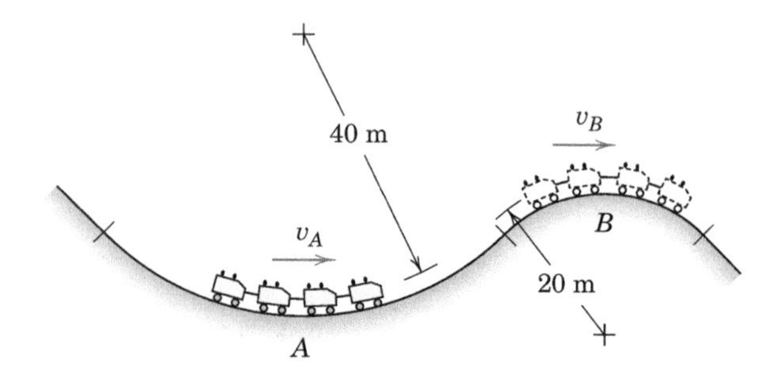

Problem 3/75

3/76 The rocket moves in a vertical plane and is being propelled by a thrust T of 32 kN. It is also subjected to an atmospheric resistance R of 9.6 kN. If the rocket has a velocity of 3 km/s and if the gravitational acceleration is 6 m/s^2 at the altitude of the rocket, calculate the radius of curvature ρ of its path for the position described and the time-rate-of-change of the magnitude v of the velocity of the rocket. The mass of the rocket at the instant considered is 2000 kg.

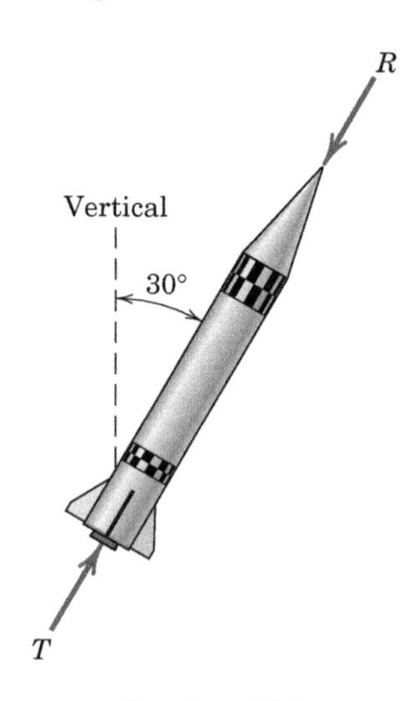

Problem 3/76

3/77 The robot arm is elevating and extending simultaneously. At a given instant, $\theta = 30°$, $\dot{\theta} = 40$ deg/s, $\ddot{\theta} = 120$ deg/s^2, $l = 0.5$ m, $\dot{l} = 0.4$ m/s, and $\ddot{l} = -0.3$ m/s^2. Compute the radial and transverse forces F_r and F_θ that the arm must exert on the gripped part P, which has a mass of 1.2 kg. Compare with the case of static equilibrium in the same position.

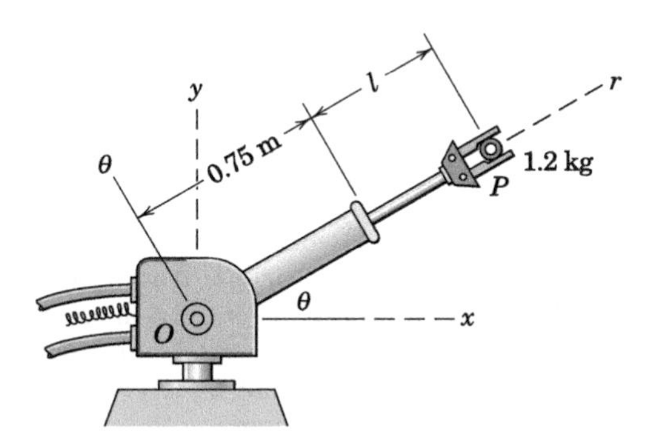

Problem 3/77

3/78 The 0.1-lb projectile A is subjected to a drag force of magnitude kv^2, where the constant $k = 0.0002$ lb-sec^2/ft^2. This drag force always opposes the velocity **v**. At the instant depicted, $v = 100$ ft/sec, $\theta = 45°$, and $r = 400$ ft. Determine the corresponding values of $\ddot{r}$ and $\ddot{\theta}$.

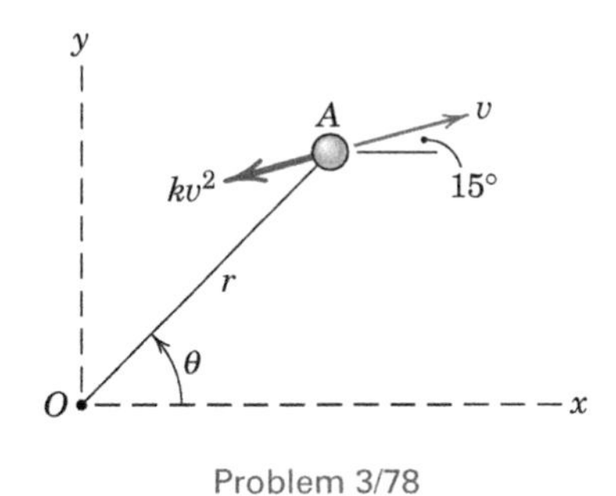

Problem 3/78

3/79 Determine the speed v at which the race car will have no tendency to slip sideways on the banked track, that is, the speed at which there is no reliance on friction. In addition, determine the minimum and maximum speeds, using the coefficient of static friction $\mu_s = 0.90$. State any assumptions.

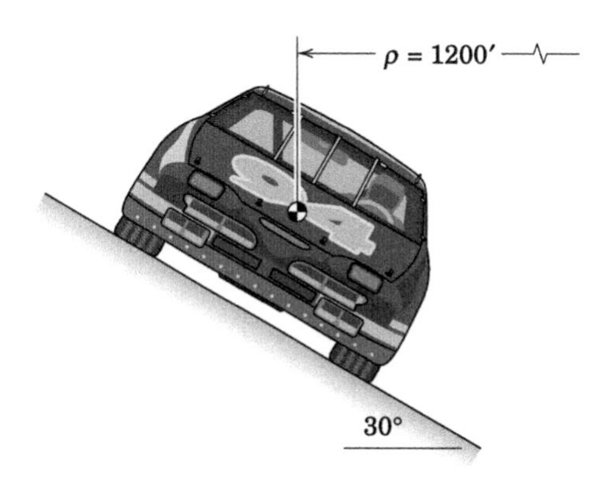

Problem 3/79

3/80 The small object is placed on the inner surface of the conical dish at the radius shown. If the coefficient of static friction between the object and the conical surface is 0.30, for what range of angular velocities ω about the vertical axis will the block remain on the dish without slipping? Assume that speed changes are made slowly so that any angular acceleration may be neglected.

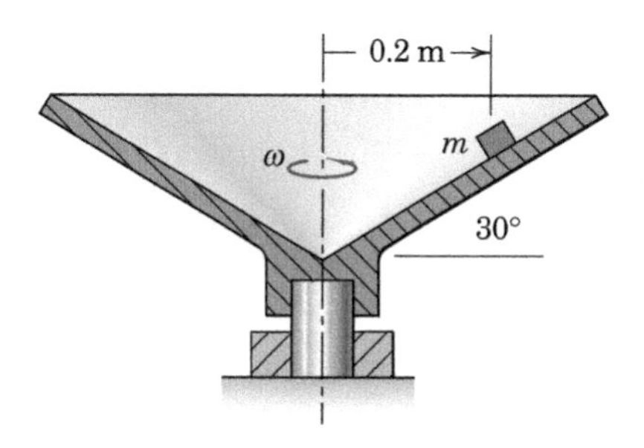

Problem 3/80

3/81 The small object of mass m is placed on the rotating conical surface at the radius shown. If the coefficient of static friction between the object and the rotating surface is 0.80, calculate the maximum angular velocity ω of the cone about the vertical axis for which the object will not slip. Assume very gradual angular-velocity changes.

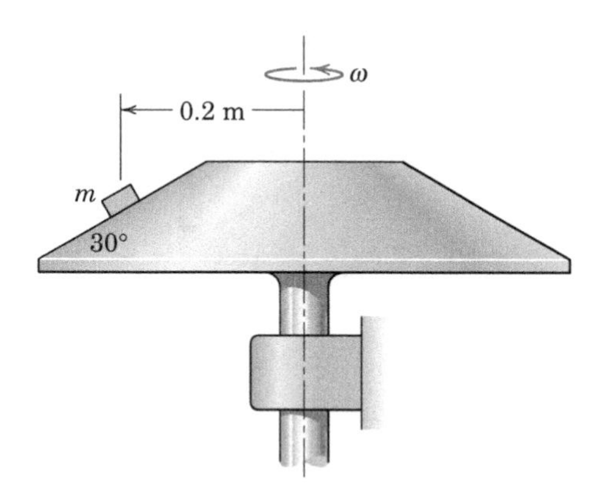

Problem 3/81

3/82 The spring-mounted 0.8-kg collar A oscillates along the horizontal rod, which is rotating at the constant angular rate $\dot{\theta} = 6$ rad/s. At a certain instant, r is increasing at the rate of 800 mm/s. If the coefficient of kinetic friction between the collar and the rod is 0.40, calculate the friction force F exerted by the rod on the collar at this instant.

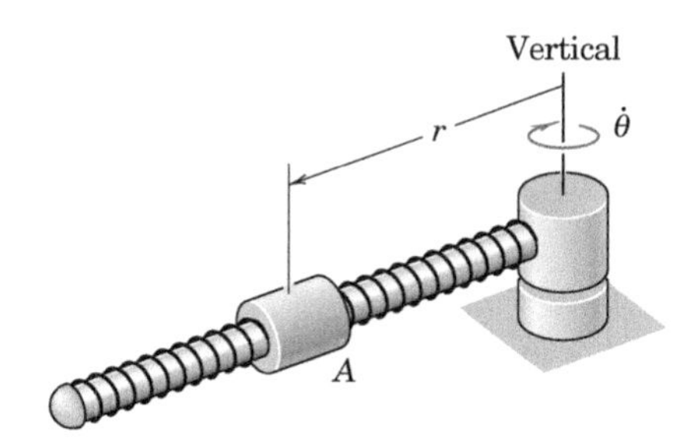

Problem 3/82

3/83 The slotted arm revolves in the horizontal plane about the fixed vertical axis through point O. The 3-lb slider C is drawn toward O at the constant rate of 2 in./sec by pulling the cord S. At the instant for which $r = 9$ in., the arm has a counterclockwise angular velocity $\omega = 6$ rad/sec and is slowing down at the rate of 2 rad/sec^2. For this instant, determine the tension T in the cord and the magnitude N of the force exerted on the slider by the sides of the smooth radial slot. Indicate which side, A or B, of the slot contacts the slider.

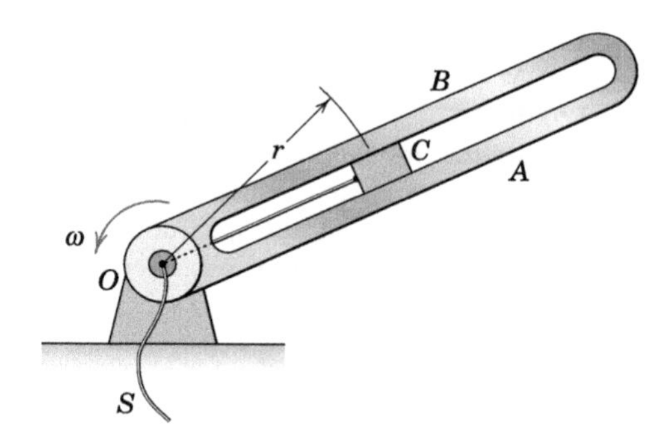

Problem 3/83

3/84 Beginning from rest when $\theta = 20°$, a 35-kg child slides with negligible friction down the sliding board which is in the shape of a 2.5-m circular arc. Determine the tangential acceleration and speed of the child, and the normal force exerted on her (*a*) when $\theta = 30°$ and (*b*) when $\theta = 90°$.

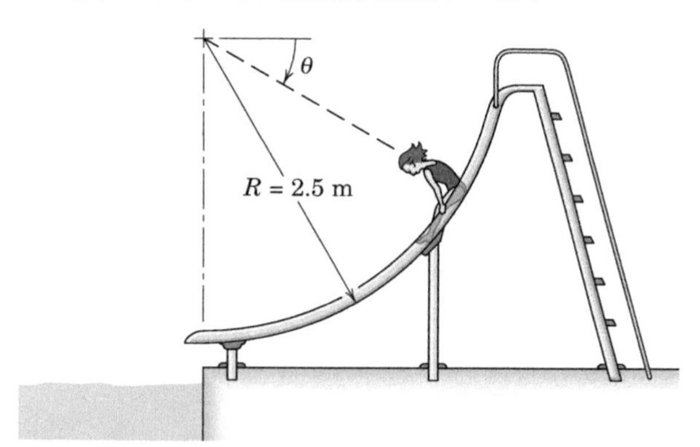

Problem 3/84

3/85 A small coin is placed on the horizontal surface of the rotating disk. If the disk starts from rest and is given a constant angular acceleration $\ddot{\theta} = \alpha$, determine an expression for the number of revolutions N through which the disk turns before the coin slips. The coefficient of static friction between the coin and the disk is μ_s.

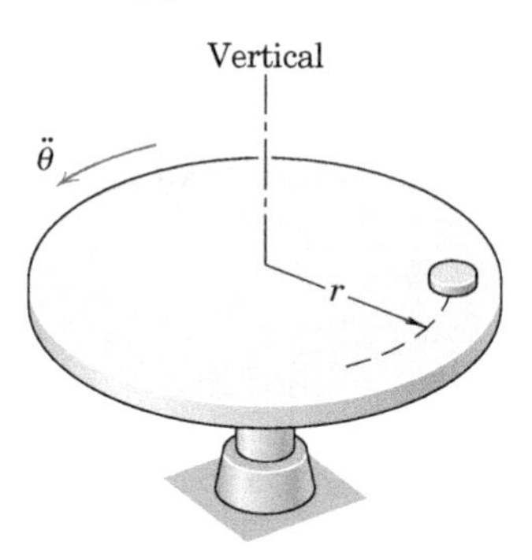

Problem 3/85

3/86 The rotating drum of a clothes dryer is shown in the figure. Determine the angular velocity Ω of the drum which results in loss of contact between the clothes and the drum at $\theta = 50°$. Assume that the small vanes prevent slipping until loss of contact.

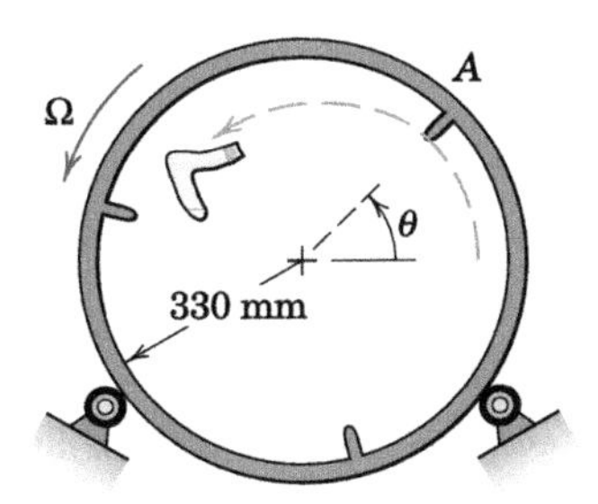

Problem 3/86

3/87 The disk spins about the fixed axis BB, which is inclined at the angle α to the vertical z-axis. A small block A is placed on the disk in its lowest position P at a distance r from the axis when the disk is at rest. The angular velocity $\omega = \dot{\theta}$ is then increased very slowly, starting from zero. At what value of ω will the block slip, and at what value of θ will the slip first occur? Use the values $\alpha = 20°$, $r = 0.4$ m, and $\mu_s = 0.60$. What is the critical value of ω if $\alpha = 0$?

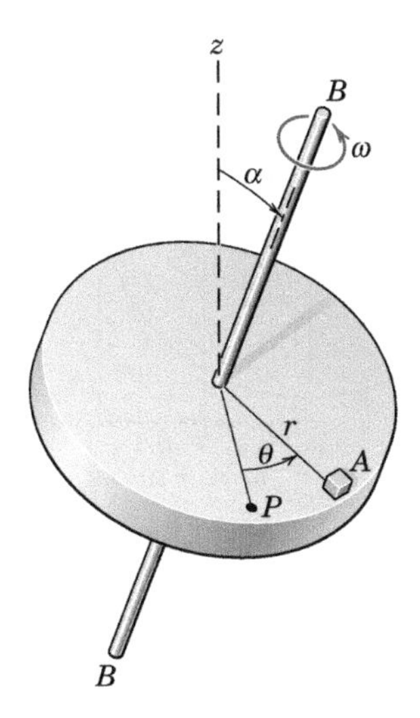

Problem 3/87

3/88 The particle P is released at time $t = 0$ from the position $r = r_0$ inside the smooth tube with no velocity relative to the tube, which is driven at the constant angular velocity ω_0 about a vertical axis. Determine the radial velocity v_r, the radial position r, and the transverse velocity v_θ as functions of time t. Explain why the radial velocity increases with time in the absence of radial forces. Plot the absolute path of the particle during the time it is inside the tube for $r_0 = 0.1$ m, $l = 1$ m, and $\omega_0 = 1$ rad/s.

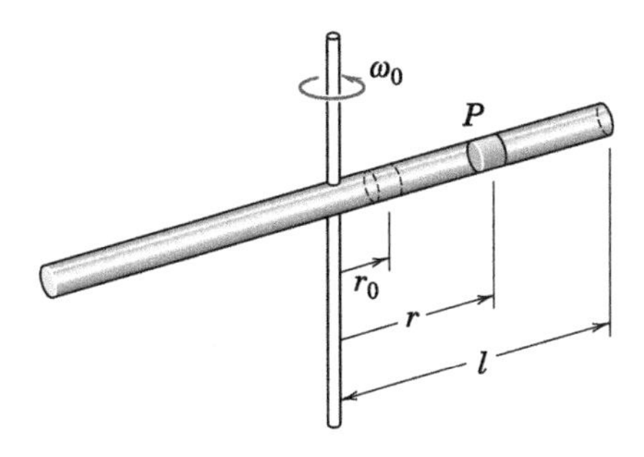

Problem 3/88

3/89 Remove the assumption of smooth surfaces as stated in Prob. 3/88 and assume a coefficient of kinetic friction μ_k between the particle and rotating tube. Determine the radial position r of the particle as a function of time t if it is released with no relative velocity at $r = r_0$ when $t = 0$. Assume that static friction is overcome.

3/90 A small vehicle enters the top A of the circular path with a horizontal velocity v_0 and gathers speed as it moves down the path. Determine an expression for the angle β which locates the point where the vehicle leaves the path and becomes a projectile. Evaluate your expression for $v_0 = 0$. Neglect friction and treat the vehicle as a particle.

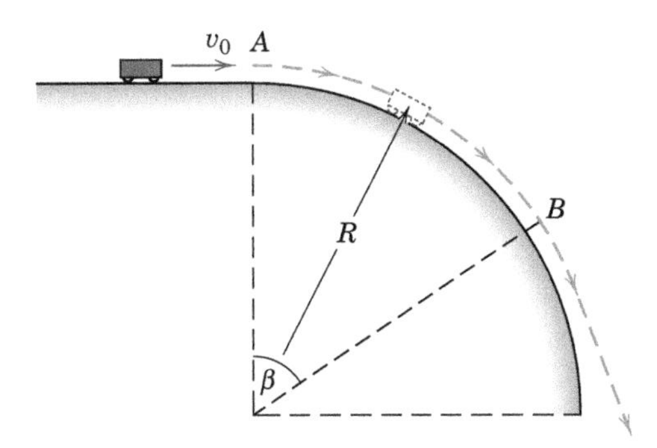

Problem 3/90

3/91 The spacecraft P is in the elliptical orbit shown. At the instant represented, its speed is $v = 13{,}244$ ft/sec. Determine the corresponding values of $\dot{r}$, $\dot{\theta}$, $\ddot{r}$, and $\ddot{\theta}$. Use $g = 32.23$ ft/sec^2 as the acceleration of gravity on the surface of the earth and $R = 3959$ mi as the radius of the earth.

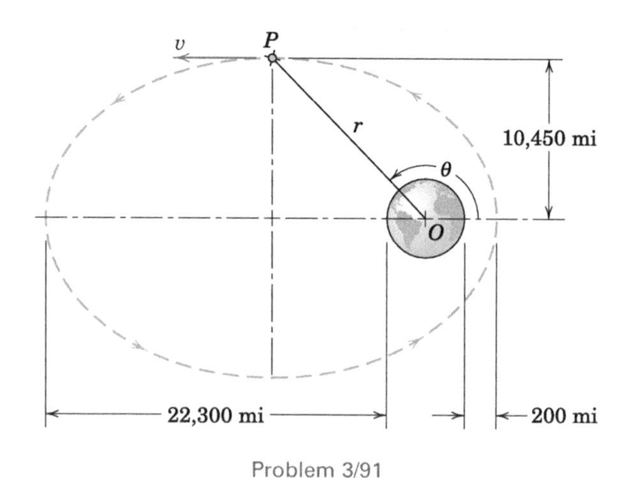

Problem 3/91

3/92 The uniform slender rod of length L, mass m, and cross-sectional area A is rotating in a horizontal plane about the vertical central axis O-O at a constant high angular velocity ω. By analyzing the horizontal forces on the accelerating differential element shown, derive an expression for the tensile stress σ in the rod as a function of r. The stress, commonly referred to as centrifugal stress, equals the tensile force divided by the cross-sectional area A.

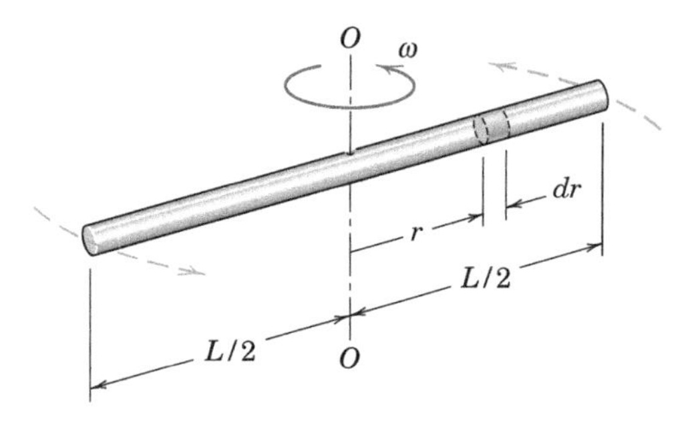

Problem 3/92

▶3/93 A small object is released from rest at A and slides with friction down the circular path. If the coefficient of friction is 0.20, determine the velocity of the object as it passes B. (*Hint:* Write the equations of motion in the n- and t-directions, eliminate N, and substitute $v\,dv = a_t r\,d\theta$. The resulting equation is a linear nonhomogeneous differential equation of the form $dy/dx + f(x)y = g(x)$, the solution of which is well known.)

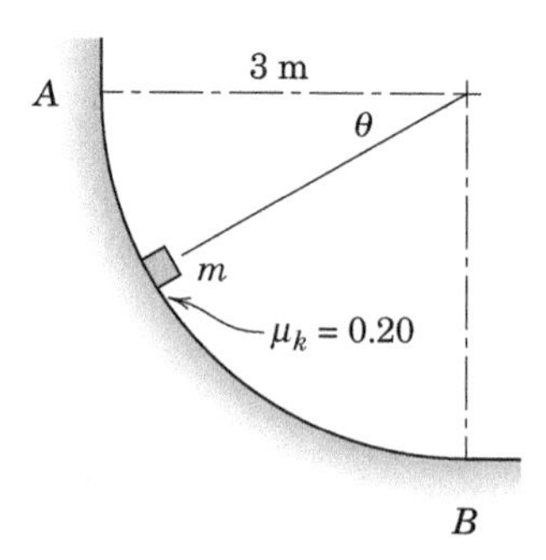

Problem 3/93

▶3/94 The slotted arm OB rotates in a horizontal plane about point O of the fixed circular cam with constant angular velocity $\dot{\theta} = 15$ rad/s. The spring has a stiffness of 5 kN/m and is uncompressed when $\theta = 0$. The smooth roller A has a mass of 0.5 kg. Determine the normal force N which the cam exerts on A and also the force R exerted on A by the sides of the slot when $\theta = 45°$. All surfaces are smooth. Neglect the small diameter of the roller.

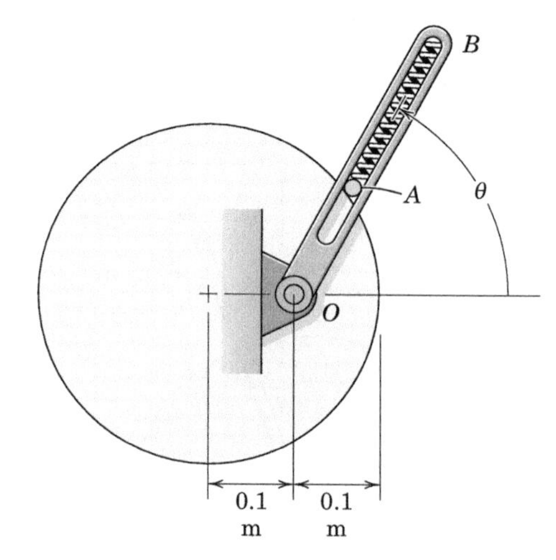

Problem 3/94

▶**3/95** Each tire on the 1350-kg car can support a maximum friction force parallel to the road surface of 2500 N. This force limit is nearly constant over all possible rectilinear and curvilinear car motions and is attainable only if the car does not skid. Under this maximum braking, determine the total stopping distance s if the brakes are first applied at point A when the car speed is 25 m/s and if the car follows the centerline of the road.

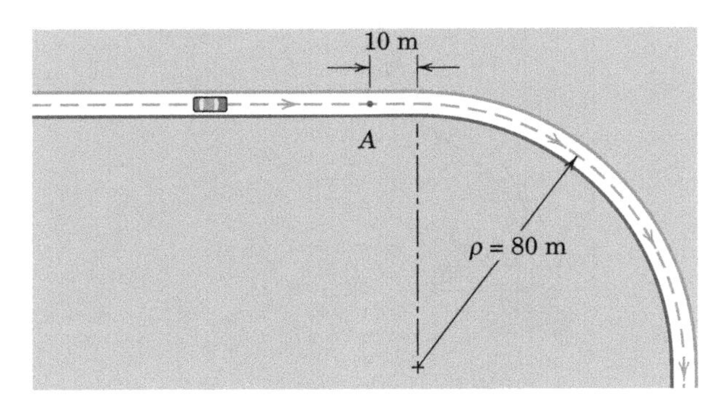

Problem 3/95

▶**3/96** A small collar of mass m is given an initial velocity of magnitude v_0 on the horizontal circular track fabricated from a slender rod. If the coefficient of kinetic friction is μ_k, determine the distance traveled before the collar comes to rest. (*Hint:* Recognize that the friction force depends on the net normal force.)

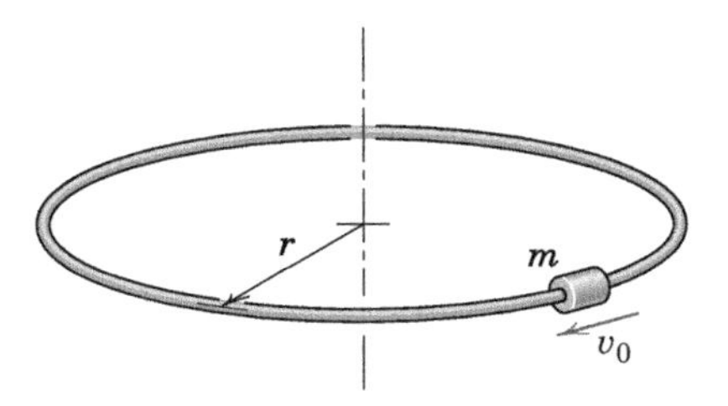

Problem 3/96

PROBLEMS

Introductory Problems

3/97 The spring is unstretched at the position $x = 0$. Under the action of a force P, the cart moves from the initial position $x_1 = -6$ in. to the final position $x_2 = 3$ in. Determine (*a*) the work done on the cart by the spring and (*b*) the work done on the cart by its weight.

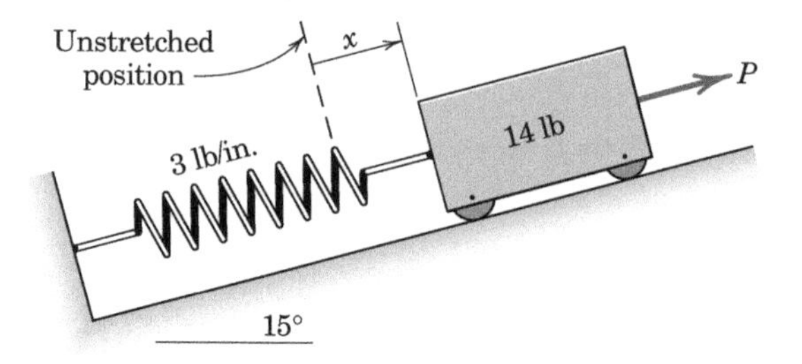

Problem 3/97

3/98 The small cart has a speed $v_A = 4$ m/s as it passes point A. It moves without appreciable friction and passes over the top hump of the track. Determine the cart speed as it passes point B. Is knowledge of the shape of the track necessary?

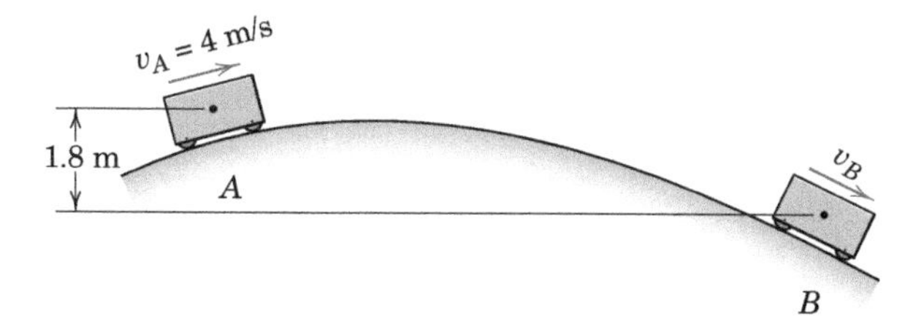

Problem 3/98

3/99 In the design of a spring bumper for a 3500-lb car, it is desired to bring the car to a stop from a speed of 5 mi/hr in a distance equal to 6 in. of spring deformation. Specify the required stiffness k for each of the two springs behind the bumper. The springs are undeformed at the start of impact.

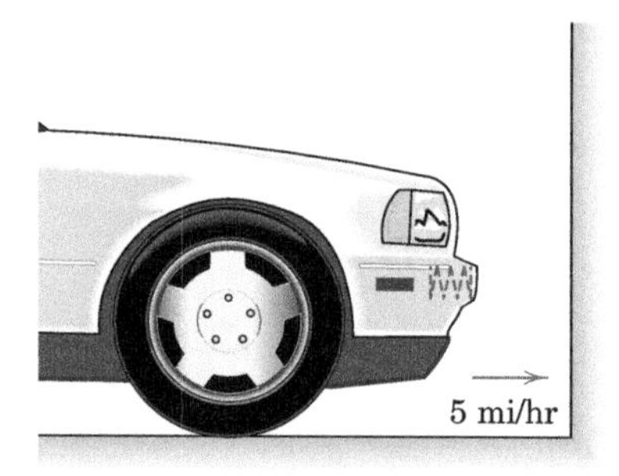

Problem 3/99

3/100 The 2-kg collar is at rest in position A when the constant force P is applied as shown. Determine the speed of the collar as it passes position B if (*a*) $P = 25$ N and (*b*) $P = 40$ N. The curved rod lies in a vertical plane, and friction is negligible.

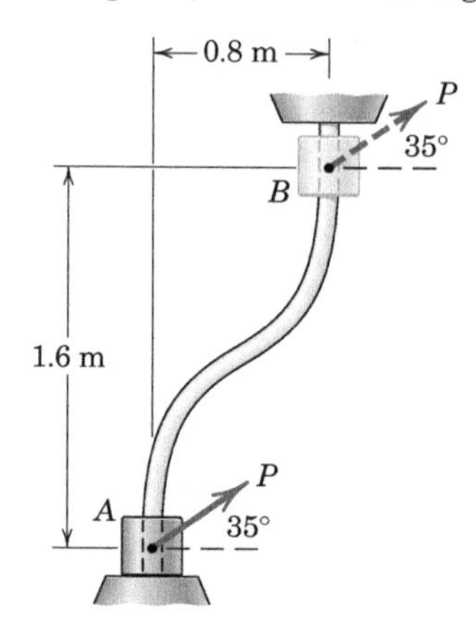

Problem 3/100

3/101 The 0.5-kg collar C starts from rest at A and slides with negligible friction on the fixed rod in the vertical plane. Determine the velocity v with which the collar strikes end B when acted upon by the 5-N force, which is constant in direction. Neglect the small dimensions of the collar.

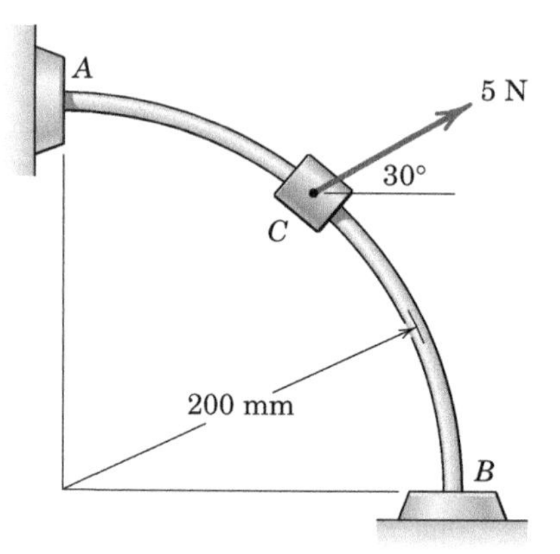

Problem 3/101

3/102 The small 0.1-kg slider enters the "loop-the-loop" with a speed $v_A = 12$ m/s as it passes point A, and it has a speed $v_B = 10$ m/s as it exits at point B. Determine the work done by friction between points A and B. The track lies in a vertical plane. Assume that contact is maintained throughout.

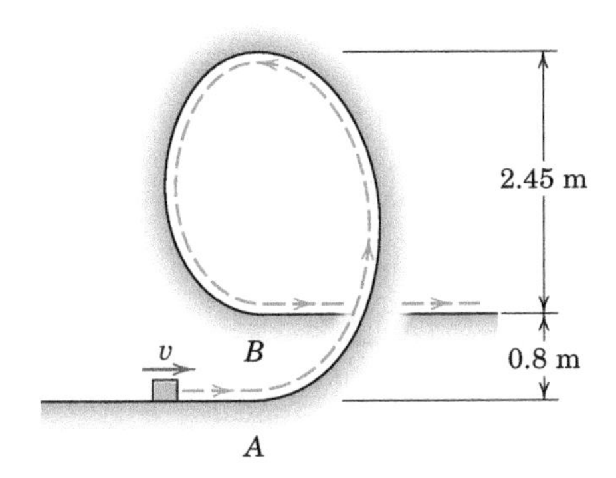

Problem 3/102

3/103 The man and his bicycle together weigh 200 lb. What power P is the man developing in riding up a 5-percent grade at a constant speed of 15 mi/hr?

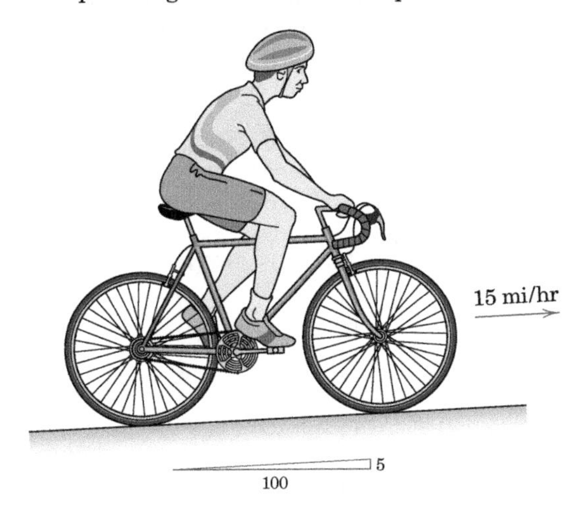

Problem 3/103

3/104 The car is moving with a speed $v_0 = 65$ mi/hr up the 6-percent grade, and the driver applies the brakes at point A, causing all wheels to skid. The coefficient of kinetic friction for the rain-slicked road is $\mu_k = 0.60$. Determine the stopping distance s_{AB}. Repeat your calculations for the case when the car is moving downhill from B to A.

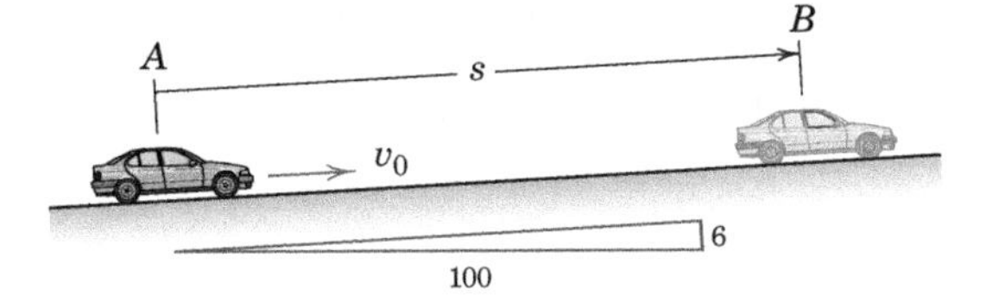

Problem 3/104

3/105 The small 0.2-kg slider is known to move from position A to position B along the vertical-plane slot. Determine (*a*) the work done on the body by its weight and (*b*) the work done on the body by the spring. The distance $R = 0.8$ m, the spring modulus $k = 180$ N/m, and the unstretched length of the spring is 0.6 m.

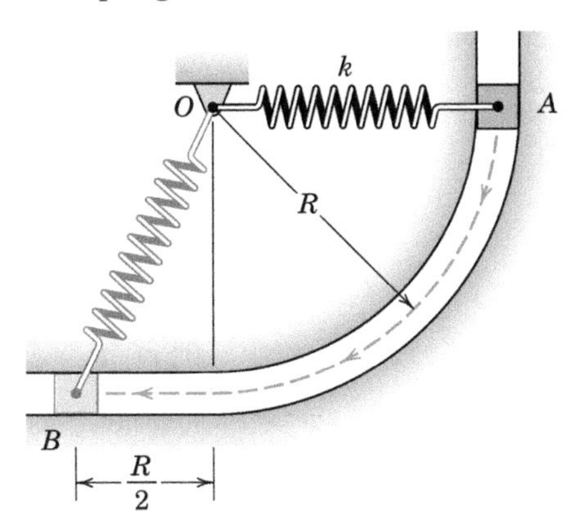

Problem 3/105

3/106 The 2-kg collar is released from rest at A and slides down the inclined fixed rod in the vertical plane. The coefficient of kinetic friction is 0.40. Calculate (*a*) the velocity v of the collar as it strikes the spring and (*b*) the maximum deflection x of the spring.

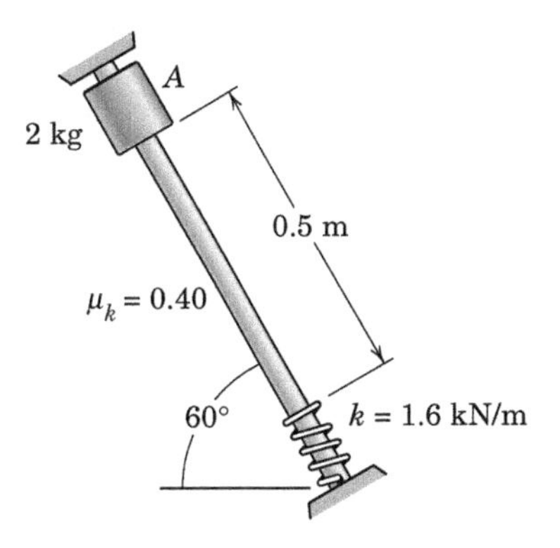

Problem 3/106

3/107 The 30-lb collar A is released from rest in the position shown and slides with negligible friction up the fixed rod inclined 30° from the horizontal under the action of a constant force $P = 50$ lb applied to the cable. Calculate the required stiffness k of the spring so that its maximum deflection equals 6 in. The position of the small pulley at B is fixed.

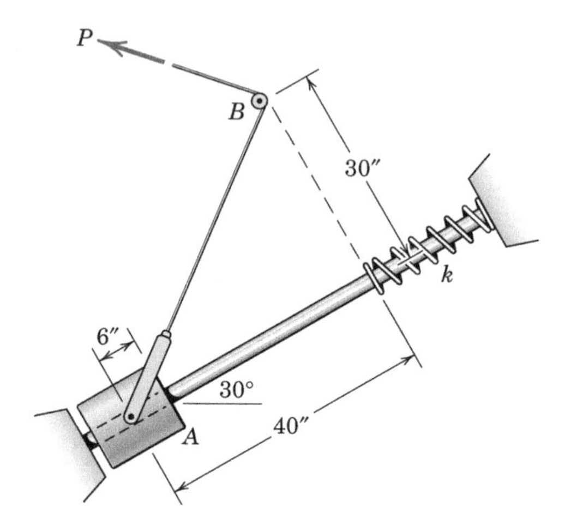

Problem 3/107

3/108 Each of the two systems is released from rest. Calculate the speed v of each 60-lb cylinder after the 40-lb cylinder has dropped 2 ft. The 30-lb cylinder of case (a) is replaced by a 30-lb force in case (b).

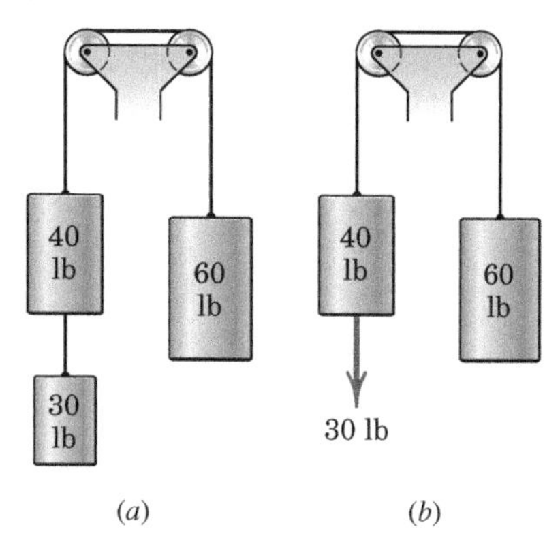

Problem 3/108

3/109 Each of the two systems is released from rest. Calculate the speed v of each 60-lb cylinder after the 40-lb cylinder has dropped 2 ft. The 30-lb cylinder of case (a) is replaced by a 30-lb force in case (b).

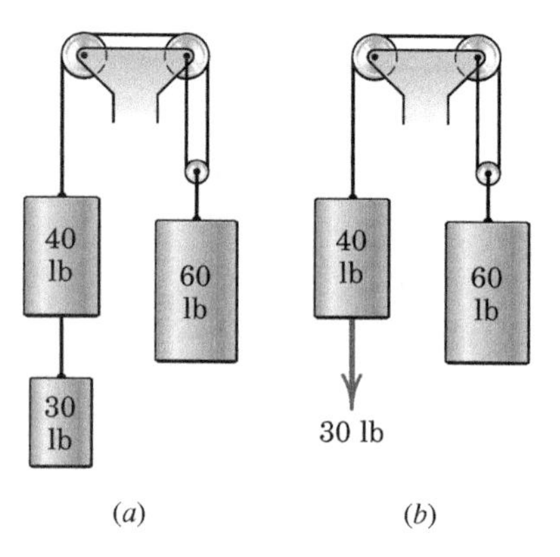

Problem 3/109

3/110 The 0.8-kg collar travels with negligible friction on the vertical rod under the action of the constant force $P = 20$ N. If the collar starts from rest at A, determine its speed as it passes point B. The value of $R = 1.6$ m.

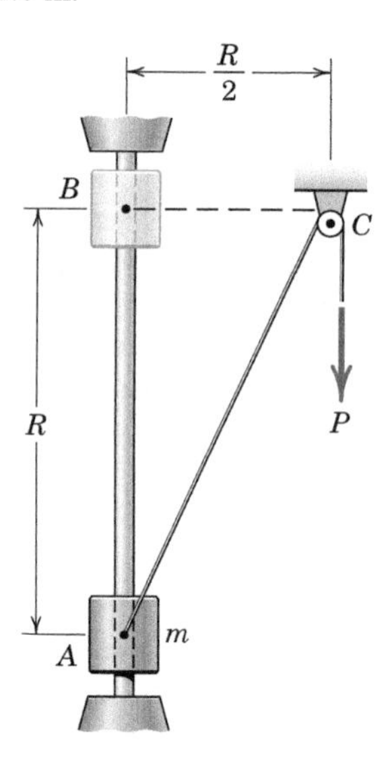

Problem 3/110

3/111 The system of the previous problem is rearranged as shown. For what constant force P will the 0.8-kg collar just reach position B with no speed after beginning from rest at position A? Friction is negligible, and $R = 1.6$ m.

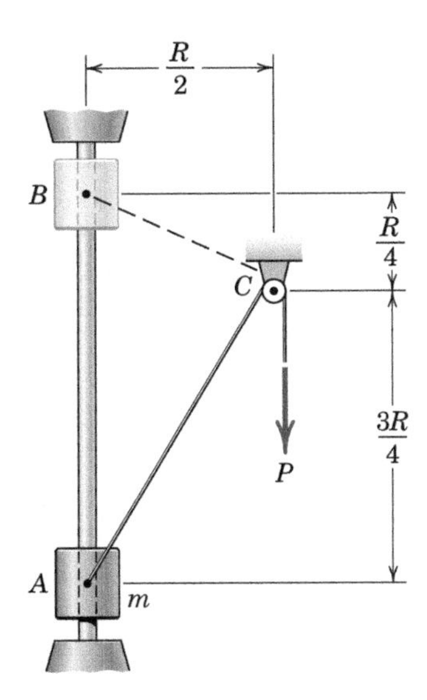

Problem 3/111

3/112 The 120-lb woman jogs up the flight of stairs in 5 seconds. Determine her average power output. Convert all given information to SI units and repeat your calculation.

Problem 3/112

Representative Problems

3/113 The 0.8-kg collar slides freely on the fixed circular rod. Calculate the velocity v of the collar as it hits the stop at B if it is elevated from rest at A by the action of the constant 40-N force in the cord. The cord is guided by the small fixed pulleys.

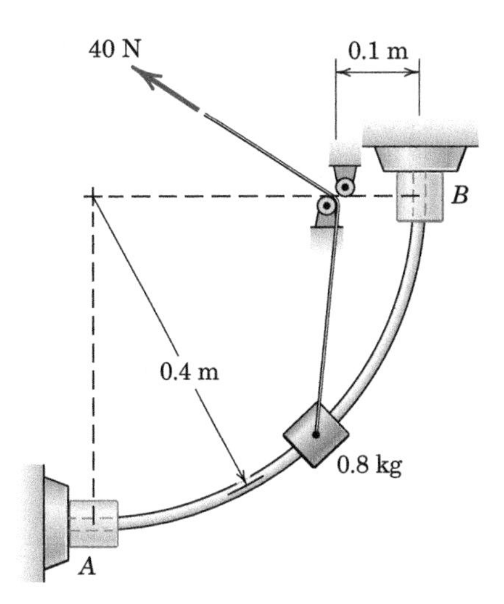

Problem 3/113

3/114 The 4-kg ball and the attached light rod rotate in the vertical plane about the fixed axis at O. If the assembly is released from rest at $\theta = 0$ and moves under the action of the 60-N force, which is maintained normal to the rod, determine the velocity v of the ball as θ approaches 90°. Treat the ball as a particle.

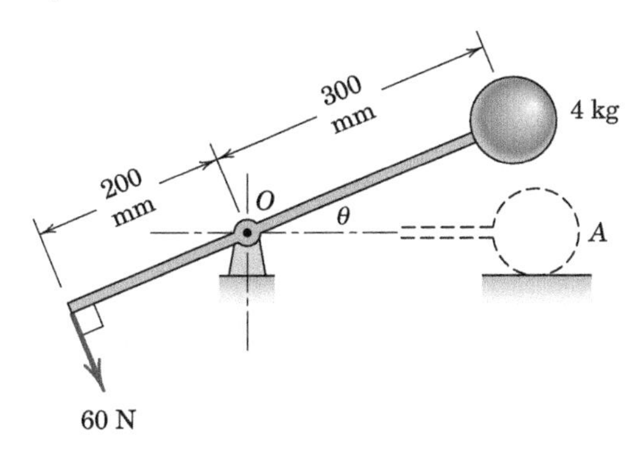

Problem 3/114

3/115 The position vector of a particle is given by $\mathbf{r} = 8t\mathbf{i} + 1.2t^2\mathbf{j} - 0.5(t^3 - 1)\mathbf{k}$, where t is the time in seconds from the start of the motion and where $\mathbf{r}$ is expressed in meters. For the condition when $t = 4$ s, determine the power P developed by the force $\mathbf{F} = 40\mathbf{i} - 20\mathbf{j} - 36\mathbf{k}$ N which acts on the particle.

3/116 An escalator handles a steady load of 30 people per minute in elevating them from the first to the second floor through a vertical rise of 24 ft. The average person weighs 140 lb. If the motor which drives the unit delivers 4 hp, calculate the mechanical efficiency e of the system.

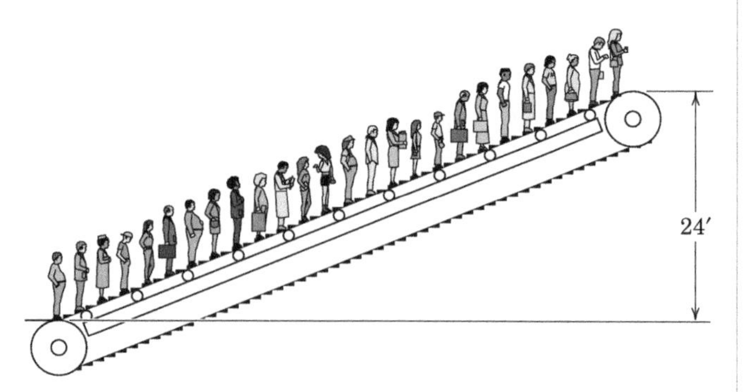

Problem 3/116

3/117 A 3600-lb car travels up the 6-percent incline shown. The car is subjected to a 60-lb aerodynamic drag force and a 50-lb force due to all other factors such as rolling resistance. Determine the power output required at a speed of 65 mi/hr if (*a*) the speed is constant and (*b*) the speed is increasing at the rate of $0.05g$.

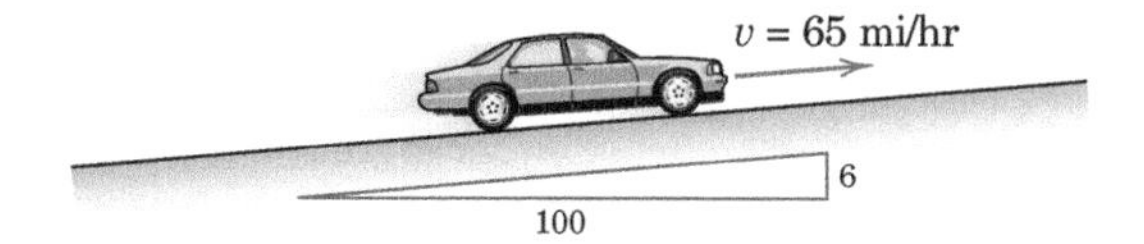

Problem 3/117

3/118 The 15-lb cylindrical collar is released from rest in the position shown and drops onto the spring. Calculate the velocity v of the cylinder when the spring has been compressed 2 in.

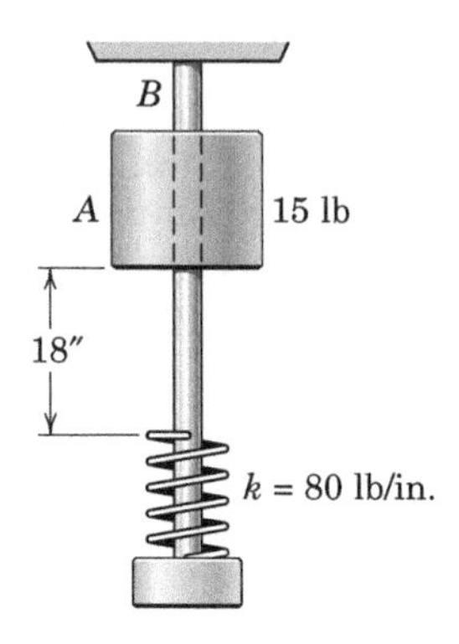

Problem 3/118

3/119 In the design of a conveyor-belt system, small metal blocks are discharged with a velocity of 0.4 m/s onto a ramp by the upper conveyor belt shown. If the coefficient of kinetic friction between the blocks and the ramp is 0.30, calculate the angle θ which the ramp must make with the horizontal so that the blocks will transfer without slipping to the lower conveyor belt moving at the speed of 0.14 m/s.

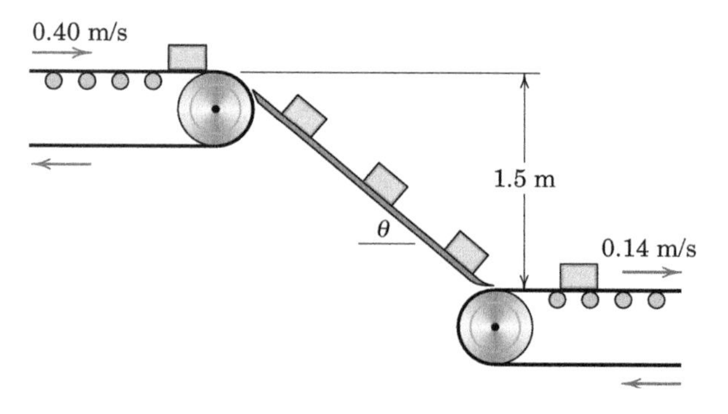

Problem 3/119

3/120 The collar of mass m is released from rest while in position A and subsequently travels with negligible friction along the vertical-plane circular guide. Determine the normal force (magnitude and direction) exerted by the guide on the collar (*a*) just before the collar passes point B, (*b*) just after the collar passes point B (i.e., the collar is now on the curved portion of the guide), (*c*) as the collar passes point C, and (*d*) just before the collar passes point D. Use the values $m = 0.4$ kg, $R = 1.2$ m, and $k = 200$ N/m. The unstretched length of the spring is $0.8R$.

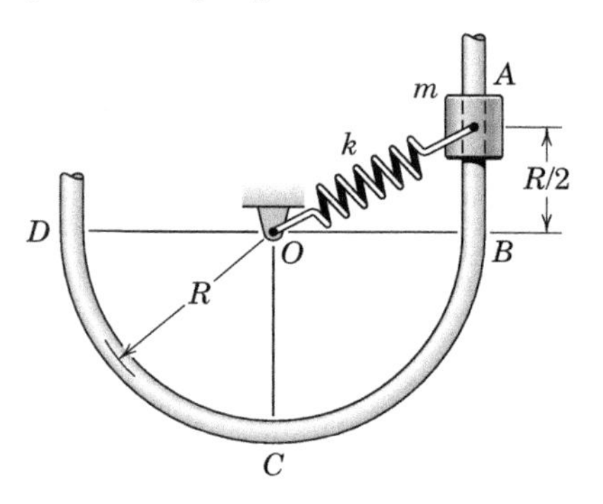

Problem 3/120

3/121 A nonlinear automobile spring is tested by having a 150-lb cylinder impact it with a speed v_0 = 12 ft/sec. The spring resistance is shown in the accompanying graph. Determine the maximum deflection δ of the spring with and without the nonlinear term present. The small platform at the top of the spring has negligible weight.

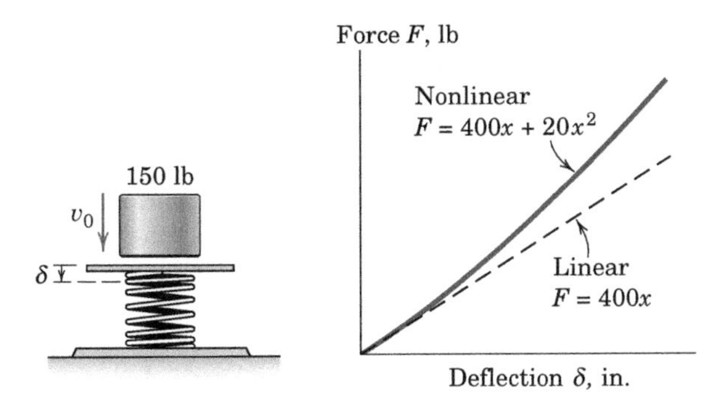

Problem 3/121

3/122 The motor unit A is used to elevate the 300-kg cylinder at a constant rate of 2 m/s. If the power meter B registers an electrical input of 2.20 kW, calculate the combined electrical and mechanical efficiency e of the system.

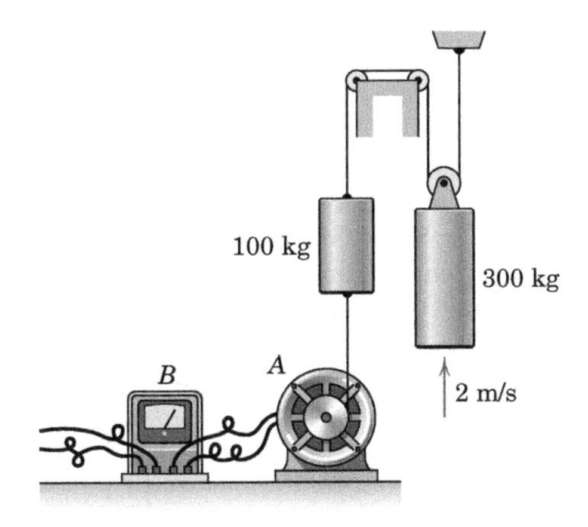

Problem 3/122

3/123 A 90-lb boy starts from rest at the bottom A of a 10-percent incline and increases his speed at a constant rate to 5 mi/hr as he passes B, 50 ft along the incline from A. Determine his power output as he approaches B.

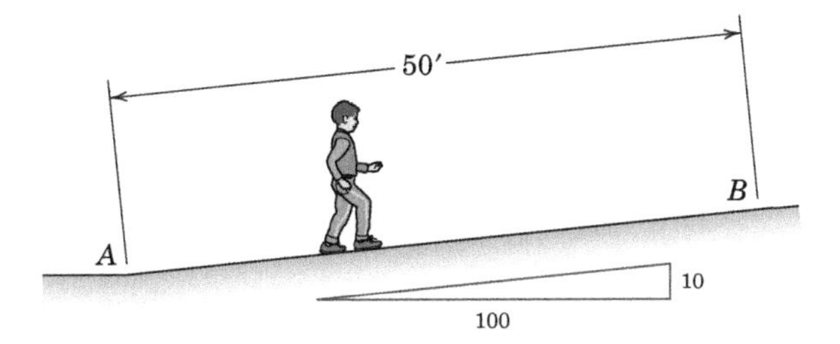

Problem 3/123

3/124 A projectile is launched from the north pole with an initial vertical velocity v_0. What value of v_0 will result in a maximum altitude of $R/3$? Neglect aerodynamic drag and use $g = 9.825$ m/s^2 as the surface-level acceleration due to gravity.

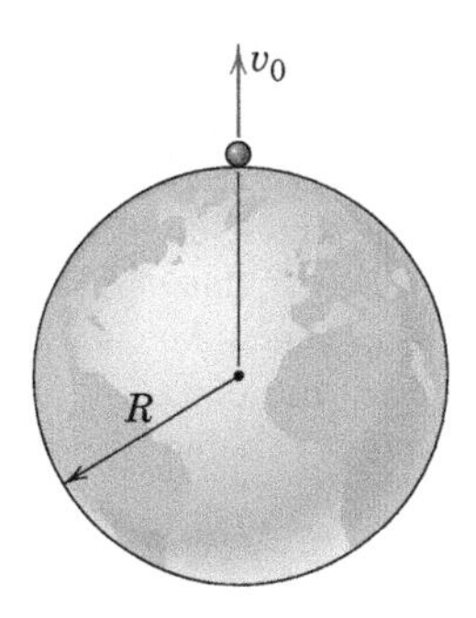

Problem 3/124

3/125 Two 425,000-lb locomotives pull fifty 200,000-lb coal hoppers. The train starts from rest and accelerates uniformly to a speed of 40 mi/hr over a distance of 8000 ft on a level track. The constant rolling resistance of each car is 0.005 times its weight. Neglect all other retarding forces and assume that each locomotive contributes equally to the tractive force. Determine (*a*) the tractive force exerted by each locomotive at 20 mi/hr, (*b*) the power required from each locomotive at 20 mi/hr, (*c*) the power required from each locomotive as the train speed approaches 40 mi/hr, and (*d*) the power required from each locomotive if the train cruises at a steady 40 mi/hr.

Problem 3/125

3/126 A car with a mass of 1500 kg starts from rest at the bottom of a 10-percent grade and acquires a speed of 50 km/h in a distance of 100 m with constant acceleration up the grade. What is the power P delivered to the drive wheels by the engine when the car reaches this speed?

3/127 The third stage of a rocket fired vertically up over the north pole coasts to a maximum altitude of 500 km following burnout of its rocket motor. Calculate the downward velocity v of the rocket when it has fallen 100 km from its position of maximum altitude. (Use the mean value of 9.825 m/s^2 for g and 6371 km for the mean radius of the earth.)

3/128 The small slider of mass m is released from rest while in position A and then slides along the vertical-plane track. The track is smooth from A to D and rough (coefficient of kinetic friction μ_k) from point D on. Determine (*a*) the normal force N_B exerted by the track on the slider just after it passes point B, (*b*) the normal force N_C exerted by the track on the slider as it passes the bottom point C, and (*c*) the distance s traveled along the incline past point D before the slider stops.

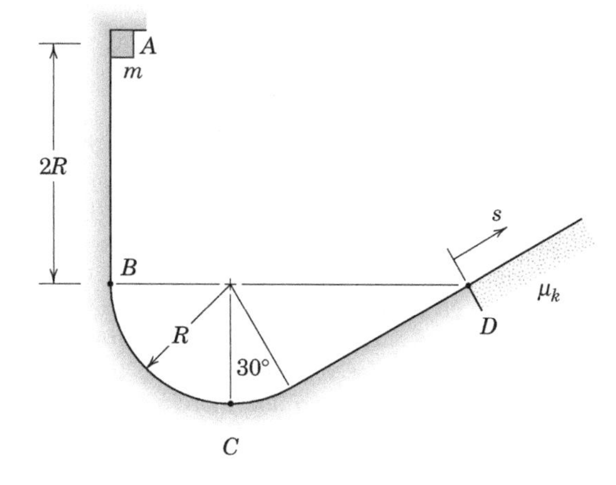

Problem 3/128

3/129 In a railroad classification yard, a 68-Mg freight car moving at 0.5 m/s at A encounters a retarder section of track at B which exerts a retarding force of 32 kN on the car in the direction opposite to motion. Over what distance x should the retarder be activated in order to limit the speed of the car to 3 m/s at C?

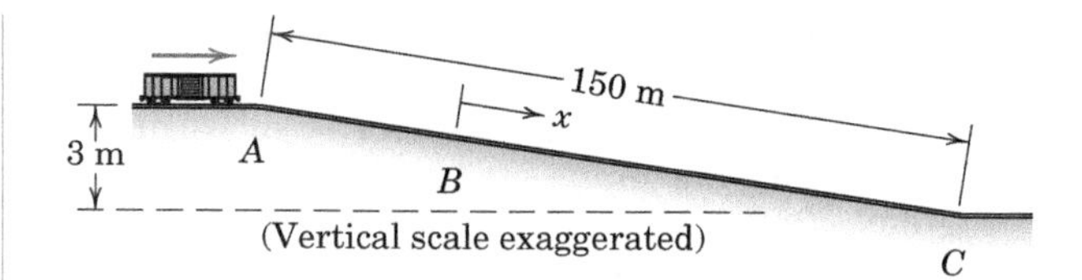

Problem 3/129

3/130 The system is released from rest with no slack in the cable and with the spring unstretched. Determine the distance s traveled by the 4-kg cart before it comes to rest (*a*) if m approaches zero and (*b*) if $m = 3$ kg. Assume no mechanical interference and no friction, and state whether the distance traveled is up or down the incline.

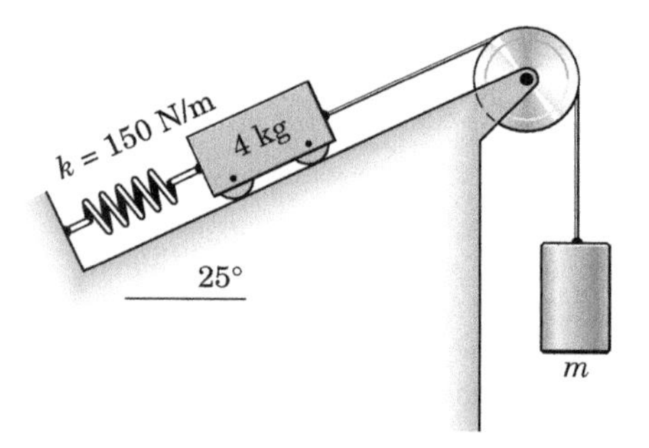

Problem 3/130

3/131 The system is released from rest with no slack in the cable and with the spring stretched 200 mm. Determine the distance s traveled by the 4-kg cart before it comes to rest (*a*) if m approaches zero and (*b*) if $m = 3$ kg. Assume no mechanical interference and no friction, and state whether the distance traveled is up or down the incline.

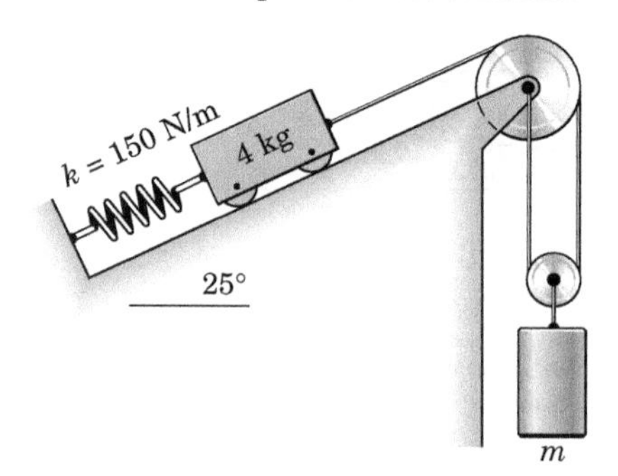

Problem 3/131

3/132 It is experimentally determined that the drive wheels of a car must exert a tractive force of 560 N on the road surface in order to maintain a steady vehicle speed of 90 km/h on a horizontal road. If it is known that the overall drivetrain efficiency is $e_m = 0.70$, determine the required motor power output P.

3/133 Once under way at a steady speed, the 1000-kg elevator A rises at the rate of 1 story (3 m) per second. Determine the power input P_{in} into the motor unit M if the combined mechanical and electrical efficiency of the system is $e = 0.8$.

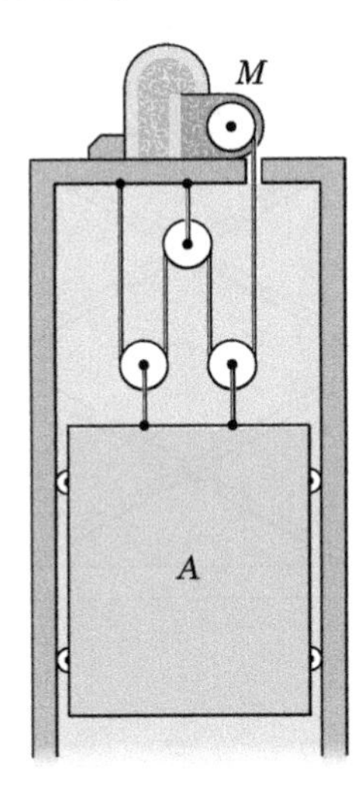

Problem 3/133

3/134 Calculate the horizontal velocity v with which the 48-lb carriage must strike the spring in order to compress it a maximum of 4 in. The spring is known as a "hardening" spring, since its stiffness increases with deflection as shown in the accompanying graph.

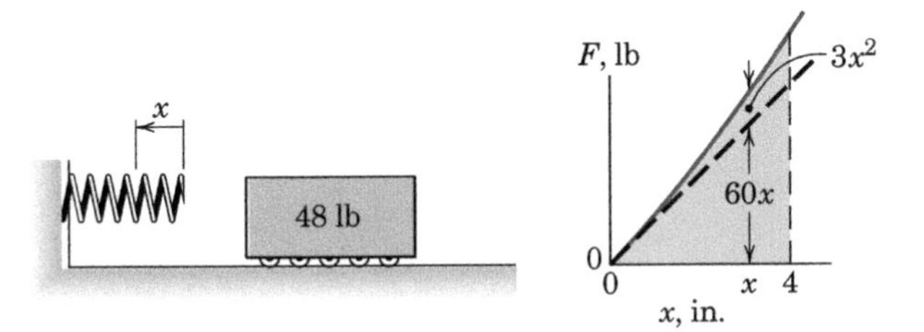

Problem 3/134

3/135 The 6-kg cylinder is released from rest in the position shown and falls on the spring, which has been initially precompressed 50 mm by the light strap and restraining wires. If the stiffness of the spring is 4 kN/m, compute the additional deflection δ of the spring produced by the falling cylinder before it rebounds.

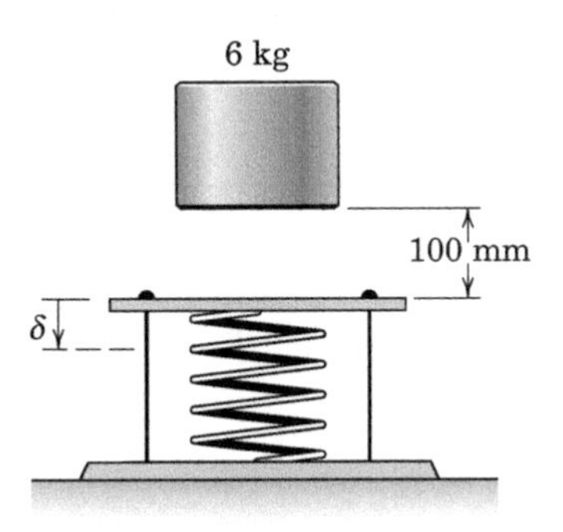

Problem 3/135

3/136 The nest of two springs is used to bring the 0.5-kg plunger A to a stop from a speed of 5 m/s and reverse its direction of motion. The inner spring increases the deceleration, and the adjustment of its position is used to control the exact point at which the reversal takes place. If this point is to correspond to a maximum deflection $\delta = 200$ mm for the outer spring, specify the adjustment of the inner spring by determining the distance s. The outer spring has a stiffness of 300 N/m and the inner one a stiffness of 150 N/m.

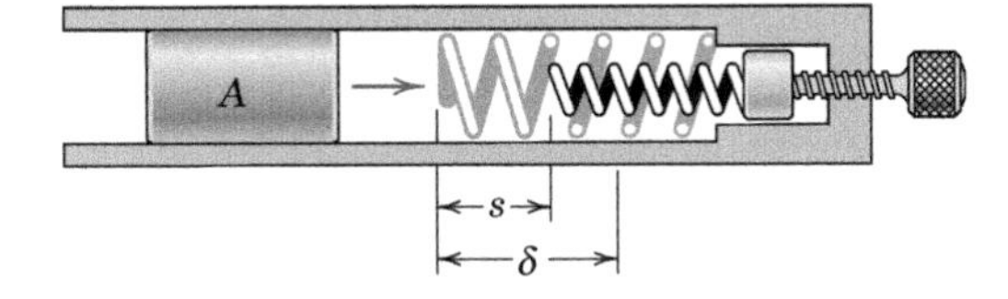

Problem 3/136

3/137 Extensive testing of an experimental 2000-lb automobile reveals the aerodynamic drag force F_D and the total nonaerodynamic rolling-resistance force F_R to be as shown in the plot. Determine (*a*) the power required for steady speeds of 30 and 60 mi/hr on a level road, (*b*) the power required for a steady speed of 60 mi/hr both up and down a 6-percent incline, and (*c*) the steady speed at which no power is required going down the 6-percent incline.

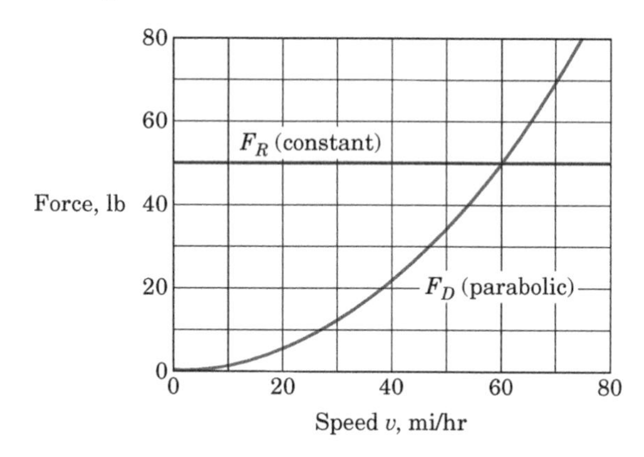

Problem 3/137

3/138 The vertical motion of the 50-lb block is controlled by the two forces P applied to the ends A and B of the linkage, where A and B are constrained to move in the horizontal guide. If forces $P = 250$ lb are applied with the linkage initially at rest with $\theta = 60°$, determine the upward velocity v of the block as θ approaches 180°. Neglect friction and the weight of the links and note that P is greater than its equilibrium value of $(5W/2) \cot 30° = 217$ lb.

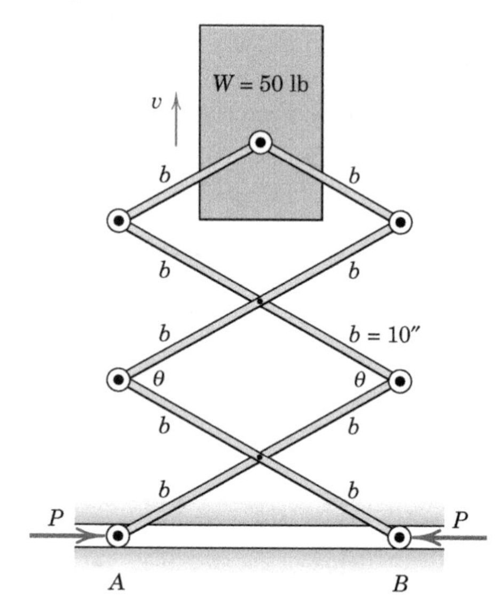

Problem 3/138

PROBLEMS

Introductory Problems

3/139 The two particles of equal mass are joined by a rod of negligible mass. If they are released from rest in the position shown and slide on the smooth guide in the vertical plane, calculate their velocity v when A reaches B's position and B is at B'.

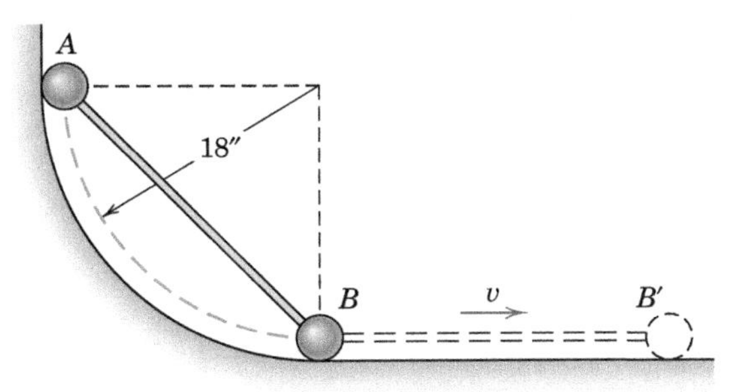

Problem 3/139

3/140 The 1.2-kg slider is released from rest in position A and slides without friction along the vertical-plane guide shown. Determine (*a*) the speed v_B of the slider as it passes position B and (*b*) the maximum deflection δ of the spring.

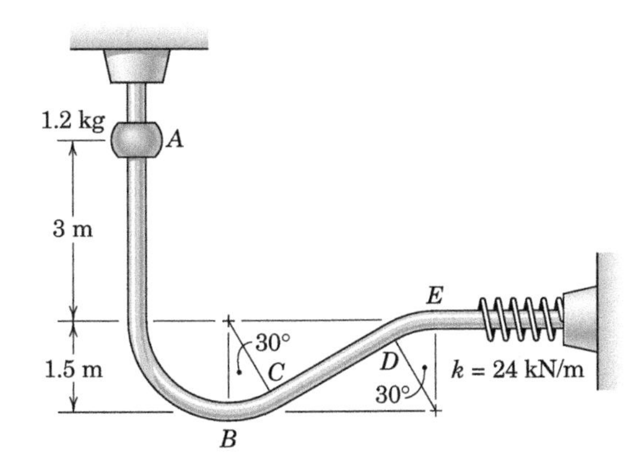

Problem 3/140

3/141 The 2-kg plunger is released from rest in the position shown where the spring of stiffness $k = 500$ N/m has been compressed to one-half its uncompressed length of 200 mm. Calculate the maximum height h above the starting position reached by the plunger.

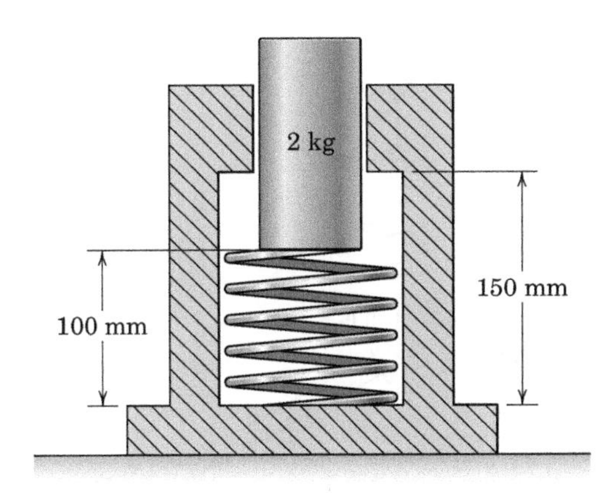

Problem 3/141

3/142 The system is released from rest with the spring initially stretched 3 in. Calculate the velocity v of the cylinder after it has dropped 0.5 in. The spring has a stiffness of 6 lb/in. Neglect the mass of the small pulley.

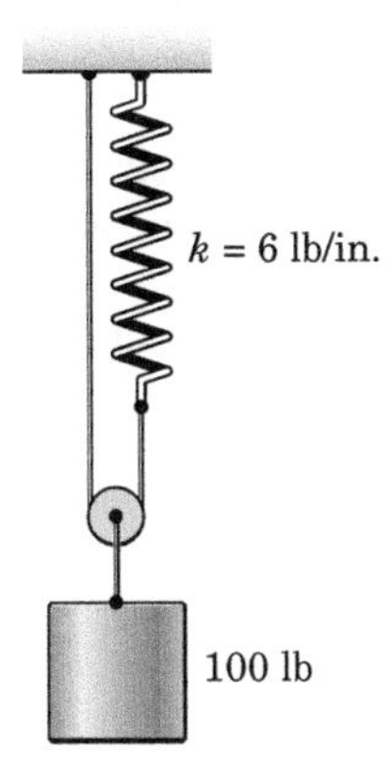

Problem 3/142

3/143 The 3-lb collar is released from rest at A and slides freely down the inclined rod. If the spring constant $k = 4$ lb/ft and the unstretched length of the spring is 50 in., determine the speed of the collar as it passes point B.

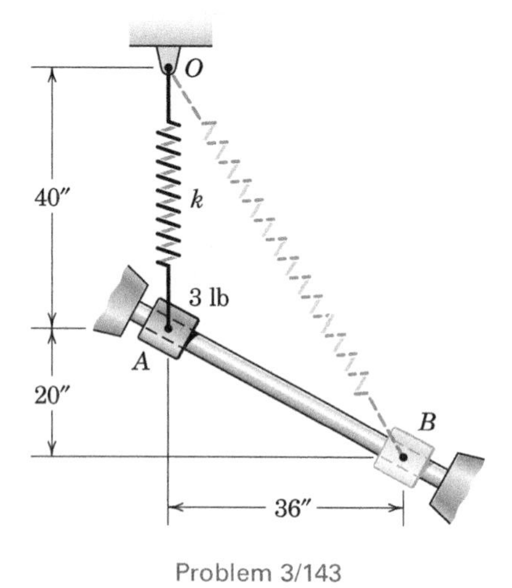

Problem 3/143

3/144 Determine the unstretched spring length which would cause the 3-lb collar of the previous problem to have no speed as it arrives at position B. All other conditions of the previous problem remain the same.

3/145 A bead with a mass of 0.25 kg is released from rest at A and slides down and around the fixed smooth wire. Determine the force N between the wire and the bead as it passes point B.

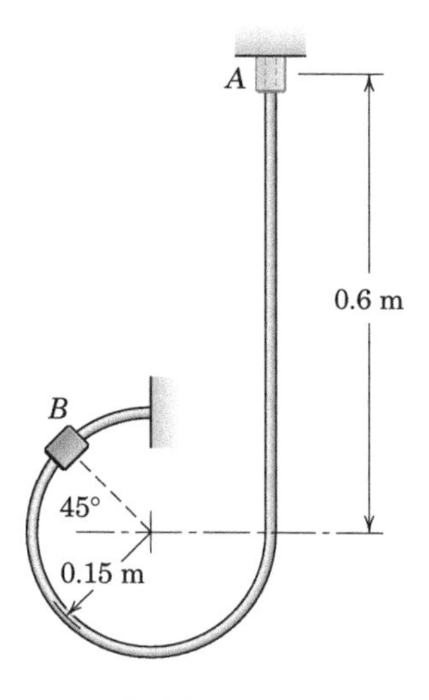

Problem 3/145

3/146 The 0.8-kg particle is attached to the system of two light rigid bars, all of which move in a vertical plane. The spring is compressed an amount $b/2$ when $\theta = 0$, and the length $b = 0.30$ m. The system is released from rest in a position slightly above that for $\theta = 0$. (a) If the maximum value of θ is observed to be 50°, determine the spring constant k. (b) For $k = 400$ N/m, determine the speed v of the particle when $\theta = 25°$. Also find the corresponding value of $\dot{\theta}$.

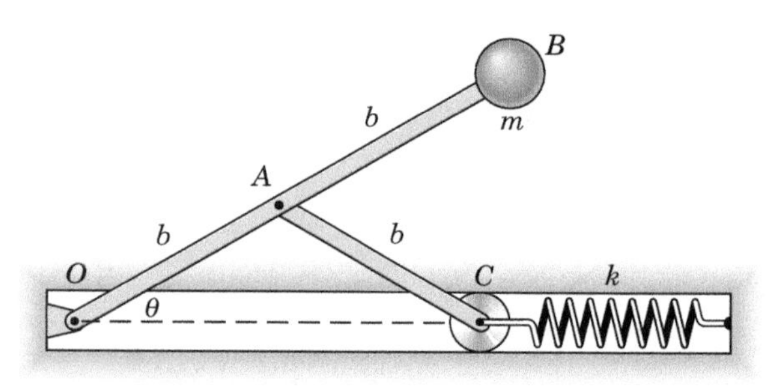

Problem 3/146

3/147 The light rod is pivoted at O and carries the 8- and 10-lb particles. If the rod is released from rest at $\theta = 30°$ and swings in the vertical plane, calculate (a) the velocity v of the 8-lb particle just before it hits the spring and (b) the maximum compression x of the spring. Assume that x is small so that the position of the rod when the spring is compressed is essentially vertical.

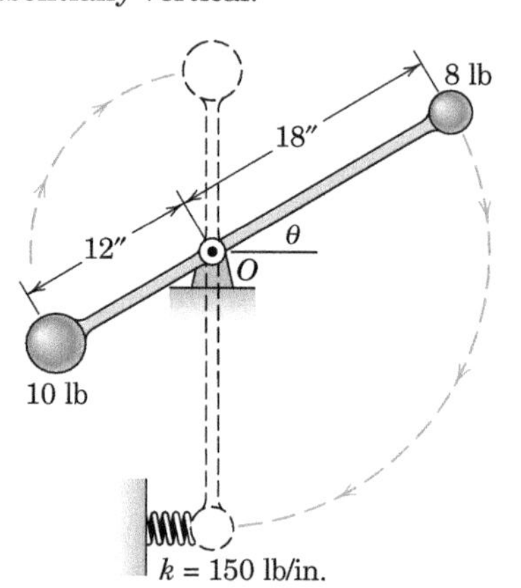

Problem 3/147

3/148 The two springs, each of stiffness $k = 1.2$ kN/m, are of equal length and undeformed when $\theta = 0$. If the mechanism is released from rest in the position $\theta = 20°$, determine its angular velocity $\dot{\theta}$ when $\theta = 0$. The mass m of each sphere is 3 kg. Treat the spheres as particles and neglect the masses of the light rods and springs.

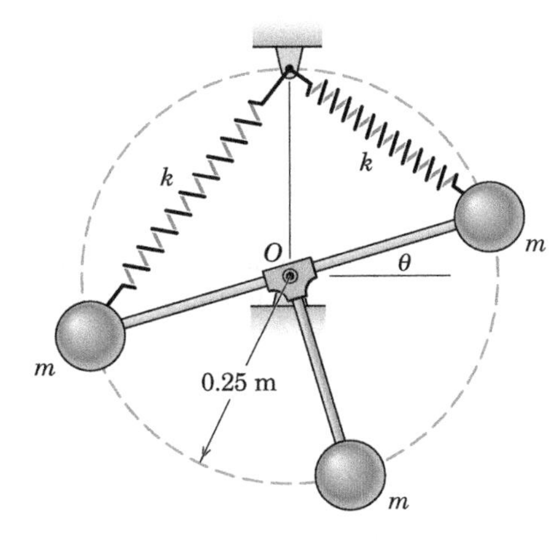

Problem 3/148

3/149 The particle of mass $m = 1.2$ kg is attached to the end of the light rigid bar of length $L = 0.6$ m. The system is released from rest while in the horizontal position shown, at which the torsional spring is undeflected. The bar is then observed to rotate 30° before stopping momentarily. (*a*) Determine the value of the torsional spring constant k_T. (*b*) For this value of k_T, determine the speed v of the particle when $\theta = 15°$.

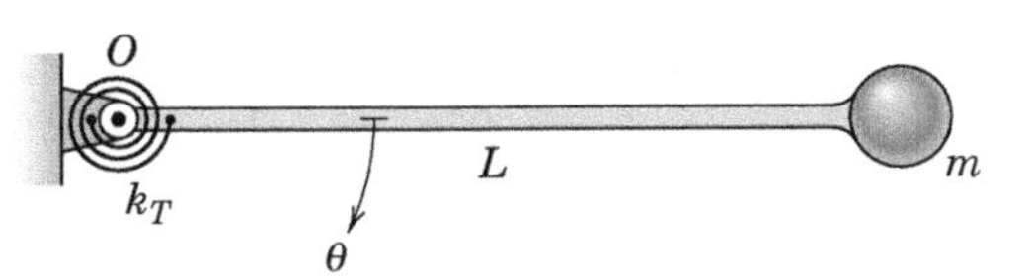

Problem 3/149

Representative Problems

3/150 The 10-kg collar slides on the smooth vertical rod and has a velocity $v_1 = 2$ m/s in position A where each spring is stretched 0.1 m. Calculate the velocity v_2 of the collar as it passes point B.

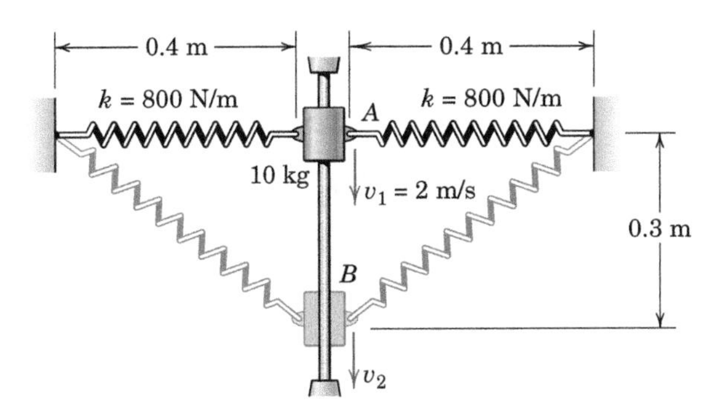

Problem 3/150

3/151 The system is released from rest with the spring initially stretched 2 in. Calculate the velocity of the 100-lb cylinder after it has dropped 6 in. Also determine the maximum drop distance of the cylinder. Neglect the mass and friction of the pulleys.

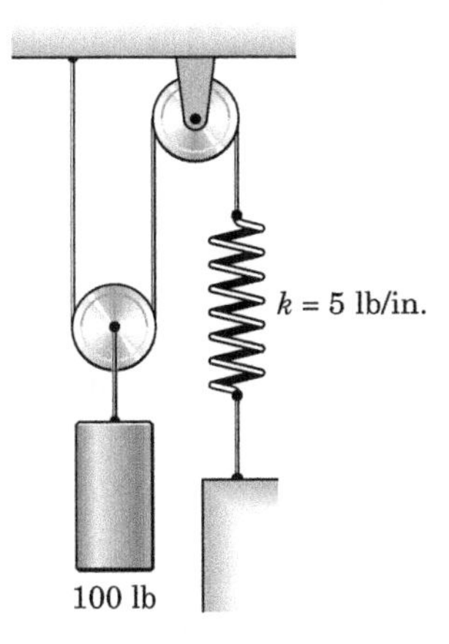

Problem 3/151

3/152 The spring has an unstretched length of 25 in. If the system is released from rest in the position shown, determine the speed v of the ball (*a*) when it has dropped a vertical distance of 10 in. and (*b*) when the rod has rotated 35°.

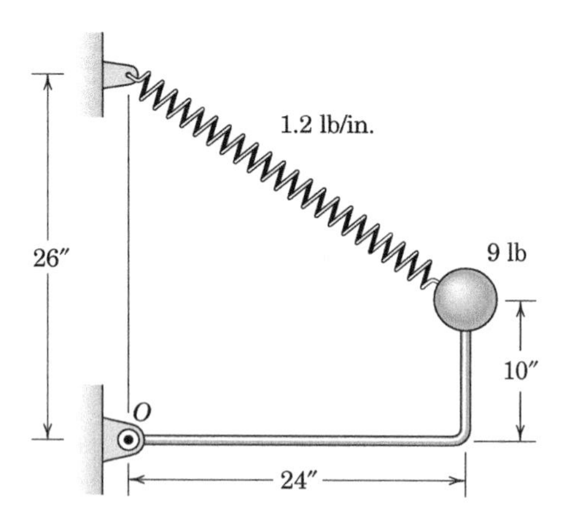

Problem 3/152

3/153 The two wheels consisting of hoops and spokes of negligible mass rotate about their respective centers and are pressed together sufficiently to prevent any slipping. The 3-lb and 2-lb eccentric masses are mounted on the rims of the wheels. If the wheels are given a slight nudge from rest in the equilibrium positions shown, compute the angular velocity $\dot{\theta}$ of the larger of the two wheels when it has revolved through a quarter of a revolution and put the eccentric masses in the dashed positions shown. Note that the angular velocity of the small wheel is twice that of the large wheel. Neglect any friction in the wheel bearings.

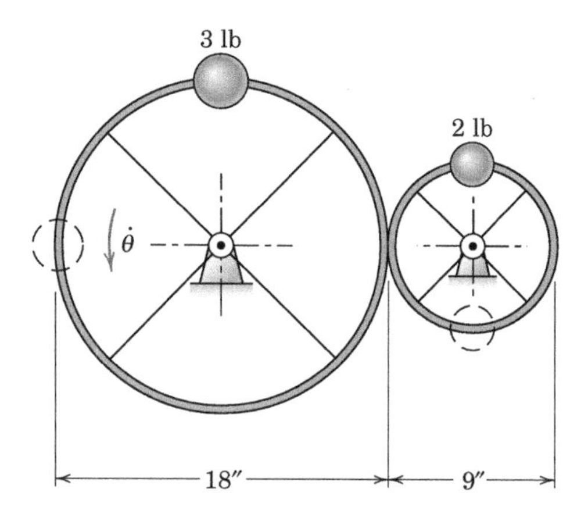

Problem 3/153

3/154 The slider of mass m is released from rest in position A and slides without friction along the vertical-plane guide shown. Determine the height h such that the normal force exerted by the guide on the slider is zero as the slider passes point C. For this value of h, determine the normal force as the slider passes point B.

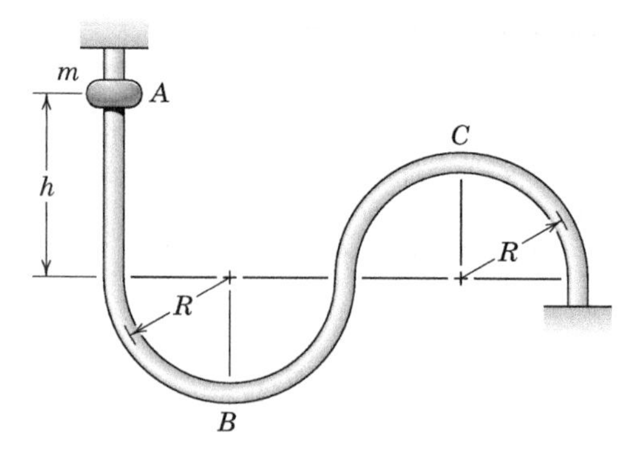

Problem 3/154

3/155 The two 1.5-kg spheres are released from rest and gently nudged outward from the position $\theta = 0$ and then rotate in a vertical plane about the fixed centers of their attached gears, thus maintaining the same angle θ for both rods. Determine the velocity v of each sphere as the rods pass the position $\theta = 30°$. The spring is unstretched when $\theta = 0$, and the masses of the two identical rods and the two gear wheels may be neglected.

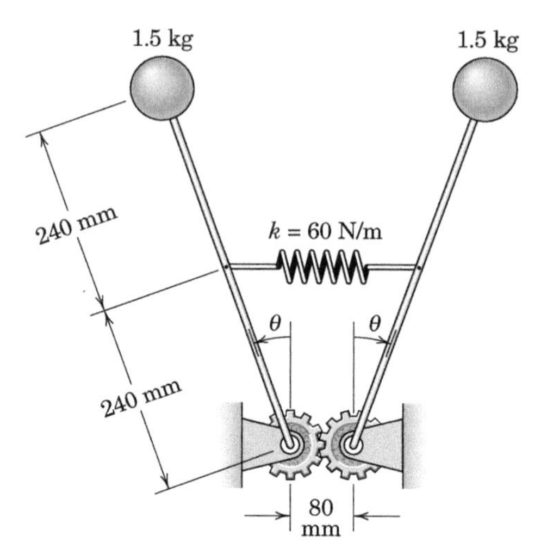

Problem 3/155

3/156 In the design of an inside loop for an amusement-park ride, it is desired to maintain the same centripetal acceleration throughout the loop. Assume negligible loss of energy during the motion and determine the radius of curvature ρ of the path as a function of the height y above the low point A, where the velocity and radius of curvature are v_0 and ρ_0, respectively. For a given value of ρ_0, what is the minimum value of v_0 for which the vehicle will not leave the track at the top of the loop?

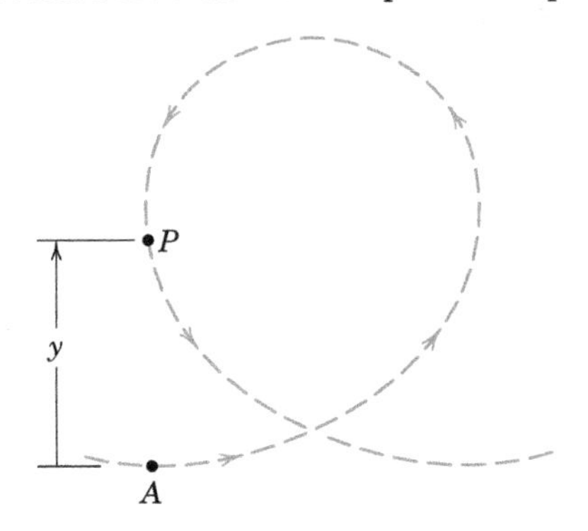

Problem 3/156

3/157 A rocket launches an unpowered space capsule at point A with an absolute velocity $v_A = 8000$ mi/hr at an altitude of 25 mi. After the capsule has traveled a distance of 250 mi measured along its absolute space trajectory, its velocity at B is 7600 mi/hr and its altitude is 50 mi. Determine the average resistance P to motion in the rarified atmosphere. The earth weight of the capsule is 48 lb, and the mean radius of the earth is 3959 mi. Consider the center of the earth fixed in space.

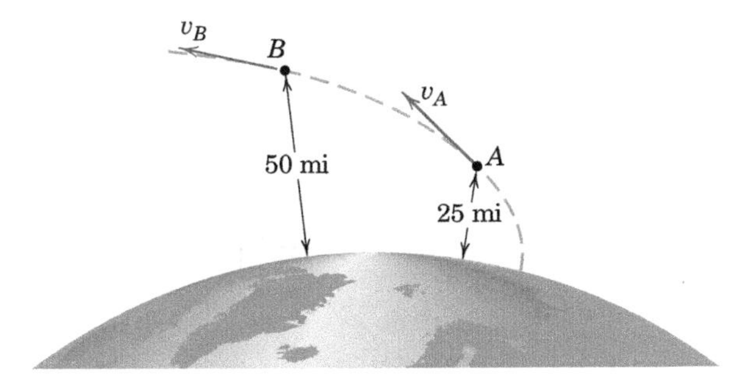

Problem 3/157

3/158 The projectile of Prob. 3/124 is repeated here. By the method of this article, determine the vertical launch velocity v_0 which will result in a maximum altitude of $R/3$. The launch is from the north pole and aerodynamic drag can be neglected. Use $g = 9.825$ m/s^2 as the surface-level acceleration due to gravity.

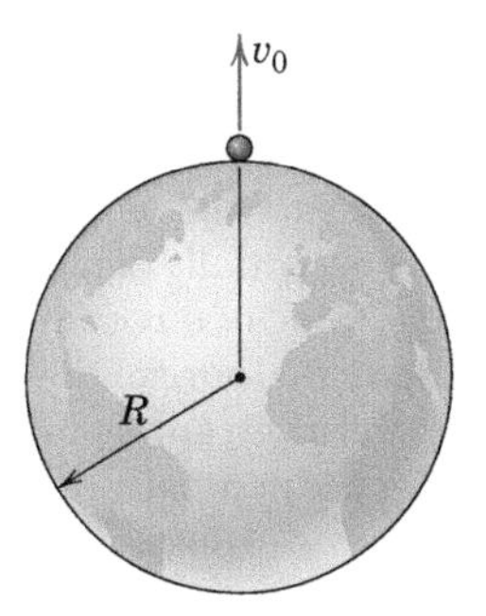

Problem 3/158

3/159 The small bodies A and B each of mass m are connected and supported by the pivoted links of negligible mass. If A is released from rest in the position shown, calculate its velocity v_A as it crosses the vertical centerline. Neglect any friction.

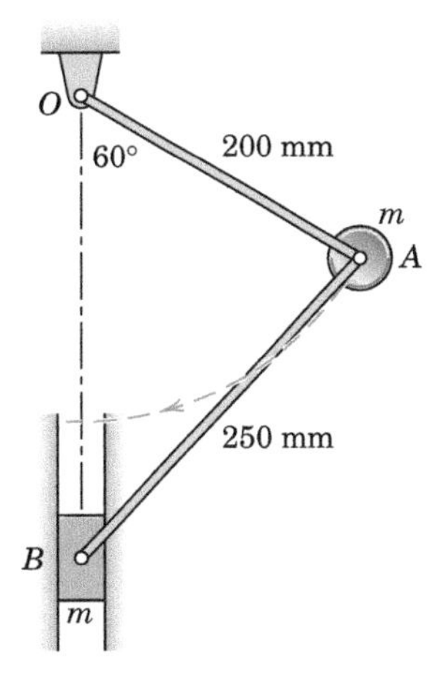

Problem 3/159

3/160 Upon its return voyage from a space mission, the spacecraft has a velocity of 24 000 km/h at point A, which is 7000 km from the center of the earth. Determine the velocity of the spacecraft when it reaches point B, which is 6500 km from the center of the earth. The trajectory between these two points is outside the effect of the earth's atmosphere.

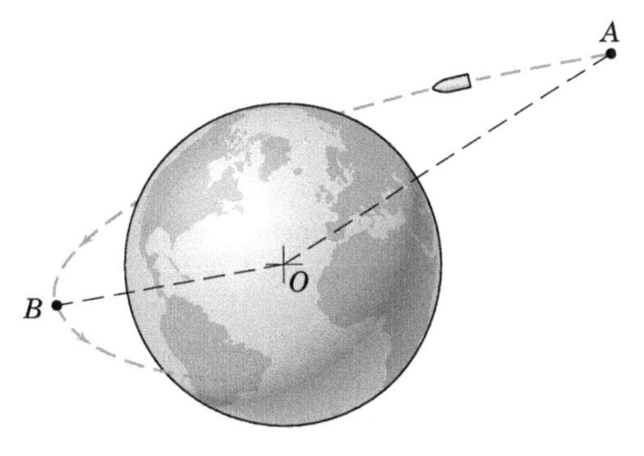

Problem 3/160

3/161 A 175-lb pole vaulter carrying a uniform 16-ft, 10-lb pole approaches the jump with a velocity v and manages to barely clear the bar set at a height of 18 ft. As he clears the bar, his velocity and that of the pole are essentially zero. Calculate the minimum possible value of v required for him to make the jump. Both the horizontal pole and the center of gravity of the vaulter are 42 in. above the ground during the approach.

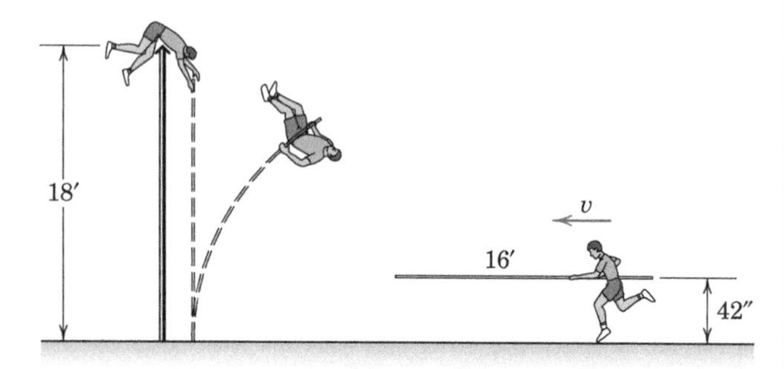

Problem 3/161

3/162 When the mechanism is released from rest in the position where $\theta = 60°$, the 4-kg carriage drops and the 6-kg sphere rises. Determine the velocity v of the sphere when $\theta = 180°$. Neglect the mass of the links and treat the sphere as a particle.

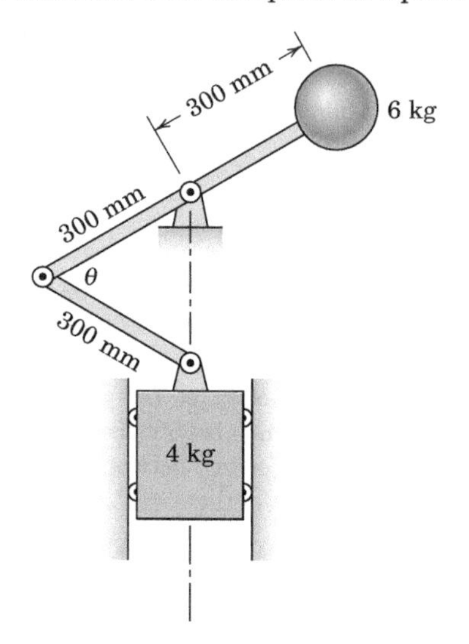

Problem 3/162

3/163 The cars of an amusement-park ride have a speed $v_1 = 90$ km/h at the lowest part of the track. Determine their speed v_2 at the highest part of the track. Neglect energy loss due to friction. (*Caution:* Give careful thought to the change in potential energy of the system of cars.)

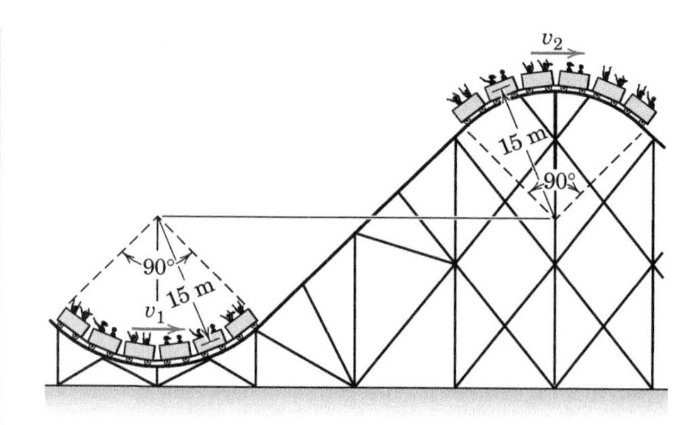

Problem 3/163

3/164 A satellite is put into an elliptical orbit around the earth and has a velocity v_P at the perigee position P. Determine the expression for the velocity v_A at the apogee position A. The radii to A and P are, respectively, r_A and r_P. Note that the total energy remains constant.

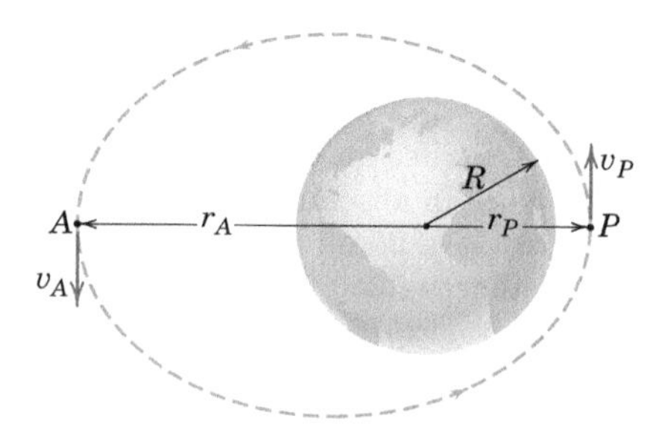

Problem 3/164

3/165 Calculate the maximum velocity of slider B if the system is released from rest with $x = y$. Motion is in the vertical plane. Assume that friction is negligible. The sliders have equal masses, and the motion is restricted to $y \geq 0$.

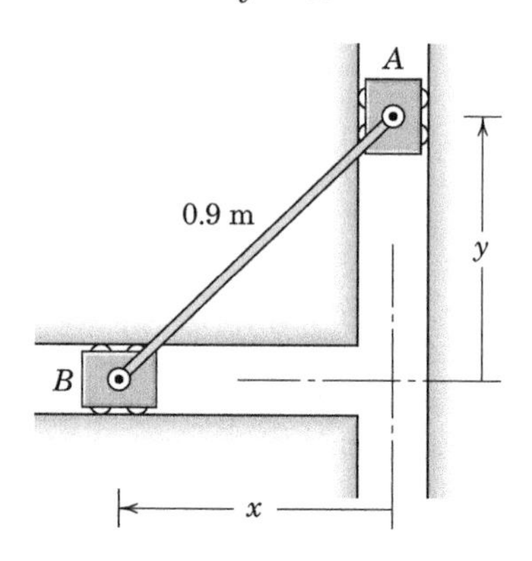

Problem 3/165

3/166 The system is initially moving with the cable taut, the 10-kg block moving down the rough incline with a speed of 0.3 m/s, and the spring stretched 25 mm. By the method of this article, (*a*) determine the velocity v of the block after it has traveled 100 mm, and (*b*) calculate the distance traveled by the block before it comes to rest.

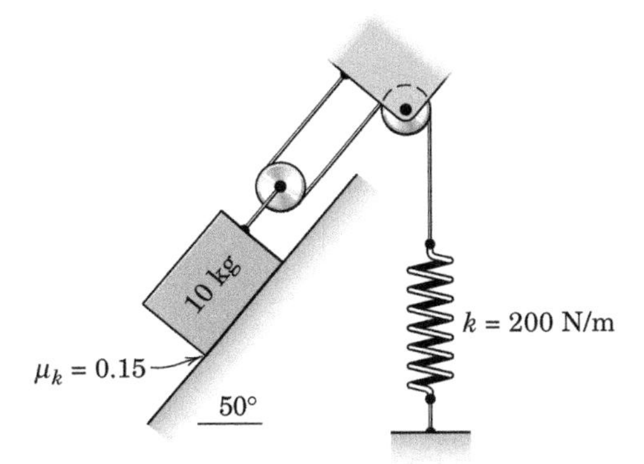

Problem 3/166

3/167 A spacecraft m is heading toward the center of the moon with a velocity of 2000 mi/hr at a distance from the moon's surface equal to the radius R of the moon. Compute the impact velocity v with the surface of the moon if the spacecraft is unable to fire its retro-rockets. Consider the moon fixed in space. The radius R of the moon is 1080 mi, and the acceleration due to gravity at its surface is 5.32 ft/sec^2.

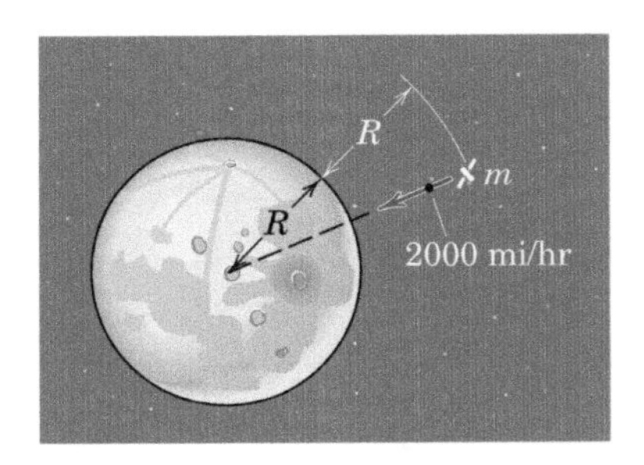

Problem 3/167

3/168 When the 10-lb plunger is released from rest in its vertical guide at $\theta = 0$, each spring of stiffness $k = 20$ lb/in. is uncompressed. The links are free to slide through their pivoted collars and compress their springs. Calculate the velocity v of the plunger when the position $\theta = 30°$ is passed.

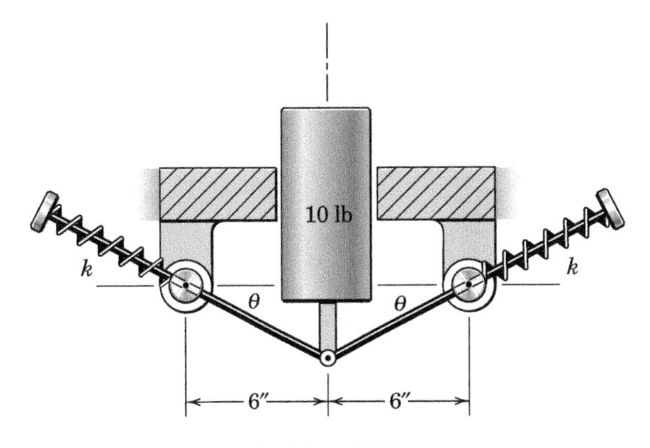

Problem 3/168

3/169 The system is released from rest with the angle $\theta = 90°$. Determine $\dot{\theta}$ when θ reaches 60°. Use the values $m_1 = 1$ kg, $m_2 = 1.25$ kg, and $b = 0.40$ m. Neglect friction and the mass of bar OB, and treat the body B as a particle.

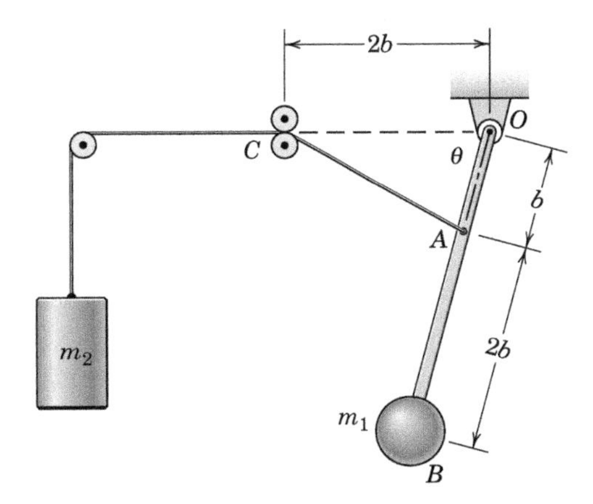

Problem 3/169

3/170 The system is at rest with the spring unstretched when $\theta = 0$. The 3-kg particle is then given a slight nudge to the right. (*a*) If the system comes to momentary rest at $\theta = 40°$, determine the spring constant k. (*b*) For the value $k = 100$ N/m, find the speed of the particle when $\theta = 25°$. Use the value $b = 0.40$ m throughout and neglect friction.

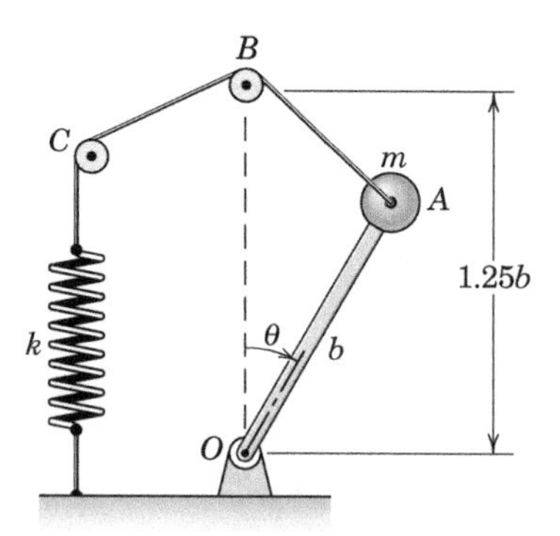

Problem 3/170

3/171 The 0.6-kg slider is released from rest at A and slides down the smooth parabolic guide (which lies in a vertical plane) under the influence of its own weight and of the spring of constant 120 N/m. Determine the speed of the slider as it passes point B and the corresponding normal force exerted on it by the guide. The unstretched length of the spring is 200 mm.

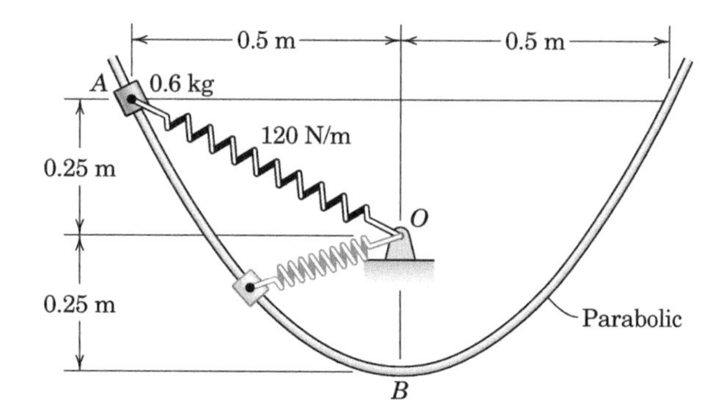

Problem 3/171

▶3/172 The two particles of mass m and $2m$, respectively, are connected by a rigid rod of negligible mass and slide with negligible friction in a circular path of radius r on the inside of the vertical circular ring. If the unit is released from rest at $\theta = 0$, determine (a) the velocity v of the particles when the rod passes the horizontal position, (b) the maximum velocity v_{max} of the particles, and (c) the maximum value of θ.

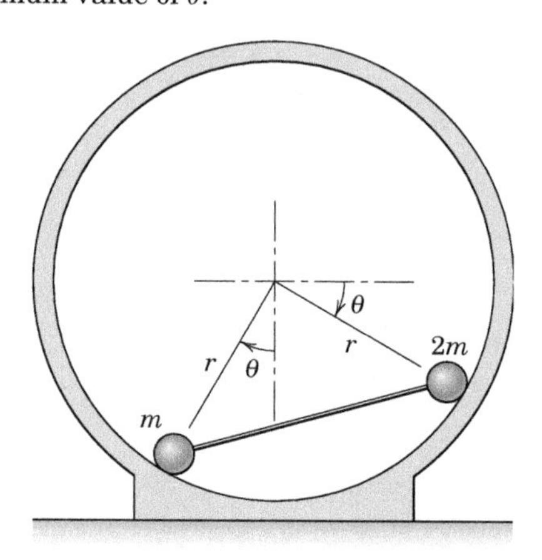

Problem 3/172

PROBLEMS

Introductory Problems

3/173 A 0.2-kg wad of clay is released from rest and drops 2 m to a concrete floor. The clay does not rebound, and the collision lasts 0.04 s. Determine the time average of the force which the floor exerts on the clay during the impact.

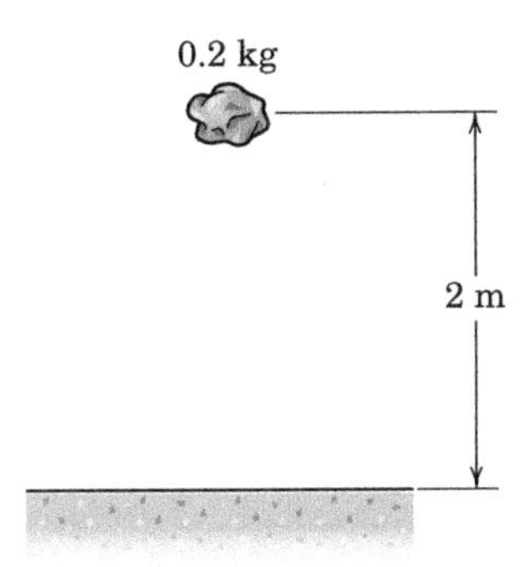

Problem 3/173

3/174 The two orbital maneuvering engines of the space shuttle develop 26 kN of thrust each. If the shuttle is traveling in orbit at a speed of 28 000 km/h, how long would it take to reach a speed of 28 100 km/h after the two engines are fired? The mass of the shuttle is 90 Mg.

3/175 A jet-propelled airplane with a mass of 10 Mg is flying horizontally at a constant speed of 1000 km/h under the action of the engine thrust T and the equal and opposite air resistance R. The pilot ignites two rocket-assist units, each of which develops a forward thrust T_0 of 8 kN for 9 s. If the velocity of the airplane in its horizontal flight is 1050 km/h at the end of the 9 s, calculate the time-average increase ΔR in air resistance. The mass of the rocket fuel used is negligible compared with that of the airplane.

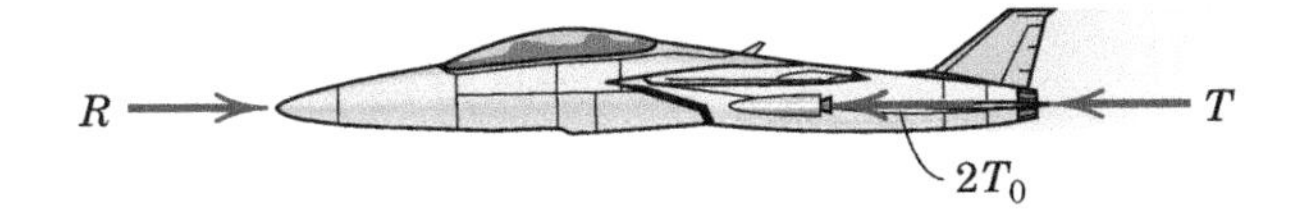

Problem 3/175

3/176 The velocity of a 1.2-kg particle is given by $\mathbf{v} = 1.5t^3\mathbf{i} + (2.4 - 3t^2)\mathbf{j} + 5\mathbf{k}$, where $\mathbf{v}$ is in meters per second and the time t is in seconds. Determine the linear momentum $\mathbf{G}$ of the particle, its magnitude G, and the net force $\mathbf{R}$ which acts on the particle when $t = 2$ s.

3/177 A 75-g projectile traveling at 600 m/s strikes and becomes embedded in the 40-kg block, which is initially stationary. Compute the energy lost during the impact. Express your answer as an absolute value $|\Delta E|$ and as a percentage n of the original system energy E.

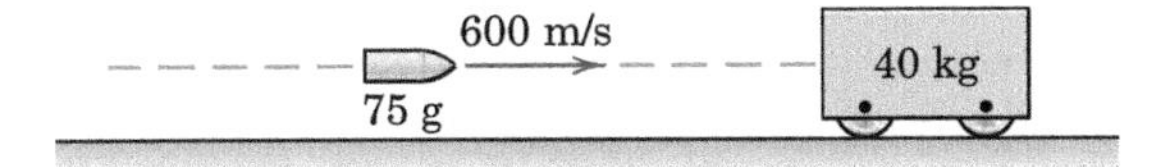

Problem 3/177

3/178 A 60-g bullet is fired horizontally with a velocity $v_1 = 600$ m/s into the 3-kg block of soft wood initially at rest on the horizontal surface. The bullet emerges from the block with the velocity $v_2 = 400$ m/s, and the block is observed to slide a distance of 2.70 m before coming to rest. Determine the coefficient of kinetic friction μ_k between the block and the supporting surface.

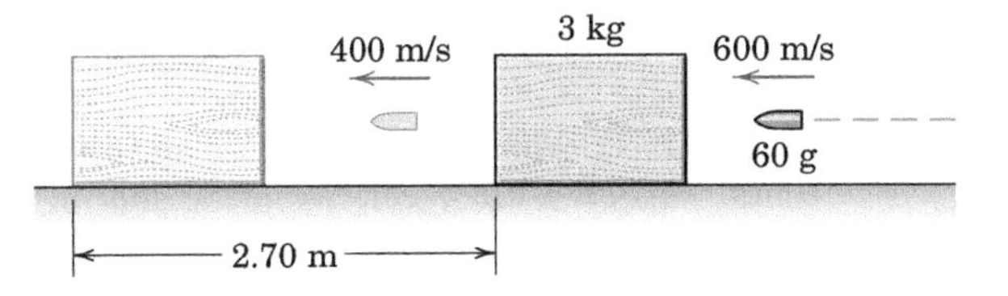

Problem 3/178

3/179 Careful measurements made during the impact of the 200-g metal cylinder with the spring-loaded plate reveal a semielliptical relation between the contact force F and the time t of impact as shown. Determine the rebound velocity v of the cylinder if it strikes the plate with a velocity of 6 m/s.

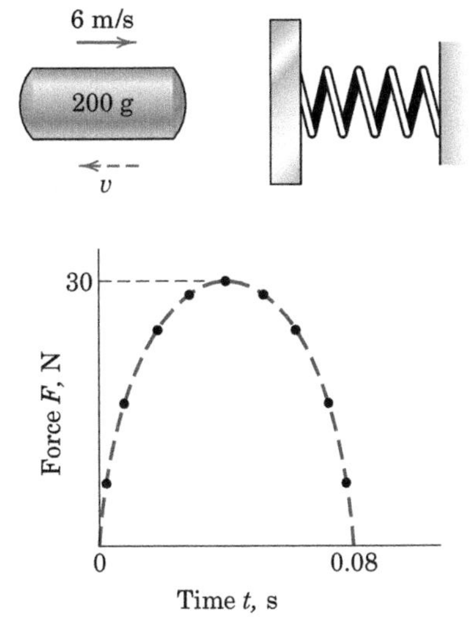

Problem 3/179

3/180 A 0.25-kg particle is moving with a velocity $\mathbf{v}_1 = 2\mathbf{i} + \mathbf{j} - \mathbf{k}$ m/s at time $t_1 = 2$ s. If the single force $\mathbf{F} = (4 + 2t)\mathbf{i} + (t^2 - 2)\mathbf{j} + 5\mathbf{k}$ N acts on the particle, determine its velocity $\mathbf{v}_2$ at time $t_2 = 4$ s.

3/181 The 90-kg man dives from the 40-kg canoe. The velocity indicated in the figure is that of the man relative to the canoe just after loss of contact. If the man, woman, and canoe are initially at rest, determine the horizontal component of the absolute velocity of the canoe just after separation. Neglect drag on the canoe, and assume that the 60-kg woman remains motionless relative to the canoe.

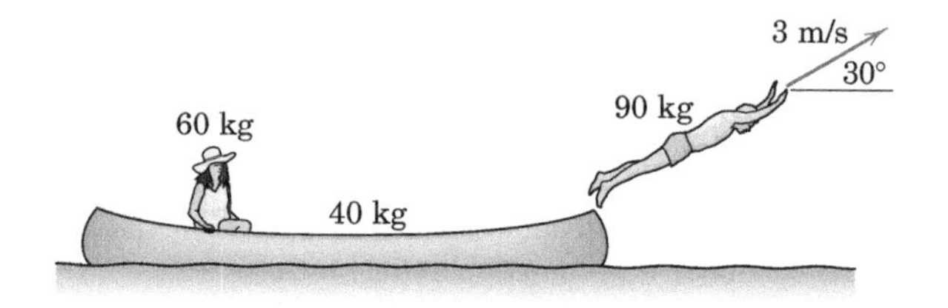

Problem 3/181

3/182 A 4-kg object, which is moving on a smooth horizontal surface with a velocity of 10 m/s in the $-x$-direction, is subjected to a force F_x which varies with time as shown. Approximate the experimental data by the dashed line and determine the velocity of the object (*a*) at $t = 0.6$ s and (*b*) at $t = 0.9$ s.

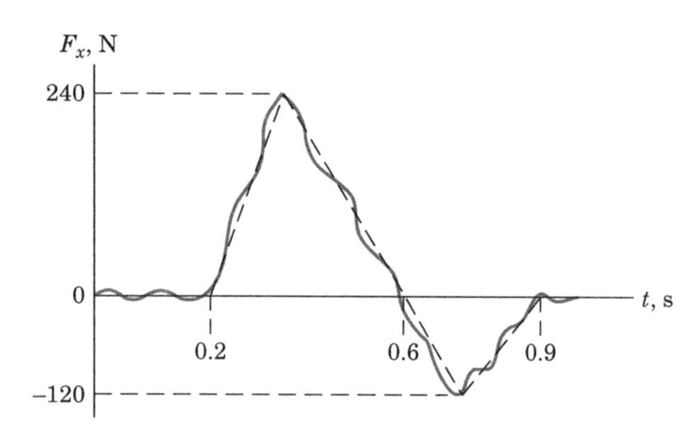

Problem 3/182

3/183 Crate *A* is traveling down the incline with a speed of 4 m/s when in the position shown. It later strikes and becomes attached to crate *B*. Determine the distance *d* moved by the pair after the collision. The coefficient of kinetic friction is $\mu_k = 0.40$ for both crates.

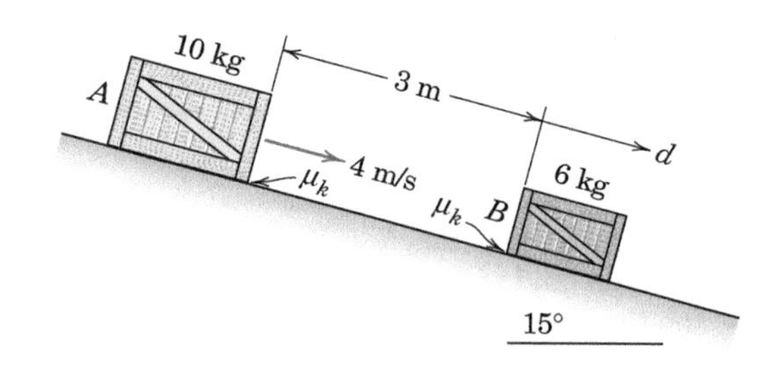

Problem 3/183

3/184 The 15 200-kg lunar lander is descending onto the moon's surface with a velocity of 2 m/s when its retro-engine is fired. If the engine produces a thrust T for 4 s which varies with time as shown and then cuts off, calculate the velocity of the lander when $t = 5$ s, assuming that it has not yet landed. Gravitational acceleration at the moon's surface is 1.62 m/s^2.

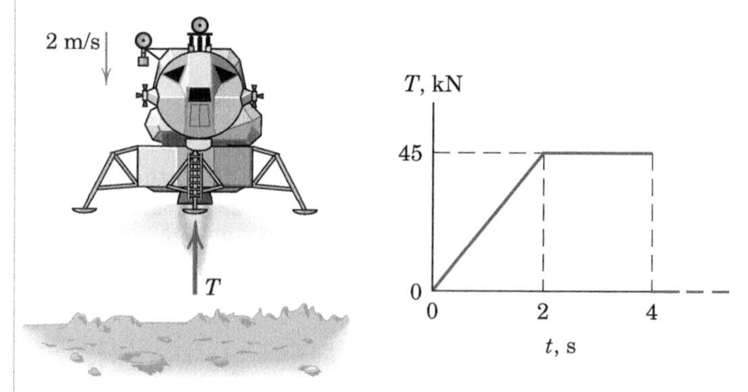

Problem 3/184

3/185 A boy weighing 100 lb runs and jumps on his 20-lb sled with a horizontal velocity of 15 ft/sec. If the sled and boy coast 80 ft on the level snow before coming to rest, compute the coefficient of kinetic friction μ_k between the snow and the runners of the sled.

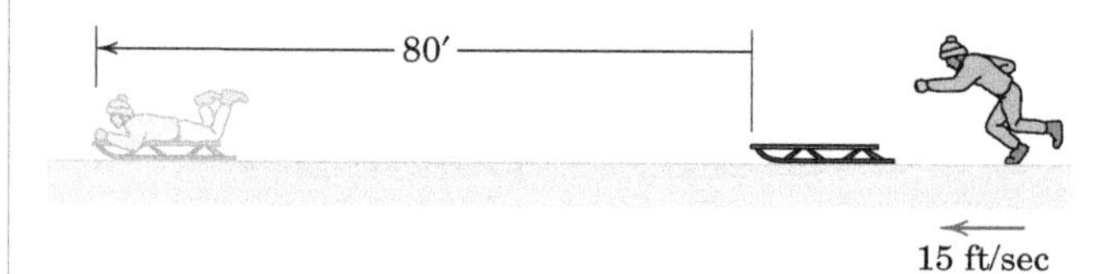

Problem 3/185

3/186 The snowboarder is traveling with a velocity of 6 m/s as shown when he lands on the incline with no rebound. If the impact has a time duration of 0.1 s, determine his speed v along the incline just after impact and the total time-average normal force exerted by the incline on the snowboard during the impact. The combined mass of the athlete and his snowboard is 60 kg.

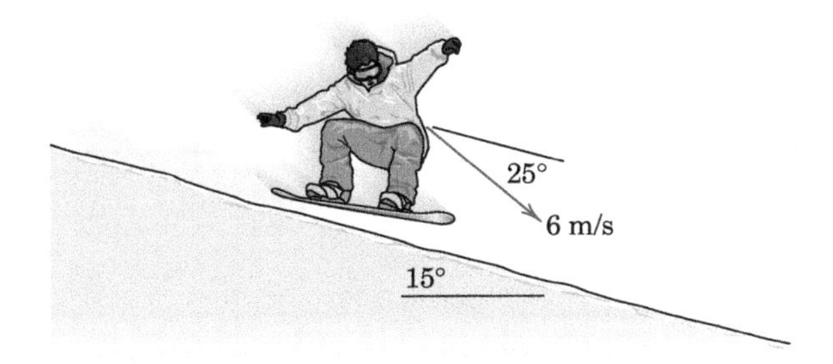

Problem 3/186

Representative Problems

3/187 The tow truck with attached 1200-kg car accelerates uniformly from 30 km/h to 70 km/h over a 15-s interval. The average rolling resistance for the car over this speed interval is 500 N. Assume that the 60° angle shown represents the time average configuration and determine the average tension in the tow cable.

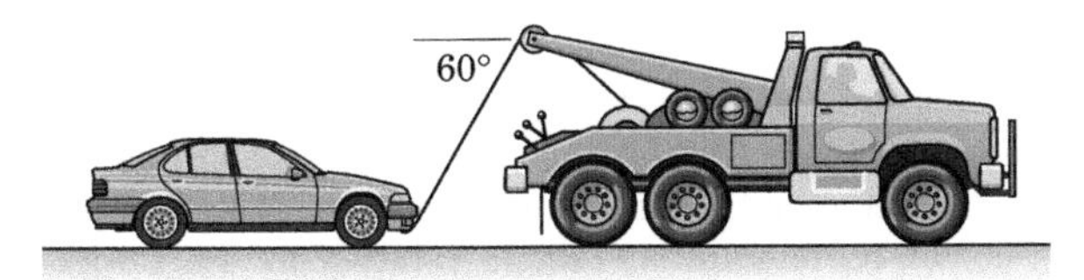

Problem 3/187

3/188 Car A weighing 3200 lb and traveling north at 20 mi/hr collides with car B weighing 3600 lb and traveling at 30 mi/hr as shown. If the two cars become entangled and move together as a unit after the crash, compute the magnitude v of their common velocity immediately after the impact and the angle θ made by the velocity vector with the north direction.

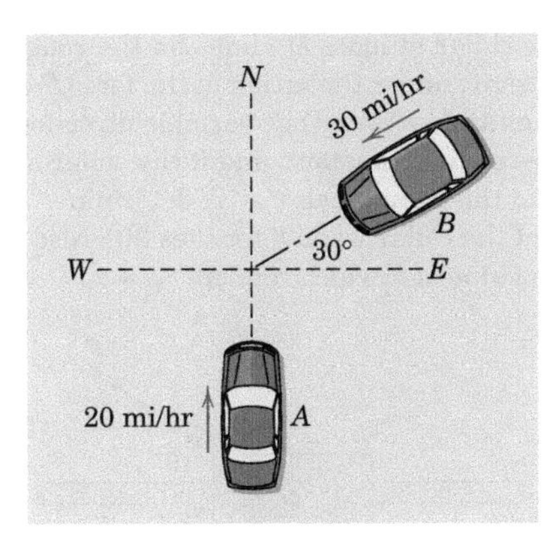

Problem 3/188

3/189 A railroad car of mass m and initial speed v collides with and becomes coupled with the two identical cars. Compute the final speed v' of the group of three cars and the fractional loss n of energy if (*a*) the initial separation distance $d = 0$ (that is, the two stationary cars are initially coupled together with no slack in the coupling) and (*b*) the distance $d \neq 0$ so that the cars are uncoupled and slightly separated. Neglect rolling resistance.

Problem 3/189

3/190 The 600,000-lb jet airliner has a touchdown velocity $v = 120$ mi/hr directed $\theta = 0.5°$ below the horizontal. The touchdown process of the eight main wheels takes 0.6 sec to complete. Treat the aircraft as a particle and estimate the average normal reaction force at each wheel during this 0.6-sec process, during which tires deflect, struts compress, etc. Assume that the aircraft lift equals the aircraft weight during the touchdown.

Problem 3/190

3/191 The collar of mass m slides on the rough horizontal shaft under the action of the force F of constant magnitude $F \leq mg$ but variable direction. If $\theta = kt$ where k is a constant, and if the collar has a speed v_1 to the right when $\theta = 0$, determine the velocity v_2 of the collar when θ reaches 90°. Also determine the value of F which renders $v_2 = v_1$.

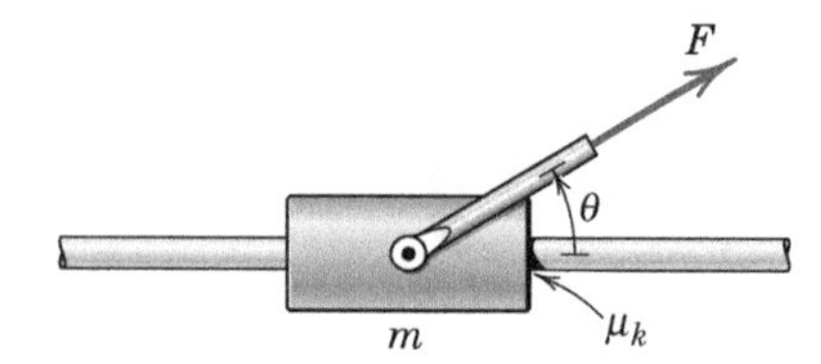

Problem 3/191

3/192 The 140-g projectile is fired with a velocity of 600 m/s and picks up three washers, each with a mass of 100 g. Find the common velocity v of the projectile and washers. Determine also the loss $|\Delta E|$ of energy during the interaction.

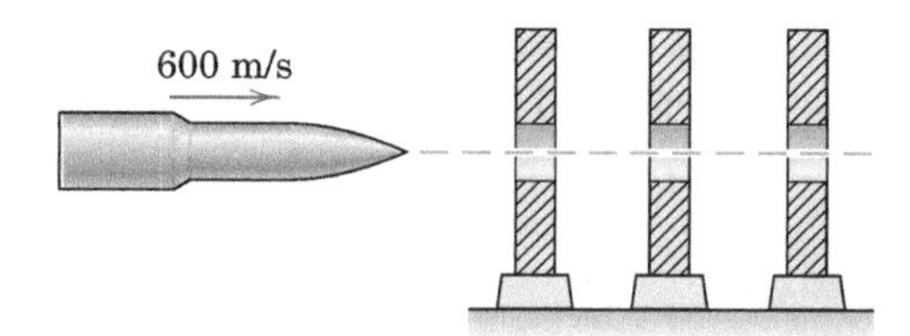

Problem 3/192

3/193 The third and fourth stages of a rocket are coasting in space with a velocity of 18 000 km/h when a small explosive charge between the stages separates them. Immediately after separation the fourth stage has increased its velocity to $v_4 = 18\ 060$ km/h. What is the corresponding velocity v_3 of the third stage? At separation the third and fourth stages have masses of 400 and 200 kg, respectively.

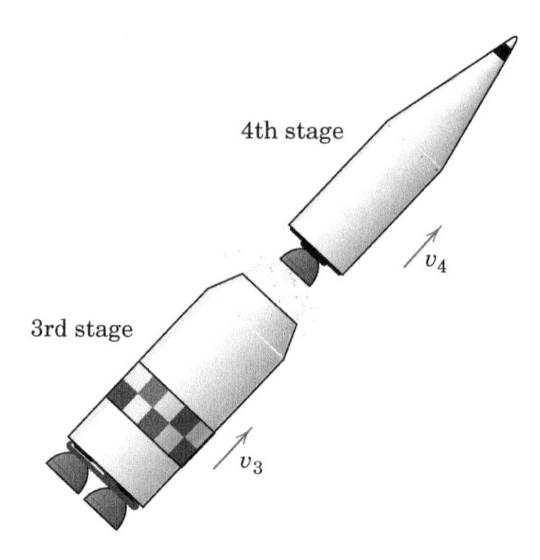

Problem 3/193

3/194 The initially stationary 20-kg block is subjected to the time-varying horizontal force whose magnitude P is shown in the plot. Note that the force is zero for all times greater than 3 s. Determine the time t_s at which the block comes to rest.

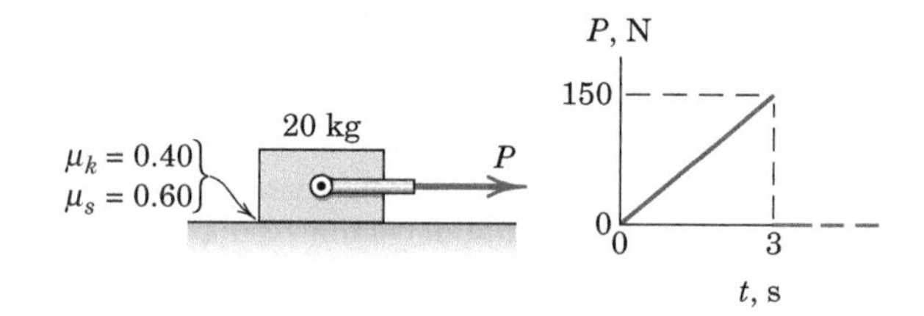

Problem 3/194

3/195 All elements of the previous problem remain unchanged, except that the force P is now held at a constant 30° angle relative to the horizontal. Determine the time t_s at which the initially stationary 20-kg block comes to rest.

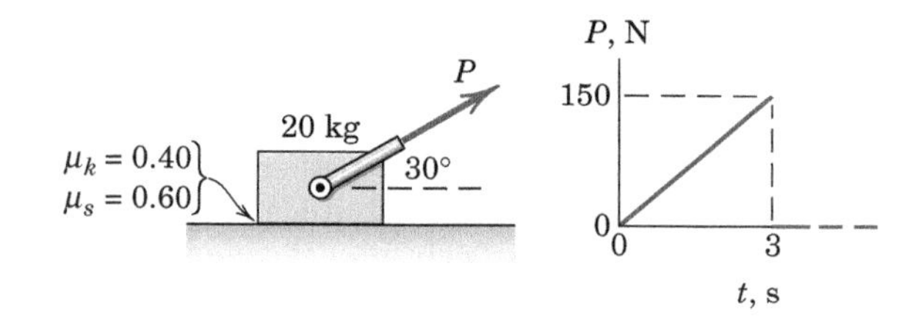

Problem 3/195

3/196 The spring of modulus $k = 200$ N/m is compressed a distance of 300 mm and suddenly released with the system at rest. Determine the absolute velocities of both masses when the spring is unstretched. Neglect friction.

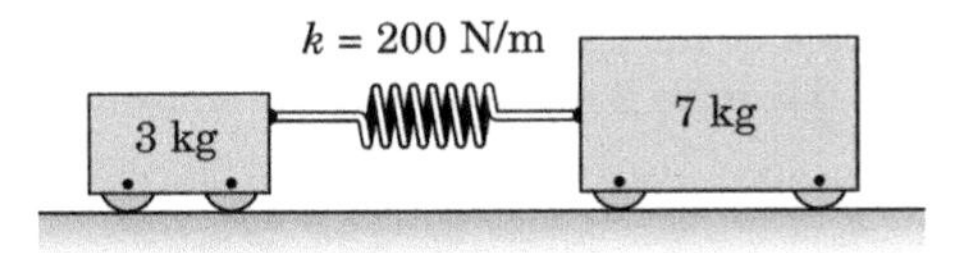

Problem 3/196

3/197 The pilot of a 90,000-lb airplane which is originally flying horizontally at a speed of 400 mi/hr cuts off all engine power and enters a 5° glide path as shown. After 120 seconds the airspeed is 360 mi/hr. Calculate the time-average drag force D (air resistance to motion along the flight path).

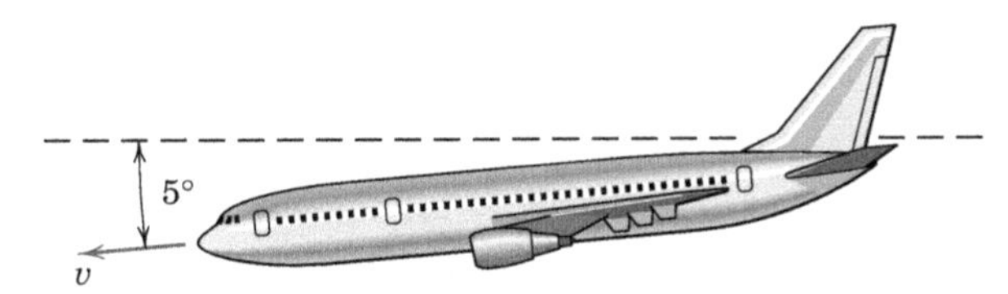

Problem 3/197

3/198 The space shuttle launches an 800-kg satellite by ejecting it from the cargo bay as shown. The ejection mechanism is activated and is in contact with the satellite for 4 s to give it a velocity of 0.3 m/s in the z-direction relative to the shuttle. The mass of the shuttle is 90 Mg. Determine the component of velocity v_f of the shuttle in the minus z-direction resulting from the ejection. Also find the time average F_{av} of the ejection force.

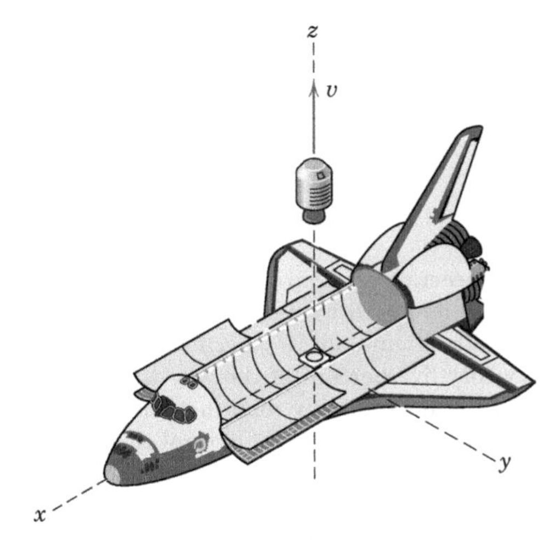

Problem 3/198

3/199 The hydraulic braking system for the truck and trailer is set to produce equal braking forces for the two units. If the brakes are applied uniformly for 5 seconds to bring the rig to a stop from a speed of 20 mi/hr down the 10-percent grade, determine the force P in the coupling between the trailer and the truck. The truck weighs 20,000 lb and the trailer weighs 15,000 lb.

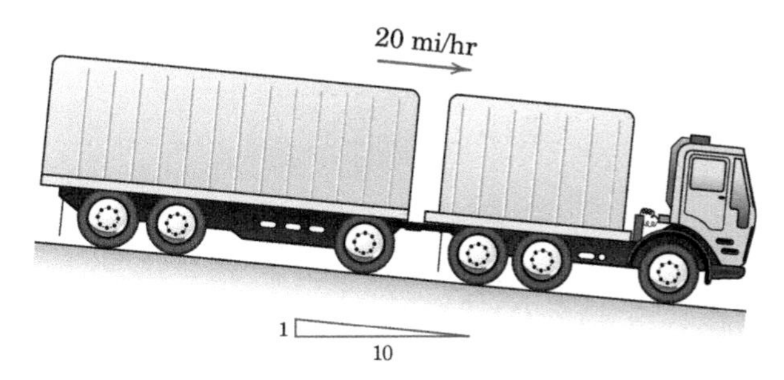

Problem 3/199

3/200 The 100-lb block is stationary at time $t = 0$, and then it is subjected to the force P shown. Note that the force is zero for all times beyond $t = 15$ sec. Determine the velocity v of the block at time $t = 15$ sec. Also calculate the time t at which the block again comes to rest.

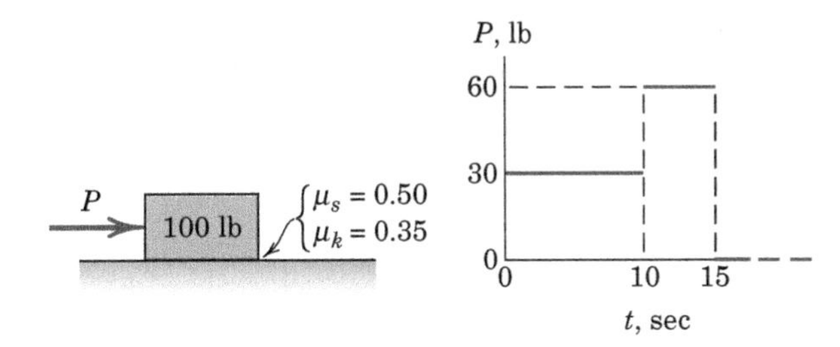

Problem 3/200

3/201 Car B is initially stationary and is struck by car A moving with initial speed $v_1 = 20$ mi/hr. The cars become entangled and move together with speed v' after the collision. If the time duration of the collision is 0.1 sec, determine (*a*) the common final speed v', (*b*) the average acceleration of each car during the collision, and (*c*) the magnitude R of the average force exerted by each car on the other car during the impact. All brakes are released during the collision.

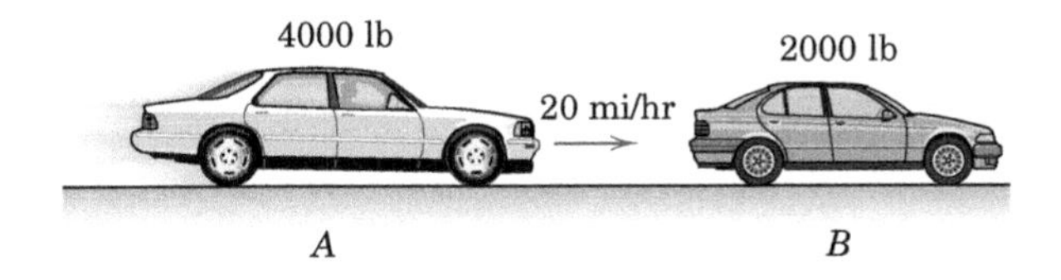

Problem 3/201

3/202 The 2.4-kg particle moves in the horizontal x-y plane and has the velocity shown at time $t = 0$. If the force $F = 2 + 3t^2/4$ newtons, where t is time in seconds, is applied to the particle in the y-direction beginning at time $t = 0$, determine the velocity v of the particle 4 seconds after F is applied and specify the corresponding angle θ measured counterclockwise from the x-axis to the direction of the velocity.

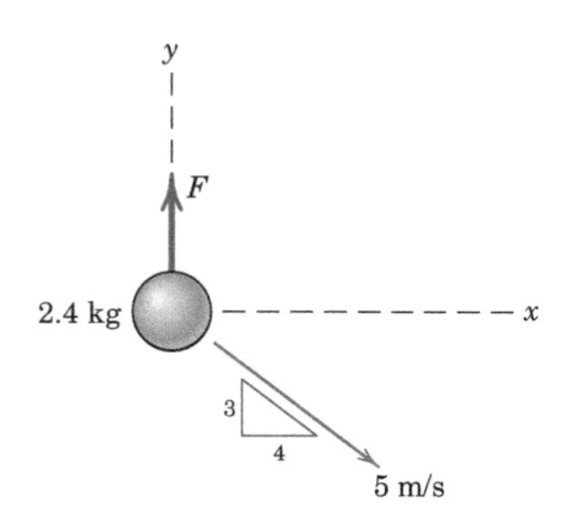

Problem 3/202

3/203 The 1.62-oz golf ball is struck by the five-iron and acquires the velocity shown in a time period of 0.001 sec. Determine the magnitude R of the average force exerted by the club on the ball. What acceleration magnitude a does this force cause, and what is the distance d over which the launch velocity is achieved, assuming constant acceleration?

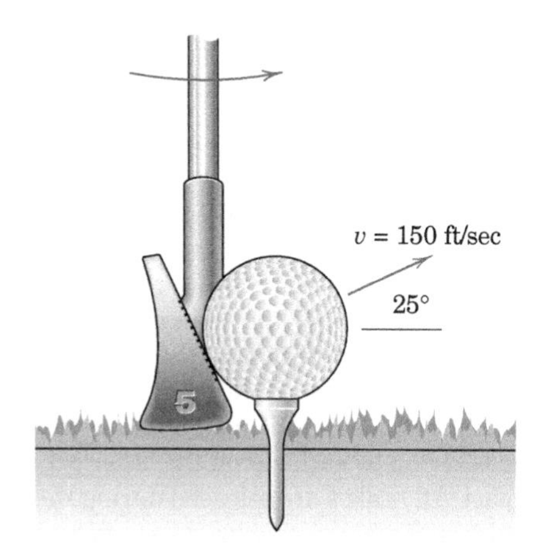

Problem 3/203

3/204 The ice-hockey puck with a mass of 0.20 kg has a velocity of 12 m/s before being struck by the hockey stick. After the impact the puck moves in the new direction shown with a velocity of 18 m/s. If the stick is in contact with the puck for 0.04 s, compute the magnitude of the average force **F** exerted by the stick on the puck during contact, and find the angle β made by **F** with the x-direction.

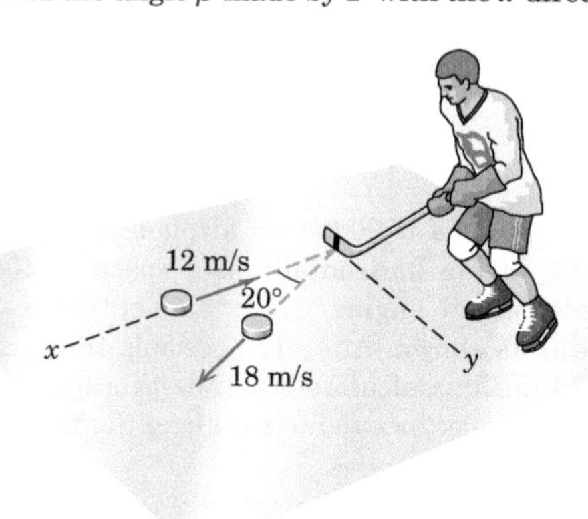

Problem 3/204

3/205 The baseball is traveling with a horizontal velocity of 85 mi/hr just before impact with the bat. Just after the impact, the velocity of the $5\frac{1}{8}$-oz ball is 130 mi/hr directed at 35° to the horizontal as shown. Determine the x- and y-components of the average force **R** exerted by the bat on the baseball during the 0.005-sec impact. Comment on the treatment of the weight of the baseball (a) during the impact and (b) over the first few seconds after impact.

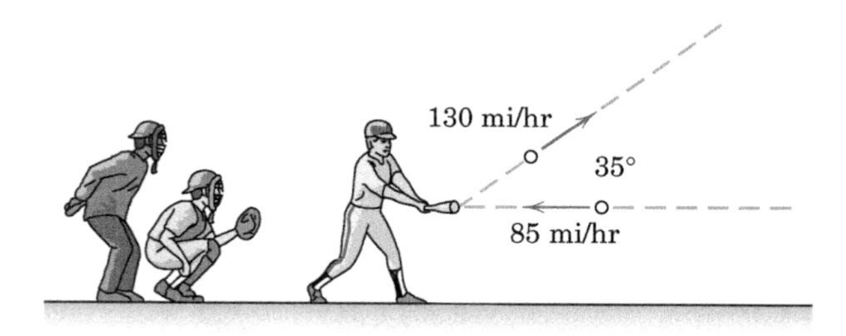

Problem 3/205

3/206 A spacecraft in deep space is programmed to increase its speed by a desired amount Δv by burning its engine for a specified time duration t. Twenty-five percent of the way through the burn, the engine suddenly malfunctions and thereafter produces only half of its normal thrust. What percent n of Δv is achieved if the rocket motor is fired for the planned time t? How much extra time t' would the rocket need to operate in order to compensate for the failure?

3/207 The slider of mass $m_1 = 0.4$ kg moves along the smooth support surface with velocity $v_1 = 5$ m/s when in the position shown. After negotiating the curved portion, it moves onto the inclined face of an initially stationary block of mass $m_2 = 2$ kg. The coefficient of kinetic friction between the slider and the block is $\mu_k = 0.30$. Determine the velocity v' of the system after the slider has come to rest relative to the block. Neglect friction at the small wheels, and neglect any effects associated with the transition.

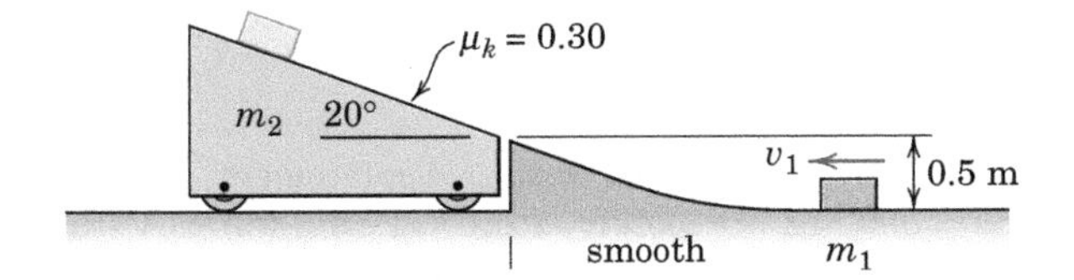

Problem 3/207

3/208 The 1.2-lb sphere is moving in the horizontal *x-y* plane with a velocity of 10 ft/sec in the direction shown and encounters a steady flow of air in the *x*-direction. If the air stream exerts an essentially constant force of 0.2 lb on the sphere in the *x*-direction, determine the time t required for the sphere to cross the *y*-axis again.

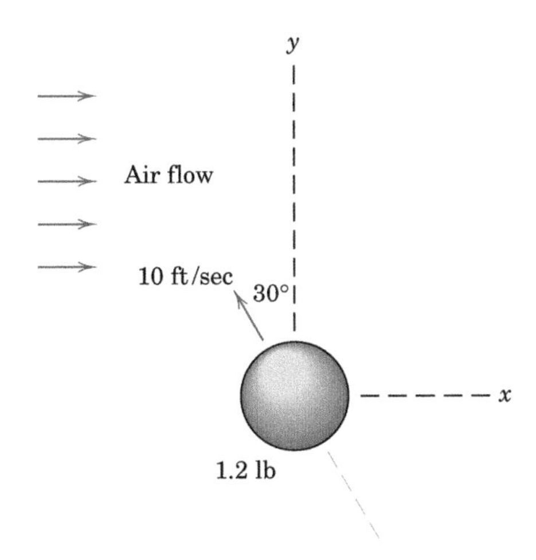

Problem 3/208

3/209 The ballistic pendulum is a simple device to measure projectile velocity v by observing the maximum angle θ to which the box of sand with embedded projectile swings. Calculate the angle θ if the 2-oz projectile is fired horizontally into the suspended 50-lb box of sand with a velocity $v =$ 2000 ft/sec. Also find the percentage of energy lost during impact.

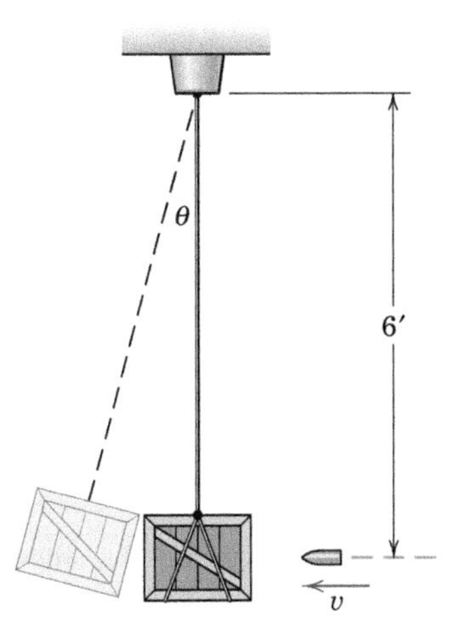

Problem 3/209

3/210 A tennis player strikes the tennis ball with her racket while the ball is still rising. The ball speed before impact with the racket is $v_1 = 15$ m/s and after impact its speed is $v_2 = 22$ m/s, with directions as shown in the figure. If the 60-g ball is in contact with the racket for 0.05 s, determine the magnitude of the average force **R** exerted by the racket on the ball. Find the angle β made by **R** with the horizontal. Comment on the treatment of the ball weight during impact.

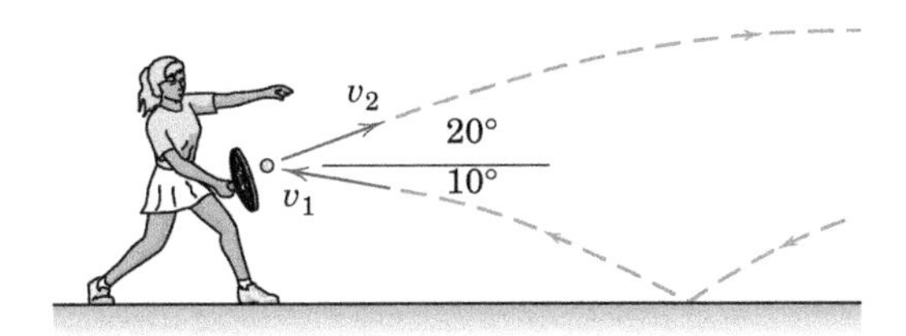

Problem 3/210

3/211 The 80-lb boy has taken a running jump from the upper surface and lands on his 10-lb skateboard with a velocity of 16 ft/sec in the plane of the figure as shown. If his impact with the skateboard has a time duration of 0.05 sec, determine the final speed v along the horizontal surface and the total normal force N exerted by the surface on the skateboard wheels during the impact.

Problem 3/211

3/212 The wad of clay A is projected as shown at the same instant that cylinder B is released. The two bodies collide and stick together at C and then ultimately strike the horizontal surface at D. Determine the horizontal distance d. Use the values $v_0 = 12$ m/s, $\theta = 40°$, $L = 6$ m, $m_A = 0.1$ kg, and $m_B = 0.2$ kg.

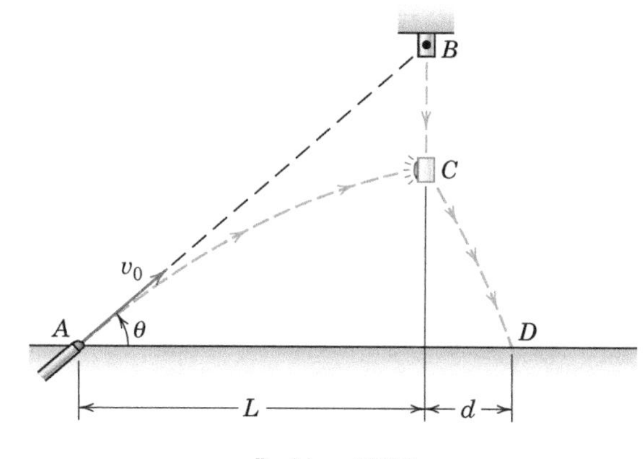

Problem 3/212

3/213 The two mine cars of equal mass are connected by a rope which is initially slack. Car A is given a shove which imparts to it a velocity of 4 ft/sec with car B initially at rest. When the slack is taken up, the rope suffers a tension impact which imparts a velocity to car B and reduces the velocity of car A. (*a*) If 40 percent of the kinetic energy of car A is lost during the rope impact, calculate the velocity v_B imparted to car B. (*b*) Following the initial impact, car B overtakes car A and the two are coupled together. Calculate their final common velocity v_C.

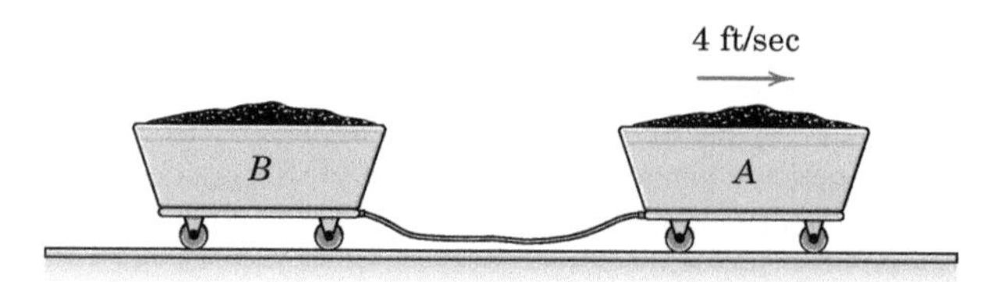

Problem 3/213

▶3/214 Two barges, each with a displacement (mass) of 500 Mg, are loosely moored in calm water. A stunt driver starts his 1500-kg car from rest at A, drives along the deck, and leaves the end of the 15° ramp at a speed of 50 km/h relative to the barge and ramp. The driver successfully jumps the gap and brings his car to rest relative to barge 2 at B. Calculate the velocity v_2 imparted to barge 2 just after the car has come to rest on the barge. Neglect the resistance of the water to motion at the low velocities involved.

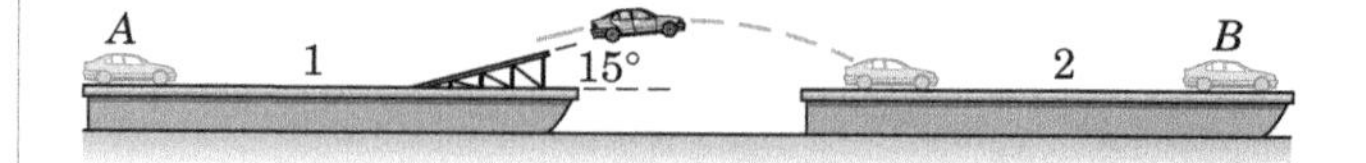

Problem 3/214

PROBLEMS

Introductory Problems

3/215 Determine the magnitude H_O of the angular momentum of the 2-kg sphere about point O (*a*) by using the vector definition of angular momentum and (*b*) by using an equivalent scalar approach. The center of the sphere lies in the x-y plane.

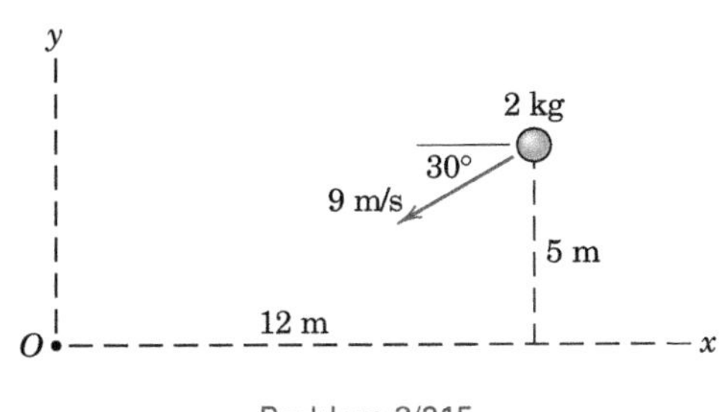

Problem 3/215

3/216 At a certain instant, the particle of mass m has the position and velocity shown in the figure, and it is acted upon by the force **F**. Determine its angular momentum about point O and the time rate of change of this angular momentum.

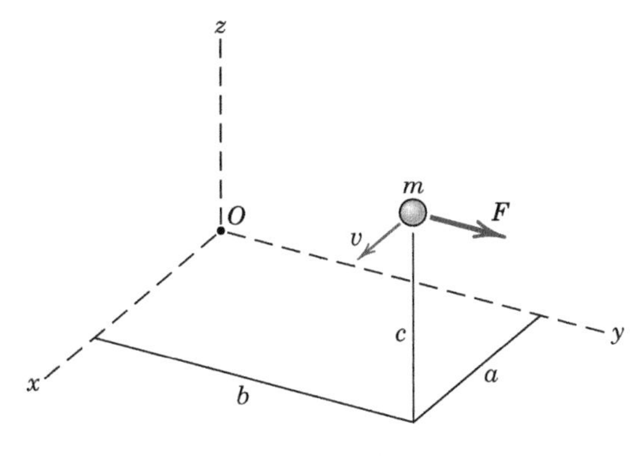

Problem 3/216

3/217 The 3-kg sphere moves in the x-y plane and has the indicated velocity at a particular instant. Determine its (*a*) linear momentum, (*b*) angular momentum about point O, and (*c*) kinetic energy.

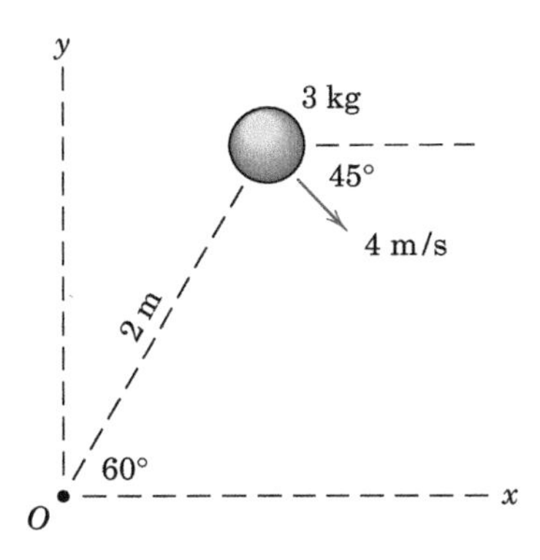

Problem 3/217

3/218 The particle of mass m is gently nudged from the equilibrium position A and subsequently slides along the smooth elliptical path which lies in a vertical plane. Determine the magnitude of its angular momentum about point O as it passes (*a*) point B and (*b*) point C. In each case, determine the time rate of change of H_O.

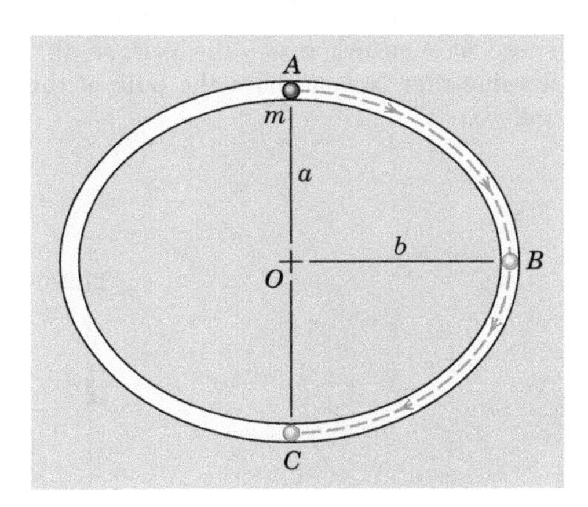

Problem 3/218

3/219 The assembly starts from rest and reaches an angular speed of 150 rev/min under the action of a 20-N force T applied to the string for t seconds. Determine t. Neglect friction and all masses except those of the four 3-kg spheres, which may be treated as particles.

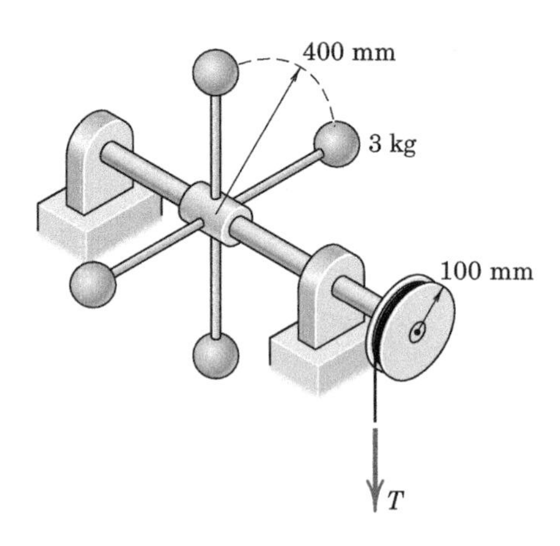

Problem 3/219

3/220 Just after launch from the earth, the space-shuttle orbiter is in the 37 × 137–mi orbit shown. At the apogee point A, its speed is 17,290 mi/hr. If nothing were done to modify the orbit, what would be its speed at the perigee P? Neglect aerodynamic drag. (Note that the normal practice is to add speed at A, which raises the perigee altitude to a value that is well above the bulk of the atmosphere.)

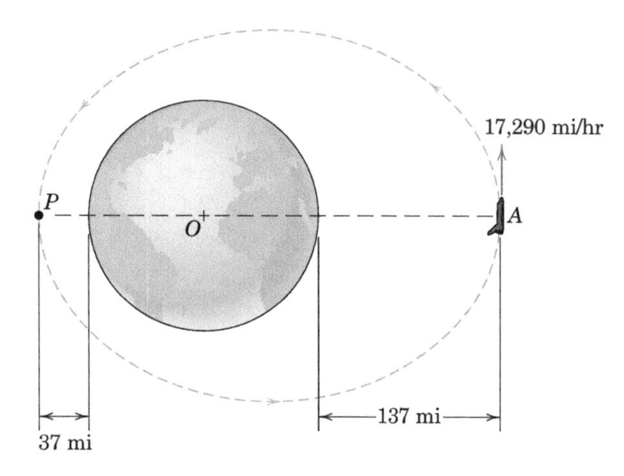

Problem 3/220

3/221 The rigid assembly which consists of light rods and two 1.2-kg spheres rotates freely about a vertical axis. The assembly is initially at rest and then a constant couple $M = 2$ N·m is applied for 5 s. Determine the final angular velocity of the assembly. Treat the small spheres as particles.

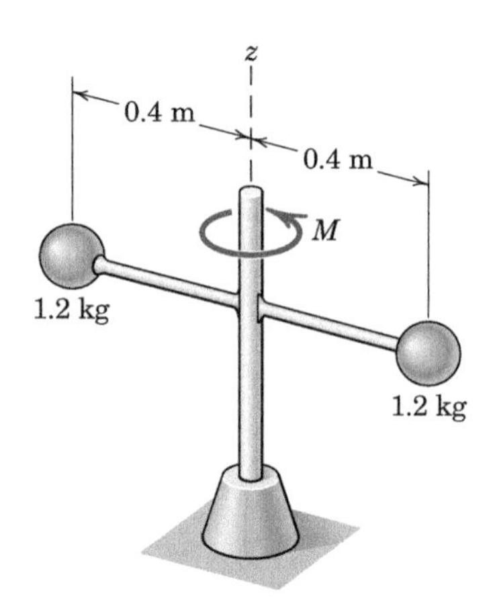

Problem 3/221

3/222 All conditions of the previous problem remain the same, except now the applied couple varies with time according to $M = 2t$, where t is in seconds and M is in newton-meters. Determine the angular velocity of the assembly at time $t = 5$ s.

Representative Problems

3/223 The small particle of mass m and its restraining cord are spinning with an angular velocity ω on the horizontal surface of a smooth disk, shown in section. As the force F is slightly relaxed, r increases and ω changes. Determine the rate of change of ω with respect to r and show that the work done by F during a movement dr equals the change in kinetic energy of the particle.

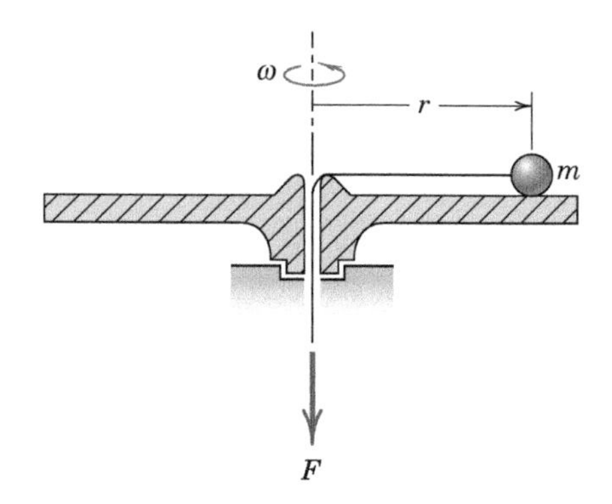

Problem 3/223

3/224 A particle with a mass of 4 kg has a position vector in meters given by $\mathbf{r} = 3t^2\mathbf{i} - 2t\mathbf{j} - 3\mathbf{k}$, where t is the time in seconds. For $t = 5$ s determine the angular momentum of the particle and the moment of all forces on the particle, both about the origin O of coordinates.

3/225 The 6-kg sphere and 4-kg block (shown in section) are secured to the arm of negligible mass which rotates in the vertical plane about a horizontal axis at O. The 2-kg plug is released from rest at A and falls into the recess in the block when the arm has reached the horizontal position. An instant before engagement, the arm has an angular velocity $\omega_0 = 2$ rad/s. Determine the angular velocity ω of the arm immediately after the plug has wedged itself in the block.

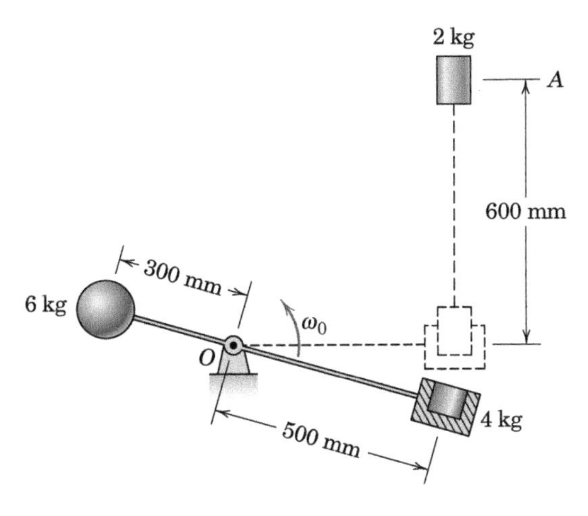

Problem 3/225

3/226 A 0.7-lb particle is located at the position $\mathbf{r}_1 = 2\mathbf{i} + 3\mathbf{j} + \mathbf{k}$ ft and has the velocity $\mathbf{v}_1 = \mathbf{i} + \mathbf{j} + 2\mathbf{k}$ ft/sec at time $t = 0$. If the particle is acted upon by a single force which has the moment $\mathbf{M}_O = (4 + 2t)\mathbf{i} + (3 - t^2)\mathbf{j} + 5\mathbf{k}$ lb-ft about the origin O of the coordinate system in use, determine the angular momentum about O of the particle when $t = 4$ sec.

3/227 The two spheres of equal mass m are able to slide along the horizontal rotating rod. If they are initially latched in position a distance r from the rotating axis with the assembly rotating freely with an angular velocity ω_0, determine the new angular velocity ω after the spheres are released and finally assume positions at the ends of the rod at a radial distance of $2r$. Also find the fraction n of the initial kinetic energy of the system which is lost. Neglect the small mass of the rod and shaft.

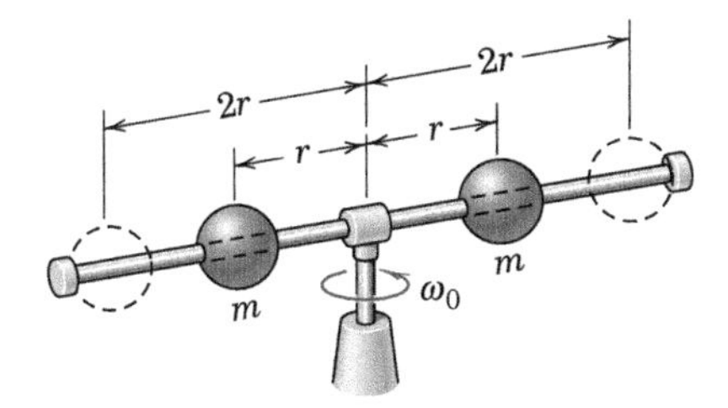

Problem 3/227

3/228 A particle of mass m moves with negligible friction on a horizontal surface and is connected to a light spring fastened at O. At position A the particle has the velocity $v_A = 4$ m/s. Determine the velocity v_B of the particle as it passes position B.

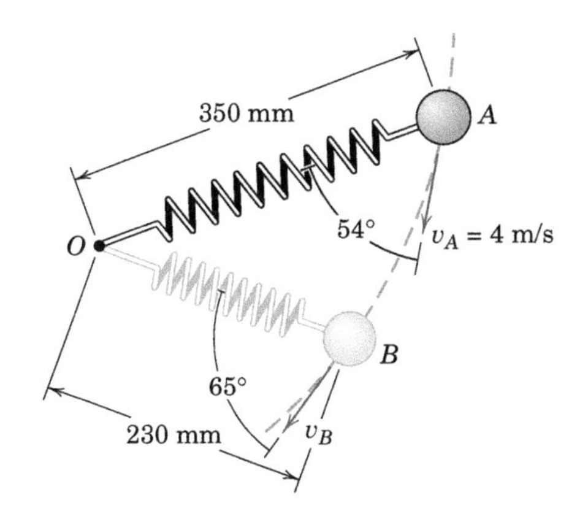

Problem 3/228

3/229 The small spheres, which have the masses and initial velocities shown in the figure, strike and become attached to the spiked ends of the rod, which is freely pivoted at O and is initially at rest. Determine the angular velocity ω of the assembly after impact. Neglect the mass of the rod.

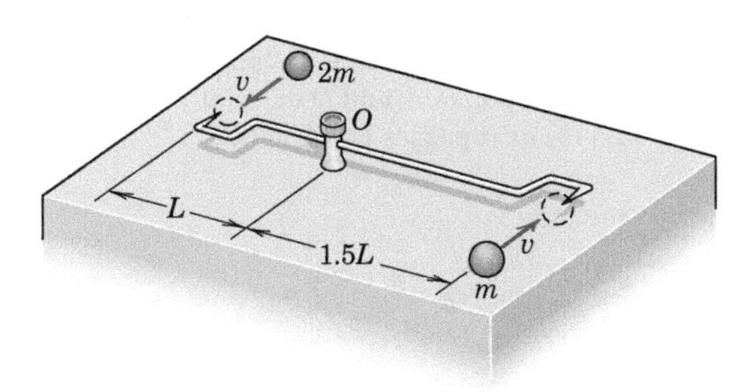

Problem 3/229

3/230 The particle of mass m is launched from point O with a horizontal velocity $\mathbf{u}$ at time $t = 0$. Determine its angular momentum $\mathbf{H}_O$ relative to point O as a function of time.

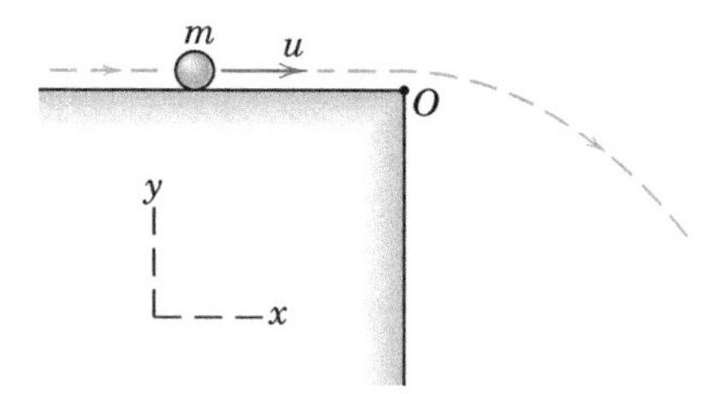

Problem 3/230

3/231 A wad of clay of mass m_1 with an initial horizontal velocity v_1 hits and adheres to the massless rigid bar which supports the body of mass m_2, which can be assumed to be a particle. The pendulum assembly is freely pivoted at O and is initially stationary. Determine the angular velocity $\dot{\theta}$ of the combined body just after impact. Why is linear momentum of the system not conserved?

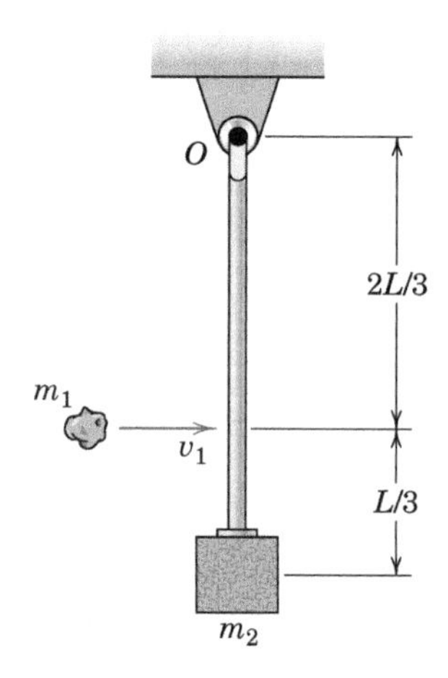

Problem 3/231

3/232 A particle of mass m is released from rest in position A and then slides down the smooth vertical-plane track. Determine its angular momentum about both points A and D (a) as it passes position B and (b) as it passes position C.

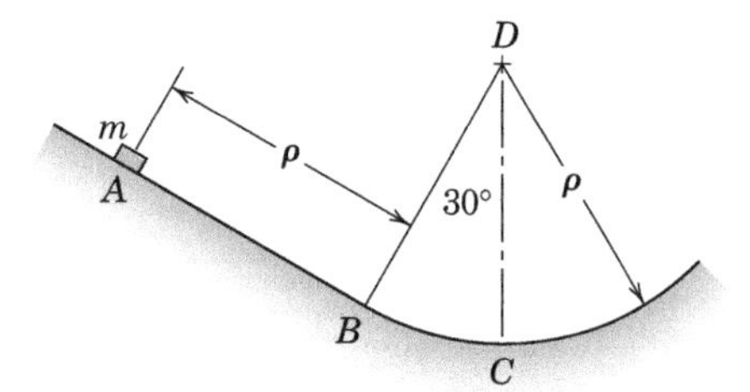

Problem 3/232

3/233 At the point A of closest approach to the sun, a comet has a velocity v_A = 188,500 ft/sec. Determine the radial and transverse components of its velocity v_B at point B, where the radial distance from the sun is $75(10^6)$ mi.

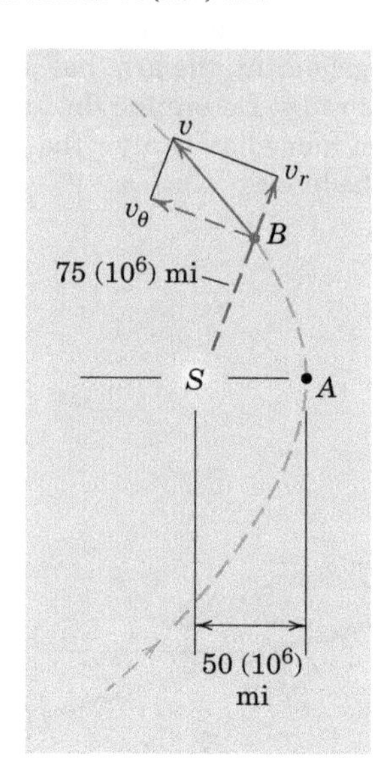

Problem 3/233

3/234 A particle moves on the inside surface of a smooth conical shell and is given an initial velocity $\mathbf{v}_0$ tangent to the horizontal rim of the surface at A. As the particle slides past point B, a distance z below A, its velocity $\mathbf{v}$ makes an angle θ with the horizontal tangent to the surface through B. Determine expressions for θ and the speed v.

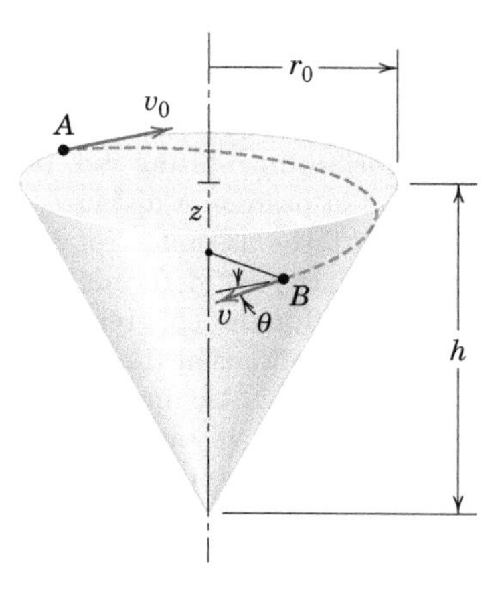

Problem 3/234

3/235 A pendulum consists of two 3.2-kg concentrated masses positioned as shown on a light but rigid bar. The pendulum is swinging through the vertical position with a clockwise angular velocity ω = 6 rad/s when a 50-g bullet traveling with velocity v = 300 m/s in the direction shown strikes the lower mass and becomes embedded in it. Calculate the angular velocity ω' which the pendulum has immediately after impact and find the maximum angular deflection θ of the pendulum.

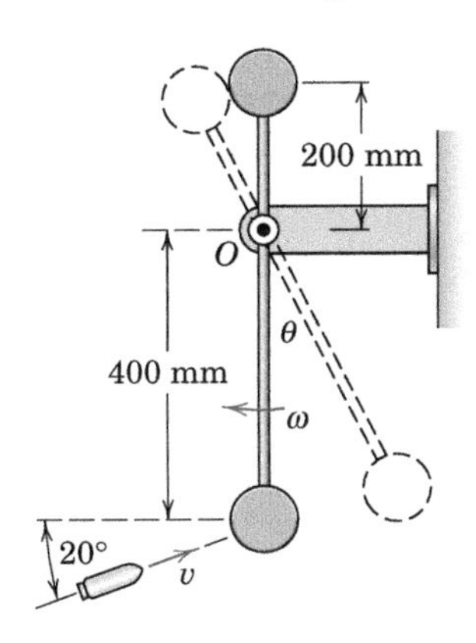

Problem 3/235

3/236 The central attractive force F on an earth satellite can have no moment about the center O of the earth. For the particular elliptical orbit with major and minor axes as shown, a satellite will have a velocity of 33 880 km/h at the perigee altitude of 390 km. Determine the velocity of the satellite at point B and at apogee A. The radius of the earth is 6371 km.

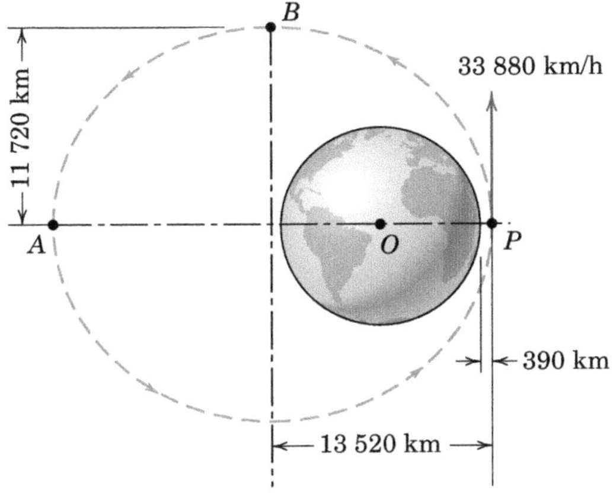

Problem 3/236

3/237 A particle is launched with a horizontal velocity v_0 = 0.55 m/s from the 30° position shown and then slides without friction along the funnel-like surface. Determine the angle θ which its velocity vector makes with the horizontal as the particle passes level O-O. The value of r is 0.9 m.

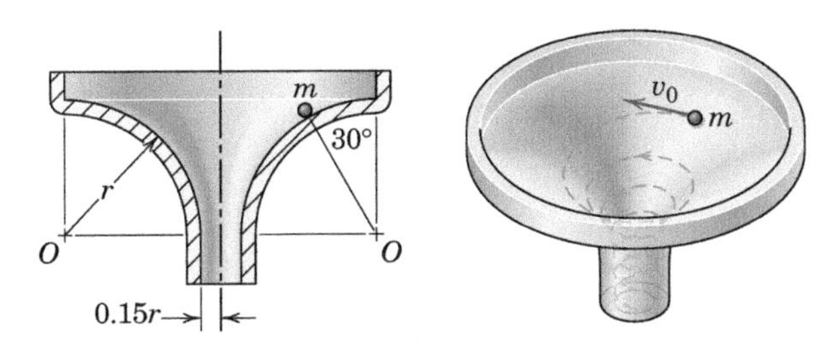

Problem 3/237

3/238 The 0.4-lb ball and its supporting cord are revolving about the vertical axis on the fixed smooth conical surface with an angular velocity of 4 rad/sec. The ball is held in the position b = 14 in. by the tension T in the cord. If the distance b is reduced to the constant value of 9 in. by increasing the tension T in the cord, compute the new angular velocity ω and the work $U'_{1\text{-}2}$ done on the system by T.

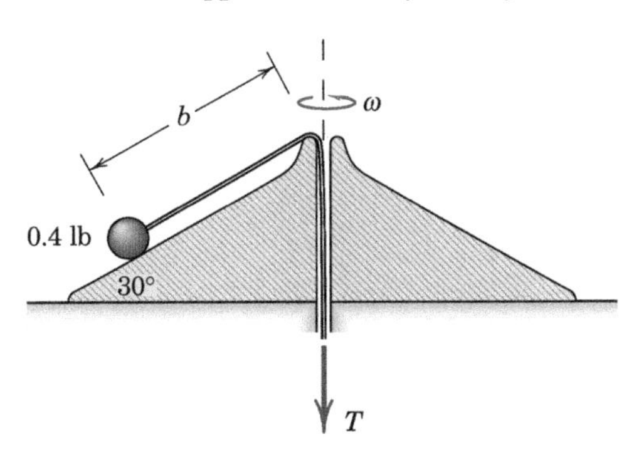

Problem 3/238

3/239 The 0.02-kg particle moves along the dashed trajectory shown and has the indicated velocities at positions A and B. Calculate the time average of the moment about O of the resultant force P acting on the particle during the 0.5 second required for it to go from A to B.

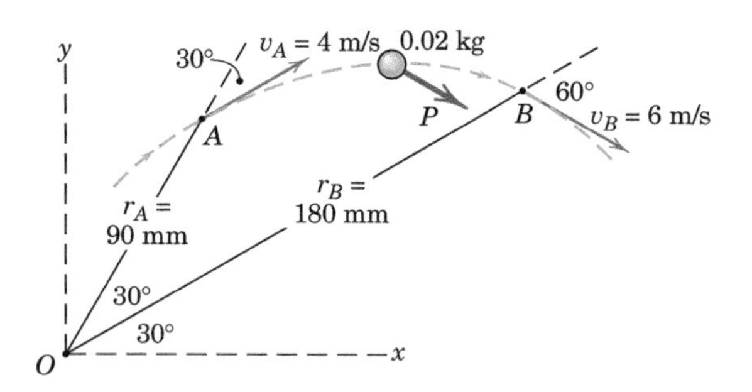

Problem 3/239

▶**3/240** The assembly of two 5-kg spheres is rotating freely about the vertical axis at 40 rev/min with $\theta = 90°$. If the force F which maintains the given position is increased to raise the base collar and reduce θ to 60°, determine the new angular velocity ω. Also determine the work U done by F in changing the configuration of the system. Assume that the mass of the arms and collars is negligible.

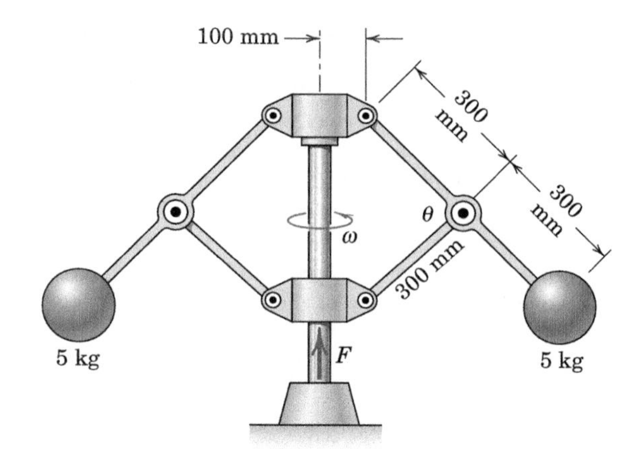

Problem 3/240

PROBLEMS

Introductory Problems

3/241 As a check of the basketball before the start of a game, the referee releases the ball from the overhead position shown, and the ball rebounds to about waist level. Determine the coefficient of restitution e and the percentage n of the original energy lost during the impact.

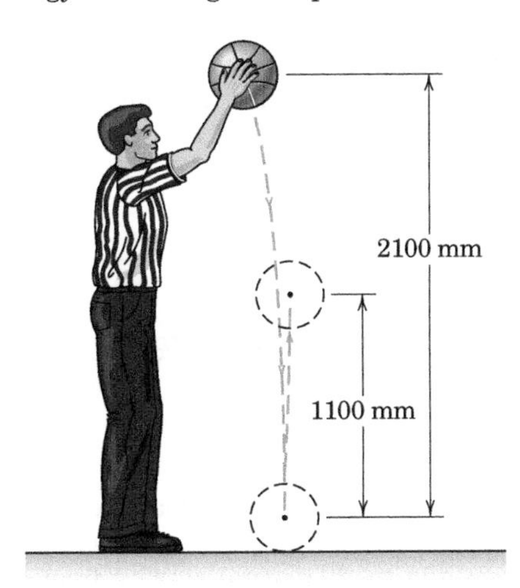

Problem 3/241

3/242 Compute the final velocities v_1' and v_2' after collision of the two cylinders which slide on the smooth horizontal shaft. The coefficient of restitution is $e = 0.8$.

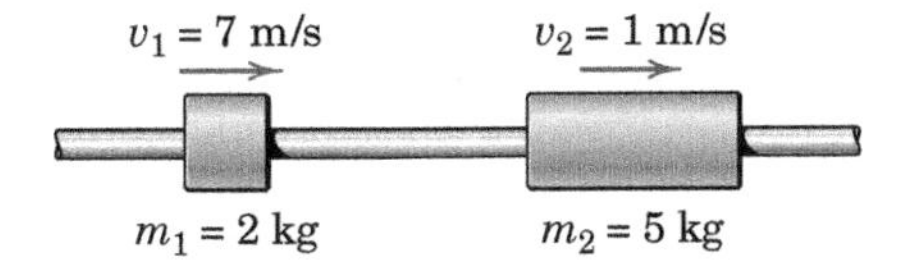

Problem 3/242

3/243 Car B is initially stationary and is struck by car A, which is moving with speed v. The mass of car B is pm, where m is the mass of car A and p is a positive constant. If the coefficient of restitution is $e = 0.1$, express the speeds v_A' and v_B' of the two cars at the end of the impact in terms of p and v. Evaluate your expressions for $p = 0.5$.

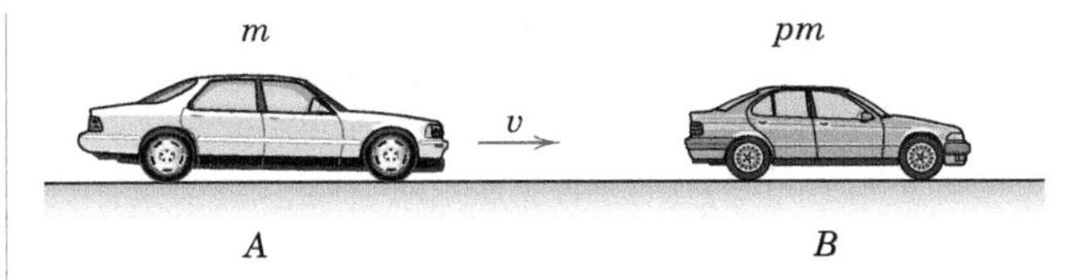

Problem 3/243

3/244 The sphere of mass m_1 travels with an initial velocity v_1 directed as shown and strikes the stationary sphere of mass m_2. For a given coefficient of restitution e, what condition on the mass ratio m_1/m_2 ensures that the final velocity of m_2 is greater than v_1?

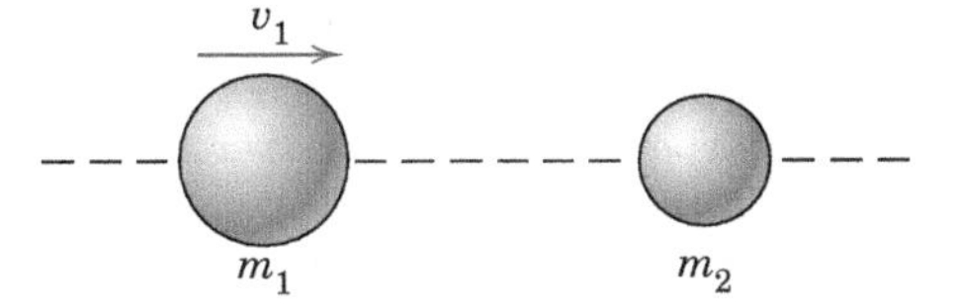

Problem 3/244

3/245 A tennis ball is projected toward a smooth surface with speed v as shown. Determine the rebound angle θ' and the final speed v'. The coefficient of restitution is 0.6.

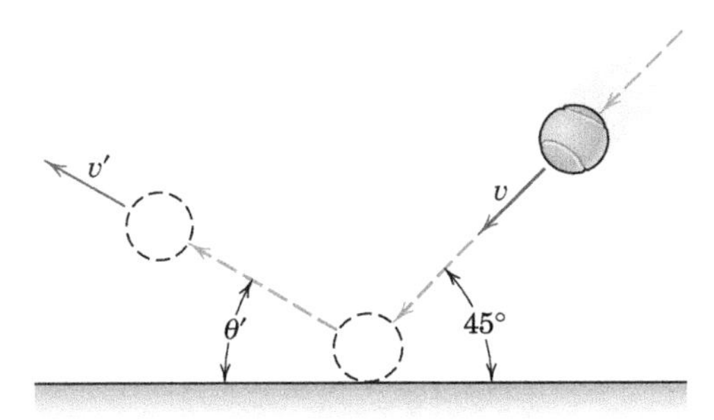

Problem 3/245

3/246 Determine the coefficient of restitution e for a steel ball dropped from rest at a height h above a heavy horizontal steel plate if the height of the second rebound is h_2.

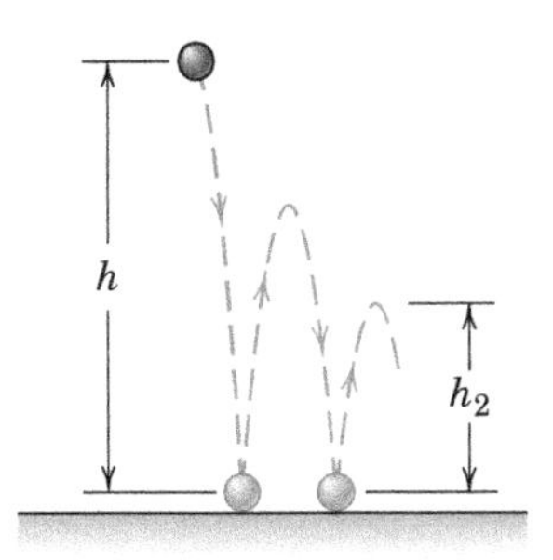

Problem 3/246

3/247 Determine the value of the coefficient of restitution e for which the outgoing angle is one-half of the incoming angle θ as shown. Evaluate your general expression for $\theta = 40°$.

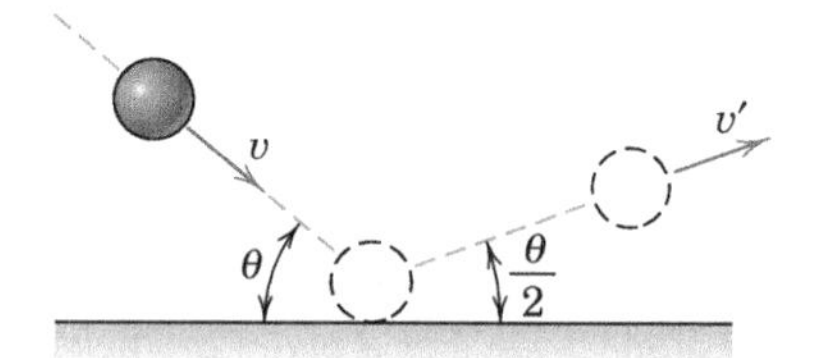

Problem 3/247

3/248 To pass inspection, steel balls designed for use in ball bearings must clear the fixed bar A at the top of their rebound when dropped from rest through the vertical distance $H = 36$ in. onto the heavy inclined steel plate. If balls which have a coefficient of restitution of less than 0.7 with the rebound plate are to be rejected, determine the position of the bar by specifying h and s. Neglect any friction during impact.

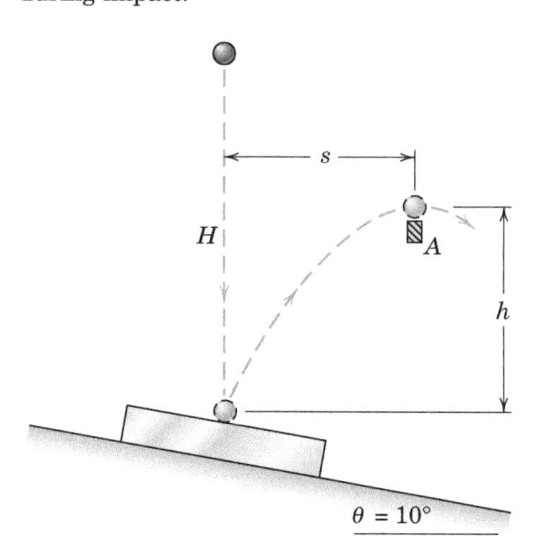

Problem 3/248

3/249 The cart of mass $m_1 = 3$ kg is moving to the right with a speed $v_1 = 6$ m/s when it collides with the initially stationary barrier of mass $m_2 = 5$ kg. The coefficient of restitution for this collision is $e = 0.75$. Determine the maximum deflection δ of the barrier, which is connected to three springs, each of which has a modulus of 4 kN/m and is undeformed before the impact.

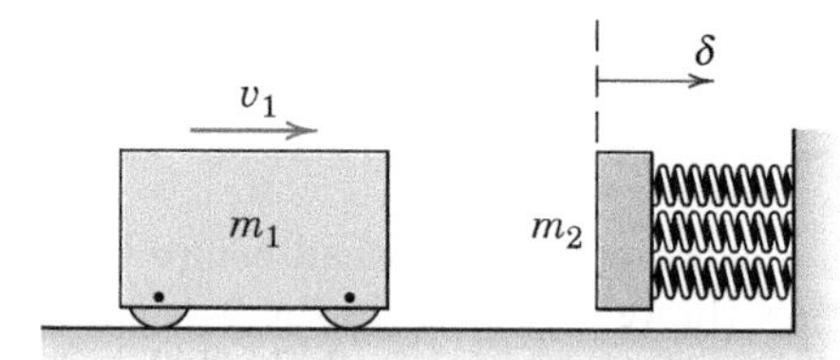

Problem 3/249

3/250 If the center of the ping-pong ball is to clear the net as shown, at what height h should the ball be horizontally served? Also determine h_2. The coefficient of restitution for the impacts between ball and table is $e = 0.9$, and the radius of the ball is $r = 0.75$ in.

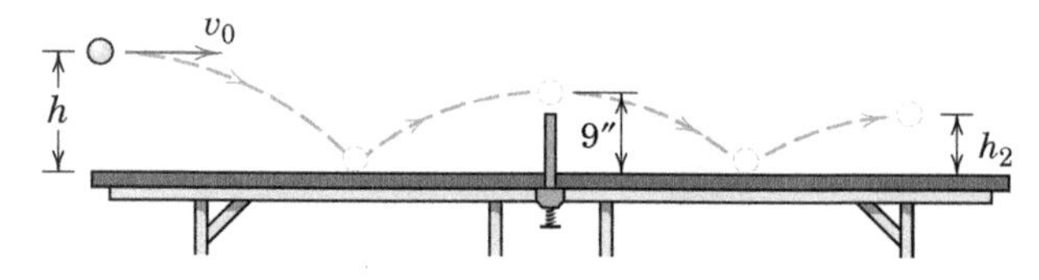

Problem 3/250

3/251 Two steel balls of the same diameter are connected by a rigid bar of negligible mass as shown and are dropped in the horizontal position from a height of 150 mm above the heavy steel and brass base plates. If the coefficient of restitution between the ball and the steel base is 0.6 and that between the other ball and the brass base is 0.4, determine the angular velocity ω of the bar immediately after impact. Assume that the two impacts are simultaneous.

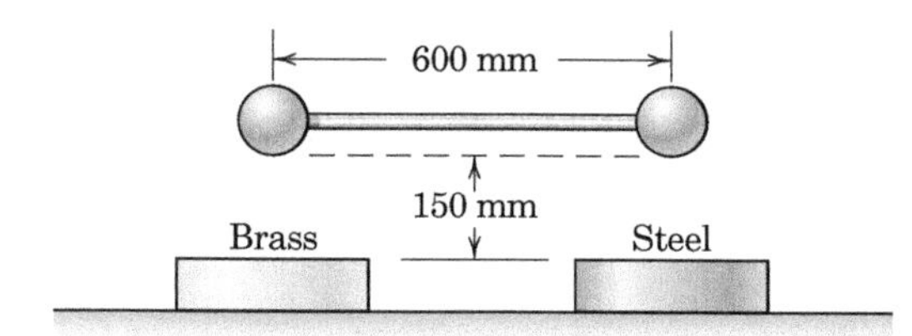

Problem 3/251

3/252 Freight car A of mass m_A is rolling to the right when it collides with freight car B of mass m_B initially at rest. If the two cars are coupled together at impact, show that the fractional loss of energy equals $m_B/(m_A + m_B)$.

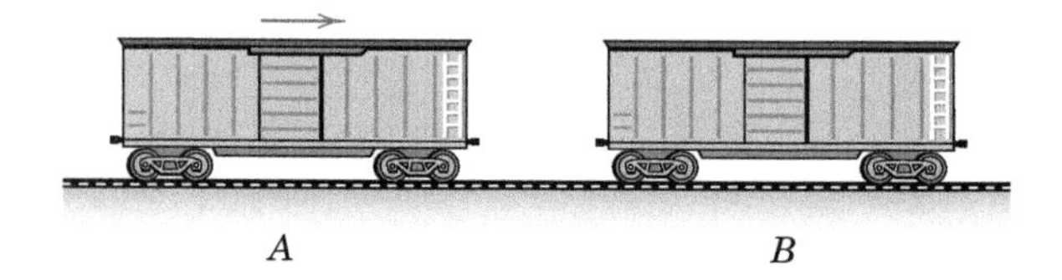

Problem 3/252

Representative Problems

3/253 A small ball is projected horizontally toward an incline as shown. Determine the slant range R. The initial speed is $v_0 = 16$ m/s, and the coefficient of restitution for the impact at A is $e = 0.6$.

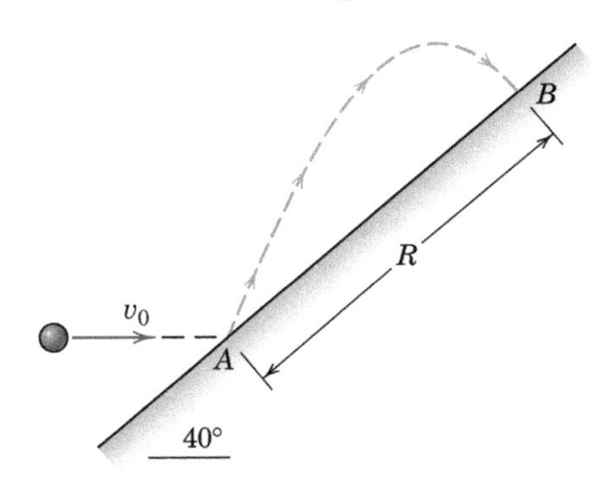

Problem 3/253

3/254 A miniature-golf shot from position A to the hole D is to be accomplished by "banking off" the 45° wall. Using the theory of this article, determine the location x for which the shot can be made. The coefficient of restitution associated with the wall collision is $e = 0.8$.

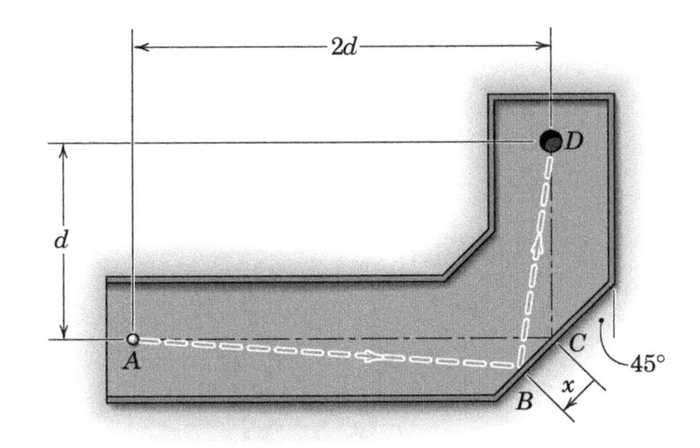

Problem 3/254

3/255 The pendulum is released from the 60° position and then strikes the initially stationary cylinder of mass m_2 when OA is vertical. Determine the maximum spring compression δ. Use the values $m_1 = 3$ kg, $m_2 = 2$ kg, $\overline{OA} = 0.8$ m, $e = 0.7$, and $k = 6$ kN/m. Assume that the bar of the pendulum is light so that the mass m_1 is effectively concentrated at point A. The rubber cushion S stops the pendulum just after the collision is over. Neglect all friction.

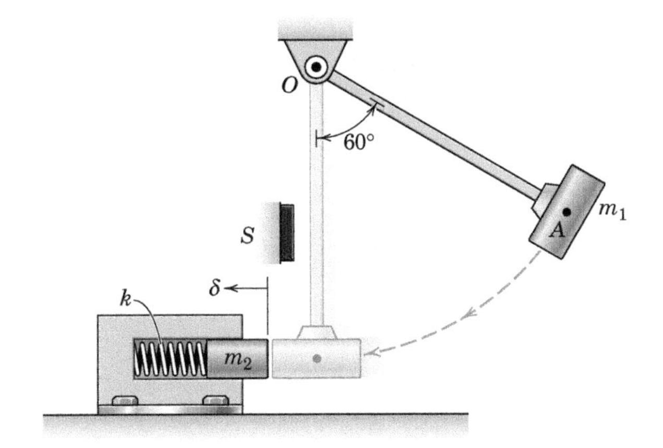

Problem 3/255

3/256 A 0.1-kg meteor and a 1000-kg spacecraft have the indicated absolute velocities just before colliding. The meteor punches a hole entirely through the spacecraft. Instruments indicate that the velocity of the meteor relative to the spacecraft just after the collision is $\mathbf{v}_{m/s}' = -1880\mathbf{i} - 6898\mathbf{j}$ m/s. Determine the direction θ of the absolute velocity of the spacecraft after the collision.

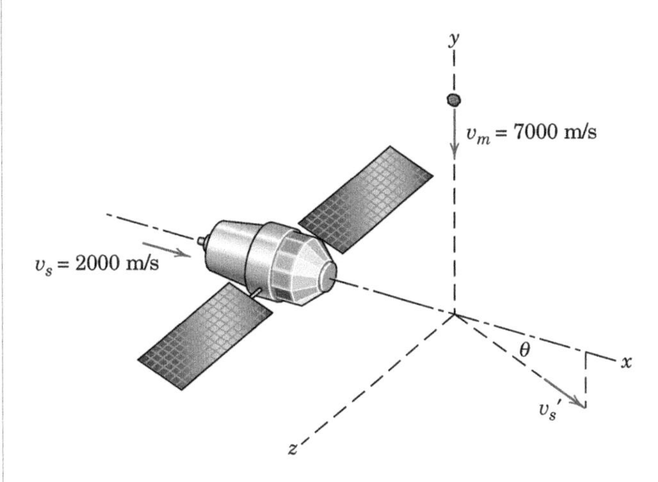

Problem 3/256

3/257 In a pool game the cue ball A must strike the eight ball in the position shown in order to send it to the pocket P with a velocity v_2'. The cue ball has a velocity v_1 before impact and a velocity v_1' after impact. The coefficient of restitution is 0.9. Both balls have the same mass and diameter. Calculate the rebound angle θ and the fraction n of the kinetic energy which is lost during the impact.

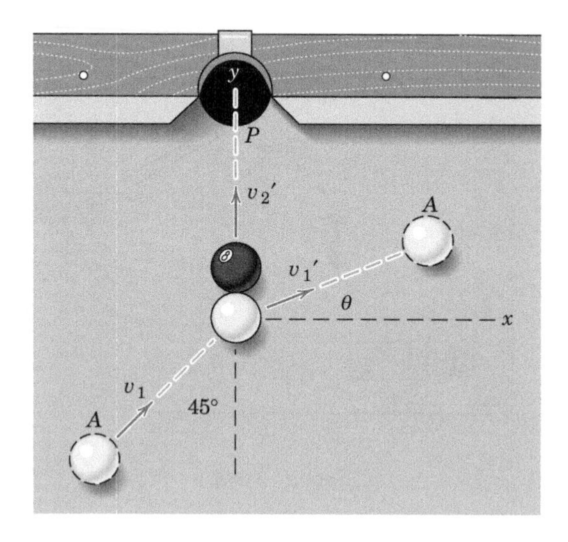

Problem 3/257

3/258 Determine the coefficient of restitution e which will allow the ball to bounce down the steps as shown. The tread and riser dimensions, d and h, respectively, are the same for every step, and the ball bounces the same distance h' above each step. What horizontal velocity v_x is required so that the ball lands in the center of each tread?

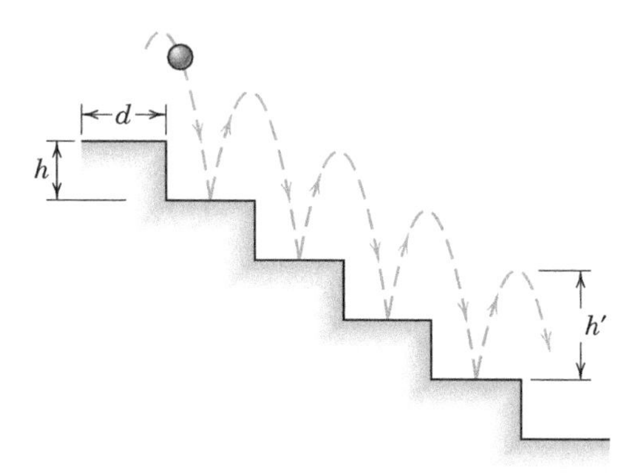

Problem 3/258

3/259 Sphere A has a mass of 23 kg and a radius of 75 mm, while sphere B has a mass of 4 kg and a radius of 50 mm. If the spheres are traveling initially along the parallel paths with the speeds shown, determine the velocities of the spheres immediately after impact. Specify the angles θ_A and θ_B with respect to the x-axis made by the rebound velocity vectors. The coefficient of restitution is 0.4 and friction is neglected.

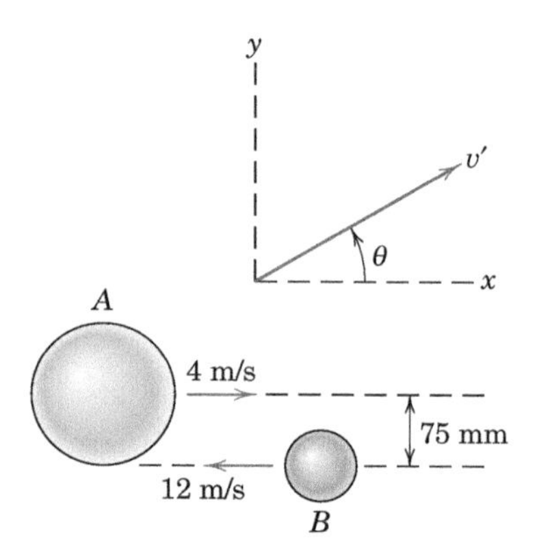

Problem 3/259

3/260 During a pregame warmup period, two basketballs collide above the hoop when in the positions shown. Just before impact, ball 1 has a velocity v_1 which makes a 30° angle with the horizontal. If the velocity v_2 of ball 2 just before impact has the same magnitude as v_1, determine the two possible values of the angle θ, measured from the horizontal, which will cause ball 1 to go directly through the center of the basket. The coefficient of restitution is $e = 0.8$.

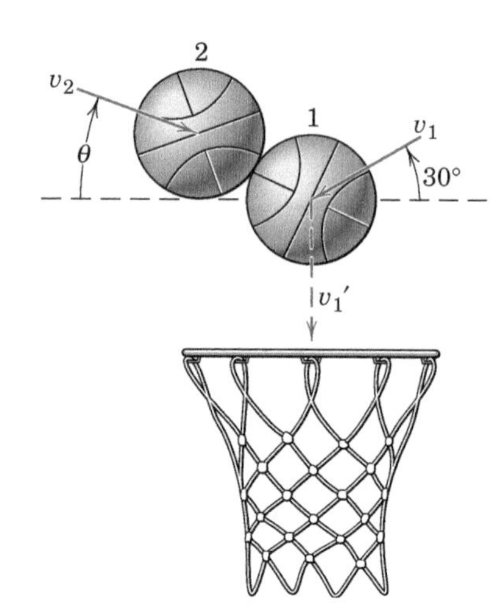

Problem 3/260

3/261 Two identical hockey pucks moving with initial velocities v_A and v_B collide as shown. If the coefficient of restitution is $e = 0.75$, determine the velocity (magnitude and direction θ with respect to the positive x-axis) of each puck just after impact. Also calculate the percentage loss n of system kinetic energy.

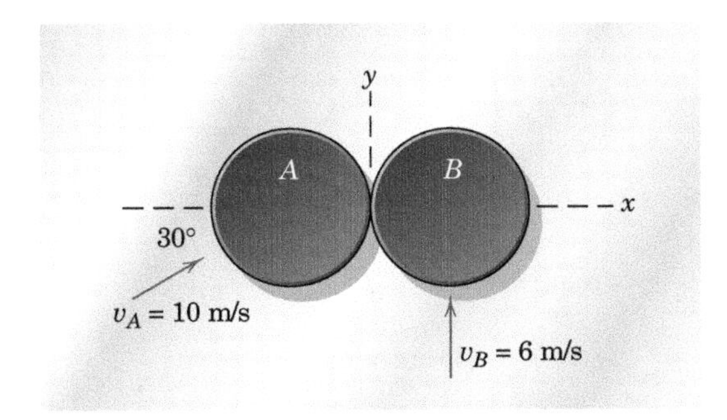

Problem 3/261

3/262 Repeat the previous problem, only now the mass of puck B is twice that of puck A.

3/263 The 3000-kg anvil A of the drop forge is mounted on a nest of heavy coil springs having a combined stiffness of $2.8(10^6)$ N/m. The 600-kg hammer B falls 500 mm from rest and strikes the anvil, which suffers a maximum downward deflection of 24 mm from its equilibrium position. Determine the height h of rebound of the hammer and the coefficient of restitution e which applies.

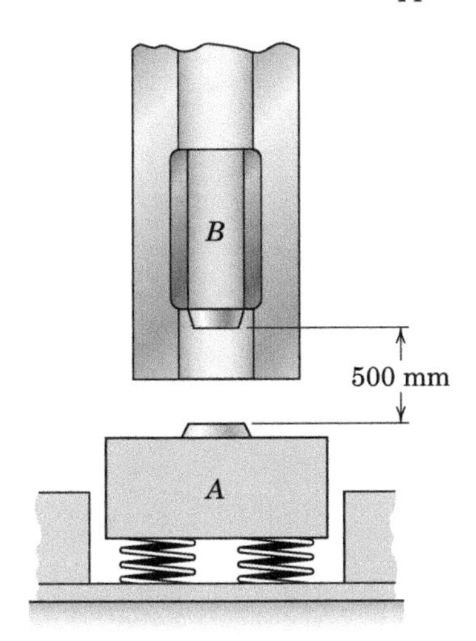

Problem 3/263

3/264 The 0.5-kg cylinder A is released from rest from the position shown and drops the distance $h_1 = 0.6$ m. It then collides with the 0.4-kg block B; the coefficient of restitution is $e = 0.8$. Determine the maximum downward displacement h_2 of block B. Neglect all friction and assume that block B is initially held in place by a hidden mechanism until the collision begins. The two springs of modulus $k = 500$ N/m are initially unstretched, and the distance $d = 0.8$ m.

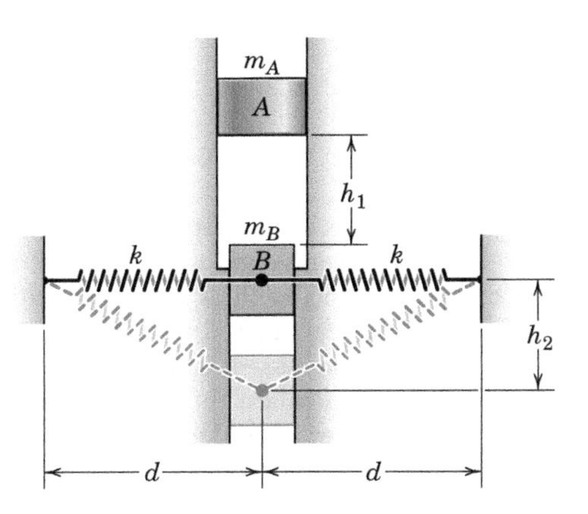

Problem 3/264

▶3/265 The elements of a device designed to measure the coefficient of restitution of bat–baseball collisions are shown. The 1-lb "bat" A is a short length of wood or aluminum which is projected to the right with a speed $v_A = 60$ ft/sec within the confines of the horizontal slot. Just before and after the moment of impact, body A is free to move horizontally. The baseball B weighs 5.125 oz and has an initial speed $v_B = 125$ ft/sec. If the coefficient of restitution is $e = 0.5$, determine the final speed of the baseball and the angle β which its final velocity makes with the horizontal.

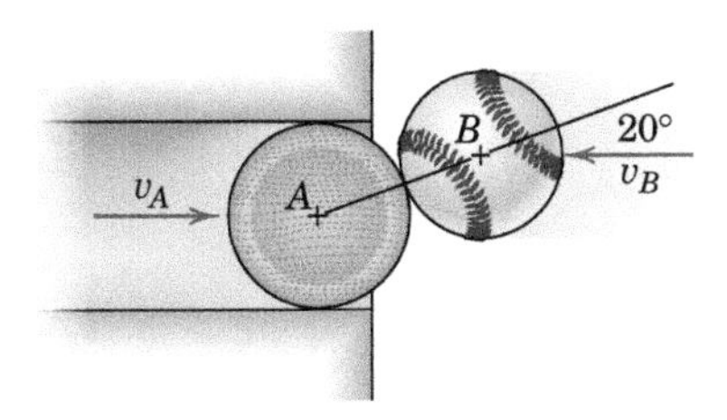

Problem 3/265

▶3/266 A child throws a ball from point A with a speed of 50 ft/sec. It strikes the wall at point B and then returns exactly to point A. Determine the necessary angle α if the coefficient of restitution in the wall impact is $e = 0.5$.

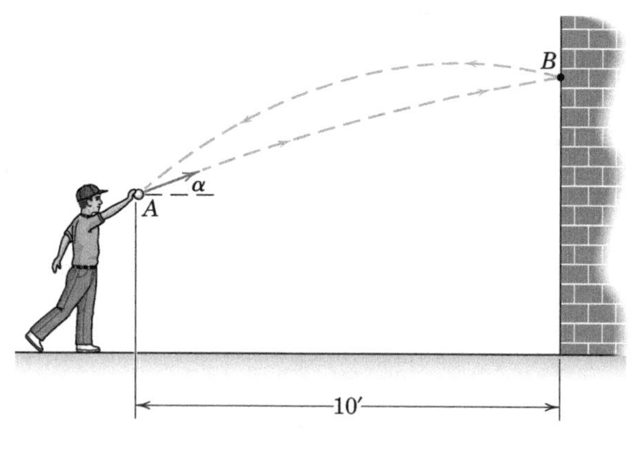

Problem 3/266

▶**3/267** The 2-kg sphere is projected horizontally with a velocity of 10 m/s against the 10-kg carriage which is backed up by the spring with stiffness of 1600 N/m. The carriage is initially at rest with the spring uncompressed. If the coefficient of restitution is 0.6, calculate the rebound velocity v', the rebound angle θ, and the maximum travel δ of the carriage after impact.

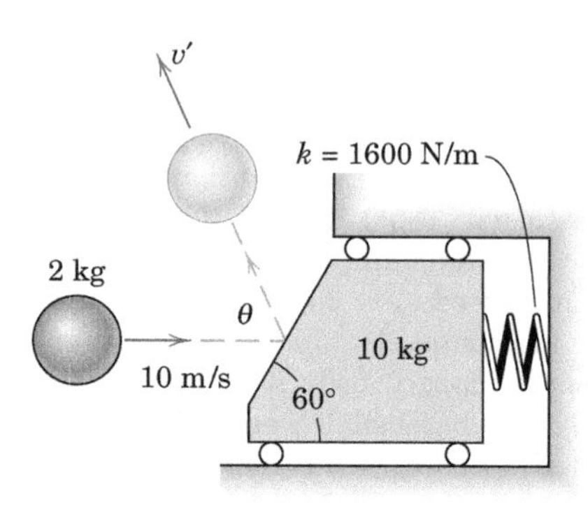

Problem 3/267

▶**3/268** A small ball is projected horizontally as shown and bounces at point A. Determine the range of initial speed v_0 for which the ball will ultimately land on the horizontal surface at B. The coefficient of restitution at A is $e = 0.8$ and the distance $d = 4$ m.

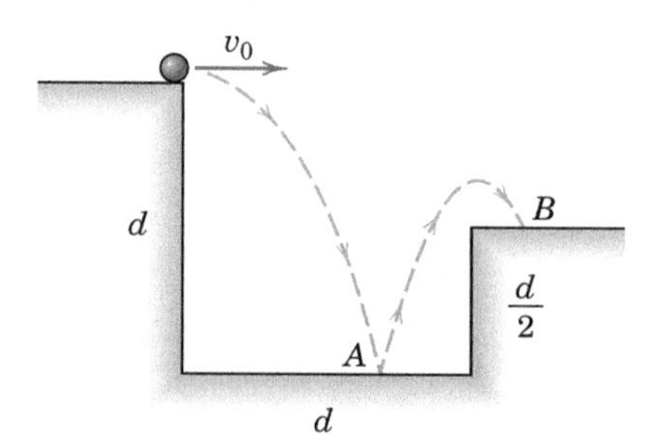

Problem 3/268

PROBLEMS

(Unless otherwise indicated, the velocities mentioned in the problems which follow are measured from a nonrotating reference frame moving with the center of the attracting body. Also, aerodynamic drag is to be neglected unless stated otherwise. Use $g = 9.825$ m/s^2 (32.23 ft/sec^2) for the absolute gravitational acceleration at the surface of the earth and treat the earth as a sphere of radius $R = 6371$ km (3959 mi).)

Introductory Problems

3/269 Calculate the velocity of a spacecraft which orbits the moon in a circular path of 80-km altitude.

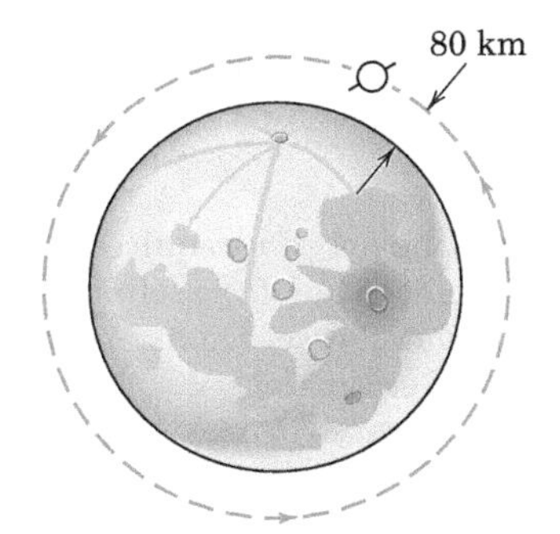

Problem 3/269

3/270 What velocity v must the space shuttle have in order to release the Hubble space telescope in a circular earth orbit 590 km above the surface of the earth?

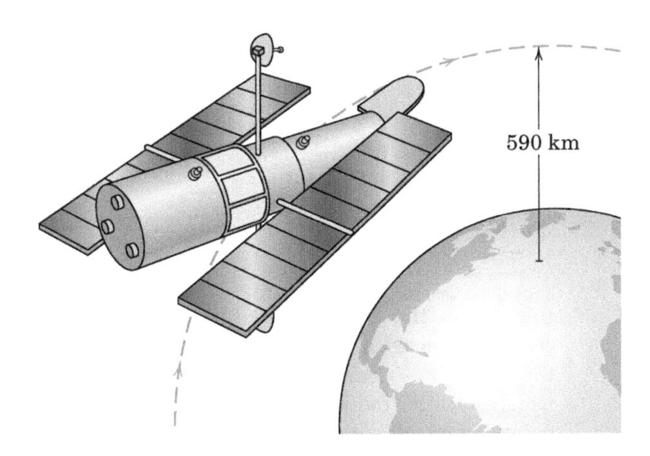

Problem 3/270

3/271 Show that the path of the moon is concave toward the sun at the position shown. Assume that the sun, earth, and moon are in the same line.

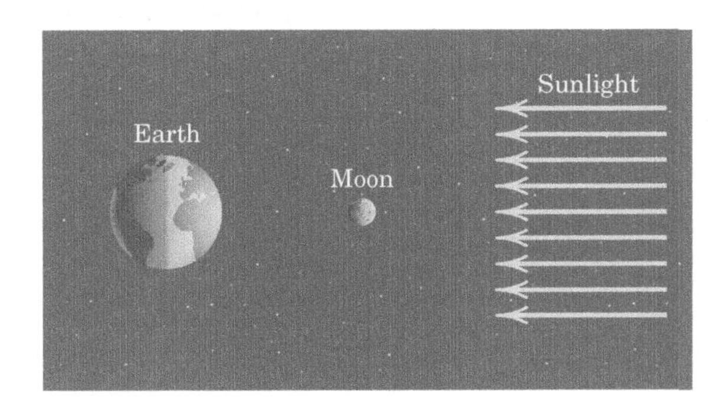

Problem 3/271

3/272 A satellite is in a circular polar orbit of altitude 300 km. Determine the separation d at the equator between the ground tracks (shown dashed) associated with two successive overhead passes of the satellite.

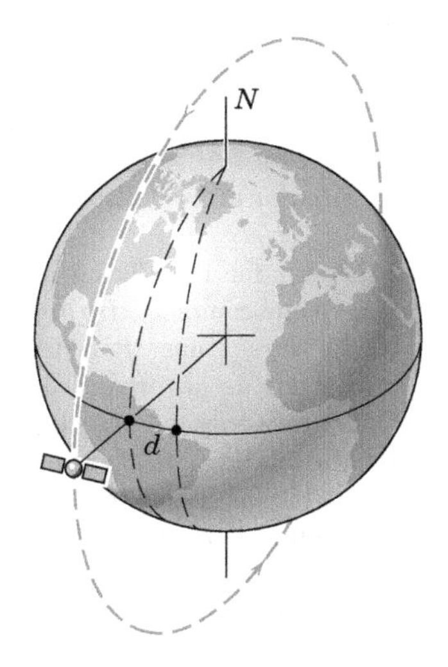

Problem 3/272

3/273 Determine the apparent velocity v_{rel} of a satellite moving in a circular equatorial orbit 200 mi above the earth as measured by an observer on the equator (*a*) for a west-to-east orbit and (*b*) for an east-to-west orbit. Why is the west-to-east orbit more easily achieved?

3/274 A spacecraft is in an initial circular orbit with an altitude of 350 km. As it passes point P, onboard thrusters give it a velocity boost of 25 m/s. Determine the resulting altitude gain Δh at point A.

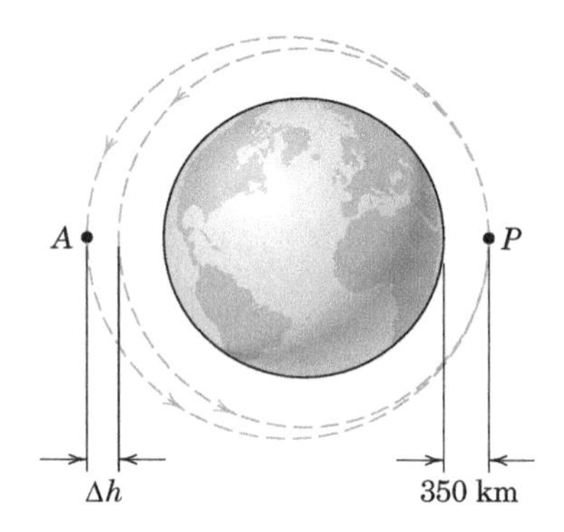

Problem 3/274

3/275 If the perigee altitude of an earth satellite is 240 km and the apogee altitude is 400 km, compute the eccentricity e of the orbit and the period τ of one complete orbit in space.

3/276 Determine the energy difference ΔE between an 80 000-kg space-shuttle orbiter on the launch pad in Cape Canaveral (latitude 28.5°) and the same orbiter in a circular orbit of altitude $h = 300$ km.

3/277 The Mars orbiter for the Viking mission was designed to make one complete trip around the planet in exactly the same time that it takes Mars to revolve once about its own axis. This time is 24 h, 37 min, 23 s. In this way, it is possible for the orbiter to pass over the landing site of the lander capsule at the same time in each Martian day at the orbiter's minimum (periapsis) altitude. For the Viking I mission, the periapsis altitude of the orbiter was 1508 km. Make use of the data in Table D/2 in Appendix D and compute the maximum (apoapsis) altitude h_a for the orbiter in its elliptical path.

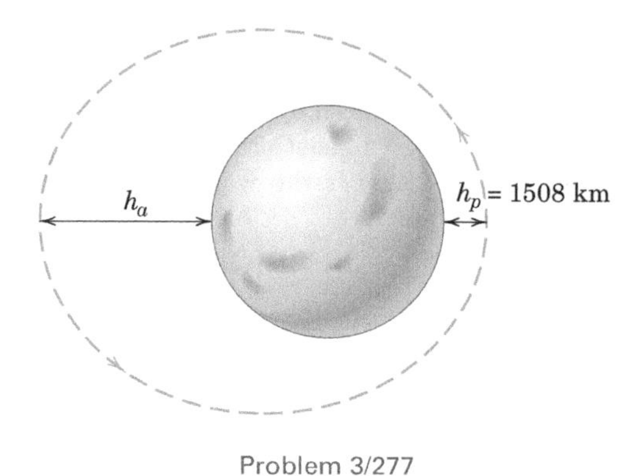

Problem 3/277

3/278 A "drag-free" satellite is one which carries a small mass inside a chamber as shown. If the satellite speed decreases because of drag, the mass speed will not, and so the mass moves relative to the chamber as indicated. Sensors detect this change in the position of the mass within the chamber, and the satellite thruster is periodically fired to re-center the mass. In this manner, compensation is made for drag. If the satellite is in a circular earth orbit of 200-km altitude and a total thruster burn time of 300 seconds occurs during 10 orbits, determine the drag force D acting on the 100-kg satellite. The thruster force T is 2 N.

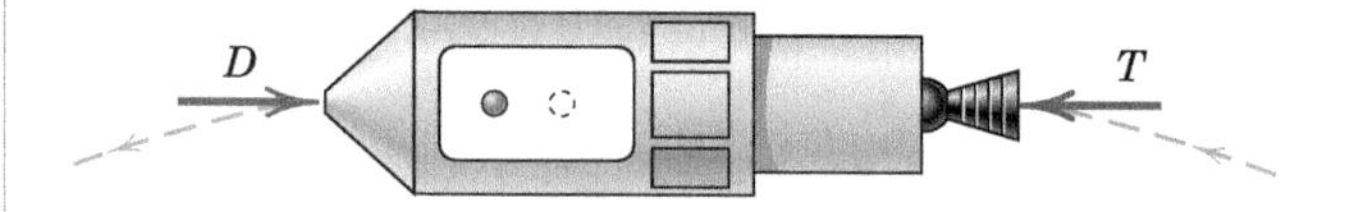

Problem 3/278

3/279 Determine the speed v required of an earth satellite at point A for (*a*) a circular orbit, (*b*) an elliptical orbit of eccentricity $e = 0.1$, (*c*) an elliptical orbit of eccentricity $e = 0.9$, and (*d*) a parabolic orbit. In cases (*b*), (*c*), and (*d*), A is the orbit perigee.

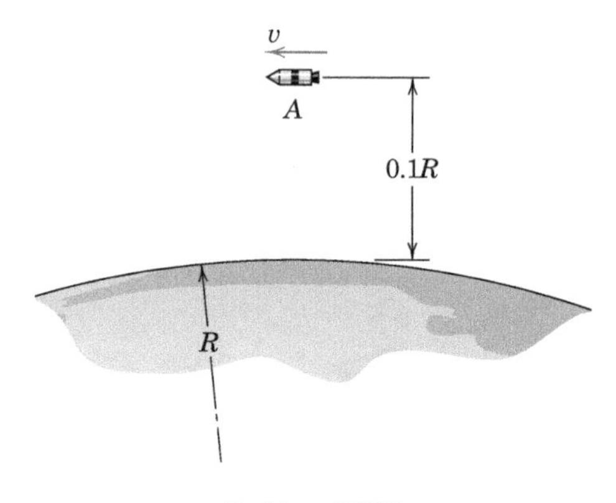

Problem 3/279

Representative Problems

3/280 Initially in the 240-km circular orbit, the spacecraft S receives a velocity boost at P which will take it to $r \rightarrow \infty$ with no speed at that point. Determine the required velocity increment Δv at point P and also determine the speed when $r = 2r_P$. At what value of θ does r become $2r_P$?

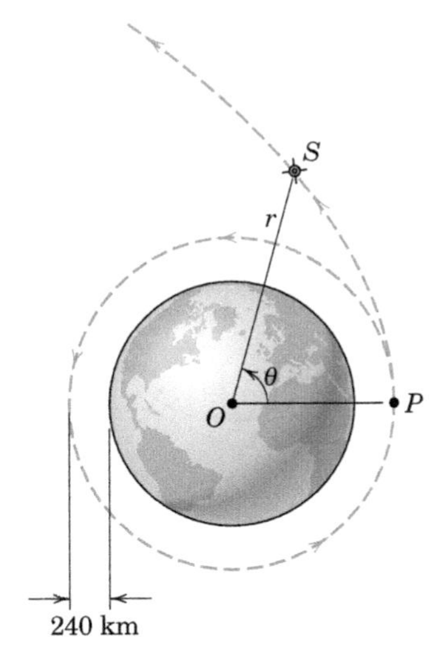

Problem 3/280

3/281 The binary star system consists of stars A and B, both of which orbit about the system mass center. Compare the orbital period τ_f calculated with the assumption of a fixed star A with the period τ_{nf} calculated without this assumption.

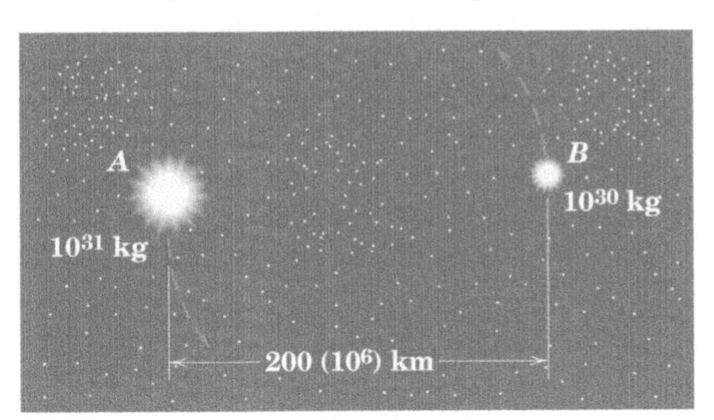

Problem 3/281

3/282 If the earth were suddenly deprived of its orbital velocity around the sun, find the time t which it would take for the earth to "fall" to the location of the center of the sun. *(Hint:* The time would be one-half the period of a degenerate elliptical orbit around the sun with the semiminor axis approaching zero.) Refer to Table D/2 for the exact period of the earth around the sun.

3/283 Just after launch from the earth, the space-shuttle orbiter is in the 37 × 137-mi orbit shown. The first time that the orbiter passes the apogee A, its two orbital-maneuvering-system (OMS) engines are fired to circularize the orbit. If the weight of the orbiter is 175,000 lb and the OMS engines have a thrust of 6000 lb each, determine the required time duration Δt of the burn.

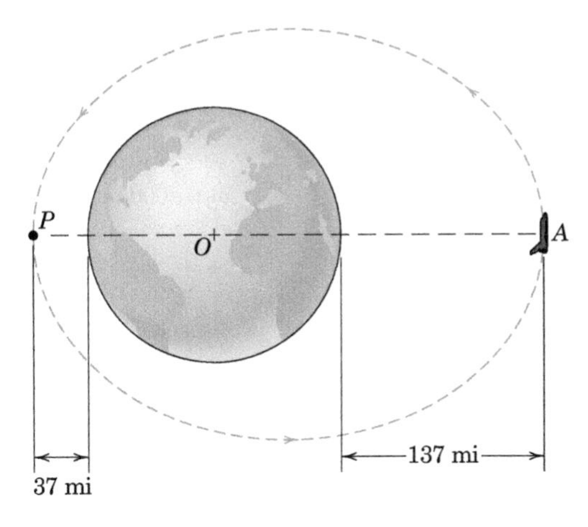

Problem 3/283

3/284 A spacecraft is in a circular orbit of radius $3R$ around the moon. At point A, the spacecraft ejects a probe which is designed to arrive at the surface of the moon at point B. Determine the necessary velocity v_r of the probe relative to the spacecraft just after ejection. Also calculate the position θ of the spacecraft when the probe arrives at point B.

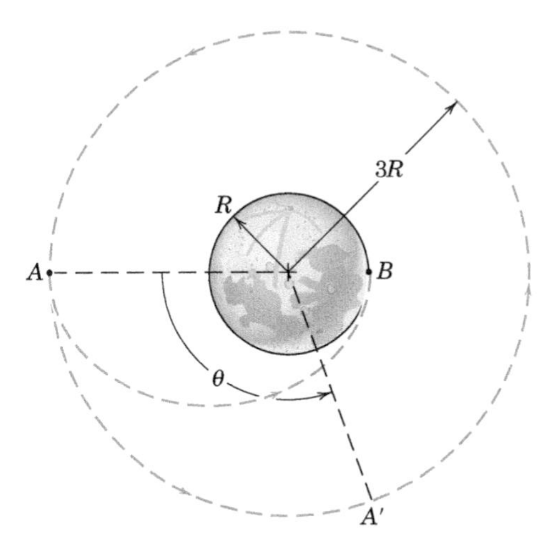

Problem 3/284

3/285 A projectile is launched from B with a speed of 2000 m/s at an angle α of 30° with the horizontal as shown. Determine the maximum altitude h_{max}.

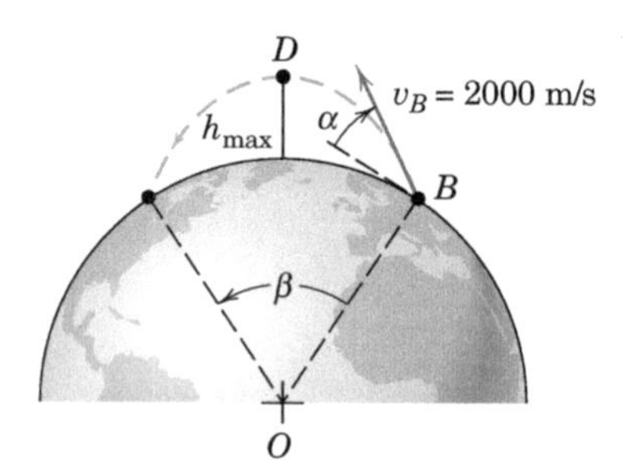

Problem 3/285

3/286 Compute the magnitude of the necessary launch velocity at B if the projectile trajectory is to intersect the earth's surface so that the angle β equals 90°. The altitude at the highest point of the trajectory is $0.5R$.

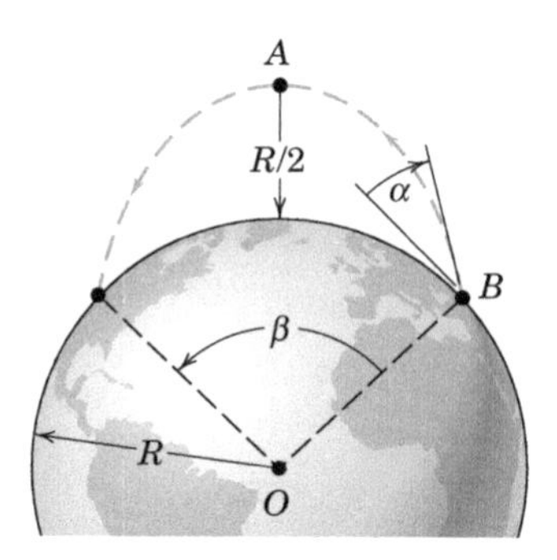

Problem 3/286

3/287 Compute the necessary launch angle α at point B for the trajectory prescribed in Prob. 3/286.

3/288 Two satellites B and C are in the same circular orbit of altitude 500 miles. Satellite B is 1000 mi ahead of satellite C as indicated. Show that C can catch up to B by "putting on the brakes." Specifically, by what amount Δv should the circular-orbit velocity of C be reduced so that it will rendezvous with B after one period in its new elliptical orbit? Check to see that C does not strike the earth in the elliptical orbit.

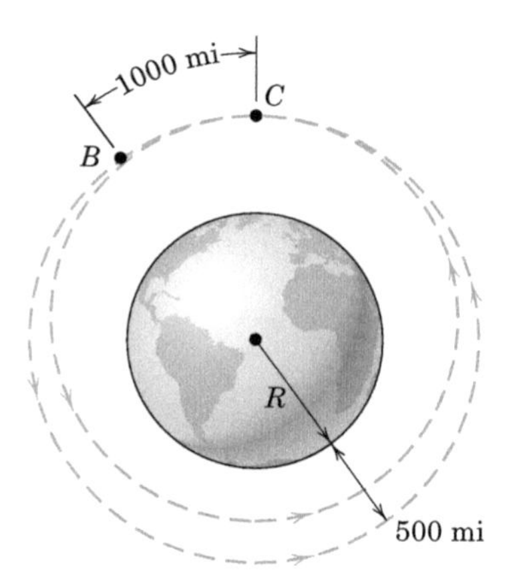

Problem 3/288

3/289 Determine the necessary amount Δv by which the circular-orbit velocity of satellite C should be reduced if the catch-up maneuver of Prob. 3/288 is to be accomplished with not one but two periods in a new elliptical orbit.

3/290 The 175,000-lb space-shuttle orbiter is in a circular orbit of altitude 200 miles. The two orbital-maneuvering-system (OMS) engines, each of which has a thrust of 6000 lb, are fired in retrothrust for 150 seconds. Determine the angle β which locates the intersection of the shuttle trajectory with the earth's surface. Assume that the shuttle position B corresponds to the completion of the OMS burn and that no loss of altitude occurs during the burn.

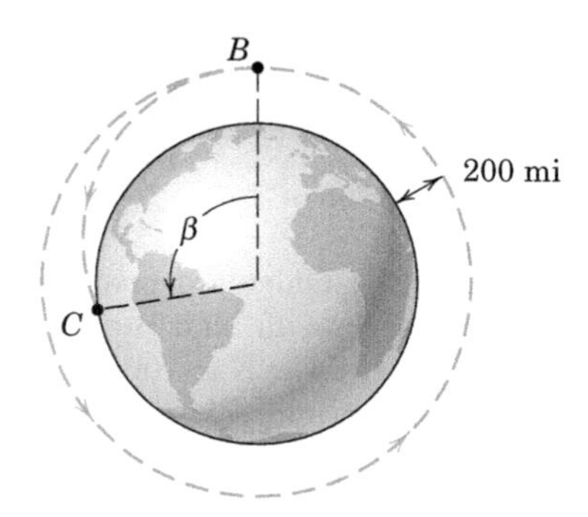

Problem 3/290

3/291 Compare the orbital period of the moon calculated with the assumption of a fixed earth with the period calculated without this assumption.

3/292 A satellite is placed in a circular polar orbit a distance H above the earth. As the satellite goes over the north pole at A, its retro-rocket is activated to produce a burst of negative thrust which reduces its velocity to a value which will ensure an equatorial landing. Derive the expression for the required reduction Δv_A of velocity at A. Note that A is the apogee of the elliptical path.

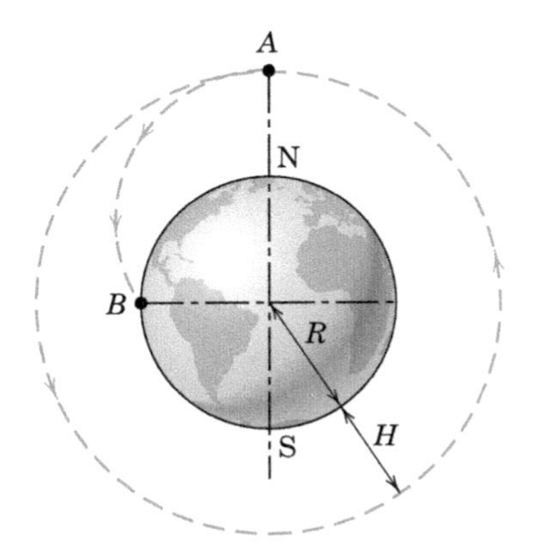

Problem 3/292

3/293 The perigee and apogee altitudes above the surface of the earth of an artificial satellite are h_p and h_a, respectively. Derive the expression for the radius of curvature ρ_p of the orbit at the perigee position. The radius of the earth is R.

3/294 A spacecraft moving in a west-to-east equatorial orbit is observed by a tracking station located on the equator. If the spacecraft has a perigee altitude $H = 150$ km and velocity v when directly over the station and an apogee altitude of 1500 km, determine an expression for the angular rate p (relative to the earth) at which the antenna dish must be rotated when the spacecraft is directly overhead. Compute p. The angular velocity of the earth is $\omega = 0.7292(10^{-4})$ rad/s.

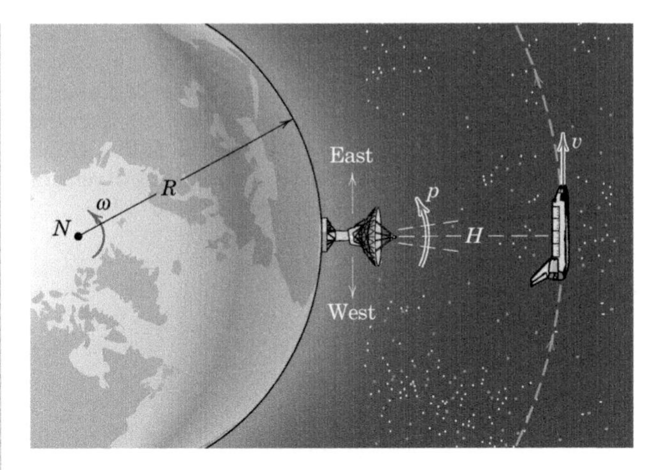

Problem 3/294

3/295 Sometime after launch from the earth, a spacecraft S is in the orbital path of the earth at some distance from the earth at position P. What velocity boost Δv at P is required so that the spacecraft arrives at the orbit of Mars at A as shown?

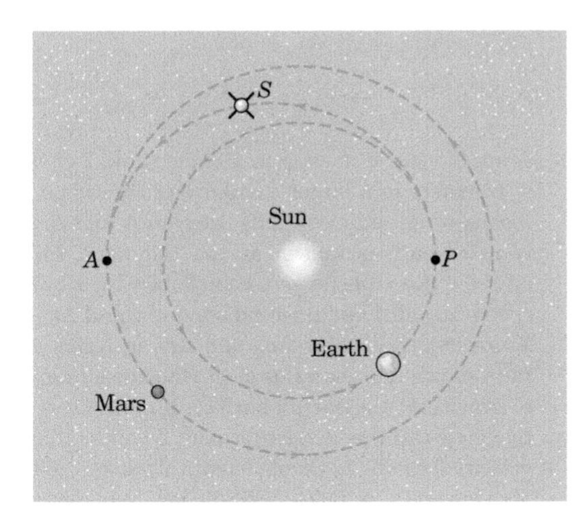

Problem 3/295

*3/296 In 1995 a spacecraft called the Solar and Heliospheric Observatory (SOHO) was placed into a circular orbit about the sun and inside that of the earth as shown. Determine the distance h so that the period of the spacecraft orbit will match that of the earth, with the result that the spacecraft will remain between the earth and the sun in a "halo" orbit.

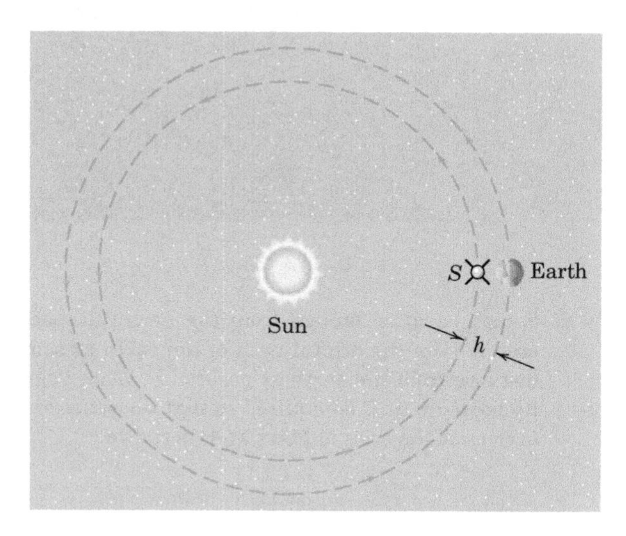

Problem 3/296

▶3/297 A space vehicle moving in a circular orbit of radius r_1 transfers to a larger circular orbit of radius r_2 by means of an elliptical path between A and B. (This transfer path is known as the Hohmann transfer ellipse.) The transfer is accomplished by a burst of speed Δv_A at A and a second burst of speed Δv_B at B. Write expressions for Δv_A and Δv_B in terms of the radii shown and the value of g of the acceleration due to gravity at the earth's surface. If each Δv is positive, how can the velocity for path 2 be less than the velocity for path 1? Compute each Δv if $r_1 = (6371 + 500)$ km and $r_2 = (6371 + 35\,800)$ km. Note that r_2 has been chosen as the radius of a geosynchronous orbit.

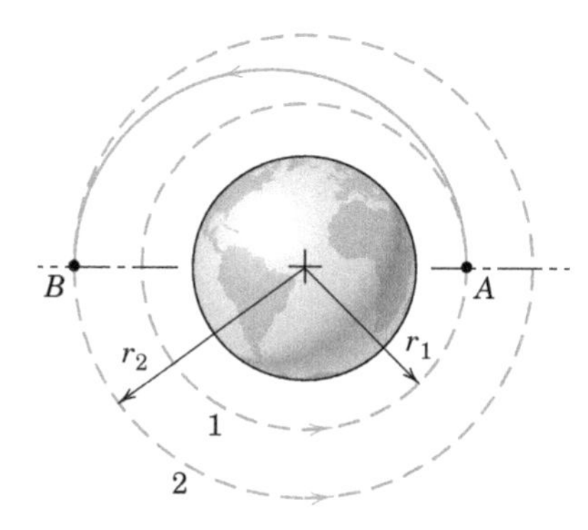

Problem 3/297

▶3/298 At the instant represented in the figure, a small experimental satellite A is ejected from the shuttle orbiter with a velocity $v_r = 100$ m/s relative to the shuttle, directed toward the center of the earth. The shuttle is in a circular orbit of altitude $h = 200$ km. For the resulting elliptical orbit of the satellite, determine the semimajor axis a and its orientation, the period τ, eccentricity e, apogee speed v_a, perigee speed v_p, r_{max}, and r_{min}. Sketch the satellite orbit.

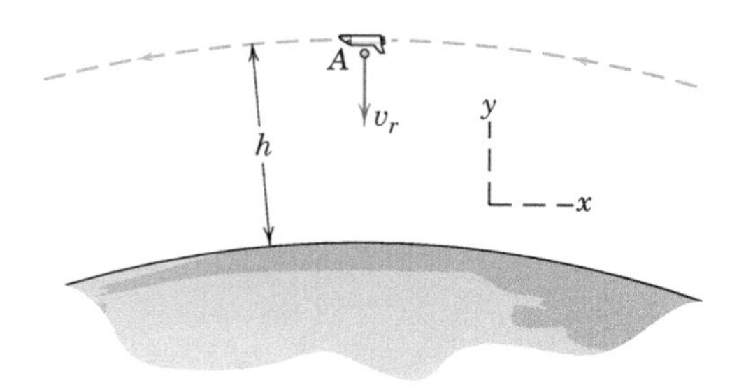

Problem 3/298

▶3/299 A spacecraft in an elliptical orbit has the position and velocity indicated in the figure at a certain instant. Determine the semimajor axis length a of the orbit and find the acute angle α between the semimajor axis and the line l. Does the spacecraft eventually strike the earth?

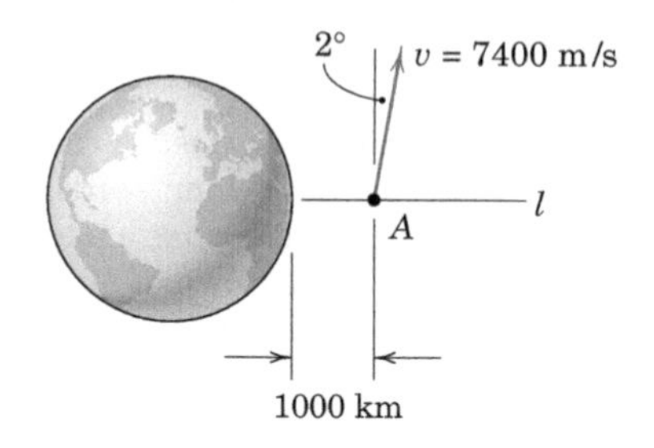

Problem 3/299

▶3/300 The satellite has a velocity at B of 3200 m/s in the direction indicated. Determine the angle β which locates the point C of impact with the earth.

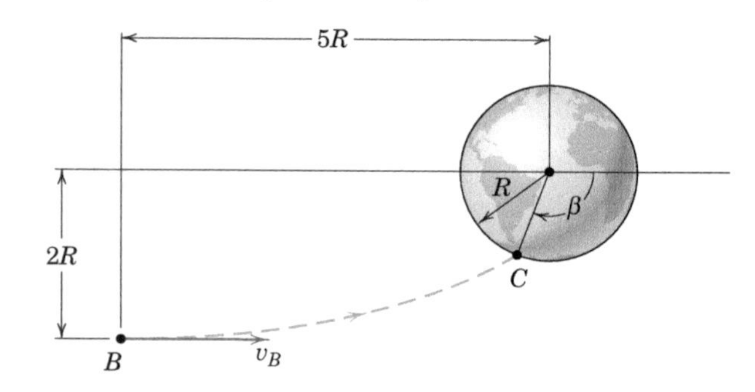

Problem 3/300

PROBLEMS

Introductory Problems

3/301 The flatbed truck is traveling at the constant speed of 60 km/h up the 15-percent grade when the 100-kg crate which it carries is given a shove which imparts to it an initial relative velocity $\dot{x} = 3$ m/s toward the rear of the truck. If the crate slides a distance $x = 2$ m measured on the truck bed before coming to rest on the bed, compute the coefficient of kinetic friction μ_k between the crate and the truck bed.

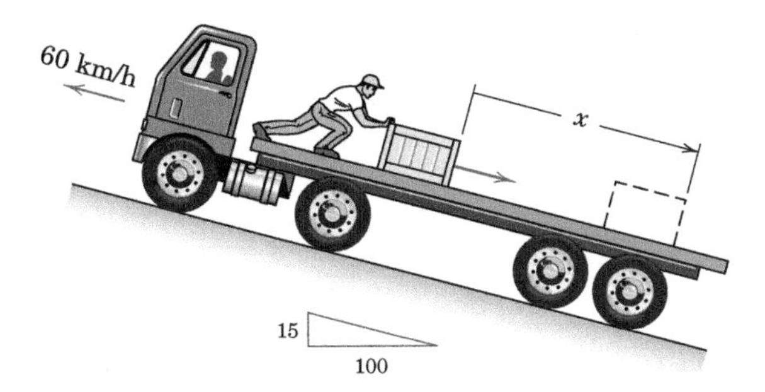

Problem 3/301

3/302 If the spring of constant k is compressed a distance δ as indicated, calculate the acceleration a_{rel} of the block of mass m_1 relative to the frame of mass m_2 upon release of the spring. The system is initially stationary.

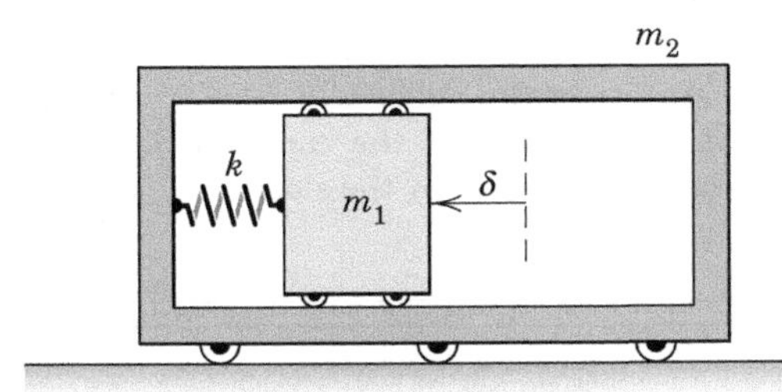

Problem 3/302

3/303 The cart with attached x-y axes moves with an absolute speed $v = 2$ m/s to the right. Simultaneously, the light arm of length $l = 0.5$ m rotates about point B of the cart with angular velocity $\dot{\theta} = 2$ rad/s. The mass of the sphere is $m = 3$ kg. Determine the following quantities for the sphere when $\theta = 0$: $\mathbf{G}$, $\mathbf{G}_{rel}$, T, T_{rel}, $\mathbf{H}_O$, $(\mathbf{H}_B)_{rel}$ where the subscript "rel" indicates measurement relative to the x-y axes. Point O is an inertially fixed point coincident with point B at the instant under consideration.

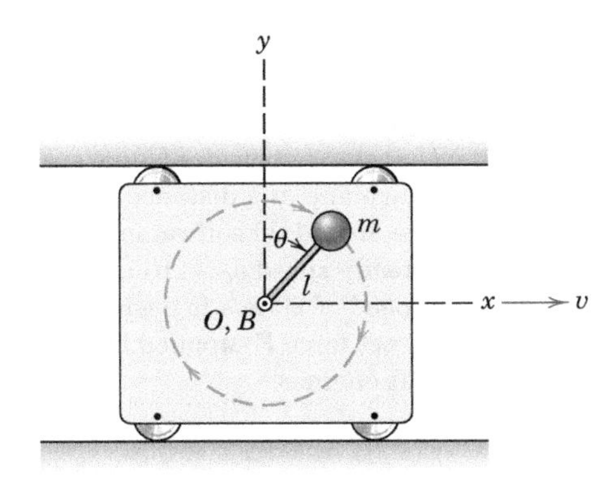

Problem 3/303

3/304 The aircraft carrier is moving at a constant speed and launches a jet plane with a mass of 3 Mg in a distance of 75 m along the deck by means of a steam-driven catapult. If the plane leaves the deck with a velocity of 240 km/h relative to the carrier and if the jet thrust is constant at 22 kN during takeoff, compute the constant force P exerted by the catapult on the airplane during the 75-m travel of the launch carriage.

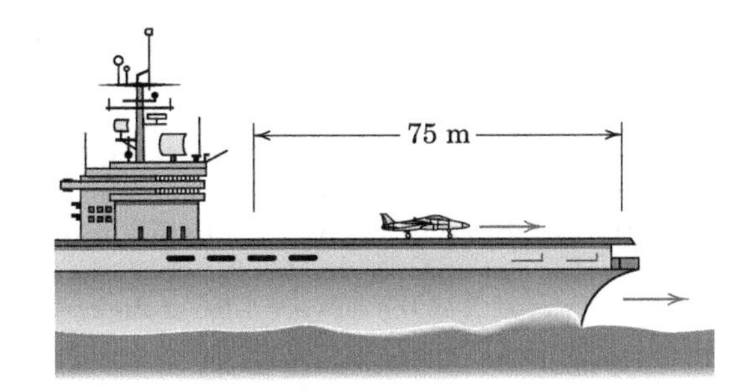

Problem 3/304

3/305 The 4000-lb van is driven from position A to position B on the barge, which is towed at a constant speed $v_0 = 10$ mi/hr. The van starts from rest relative to the barge at A, accelerates to $v = 15$ mi/hr relative to the barge over a distance of 80 ft, and then stops with a deceleration of the same magnitude. Determine the magnitude of the net force F between the tires of the van and the barge during this maneuver.

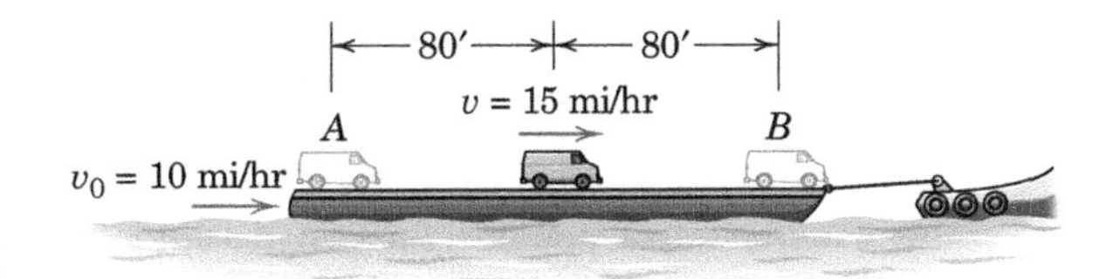

Problem 3/305

Representative Problems

3/306 The launch catapult of the aircraft carrier gives the 7-Mg jet airplane a constant acceleration and launches the airplane in a distance of 100 m measured along the angled takeoff ramp. The carrier is moving at a steady speed v_C = 16 m/s. If an absolute aircraft speed of 90 m/s is desired for takeoff, determine the net force F supplied by the catapult and the aircraft engines.

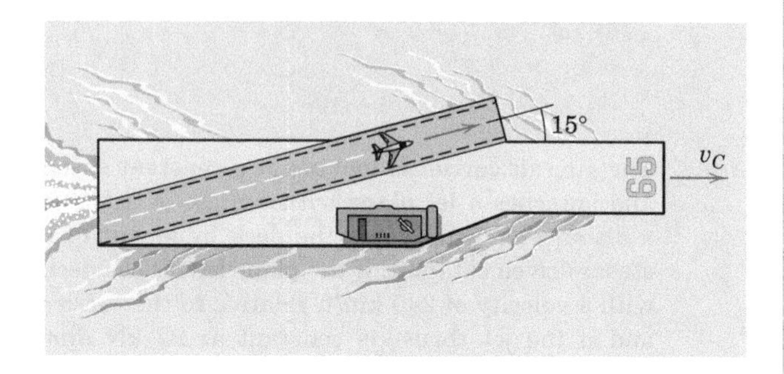

Problem 3/306

3/307 The coefficients of friction between the flatbed of the truck and crate are $\mu_s = 0.80$ and $\mu_k = 0.70$. The coefficient of kinetic friction between the truck tires and the road surface is 0.90. If the truck stops from an initial speed of 15 m/s with maximum braking (wheels skidding), determine where on the bed the crate finally comes to rest or the velocity v_{rel} relative to the truck with which the crate strikes the wall at the forward edge of the bed.

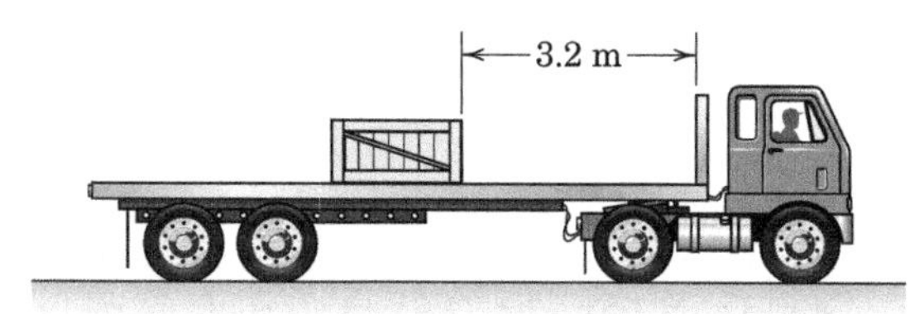

Problem 3/307

3/308 A boy of mass m is standing initially at rest relative to the moving walkway, which has a constant horizontal speed u. He decides to accelerate his progress and starts to walk from point A with a steadily increasing speed and reaches point B with a speed $\dot{x} = v$ relative to the walkway. During his acceleration he generates an average horizontal force F between his shoes and the walkway. Write the work-energy equations for his absolute and relative motions and explain the meaning of the term muv.

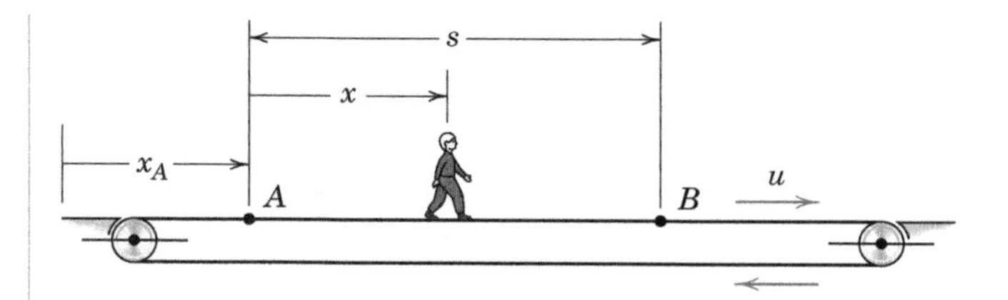

Problem 3/308

3/309 The block of mass m is attached to the frame by the spring of stiffness k and moves horizontally with negligible friction within the frame. The frame and block are initially at rest with $x = x_0$, the uncompressed length of the spring. If the frame is given a constant acceleration a_0, determine the maximum velocity $\dot{x}_{max} = (v_{rel})_{max}$ of the block relative to the frame.

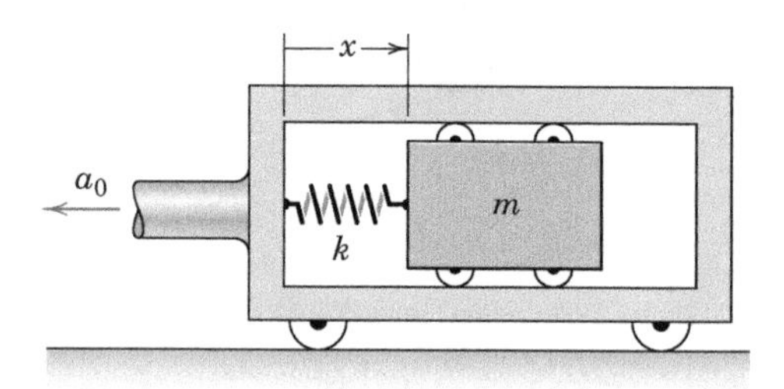

Problem 3/309

3/310 The slider A has a mass of 2 kg and moves with negligible friction in the 30° slot in the vertical sliding plate. What horizontal acceleration a_0 should be given to the plate so that the absolute acceleration of the slider will be vertically down? What is the value of the corresponding force R exerted on the slider by the slot?

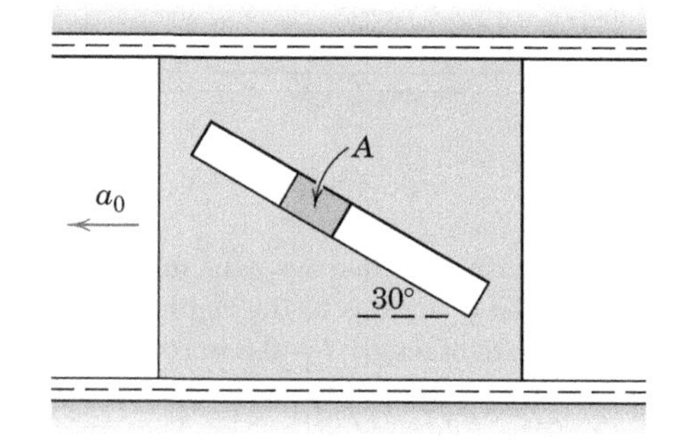

Problem 3/310

3/311 The ball A of mass 10 kg is attached to the light rod of length $l = 0.8$ m. The mass of the carriage alone is 250 kg, and it moves with an acceleration a_O as shown. If $\dot{\theta} = 3$ rad/s when $\theta = 90°$, find the kinetic energy T of the system if the carriage has a velocity of 0.8 m/s (*a*) in the direction of a_O and (*b*) in the direction opposite to a_O. Treat the ball as a particle.

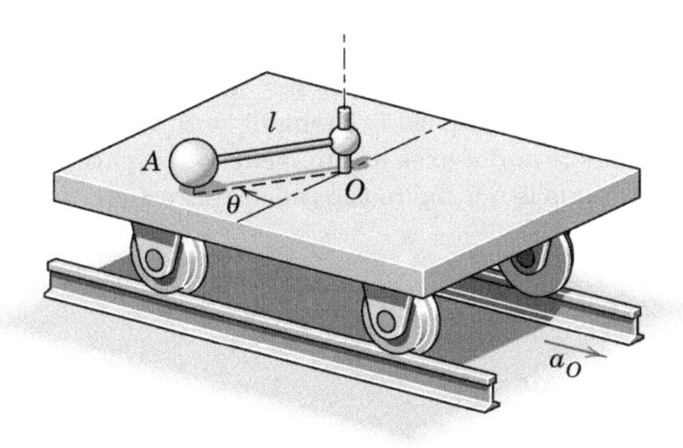

Problem 3/311

3/312 Consider the system of Prob. 3/311 where the mass of the ball is $m = 10$ kg and the length of the light rod is $l = 0.8$ m. The ball–rod assembly is free to rotate about a vertical axis through O. The carriage, rod, and ball are initially at rest with $\theta = 0$ when the carriage is given a constant acceleration $a_O = 3$ m/s^2. Write an expression for the tension T in the rod as a function of θ and calculate T for the position $\theta = \pi/2$.

3/313 A simple pendulum is placed on an elevator, which accelerates upward as shown. If the pendulum is displaced an amount θ_0 and released from rest relative to the elevator, find the tension T_0 in the supporting light rod when $\theta = 0$. Evaluate your result for $\theta_0 = \pi/2$.

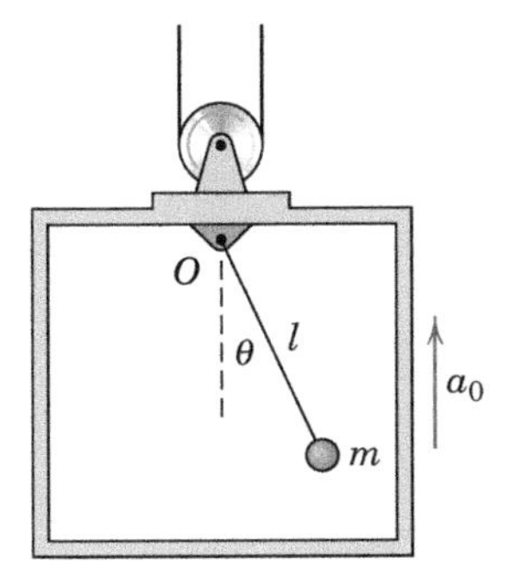

Problem 3/313

3/314 A boy of mass m is standing initially at rest relative to the moving walkway inclined at the angle θ and moving with a constant speed u. He decides to accelerate his progress and starts to walk from point A with a steadily increasing speed and reaches point B with a speed v_r relative to the walkway. During his acceleration he generates a constant average force F tangent to the walkway between his shoes and the walkway surface. Write the work-energy equations for the motion between A and B for his absolute motion and his relative motion and explain the meaning of the term muv_r. If the boy weighs 150 lb and if $u = 2$ ft/sec, $s = 30$ ft, and $\theta = 10°$, calculate the power P_{rel} developed by the boy as he reaches the speed of 2.5 ft/sec relative to the walkway.

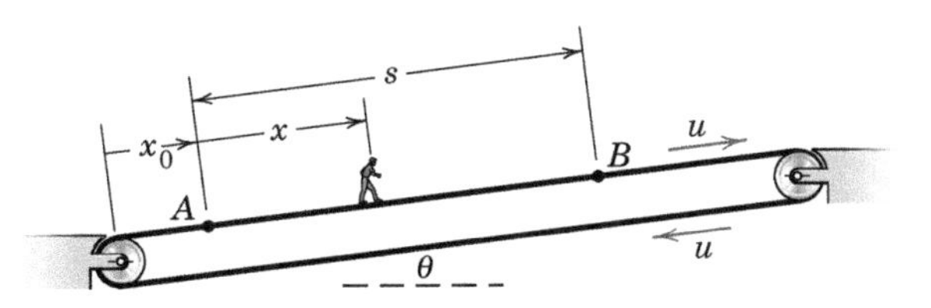

Problem 3/314

▶3/315 A ball is released from rest relative to the elevator at a distance h_1 above the floor. The speed of the elevator at the time of ball release is v_0. Determine the bounce height h_2 of the ball (*a*) if v_0 is constant and (*b*) if an upward elevator acceleration $a = g/4$ begins at the instant the ball is released. The coefficient of restitution for the impact is e.

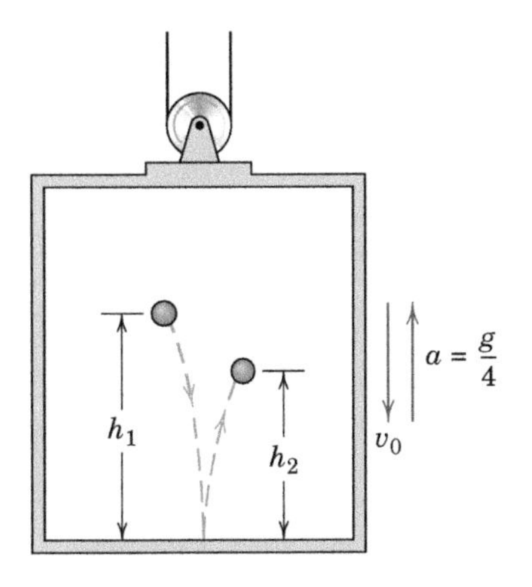

Problem 3/315

▶**3/316** The small slider A moves with negligible friction down the tapered block, which moves to the right with constant speed $v = v_0$. Use the principle of work-energy to determine the magnitude v_A of the absolute velocity of the slider as it passes point C if it is released at point B with no velocity relative to the block. Apply the equation, both as an observer fixed to the block and as an observer fixed to the ground, and reconcile the two relations.

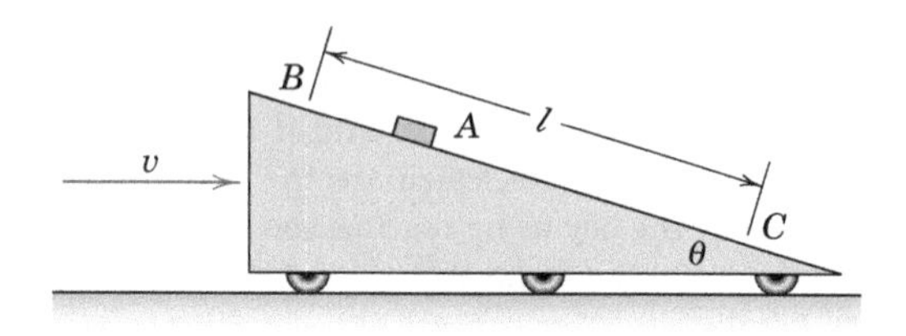

Problem 3/316

▶**3/317** When a particle is dropped from rest relative to the surface of the earth at a latitude γ, the initial apparent acceleration is the relative acceleration due to gravity g_{rel}. The absolute acceleration due to gravity g is directed toward the center of the earth. Derive an expression for g_{rel} in terms of g, R, ω, and γ, where R is the radius of the earth treated as a sphere and ω is the constant angular velocity of the earth about the polar axis considered fixed. Although axes x-y-z are attached to the earth and hence rotate, we may use Eq. 3/50 as long as the particle has no velocity relative to x-y-z. (*Hint:* Use the first two terms of the binomial expansion for the approximation.)

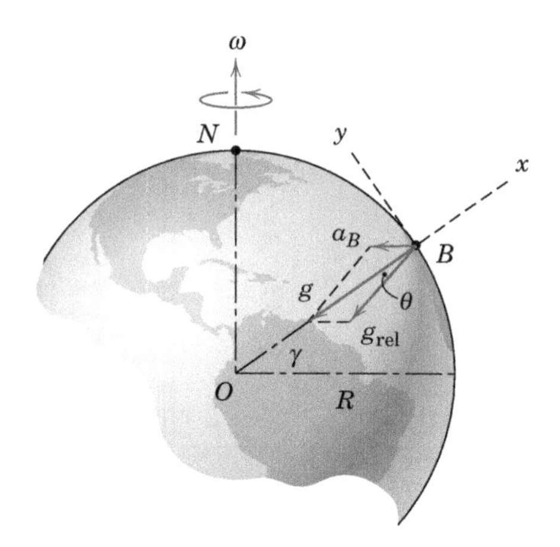

Problem 3/317

▶**3/318** The figure represents the space shuttle S, which is (*a*) in a circular orbit about the earth and (*b*) in an elliptical orbit where P is its perigee position. The exploded views on the right represent the cabin space with its x-axis oriented in the direction of the orbit. The astronauts conduct an experiment by applying a known force F in the x-direction to a small mass m. Explain why $F = m\ddot{x}$ does or does not hold in each case, where x is measured within the spacecraft. Assume that the shuttle is between perigee and apogee in the elliptical orbit so that the orbital speed is changing with time. Note that the t- and x-axes are tangent to the path, and the θ-axis is normal to the radial r-direction.

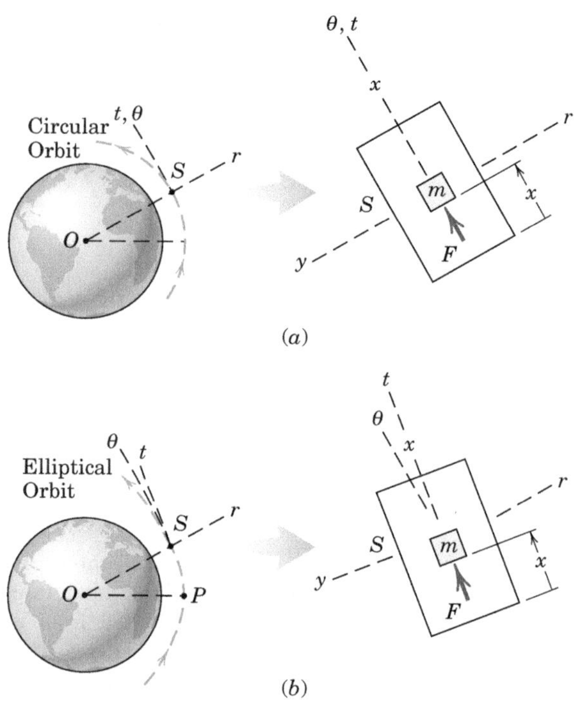

Problem 3/318

PROBLEMS

Introductory Problems

4/1 The system of three particles has the indicated particle masses, velocities, and external forces. Determine $\bar{\mathbf{r}}$, $\dot{\bar{\mathbf{r}}}$, $\ddot{\bar{\mathbf{r}}}$, T, $\mathbf{H}_O$, and $\dot{\mathbf{H}}_O$ for this two-dimensional system.

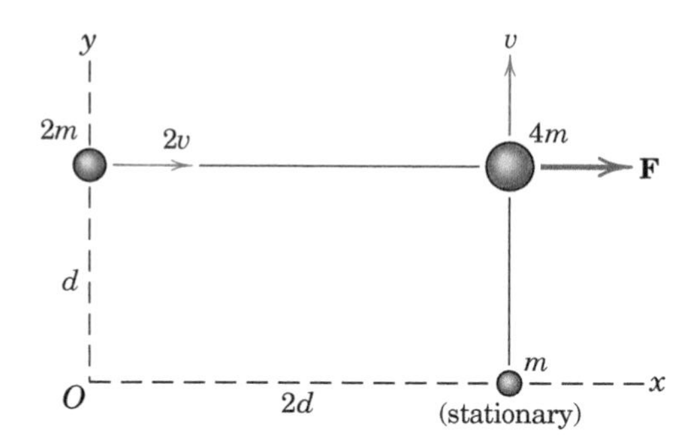

Problem 4/1

4/2 For the particle system of Prob. 4/1, determine $\mathbf{H}_G$ and $\dot{\mathbf{H}}_G$.

4/3 The system of three particles has the indicated particle masses, velocities, and external forces. Determine $\bar{\mathbf{r}}$, $\dot{\bar{\mathbf{r}}}$, $\ddot{\bar{\mathbf{r}}}$, T, $\mathbf{H}_O$, and $\dot{\mathbf{H}}_O$ for this three-dimensional system.

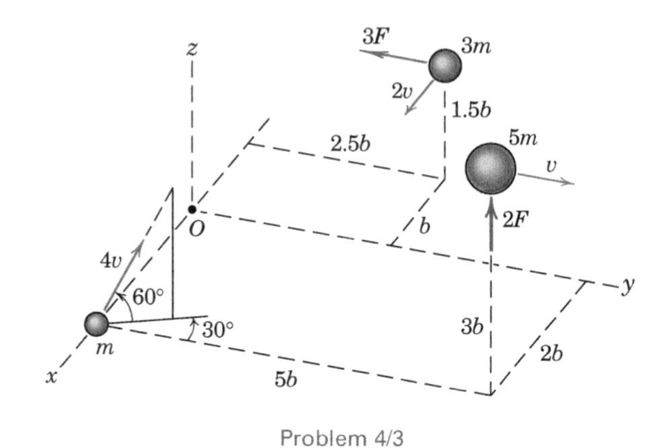

Problem 4/3

4/4 For the particle system of Prob. 4/3, determine $\mathbf{H}_G$ and $\dot{\mathbf{H}}_G$.

4/5 The system consists of the two smooth spheres, each weighing 3 lb and connected by a light spring, and the two bars of negligible weight hinged freely at their ends and hanging in the vertical plane. The spheres are confined to slide in the smooth horizontal guide. If a horizontal force $F = 10$ lb is applied to the one bar at the position shown, what is the acceleration of the center C of the spring? Why does the result not depend on the dimension b?

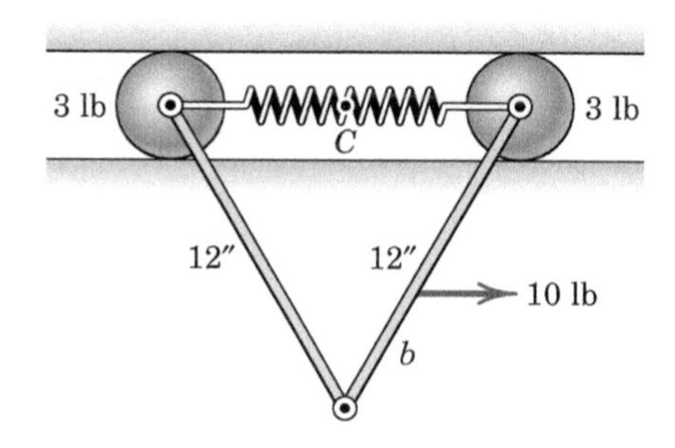

Problem 4/5

4/6 The two small spheres, each of mass m, are rigidly connected by a rod of negligible mass and are released from rest in the position shown and slide down the smooth circular guide in the vertical plane. Determine their common velocity v as they reach the horizontal dashed position. Also find the force N between sphere 1 and the supporting surface an instant before the sphere reaches the bottom position A.

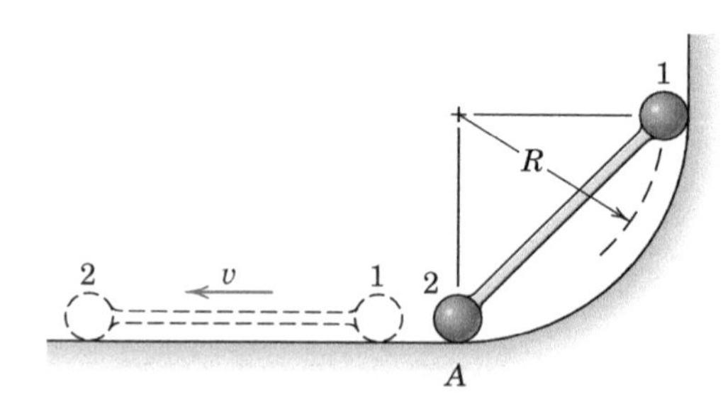

Problem 4/6

4/7 The total linear momentum of a system of five particles at time $t = 2.2$ s is given by $\mathbf{G}_{2.2} = 3.4\mathbf{i} - 2.6\mathbf{j} + 4.6\mathbf{k}$ kg·m/s. At time $t = 2.4$ s, the linear momentum has changed to $\mathbf{G}_{2.4} = 3.7\mathbf{i} - 2.2\mathbf{j} + 4.9\mathbf{k}$ kg·m/s. Calculate the magnitude F of the time average of the resultant of the external forces acting on the system during the interval.

4/8 The angular momentum of a system of six particles about a fixed point O at time $t = 4$ s is $\mathbf{H}_4 = 3.65\mathbf{i} + 4.27\mathbf{j} - 5.36\mathbf{k}$ kg·m²/s. At time $t = 4.1$ s, the angular momentum is $\mathbf{H}_{4.1} = 3.67\mathbf{i} + 4.30\mathbf{j} - 5.20\mathbf{k}$ kg·m²/s. Determine the average value of the resultant moment about point O of all forces acting on all particles during the 0.1-s interval.

4/9 Three monkeys A, B, and C weighing 20, 25, and 15 lb, respectively, are climbing up and down the rope suspended from D. At the instant represented, A is descending the rope with an acceleration of 5 ft/sec^2, and C is pulling himself up with an acceleration of 3 ft/sec^2. Monkey B is climbing up with a constant speed of 2 ft/sec. Treat the rope and monkeys as a complete system and calculate the tension T in the rope at D.

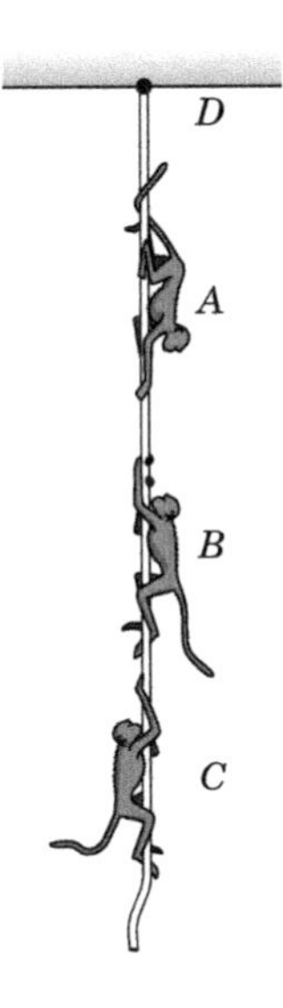

Problem 4/9

4/10 The monkeys of Prob. 4/9 are now climbing along the heavy rope wall suspended from the uniform beam. If monkeys A, B, and C have velocities of 5, 3, and 2 ft/sec, and accelerations of 1.5, 0.5, and 2 ft/sec^2, respectively, determine the changes in the reactions at D and E caused by the motion and weight of the monkeys. The support at E makes contact with only one side of the beam at a time. Assume for this analysis that the rope wall remains rigid.

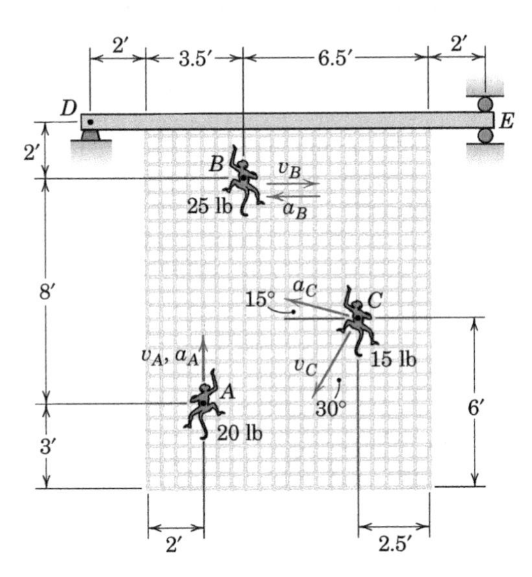

Problem 4/10

4/11 The two spheres, each of mass m, are connected by the spring and hinged bars of negligible mass. The spheres are free to slide in the smooth guide up the incline θ. Determine the acceleration a_C of the center C of the spring.

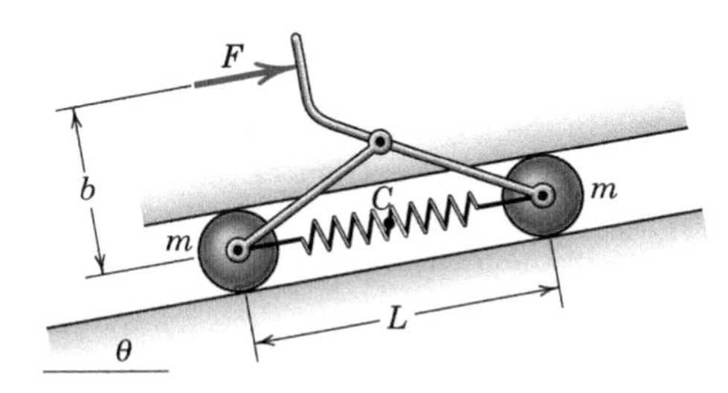

Problem 4/11

4/12 Each of the five connected particles has a mass of 0.5 kg, and G is the mass center of the system. At a certain instant the angular velocity of the body is $\omega = 2$ rad/s and the linear velocity of G is $v_G = 4$ m/s in the direction shown. Determine the linear momentum of the body and its angular momentum about G and about O.

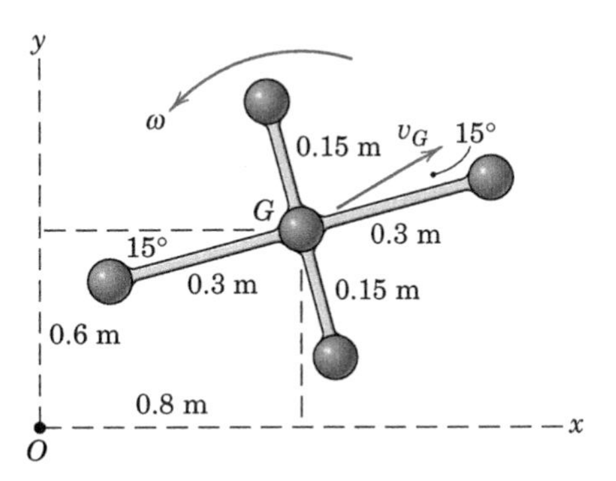

Problem 4/12

4/13 Calculate the acceleration of the center of mass of the system of the four 10-kg cylinders. Neglect friction and the mass of the pulleys and cables.

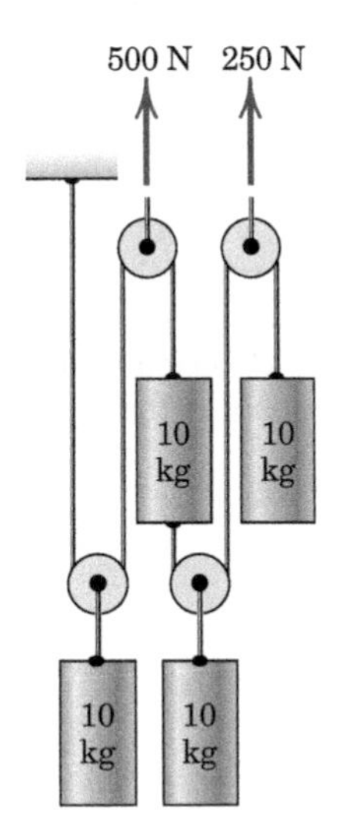

Problem 4/13

4/14 The four systems slide on a smooth horizontal surface and have the same mass m. The configurations of mass in the two pairs are identical. What can be said about the acceleration of the mass center for each system? Explain any difference in the accelerations of the members.

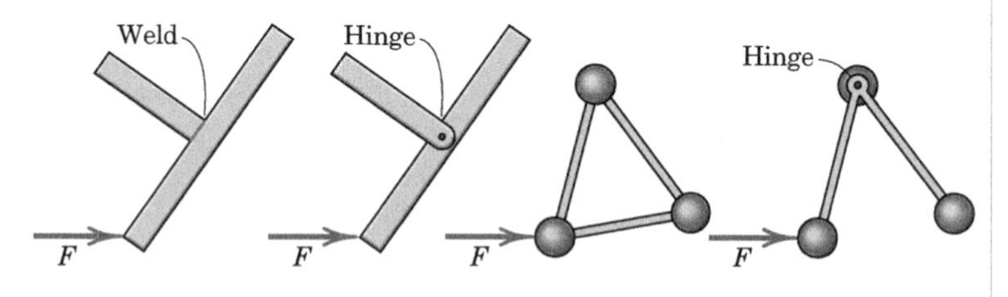

Problem 4/14

Representative Problems

4/15 Calculate the vertical acceleration of the system mass center and the individual vertical accelerations of spheres 1 and 2 for the cases (*a*) $\alpha = \beta$ and (*b*) $\alpha \neq \beta$. Each sphere has a mass of 2 kg, and the 50-N force is applied vertically to the junction O of the two light wires. The spheres are initially at rest. State any assumptions.

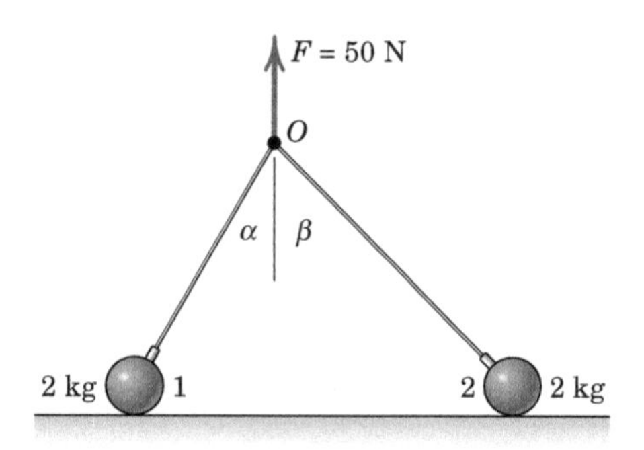

Problem 4/15

4/16 The two small spheres, each of mass m, and their connecting rod of negligible mass are rotating about their mass center G with an angular velocity ω. At the same instant the mass center has a velocity v in the x-direction. Determine the angular momentum $\mathbf{H}_O$ of the assembly at the instant when G has coordinates x and y.

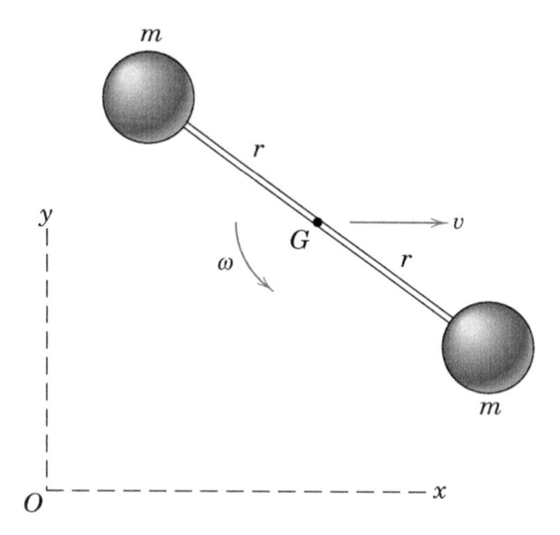

Problem 4/16

4/17 A department-store escalator makes an angle of 30° with the horizontal and takes 40 seconds to transport a person from the first to the second floor with a vertical rise of 20 ft. At a certain instant, there are 10 people on the escalator averaging 150 lb per person and standing at rest relative to the moving steps. Additionally, three boys averaging 120 lb each are running down the escalator at a speed of 2 ft/sec relative to the moving steps. Calculate the power output P of the driving motor to maintain the constant speed of the escalator. The no-load power without passengers is 2.2 hp to overcome friction in the mechanism.

4/18 A centrifuge consists of four cylindrical containers, each of mass m, at a radial distance r from the rotation axis. Determine the time t required to bring the centrifuge to an angular velocity ω from rest under a constant torque M applied to the shaft. The diameter of each container is small compared with r, and the mass of the shaft and supporting arms is small compared with m.

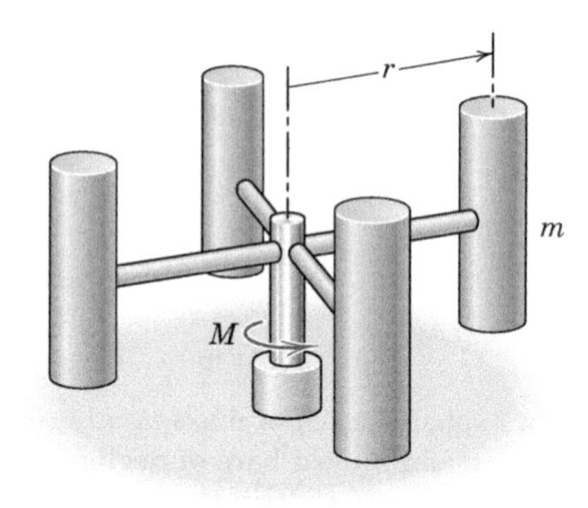

Problem 4/18

4/19 The three small spheres are welded to the light rigid frame which is rotating in a horizontal plane about a vertical axis through O with an angular velocity $\dot{\theta} = 20$ rad/s. If a couple $M_O = 30$ N·m is applied to the frame for 5 seconds, compute the new angular velocity $\dot{\theta}'$.

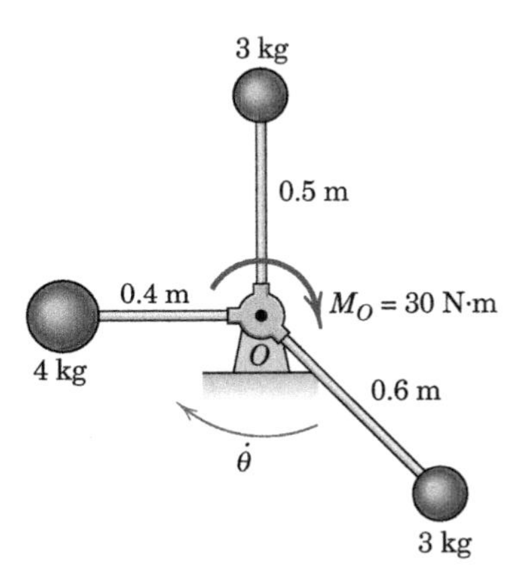

Problem 4/19

4/20 Billiard ball A is moving in the y-direction with a velocity of 2 m/s when it strikes ball B of identical size and mass initially at rest. Following the impact, the balls are observed to move in the directions shown. Calculate the velocities v_A and v_B which the balls have immediately after the impact. Treat the balls as particles and neglect any friction forces acting on the balls compared with the force of impact.

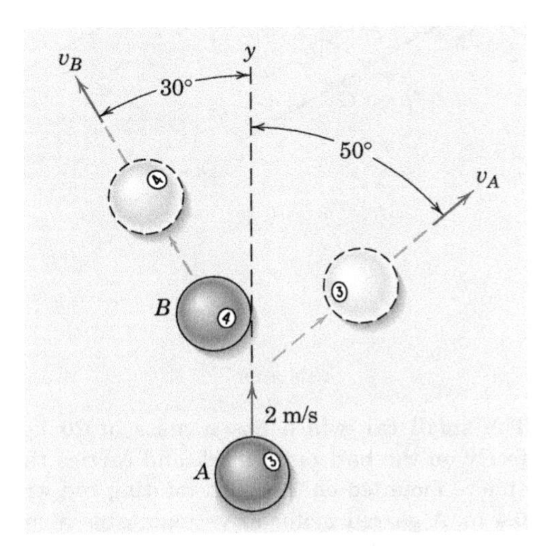

Problem 4/20

4/21 The 300-kg and 400-kg mine cars are rolling in opposite directions along the horizontal track with the respective speeds of 0.6 m/s and 0.3 m/s. Upon impact the cars become coupled together. Just prior to impact, a 100-kg boulder leaves the delivery chute with a velocity of 1.2 m/s in the direction shown and lands in the 300-kg car. Calculate the velocity v of the system after the boulder has come to rest relative to the car. Would the final velocity be the same if the cars were coupled before the boulder dropped?

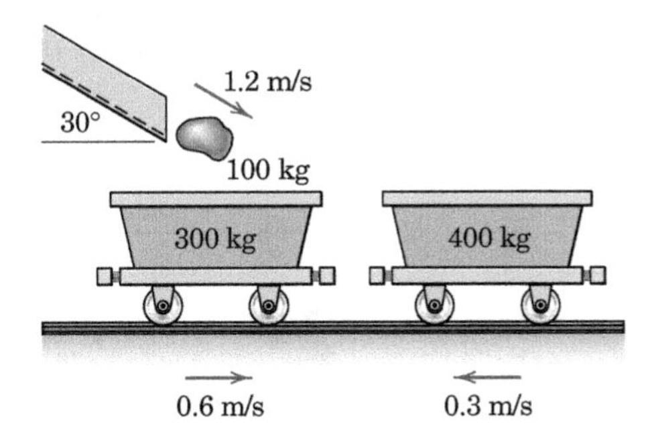

Problem 4/21

4/22 The three freight cars are rolling along the horizontal track with the velocities shown. After the impacts occur, the three cars become coupled together and move with a common velocity v. The weights of the loaded cars A, B, and C are 130,000, 100,000, and 150,000 lb, respectively. Determine v and calculate the percentage loss n of energy of the system due to coupling.

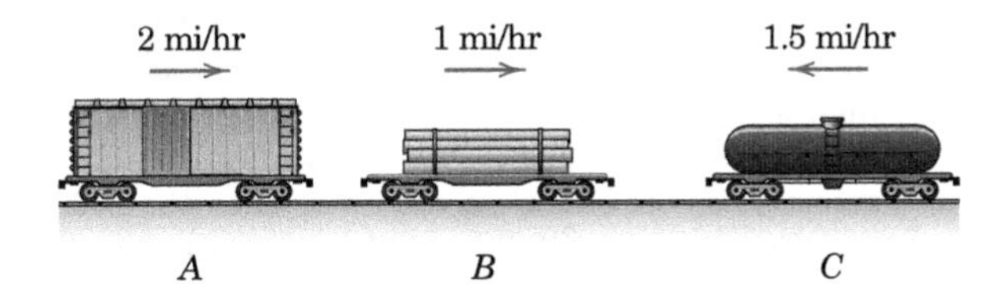

Problem 4/22

4/23 The man of mass m_1 and the woman of mass m_2 are standing on opposite ends of the platform of mass m_0 which moves with negligible friction and is initially at rest with $s = 0$. The man and woman begin to approach each other. Derive an expression for the displacement s of the platform when the two meet in terms of the displacement x_1 of the man relative to the platform.

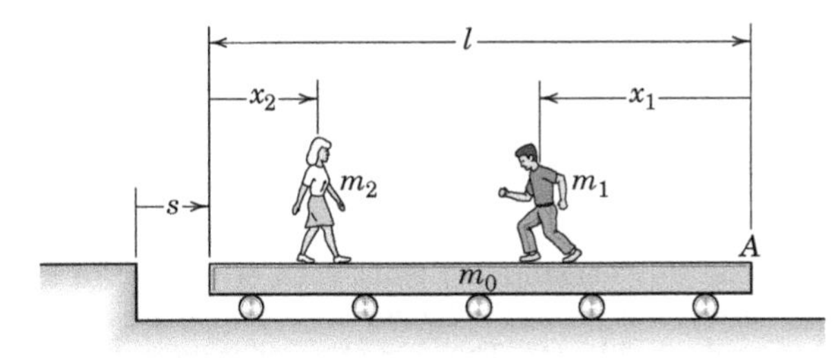

Problem 4/23

4/24 The woman A, the captain B, and the sailor C weigh 120, 180, and 160 lb, respectively, and are sitting in the 300-lb skiff, which is gliding through the water with a speed of 1 knot. If the three people change their positions as shown in the second figure, find the distance x from the skiff to the position where it would have been if the people had not moved. Neglect any resistance to motion afforded by the water. Does the sequence or timing of the change in positions affect the final result?

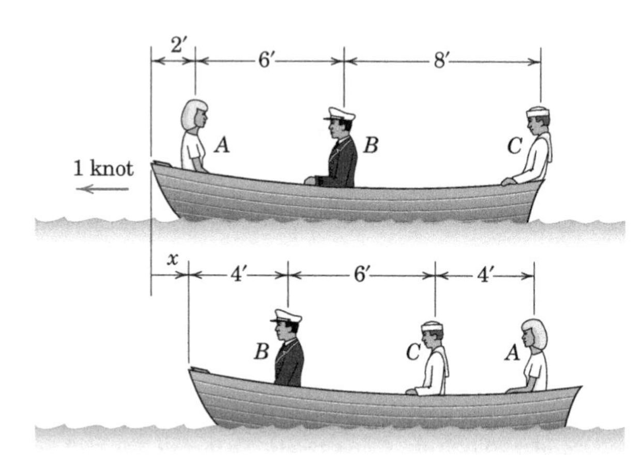

Problem 4/24

4/25 Each of the bars A and B has a mass of 10 kg and slides in its horizontal guideway with negligible friction. Motion is controlled by the lever of negligible mass connected to the bars as shown. Calculate the acceleration of point C on the lever when the 200-N force is applied as indicated. To verify your result, analyze the kinetics of each member separately and determine a_C by kinematic considerations from the calculated accelerations of the two bars.

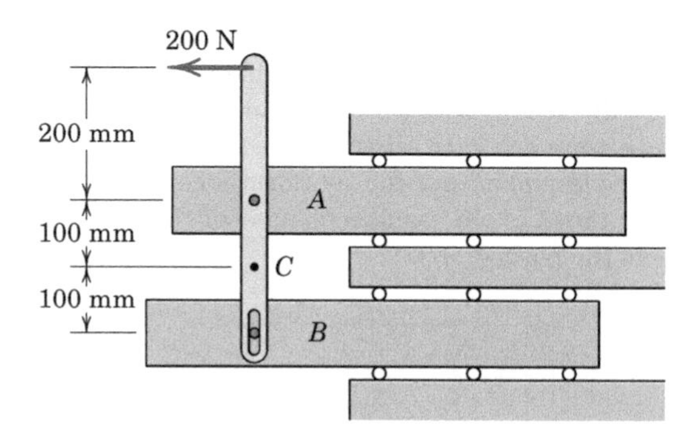

Problem 4/25

4/26 The three small steel balls, each of mass 2.75 kg, are connected by the hinged links of negligible mass and equal length. They are released from rest in the positions shown and slide down the quarter-circular guide in the vertical plane. When the upper sphere reaches the bottom position, the spheres have a horizontal velocity of 1.560 m/s. Calculate the energy loss ΔQ due to friction and the total impulse I_x on the system of three spheres during this interval.

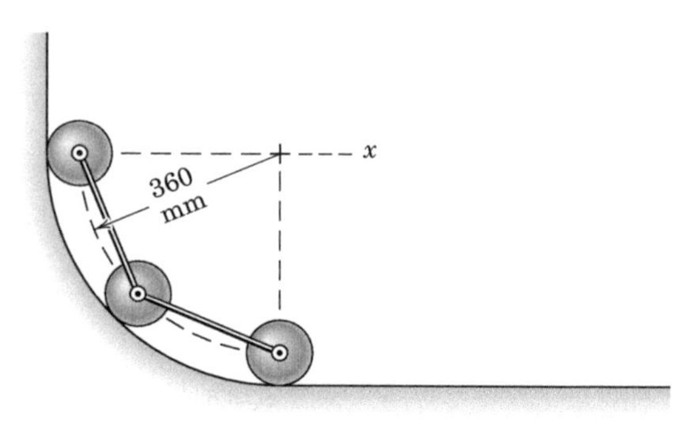

Problem 4/26

4/27 Two steel balls, each of mass m, are welded to a light rod of length L and negligible mass and are initially at rest on a smooth horizontal surface. A horizontal force of magnitude F is suddenly applied to the rod as shown. Determine (*a*) the instantaneous acceleration $\bar{a}$ of the mass center G and (*b*) the corresponding rate $\ddot{\theta}$ at which the angular velocity of the assembly about G is changing with time.

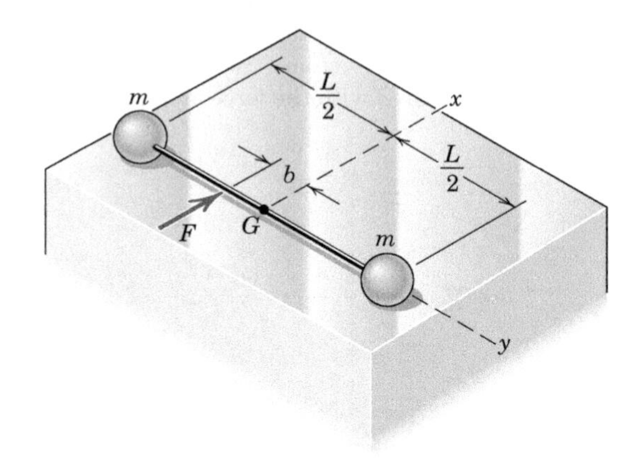

Problem 4/27

4/28 The small car, which has a mass of 20 kg, rolls freely on the horizontal track and carries the 5-kg sphere mounted on the light rotating rod with r = 0.4 m. A geared motor drive maintains a constant angular speed $\dot{\theta}$ = 4 rad/s of the rod. If the car has a velocity v = 0.6 m/s when θ = 0, calculate v when θ = 60°. Neglect the mass of the wheels and any friction.

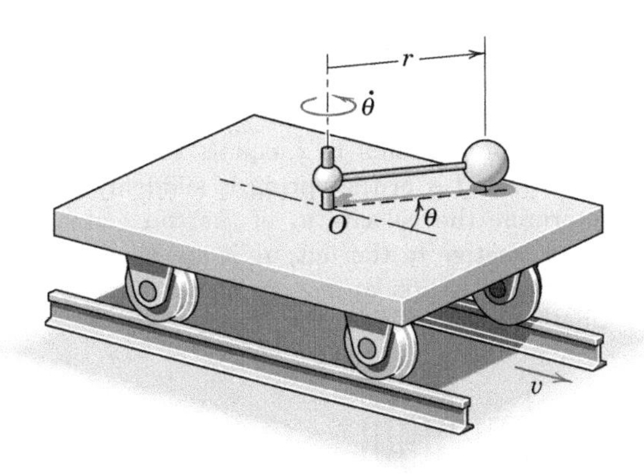

Problem 4/28

4/29 The cars of a roller-coaster ride have a speed of 30 km/h as they pass over the top of the circular track. Neglect any friction and calculate their speed v when they reach the horizontal bottom position. At the top position, the radius of the circular path of their mass centers is 18 m, and all six cars have the same mass.

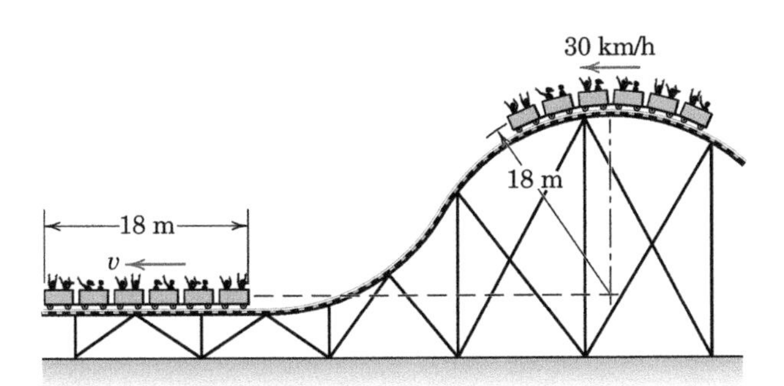

Problem 4/29

4/30 The carriage of mass $2m$ is free to roll along the horizontal rails and carries the two spheres, each of mass m, mounted on rods of length l and negligible mass. The shaft to which the rods are secured is mounted in the carriage and is free to rotate. If the system is released from rest with the rods in the vertical position where $\theta = 0$, determine the velocity v_x of the carriage and the angular velocity $\dot{\theta}$ of the rods for the instant when $\theta = 180°$. Treat the carriage and the spheres as particles and neglect any friction.

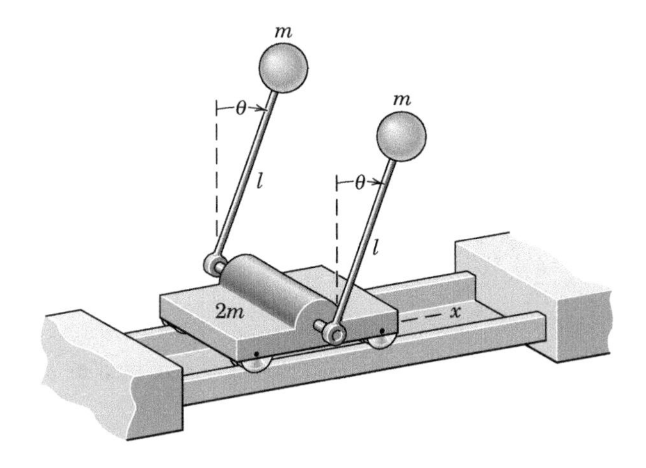

Problem 4/30

▶4/31 The 50,000-lb flatcar supports a 15,000-lb vehicle on a 5° ramp built on the flatcar. If the vehicle is released from rest with the flatcar also at rest, determine the velocity v of the flatcar when the vehicle has rolled $s = 40$ ft down the ramp just before hitting the stop at B. Neglect all friction and treat the vehicle and the flatcar as particles.

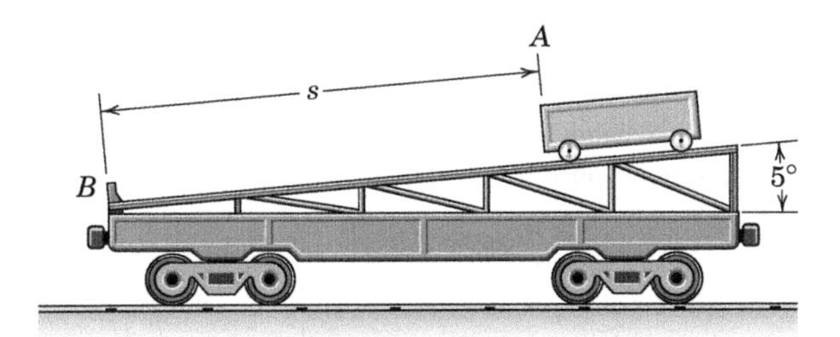

Problem 4/31

▶4/32 A 60-kg rocket is fired from O with an initial velocity $v_0 = 125$ m/s along the indicated trajectory. The rocket explodes 7 seconds after launch and breaks into three pieces A, B, and C having masses of 10, 30, and 20 kg, respectively. Pieces B and C are recovered at the impact coordinates shown. Instrumentation records reveal that piece B reached a maximum altitude of 1500 m after the explosion and that piece C struck the ground 6 seconds after the explosion. What are the impact coordinates for piece A? Neglect air resistance.

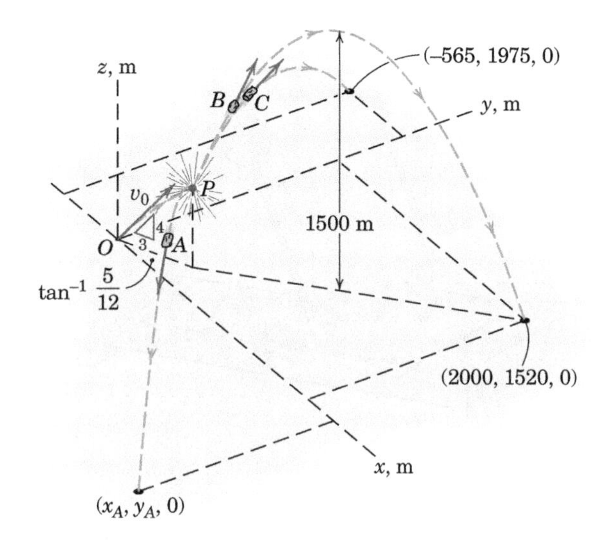

Problem 4/32

▶**4/33** A horizontal bar of mass m_1 and small diameter is suspended by two wires of length l from a carriage of mass m_2 which is free to roll along the horizontal rails. If the bar and carriage are released from rest with the wires making an angle θ with the vertical, determine the velocity $v_{b/c}$ of the bar relative to the carriage and the velocity v_c of the carriage at the instant when $\theta = 0$. Neglect all friction and treat the carriage and the bar as particles in the vertical plane of motion.

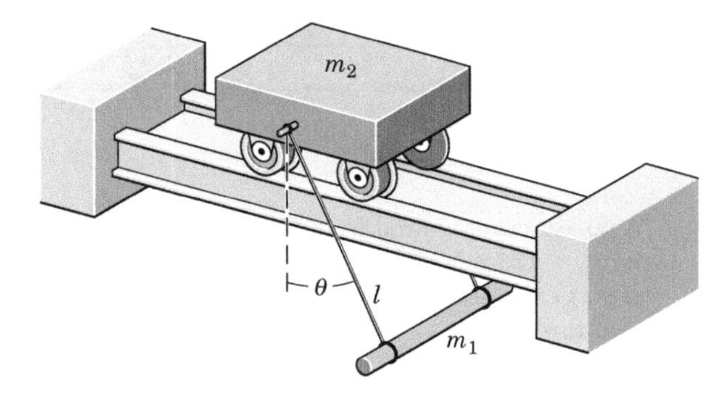

Problem 4/33

▶**4/34** In the unstretched position the coils of the 3-lb spring are just touching one another, as shown in part a of the figure. In the stretched position the force P, proportional to x, equals 200 lb when x = 20 in. If end A of the spring is suddenly released, determine the velocity v_A of the coil end A, measured positive to the left, as it approaches its unstretched position at $x = 0$. What happens to the kinetic energy of the spring?

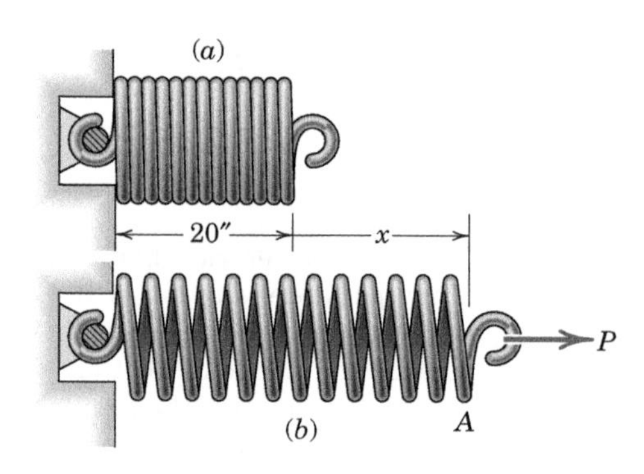

Problem 4/34

PROBLEMS

Introductory Problems

4/35 The experimental race car is propelled by a rocket motor and is designed to reach a maximum speed $v = 300$ mi/hr under the thrust T of its motor. Prior wind-tunnel tests disclose that the wind resistance at this speed is 225 lb. If the rocket motor is burning fuel at the rate of 3.5 lb/sec, determine the velocity u of the exhaust gases relative to the car.

Problem 4/35

4/36 The jet aircraft has a mass of 4.6 Mg and a drag (air resistance) of 32 kN at a speed of 1000 km/h at a particular altitude. The aircraft consumes air at the rate of 106 kg/s through its intake scoop and uses fuel at the rate of 4 kg/s. If the exhaust has a rearward velocity of 680 m/s relative to the exhaust nozzle, determine the maximum angle of elevation α at which the jet can fly with a constant speed of 1000 km/h at the particular altitude in question.

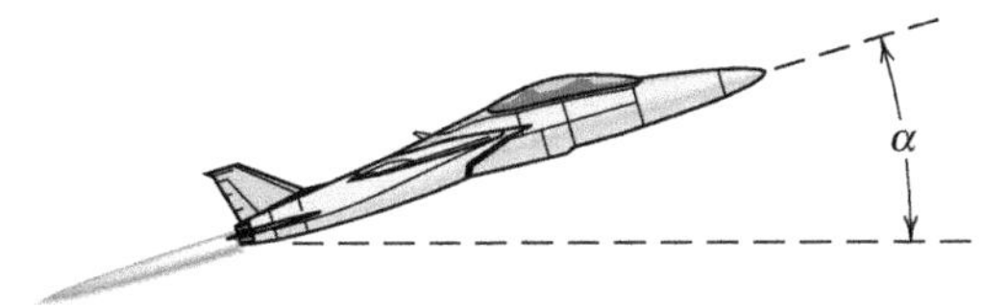

Problem 4/36

4/37 Fresh water issues from the nozzle with a velocity of 30 m/s at the rate of 0.05 m^3/s and is split into two equal streams by the fixed vane and deflected through 60° as shown. Calculate the force F required to hold the vane in place. The density of water is 1000 kg/m^3.

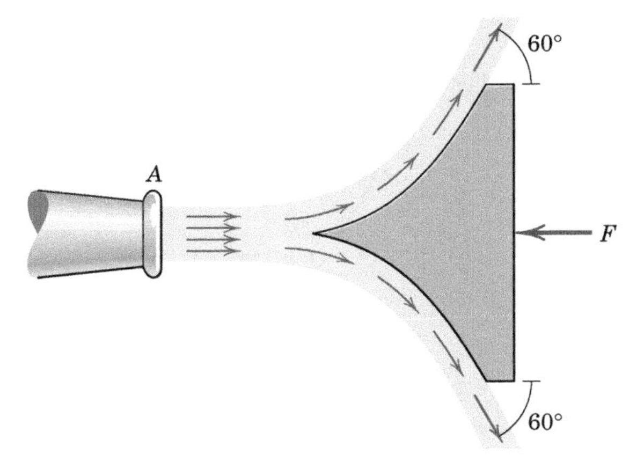

Problem 4/37

4/38 In an unwise effort to remove debris, a homeowner directs the nozzle of his backpack blower directly toward the garage door. The nozzle velocity is 130 mi/hr and the flow rate is 410 ft^3/min. Estimate the force F exerted by the airflow on the door. The specific weight of air is 0.0753 lb/ft^3.

Problem 4/38

4/39 The jet water ski has reached its maximum velocity of 70 km/h when operating in salt water. The water intake is in the horizontal tunnel in the bottom of the hull, so the water enters the intake at the velocity of 70 km/h relative to the ski. The motorized pump discharges water from the horizontal exhaust nozzle of 50-mm diameter at the rate of 0.082 m^3/s. Calculate the resistance R of the water to the hull at the operating speed.

Problem 4/39

4/40 The 25-mm steel slab 1.2 m wide enters the rolls at the speed of 0.4 m/s and is reduced in thickness to 19 mm. Calculate the small horizontal thrust T on the bearings of each of the two rolls.

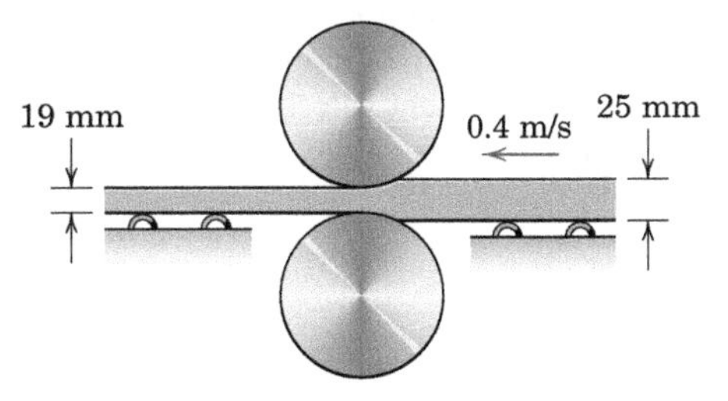

Problem 4/40

4/41 The fire tug discharges a stream of salt water (density 1030 kg/m^3) with a nozzle velocity of 40 m/s at the rate of 0.080 m^3/s. Calculate the propeller thrust T which must be developed by the tug to maintain a fixed position while pumping.

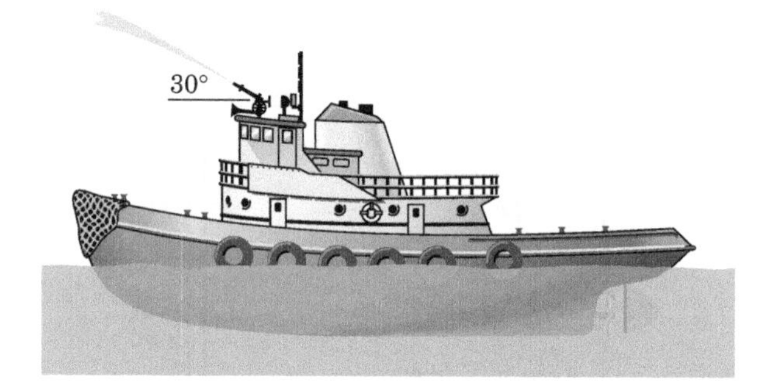

Problem 4/41

4/42 The pump shown draws air with a density ρ through the fixed duct A of diameter d with a velocity u and discharges it at high velocity v through the two outlets B. The pressure in the airstreams at A and B is atmospheric. Determine the expression for the tension T exerted on the pump unit through the flange at C.

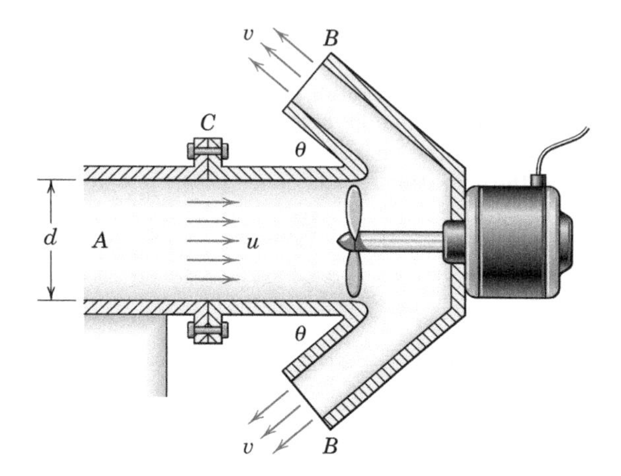

Problem 4/42

4/43 A jet-engine noise suppressor consists of a movable duct which is secured directly behind the jet exhaust by cable A and deflects the blast directly upward. During a ground test, the engine sucks in air at the rate of 43 kg/s and burns fuel at the rate of 0.8 kg/s. The exhaust velocity is 720 m/s. Determine the tension T in the cable.

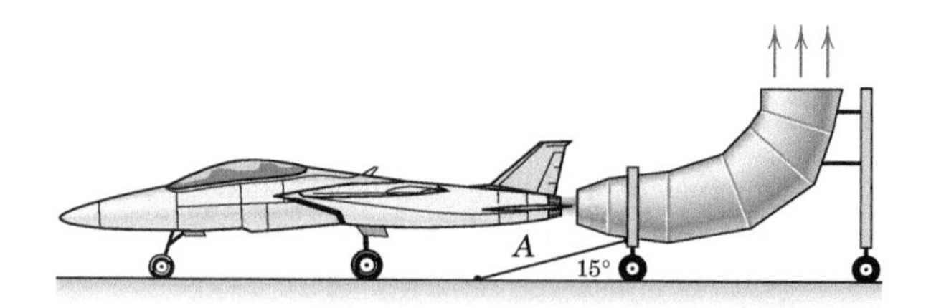

Problem 4/43

4/44 The 90° vane moves to the left with a constant velocity of 10 m/s against a stream of fresh water issuing with a velocity of 20 m/s from the 25-mm-diameter nozzle. Calculate the forces F_x and F_y on the vane required to support the motion.

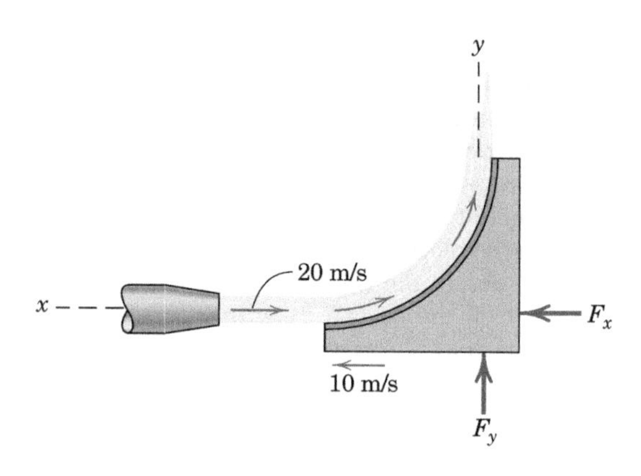

Problem 4/44

Representative Problems

4/45 The pipe bend shown has a cross-sectional area A and is supported in its plane by the tension T applied to its flanges by the adjacent connecting pipes (not shown). If the velocity of the liquid is v, its density ρ, and its static pressure p, determine T and show that it is independent of the angle θ.

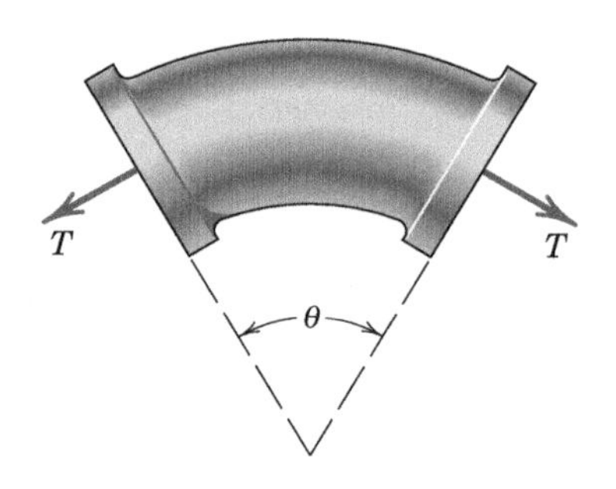

Problem 4/45

4/46 A jet of fluid with cross-sectional area A and mass density ρ issues from the nozzle with a velocity v and impinges on the inclined trough shown in section. Some of the fluid is diverted in each of the two directions. If the trough is smooth, the velocity of both diverted streams remains v, and the only force which can be exerted on the trough is normal to the bottom surface. Hence, the trough will be held in position by forces whose resultant is F normal to the trough. By writing impulse-momentum equations for the directions along and normal to the trough, determine the force F required to support the trough. Also find the volume rates of flow Q_1 and Q_2 for the two streams.

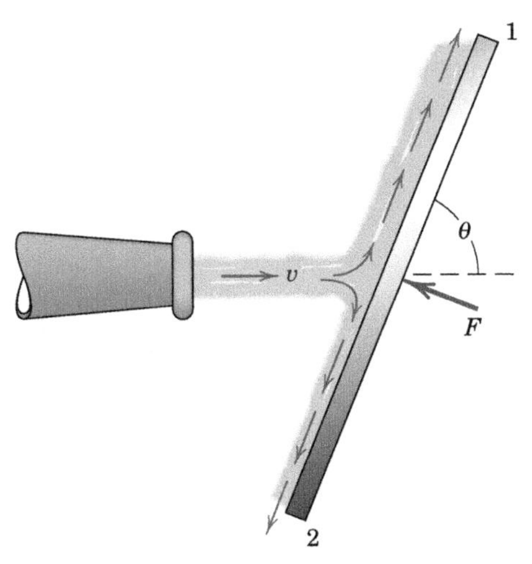

Problem 4/46

4/47 The 8-oz ball is supported by the vertical stream of fresh water which issues from the 1/2-in.-diameter nozzle with a velocity of 35 ft/sec. Calculate the height h of the ball above the nozzle. Assume that the stream remains intact and there is no energy lost in the jet stream.

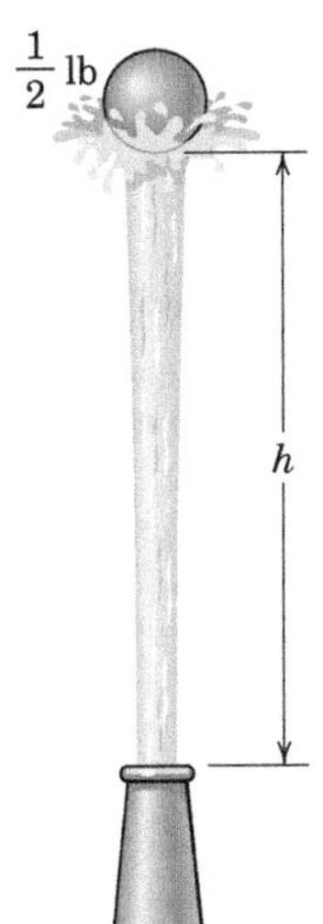

Problem 4/47

4/48 A jet-engine thrust reverser to reduce an aircraft speed of 200 km/h after landing employs folding vanes which deflect the exhaust gases in the direction indicated. If the engine is consuming 50 kg of air and 0.65 kg of fuel per second, calculate the braking thrust as a fraction n of the engine thrust without the deflector vanes. The exhaust gases have a velocity of 650 m/s relative to the nozzle.

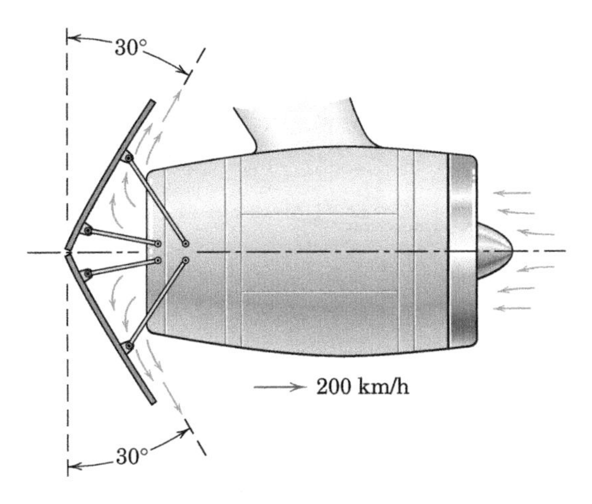

Problem 4/48

4/49 Water issues from a nozzle with an initial velocity v and supports a thin plate of mass m at a height h above the nozzle exit. A hole in the center of the plate allows some of the water to travel upward to a maximum altitude $2h$ above the plate. Determine the mass m of the plate. Neglect any effects of water falling onto the plate after reaching maximum altitude. Let ρ be the density of water.

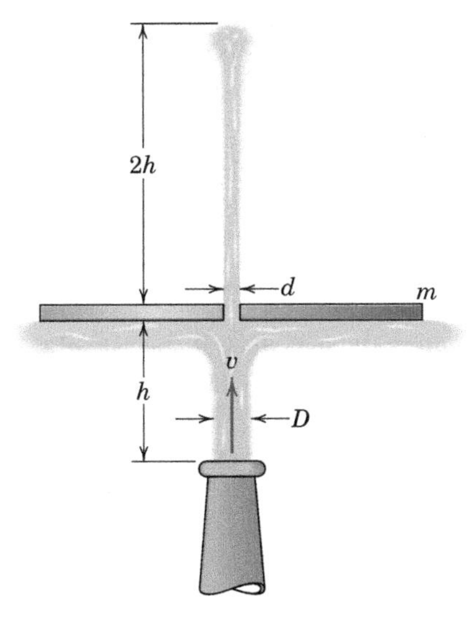

Problem 4/49

4/50 The axial-flow fan C pumps air through the duct of circular cross section and exhausts it with a velocity v at B. The air densities at A and B are ρ_A and ρ_B, respectively, and the corresponding pressures are p_A and p_B. The fixed deflecting blades at D restore axial flow to the air after it passes through the propeller blades C. Write an expression for the resultant horizontal force R exerted on the fan unit by the flange and bolts at A.

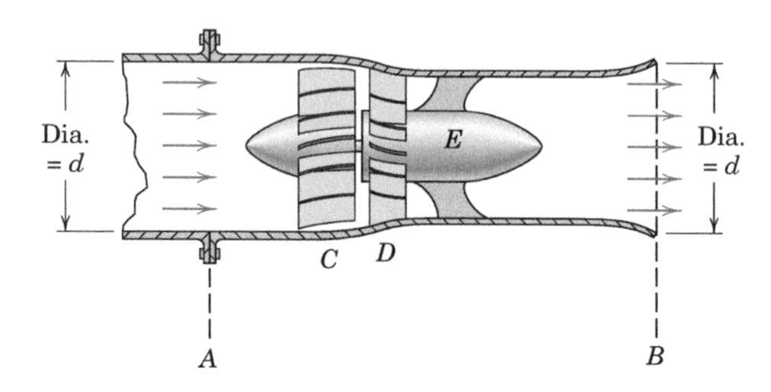

Problem 4/50

4/51 Air is pumped through the stationary duct A with a velocity of 50 ft/sec and exhausted through an experimental nozzle section BC. The average static pressure across section B is 150 lb/in.^{2} gage, and the specific weight of air at this pressure and at the temperature prevailing is 0.840 lb/ft^3. The average static pressure across the exit section C is measured to be 2 lb/in.^{2} gage, and the corresponding specific weight of air is 0.0760 lb/ft^3. Calculate the force T exerted on the nozzle flange at B by the bolts and the gasket to hold the nozzle in place.

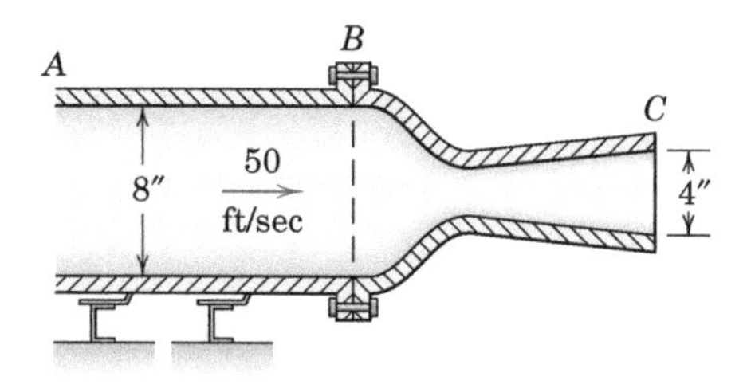

Problem 4/51

4/52 Air enters the pipe at A at the rate of 6 kg/s under a pressure of 1400 kPa gage and leaves the whistle at atmospheric pressure through the opening at B. The entering velocity of the air at A is 45 m/s, and the exhaust velocity at B is 360 m/s. Calculate the tension T, shear V, and bending moment M in the pipe at A. The net flow area at A is 7500 mm^2.

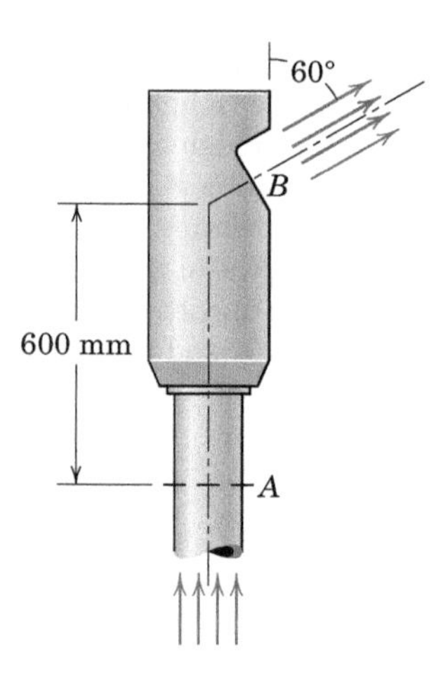

Problem 4/52

4/53 The sump pump has a net mass of 310 kg and pumps fresh water against a 6-m head at the rate of 0.125 m^3/s. Determine the vertical force R between the supporting base and the pump flange at A during operation. The mass of water in the pump may be taken as the equivalent of a 200-mm-diameter column 6 m in height.

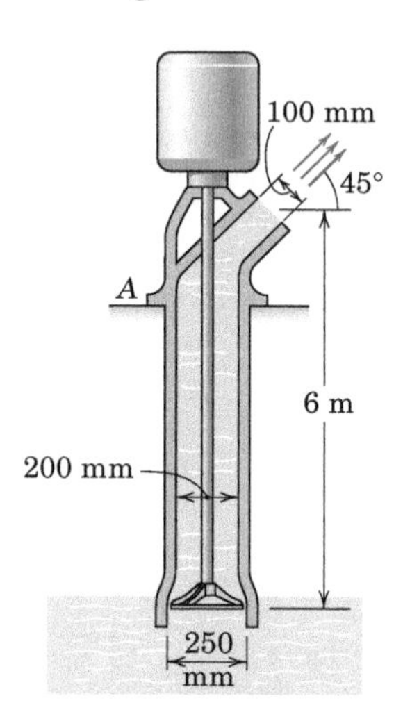

Problem 4/53

4/54 The experimental ground-effect machine has a total weight of 4200 lb. It hovers 1 or 2 ft off the ground by pumping air at atmospheric pressure through the circular intake duct at B and discharging it horizontally under the periphery of the skirt C. For an intake velocity v of 150 ft/sec, calculate the average air pressure p under the 18-ft-diameter machine at ground level. The specific weight of the air is 0.076 lb/ft^3.

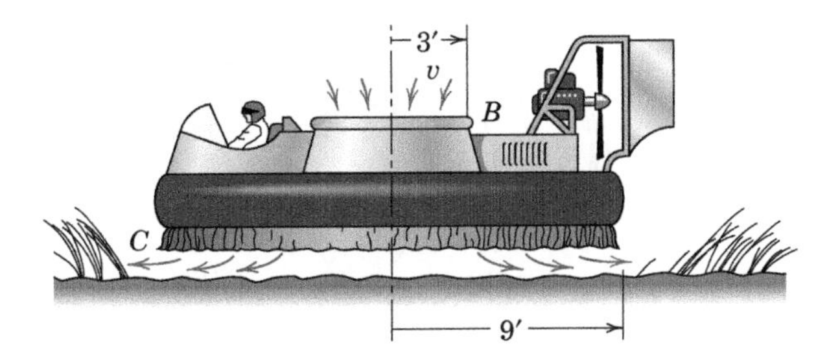

Problem 4/54

4/55 The leaf blower draws in air at a rate of 400 ft^3/min and discharges it at a speed v = 240 mi/hr. If the specific weight of the air being drawn into the blower is 7.53(10^{-2}) lb/ft^3, determine the added torque which the man must exert on the handle of the blower when it is running, compared with that when it is off, to maintain a steady orientation.

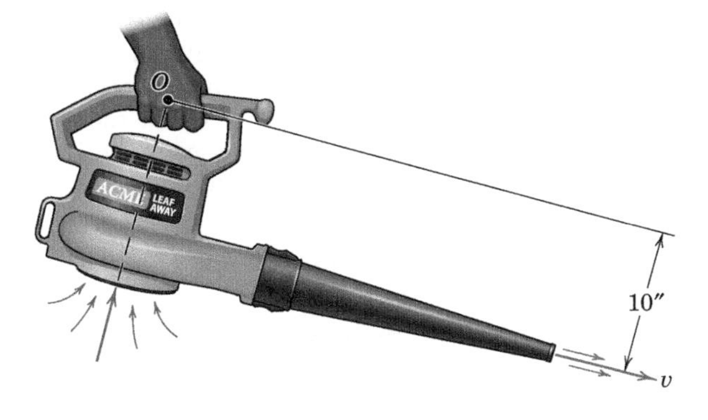

Problem 4/55

4/56 The ducted fan unit of mass m is supported in the vertical position on its flange at A. The unit draws in air with a density ρ and a velocity u through section A and discharges it through section B with a velocity v. Both inlet and outlet pressures are atmospheric. Write an expression for the force R applied to the flange of the fan unit by the supporting slab.

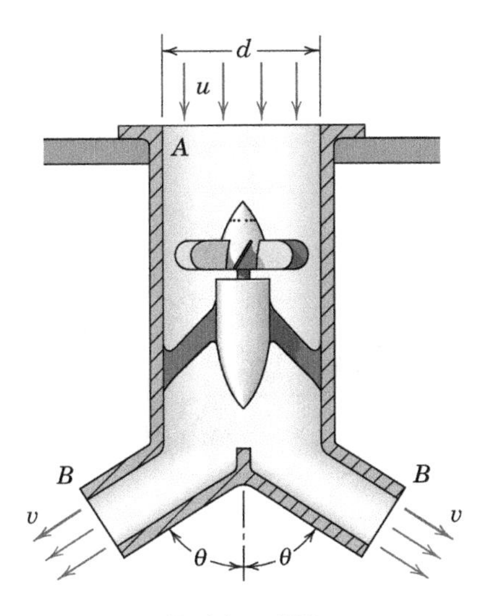

Problem 4/56

4/57 The fire hydrant is tested under a high standpipe pressure. The total flow of 10 ft^3/sec is divided equally between the two outlets, each of which has a cross-sectional area of 0.040 ft^2. The inlet cross-sectional area at the base is 0.75 ft^2. Neglect the weight of the hydrant and water within it and compute the tension T, the shear V, and the bending moment M in the base of the standpipe at B. The specific weight of water is 62.4 lb/ft^3. The static pressure of the water as it enters the base at B is 120 lb/in.^{2}

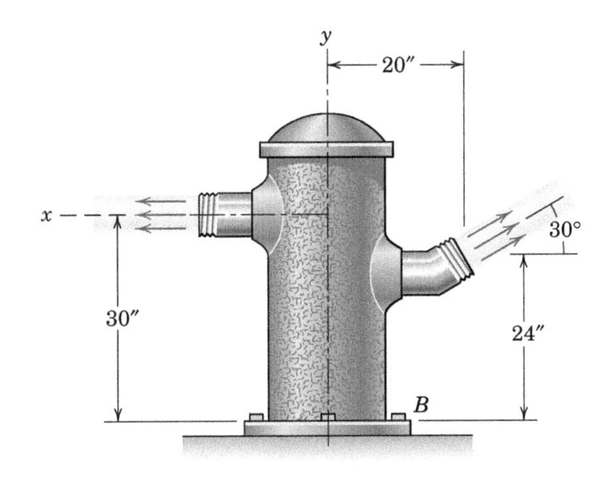

Problem 4/57

4/58 A rotary snow plow mounted on a large truck eats its way through a snow drift on a level road at a constant speed of 20 km/h. The plow discharges 60 Mg of snow per minute from its 45° chute with a velocity of 12 m/s relative to the plow. Calculate the tractive force P on the tires in the direction of motion necessary to move the plow and find the corresponding lateral force R between the tires and the road.

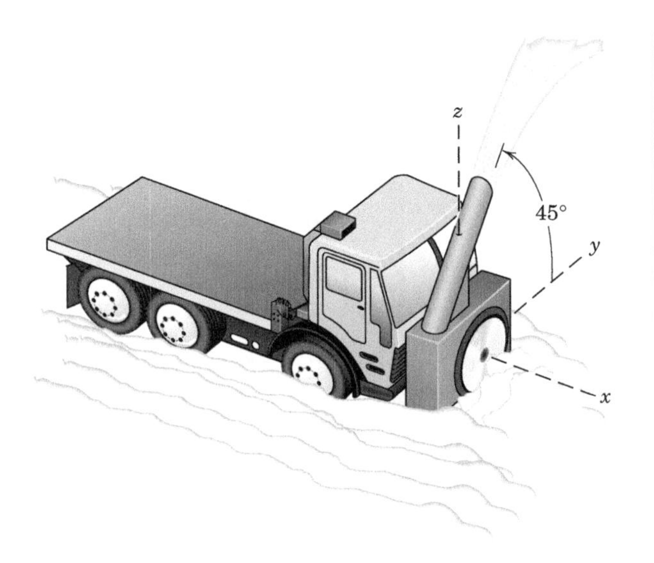

Problem 4/58

4/59 Salt water flows through the fixed 12-in.-inside-diameter pipe at a speed $v_0 = 4$ ft/sec and enters the 150° bend with inside radius of 24 in. The water exits to the atmosphere through the 6-in.-diameter nozzle C. Determine the shear force V, axial force P, and bending moment M at flanges A and B which result from the flow of the salt water. The gage pressure in the pipe at flange A is 250 lb/in.^{2} and the pressure drop between A and B due to head loss may be neglected.

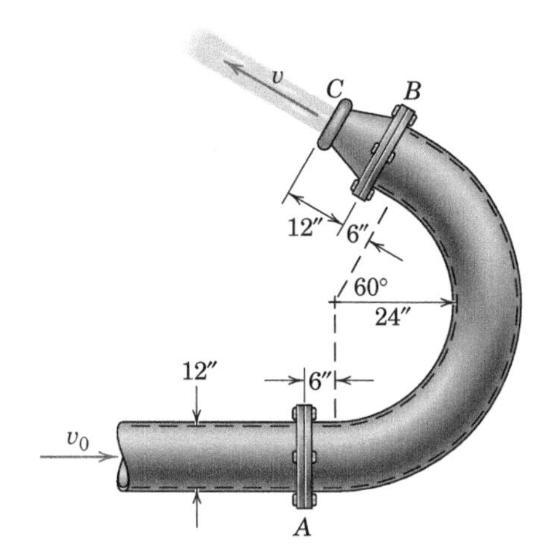

Problem 4/59

4/60 The industrial blower sucks in air through the axial opening A with a velocity v_1 and discharges it at atmospheric pressure and temperature through the 150-mm-diameter duct B with a velocity v_2. The blower handles 16 m^3 of air per minute with the motor and fan running at 3450 rev/min. If the motor requires 0.32 kW of power under no load (both ducts closed), calculate the power P consumed while air is being pumped.

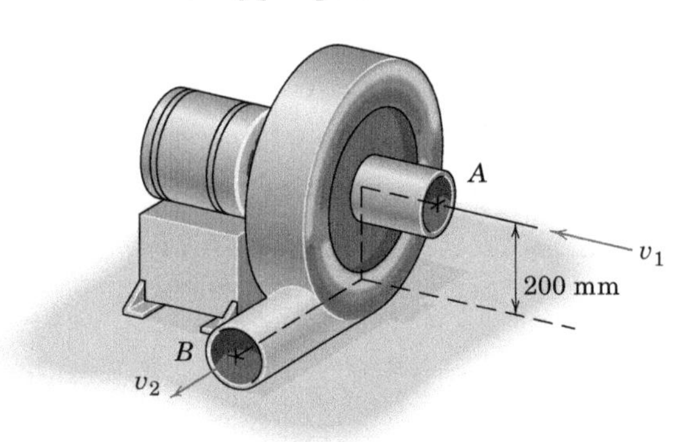

Problem 4/60

4/61 The feasibility of a one-passenger VTOL (vertical takeoff and landing) craft is under review. The preliminary design calls for a small engine with a high power-to-weight ratio driving an air pump that draws in air through the 70° ducts with an inlet velocity $v = 40$ m/s at a static gage pressure of -1.8 kPa across the inlet areas totaling 0.1320 m^2. The air is exhausted vertically down with a velocity $u = 420$ m/s. For a 90-kg passenger, calculate the maximum net mass m of the machine for which it can take off and hover. (See Table D/1 for air density.)

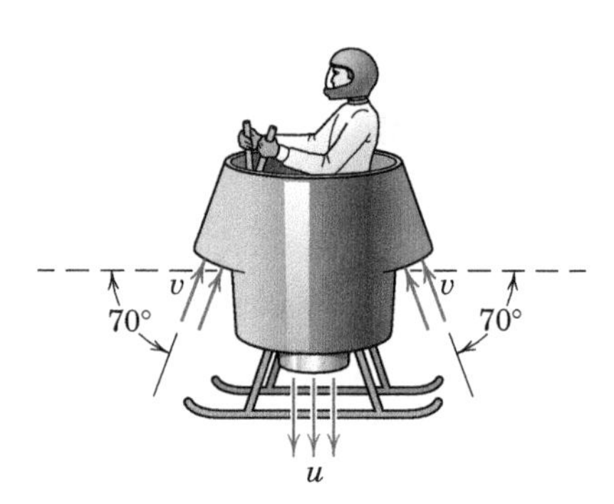

Problem 4/61

4/62 The helicopter shown has a mass m and hovers in position by imparting downward momentum to a column of air defined by the slipstream boundary shown. Find the downward velocity v given to the air by the rotor at a section in the stream below the rotor, where the pressure is atmospheric and the stream radius is r. Also find the power P required of the engine. Neglect the rotational energy of the air, any temperature rise due to air friction, and any change in air density ρ.

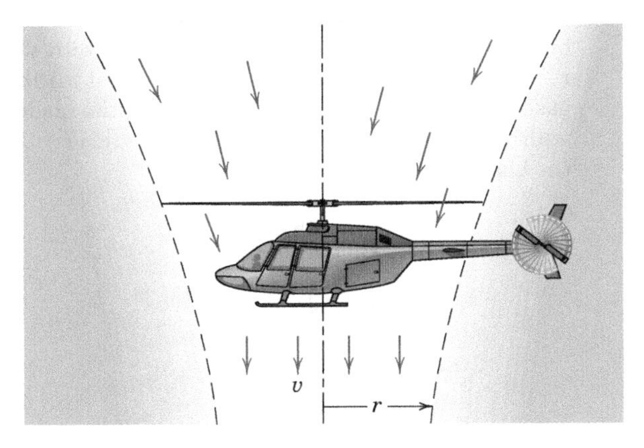

Problem 4/62

4/63 The sprinkler is made to rotate at the constant angular velocity ω and distributes water at the volume rate Q. Each of the four nozzles has an exit area A. Water is ejected from each nozzle at an angle ϕ that is measured in the horizontal plane as shown. Write an expression for the torque M on the shaft of the sprinkler necessary to maintain the given motion. For a given pressure and thus flow rate Q, at what speed ω_0 will the sprinkler operate with no applied torque? Let ρ be the density of water.

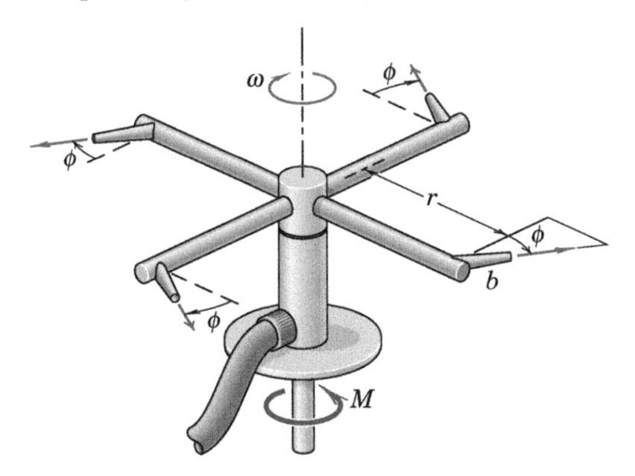

Problem 4/63

4/64 The VTOL (vertical takeoff and landing) military aircraft is capable of rising vertically under the action of its jet exhaust, which can be "vectored" from $\theta \cong 0$ for takeoff and hovering to $\theta = 90°$ for forward flight. The loaded aircraft has a mass of 8600 kg. At full takeoff power, its turbo-fan engine consumes air at the rate of 90 kg/s and has an air–fuel ratio of 18. Exhaust-gas velocity is 1020 m/s with essentially atmospheric pressure across the exhaust nozzles. Air with a density of 1.206 kg/m^3 is sucked into the intake scoops at a pressure of -2 kPa (gage) over the total inlet area of 1.10 m^2. Determine the angle θ for vertical takeoff and the corresponding vertical acceleration a_y of the aircraft.

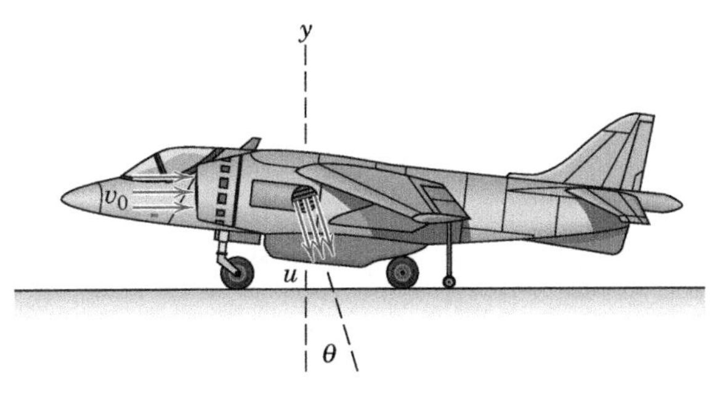

Problem 4/64

4/65 A marine terminal for unloading bulk wheat from a ship is equipped with a vertical pipe with a nozzle at A which sucks wheat up the pipe and transfers it to the storage building. Calculate the x- and y-components of the force **R** required to change the momentum of the flowing mass in rounding the bend. Identify all forces applied externally to the bend and mass within it. Air flows through the 14-in.-diameter pipe at the rate of 18 tons per hour under a vacuum of 9 in. of mercury ($p = -4.42$ lb/in.^{2}) and carries with it 150 tons of wheat per hour at a speed of 124 ft/sec.

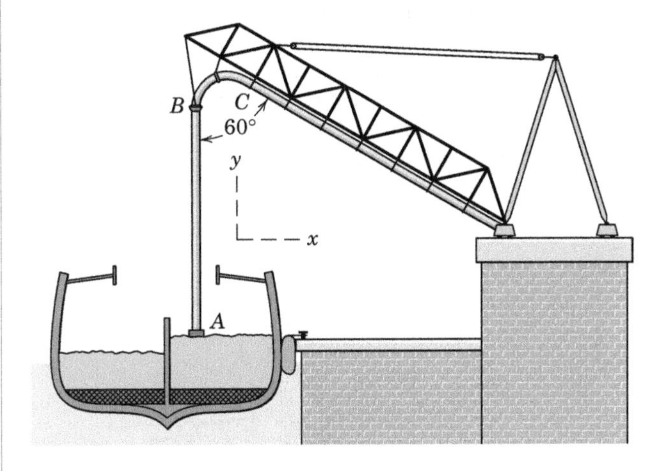

Problem 4/65

▸4/66 An axial section of the suction nozzle A for a bulk wheat unloader is shown here. The outer pipe is secured to the inner pipe by several longitudinal webs which do not restrict the flow of air. A vacuum of 9 in. of mercury ($p = -4.42$ lb/in.^{2} gage) is maintained in the inner pipe, and the pressure across the bottom of the outer pipe is atmospheric ($p = 0$). Air at 0.075 lb/ft^3 is drawn in through the space between the pipes at a rate of 18 tons/hr at atmospheric pressure and draws with it 150 tons of wheat per hour up the pipe at a velocity of 124 ft/sec. If the nozzle unit below section A-A weighs 60 lb, calculate the compression C in the connection at A-A.

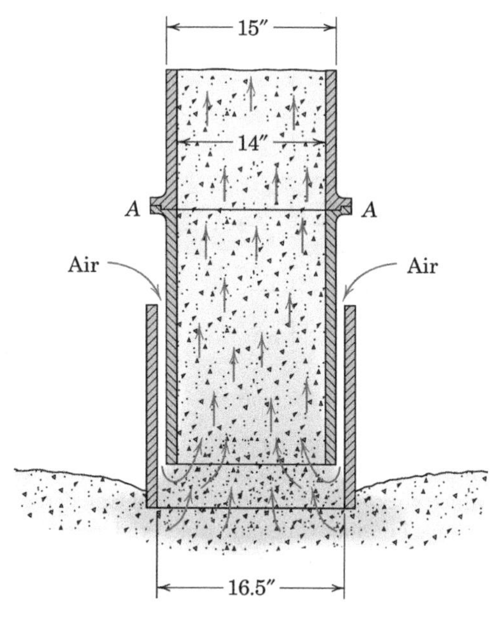

Problem 4/66

▶**4/67** The valve, which is screwed into the fixed pipe at section A-A, is designed to discharge fresh water at the rate of 340 gal/min into the atmosphere in the x-y plane as shown. Water pressure at A-A is 150 lb/in.^{2} gage. The flow area at A-A has a diameter of 2 in., and the diameter of the discharge area at B is 1 in. Neglect the weight of the valve and water within it and compute the shear V, tension F, torsion T, and bending moment M at section A-A. (1 gallon contains 231 in.^{3})

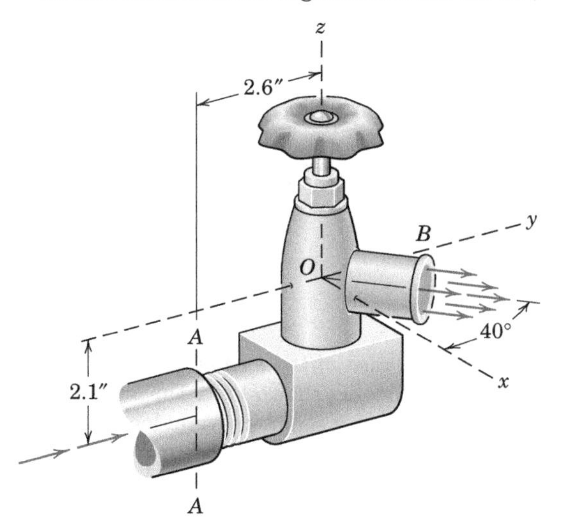

Problem 4/67

▶**4/68** In the figure is shown a detail of the stationary nozzle diaphragm A and the rotating blades B of a gas turbine. The products of combustion pass through the fixed diaphragm blades at the 27° angle and impinge on the moving rotor blades. The angles shown are selected so that the velocity of the gas relative to the moving blade at entrance is at the 20° angle for minimum turbulence, corresponding to a mean blade velocity of 315 m/s at a radius of 375 mm. If gas flows past the blades at the rate of 15 kg/s, determine the theoretical power output P of the turbine. Neglect fluid and mechanical friction with the resulting heat-energy loss and assume that all the gases are deflected along the surfaces of the blades with a velocity of constant magnitude relative to the blade.

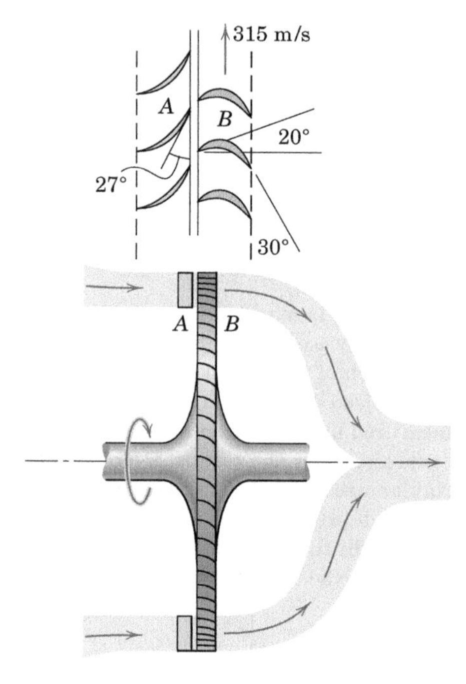

Problem 4/68

PROBLEMS

Introductory Problems

4/69 When the rocket reaches the position in its trajectory shown, it has a mass of 3 Mg and is beyond the effect of the earth's atmosphere. Gravitational acceleration is 9.60 m/s^2. Fuel is being consumed at the rate of 130 kg/s, and the exhaust velocity relative to the nozzle is 600 m/s. Compute the n- and t-components of acceleration of the rocket.

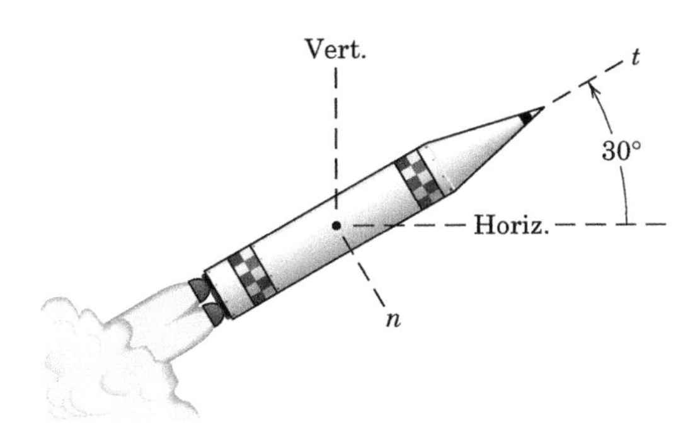

Problem 4/69

4/70 At the instant of vertical launch the rocket expels exhaust at the rate of 220 kg/s with an exhaust velocity of 820 m/s. If the initial vertical acceleration is 6.80 m/s^2, calculate the total mass of the rocket and fuel at launch.

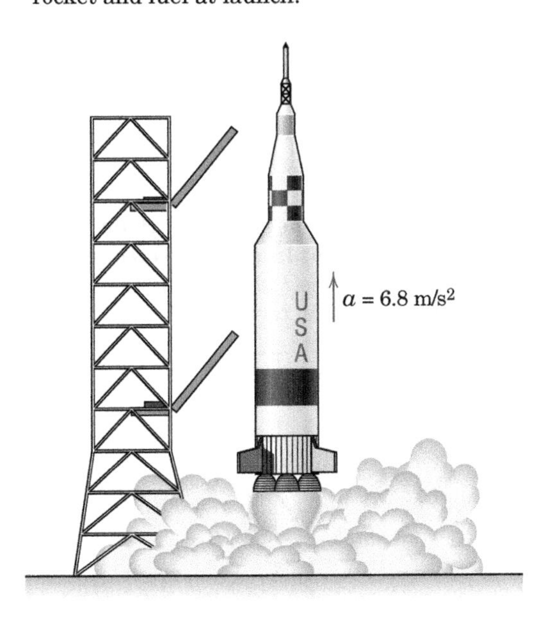

Problem 4/70

4/71 The space shuttle, together with its central fuel tank and two booster rockets, has a total mass of 2.04(10^6) kg at liftoff. Each of the two booster rockets produces a thrust of 11.80(10^6) N, and each of the three main engines of the shuttle produces a thrust of 2.00(10^6) N. The specific impulse (ratio of exhaust velocity to gravitational acceleration) for each of the three main engines of the shuttle is 455 s. Calculate the initial vertical acceleration a of the assembly with all five engines operating and find the rate at which fuel is being consumed by each of the shuttle's three engines.

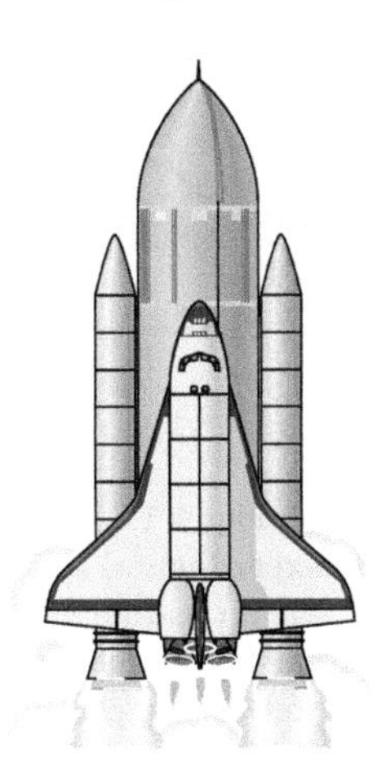

Problem 4/71

4/72 A small rocket of initial mass m_0 is fired vertically upward near the surface of the earth (g constant). If air resistance is neglected, determine the manner in which the mass m of the rocket must vary as a function of the time t after launching in order that the rocket may have a constant vertical acceleration a, with a constant relative velocity u of the escaping gases with respect to the nozzle.

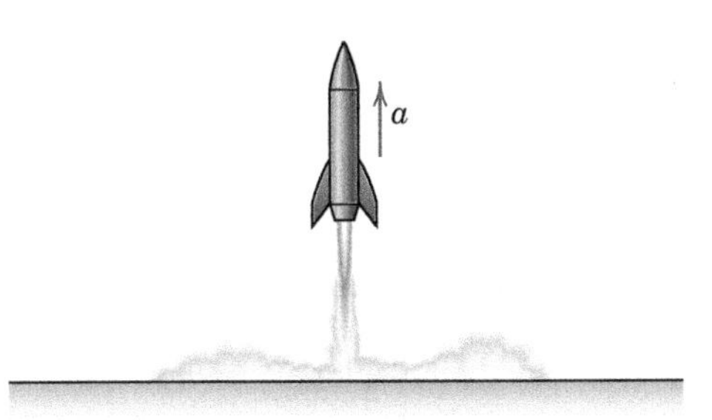

Problem 4/72

4/73 The mass m of a raindrop increases as it picks up moisture during its vertical descent through still air. If the air resistance to motion of the drop is R and its downward velocity is v, write the equation of motion for the drop and show that the relation $\Sigma F = d(mv)/dt$ is obeyed as a special case of the variable-mass equation.

4/74 A tank truck for washing down streets has a total weight of 20,000 lb when its tank is full. With the spray turned on, 80 lb of water per second issue from the nozzle with a velocity of 60 ft/sec relative to the truck at the 30° angle shown. If the truck is to accelerate at the rate of 2 ft/sec^2 when starting on a level road, determine the required tractive force P between the tires and the road when (*a*) the spray is turned on and (*b*) the spray is turned off.

Problem 4/74

4/75 A model rocket weighs 1.5 lb just before its vertical launch. Its experimental solid-fuel motor carries 0.1 lb of fuel, has an escape velocity of 3000 ft/sec, and burns the fuel for 0.9 sec. Determine the acceleration of the rocket at launch and its burnout velocity. Neglect aerodynamic drag and state any other assumptions.

Problem 4/75

4/76 The magnetometer boom for a spacecraft consists of a large number of triangular-shaped units which spring into their deployed configuration upon release from the canister in which they were folded and packed prior to release. Write an expression for the force F which the base of the canister must exert on the boom during its deployment in terms of the increasing length x and its time derivatives. The mass of the boom per unit of deployed length is ρ. Treat the supporting base on the spacecraft as a fixed platform and assume that the deployment takes place outside of any gravitational field. Neglect the dimension b compared with x.

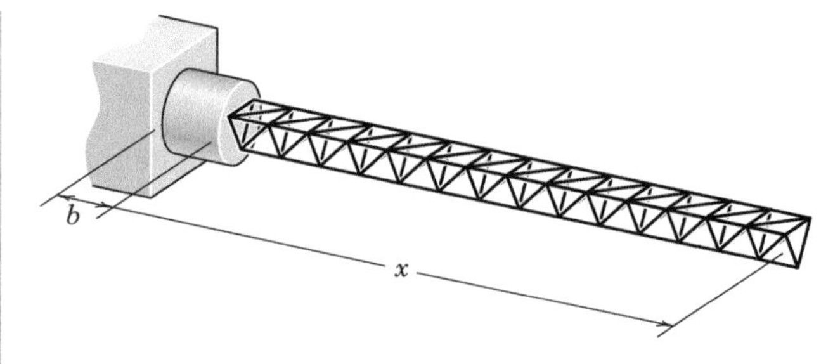

Problem 4/76

4/77 Fresh water issues from the two 30-mm-diameter holes in the bucket with a velocity of 2.5 m/s in the directions shown. Calculate the force P required to give the bucket an upward acceleration of 0.5 m/s^2 from rest if it contains 20 kg of water at that time. The empty bucket has a mass of 0.6 kg.

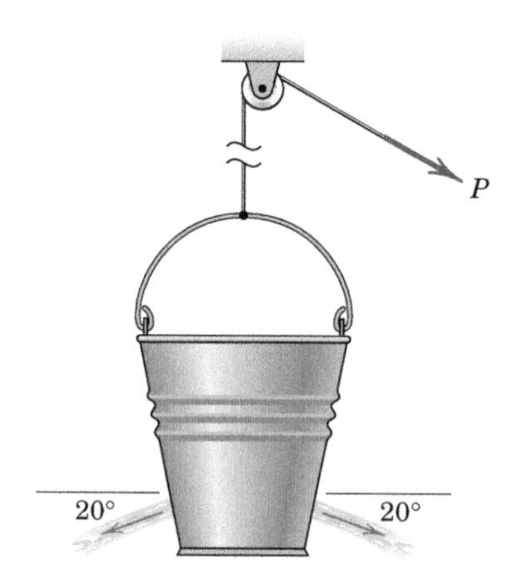

Problem 4/77

Representative Problems

4/78 The upper end of the open-link chain of length L and mass ρ per unit length is lowered at a constant speed v by the force P. Determine the reading R of the platform scale in terms of x.

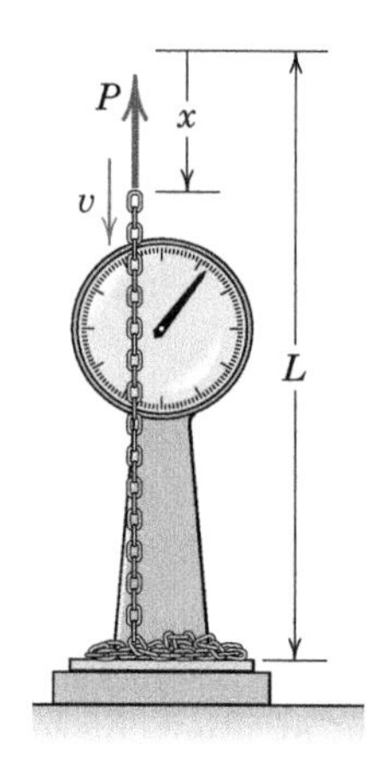

Problem 4/78

4/79 A rocket stage designed for deep-space missions consists of 200 kg of fuel and 300 kg of structure and payload combined. In terms of burnout velocity, what would be the advantage of reducing the structural/payload mass by 1 percent (3 kg) and using that mass for additional fuel? Express your answer in terms of a percent increase in burnout velocity. Repeat your calculation for a 5 percent reduction in the structural/payload mass.

4/80 At a bulk loading station, gravel leaves the hopper at the rate of 220 lb/sec with a velocity of 10 ft/sec in the direction shown and is deposited on the moving flatbed truck. The tractive force between the driving wheels and the road is 380 lb, which overcomes the 200 lb of frictional road resistance. Determine the acceleration a of the truck 4 seconds after the hopper is opened over the truck bed, at which instant the truck has a forward speed of 1.5 mi/hr. The empty weight of the truck is 12,000 lb.

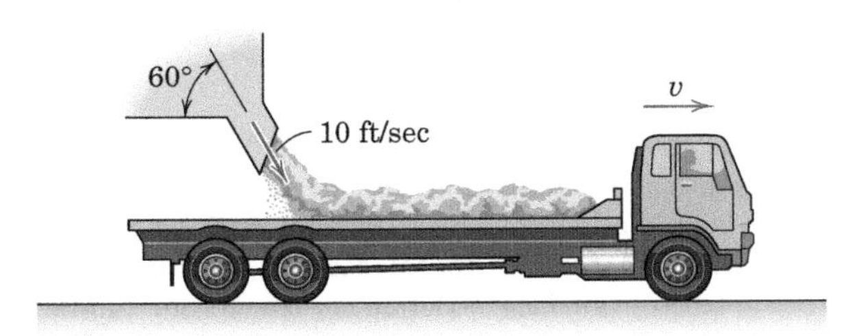

Problem 4/80

4/81 A railroad coal car weighs 54,600 lb empty and carries a total load of 180,000 lb of coal. The bins are equipped with bottom doors which permit discharging coal through an opening between the rails. If the car dumps coal at the rate of 20,000 lb/sec in a downward direction relative to the car, and if frictional resistance to motion is 4 lb per ton of total remaining weight, determine the coupler force P required to give the car an acceleration of 0.15 ft/sec^2 in the direction of P at the instant when half the coal has been dumped.

Problem 4/81

4/82 A coil of heavy flexible cable with a total length of 100 m and a mass of 1.2 kg/m is to be laid along a straight horizontal line. The end is secured to a post at A, and the cable peels off the coil and emerges through the horizontal opening in the cart as shown. The cart and drum together have a mass of 40 kg. If the cart is moving to the right with a velocity of 2 m/s when 30 m of cable remain in the drum and the tension in the rope at the post is 2.4 N, determine the force P required to give the cart and drum an acceleration of 0.3 m/s^2. Neglect all friction.

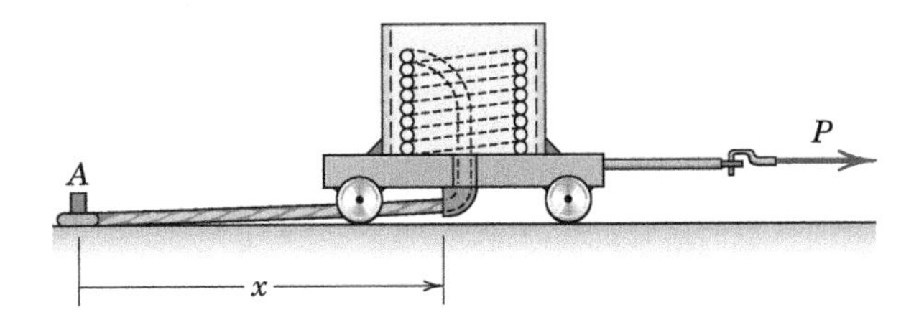

Problem 4/82

4/83 By lowering a scoop as it skims the surface of a body of water, the aircraft (nicknamed the "Super Scooper") is able to ingest 4.5 m^3 of fresh water during a 12-second run. The plane then flies to a fire area and makes a massive water drop with the ability to repeat the procedure as many times as necessary. The plane approaches its run with a velocity of 280 km/h and an initial mass of 16.4 Mg. As the scoop enters the water, the pilot advances the throttle to provide an additional 300 hp (223.8 kW) needed to prevent undue deceleration. Determine the initial deceleration when the scooping action starts. (Neglect the difference between the average and the initial rates of water intake.)

Problem 4/83

4/84 A small rocket-propelled vehicle weighs 125 lb, including 20 lb of fuel. Fuel is burned at the constant rate of 2 lb/sec with an exhaust velocity relative to the nozzle of 400 ft/sec. Upon ignition the vehicle is released from rest on the 10° incline. Calculate the maximum velocity v reached by the vehicle. Neglect all friction.

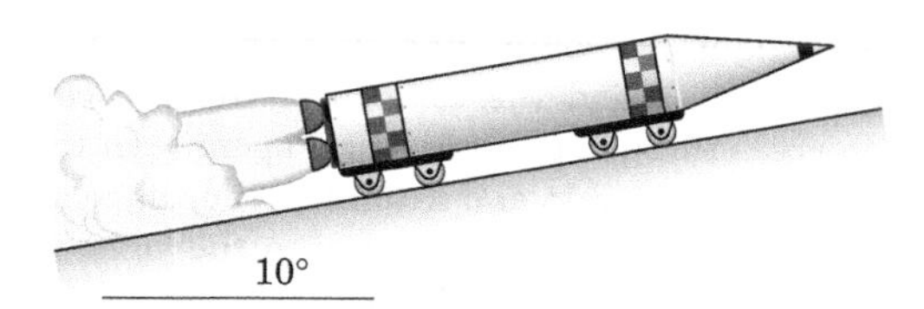

Problem 4/84

4/85 The end of a pile of loose-link chain of mass ρ per unit length is being pulled horizontally along the surface by a constant force P. If the coefficient of kinetic friction between the chain and the surface is μ_k, determine the acceleration a of the chain in terms of x and $\dot{x}$.

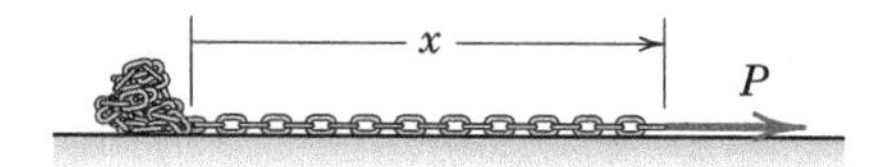

Problem 4/85

4/86 A coal car with an empty mass of 25 Mg is moving freely with a speed of 1.2 m/s under a hopper which opens and releases coal into the moving car at the constant rate of 4 Mg per second. Determine the distance x moved by the car during the time that 32 Mg of coal are deposited in the car. Neglect any frictional resistance to rolling along the horizontal track.

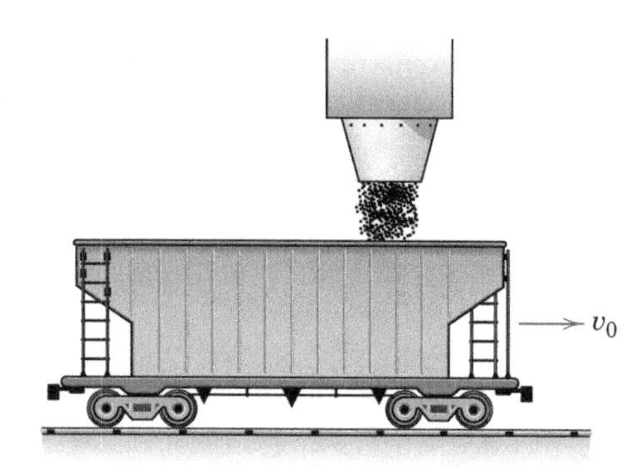

Problem 4/86

4/87 Sand is released from the hopper H with negligible velocity and then falls a distance h to the conveyor belt. The mass flow rate from the hopper is m'. Develop an expression for the steady-state belt speed v for the case $h = 0$. Assume that the sand quickly acquires the belt velocity with no rebound, and neglect friction at the pulleys A and B.

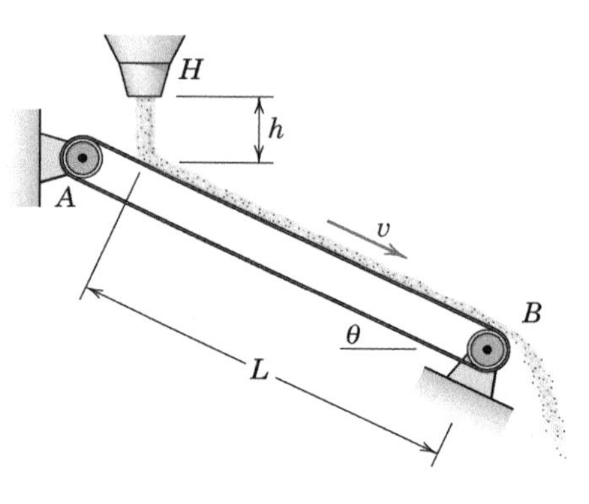

Problem 4/87

4/88 Repeat the previous problem, but now let $h \neq 0$. Then evaluate your expression for the conditions $h = 2$ m, $L = 10$ m, and $\theta = 25°$.

4/89 The open-link chain of length L and mass ρ per unit length is released from rest in the position shown, where the bottom link is almost touching the platform and the horizontal section is supported on a smooth surface. Friction at the corner guide is negligible. Determine (*a*) the velocity v_1 of end A as it reaches the corner and (*b*) its velocity v_2 as it strikes the platform. (*c*) Also specify the total loss Q of energy.

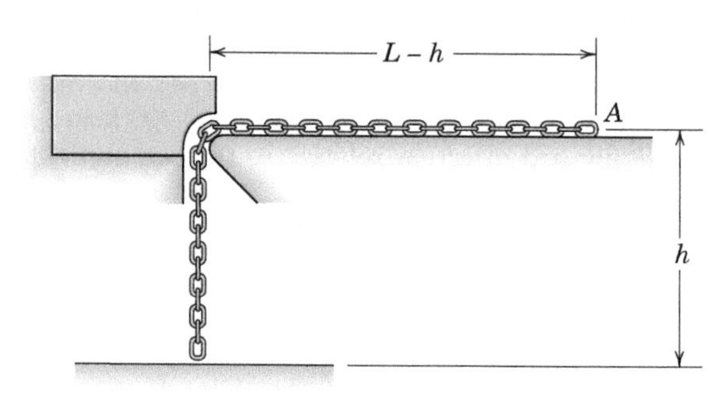

Problem 4/89

4/90 In the figure is shown a system used to arrest the motion of an airplane landing on a field of restricted length. The plane of mass m rolling freely with a velocity v_0 engages a hook which pulls the ends of two heavy chains, each of length L and mass ρ per unit length, in the manner shown. A conservative calculation of the effectiveness of the device neglects the retardation of chain friction on the ground and any other resistance to the motion of the airplane. With these assumptions, compute the velocity v of the airplane at the instant when the last link of each chain is put in motion. Also determine the relation between the displacement x and the time t after contact with the chain. Assume each link of the chain acquires its velocity v suddenly upon contact with the moving links.

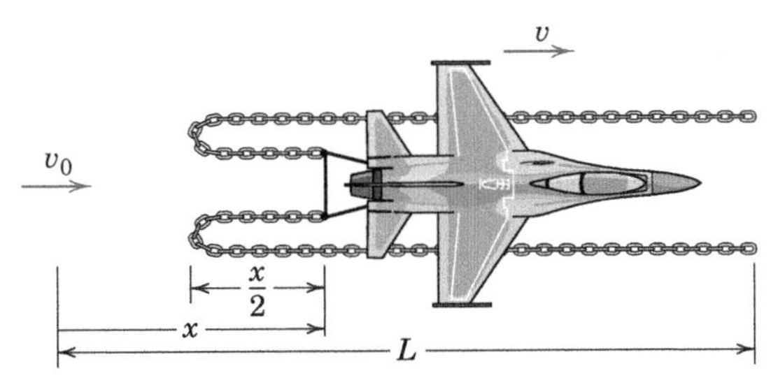

Problem 4/90

▶4/91 The free end of the flexible and inextensible rope of mass ρ per unit length and total length L is given a constant upward velocity v. Write expressions for P, the force R supporting the fixed end, and the tension T_1 in the rope at the loop in terms of x. (For the loop of negligible size, the tension is the same on both sides.)

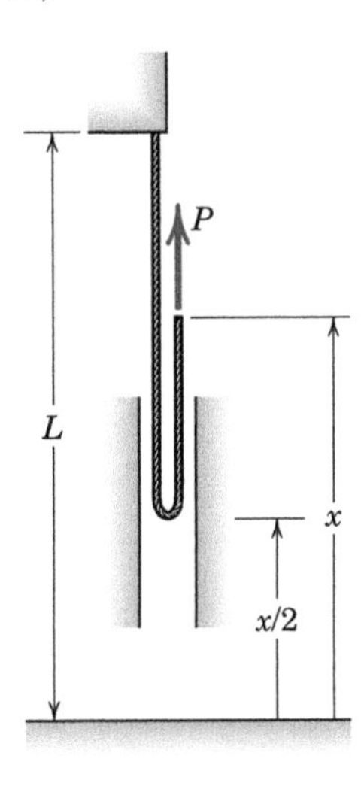

Problem 4/91

▶4/92 Replace the rope of Prob. 4/91 by an open-link chain with the same mass ρ per unit length. The free end is given a constant upward velocity v. Write expressions for P, the tension T_1 at the bottom of the moving part, and the force R supporting the fixed end in terms of x. Also find the energy loss Q in terms of x.

PART II

DYNAMICS OF RIGID BODIES

PROBLEMS

Introductory Problems

5/1 A torque applied to a flywheel causes it to accelerate uniformly from a speed of 300 rev/min to a speed of 900 rev/min in 6 seconds. Determine the number of revolutions N through which the wheel turns during this interval. (*Suggestion:* Use revolutions and minutes for units in your calculations.)

5/2 The circular sector rotates about a fixed axis through point O with angular velocity $\omega = 2$ rad/s and angular acceleration $\alpha = 4$ rad/s^2 with directions as indicated in the figure. Determine the instantaneous velocity and acceleration of point A.

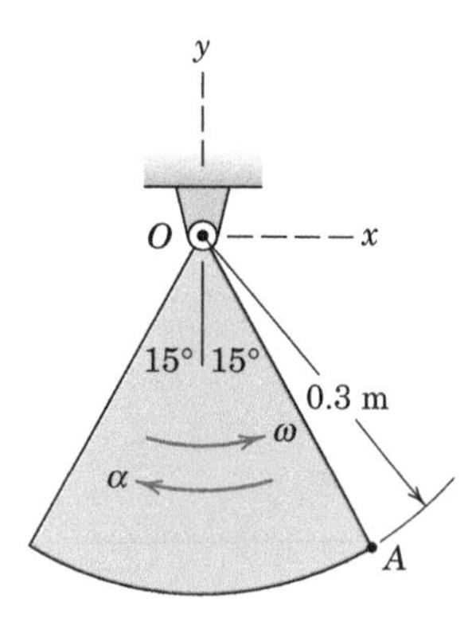

Problem 5/2

5/3 The angular velocity of a gear is controlled according to $\omega = 12 - 3t^2$ where ω, in radians per second, is positive in the clockwise sense and where t is the time in seconds. Find the net angular displacement $\Delta\theta$ from the time $t = 0$ to $t = 3$ s. Also find the total number of revolutions N through which the gear turns during the 3 seconds.

5/4 Magnetic tape is fed over and around the light pulleys mounted in a computer frame. If the speed v of the tape is constant and if the ratio of the magnitudes of the acceleration of points A and B is 2/3, determine the radius r of the larger pulley.

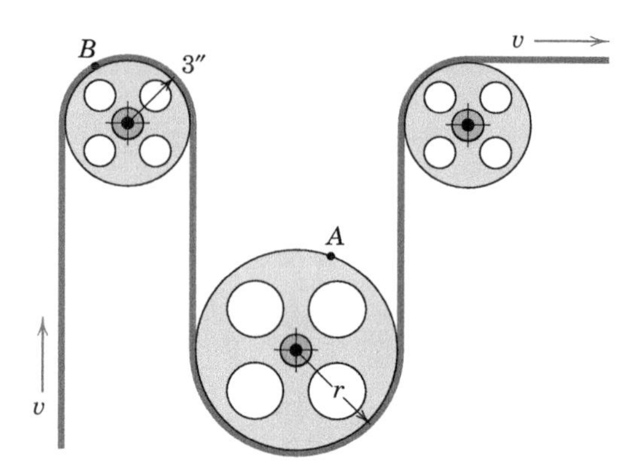

Problem 5/4

5/5 When switched on, the grinding machine accelerates from rest to its operating speed of 3450 rev/min in 6 seconds. When switched off, it coasts to rest in 32 seconds. Determine the number of revolutions turned during both the startup and shutdown periods. Assume uniform angular acceleration in both cases.

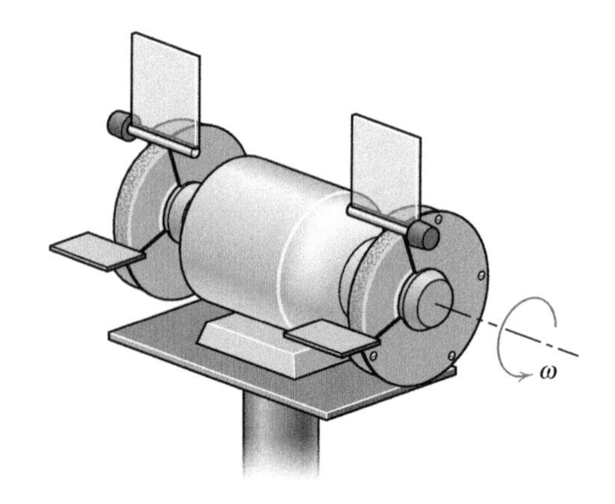

Problem 5/5

5/6 The small cart is released from rest in position 1 and requires 0.638 seconds to reach position 2 at the bottom of the path, where its center G has a velocity of 14.20 ft/sec. Determine the angular velocity ω of line AB in position 2 and the average angular velocity ω_{av} of AB during the interval.

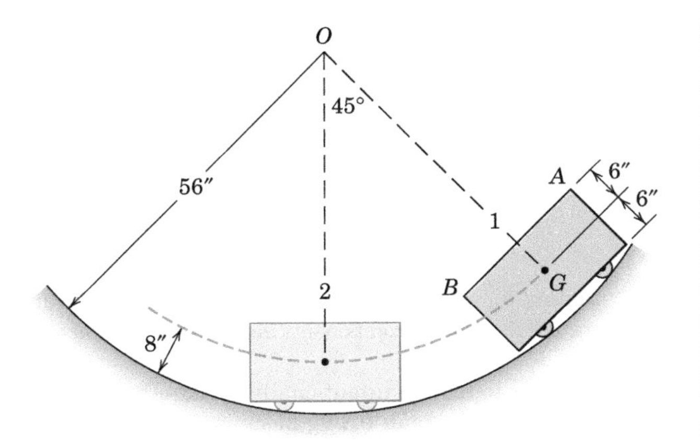

Problem 5/5

5/7 The flywheel has a diameter of 600 mm and rotates with increasing speed about its z-axis shaft. When point P on the rim crosses the y-axis with $\theta = 90°$, it has an acceleration given by $\mathbf{a} = -1.8\mathbf{i} - 4.8\mathbf{j}$ m/s^2. For this instant, determine the angular velocity ω and the angular acceleration α of the flywheel.

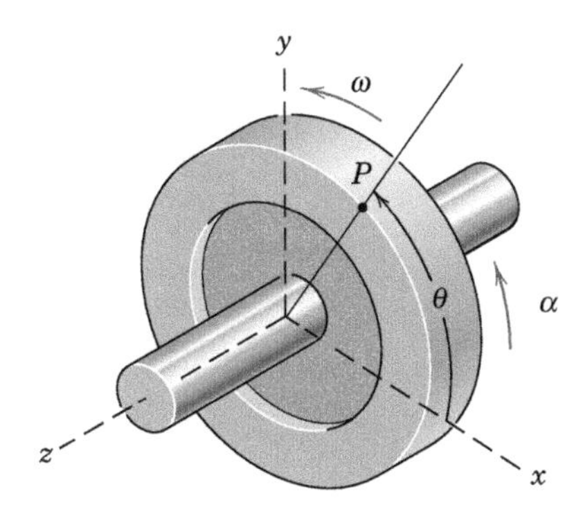

Problem 5/7

5/8 The drive mechanism imparts to the semicircular plate simple harmonic motion of the form $\theta = \theta_0 \sin \omega_0 t$, where θ_0 is the amplitude of the oscillation and ω_0 is its circular frequency. Determine the amplitudes of the angular velocity and angular acceleration and state where in the motion cycle these maxima occur. Note that this motion is *not* that of a freely pivoted and undriven body undergoing arbitrarily large-amplitude angular motion.

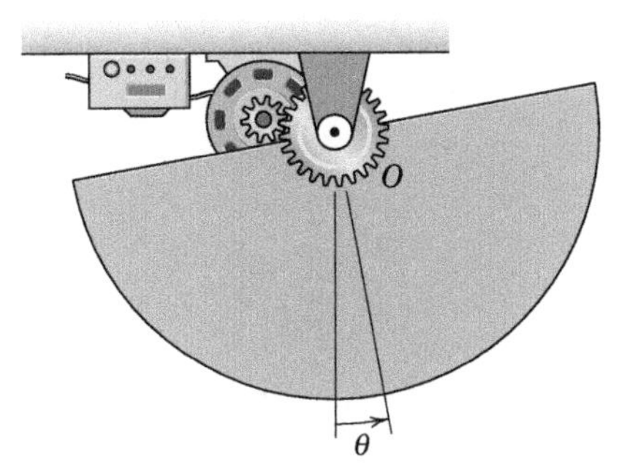

Problem 5/8

5/9 The cylinder rotates about the fixed z-axis in the direction indicated. If the speed of point A is $v_A = 2$ ft/sec and the magnitude of its acceleration is $a_A = 12$ ft/sec^2, determine the angular velocity and angular acceleration of the cylinder. Is knowledge of the angle θ necessary?

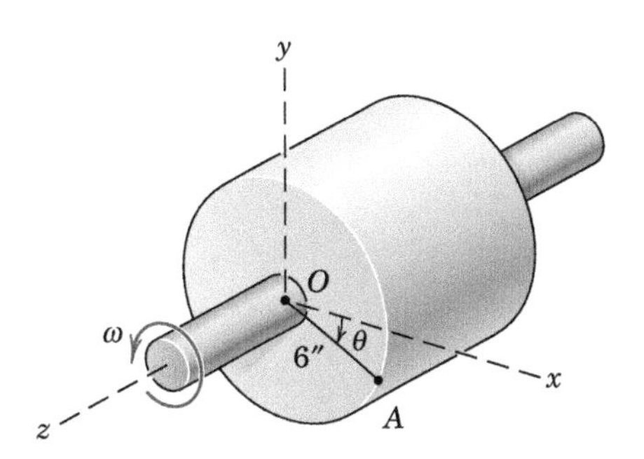

Problem 5/9

Representative Problems

5/10 The angular acceleration of a body which is rotating about a fixed axis is given by $\alpha = -k\omega^2$, where the constant $k = 0.1$ (no units). Determine the angular displacement and time elapsed when the angular velocity has been reduced to one-third its initial value $\omega_0 = 12$ rad/s.

5/11 The device shown rotates about the fixed z-axis with angular velocity $\omega = 20$ rad/s and angular acceleration $\alpha = 40$ rad/s^2 in the directions indicated. Determine the instantaneous velocity and acceleration of point B.

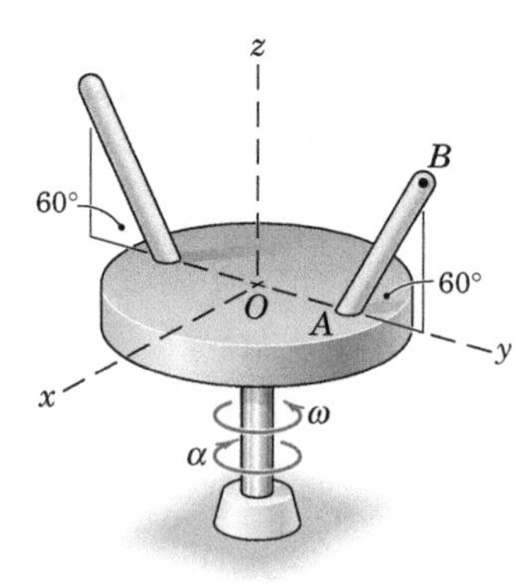

$\overline{OA}$ = 300 mm, $\overline{AB}$ = 500 mm

Problem 5/11

5/12 The circular disk rotates with a constant angular velocity $\omega = 40$ rad/sec about its axis, which is inclined in the y-z plane at the angle $\theta = \tan^{-1}\frac{3}{4}$. Determine the vector expressions for the velocity and acceleration of point P, whose position vector at the instant shown is $\mathbf{r} = 15\mathbf{i} + 16\mathbf{j} - 12\mathbf{k}$ in. (Check the magnitudes of your results from the scalar values $v = r\omega$ and $a_n = r\omega^2$.)

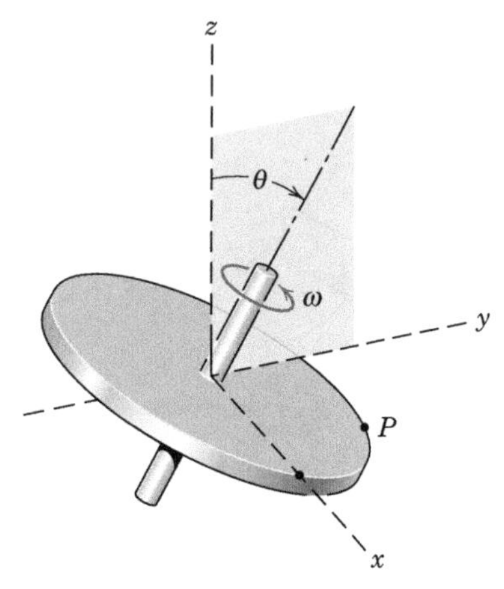

Problem 5/12

5/13 The T-shaped body rotates about a horizontal axis through point O. At the instant represented, its angular velocity is $\omega = 5$ rad/sec and its angular acceleration is $\alpha = 10$ rad/sec^2 in the directions indicated. Determine the velocity and acceleration of (a) point A and (b) point B. Express your results in terms of components along the x- and y-axes shown.

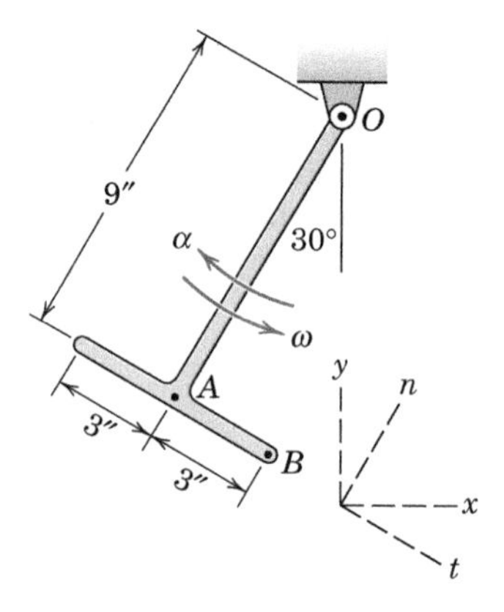

Problem 5/13

5/14 Repeat the previous problem, but now express your results in terms of components along the n- and t-axes.

5/15 In order to test an intentionally weak adhesive, the bottom of the small 0.3-kg block is coated with adhesive and then the block is pressed onto the turntable with a known force. The turntable starts from rest at time $t = 0$ and uniformly accelerates with $\alpha = 2$ rad/s^2. If the adhesive fails at exactly $t = 3$ s, determine the ultimate shear force which the adhesive supports. What is the angular displacement of the turntable at the time of failure?

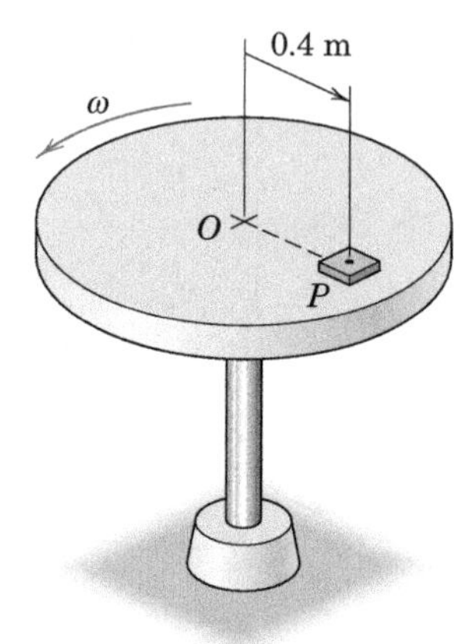

Problem 5/15

5/16 The two attached pulleys are driven by the belt with increasing speed. When the belt reaches a speed $v = 2$ ft/sec, the total acceleration of point P is 26 ft/sec^2. For this instant determine the angular acceleration α of the pulleys and the acceleration of point B on the belt.

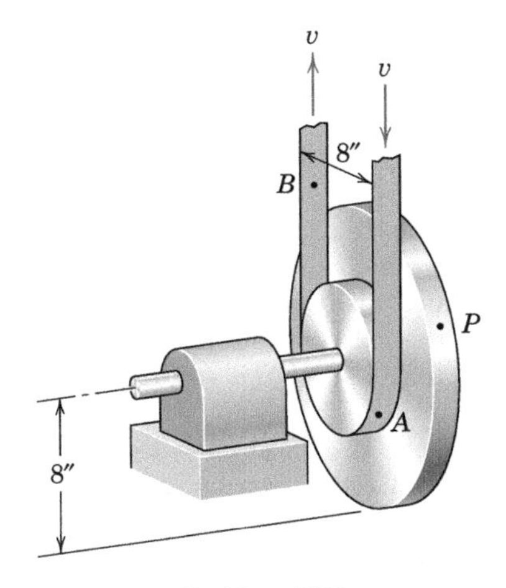

Problem 5/16

5/17 The bent flat bar rotates about a fixed axis through point O with the instantaneous angular properties indicated in the figure. Determine the velocity and acceleration of point A.

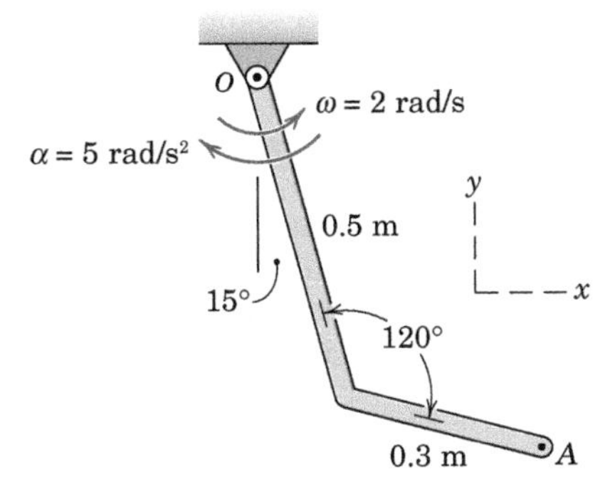

Problem 5/17

5/18 At time $t = 0$, the arm is rotating about the fixed z-axis with an angular velocity $\omega = 200$ rad/s in the direction shown. At that time, a constant angular deceleration begins and the arm comes to a stop in 10 seconds. At what time t does the acceleration of point P make a 15° angle with the arm AB?

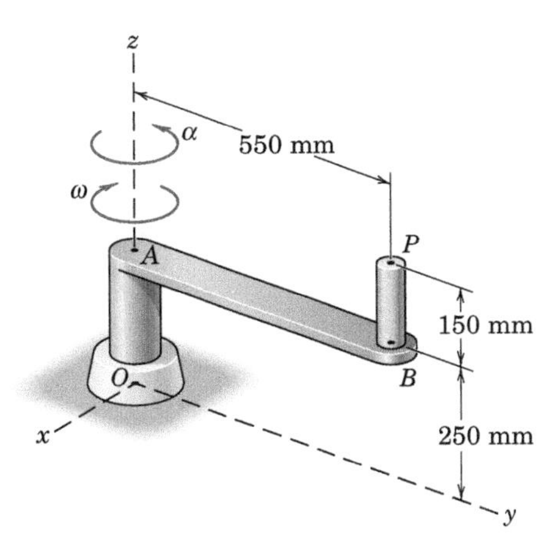

Problem 5/18

5/19 A variable torque is applied to a rotating wheel at time $t = 0$ and causes the clockwise angular acceleration to increase linearly with the clockwise angular displacement θ of the wheel during the next 30 revolutions. When the wheel has turned the additional 30 revolutions, its angular velocity is 90 rad/s. Determine its angular velocity ω_0 at the start of the interval at $t = 0$.

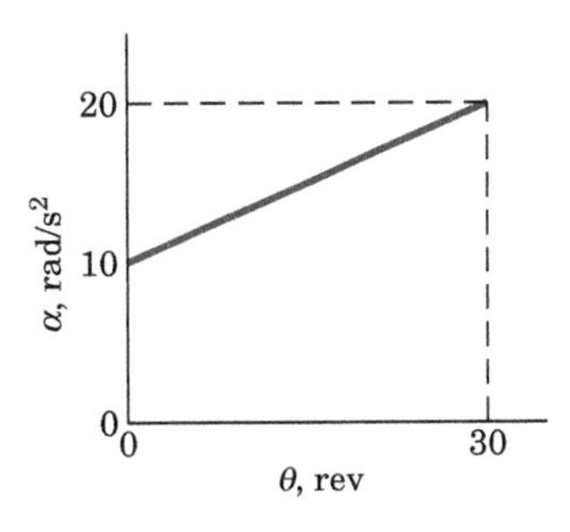

Problem 5/19

5/20 Develop general expressions for the instantaneous velocity and acceleration of point A of the square plate, which rotates about a fixed axis through point O. Take all variables to be positive. Then evaluate your expressions for $\theta = 30°$, $b = 0.2$ m, $\omega = 1.4$ rad/s, and $\alpha = 2.5$ rad/s^2.

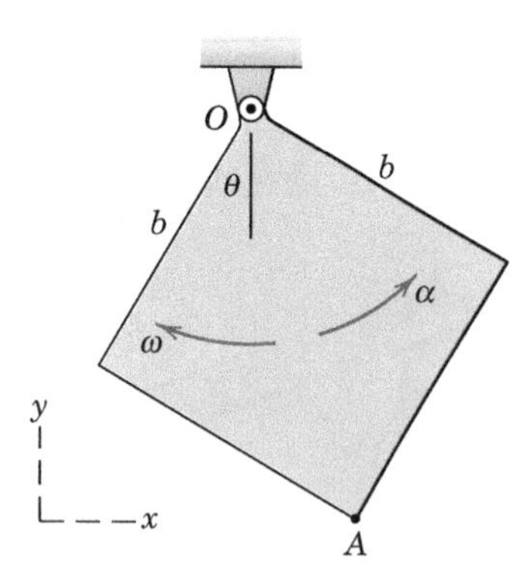

Problem 5/20

5/21 The motor A accelerates uniformly from zero to 3600 rev/min in 8 seconds after it is turned on at time $t = 0$. It drives a fan (not shown) which is attached to drum B. The effective pulley radii are shown in the figure. Determine (a) the number of revolutions turned by drum B during the 8-second startup period, (b) the angular velocity of drum B at time $t = 4$ s, and (c) the number of revolutions turned by drum B during the first 4 seconds of motion. Assume no belt slippage.

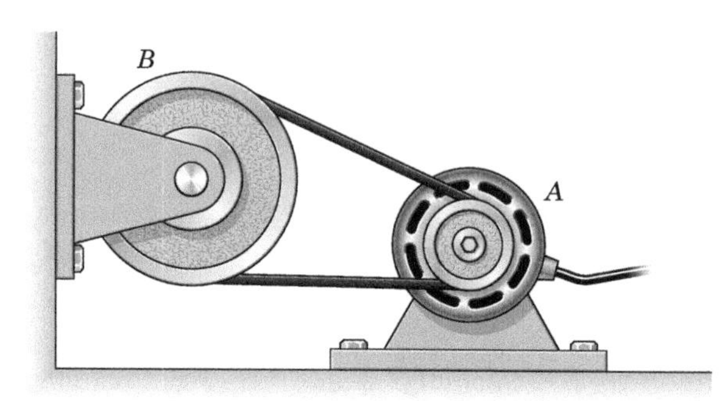

$r_A = 75$ mm, $r_B = 200$ mm

Problem 5/21

5/22 Point A of the circular disk is at the angular position $\theta = 0$ at time $t = 0$. The disk has angular velocity $\omega_0 = 0.1$ rad/s at $t = 0$ and subsequently experiences a constant angular acceleration $\alpha = 2$ rad/s^2. Determine the velocity and acceleration of point A in terms of fixed **i** and **j** unit vectors at time $t = 1$ s.

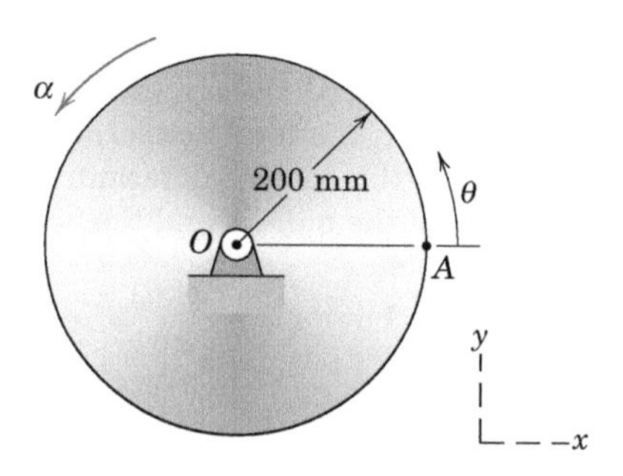

Problem 5/22

5/23 Repeat Prob. 5/22, except now the angular acceleration of the disk is given by $\alpha = 2t$, where t is in seconds and α is in radians per second squared. Determine the velocity and acceleration of point A in terms of fixed **i** and **j** unit vectors at time $t = 2$ s.

5/24 Repeat Prob. 5/22, except now the angular acceleration of the disk is given by $\alpha = 2\omega$, where ω is in radians per second and α is in radians per second squared. Determine the velocity and acceleration of point A in terms of fixed **i** and **j** unit vectors at time $t = 1$ s.

5/25 The disk of Prob. 5/22 is at the angular position $\theta = 0$ at time $t = 0$. Its angular velocity at $t = 0$ is $\omega_0 = 0.1$ rad/s, and then it experiences an angular acceleration given by $\alpha = 2\theta$, where θ is in radians and α is in radians per second squared. Determine the angular position of point A at time $t = 2$ s.

5/26 During its final spin cycle, a front-loading washing machine has a spin rate of 1200 rev/min. Once power is removed, the drum is observed to uniformly decelerate to rest in 25 s. Determine the number of revolutions made during this period as well as the number of revolutions made during the first half of it.

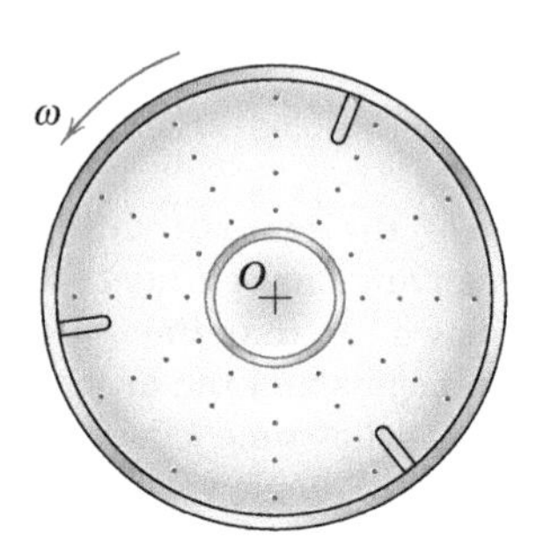

Problem 5/26

5/27 The design characteristics of a gear-reduction unit are under review. Gear B is rotating clockwise with a speed of 300 rev/min when a torque is applied to gear A at time $t = 2$ s to give gear A a counterclockwise angular acceleration α which varies with time for a duration of 4 seconds as shown. Determine the speed N_B of gear B when $t = 6$ s.

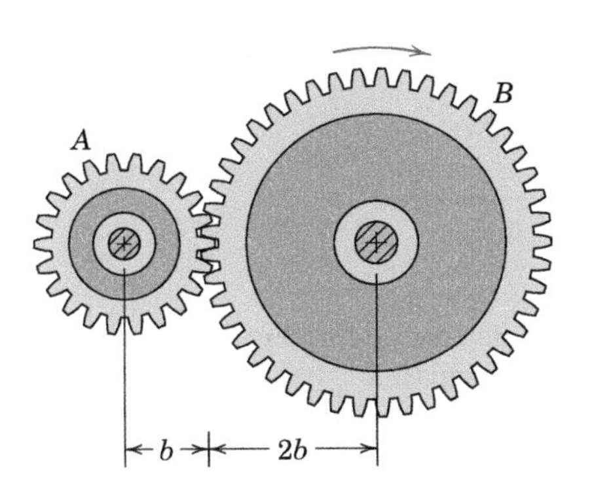

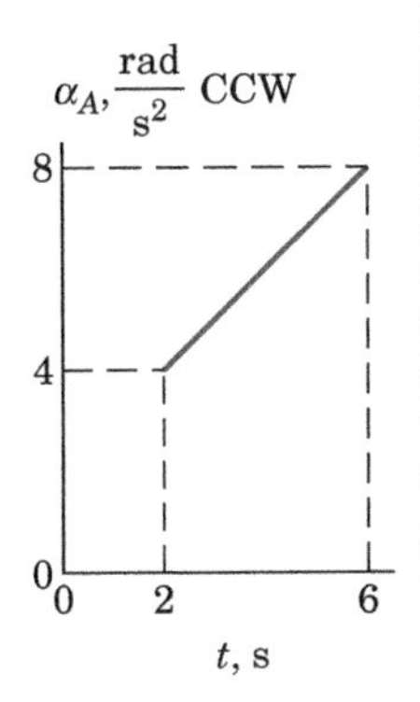

Problem 5/27

▶5/28 A V-belt speed-reduction drive is shown where pulley A drives the two integral pulleys B which in turn drive pulley C. If A starts from rest at time $t = 0$ and is given a constant angular acceleration α_1, derive expressions for the angular velocity of C and the magnitude of the acceleration of a point P on the belt, both at time t.

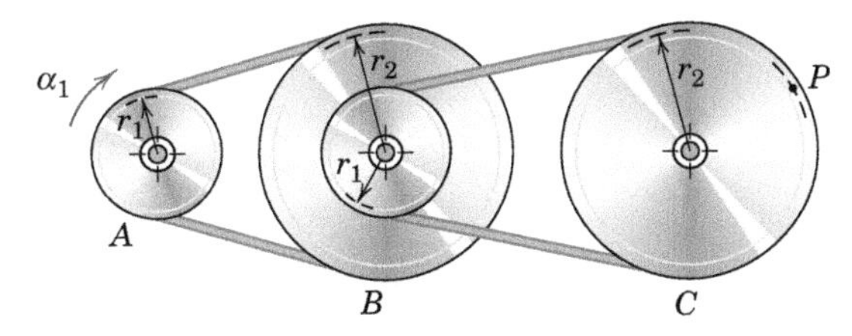

Problem 5/28

PROBLEMS

Introductory Problems

5/29 Slider A moves in the horizontal slot with a constant speed v for a short interval of motion. Determine the angular velocity ω of bar AB in terms of the displacement x_A.

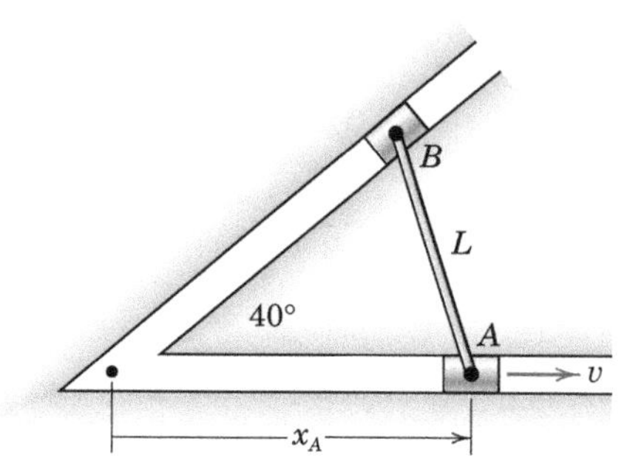

Problem 5/29

5/30 The fixed hydraulic cylinder C imparts a constant upward velocity v to the collar B, which slips freely on rod OA. Determine the resulting angular velocity ω_{OA} in terms of v, the displacement s of point B, and the fixed distance d.

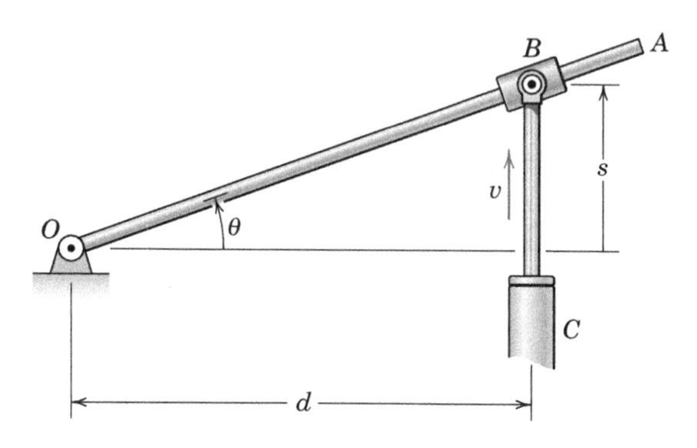

Problem 5/30

5/31 The concrete pier P is being lowered by the pulley and cable arrangement shown. If points A and B have velocities of 0.4 m/s and 0.2 m/s, respectively, compute the velocity of P, the velocity of point C for the instant represented, and the angular velocity of the pulley.

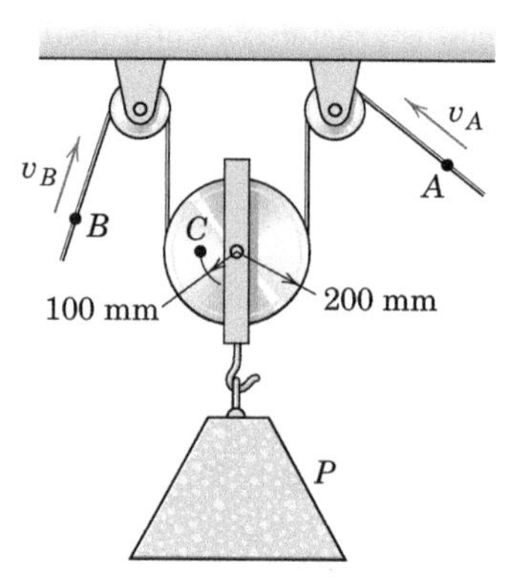

Problem 5/31

5/32 At the instant under consideration, the hydraulic cylinder AB has a length $L = 0.75$ m, and this length is momentarily increasing at a constant rate of 0.2 m/s. If $v_A = 0.6$ m/s and $\theta = 35°$, determine the velocity of slider B.

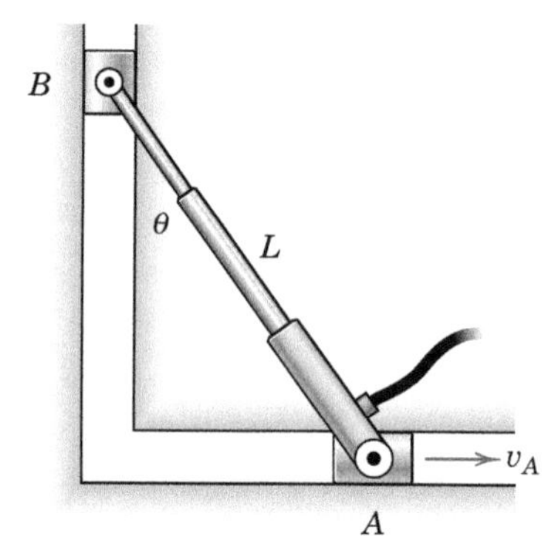

Problem 5/32

5/33 The hydraulic cylinder D is causing the distance OA to increase at the rate of 2 in./sec. Calculate the velocity of the pin at C in its horizontal guide for the instant when $\theta = 50°$.

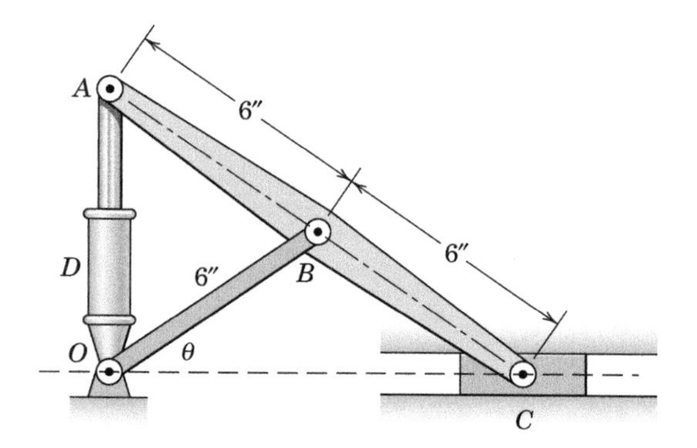

Problem 5/33

5/34 The Scotch-yoke mechanism converts rotational motion of the disk to oscillatory translation of the shaft. For given values of θ, ω, α, r, and d, determine the velocity and acceleration of point P of the shaft.

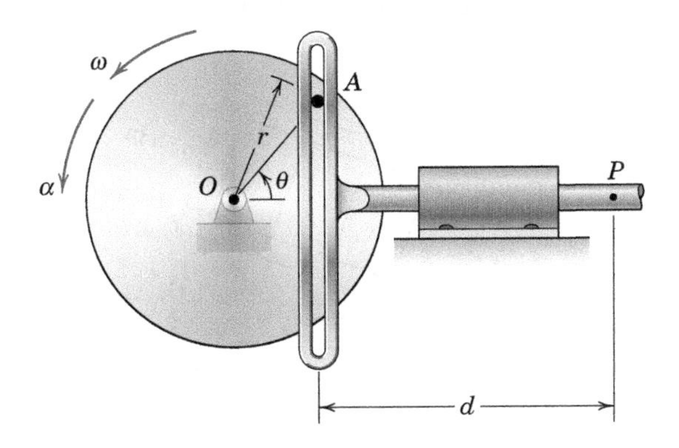

Problem 5/34

5/35 The Scotch-yoke mechanism of Prob. 5/34 is modified as shown in the figure. For given values of ω, α, r, θ, d, and β, determine the velocity and acceleration of point P of the shaft.

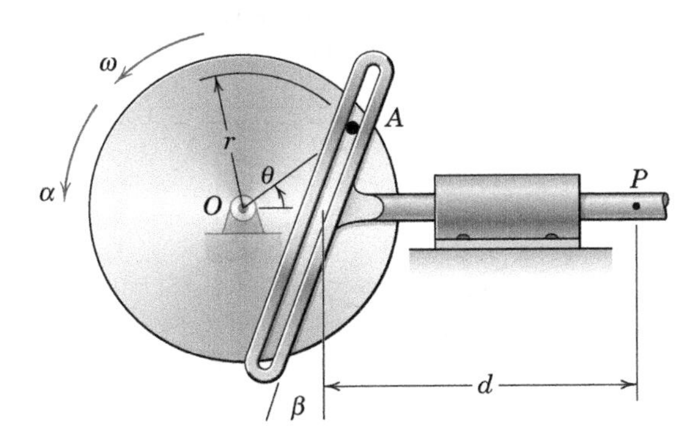

Problem 5/35

5/36 The wheel of radius r rolls without slipping, and its center O has a constant velocity v_O to the right. Determine expressions for the magnitudes of the velocity $\mathbf{v}$ and acceleration $\mathbf{a}$ of point A on the rim by differentiating its x- and y-coordinates. Represent your results graphically as vectors on your sketch and show that $\mathbf{v}$ is the vector sum of two vectors, each of which has a magnitude v_O.

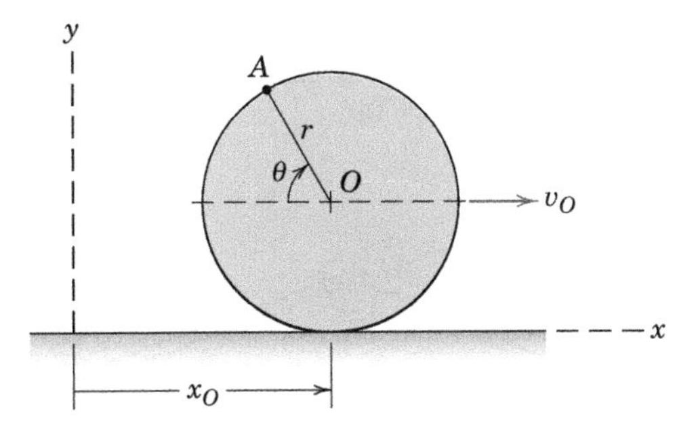

Problem 5/36

5/37 Link OA rotates with a clockwise angular velocity $\omega = 7$ rad/s. Determine the velocity of point B for the position $\theta = 30°$. Use the values $b = 80$ mm, $d = 100$ mm, and $h = 30$ mm.

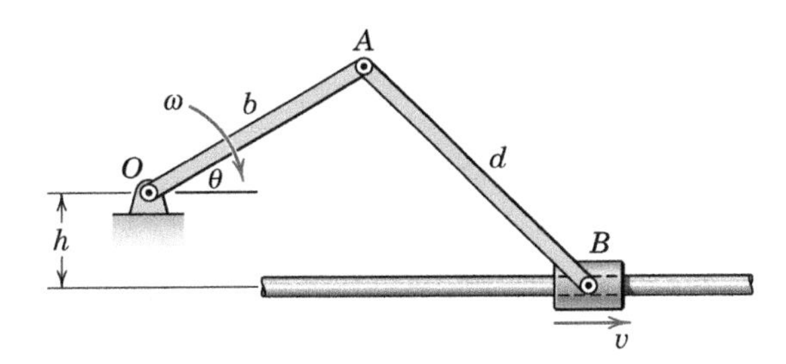

Problem 5/37

5/38 Determine the acceleration of the shaft B for $\theta = 60°$ if the crank OA has an angular acceleration $\ddot{\theta} = 8$ rad/s^2 and an angular velocity $\dot{\theta} = 4$ rad/s at this position. The spring maintains contact between the roller and the surface of the plunger.

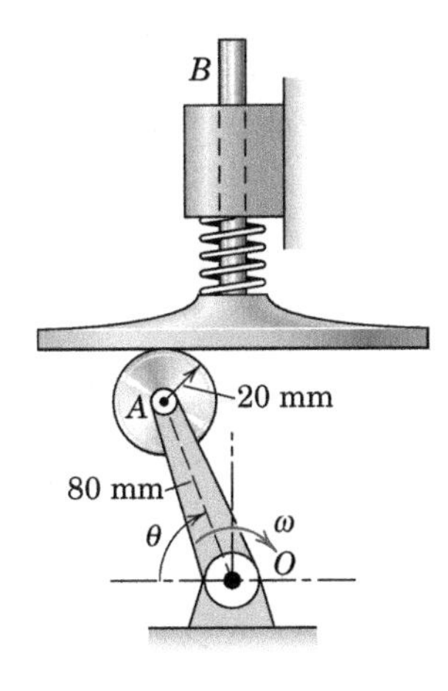

Problem 5/38

5/39 Link OA rotates with a counterclockwise angular velocity $\omega = 3$ rad/s. Determine the angular velocity of bar BC when $\theta = 20°$.

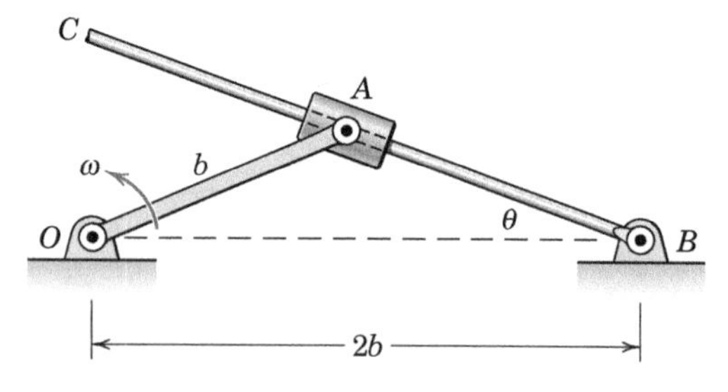

Problem 5/39

5/40 The collar C moves to the left on the fixed guide with speed v. Determine the angular velocity ω_{OA} as a function of v, the collar position s, and the height h.

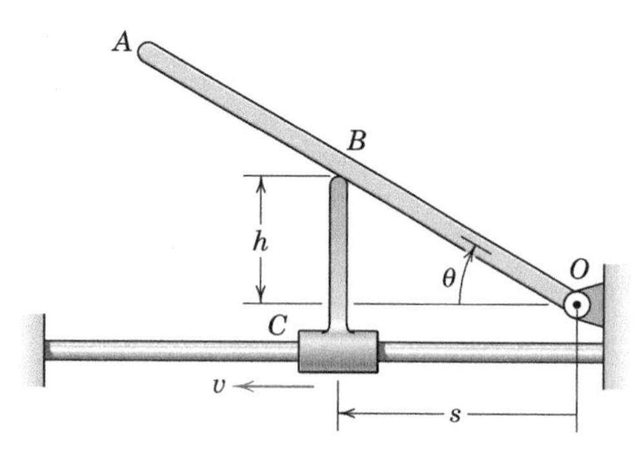

Problem 5/40

Representative Problems

5/41 Boom OA is being elevated by the rope-and-pulley arrangement shown. If point B on the rope is given a constant velocity $v_B = 9$ ft/sec, determine the angular velocity ω and angular acceleration α of the boom for $\theta = 30°$.

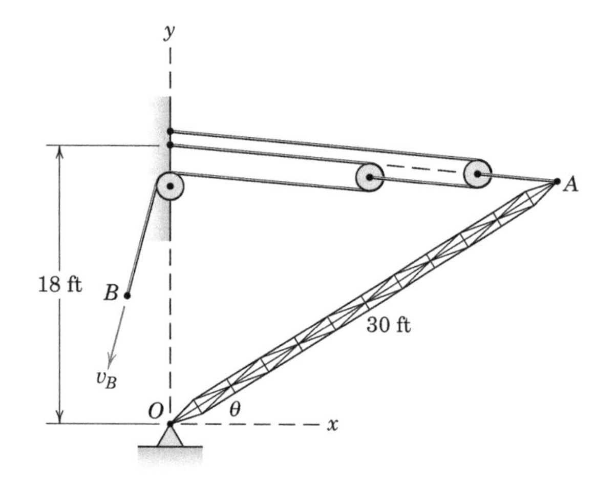

Problem 5/41

5/42 The hydraulic cylinder imparts a constant upward velocity $v_A = 0.2$ m/s to corner A of the rectangular container during an interval of its motion. For the instant when $\theta = 20°$, determine the velocity and acceleration of roller B. Also, determine the corresponding angular velocity of edge CD.

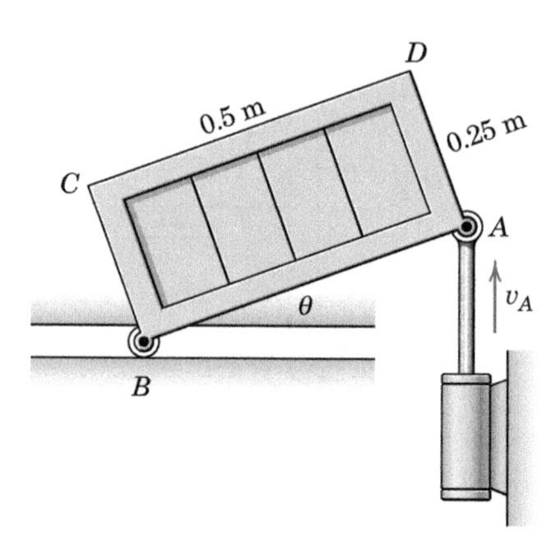

Problem 5/42

5/43 Vertical motion of the work platform is controlled by the horizontal motion of pin A. If A has a velocity v_0 to the left, determine the vertical velocity v of the platform for any value of θ.

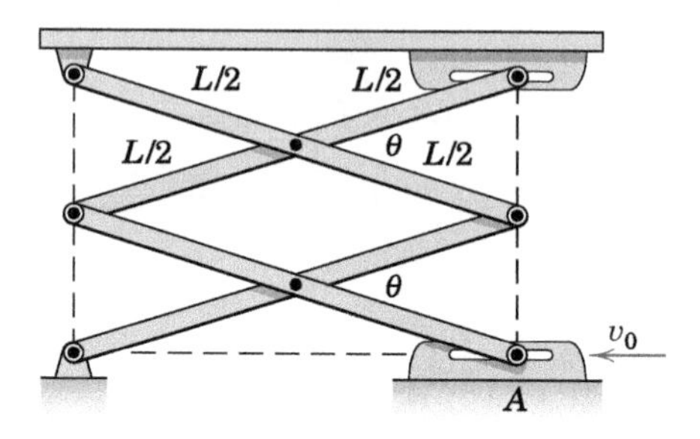

Problem 5/43

5/44 The rod OB slides through the collar pivoted to the rotating link at A. If CA has an angular velocity $\omega = 3$ rad/s for an interval of motion, calculate the angular velocity of OB when $\theta = 45°$.

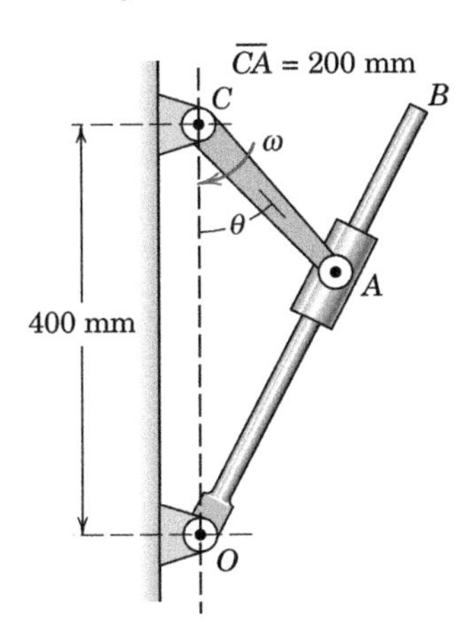

Problem 5/44

5/45 A roadway speed bump is being installed on a level road to remind motorists of the existing speed limit. If the driver of the car experiences at G a vertical acceleration of as much as g, up or down, he is expected to realize that his speed is bordering on being excessive. For the speed bump with the cosine contour shown, derive an expression for the height h of the bump which will produce a vertical component of acceleration at G of g at a car speed v. Compute h if $b = 1$ m and $v = 20$ km/h. Neglect the effects of suspension-spring flexing and finite wheel diameter.

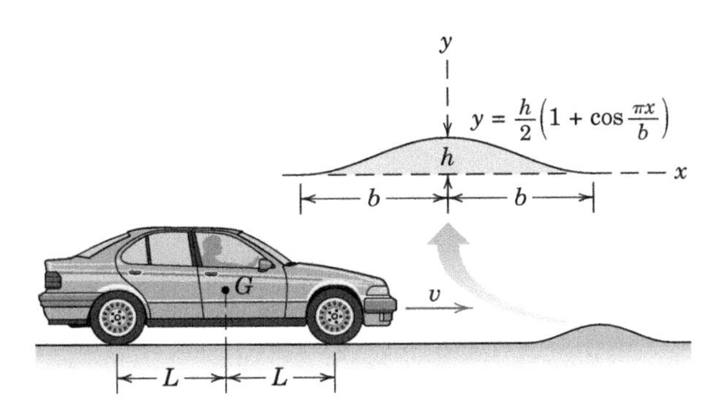

Problem 5/45

5/46 Motion of the wheel as it rolls up the fixed rack on its geared hub is controlled through the peripheral cable by the driving wheel D, which turns counterclockwise at the constant rate $\omega_0 = 4$ rad/s for a short interval of motion. By examining the geometry of a small (differential) rotation of line $AOCB$ as it pivots momentarily about the contact point C, determine the angular velocity ω of the wheel and the velocities of point A and the center O. Also find the acceleration of point C.

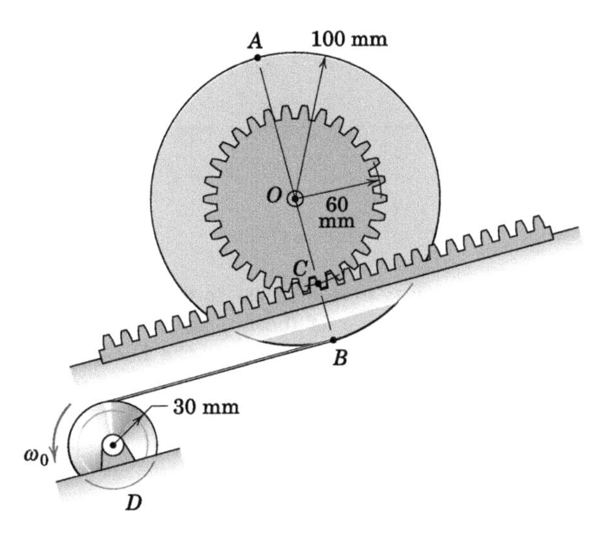

Problem 5/46

5/47 Link OA is given a clockwise angular velocity $\omega = 2$ rad/sec as indicated. Determine the velocity v of point C for the position $\theta = 30°$ if $b = 8$ in.

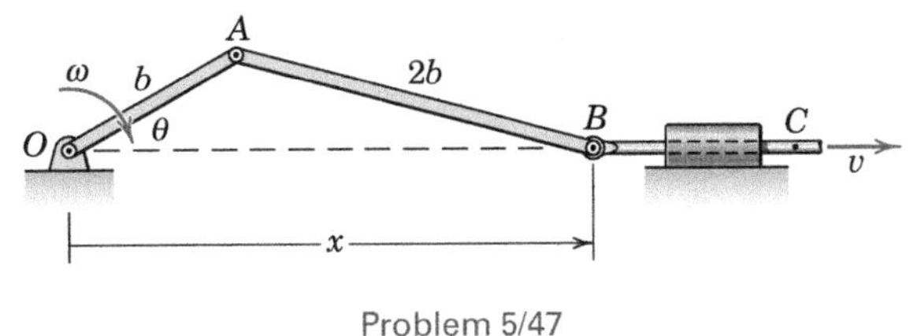

Problem 5/47

5/48 Determine the acceleration of point C of the previous problem if the clockwise angular velocity of link OA is constant at $\omega = 2$ rad/sec.

5/49 Derive an expression for the upward velocity v of the car hoist in terms of θ. The piston rod of the hydraulic cylinder is extending at the rate $\dot{s}$.

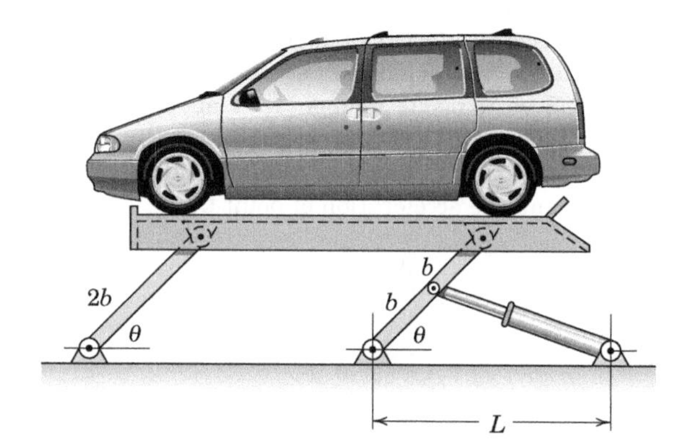

Problem 5/49

5/50 It is desired to design a system for controlling the rate of extension $\dot{x}$ of the fire-truck ladder during elevation of the ladder so that the bucket B will have vertical motion only. Determine $\dot{x}$ in terms of the elongation rate $\dot{c}$ of the hydraulic cylinder for given values of θ and x.

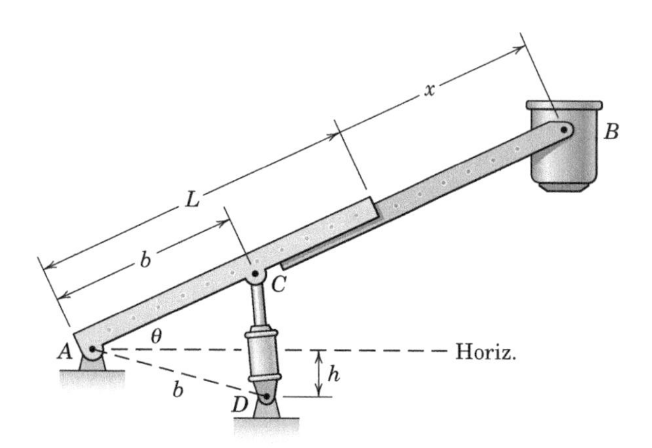

Problem 5/50

5/51 Show that the expressions $v = r\omega$ and $a_t = r\alpha$ hold for the motion of the center O of the wheel which rolls on the concave or convex circular arc, where ω and α are the absolute angular velocity and acceleration, respectively, of the wheel. (*Hint:* Follow the example of Sample Problem 5/4 and allow the wheel to roll a small distance. Be very careful to identify the correct *absolute* angle through which the wheel turns in each case in determining its angular velocity and angular acceleration.)

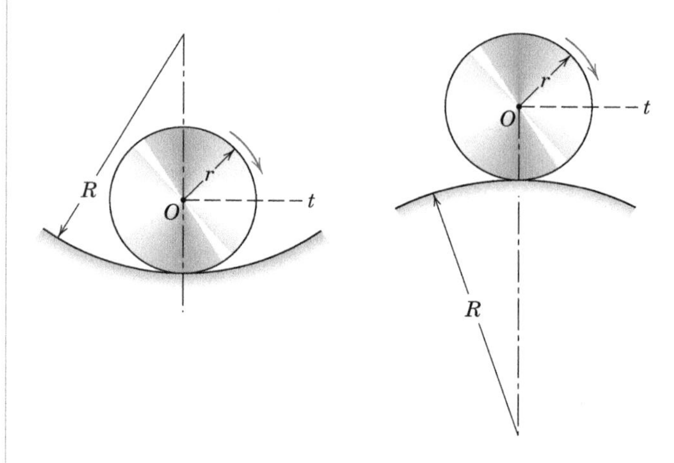

Problem 5/51

5/52 The rotation of link AO is controlled by the piston rod of hydraulic cylinder BC, which is elongating at the constant rate $\dot{s} = k$ for an interval of motion. Write the vector expression for the acceleration of end A for a given value of θ using unit vectors $\mathbf{e}_n$ and $\mathbf{e}_t$ with $n-t$ coordinates.

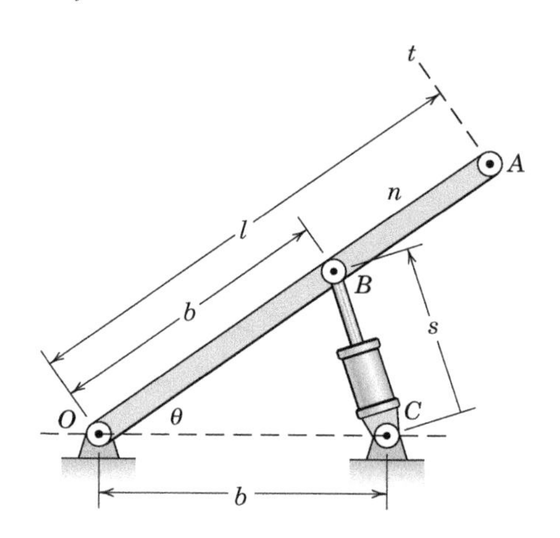

Problem 5/52

5/53 A variable-speed belt drive consists of the two pulleys, each of which is constructed of two cones which turn as a unit but are capable of being drawn together or separated so as to change the effective radius of the pulley. If the angular velocity ω_1 of pulley 1 is constant, determine the expression for the angular acceleration $\alpha_2 = \dot{\omega}_2$ of pulley 2 in terms of the rates of change $\dot{r}_1$ and $\dot{r}_2$ of the effective radii.

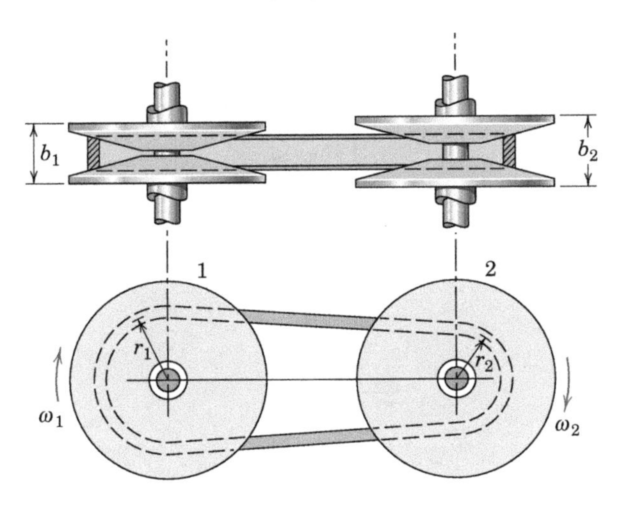

Problem 5/53

5/54 Angular oscillation of the slotted link is achieved by the crank OA, which rotates clockwise at the steady speed $N = 120$ rev/min. Determine an expression for the angular velocity $\dot{\beta}$ of the slotted link in terms of θ.

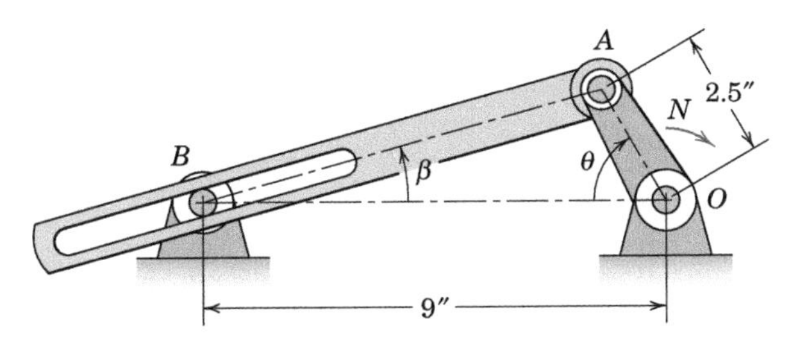

Problem 5/54

5/55 The two gears form an integral unit and roll on the fixed rack. The large gear has 48 teeth, and the worm turns with a speed of 120 rev/min. Find the velocity v_O of the center O of the gear.

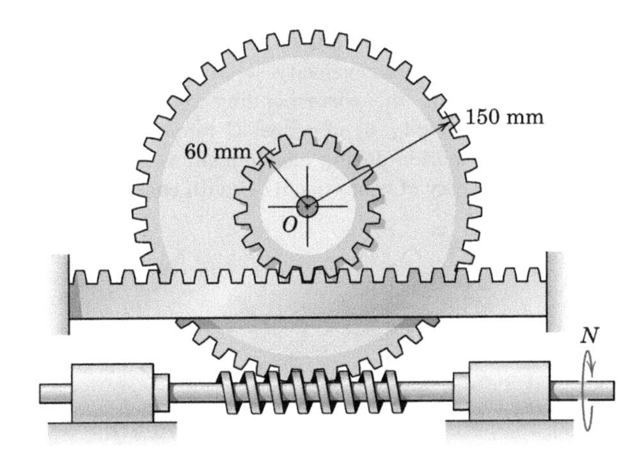

Problem 5/55

▶5/56 One of the most common mechanisms is the slider-crank. Express the angular velocity ω_{AB} and angular acceleration α_{AB} of the connecting rod AB in terms of the crank angle θ for a given constant crank speed ω_0. Take ω_{AB} and α_{AB} to be positive counterclockwise.

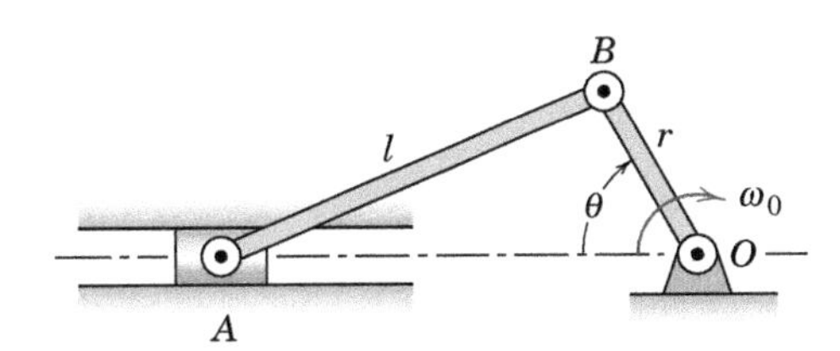

Problem 5/56

▶**5/57** The Geneva wheel is a mechanism for producing intermittent rotation. Pin P in the integral unit of wheel A and locking plate B engages the radial slots in wheel C, thus turning wheel C one-fourth of a revolution for each revolution of the pin. At the engagement position shown, $\theta = 45°$. For a constant clockwise angular velocity $\omega_1 = 2$ rad/s of wheel A, determine the corresponding counterclockwise angular velocity ω_2 of wheel C for $\theta = 20°$. (Note that the motion during engagement is governed by the geometry of triangle O_1O_2P with changing θ.)

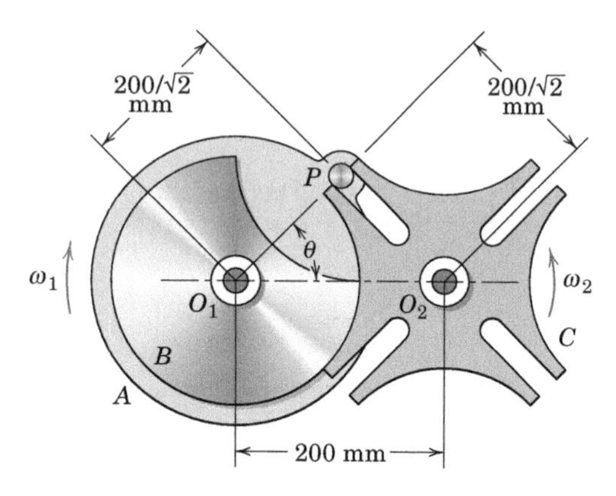

Problem 5/57

▶**5/58** The punch is operated by a simple harmonic oscillation of the pivoted sector given by $\theta = \theta_0 \sin 2\pi t$ where the amplitude is $\theta_0 = \pi/12$ rad (15°) and the time for one complete oscillation is 1 second. Determine the acceleration of the punch when (*a*) $\theta = 0$ and (*b*) $\theta = \pi/12$.

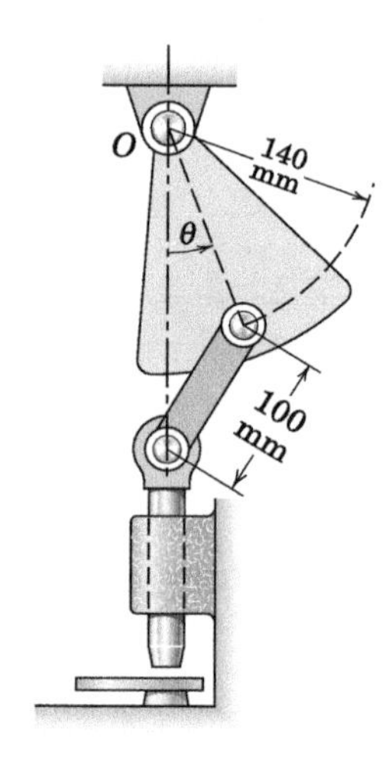

Problem 5/58

PROBLEMS

Introductory Problems

5/59 The right-angle link AB has a clockwise angular velocity $\omega = 2$ rad/sec when in the position shown. Determine the velocity of B with respect to A for this instant.

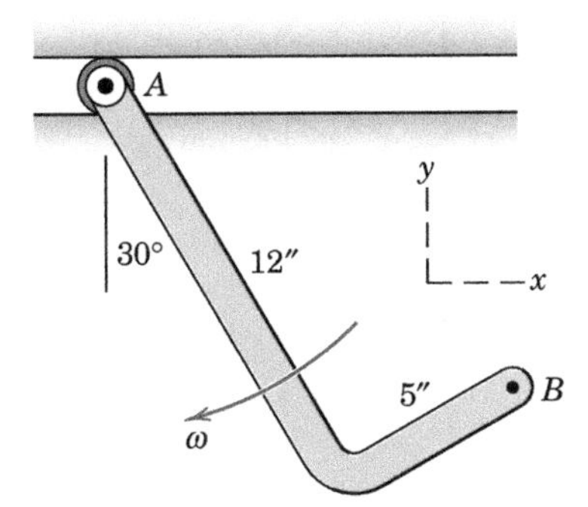

Problem 5/59

5/60 The uniform rectangular plate moves on the horizontal surface. Its mass center has a velocity $v_G =$ 3 m/s directed parallel to the x-axis and the plate has a counterclockwise (as seen from above) angular velocity $\omega = 4$ rad/s. Determine the velocities of points A and B.

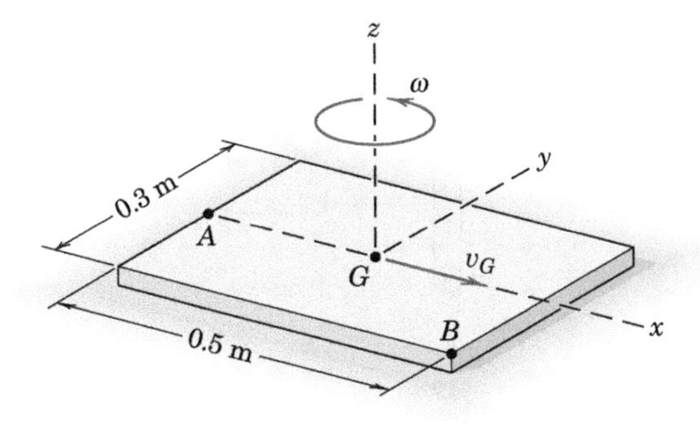

Problem 5/60

5/61 The cart has a velocity of 4 ft/sec to the right. Determine the angular speed N of the wheel so that point A on the top of the rim has a velocity (a) equal to 4 ft/sec to the left, (b) equal to zero, and (c) equal to 8 ft/sec to the right.

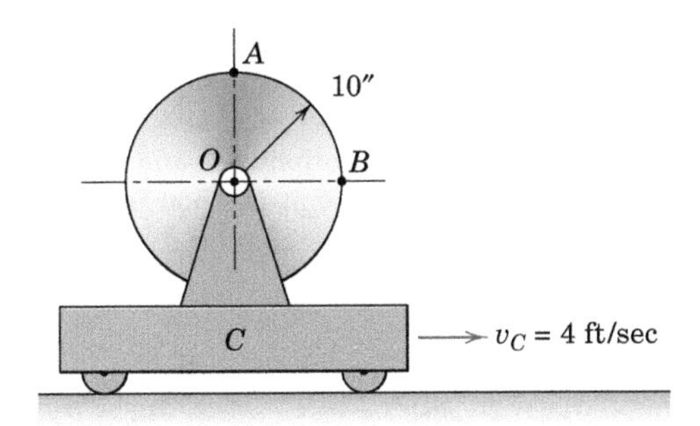

Problem 5/61

5/62 End A of the 2-ft link has a velocity of 5 ft/sec in the direction shown. At the same instant, end B has a velocity whose magnitude is 7 ft/sec as indicated. Find the angular velocity ω of the link in two ways.

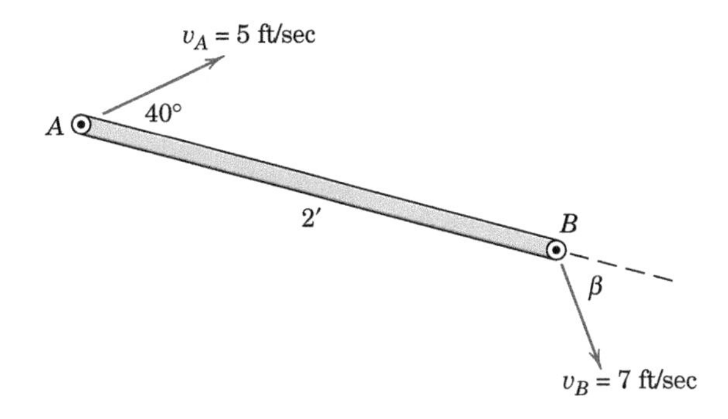

Problem 5/62

5/63 The speed of the center of the earth as it orbits the sun is $v = 107\,257$ km/h, and the absolute angular velocity of the earth about its north–south spin axis is $\omega = 7.292(10^{-5})$ rad/s. Use the value $R = 6371$ km for the radius of the earth and determine the velocities of points A, B, C, and D, all of which are on the equator. The inclination of the axis of the earth is neglected.

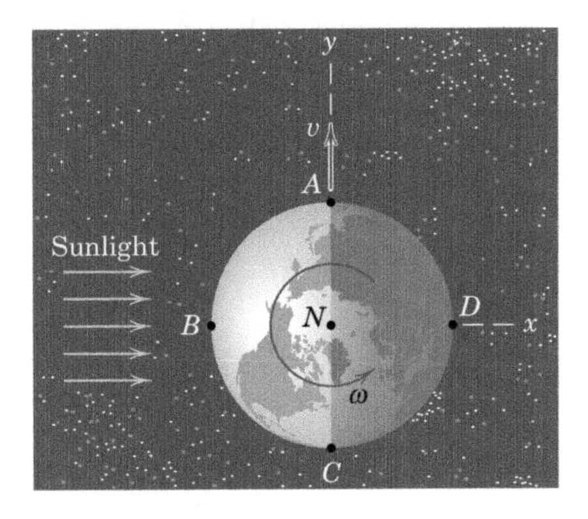

Problem 5/63

5/64 The center C of the smaller wheel has a velocity $v_C = 0.4$ m/s in the direction shown. The cord which connects the two wheels is securely wrapped around the respective peripheries and does not slip. Calculate the speed of point D when in the position shown. Also compute the change Δx which occurs per second if v_C is constant.

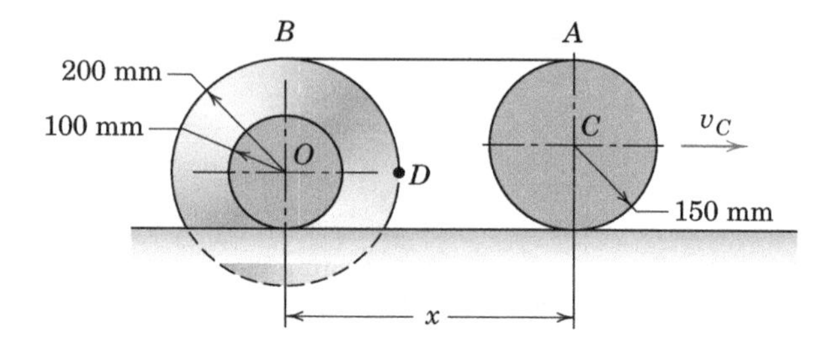

Problem 5/64

5/65 The circular disk of radius 8 in. is released very near the horizontal surface with a velocity of its center $v_O = 27$ in./sec to the right and a clockwise angular velocity $\omega = 2$ rad/sec. Determine the velocities of points A and P of the disk. Describe the motion upon contact with the ground.

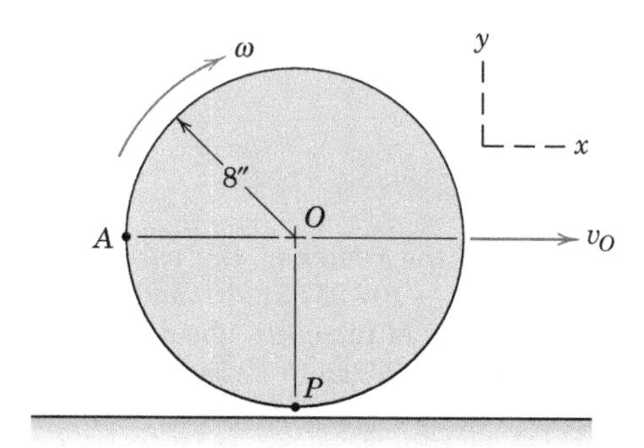

Problem 5/65

5/66 For a short interval, collars A and B are sliding along the fixed vertical shaft with velocities $v_A = 2$ m/s and $v_B = 3$ m/s in the directions shown. Determine the magnitude of the velocity of point C for the position $\theta = 60°$.

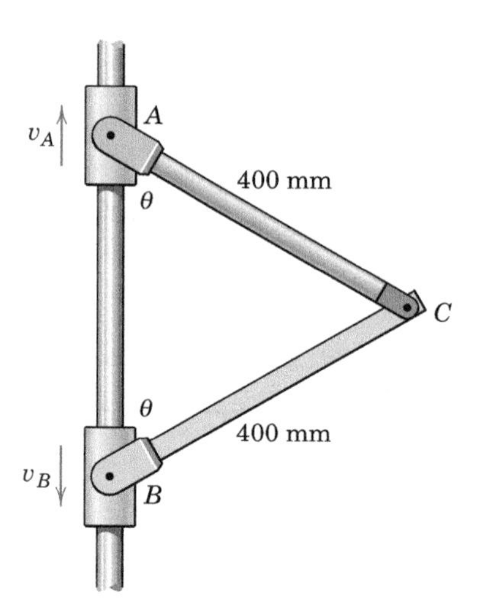

Problem 5/66

5/67 The right-angle link has a counterclockwise angular velocity of 3 rad/s at the instant represented, and point B has a velocity $\mathbf{v}_B = 2\mathbf{i} - 0.3\mathbf{j}$ m/s. Determine the velocity of A using vector notation. Sketch the vector polygon which corresponds to the terms in the relative-velocity equation and estimate or measure the magnitude of $\mathbf{v}_A$.

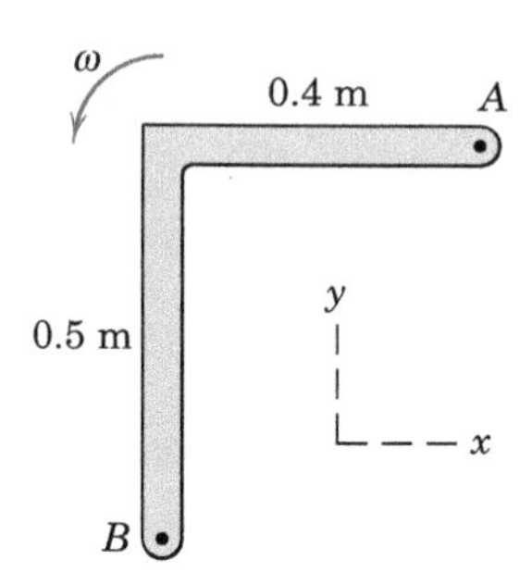

Problem 5/67

5/68 The magnitude of the absolute velocity of point A on the automobile tire is 12 m/s when A is in the position shown. What are the corresponding velocity v_O of the car and the angular velocity ω of the wheel? (The wheel rolls without slipping.)

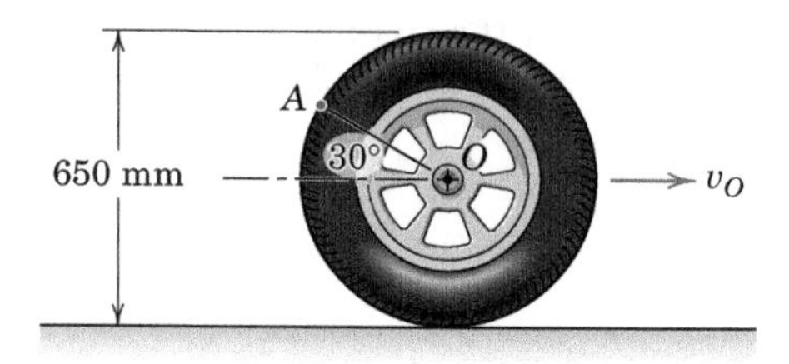

Problem 5/68

5/69 The two pulleys are riveted together to form a single rigid unit, and each of the two cables is securely wrapped around its respective pulley. If point A on the hoisting cable has a velocity $v = 3$ ft/sec, determine the magnitudes of the velocity of point O and the velocity of point B on the larger pulley for the position shown.

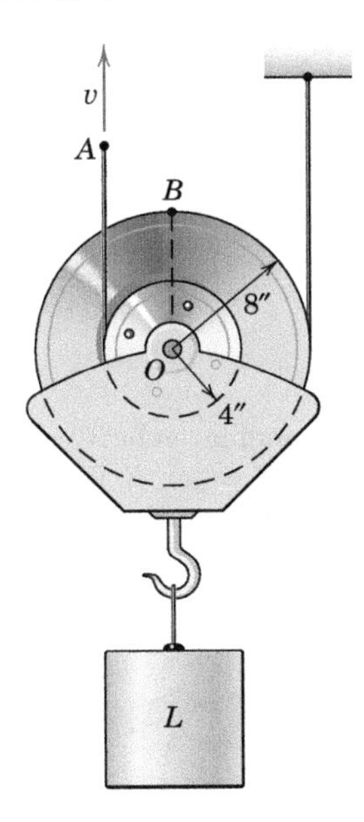

Problem 5/69

5/70 The rider of the bicycle shown pumps steadily to maintain a constant speed of 16 km/h against a slight head wind. Calculate the maximum and minimum magnitudes of the absolute velocity of the pedal A.

Problem 5/70

5/71 Determine the angular velocity of bar AB just after roller B has begun moving up the 15° incline. At the instant under consideration, the velocity of roller A is v_A.

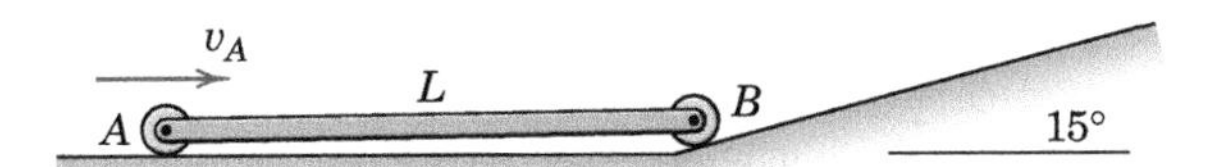

Problem 5/71

5/72 For the instant represented, point B crosses the horizontal axis through point O with a downward velocity $v = 0.6$ m/s. Determine the corresponding value of the angular velocity ω_{OA} of link OA.

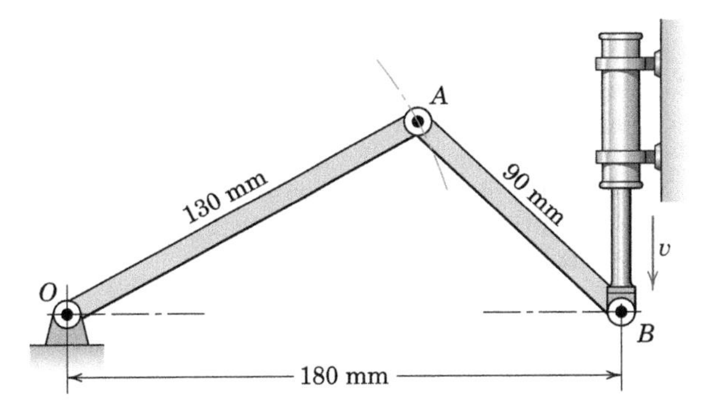

Problem 5/72

Representative Problems

5/73 The spoked wheel of radius r is made to roll up the incline by the cord wrapped securely around a shallow groove on its outer rim. For a given cord speed v at point P, determine the velocities of points A and B. No slipping occurs.

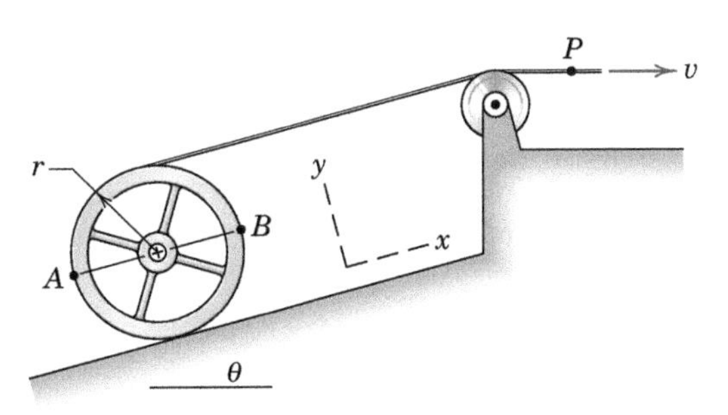

Problem 5/73

5/74 At the instant represented, the velocity of point A of the 1.2-m bar is 3 m/s to the right. Determine the speed v_B of point B and the angular velocity ω of the bar. The diameter of the small end wheels may be neglected.

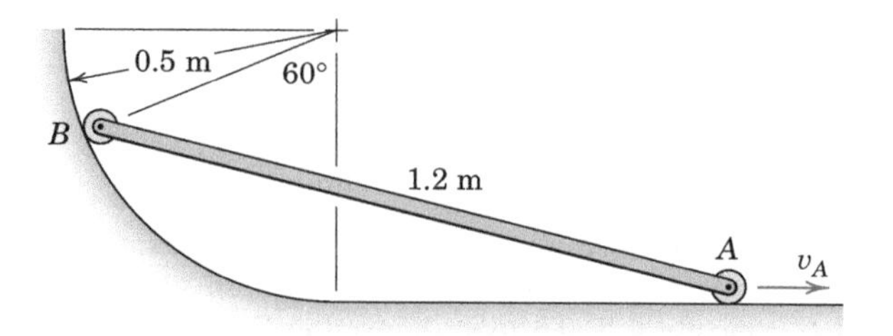

Problem 5/74

5/75 Determine the angular velocity of link BC for the instant indicated. In case (a), the center O of the disk is a fixed pivot, while in case (b), the disk rolls without slipping on the horizontal surface. In both cases, the disk has clockwise angular velocity ω. Neglect the small distance of pin A from the edge of the disk.

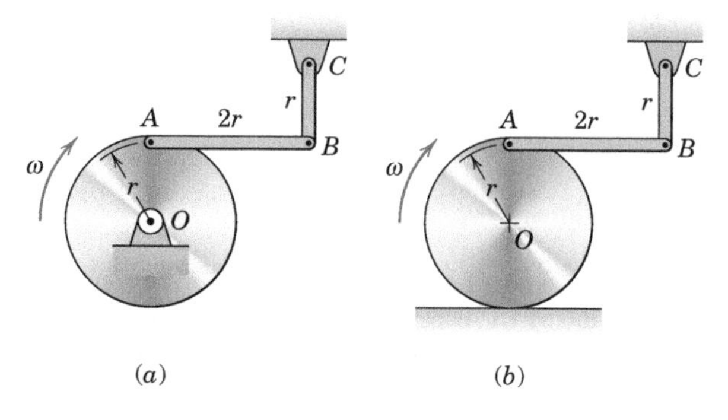

Problem 5/75

5/76 The elements of a switching device are shown. If the vertical control rod has a downward velocity v = 2 ft/sec when the device is in the position shown, determine the corresponding speed of point A. Roller C is in continuous contact with the inclined surface.

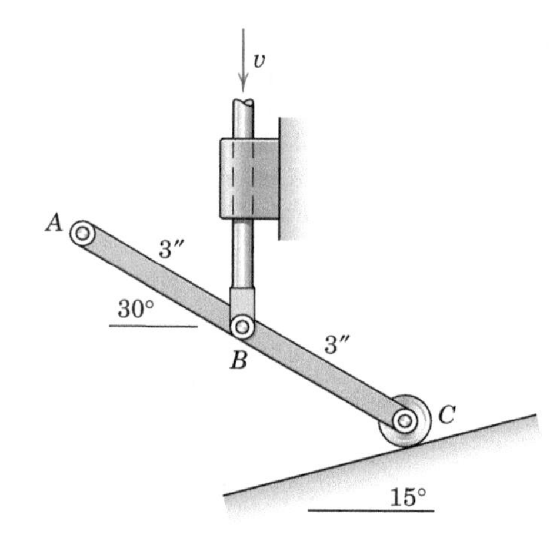

Problem 5/76

5/77 Determine the angular velocity ω_{AB} of link AB and the velocity v_B of collar B for the instant represented. Assume the quantities ω_0 and r to be known.

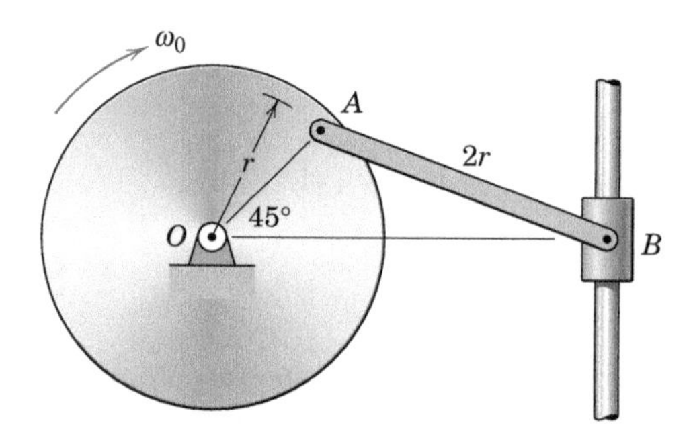

Problem 5/77

5/78 Determine the angular velocity ω_{AB} of link AB and the velocity v_B of collar B for the instant represented. Assume the quantities ω_0 and r to be known.

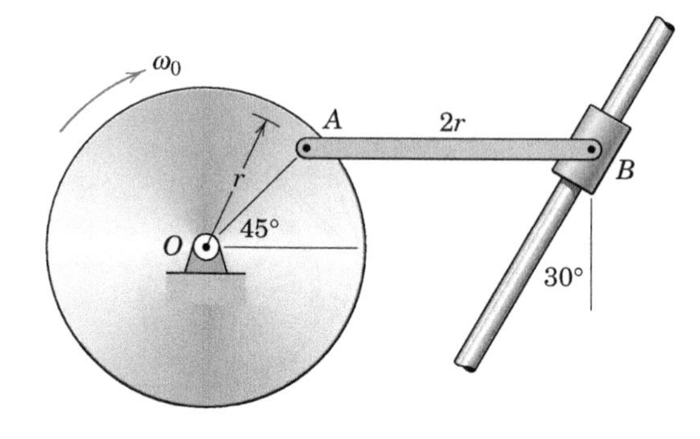

Problem 5/78

5/79 The rotation of the gear is controlled by the horizontal motion of end A of the rack AB. If the piston rod has a constant velocity $\dot{x} = 300$ mm/s during a short interval of motion, determine the angular velocity ω_0 of the gear and the angular velocity ω_{AB} of AB at the instant when $x = 800$ mm.

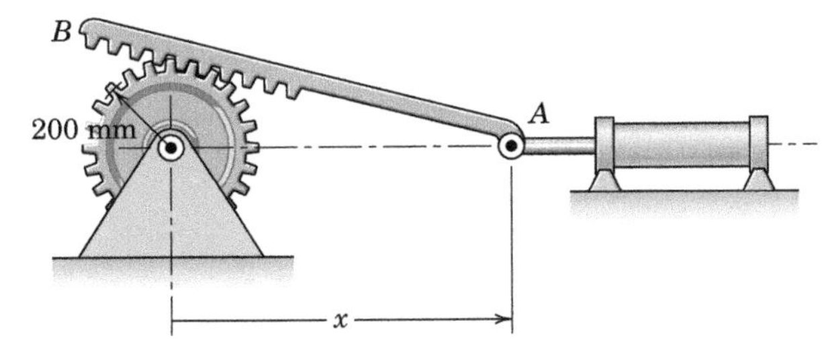

Problem 5/79

5/80 Motion of the rectangular plate P is controlled by the two links which cross without touching. For the instant represented where the links are perpendicular to each other, the plate has a counterclockwise angular velocity $\omega_P = 2$ rad/s. Determine the corresponding angular velocities of the two links.

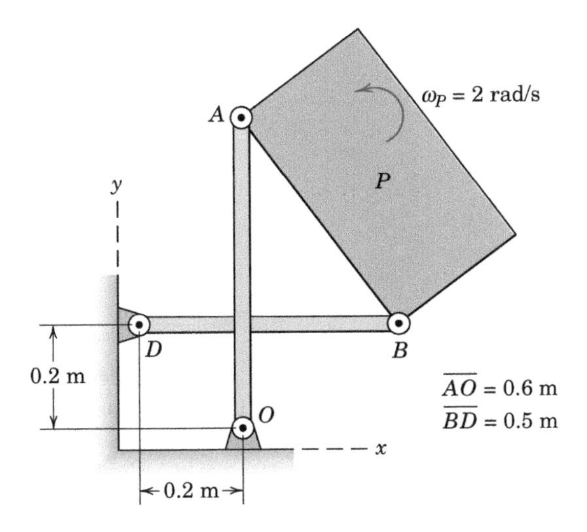

Problem 5/80

5/81 The elements of a simplified clam-shell bucket for a dredge are shown. The cable which opens and closes the bucket passes through the block at O. With O as a fixed point, determine the angular velocity ω of the bucket jaws when $\theta = 45°$ as they are closing. The upward velocity of the control cable is 0.5 m/s as it passes through the block.

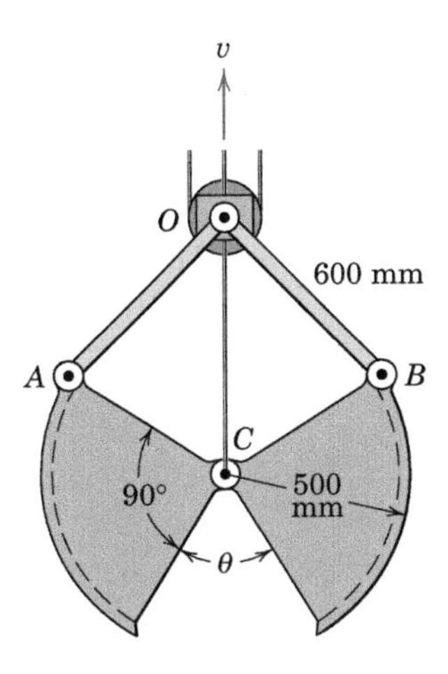

Problem 5/81

5/82 The ends of the 0.4-m slender bar remain in contact with their respective support surfaces. If end B has a velocity $v_B = 0.5$ m/s in the direction shown, determine the angular velocity of the bar and the velocity of end A.

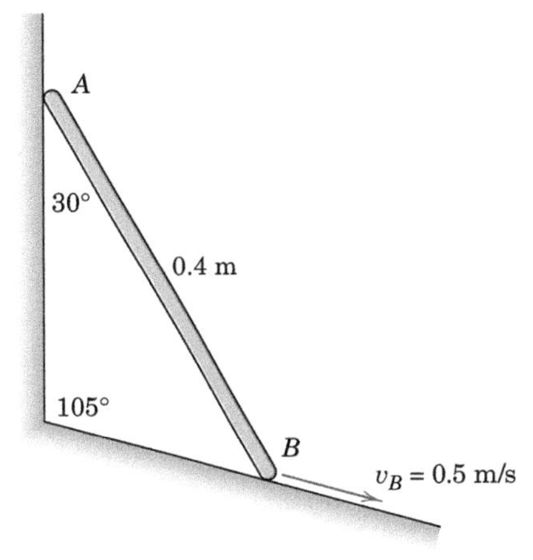

Problem 5/82

5/83 In the design of a produce-processing plant, roller trays of produce are to be oscillated under water spray by the action of the connecting link AB and crank OB. For the instant when $\theta = 15°$, the angular velocity of AB is 0.086 rad/s clockwise. Find the corresponding angular velocity $\dot{\theta}$ of the crank and the velocity v_A of the tray. Solve the relative-velocity equation by either vector algebra or vector geometry.

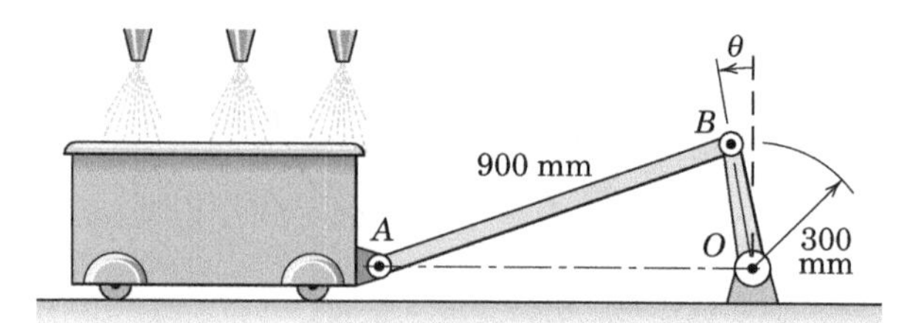

Problem 5/83

5/84 The vertical rod has a downward velocity $v = 0.8$ m/s when link AB is in the 30° position shown. Determine the corresponding angular velocity of AB and the speed of roller B if $R = 0.4$ m.

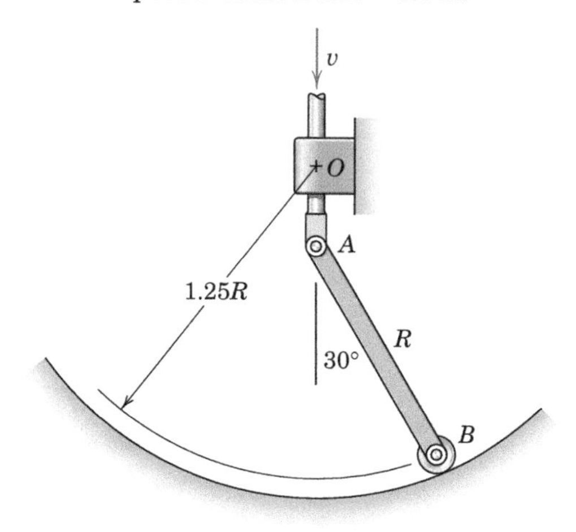

Problem 5/84

5/85 Pin P on the end of the horizontal rod slides freely in the slotted gear. The gear engages the moving rack A and the fixed rack B (teeth not shown) so it rolls without slipping. If A has a velocity of 0.4 ft/sec to the left for the instant shown, determine the velocity v_P of the rod for this position.

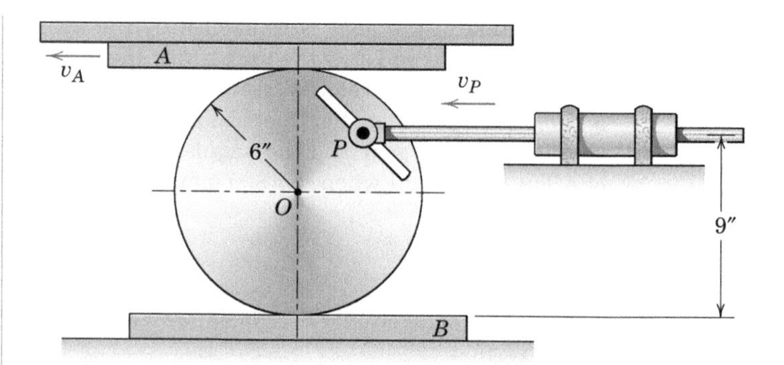

Problem 5/85

5/86 A four-bar linkage is shown in the figure (the ground "link" OC is considered the fourth bar). If the drive link OA has a counterclockwise angular velocity $\omega_0 = 10$ rad/s, determine the angular velocities of links AB and BC.

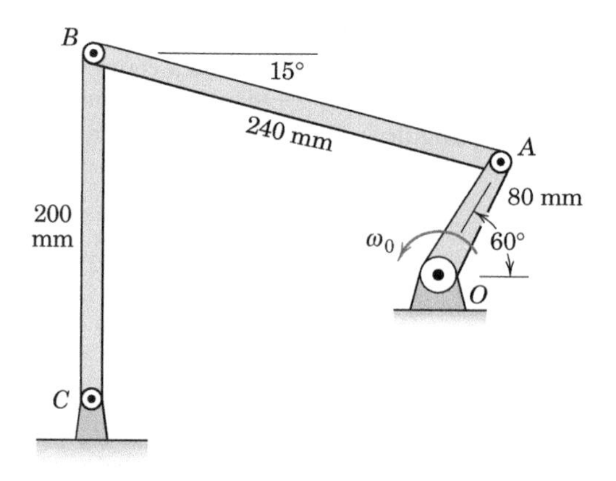

Problem 5/86

5/87 The mechanism is part of a latching device where rotation of link AOB is controlled by the rotation of slotted link D about C. If member D has a clockwise angular velocity of 1.5 rad/s when the slot is parallel to OC, determine the corresponding angular velocity of AOB. Solve graphically or geometrically.

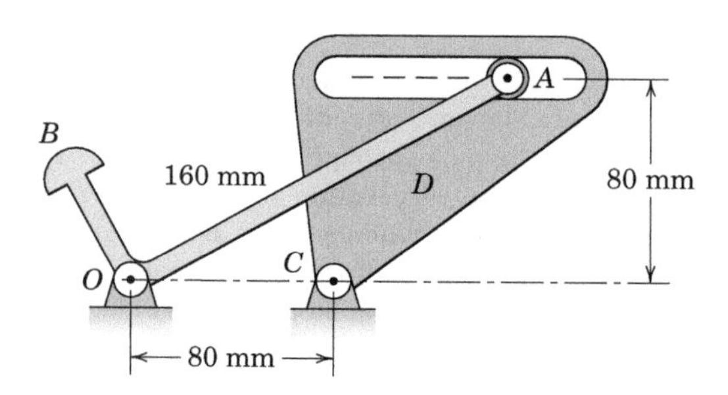

Problem 5/87

5/88 The elements of the mechanism for deployment of a spacecraft magnetometer boom are shown. Determine the angular velocity of the boom when the driving link OB crosses the y-axis with an angular velocity $\omega_{OB} = 0.5$ rad/sec if $\tan \theta = 4/3$ at this instant.

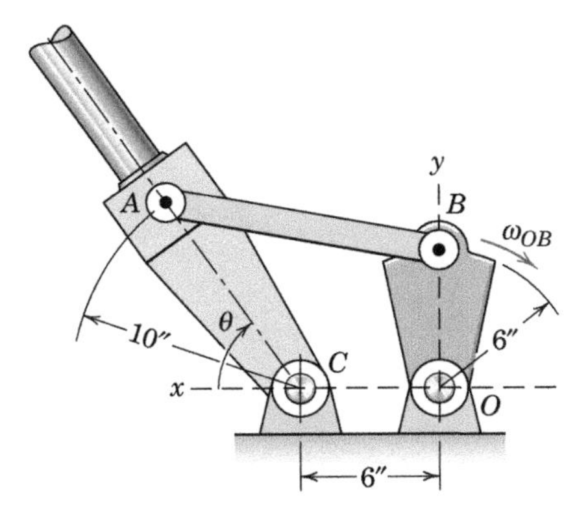

Problem 5/88

5/89 A mechanism for pushing small boxes from an assembly line onto a conveyor belt is shown with arm OD and crank CB in their vertical positions. The crank revolves clockwise at a constant rate of 1 revolution every 2 seconds. For the position shown, determine the speed at which the box is being shoved horizontally onto the conveyor belt.

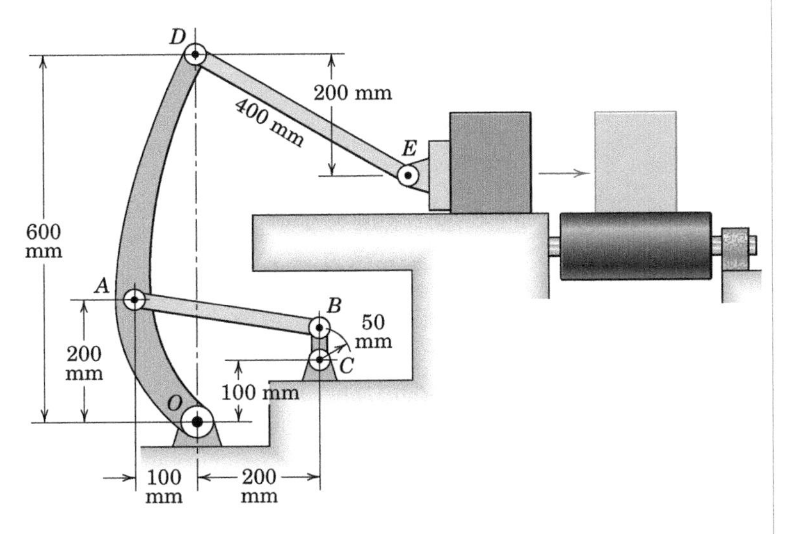

Problem 5/89

▶**5/90** The wheel rolls without slipping. For the instant portrayed, when O is directly under point C, link OA has a velocity $v = 1.5$ m/s to the right and $\theta = 30°$. Determine the angular velocity ω of the slotted link.

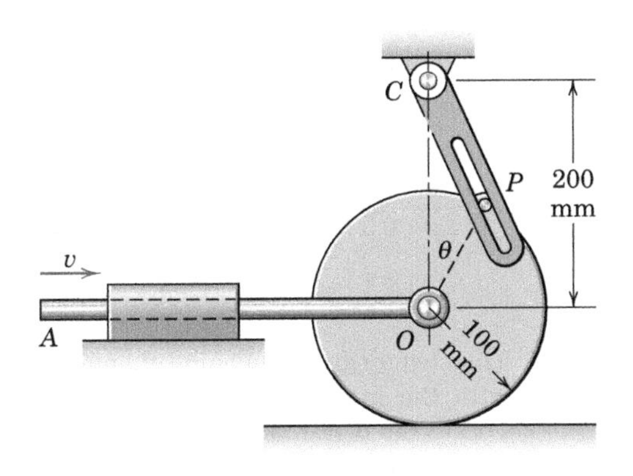

Problem 5/90

PROBLEMS

Introductory Problems

5/91 The slender bar is moving in general plane motion with the indicated linear and angular properties. Locate the instantaneous center of zero velocity and determine the speeds of points A and B.

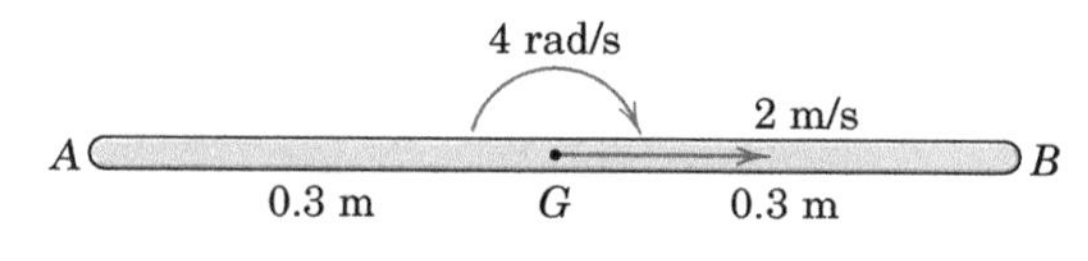

Problem 5/91

5/92 The slender bar is moving in general plane motion with the indicated linear and angular properties. Locate the instantaneous center of zero velocity and determine the velocities of points A and B.

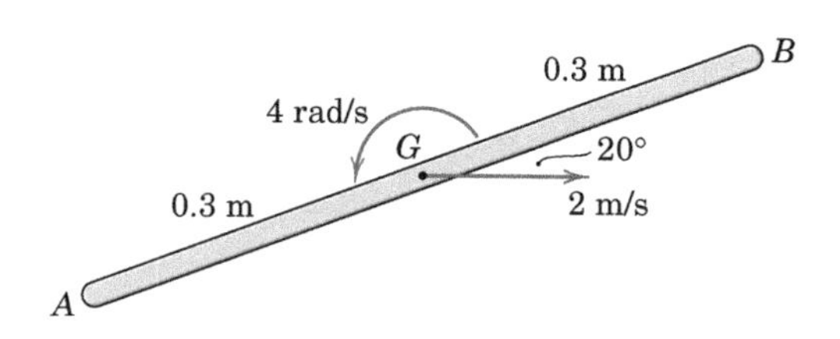

Problem 5/92

5/93 For the instant represented, corner A of the rectangular plate has a velocity $v_A = 2.8$ m/s and the plate has a clockwise angular velocity $\omega = 12$ rad/s. Determine the magnitude of the corresponding velocity of point B.

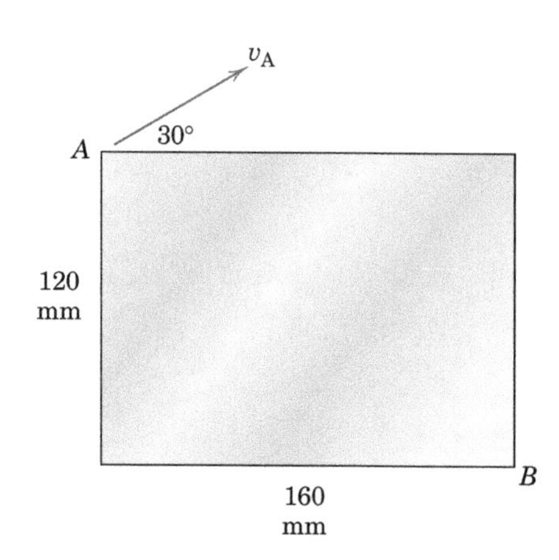

Problem 5/93

5/94 Roller B of the quarter-circular link has a velocity $v_B = 3$ ft/sec directed down the 15° incline. The link has a counterclockwise angular velocity $\omega = 2$ rad/sec. By the method of this article, determine the velocity of roller A.

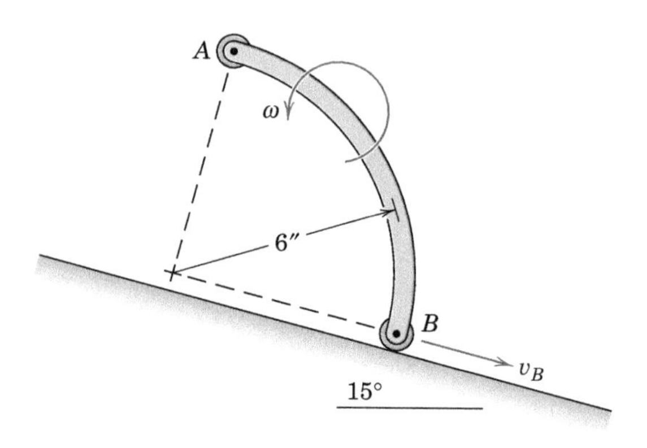

Problem 5/94

5/95 The bar of Prob. 5/82 is repeated here. By the method of this article, determine the velocity of end A. Both ends remain in contact with their respective support surfaces.

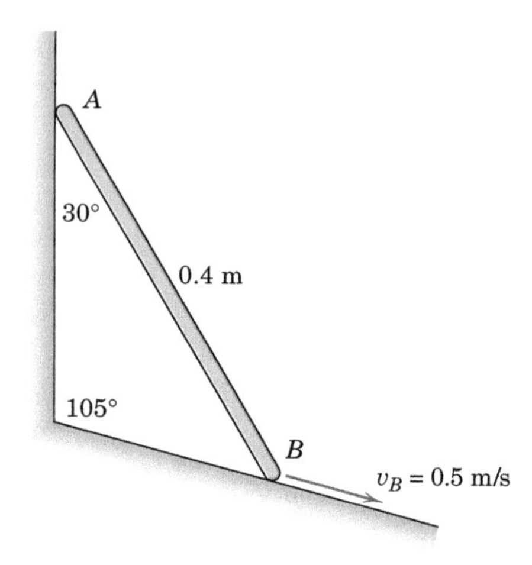

Problem 5/95

5/96 The bar AB has a counterclockwise angular velocity of 6 rad/sec. Construct the velocity vectors for points A and G of the bar and specify their magnitudes if the instantaneous center of zero velocity for the bar is (*a*) at C_1, (*b*) at C_2, and (*c*) at C_3.

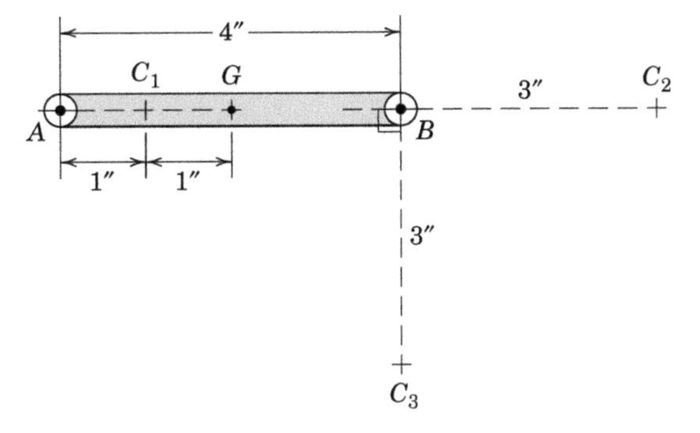

Problem 5/96

5/97 A car mechanic "walks" two wheel/tire units across a horizontal floor as shown. He walks with constant speed v and keeps the tires in the configuration shown with the same position relative to his body. If there is no slipping at any interface, determine (*a*) the angular velocity of the lower tire, (*b*) the angular velocity of the upper tire, and (*c*) the velocities of points A, B, C, and D. The radius of both tires is r.

Problem 5/97

5/98 At a certain instant vertex B of the right-triangular plate has a velocity of 200 mm/s in the direction shown. If the instantaneous center of zero velocity for the plate is 40 mm from point B and if the angular velocity of the plate is clockwise, determine the speed of point D.

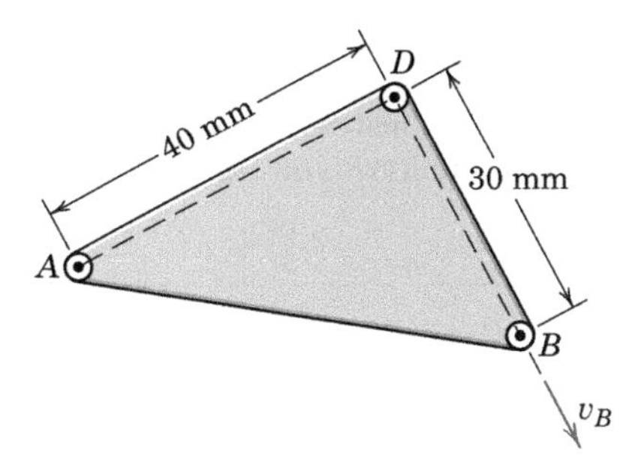

Problem 5/98

5/99 At the instant represented, crank OB has a clockwise angular velocity $\omega = 0.8$ rad/sec and is passing the horizontal position. By the method of this article, determine the corresponding speed of the guide roller A in the 20° slot and the speed of point C midway between A and B.

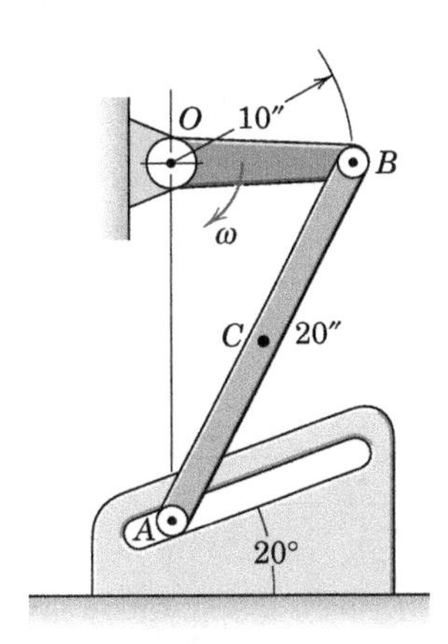

Problem 5/99

5/100 Crank OA rotates with a counterclockwise angular velocity of 9 rad/s. By the method of this article, determine the angular velocity ω of link AB and the velocity of roller B for the position illustrated. Also, find the velocity of the center G of link AB.

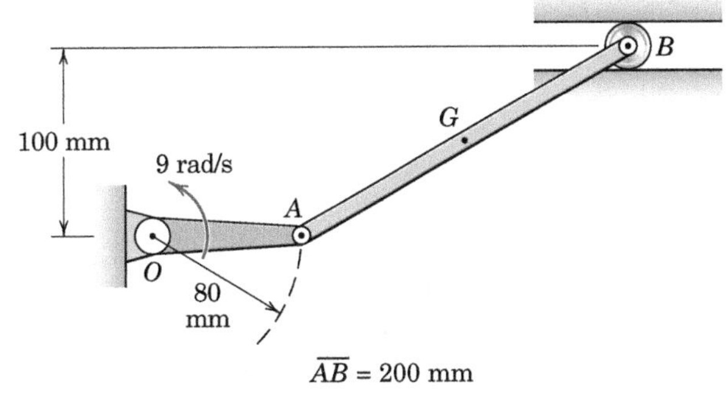

Problem 5/100

5/101 The mechanism of Prob. 5/100 is now shown in a different position, with the crank OA 30° below the horizontal as illustrated. Determine the angular velocity ω of link AB and the velocity of roller B.

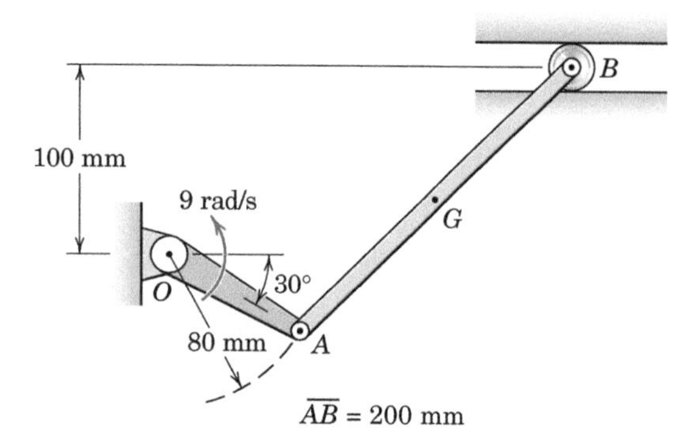

Problem 5/101

5/102 If link OA has a clockwise angular velocity of 2 rad/s in the position for which $x = 75$ mm, determine the velocity of the slider at B by the method of this article.

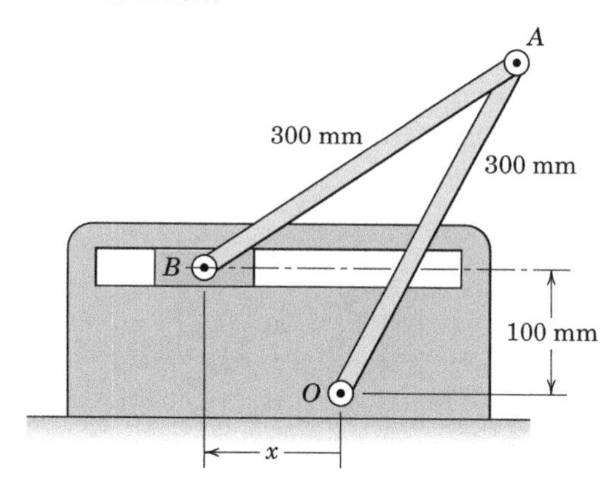

Problem 5/102

5/103 Motion of the bar is controlled by the constrained paths of A and B. If the angular velocity of the bar is 2 rad/s counterclockwise as the position $\theta = 45°$ is passed, determine the speeds of points A and P.

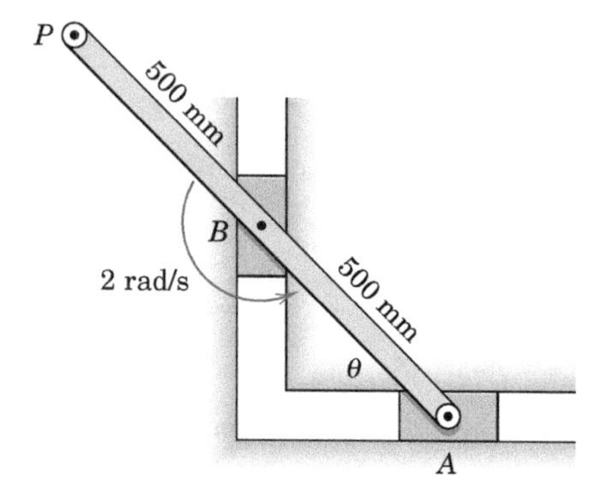

Problem 5/103

5/104 The switching device of Prob. 5/76 is repeated here. If the vertical control rod has a downward velocity $v = 2$ ft/sec when the device is in the position shown, determine the corresponding speed of point A by the method of this article. Roller C is in continuous contact with the inclined surface.

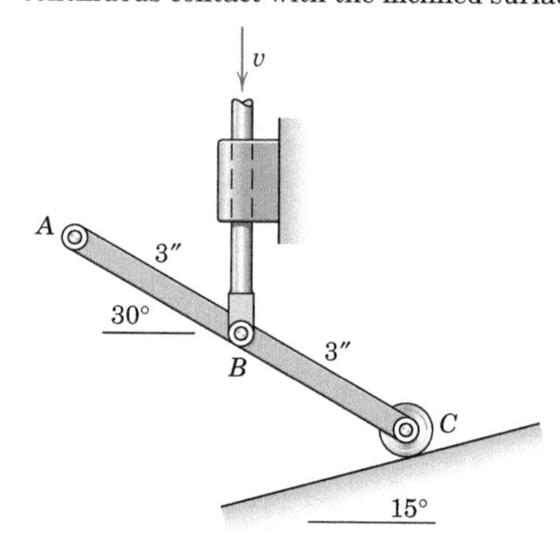

Problem 5/104

5/105 The shaft of the wheel unit rolls without slipping on the fixed horizontal surface, and point O has a velocity of 3 ft/sec to the right. By the method of this article, determine the velocities of points A, B, C, and D.

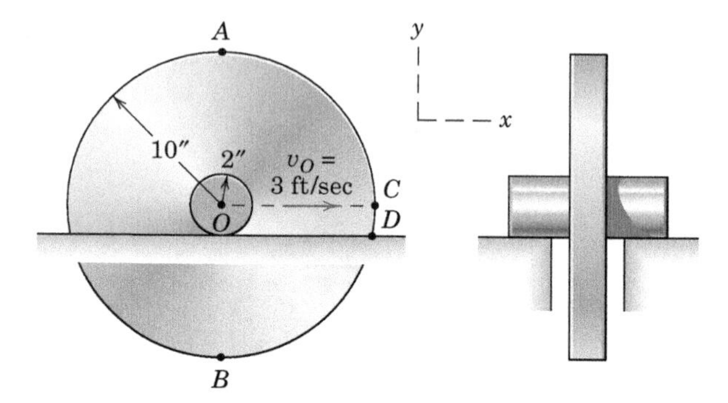

Problem 5/105

5/106 The center D of the car follows the centerline of the 100-ft skidpad. The speed of point D is v = 45 ft/sec. Determine the angular velocity of the car and the speeds of points A and B of the car.

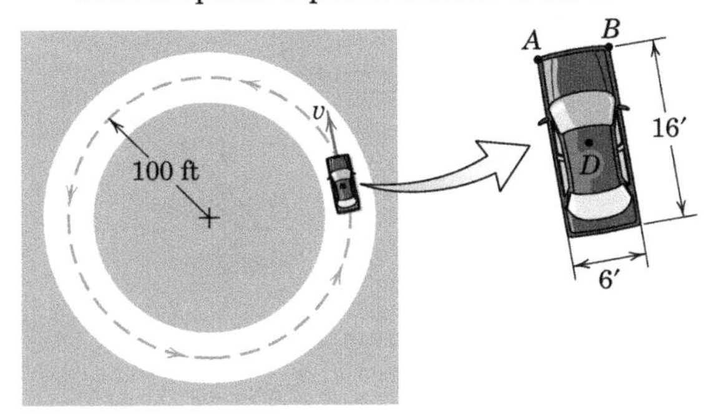

Problem 5/106

Representative Problems

5/107 The attached wheels roll without slipping on the plates A and B, which are moving in opposite directions as shown. If v_A = 60 mm/s to the right and v_B = 200 mm/s to the left, determine the speeds of the center O and the point P for the position shown.

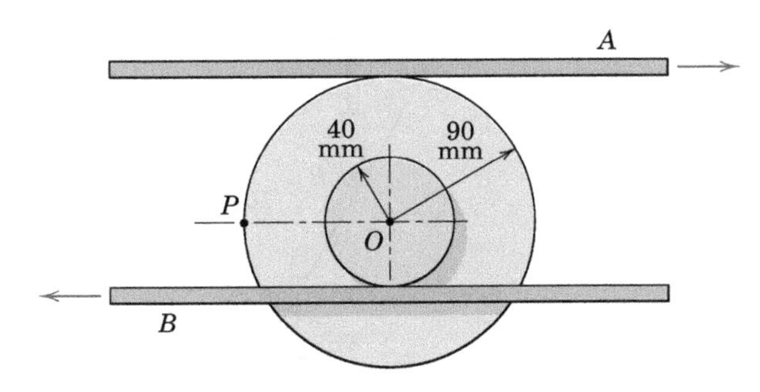

Problem 5/107

5/108 The mechanism of Prob. 5/77 is repeated here. By the method of this article, determine the angular velocity of link AB and the velocity of collar B for the position shown. Assume the quantities ω_0 and r to be known.

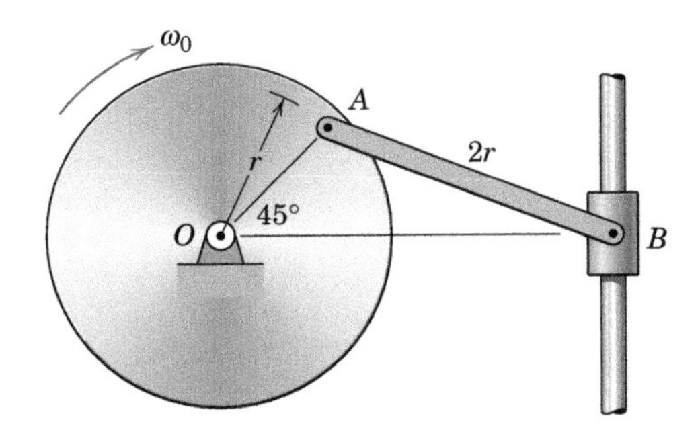

Problem 5/108

5/109 The mechanism of Prob. 5/78 is repeated here. By the method of this article, determine the angular velocity of link AB and the velocity of collar B for the instant depicted. Assume the quantities ω_0 and r to be known.

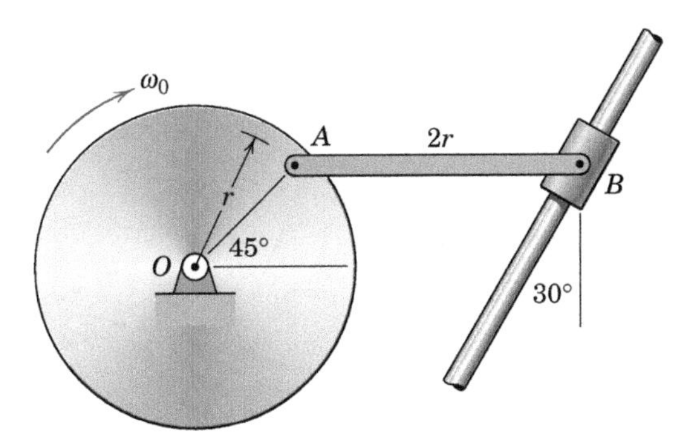

Problem 5/109

5/110 At the instant under consideration, the rod of the hydraulic cylinder is extending at the rate v_A = 2 m/s. Determine the corresponding angular velocity ω_{OB} of link OB.

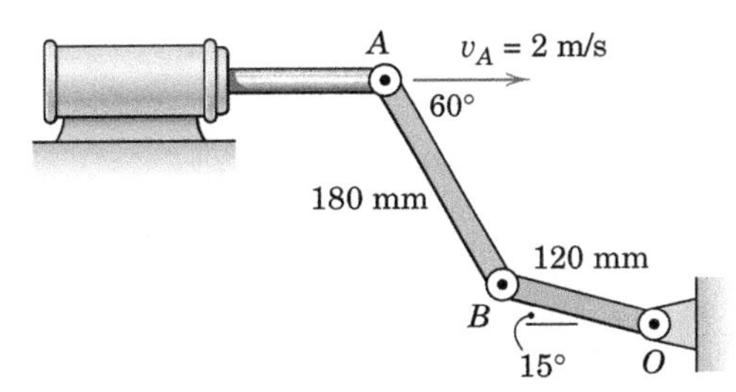

Problem 5/110

5/111 End A of the slender pole is given a velocity v_A to the right along the horizontal surface. Show that the magnitude of the velocity of end B equals v_A when the midpoint M of the pole comes in contact with the semicircular obstruction.

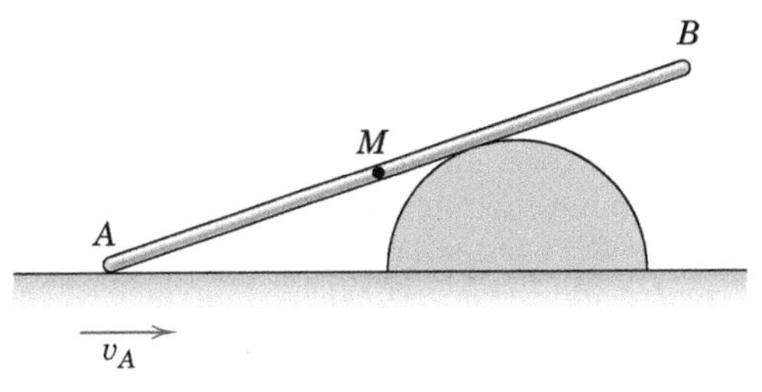

Problem 5/111

5/112 The flexible band F is attached at E to the rotating sector and leads over the guide pulley G. Determine the angular velocities of links AB and BD for the position shown if the band has a speed of 2 m/s.

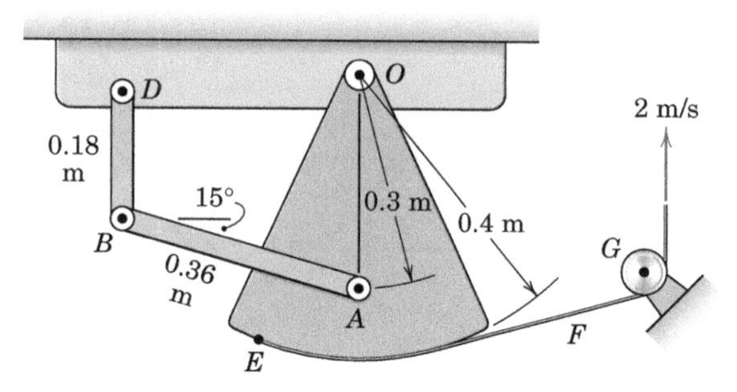

Problem 5/112

5/113 The rear driving wheel of a car has a diameter of 26 in. and has an angular speed N of 200 rev/min on an icy road. If the instantaneous center of zero velocity is 4 in. above the point of contact of the tire with the road, determine the velocity v of the car and the slipping velocity v_s of the tire on the ice.

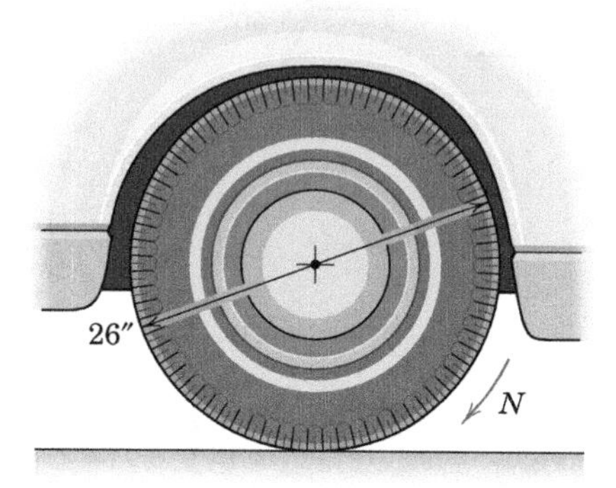

Problem 5/113

5/114 Solve for the speed of point D in Prob. 5/64 by the method of Art. 5/5.

5/115 Link OA has a counterclockwise angular velocity $\dot{\theta} = 4$ rad/sec during an interval of its motion. Determine the angular velocity of link AB and of sector BD for $\theta = 45°$ at which instant AB is horizontal and BD is vertical.

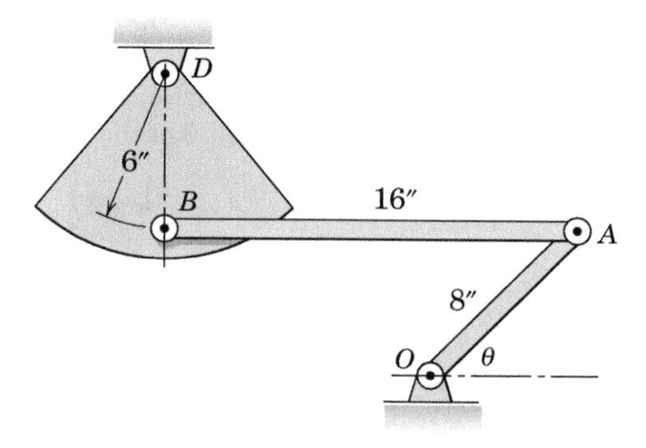

Problem 5/115

5/116 Vertical oscillation of the spring-loaded plunger F is controlled by a periodic change in pressure in the vertical hydraulic cylinder E. For the position $\theta = 60°$, determine the angular velocity of AD and the velocity of the roller A in its horizontal guide if the plunger F has a downward velocity of 2 m/s.

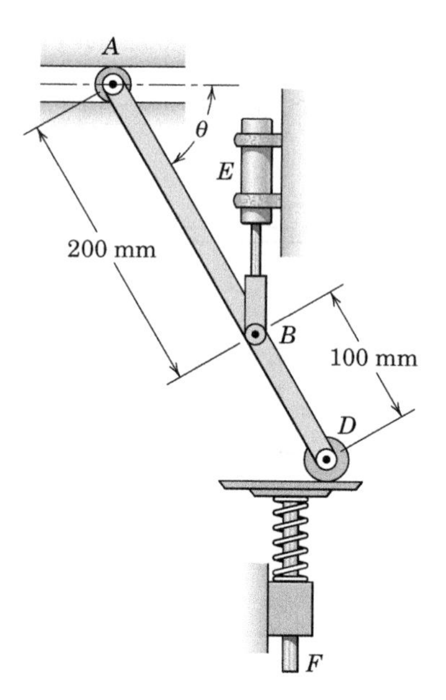

Problem 5/116

5/117 A device which tests the resistance to wear of two materials A and B is shown. If the link EO has a velocity of 4 ft/sec to the right when $\theta = 45°$, determine the rubbing velocity v_A.

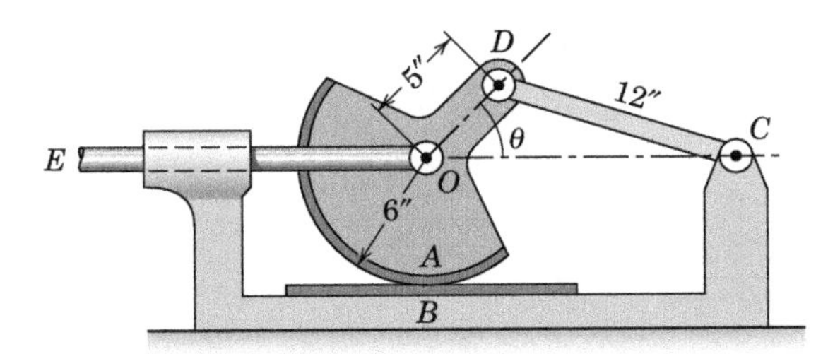

Problem 5/117

5/118 Motion of the roller A against its restraining spring is controlled by the downward motion of the plunger E. For an interval of motion the velocity of E is $v = 0.2$ m/s. Determine the velocity of A when θ becomes 90°.

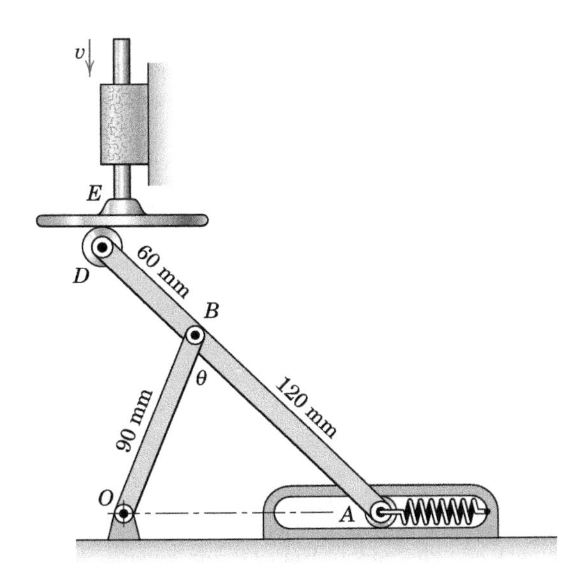

Problem 5/118

5/119 In the design of the mechanism shown, collar A is to slide along the fixed shaft as angle θ increases. When $\theta = 30°$, the control link at D is to have a downward component of velocity of 0.60 m/s. Determine the corresponding velocity of collar A by the method of this article.

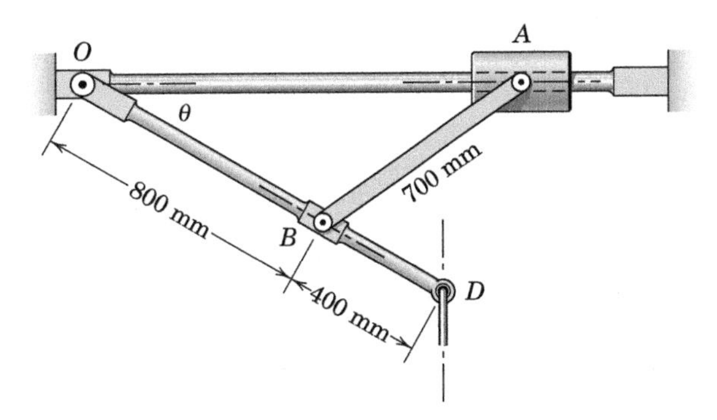

Problem 5/119

▶5/120 Determine the angular velocity ω of the ram head AE of the rock crusher in the position for which $\theta = 60°$. The crank OB has an angular speed of 60 rev/min. When B is at the bottom of its circle, D and E are on a horizontal line through F, and lines BD and AE are vertical. The dimensions are $\overline{OB} = 4$ in., $\overline{BD} = 30$ in., and $\overline{AE} = \overline{ED} = \overline{DF} = 15$ in. Carefully construct the configuration graphically, and use the method of this article.

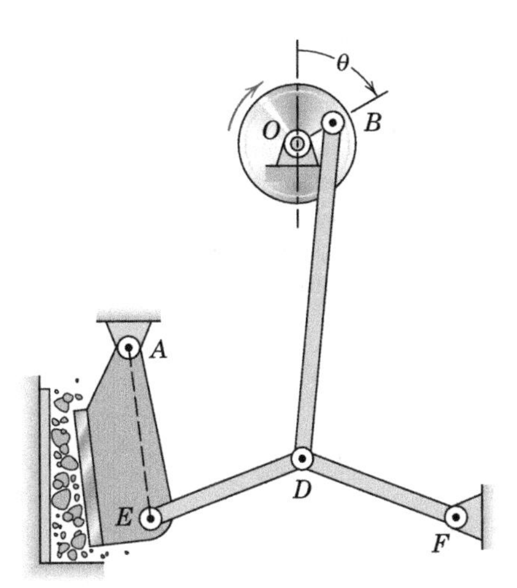

Problem 5/120

PROBLEMS

Introductory Problems

5/121 For the instant represented, corner C of the rectangular plate has an acceleration of 5 m/s^2 in the negative y-direction, and the plate has a clockwise angular velocity of 4 rad/s which is decreasing by 12 rad/s each second. Determine the magnitude of the acceleration of A at this instant. Solve by scalar-geometric and by vector-algebraic methods.

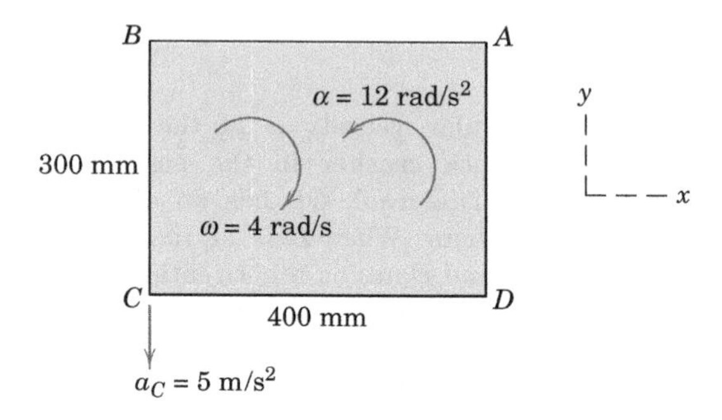

Problem 5/121

5/122 The two rotor blades of 800-mm radius rotate counterclockwise with a constant angular velocity $\omega = \dot{\theta} = 2$ rad/s about the shaft at O mounted in the sliding block. The acceleration of the block is $a_O = 3$ m/s^2. Determine the magnitude of the acceleration of the tip A of the blade when (*a*) $\theta = 0$, (*b*) $\theta = 90°$, and (*c*) $\theta = 180°$. Does the velocity of O or the sense of ω enter into the calculation?

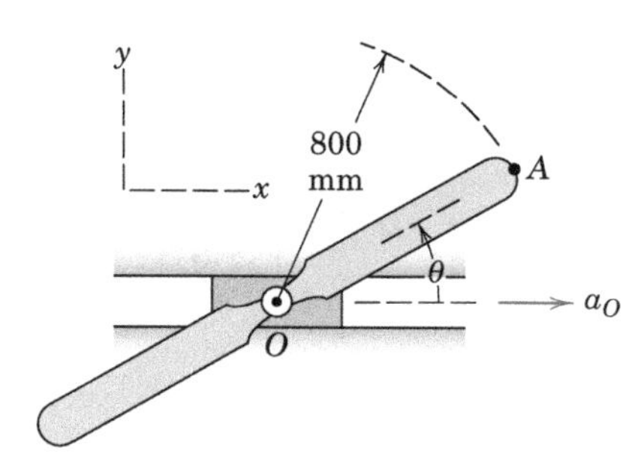

Problem 5/122

5/123 A container for waste materials is dumped by the hydraulically-activated linkage shown. If the piston rod starts from rest in the position indicated and has an acceleration of 1.5 ft/sec^2 in the direction shown, compute the initial angular acceleration of the container.

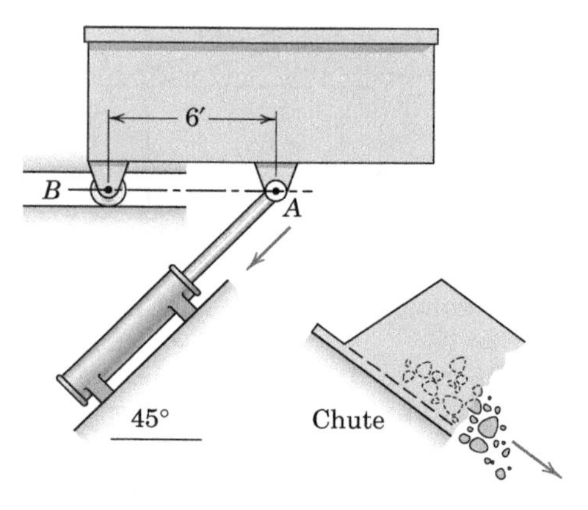

Problem 5/123

5/124 Determine the angular velocity and angular acceleration of the slender bar AB just after roller B passes point C and enters the circular portion of the support surface.

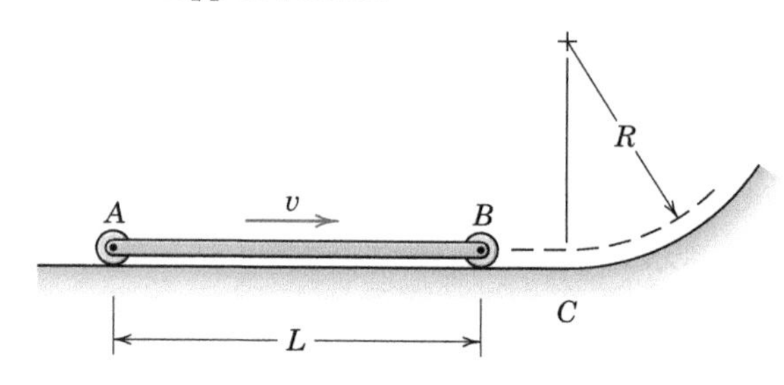

Problem 5/124

5/125 The wheel of radius R rolls without slipping, and its center O has an acceleration a_O. A point P on the wheel is a distance r from O. For given values of a_O, R, and r, determine the angle θ and the velocity v_O of the wheel for which P has no acceleration in this position.

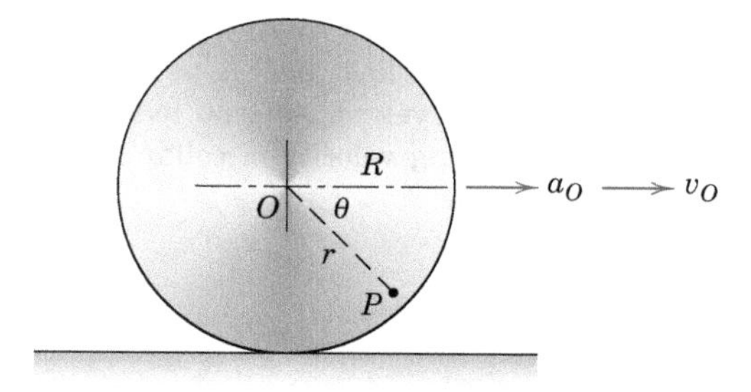

Problem 5/125

5/126 The 9-m steel beam is being hoisted from its horizontal position by the two cables attached at A and B. If the initial angular accelerations of the hoisting drums are $\alpha_1 = 0.5\ \text{rad/s}^2$ and $\alpha_2 = 0.2\ \text{rad/s}^2$ in the directions shown, determine the corresponding angular acceleration α of the beam, the acceleration of C, and the distance b from B to a point P on the beam centerline which has no acceleration.

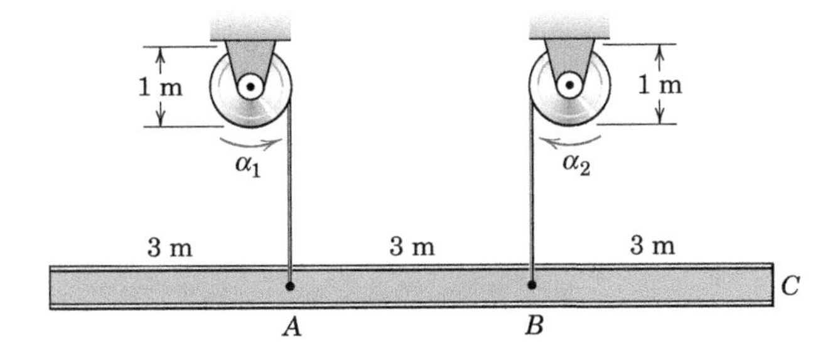

Problem 5/126

5/127 The bar of Prob. 5/82 is repeated here. The ends of the 0.4-m bar remain in contact with their respective support surfaces. End B has a velocity of 0.5 m/s and an acceleration of 0.3 m/s² in the directions shown. Determine the angular acceleration of the bar and the acceleration of end A.

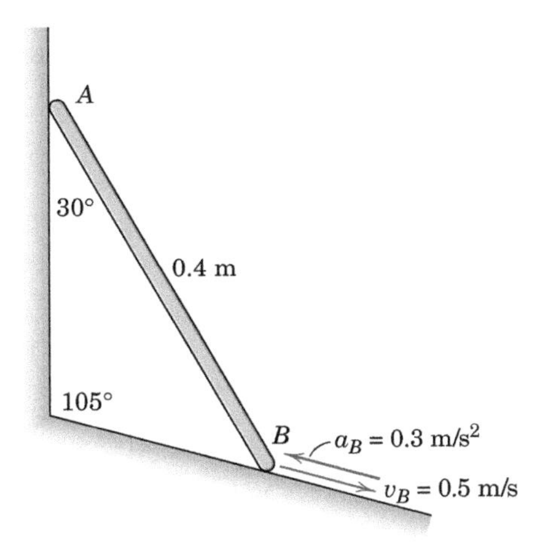

Problem 5/127

5/128 Determine the acceleration of point B on the equator of the earth, repeated here from Prob. 5/63. Use the data given with that problem and assume that the earth's orbital path is circular, consulting Table D/2 as necessary. Consider the center of the sun fixed and neglect the tilt of the axis of the earth.

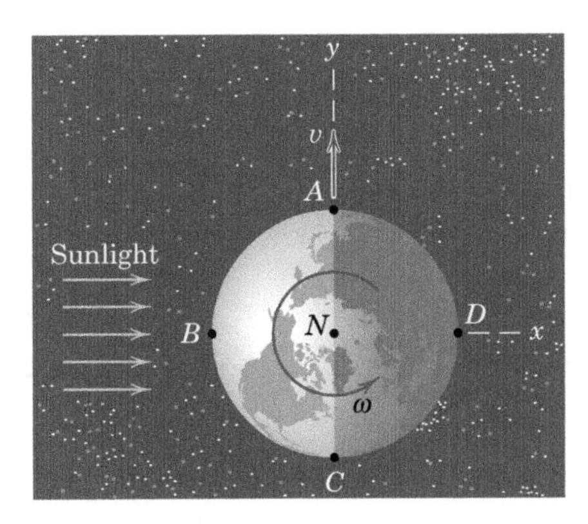

Problem 5/128

5/129 The spoked wheel of Prob. 5/73 is repeated here with additional information supplied. For a given cord speed v and acceleration a at point P and wheel radius r, determine the acceleration of point B with respect to point A.

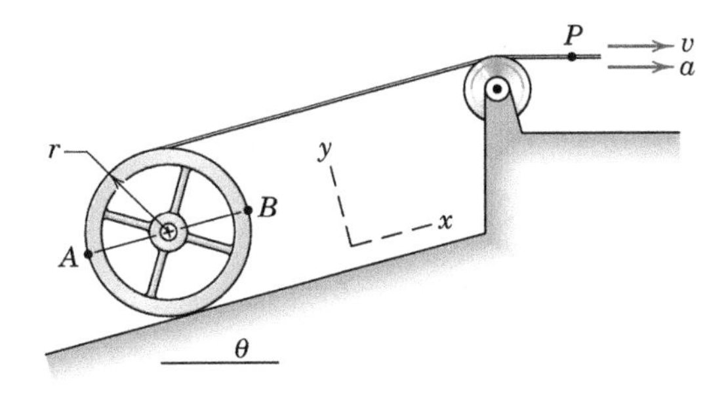

Problem 5/129

5/130 Calculate the angular acceleration of the plate in the position shown, where control link AO has a constant angular velocity $\omega_{OA} = 4$ rad/sec and $\theta = 60°$ for both links.

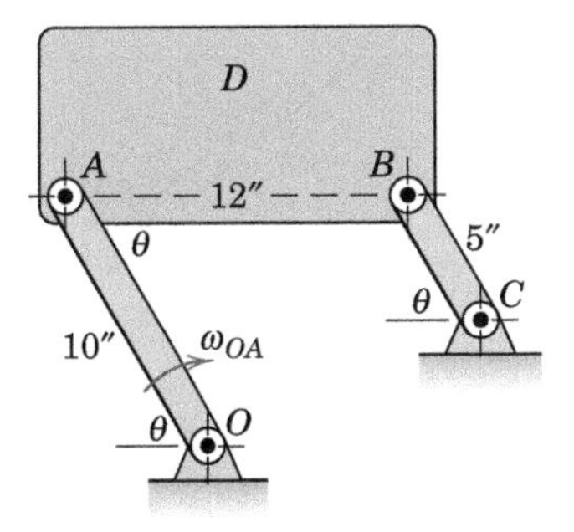

Problem 5/130

5/131 The bar AB of Prob. 5/71 is repeated here. At the instant under consideration, roller B has just begun moving on the 15° incline, and the velocity and acceleration of roller A are given. Determine the angular acceleration of bar AB and the acceleration of roller B.

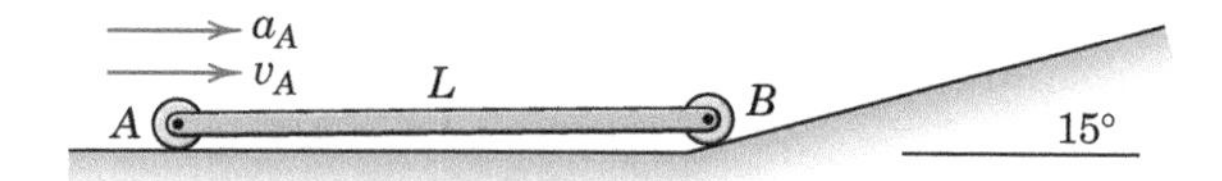

Problem 5/131

5/132 Determine the angular acceleration α_{AB} of AB for the position shown if link OB has a constant angular velocity ω.

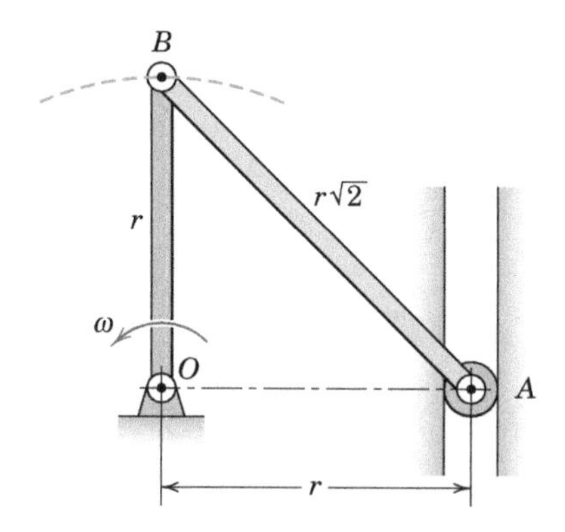

Problem 5/132

Representative Problems

5/133 Determine the angular acceleration of AB and the linear acceleration of A for the position $\theta = 90°$ if $\dot{\theta} = 4$ rad/s and $\ddot{\theta} = 0$ at that position.

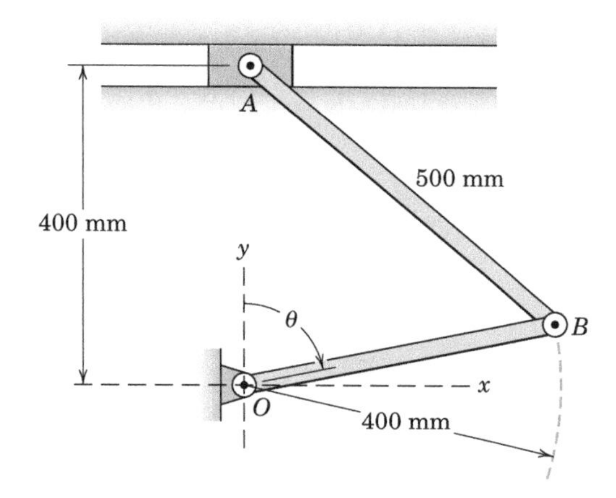

Problem 5/133

5/134 The switching device of Prob. 5/76 is repeated here. If the vertical control rod has a downward velocity $v = 2$ ft/sec and an upward acceleration $a = 1.2$ ft/sec^2 when the device is in the position shown, determine the magnitude of the acceleration of point A. Roller C is in continuous contact with the inclined surface.

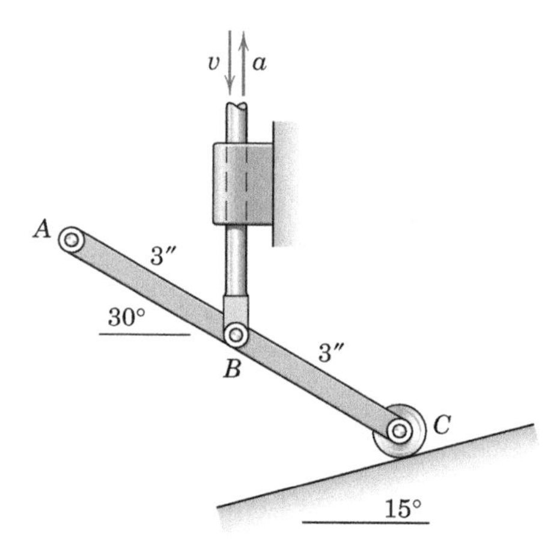

Problem 5/134

5/135 The two connected wheels of Prob. 5/64 are shown again here. Determine the magnitude of the acceleration of point D in the position shown if the center C of the smaller wheel has an acceleration to the right of 0.8 m/s^2 and has reached a velocity of 0.4 m/s at this instant.

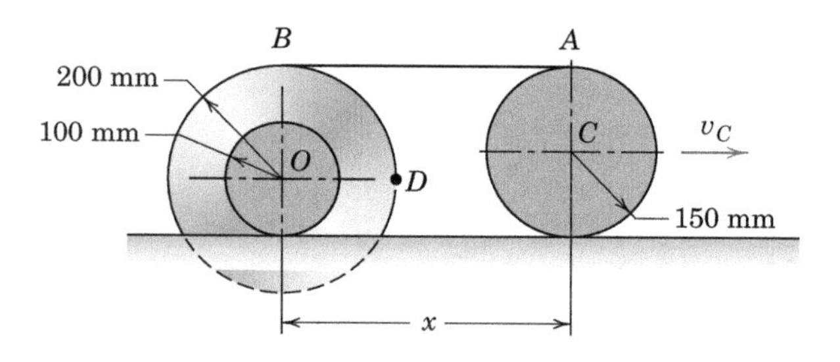

Problem 5/135

5/136 The end rollers of bar AB are constrained to the slot shown. If roller A has a downward velocity of 1.2 m/s and this speed is constant over a small motion interval, determine the tangential acceleration of roller B as it passes the topmost position. The value of R is 0.5 m.

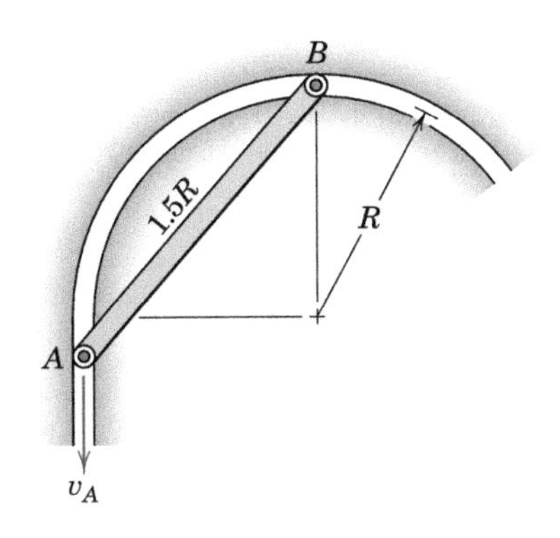

Problem 5/136

5/137 If the wheel in each case rolls on the circular surface without slipping, determine the acceleration of point C on the wheel momentarily in contact with the circular surface. The wheel has an angular velocity ω and an angular acceleration α.

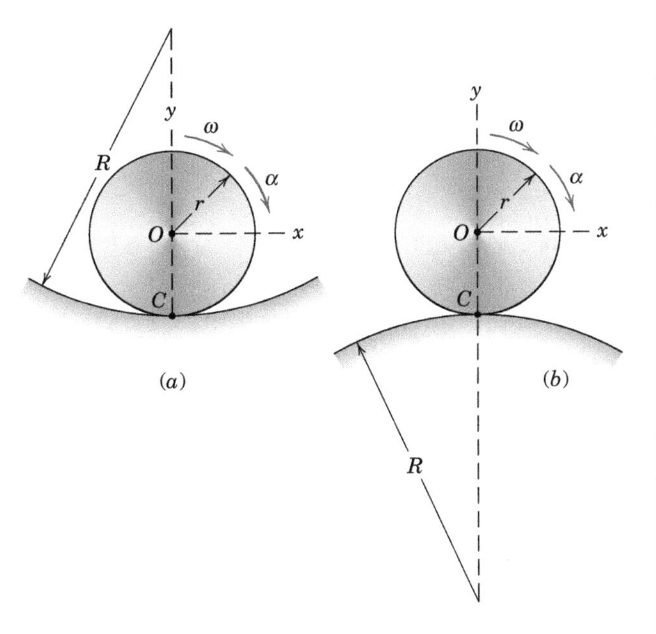

Problem 5/137

5/138 The system of Prob. 5/100 is repeated here. Crank OA rotates with a constant counterclockwise angular velocity of 9 rad/s. Determine the angular acceleration α_{AB} of link AB for the position shown.

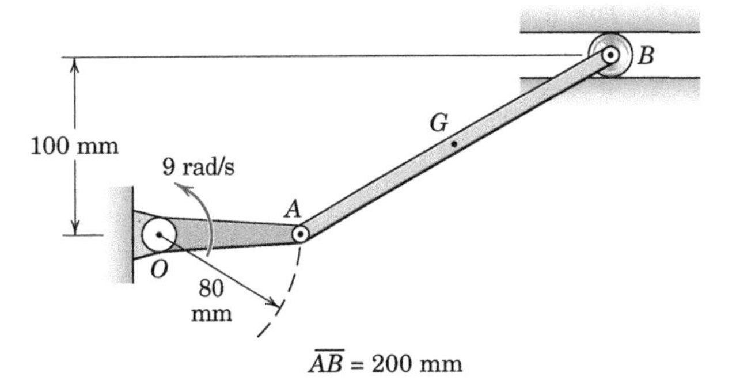

Problem 5/138

5/139 The system of Prob. 5/101 is repeated here. Crank OA is rotating at a counterclockwise angular rate of 9 rad/s, and this rate is decreasing at 5 rad/s^2. Determine the angular acceleration α_{AB} of link AB for the position shown.

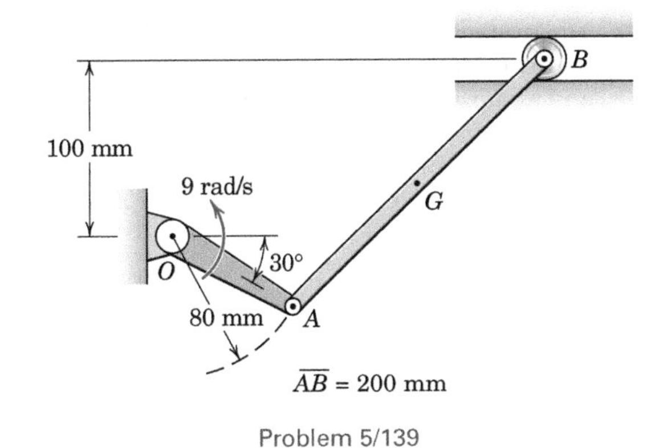

Problem 5/139

5/140 The triangular plate ABD has a clockwise angular velocity of 3 rad/sec and link OA has zero angular acceleration for the instant represented. Determine the angular accelerations of plate ABD and link BC for this instant.

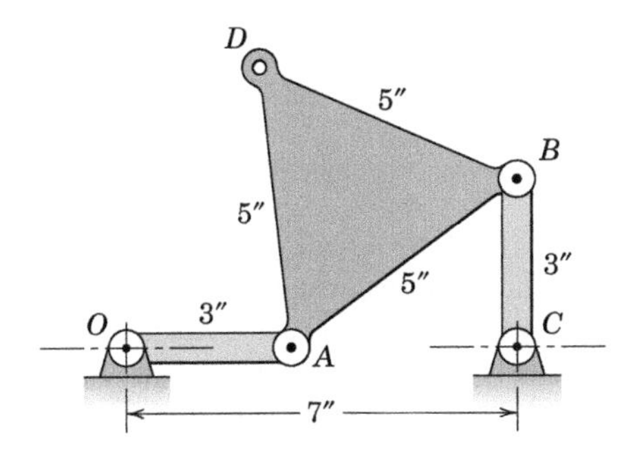

Problem 5/140

5/141 The mechanism of Prob. 5/77 is repeated here. The angular velocity ω_0 of the disk is constant. For the instant represented, determine the angular acceleration α_{AB} of link AB and the acceleration a_B of collar B. Assume the quantities ω_0 and r to be known.

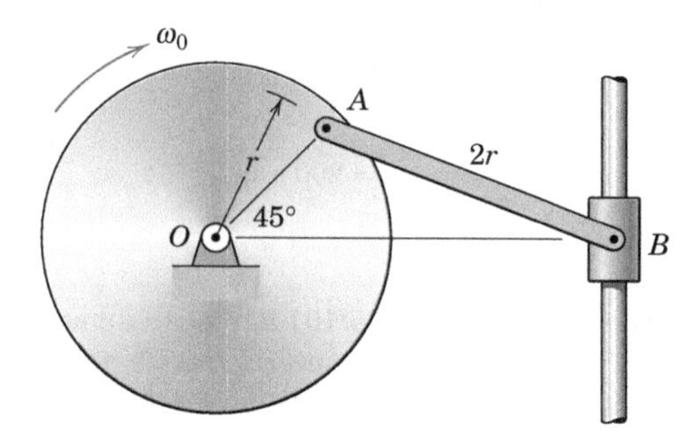

Problem 5/141

5/142 The system of Prob. 5/84 is repeated here. If the vertical rod has a downward velocity $v = 0.8$ m/s and an upward acceleration $a = 1.2$ m/s^2 when the device is in the position shown, determine the corresponding angular acceleration α of bar AB and the magnitude of the acceleration of roller B. The value of R is 0.4 m.

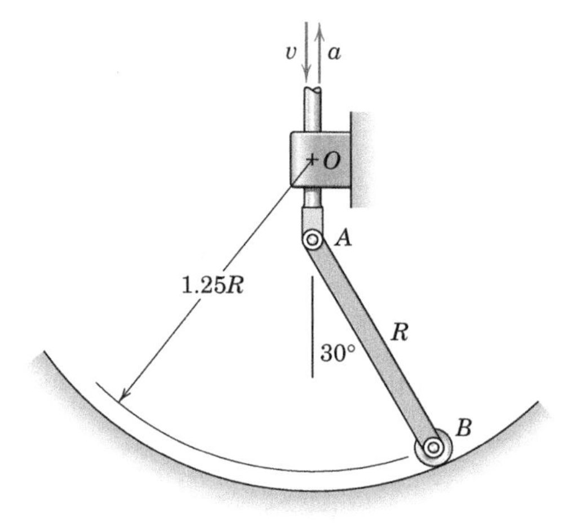

Problem 5/142

5/143 The shaft of the wheel unit rolls without slipping on the fixed horizontal surface. If the velocity and acceleration of point O are 3 ft/sec to the right and 4 ft/sec^2 to the left, respectively, determine the accelerations of points A and D.

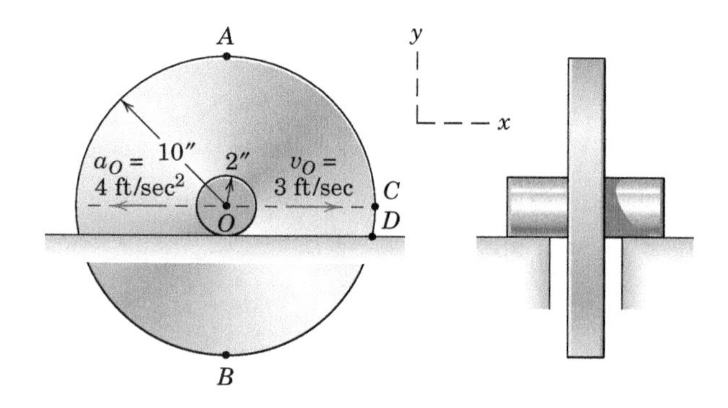

Problem 5/143

5/144 Plane motion of the triangular plate ABC is controlled by crank OA and link DB. For the instant represented, when OA and DB are vertical, OA has a clockwise angular velocity of 3 rad/s and a counterclockwise angular acceleration of 10 rad/s^2. Determine the angular acceleration of DB for this instant.

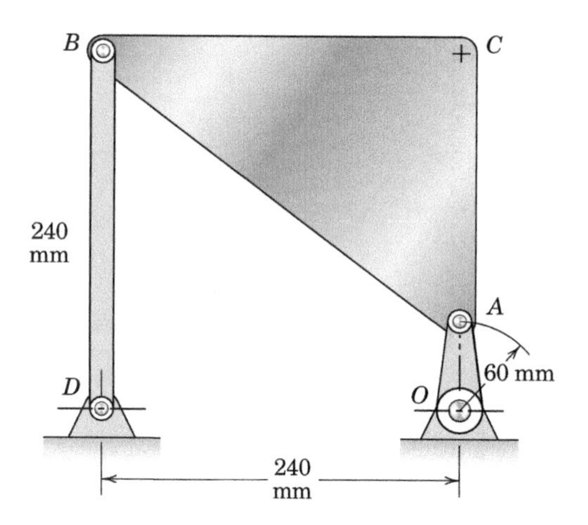

Problem 5/144

5/145 The system of Prob. 5/110 is repeated here. At the instant under consideration, the rod of the hydraulic cylinder is extending at the constant rate $v_A = 2$ m/s. Determine the angular acceleration α_{OB} of link OB.

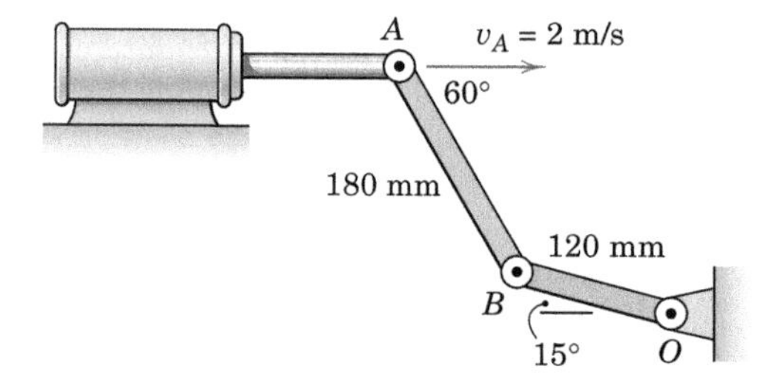

Problem 5/145

5/146 The velocity of roller A is $v_A = 0.5$ m/s to the right as shown, and this velocity is momentarily decreasing at a rate of 2 m/s^2. Determine the corresponding value of the angular acceleration α of bar AB as well as the tangential acceleration of roller B along the circular guide. The value of R is 0.6 m.

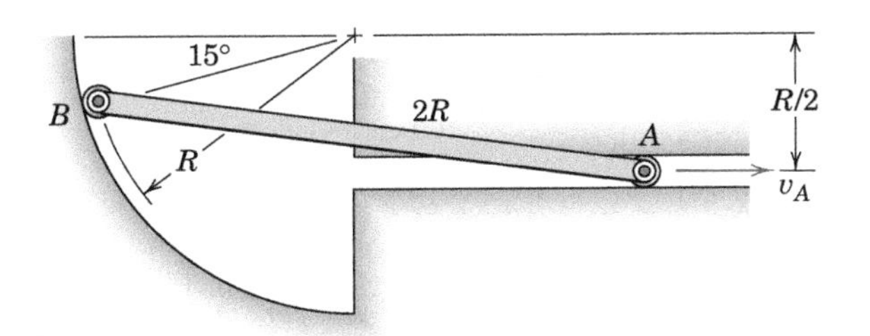

Problem 5/146

5/147 In the design of this linkage, motion of the square plate is controlled by the two pivoted links. Link OA has a constant angular velocity $\omega = 4$ rad/s during a short interval of motion. For the instant represented, $\theta = \tan^{-1} 4/3$ and AB is parallel to the x-axis. For this instant, determine the angular acceleration of both the plate and link CB.

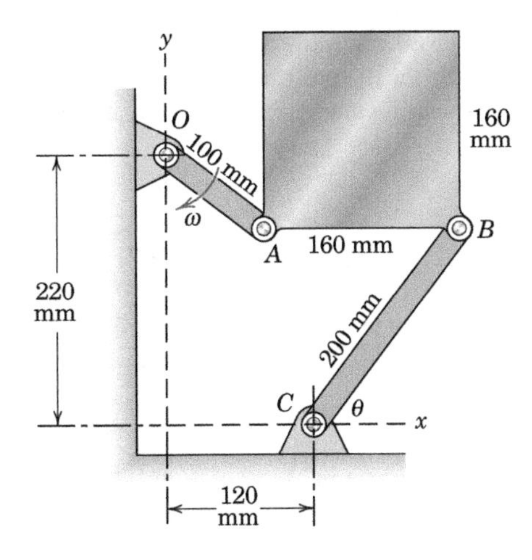

Problem 5/147

5/148 The mechanism of Prob. 5/112 is repeated here. If the band has a constant speed of 2 m/s as indicated in the figure, determine the angular acceleration α_{AB} of link AB.

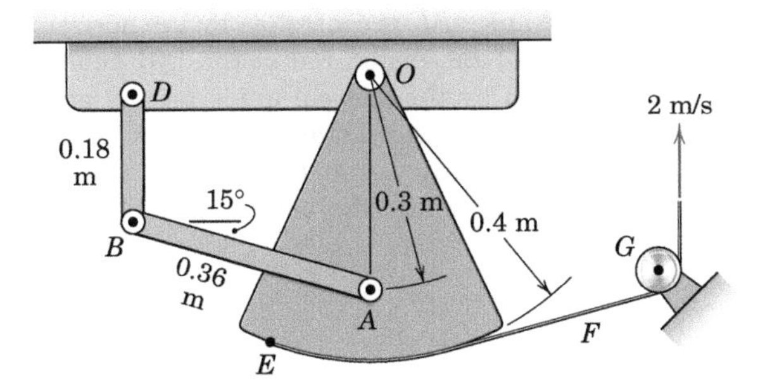

Problem 5/148

5/149 The bar AB from Prob. 5/74 is repeated here. If the velocity of point A is 3 m/s to the right and is constant for an interval including the position shown, determine the tangential acceleration of point B along its path and the angular acceleration of the bar.

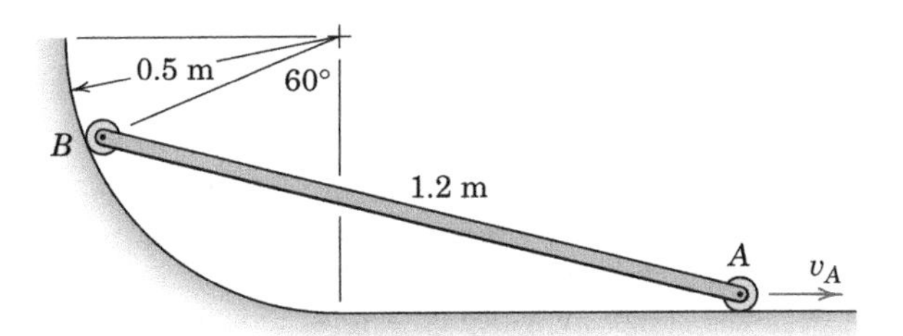

Problem 5/149

5/150 If the piston rod of the hydraulic cylinder C has a constant upward velocity of 0.5 m/s, calculate the acceleration of point D for the position where θ is 45°.

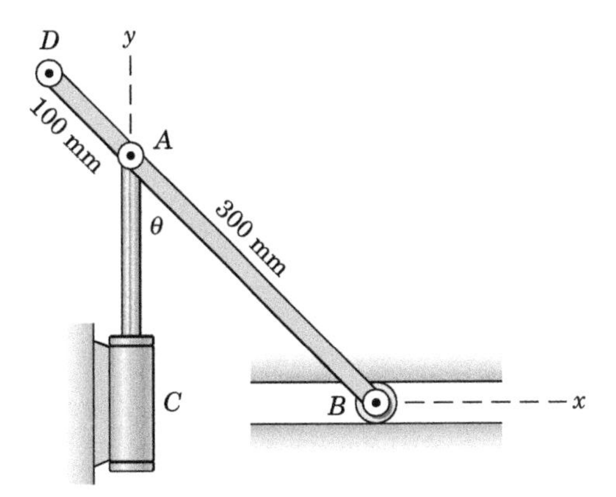

Problem 5/150

5/151 Motion of link ABC is controlled by the horizontal movement of the piston rod of the hydraulic cylinder D and by the vertical guide for the pinned slider at B. For the instant when $\theta = 45°$, the piston rod is retracting at the constant rate $v_C = 0.6$ ft/sec. Determine the acceleration of point A for this instant.

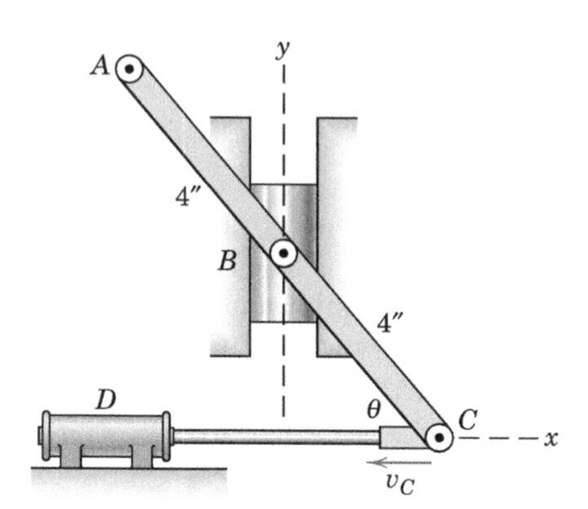

Problem 5/151

5/152 The deployment mechanism for the spacecraft magnetometer boom of Prob. 5/88 is shown again here. The driving link OB has a constant clockwise angular velocity ω_{OB} of 0.5 rad/sec as it crosses the vertical position. Determine the angular acceleration $\boldsymbol{\alpha}_{CA}$ of the boom for the position shown where $\tan\theta = 4/3$.

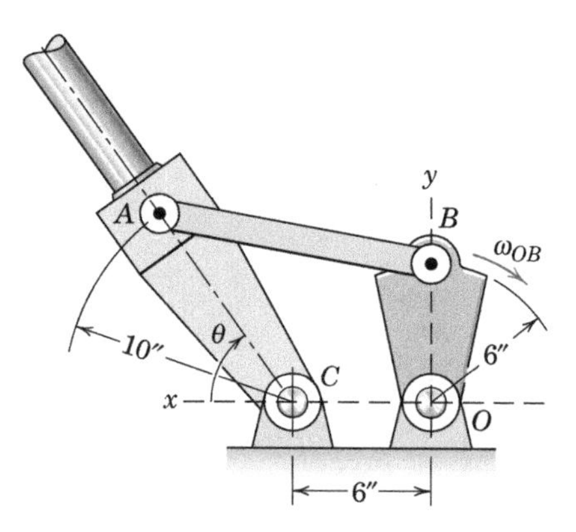

Problem 5/152

5/153 The four-bar linkage of Prob. 5/86 is repeated here. If the angular velocity and angular acceleration of drive link OA are 10 rad/s and 5 rad/s^2, respectively, both counterclockwise, determine the angular accelerations of bars AB and BC for the instant represented.

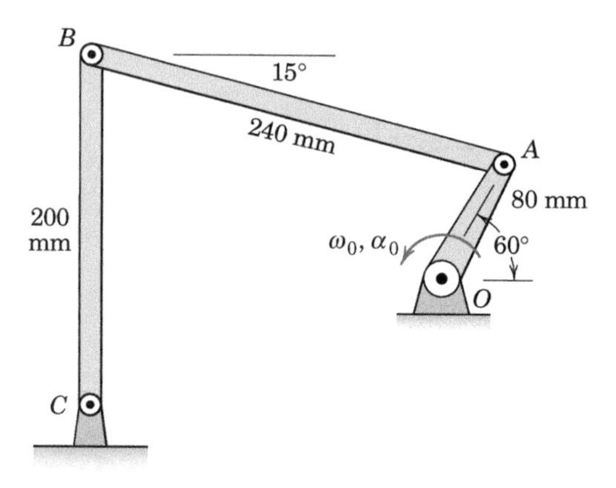

Problem 5/153

5/154 The elements of a power hacksaw are shown in the figure. The saw blade is mounted in a frame which slides along the horizontal guide. If the motor turns the flywheel at a constant counterclockwise speed of 60 rev/min, determine the acceleration of the blade for the position where $\theta = 90°$, and find the corresponding angular acceleration of the link AB.

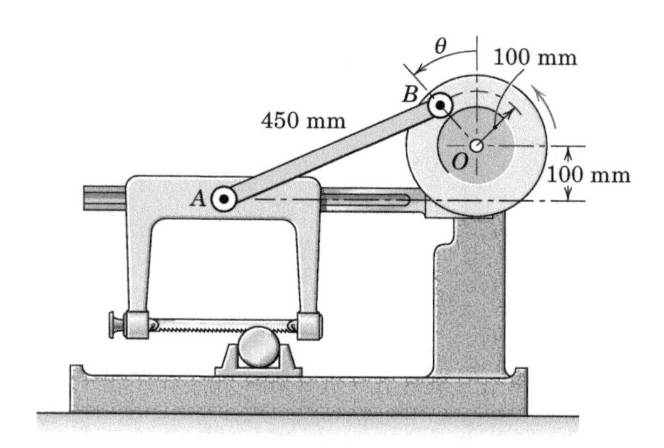

Problem 5/154

▶5/155 A mechanism for pushing small boxes from an assembly line onto a conveyor belt, repeated from Prob. 5/89, is shown with arm OD and crank CB in their vertical positions. For the configuration shown, crank CB has a constant clockwise angular velocity of π rad/s. Determine the acceleration of E.

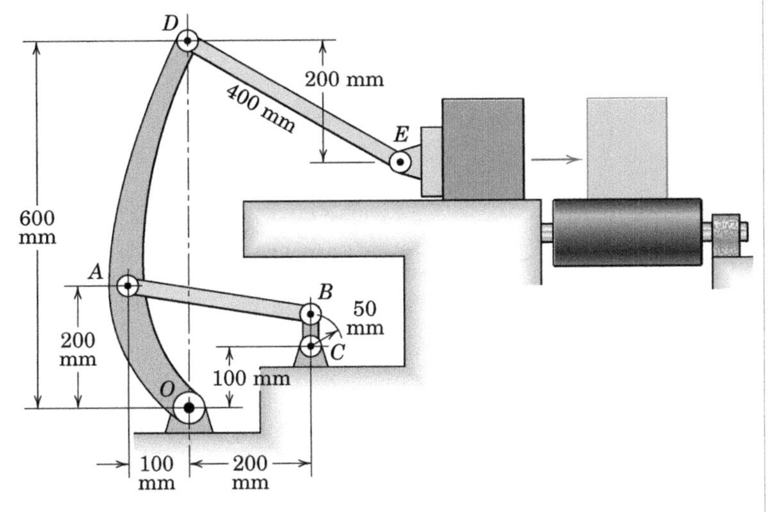

Problem 5/155

▶5/156 An intermittent-drive mechanism for perforated tape F consists of the link DAB driven by the crank OB. The trace of the motion of the finger at D is shown by the dashed line. Determine the magnitude of the acceleration of D at the instant represented when both OB and CA are horizontal if OB has a constant clockwise rotational velocity of 120 rev/min.

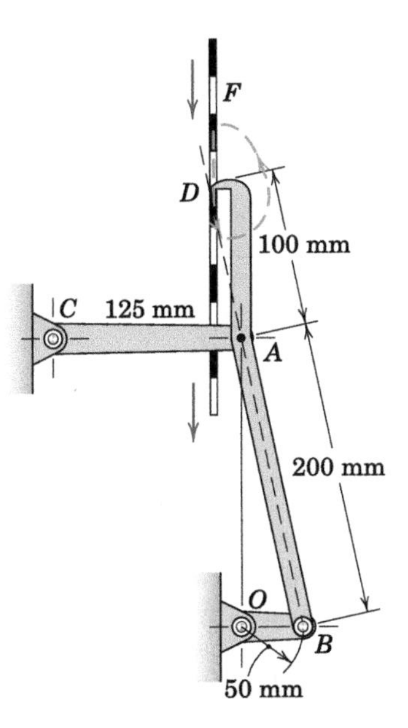

Problem 5/156

PROBLEMS

Introductory Problems

5/157 The disk rotates about a fixed axis through O with angular velocity $\omega = 5$ rad/s and angular acceleration $\alpha = 3$ rad/s^2 at the instant represented, in the directions shown. The slider A moves in the straight slot. Determine the absolute velocity and acceleration of A for the same instant, when $x = 36$ mm, $\dot{x} = -100$ mm/s, and $\ddot{x} = 150$ mm/s^2.

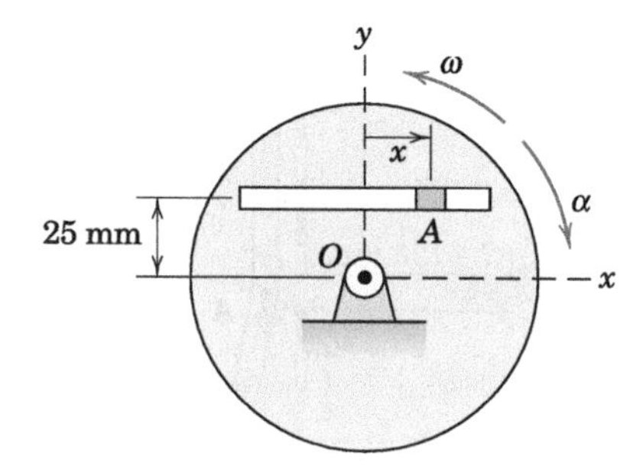

Problem 5/157

5/158 The sector rotates with the indicated angular quantities about a fixed axis through point B. Simultaneously, the particle A moves in the curved slot with constant speed u relative to the sector. Determine the absolute velocity and acceleration of particle A, and identify the Coriolis acceleration.

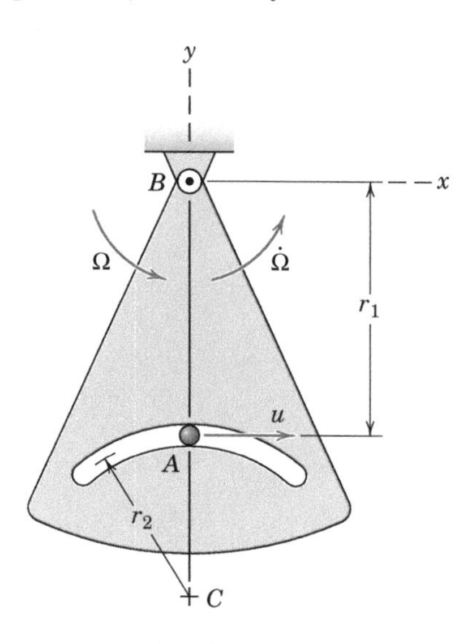

Problem 5/158

5/159 The slotted wheel rolls to the right without slipping, with a constant speed $v = 2$ ft/sec of its center O. Simultaneously, motion of the sliding block A is controlled by a mechanism not shown so that $\dot{x} = 1.5$ ft/sec with $\ddot{x} = 0$. Determine the magnitude of the acceleration of A for the instant when $x = 6$ in. and $\theta = 30°$.

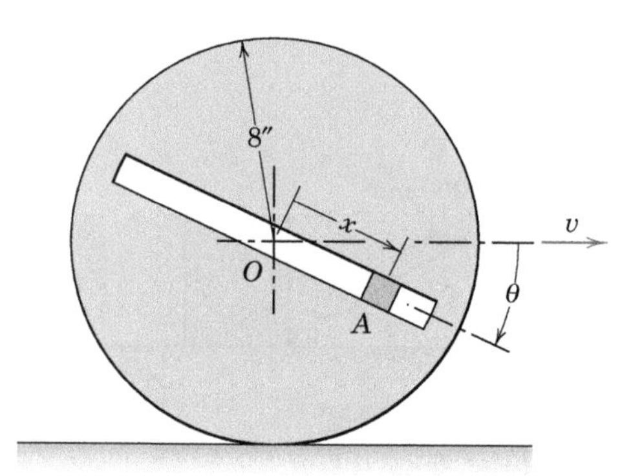

Problem 5/159

5/160 The disk rolls without slipping on the horizontal surface, and at the instant represented, the center O has the velocity and acceleration shown in the figure. For this instant, the particle A has the indicated speed u and time rate of change of speed $\dot{u}$, both relative to the disk. Determine the absolute velocity and acceleration of particle A.

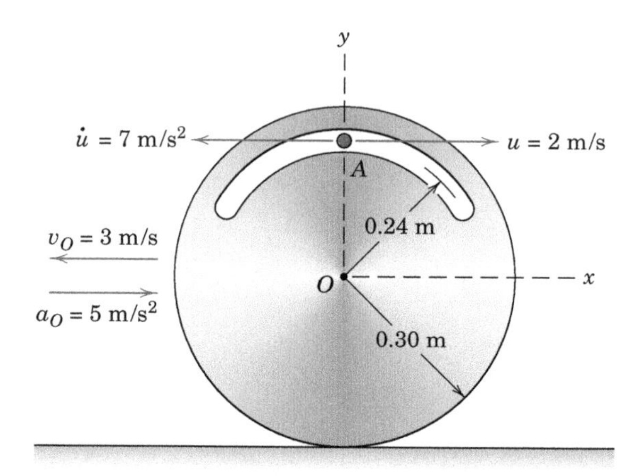

Problem 5/160

5/161 The cars of the roller coaster have a speed $v = 25$ ft/sec at the instant under consideration. As rider B passes the topmost point, she observes a stationary friend A. What velocity of A does she observe? At the position under consideration, the center of curvature of the path of rider B is point C.

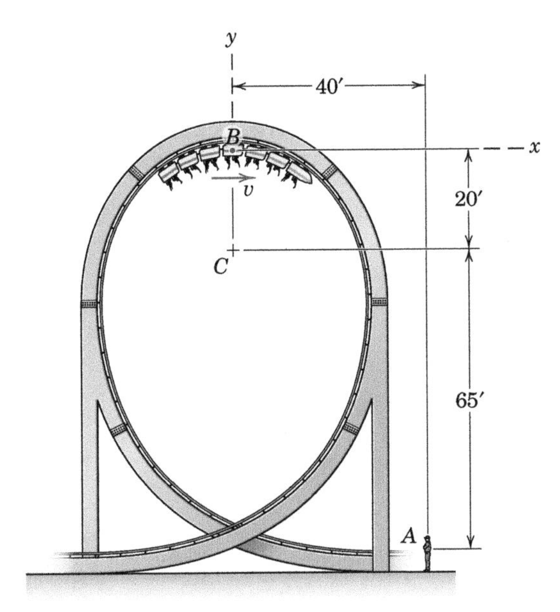

Problem 5/161

5/162 An experimental vehicle A travels with constant speed v relative to the earth along a north–south track. Determine the Coriolis acceleration $\mathbf{a}_{Cor}$ as a function of the latitude θ. Assume an earth-fixed rotating frame $Bxyz$ and a spherical earth. If the vehicle speed is $v = 500$ km/h, determine the magnitude of the Coriolis acceleration at (*a*) the equator and (*b*) the north pole.

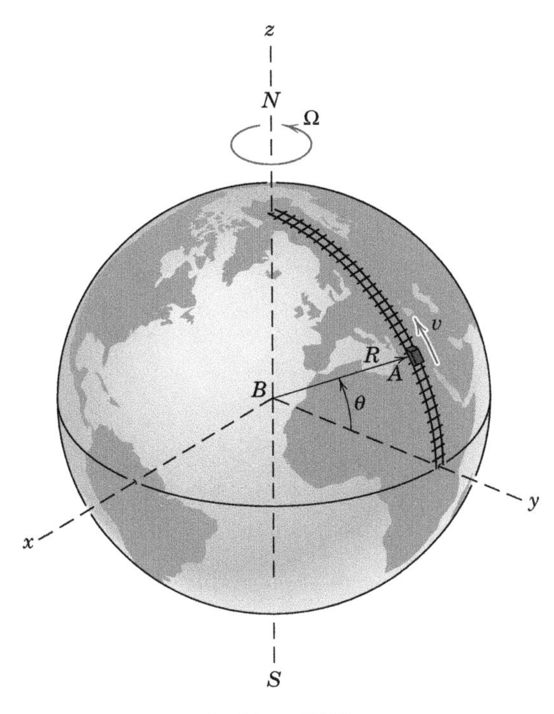

Problem 5/162

5/163 Car B is rounding the curve with a constant speed of 54 km/h, and car A is approaching car B in the intersection with a constant speed of 72 km/h. Determine the velocity which car A appears to have to an observer riding in and turning with car B. The x-y axes are attached to car B. Is this apparent velocity the negative of the velocity which B appears to have to a nonrotating observer in car A? The distance separating the two cars at the instant depicted is 40 m.

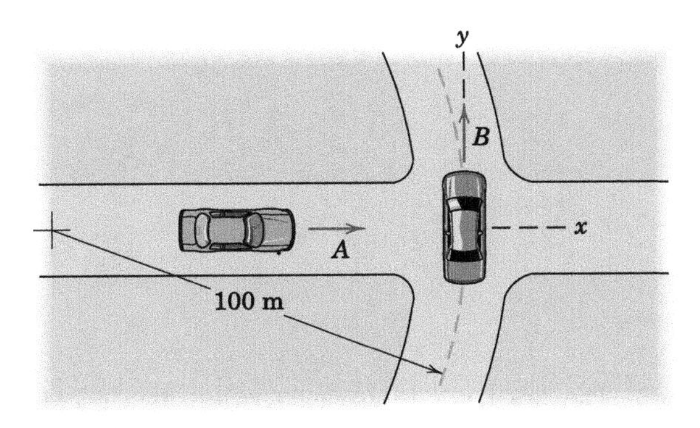

Problem 5/163

5/164 For the cars of Prob. 5/163 traveling with constant speed, determine the acceleration which car A appears to have to an observer riding in and turning with car B.

5/165 The small collar A is sliding on the bent bar with speed u relative to the bar as shown. Simultaneously, the bar is rotating with angular velocity ω about the fixed pivot B. Take the x-y axes to be fixed to the bar and determine the Coriolis acceleration of the slider for the instant represented. Interpret your result.

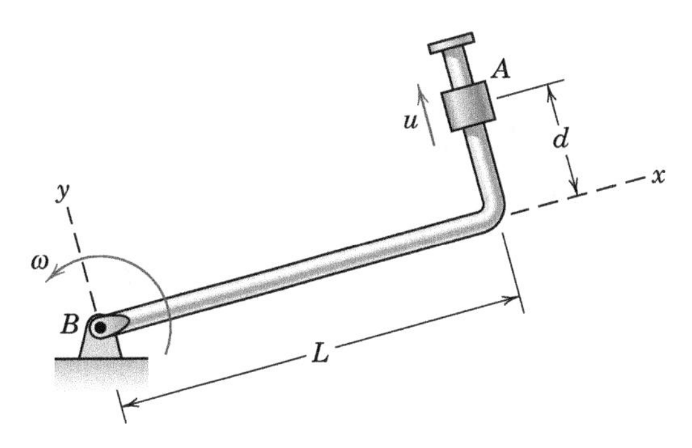

Problem 5/165

5/166 A train traveling at a constant speed $v = 25$ mi/hr has entered a circular portion of track with a radius $R = 200$ ft. Determine the velocity and acceleration of point A of the train as observed by the engineer B, who is fixed to the locomotive. Use the axes given in the figure.

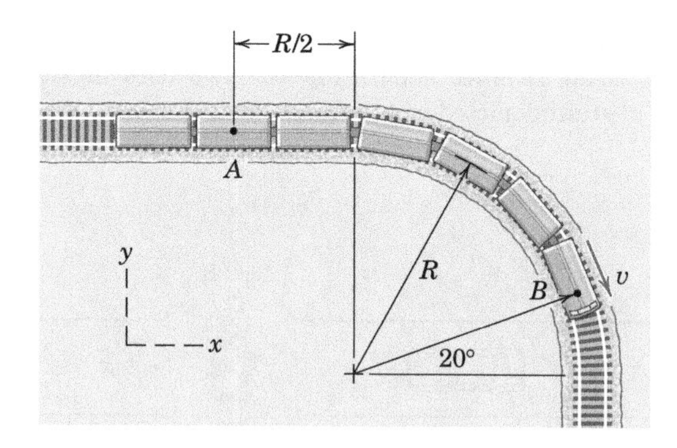

Problem 5/166

5/167 The fire truck is moving forward at a speed of 35 mi/hr and is decelerating at the rate of 10 ft/sec^2. Simultaneously, the ladder is being raised and extended. At the instant considered the angle θ is 30° and is increasing at the constant rate of 10 deg/sec. Also at this instant the extension b of the ladder is 5 ft, with $\dot{b} = 2$ ft/sec and $\ddot{b} = -1$ ft/sec^2. For this instant determine the acceleration of the end A of the ladder (*a*) with respect to the truck and (*b*) with respect to the ground.

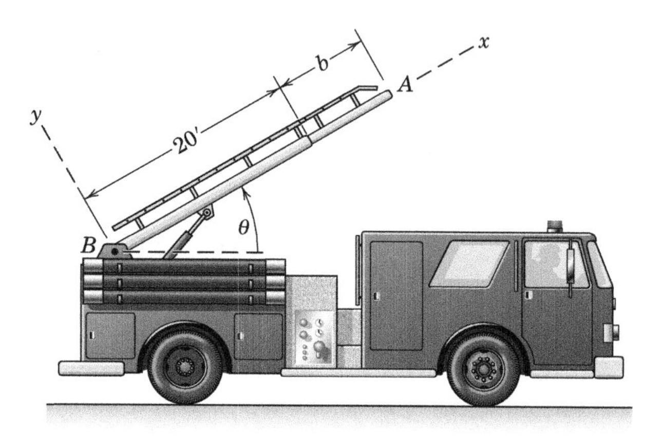

Problem 5/167

Representative Problems

5/168 Vehicle A travels west at high speed on a perfectly straight road B which is tangent to the surface of the earth at the equator. The road has no curvature whatsoever in the vertical plane. Determine the necessary speed v_{rel} of the vehicle relative to the road which will give rise to zero acceleration of the vehicle in the vertical direction. Assume that the center of the earth has no acceleration.

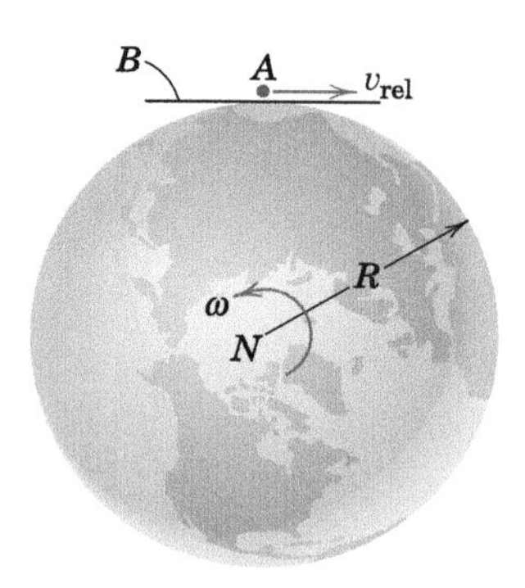

Problem 5/168

5/169 Aircraft B has a constant speed of 540 km/h at the bottom of a circular loop of 400-m radius. Aircraft A flying horizontally in the plane of the loop passes 100 m directly under B at a constant speed of 360 km/h. With coordinate axes attached to B as shown, determine the acceleration which A appears to have to the pilot of B for this instant.

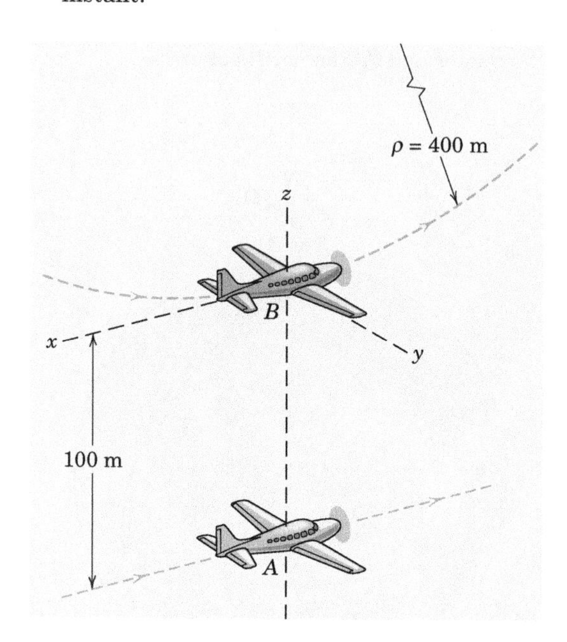

Problem 5/169

5/170 Bar OC rotates with a clockwise angular velocity $\omega_{OC} = 2$ rad/s. The pin A attached to bar OC engages the straight slot of the sector. Determine the angular velocity ω of the sector and the velocity of pin A relative to the sector for the instant represented.

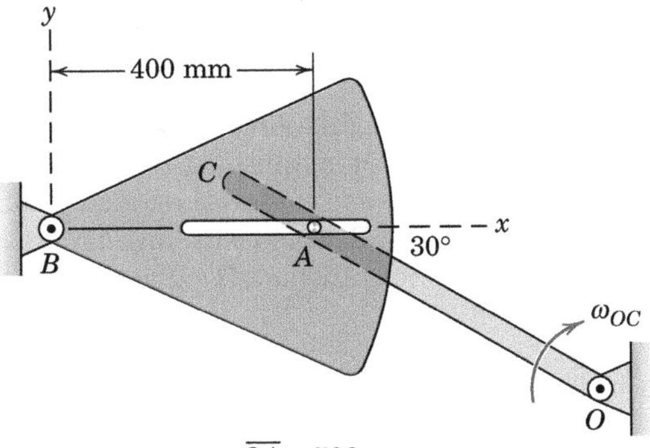

Problem 5/170

5/171 If the bar OC of the previous problem rotates with a clockwise angular velocity $\omega_{OC} = 2$ rad/s and a counterclockwise angular acceleration $\alpha_{OC} = 4$ rad/s^2, determine the angular acceleration α of the sector and the acceleration of point A relative to the sector.

5/172 The system of Prob. 5/170 is modified in that OC is now a slotted member which accommodates the pin A attached to the sector. If bar OC rotates with a clockwise angular velocity $\omega_{OC} = 2$ rad/s and a counterclockwise angular acceleration $\alpha_{OC} = 4$ rad/s^2, determine the angular velocity ω and the angular acceleration α of the sector.

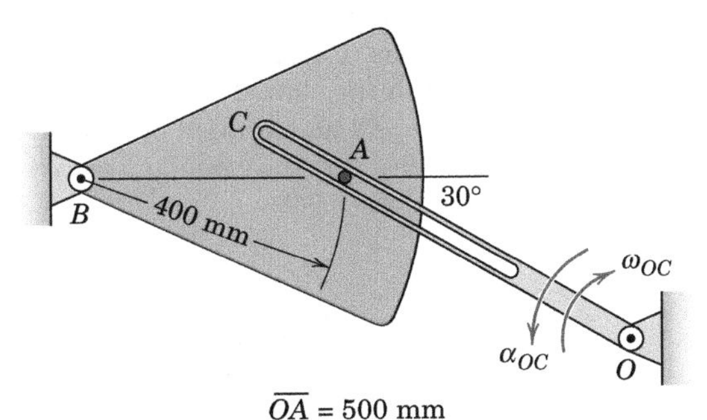

Problem 5/172

5/173 A smooth bowling alley is oriented north–south as shown. A ball A is released with speed v along the lane as shown. Because of the Coriolis effect, it will deflect a distance δ as shown. Develop a general expression for δ. The bowling alley is located at a latitude θ in the northern hemisphere. Evaluate your expression for the conditions $L = 60$ ft, $v = 15$ ft/sec, and $\theta = 40°$. Should bowlers prefer east–west alleys? State any assumptions.

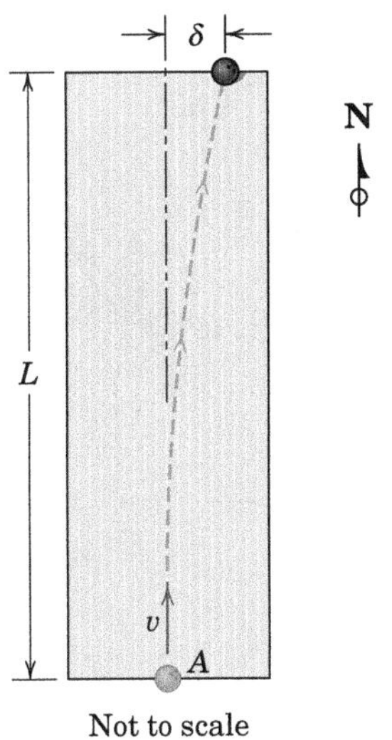

Problem 5/173

5/174 Under the action of its stern and starboard bow thrusters, the cruise ship has the velocity $v_B = 1$ m/s of its mass center B and angular velocity $\omega = 1$ deg/s about a vertical axis. The velocity of B is constant, but the angular rate ω is decreasing at 0.5 deg/s^2. Person A is stationary on the dock. What velocity and acceleration of A are observed by a passenger fixed to and rotating with the ship? Treat the problem as two-dimensional.

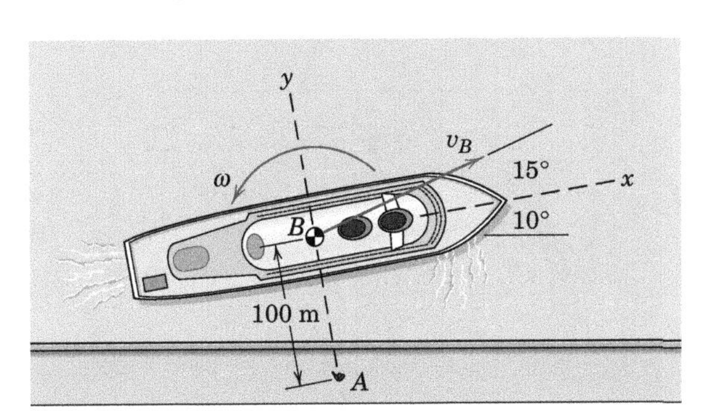

Problem 5/174

5/175 The air transport B is flying with a constant speed of 480 mi/hr in a horizontal arc of 9-mi radius. When B reaches the position shown, aircraft A, flying southwest at a constant speed of 360 mi/hr, crosses the radial line from B to the center of curvature C of its path. Write the vector expression, using the x-y axes attached to B, for the velocity of A as measured by an observer in and turning with B.

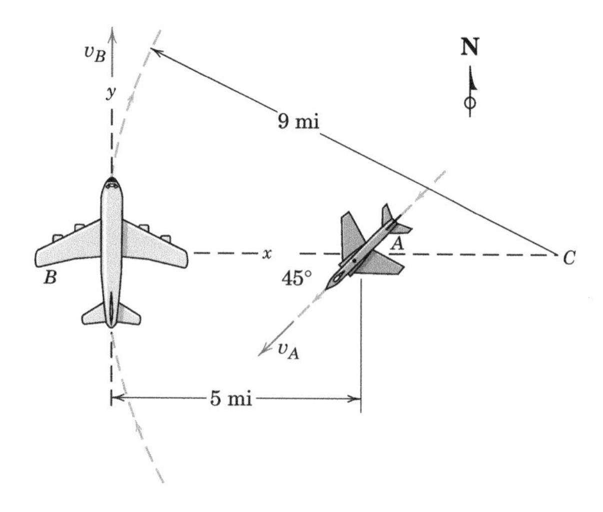

Problem 5/175

5/176 For the conditions of Prob. 5/175, obtain the vector expression for the acceleration which aircraft A appears to have to an observer in and turning with aircraft B, to which axes x-y are attached.

5/177 Car A is traveling along the straightaway with constant speed v. Car B is moving along the circular on-ramp with constant speed $v/2$. Determine the velocity and acceleration of car A as seen by an observer fixed to car B. Use the values v = 60 mi/hr and R = 200 ft, and utilize the x-y coordinates shown in the figure.

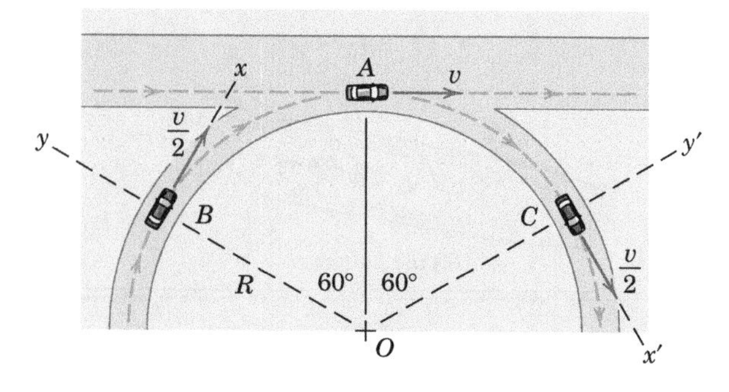

Problem 5/177

5/178 Refer to the figure for Prob. 5/177. Car A is traveling along the straightaway with speed v, and this speed is decreasing at a rate a. Car C is moving along the circular off-ramp with speed $v/2$, and this speed is decreasing at a rate $a/2$. Determine the velocity and acceleration which car A appears to have to an observer fixed to car C. Use the values v = 60 mi/hr, a = 10 ft/sec^2, and R = 200 ft, and utilize the x'-y' coordinates shown in the figure.

5/179 For the instant represented, link CB is rotating counterclockwise at a constant rate N = 4 rad/s, and its pin A causes a clockwise rotation of the slotted member ODE. Determine the angular velocity ω and angular acceleration α of ODE for this instant.

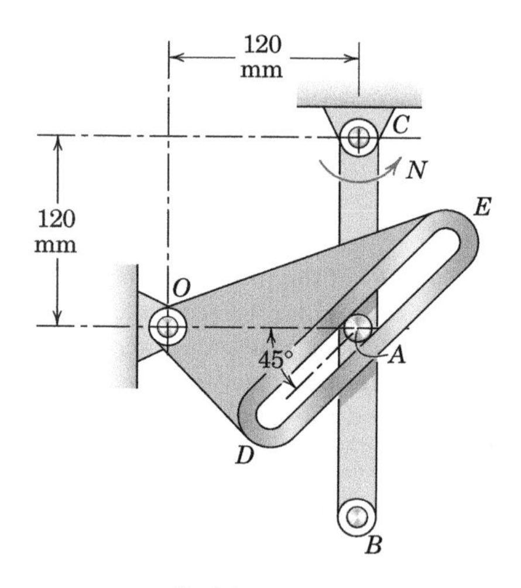

Problem 5/179

5/180 The disk rotates about a fixed axis through point O with a clockwise angular velocity ω_0 = 20 rad/s and a counterclockwise angular acceleration α_0 = 5 rad/s^2 at the instant under consideration. The value of r is 200 mm. Pin A is fixed to the disk but slides freely within the slotted member BC. Determine the velocity and acceleration of A relative to slotted member BC and the angular velocity and angular acceleration of BC.

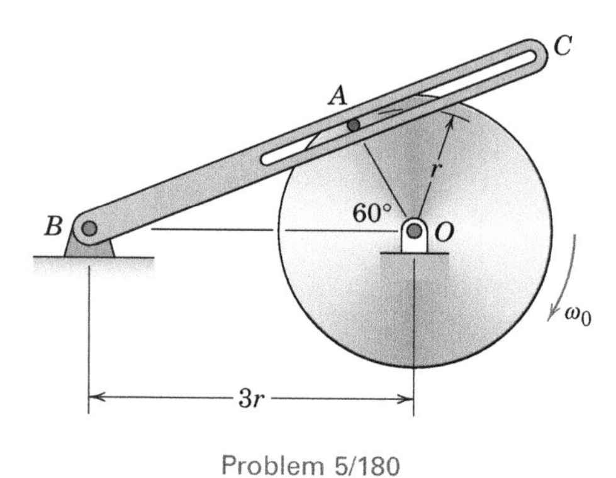

Problem 5/180

5/181 All conditions of the previous problem remain the same, except now, rather than rotating about a fixed center, the disk rolls without slipping on the horizontal surface. If the disk has a clockwise angular velocity of 20 rad/s and a counterclockwise angular acceleration of 5 rad/s^2, determine the velocity and acceleration of pin A relative to the slotted member BC and the angular velocity and angular acceleration of BC. The value of r is 200 mm. Neglect the distance from the center of pin A to the edge of the disk.

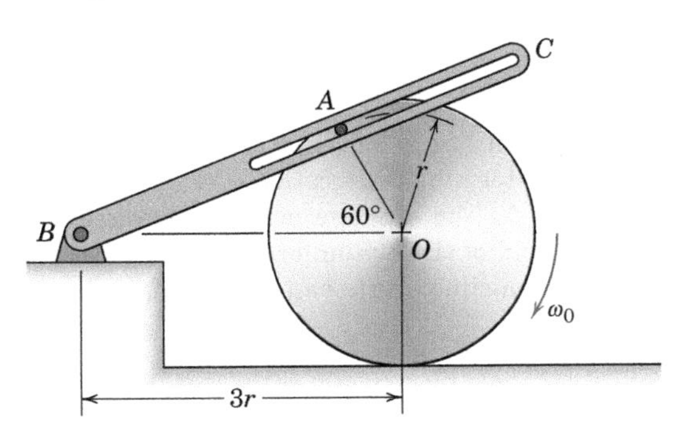

Problem 5/181

▶5/182 The space shuttle A is in an equatorial circular orbit of 240-km altitude and is moving from west to east. Determine the velocity and acceleration which it appears to have to an observer B fixed to and rotating with the earth at the equator as the shuttle passes overhead. Use $R = 6378$ km for the radius of the earth. Also use Fig. 1/1 for the appropriate value of g and carry out your calculations to four-figure accuracy.

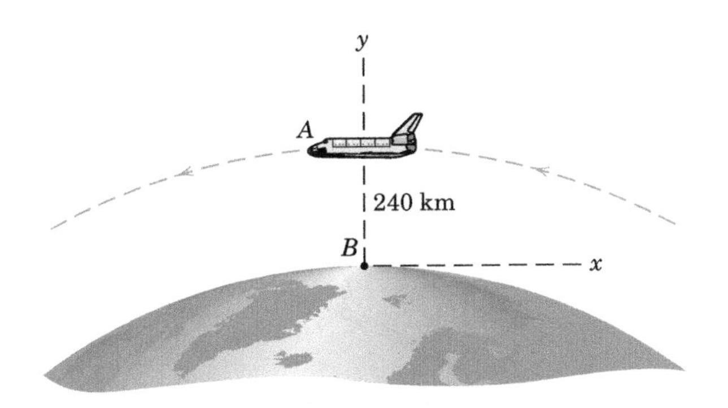

Problem 5/182

▶5/183 Determine the angular acceleration of link EC in the position shown, where $\omega = \dot{\beta} = 2$ rad/sec and $\ddot{\beta} = 6$ rad/sec^2 when $\theta = \beta = 60°$. Pin A is fixed to link EC. The circular slot in link DO has a radius of curvature of 6 in. In the position shown, the tangent to the slot at the point of contact is parallel to AO.

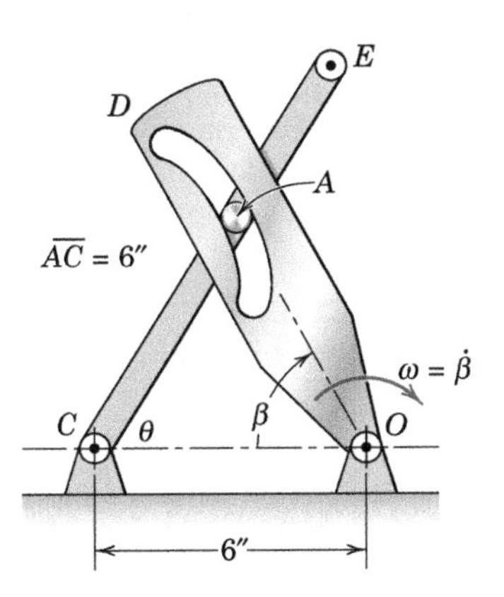

Problem 5/183

▶5/184 One wheel of an experimental vehicle F, which has a constant velocity $v = 36$ km/h, is shown. The wheel rolls without slipping and causes an oscillation of the slotted arm through the action of its pin A. Control rod DB, in turn, moves back and forth relative to the vehicle by virtue of the motion imparted to pin B. For the position shown, determine the acceleration a_B of the control rod DB. (*Suggestion:* Consider the justification and advantage of using a reference frame attached to the vehicle.)

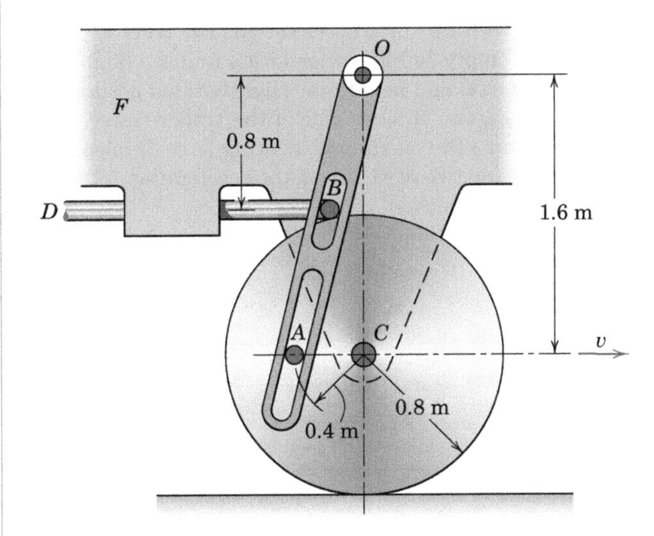

Problem 5/184

PROBLEMS

Introductory Problems

6/1 The right-angle bar with equal legs weighs 6 lb and is freely hinged to the vertical plate at C. The bar is prevented from rotating by the two pegs A and B fixed to the plate. Determine the acceleration a of the plate for which no force is exerted on the bar by either peg A or B.

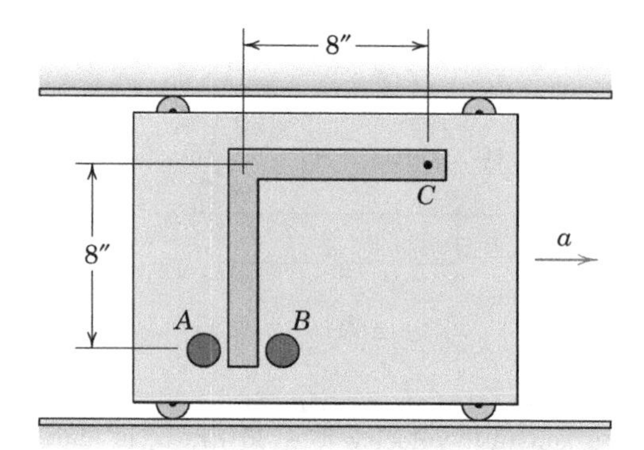

Problem 6/1

6/2 In Prob. 6/1, if the plate is given a horizontal acceleration $a = 2g$, calculate the force exerted on the bar by either peg A or B.

6/3 The driver of a pickup truck accelerates from rest to a speed of 45 mi/hr over a horizontal distance of 225 ft with constant acceleration. The truck is hauling an empty 500-lb trailer with a uniform 60-lb gate hinged at O and held in the slightly tilted position by two pegs, one on each side of the trailer frame at A. Determine the maximum shearing force developed in each of the two pegs during the acceleration.

Problem 6/3

6/4 A passenger car of an overhead monorail system is driven by one of its two small wheels A or B. Select the one for which the car can be given the greater acceleration without slipping the driving wheel and compute the maximum acceleration if the effective coefficient of friction is limited to 0.25 between the wheels and the rail. Neglect the small mass of the wheels.

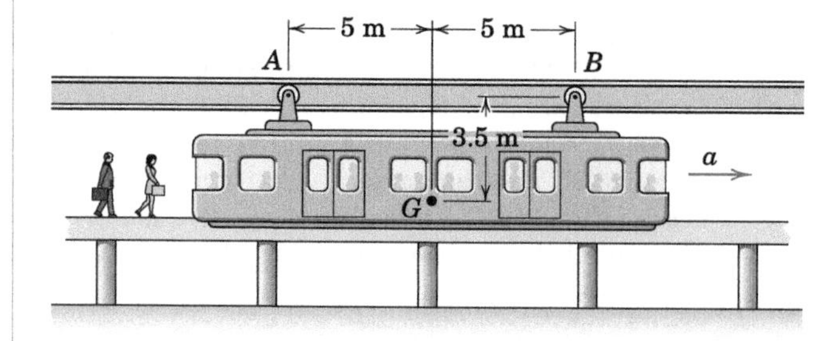

Problem 6/4

6/5 The uniform box of mass m slides down the rough incline. Determine the location d of the effective normal force N. The effective normal force is located at the centroid of the nonuniform pressure distribution which the incline exerts on the bottom surface of the block.

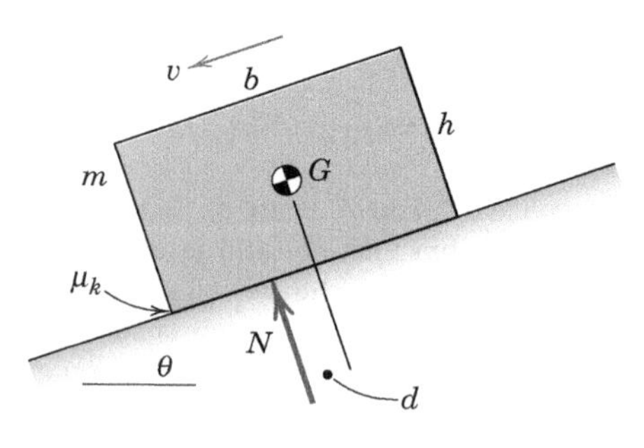

Problem 6/5

6/6 The uniform slender bar of mass m and length L is held in the position shown by the stop at A. What acceleration a will cause the normal force acting on the roller at B to become (a) one-half of the static value, (b) one-fourth of the static value, and (c) zero?

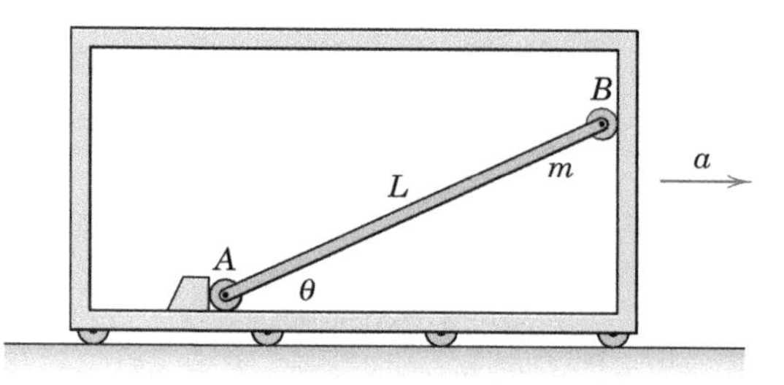

Problem 6/6

6/7 The homogeneous crate of mass m is mounted on small wheels as shown. Determine the maximum force P which can be applied without overturning the crate about (a) its lower front edge with $h = b$ and (b) its lower back edge with $h = 0$.

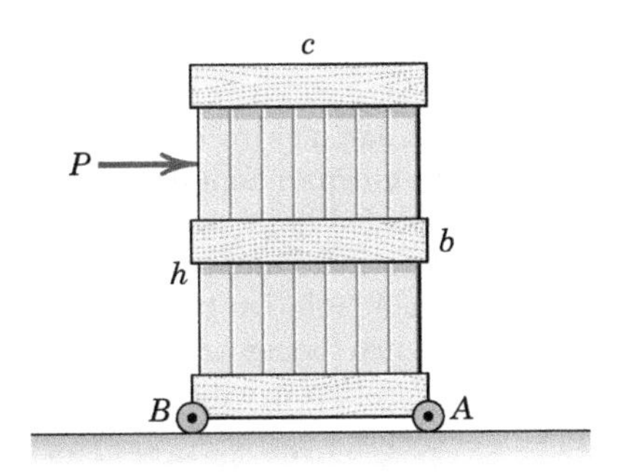

Problem 6/7

6/8 The frame is made from uniform rod which has a mass ρ per unit length. A smooth recessed slot constrains the small rollers at A and B to travel horizontally. Force P is applied to the frame through a cable attached to an adjustable collar C. Determine the magnitudes and directions of the normal forces which act on the rollers if (*a*) $h = 0.3L$, (*b*) $h = 0.5L$, and (*c*) $h = 0.9L$. Evaluate your results for $\rho = 2$ kg/m, $L = 500$ mm, and $P = 60$ N. What is the acceleration of the frame in each case?

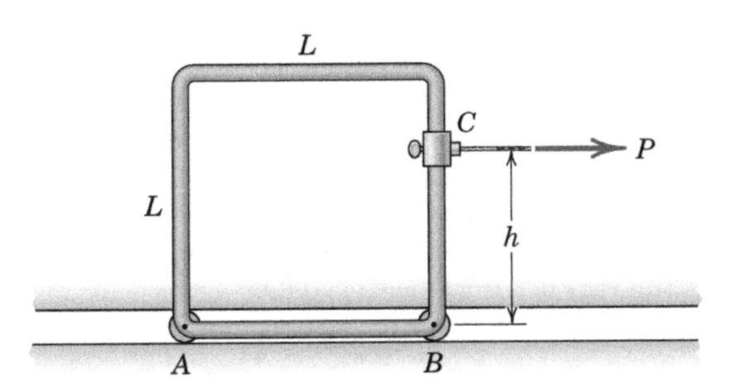

Problem 6/8

6/9 A uniform slender rod rests on a car seat as shown. Determine the deceleration a for which the rod will begin to tip forward. Assume that friction at B is sufficient to prevent slipping.

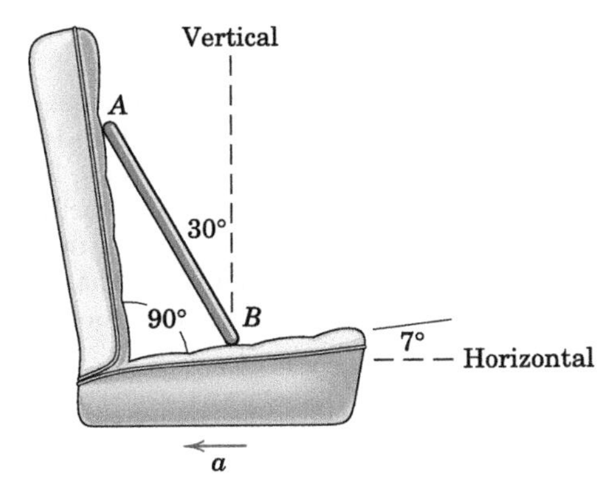

Problem 6/9

6/10 Determine the value of P which will cause the homogeneous cylinder to begin to roll up out of its rectangular recess. The mass of the cylinder is m and that of the cart is M. The cart wheels have negligible mass and friction.

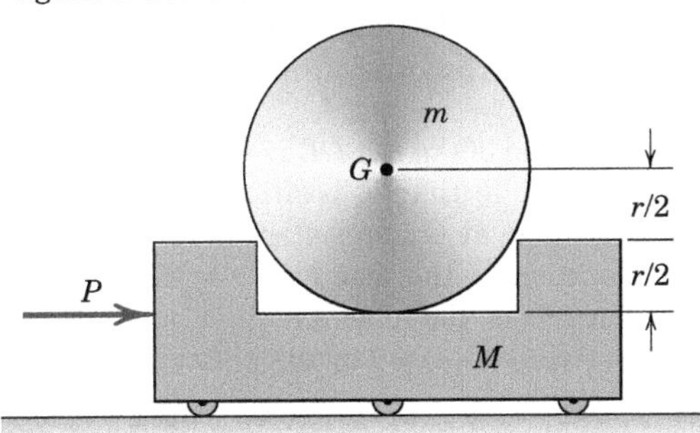

Problem 6/10

6/11 The uniform 5-kg bar AB is suspended in a vertical position from an accelerating vehicle and restrained by the wire BC. If the acceleration is $a = 0.6g$, determine the tension T in the wire and the magnitude of the total force supported by the pin at A.

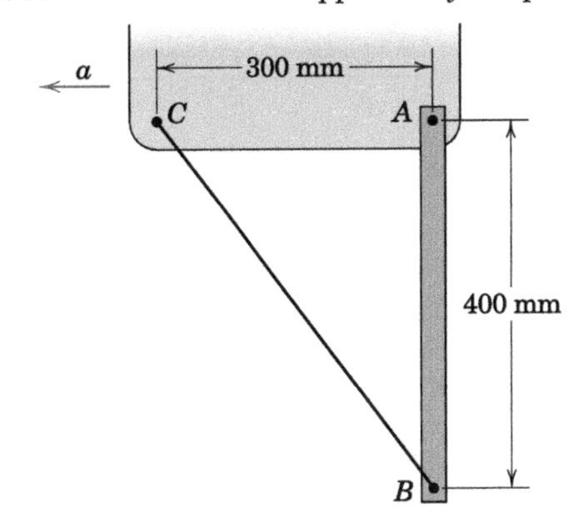

Problem 6/11

6/12 If the collar P is given a constant acceleration $a = 3g$ to the right, the pendulum will assume a steady-state deflection $\theta = 30°$. Determine the stiffness k_T of the torsional spring which will allow this to happen. The torsional spring is undeformed when the pendulum is in the vertical position.

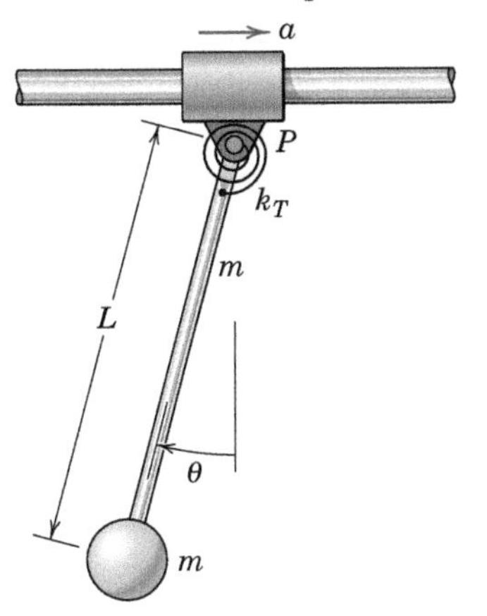

Problem 6/12

*6/13 If the collar P of the pendulum of Prob. 6/12 is given a constant acceleration $a = 5g$, what will be the steady-state deflection of the pendulum from the vertical? Use the value $k_T = 7mgL$.

6/14 The uniform 30-kg bar OB is secured to the accelerating frame in the 30° position from the horizontal by the hinge at O and roller at A. If the horizontal acceleration of the frame is $a = 20$ m/s^2, compute the force F_A on the roller and the x- and y-components of the force supported by the pin at O.

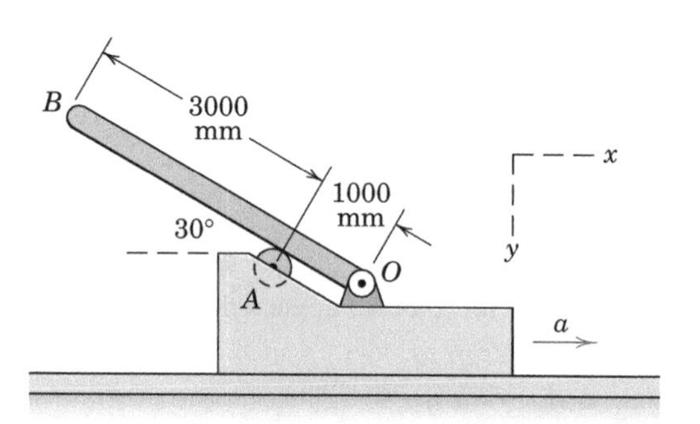

Problem 6/14

6/15 The bicyclist applies the brakes as he descends the 10° incline. What deceleration a would cause the dangerous condition of tipping about the front wheel A? The combined center of mass of the rider and bicycle is at G.

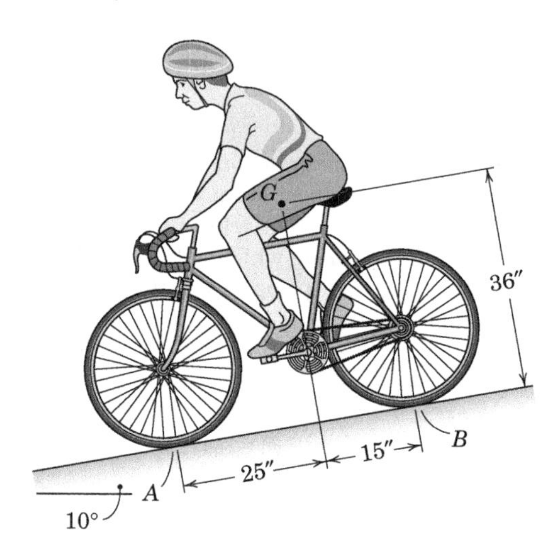

Problem 6/15

Representative Problems

6/16 The right-angle bar acts to control the maximum acceleration of an experimental vehicle by depressing the spring-loaded limit switch A with a vertical force. If a 60-percent reduction in the force applied to the button at A (relative to the static value) results in the switch being tripped, determine the maximum acceleration a which the vehicle is permitted to experience.

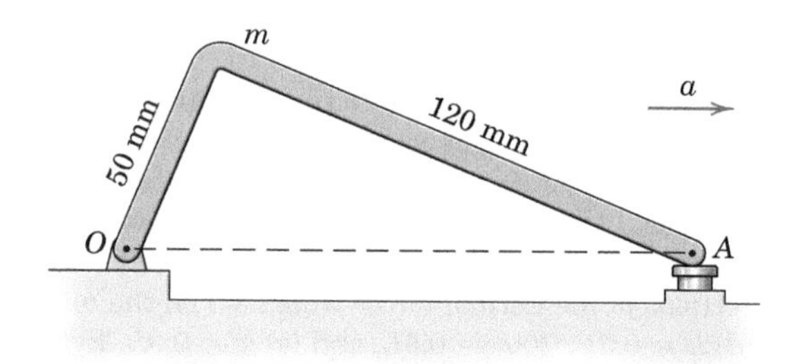

Problem 6/16

6/17 The 1650-kg car has its mass center at G. Calculate the normal forces N_A and N_B between the road and the front and rear pairs of wheels under conditions of maximum acceleration. The mass of the wheels is small compared with the total mass of the car. The coefficient of static friction between the road and the rear driving wheels is 0.80.

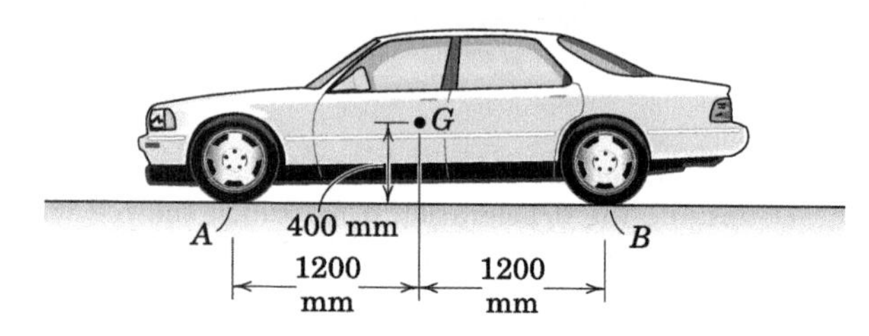

Problem 6/17

6/18 The four-wheel-drive all-terrain vehicle has a mass of 300 kg with center of mass G_2. The driver has a mass of 85 kg with center of mass G_1. If all four wheels are observed to spin momentarily as the driver attempts to go forward, what is the forward acceleration of the driver and ATV? The coefficient of friction between the tires and the ground is 0.40.

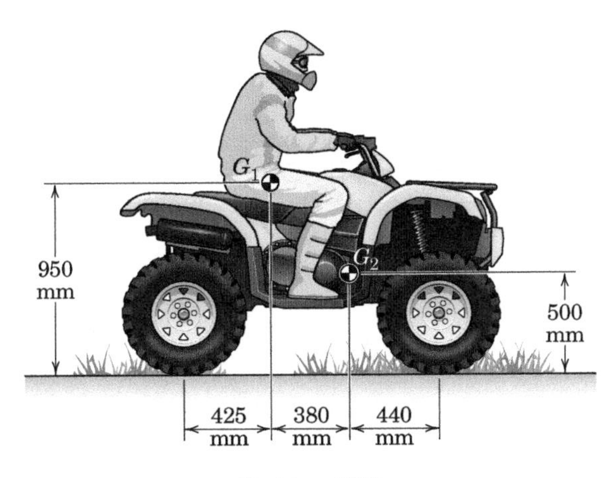

Problem 6/18

6/19 A cleated conveyor belt transports solid homogeneous cylinders up a 15° incline. The diameter of each cylinder is half its height. Determine the maximum acceleration which the belt may have without tipping the cylinders as it starts.

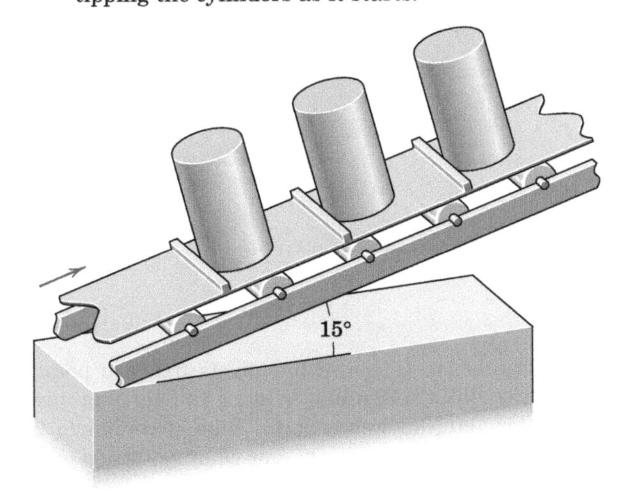

Problem 6/19

6/20 The thin hoop of negligible mass and radius r contains a homogeneous semicylinder of mass m which is rigidly attached to the hoop and positioned such that its diametral face is vertical. The assembly is centered on the top of a cart of mass M which rolls freely on the horizontal surface. If the system is released from rest, what x-directed force P must be applied to the cart to keep the hoop and semicylinder stationary with respect to the cart, and what is the resulting acceleration a of the cart? Motion takes place in the x-y plane. Neglect the mass of the cart wheels and any friction in the wheel bearings. What is the requirement on the coefficient of static friction between the hoop and cart?

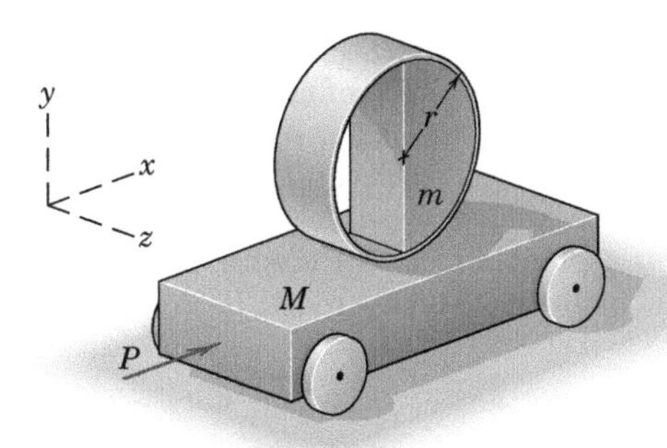

Problem 6/20

6/21 Determine the magnitude P and direction θ of the force required to impart a rearward acceleration $a = 5\text{ ft/sec}^2$ to the loaded wheelbarrow with no rotation from the position shown. The combined weight of the wheelbarrow and its load is 500 lb with center of gravity at G. Compare the normal force at B under acceleration with that for static equilibrium in the position shown. Neglect the friction and mass of the wheel.

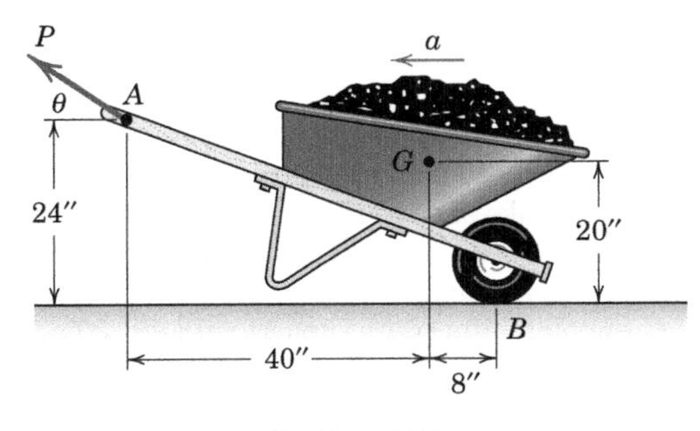

Problem 6/21

6/22 The mine skip has a loaded mass of 2000 kg and is attached to the towing vehicle by the light hinged link CD. If the towing vehicle has an acceleration of 3 m/s^2, calculate the corresponding reactions under the small wheels at A and B.

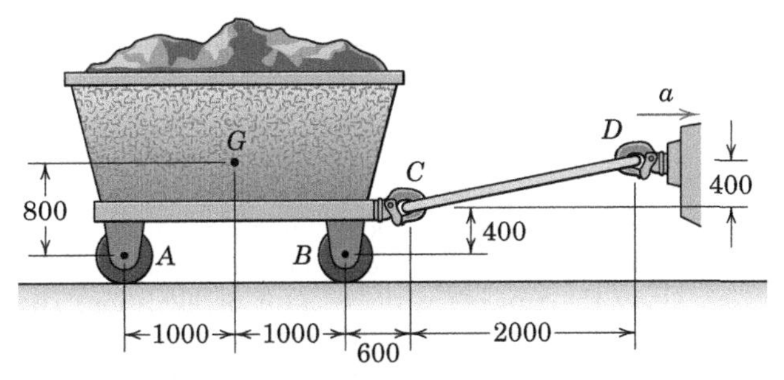

Problem 6/22

6/23 The block A and attached rod have a combined mass of 60 kg and are confined to move along the 60° guide under the action of the 800-N applied force. The uniform horizontal rod has a mass of 20 kg and is welded to the block at B. Friction in the guide is negligible. Compute the bending moment M exerted by the weld on the rod at B.

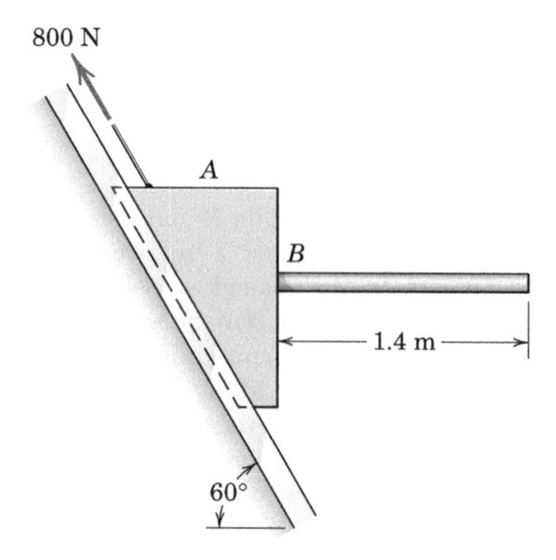

Problem 6/23

6/24 The homogeneous rectangular plate weighs 40 lb and is supported in the vertical plane by the light parallel links shown. If a couple $M = 80$ lb-ft is applied to the end of link AB with the system initially at rest, calculate the force supported by the pin at C as the plate lifts off its support with $\theta = 30°$.

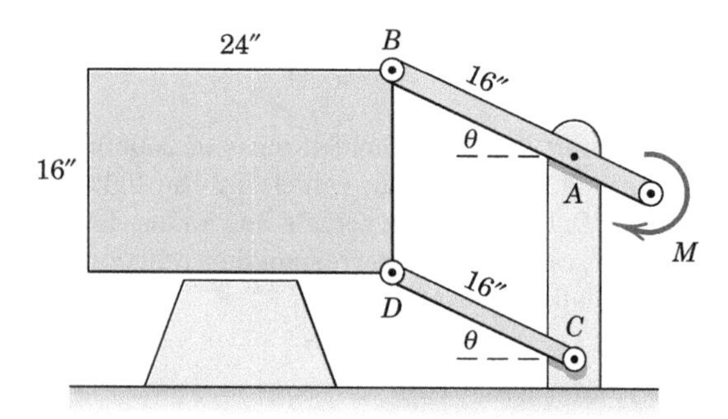

Problem 6/24

6/25 A jet transport with a landing speed of 200 km/h reduces its speed to 60 km/h with a negative thrust R from its jet thrust reversers in a distance of 425 m along the runway with constant deceleration. The total mass of the aircraft is 140 Mg with mass center at G. Compute the reaction N under the nose wheel B toward the end of the braking interval and prior to the application of mechanical braking. At the lower speed, aerodynamic forces on the aircraft are small and may be neglected.

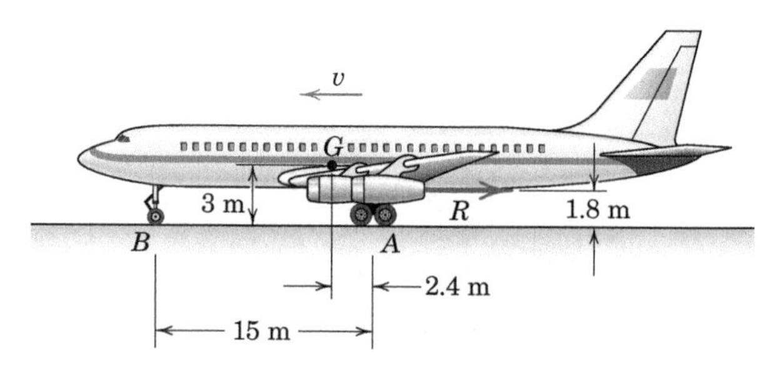

Problem 6/25

6/26 The 1300-lb homogeneous plate is suspended from the overhead carriage by the two parallel steel cables. What acceleration a will cause the tensions in the two cables to be equal? What is the resulting steady-state deflection θ of the cables from the vertical? Evaluate your results for the case where $b = 2h$.

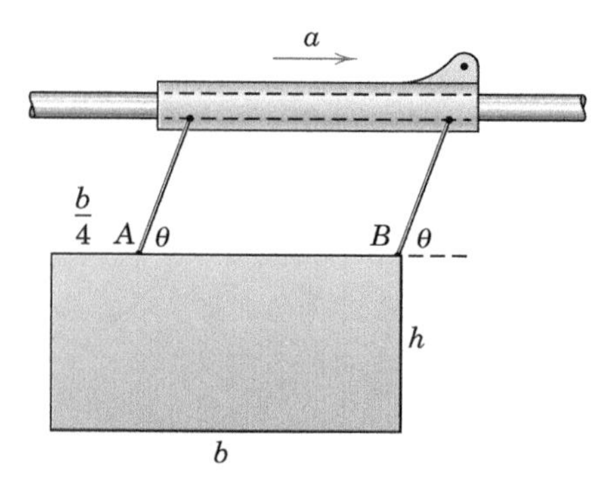

Problem 6/26

6/27 The uniform L-shaped bar pivots freely at point P of the slider, which moves along the horizontal rod. Determine the steady-state value of the angle θ if (a) $a = 0$ and (b) $a = g/2$. For what value of a would the steady-state value of θ be zero?

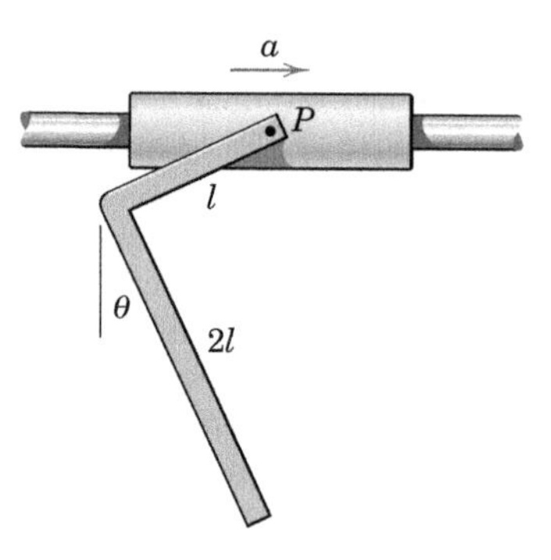

Problem 6/27

►6/28 The 30,000-lb concrete pipe section is being transported on a flatbed truck. Five inextensible cables are passed across the top of the pipe and tightened securely to the flatbed with an initial tension of 2000 lb. What is the maximum deceleration a which the truck can experience if the pipe is to remain stationary relative to the truck? The coefficient of static friction between the concrete and the flatbed is 0.80, and that between the cables and the concrete is 0.75.

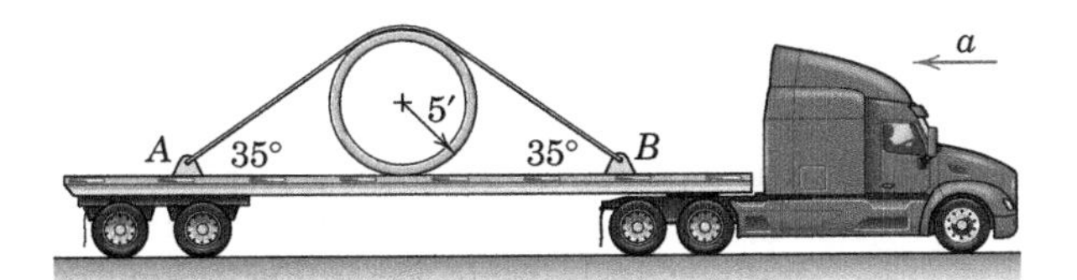

Problem 6/28

6/29 Determine the maximum counterweight W for which the loaded 4000-lb coal car will not overturn about the rear wheels B. Neglect the mass of all pulleys and wheels. (Note that the tension in the cable at C is not $2W$.)

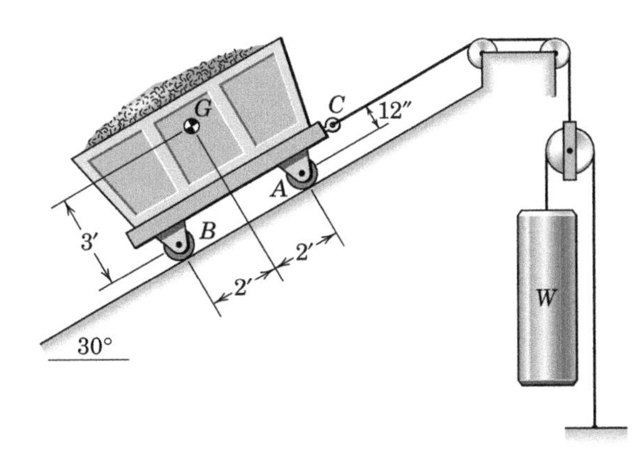

Problem 6/29

6/30 The 1800-kg rear-wheel-drive car accelerates forward at a rate of $g/2$. If the modulus of each of the rear and front springs is 35 kN/m, estimate the resulting momentary nose-up pitch angle θ. (This upward pitch angle during acceleration is called *squat*, while the downward pitch angle during braking is called *dive*!) Neglect the unsprung mass of the wheels and tires. (*Hint:* Begin by assuming a rigid vehicle.)

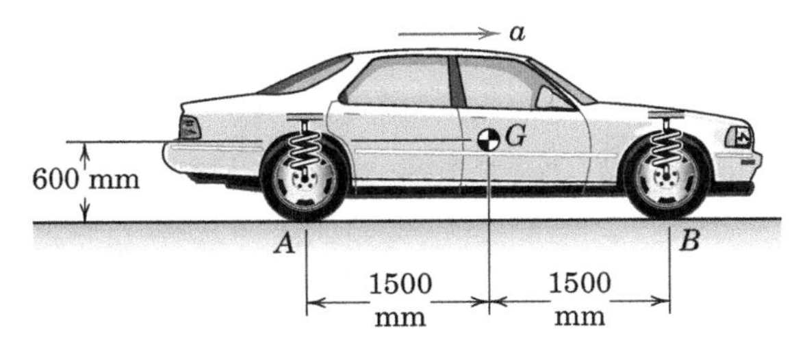

Problem 6/30

6/31 The experimental Formula One race car is traveling at 300 km/h when the driver begins braking to investigate the behavior of the extreme-grip tires. An accelerometer in the car records a maximum deceleration of $4g$ when both the front and rear tires are on the verge of slipping. The car and driver have a combined mass of 690 kg with mass center G. The horizontal drag acting on the car at this speed is 4 kN and may be assumed to pass through the mass center G. The downforce acting over the body of the car at this speed is 13 kN. For simplicity, assume that 35% of this force acts directly over the front wheels, 40% acts directly over the rear wheels, and the remaining portion acts at the mass center. What is the necessary coefficient of friction μ between the tires and the road for this condition? Compare your results with those for passenger car tires.

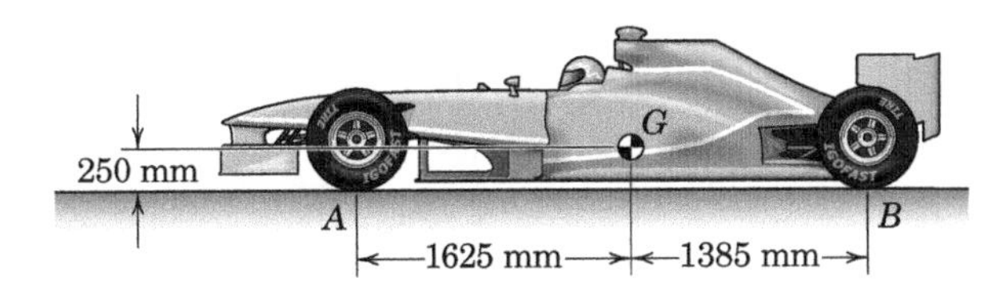

Problem 6/31

▸6/32 The uniform 225-lb crate is supported by the thin homogeneous 40-lb platform BF and light support links whose motion is controlled by the hydraulic cylinder CD. If the cylinder is extending at a constant rate of 6 in./sec when $\theta = 75°$, determine the magnitudes of the forces supported by the pins at B and F. Additionally, determine the total friction force acting on the crate. The crate is centered on the 6-ft platform, and friction is sufficient to keep the crate motionless relative to the platform. (*Hint:* Be careful with the location of the resultant normal force beneath the crate.)

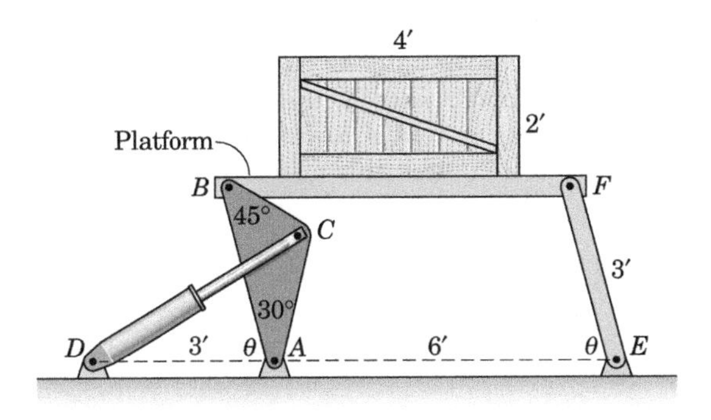

Problem 6/32

PROBLEMS

Introductory Problems

6/33 Two pulleys are fastened together to form an integral unit. At a certain instant, the indicated belt tensions act on the unit and the unit is turning counterclockwise. Determine the angular acceleration of the unit for this instant if the moment due to friction in the bearing at O is 2.5 N·m.

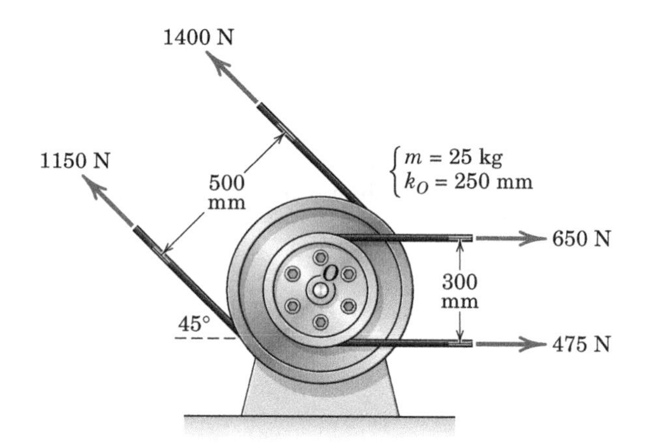

Problem 6/33

6/34 The uniform 20-kg slender bar is pivoted at O and swings freely in the vertical plane. If the bar is released from rest in the horizontal position, calculate the initial value of the force R exerted by the bearing on the bar an instant after release.

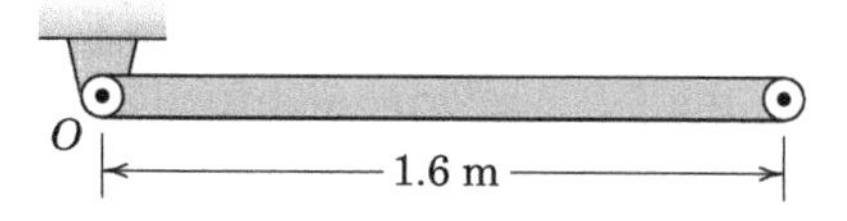

Problem 6/34

6/35 The figure shows an overhead view of a hydraulically-operated gate. As fluid enters the piston side of the cylinder near A, the rod at B extends causing the gate to rotate about a vertical axis through O. For a 2-in.-diameter piston, what fluid pressure p will give the gate an initial counterclockwise angular acceleration of 4 rad/sec? The radius of gyration about O for the 500-lb gate is k_O = 38 in.

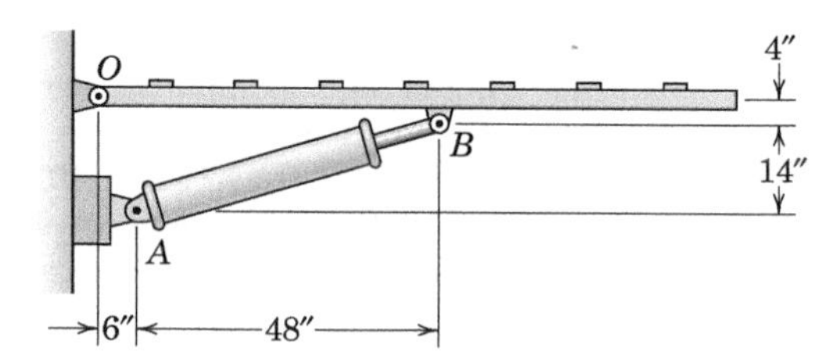

Problem 6/35

6/36 The uniform 100-kg beam is freely hinged about its upper end A and is initially at rest in the vertical position with $\theta = 0$. Determine the initial angular acceleration α of the beam and the magnitude F_A of the force supported by the pin at A due to the application of the force P = 300 N on the attached cable.

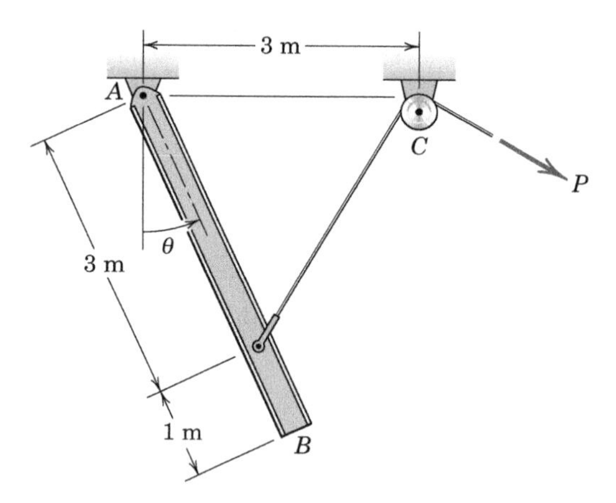

Problem 6/36

6/37 The motor M is used to hoist the 12,000-lb stadium panel (centroidal radius of gyration $\bar{k}$ = 6.5 ft) into position by pivoting the panel about its corner A. If the motor is capable of producing 5000 lb-ft of torque, what pulley diameter d will give the panel an initial counterclockwise angular acceleration of 1.5 deg/sec^2? Neglect all friction.

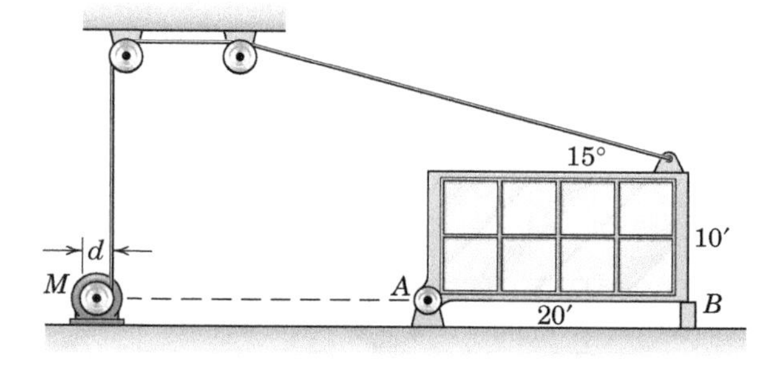

Problem 6/37

6/38 A momentum wheel for dynamics-class demonstrations is shown. It is basically a bicycle wheel modified with rim band-weighting, handles, and a pulley for cord startup. The heavy rim band causes the radius of gyration of the 7-lb wheel to be 11 in. If a steady 10-lb pull T is applied to the cord, determine the angular acceleration of the wheel. Neglect bearing friction.

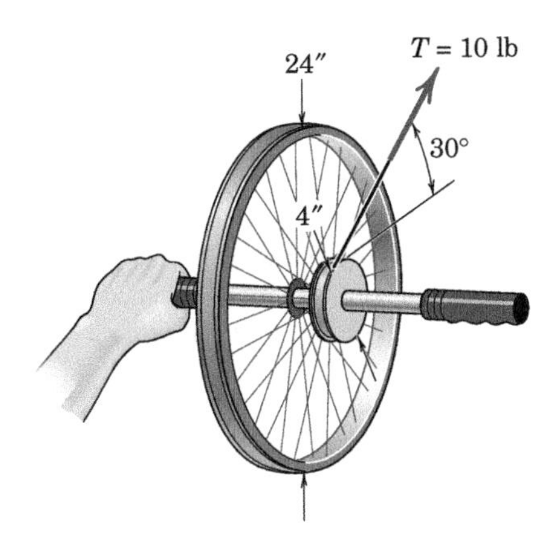

Problem 6/38

6/39 Each of the two drums and connected hubs of 8-in. radius weighs 200 lb and has a radius of gyration about its center of 15 in. Calculate the angular acceleration of each drum. Friction in each bearing is negligible.

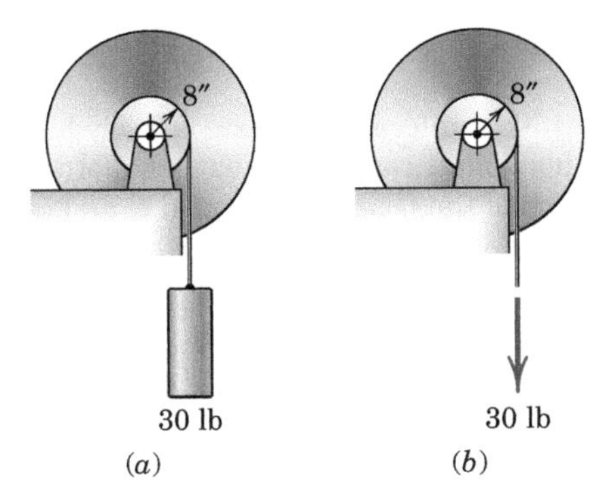

Problem 6/39

6/40 Determine the angular acceleration and the force on the bearing at O for (a) the narrow ring of mass m and (b) the flat circular disk of mass m immediately after each is released from rest in the vertical plane with OC horizontal.

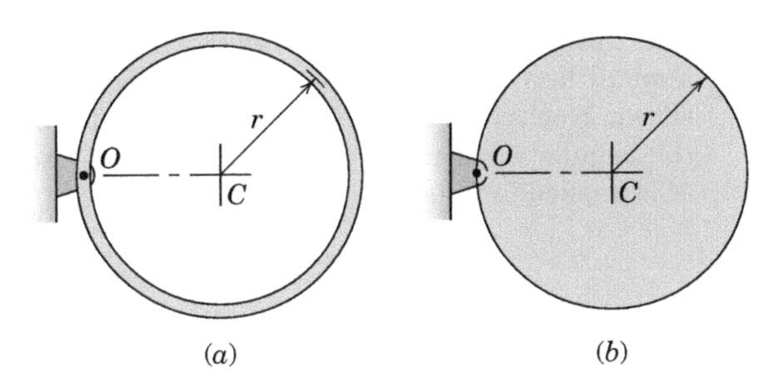

Problem 6/40

6/41 The uniform 5-kg portion of a circular hoop is released from rest while in the position shown where the torsional spring of stiffness k_T = 15 N·m/rad has been twisted 90° clockwise from its undeformed position. Determine the magnitude of the pin force at O at the instant of release. Motion takes place in a vertical plane and the hoop radius is r = 150 mm.

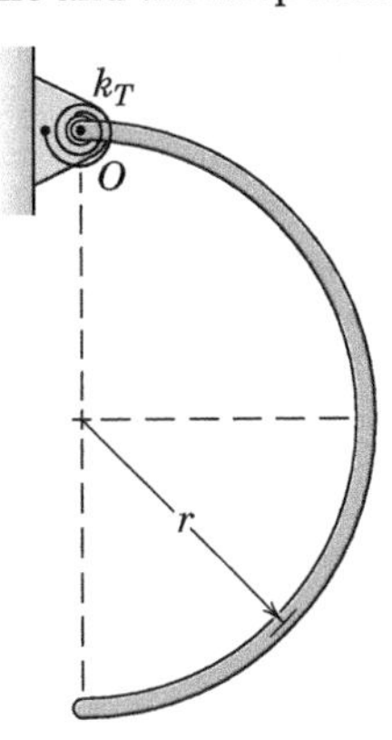

Problem 6/41

6/42 The 30-in. slender bar weighs 20 lb and is mounted on a vertical shaft at O. If a torque M = 100 lb-in. is applied to the bar through its shaft, calculate the horizontal force R on the bearing as the bar starts to rotate.

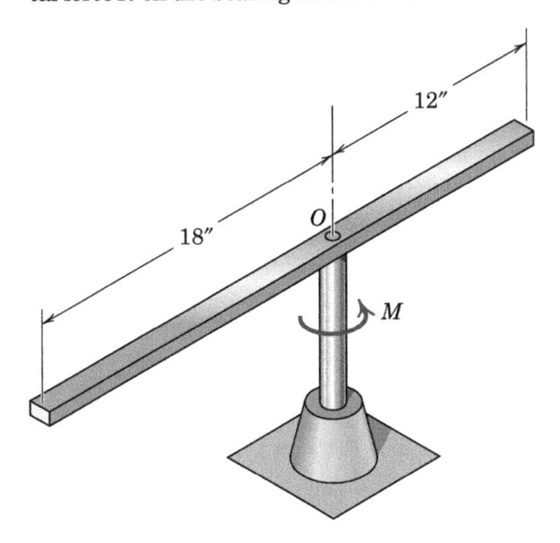

Problem 6/42

6/43 The half ring of mass m and radius r is welded to a small horizontal shaft mounted in a bearing as shown. Neglect the mass of the shaft and determine the angular acceleration of the ring when a torque M is applied to the shaft.

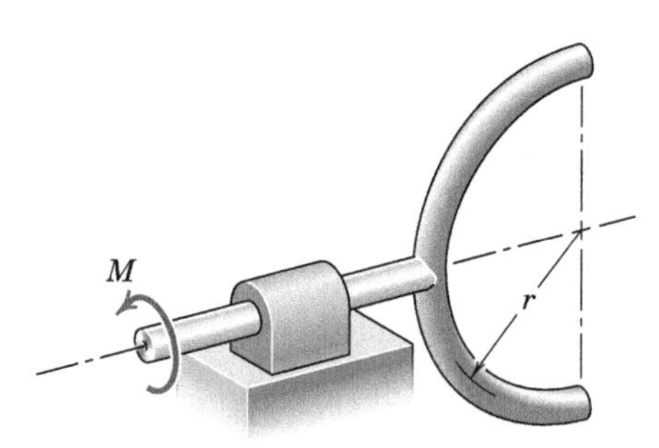

Problem 6/43

6/44 The uniform plate of mass m is released from rest while in the position shown. Determine the initial angular acceleration α of the plate and the magnitude of the force supported by the pin at O. The axis of rotation is horizontal.

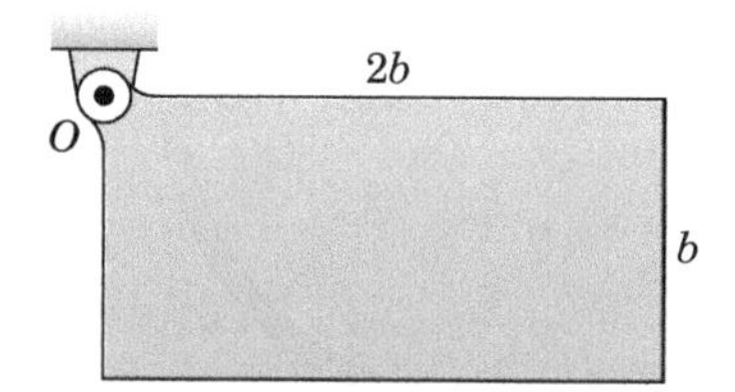

Problem 6/44

6/45 The uniform slender bar AB has a mass of 8 kg and swings in a vertical plane about the pivot at A. If $\dot{\theta} = 2$ rad/s when $\theta = 30°$, compute the force supported by the pin at A at that instant.

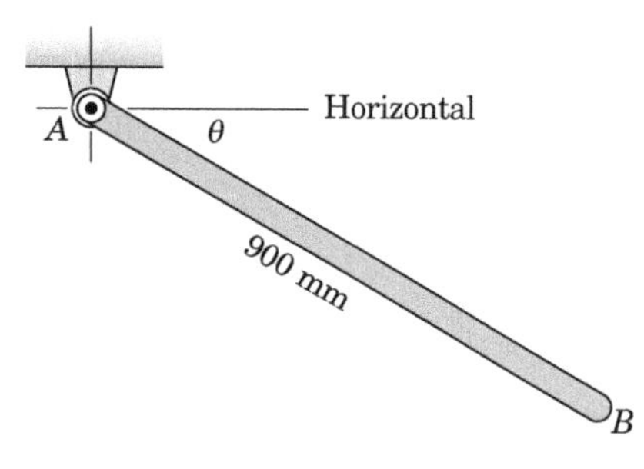

Problem 6/45

6/46 The uniform 16.1-lb slender bar is hinged about a horizontal axis through O and released from rest in the horizontal position. Determine the distance b from the mass center to O which will result in an initial angular acceleration of 16.1 rad/sec^2, and find the force R on the bar at O just after release.

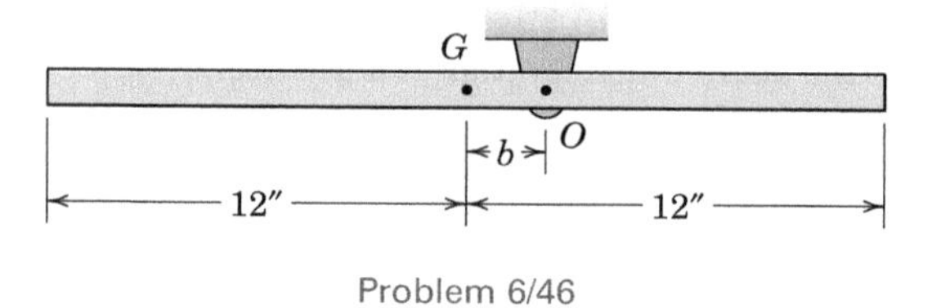

Problem 6/46

6/47 The uniform quarter-circular sector of mass m is released from rest with one straight edge vertical as shown. Determine the initial angular acceleration and the horizontal and vertical components of the reaction at the ideal pivot at O.

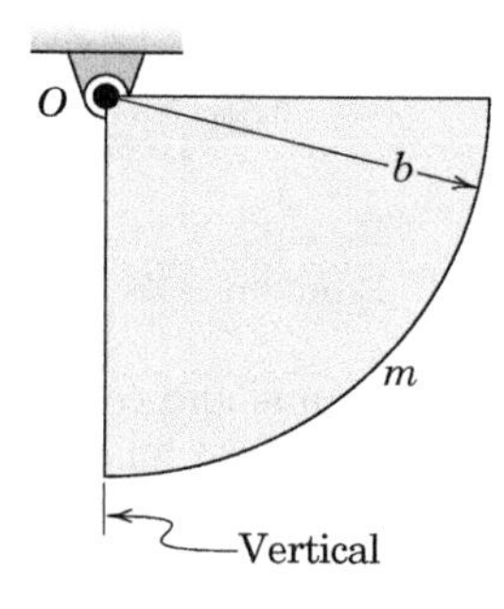

Problem 6/47

Representative Problems

6/48 The 15-kg uniform steel plate is freely hinged about the horizontal z-axis. Calculate the force supported by each of the bearings at A and B an instant after the plate is released from rest while in the horizontal position shown.

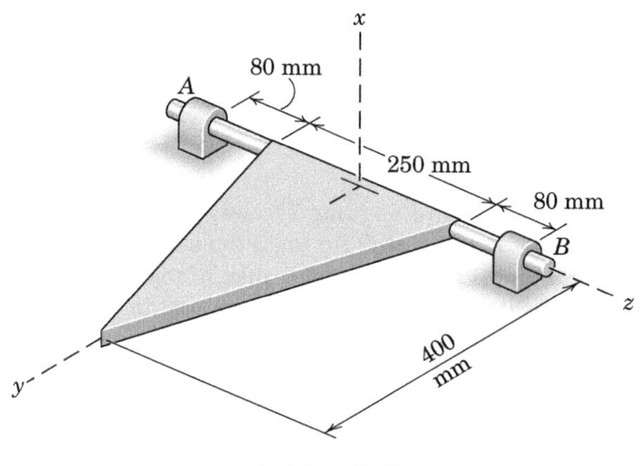

Problem 6/48

6/49 The square frame is composed of four equal lengths of uniform slender rod, and the ball attachment at O is suspended in a socket (not shown). Beginning from the position shown, the assembly is rotated 45° about axis A-A and released. Determine the initial angular acceleration of the frame. Repeat for a 45° rotation about axis B-B. Neglect the small mass, offset, and friction of the ball.

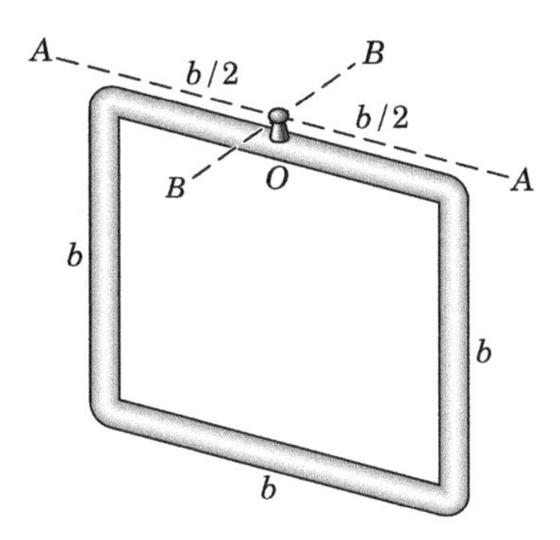

Problem 6/49

6/50 A uniform torus and a cylindrical ring, each solid and of mass m, are released from rest in the positions shown. Determine the magnitude of the pin reaction at O and the angular acceleration of each body an instant after release. Neglect friction in the pivot at O for each case.

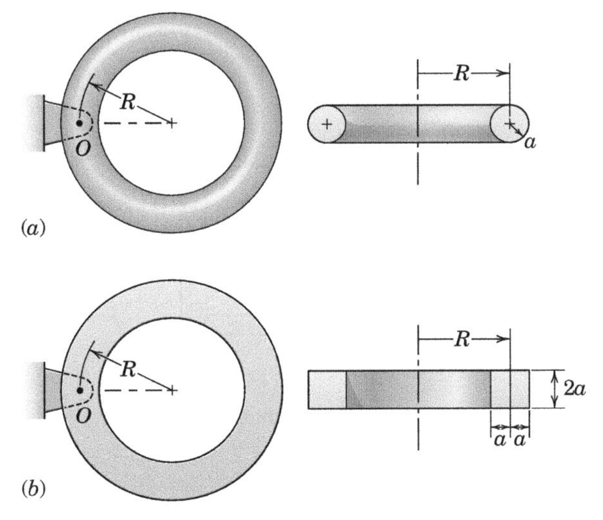

Problem 6/50

6/51 A reel of flexible power cable is mounted on the dolly, which is fixed in position. There are 200 ft of cable weighing 0.436 lb per foot of length wound on the reel at a radius of 15 in. The empty spool weighs 62 lb and has a radius of gyration about its axis of 12 in. A tension T of 20 lb is required to overcome frictional resistance to turning. Calculate the angular acceleration α of the reel if a tension of 40 lb is applied to the free end of the cable.

Problem 6/51

6/52 The uniform bar of mass m is supported by the smooth pin at O and is connected to the cylinder of mass m_1 by the light cable which passes over the light pulley at C. If the system is released from rest while in the position shown, determine the tension in the cable. Use the values $m = 30$ kg, $m_1 = 20$ kg, and $L = 6$ m.

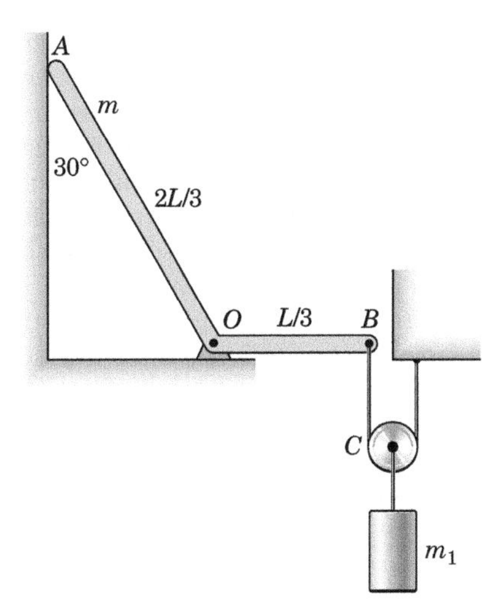

Problem 6/52

6/53 An *air table* is used to study the elastic motion of flexible spacecraft models. Pressurized air escaping from numerous small holes in the horizontal surface provides a supporting air cushion which largely eliminates friction. The model shown consists of a cylindrical hub of radius r and four appendages of length l and small thickness t. The hub and the four appendages all have the same depth d and are constructed of the same material of density ρ. Assume that the spacecraft is rigid and determine the moment M which must be applied to the hub to spin the model from rest to an angular velocity ω in a time period of τ seconds. (Note that for a spacecraft with highly flexible appendages, the moment must be judiciously applied to the rigid hub to avoid undesirable large elastic deflections of the appendages.)

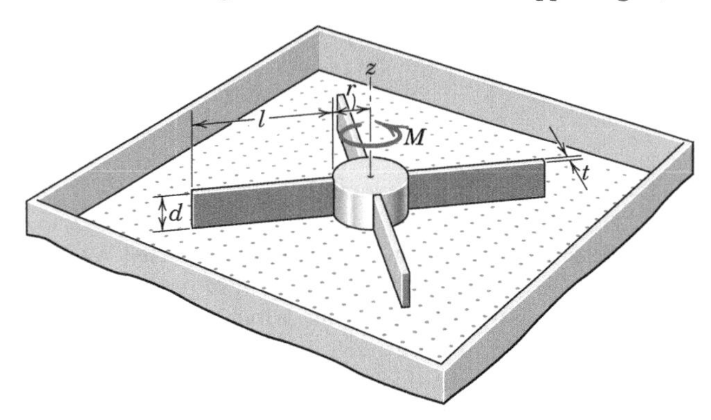

Problem 6/53

6/54 A vibration test is run to check the design adequacy of bearings A and B. The unbalanced rotor and attached shaft have a combined mass of 2.8 kg. To locate the mass center, a torque of 0.660 N·m is applied to the shaft to hold it in equilibrium in a position rotated 90° from that shown. A constant torque $M = 1.5$ N·m is then applied to the shaft, which reaches a speed of 1200 rev/min in 18 revolutions starting from rest. (During each revolution the angular acceleration varies, but its average value is the same as for constant acceleration.) Determine (*a*) the radius of gyration k of the rotor and shaft about the rotation axis, (*b*) the force F which each bearing exerts on the shaft immediately after M is applied, and (*c*) the force R exerted by each bearing when the speed of 1200 rev/min is reached and M is removed. Neglect any frictional resistance and the bearing forces due to static equilibrium.

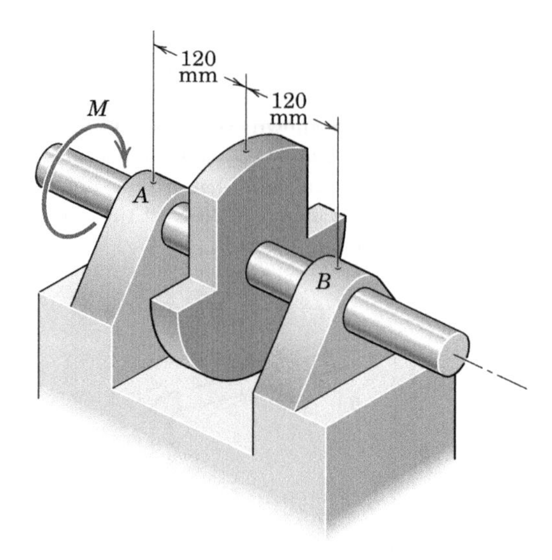

Problem 6/54

6/55 The solid cylindrical rotor B has a mass of 43 kg and is mounted on its central axis C-C. The frame A rotates about the fixed vertical axis O-O under the applied torque $M = 30$ N·m. The rotor may be unlocked from the frame by withdrawing the locking pin P. Calculate the angular acceleration α of the frame A if the locking pin is (*a*) in place and (*b*) withdrawn. Neglect all friction and the mass of the frame.

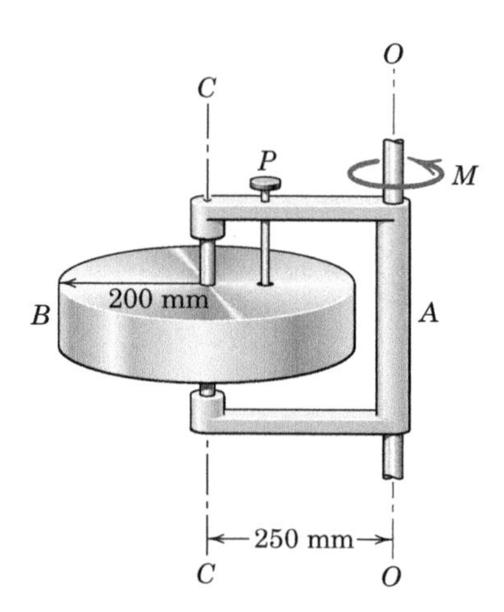

Problem 6/55

6/56 The right-angle body is made of uniform slender bar of mass m and length L. It is released from rest while in the position shown. Determine the initial angular acceleration α of the body and the magnitude of the force supported by the pivot at O.

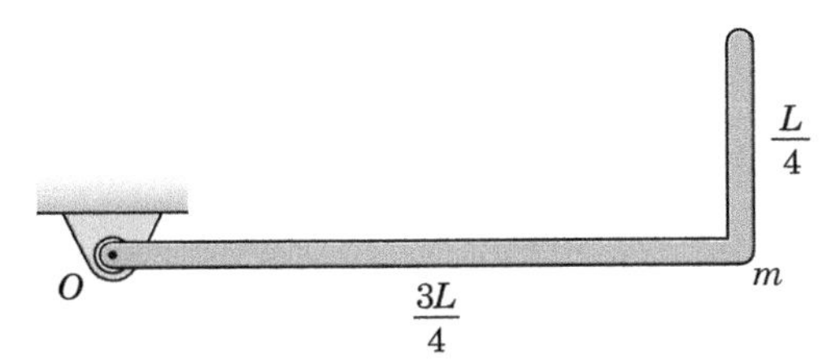

Problem 6/56

6/57 Each of the two grinding wheels has a diameter of 6 in., a thickness of 3/4 in., and a specific weight of 425 lb/ft^3. When switched on, the machine accelerates from rest to its operating speed of 3450 rev/min in 5 sec. When switched off, it comes to rest in 35 sec. Determine the motor torque and frictional moment, assuming that each is constant. Neglect the effects of the inertia of the rotating motor armature.

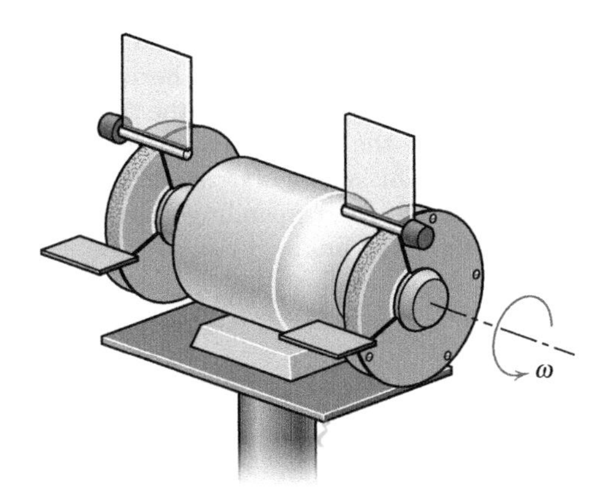

Problem 6/57

6/58 The uniform slender bar is released from rest in the horizontal position shown. Determine the value of x for which the angular acceleration is a maximum, and determine the corresponding angular acceleration α.

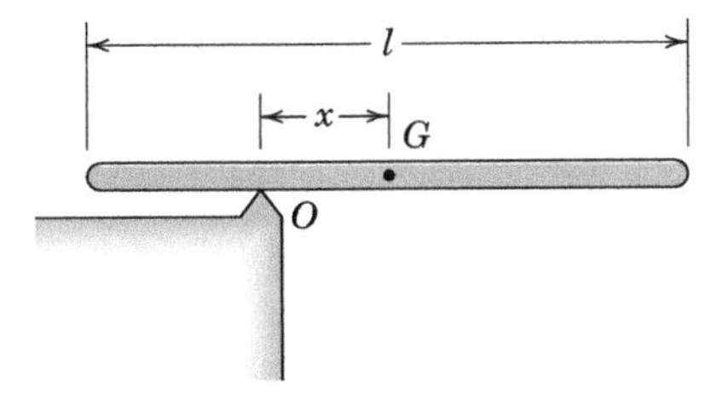

Problem 6/58

6/59 The assembly from Prob. 3/219 is repeated here with the following additional information. The 2-kg collar at C has an outer diameter of 80 mm and is press fitted to the light 50-mm-diameter shaft. Each spoke has a mass of 1.5 kg and carries a 3-kg sphere with a radius of 40 mm attached to its end. The pulley at D has a mass of 5 kg with a centroidal radius of gyration of 60 mm. If a tension $T = 20$ N is applied to the end of the securely wrapped cable with the assembly initially at rest, determine the initial angular acceleration of the assembly. Neglect friction in the bearings at A and B and state any assumptions.

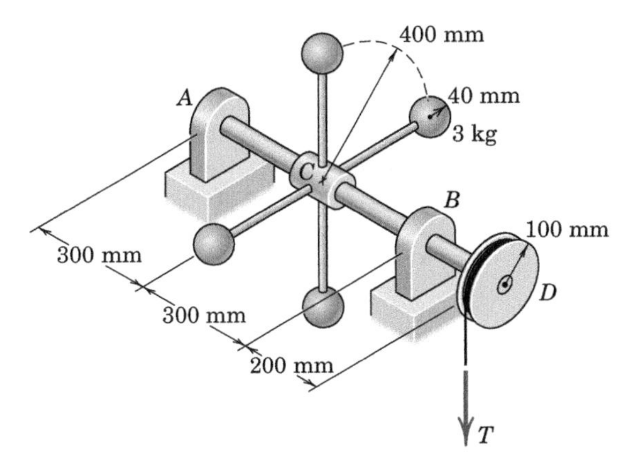

Problem 6/59

6/60 The right-angle plate is formed from a flat plate having a mass ρ per unit area and is welded to the horizontal shaft mounted in the bearing at O. If the shaft is free to rotate, determine the initial angular acceleration α of the plate when it is released from rest with the upper surface in the horizontal plane. Also determine the y- and z-components of the resultant force on the shaft at O.

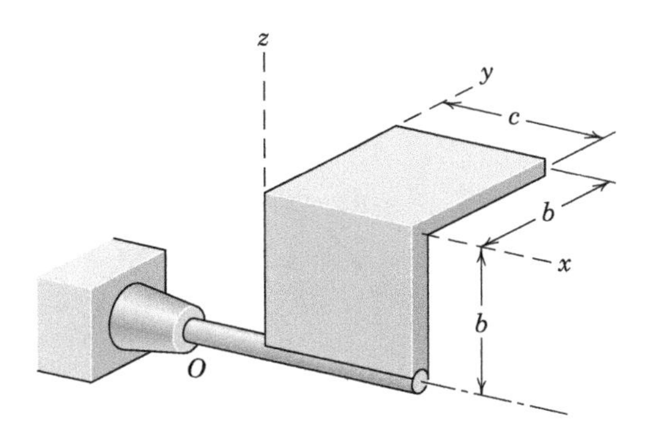

Problem 6/60

6/61 The semicircular disk of mass m and radius r is released from rest at $\theta = 0$ and rotates freely in the vertical plane about its fixed bearing at O. Derive expressions for the n- and t-components of the force F on the bearing as functions of θ.

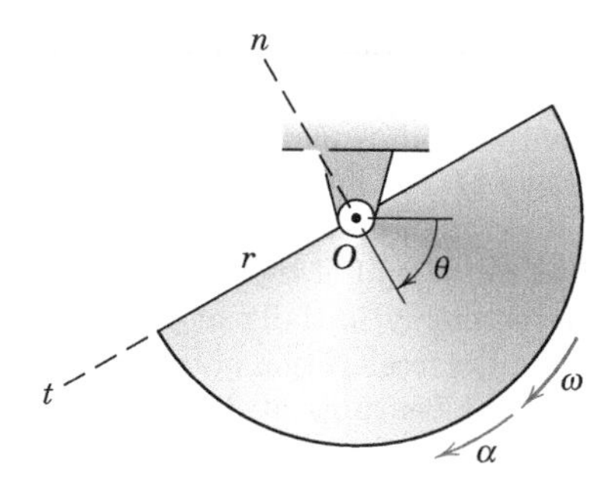

Problem 6/61

6/62 The uniform steel I-beam has a mass of 300 kg and is supported in the vertical plane as shown. Calculate the force R supported by the pin at O for the condition immediately after the support at B is suddenly removed. The mass of the bracket on the left end is small and may be neglected. Also treat the beam as a uniform slender bar.

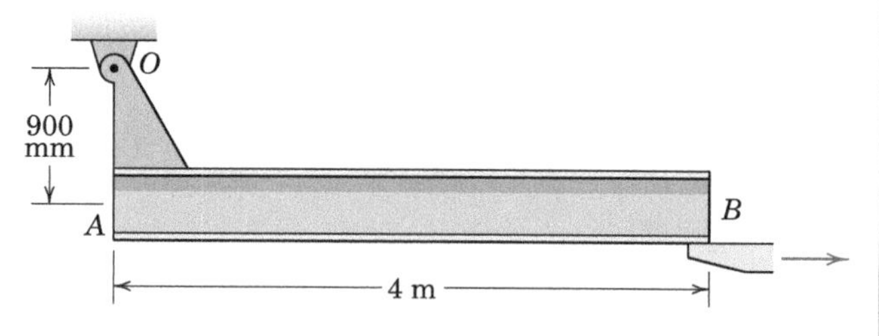

Problem 6/62

6/63 The gear train shown operates in a horizontal plane and is used to transmit motion to the rack D of mass m_D. If an input torque M is applied to gear A, what will be the resulting acceleration a of the unloaded rack? (The mechanism which it normally drives has been disengaged.) Gear C is keyed to the same shaft as gear B. Gears A, B, and C have pitch diameters d_A, d_B, and d_C, and centroidal mass moments of inertia I_A, I_B, and I_C, respectively. All friction is negligible.

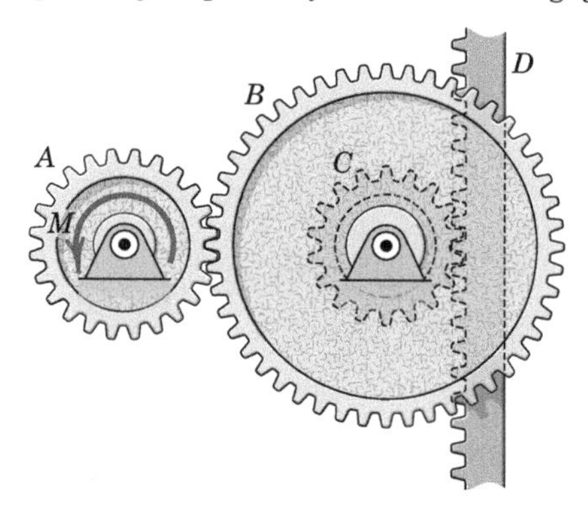

Problem 6/63

6/64 The uniform semicircular ring of mass $m = 2.5$ kg and mean radius $r = 200$ mm is mounted on spokes of negligible mass and pivoted about a horizontal axis through O. If the ring is released from rest in the position $\theta = 30°$, determine the force R supported by the bearing O just after release.

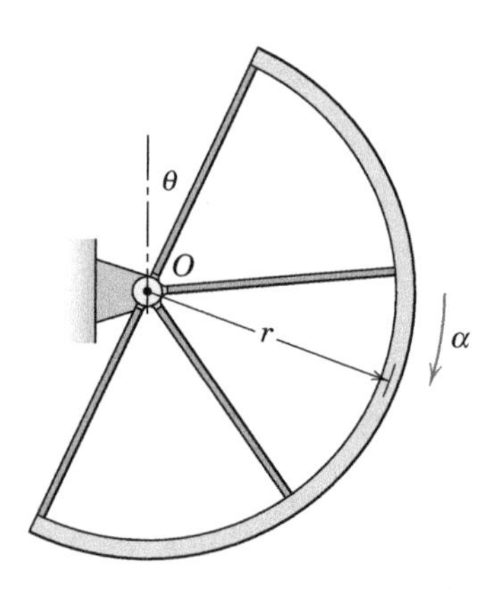

Problem 6/64

6/65 The link B weighs 0.80 lb with center of mass 2.20 in. from O-O and has a radius of gyration about O-O of 2.76 in. The link is welded to the steel tube and is free to rotate about the fixed horizontal shaft at O-O. The tube weighs 1.84 lb. If the tube is released from rest with the link in the horizontal position, calculate the initial angular acceleration α of the assembly and the corresponding reaction O exerted by the shaft on the link.

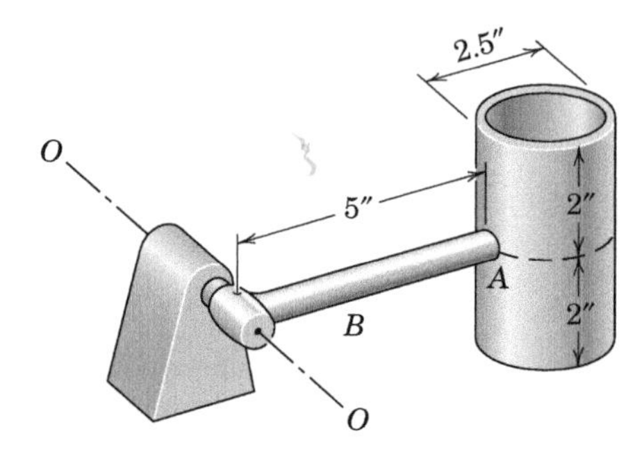

Problem 6/65

6/66 A flexible cable 60 meters long with a mass of 0.160 kg per meter of length is wound around the reel. With $y = 0$, the weight of the 4-kg cylinder is required to start turning the reel to overcome friction in its bearings. Determine the downward acceleration a in meters per second squared of the cylinder as a function of y in meters. The empty reel has a mass of 16 kg with a radius of gyration about its bearing of 200 mm.

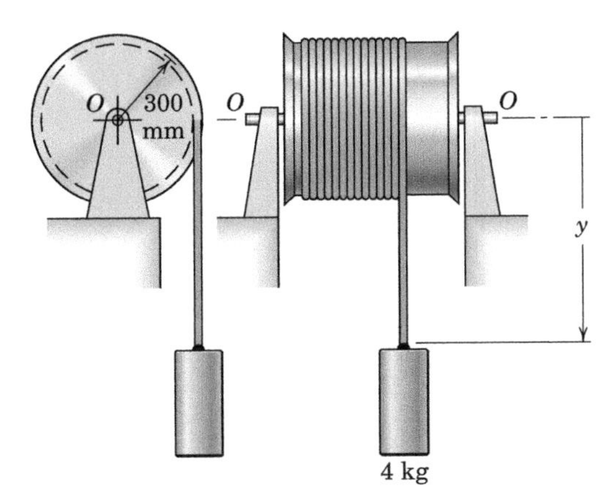

Problem 6/66

6/67 The uniform 72-ft mast weighs 600 lb and is hinged at its lower end to a fixed support at O. If the winch C develops a starting torque of 900 lb-ft, calculate the total force supported by the pin at O as the mast begins to lift off its support at B. Also find the corresponding angular acceleration α of the mast. The cable at A is horizontal, and the mass of the pulleys and winch is negligible.

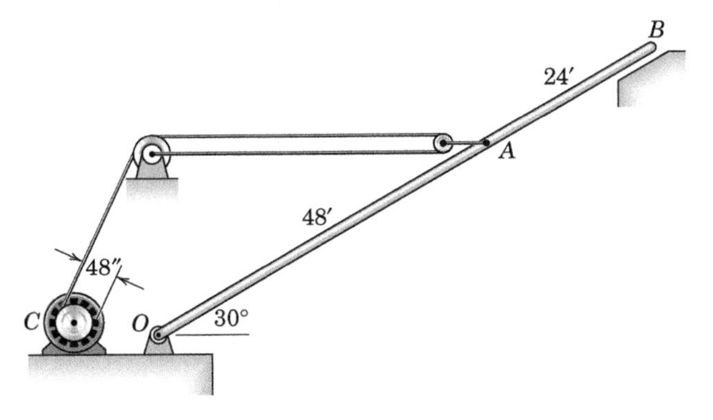

Problem 6/67

6/68 The robotic device consists of the stationary pedestal OA, arm AB pivoted at A, and arm BC pivoted at B. The rotation axes are normal to the plane of the figure. Estimate (a) the moment M_A applied to arm AB required to rotate it about joint A at 4 rad/s^2 counterclockwise from the position shown with joint B locked and (b) the moment M_B applied to arm BC required to rotate it about joint B at the same rate with joint A locked. The mass of arm AB is 25 kg and that of BC is 4 kg, with the stationary portion of joint A excluded entirely and the mass of joint B divided equally between the two arms. Assume that the centers of mass G_1 and G_2 are in the geometric centers of the arms and model the arms as slender rods.

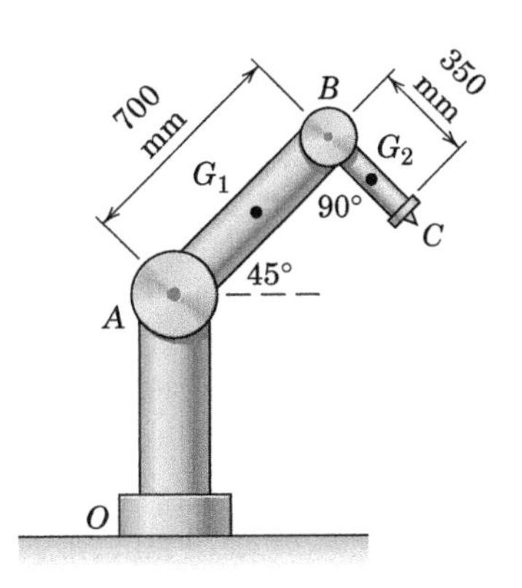

Problem 6/68

▶6/69 For the beam described in Prob. 6/62, determine the maximum angular velocity ω reached by the beam as it rotates in the vertical plane about the bearing at O. Also determine the corresponding force R supported by the pin at O for this condition. Again, treat the beam as a slender 300-kg bar and neglect the mass of the supporting bracket.

▶6/70 The uniform parabolic plate has a mass per unit area of 225 kg/m^2. If the plate is released from rest while in the horizontal position shown, determine the initial angular acceleration $\boldsymbol{\alpha}$ about the bearing axis AB.

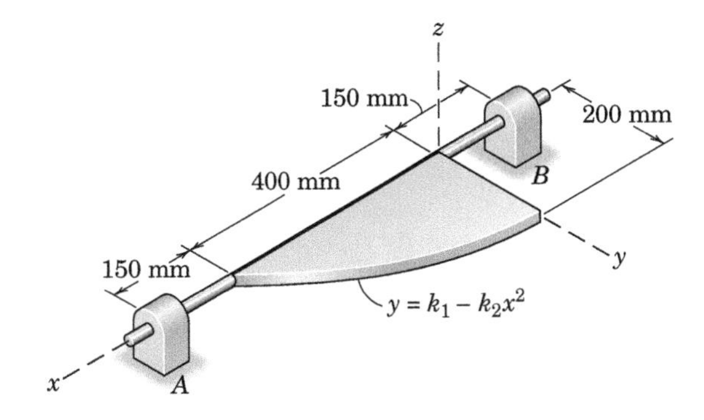

Problem 6/70

PROBLEMS

Introductory Problems

6/71 The uniform slender bar rests on a smooth horizontal surface when a force F is applied normal to the bar at point A. Point A is observed to have an initial acceleration a_A of 20 m/s^2, and the bar has a corresponding angular acceleration α of 18 rad/s^2. Determine the distance b.

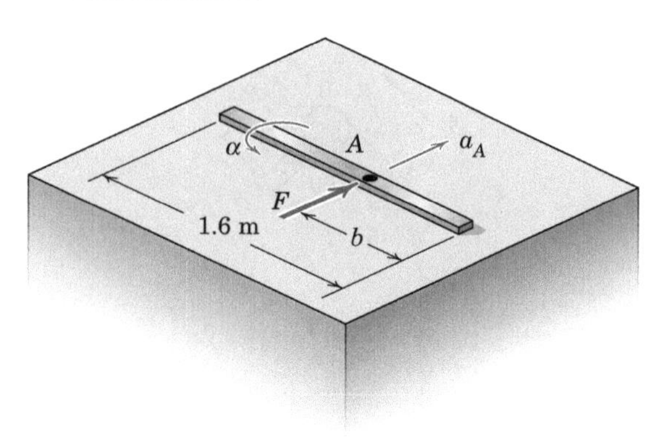

Problem 6/71

6/72 The 64.4-lb solid circular disk is initially at rest on the horizontal surface when a 3-lb force P, constant in magnitude and direction, is applied to the cord wrapped securely around its periphery. Friction between the disk and the surface is negligible. Calculate the angular velocity ω of the disk after the 3-lb force has been applied for 2 seconds and find the linear velocity v of the center of the disk after it has moved 3 feet from rest.

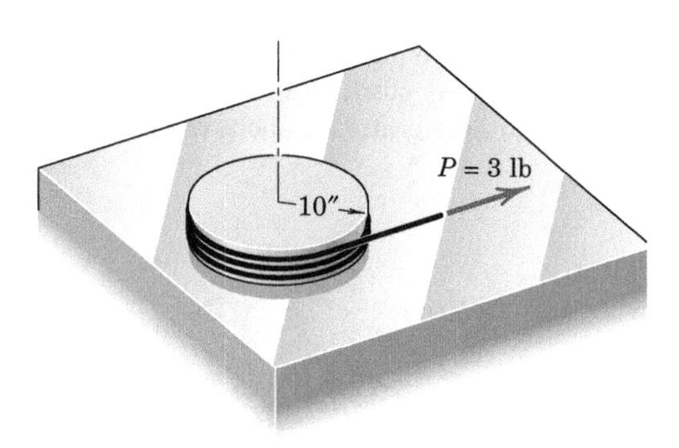

Problem 6/72

6/73 A long cable of length L and mass ρ per unit length is wrapped around the periphery of a spool of negligible mass. One end of the cable is fixed, and the spool is released from rest in the position shown. Find the initial acceleration a of the center of the spool.

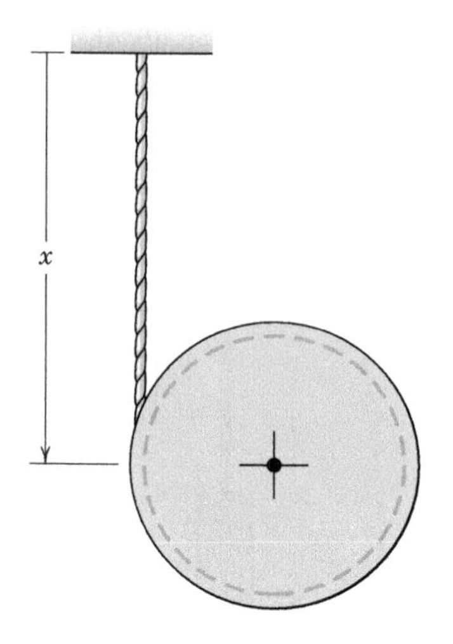

Problem 6/73

6/74 The uniform semicircular rod of mass m is lying motionless on the smooth horizontal surface when the force F is applied at B as shown. Determine the resulting initial acceleration of point A.

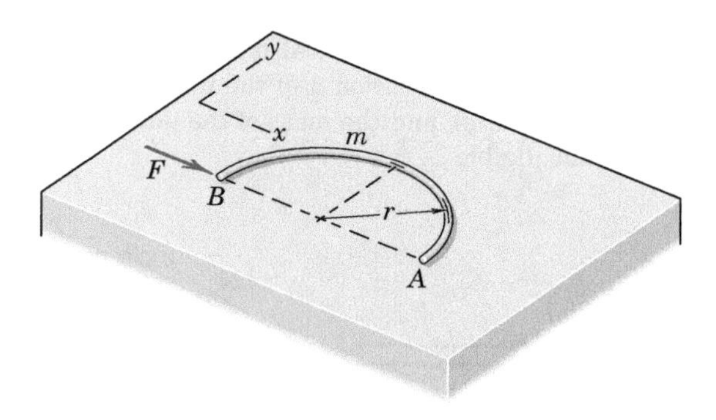

Problem 6/74

6/75 Repeat Prob. 6/74, except that the direction of the applied force has been changed as shown in the figure.

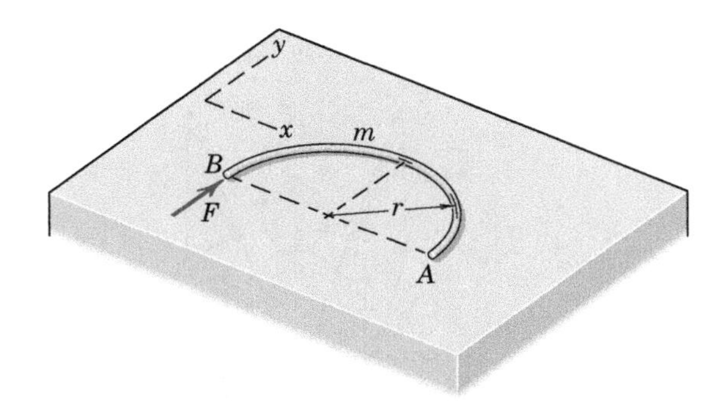

Problem 6/75

6/76 The spacecraft is spinning with a constant angular velocity ω about the z-axis at the same time that its mass center O is traveling with a velocity v_O in the y-direction. If a tangential hydrogen-peroxide jet is fired when the craft is in the position shown, determine the expression for the absolute acceleration of point A on the spacecraft rim at the instant the jet force is F. The radius of gyration of the craft about the z-axis is k, and its mass is m.

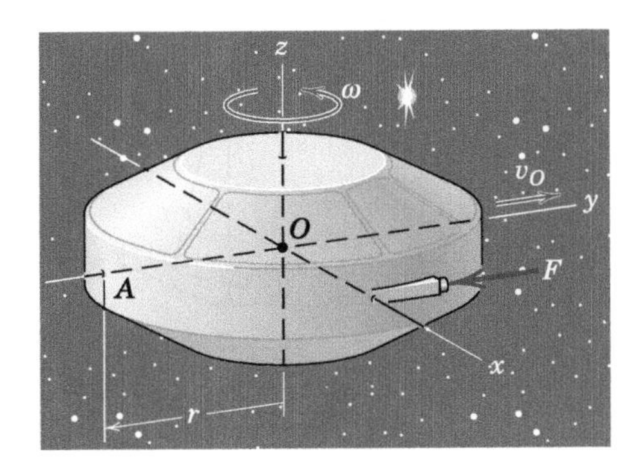

Problem 6/76

6/77 The body consists of a uniform slender bar and a uniform disk, each of mass $m/2$. It rests on a smooth surface. Determine the angular acceleration α and the acceleration of the mass center of the body when the force $P = 6$ N is applied as shown. The value of the mass m of the entire body is 1.2 kg.

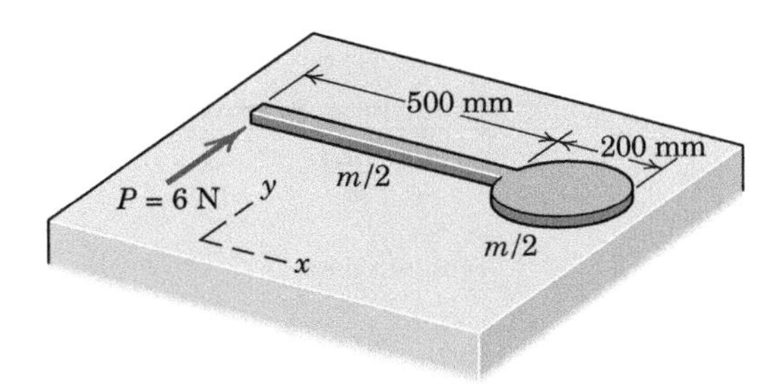

Problem 6/77

6/78 Determine the angular acceleration of each of the two wheels as they roll without slipping down the inclines. For wheel A investigate the case where the mass of the rim and spokes is negligible and the mass of the bar is concentrated along its centerline. For wheel B assume that the thickness of the rim is negligible compared with its radius so that all of the mass is concentrated in the rim. Also specify the minimum coefficient of static friction μ_s required to prevent each wheel from slipping.

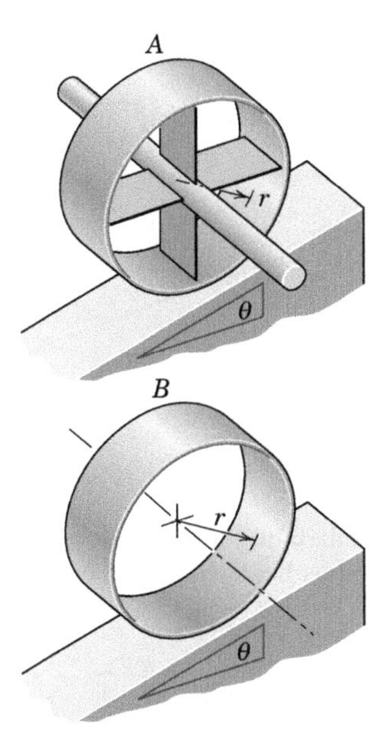

Problem 6/78

6/79 The solid homogeneous cylinder is released from rest on the ramp. If $\theta = 40°$, $\mu_s = 0.30$, and $\mu_k = 0.20$, determine the acceleration of the mass center G and the friction force exerted by the ramp on the cylinder.

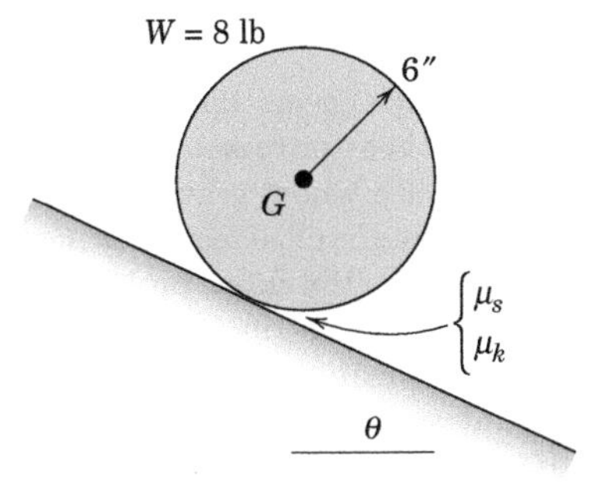

Problem 6/79

6/80 The 30-kg spool of outer radius r_o = 450 mm has a centroidal radius of gyration $\bar{k}$ = 275 mm and a central shaft of radius r_i = 200 mm. The spool is at rest on the incline when a tension T = 300 N is applied to the end of a cable which is wrapped securely around the central shaft as shown. Determine the acceleration of the spool center G and the magnitude and direction of the friction force acting at the interface of the spool and incline. The friction coefficients there are μ_s = 0.45 and μ_k = 0.30. The tension T is applied parallel to the incline and the angle θ = 20°.

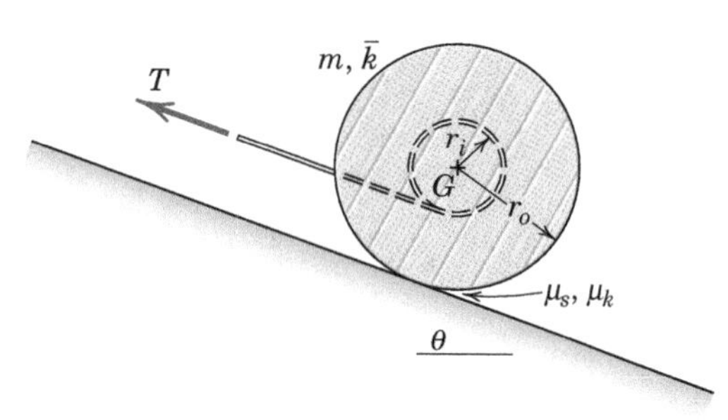

Problem 6/80

6/81 Repeat Prob. 6/80 for the case where the cable configuration has been changed as shown in the figure.

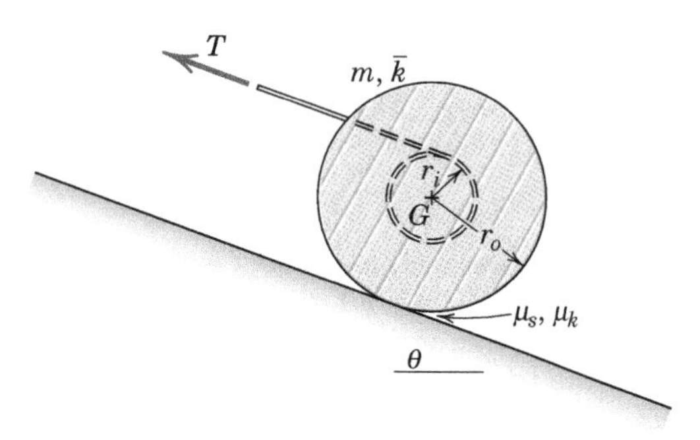

Problem 6/81

6/82 The fairing which covers the spacecraft package in the nose of the booster rocket is jettisoned when the rocket is in space where gravitational attraction is negligible. A mechanical actuator moves the two halves slowly from the closed position I to position II at which point the fairings are released to rotate freely about their hinges at O under the influence of the constant acceleration a of the rocket. When position III is reached, the hinge at O is released and the fairings drift away from the rocket. Determine the angular velocity ω of the fairing at the 90° position. The mass of each fairing is m with center of mass at G and radius of gyration k_O about O.

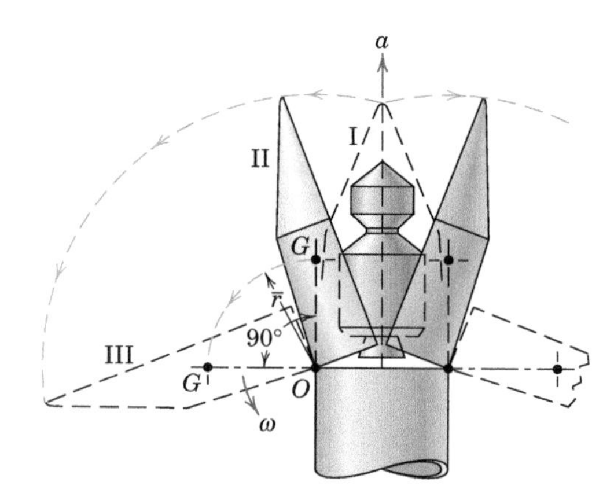

Problem 6/82

6/83 The uniform heavy bar AB of mass m is moving on its light end rollers along the horizontal with a velocity v when end A passes point C and begins to move on the curved portion of the path with radius r. Determine the force exerted by the path on roller A immediately after it passes C.

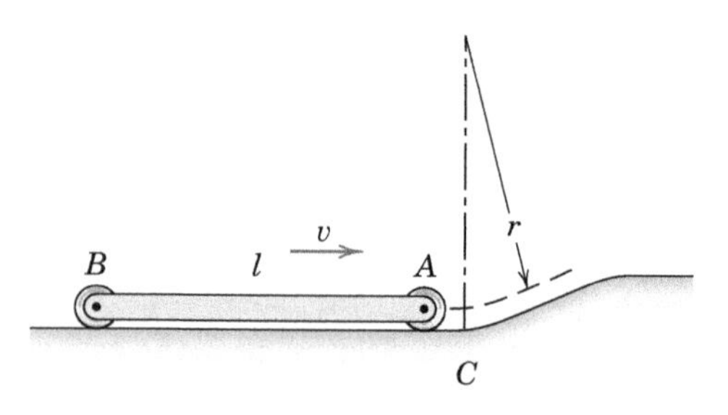

Problem 6/83

6/84 The uniform slender bar of mass m and total length L is released from rest in the position shown. Determine the force supported by the small roller at A and the acceleration of roller A along the smooth guide. Evaluate your results for θ = 15°.

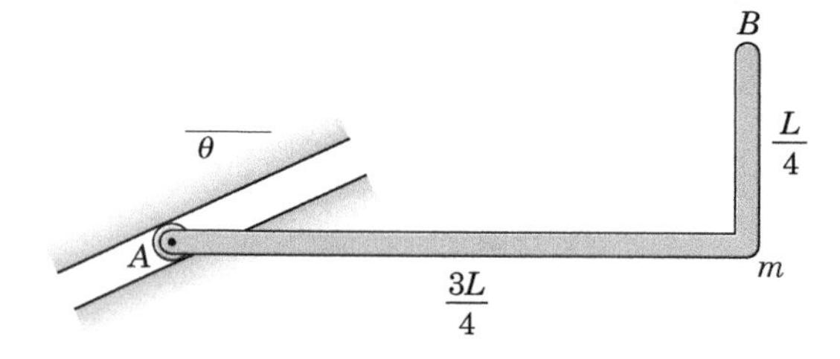

Problem 6/84

6/85 During a test, the car travels in a horizontal circle of radius R and has a forward tangential acceleration a. Determine the lateral reactions at the front and rear wheel pairs if (a) the car speed $v = 0$ and (b) the speed $v \neq 0$. The car mass is m and its polar moment of inertia (about a vertical axis through G) is $\bar{I}$. Assume that $R \gg d$.

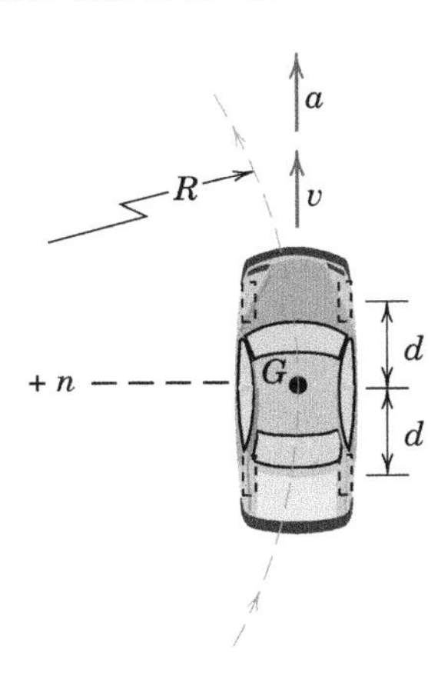

Problem 6/85

6/86 The system of Prob. 6/20 is repeated here. If the hoop- and semicylinder-assembly is centered on the top of the stationary cart and the system is released from rest, determine the initial acceleration a of the cart and the angular acceleration α of the hoop and semicylinder. Friction between the hoop and cart is sufficient to prevent slip. Motion takes place in the x-y plane. Neglect the mass of the cart wheels and any friction in the wheel bearings.

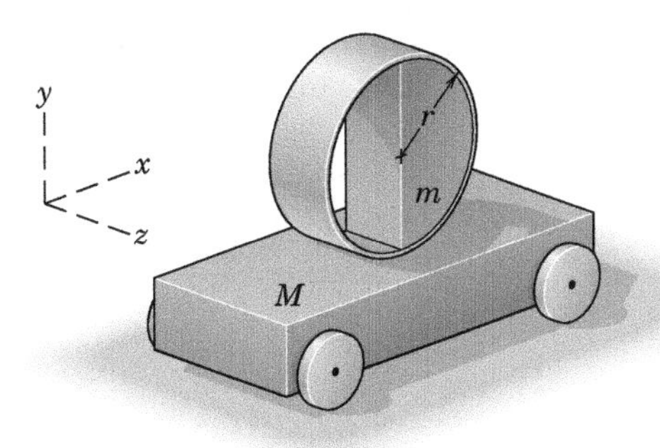

Problem 6/86

6/87 The 9-ft steel beam weighs 280 lb and is hoisted from rest where the tension in each of the cables is 140 lb. If the hoisting drums are given initial angular accelerations $\alpha_1 = 4$ rad/sec^2 and $\alpha_2 = 6$ rad/sec^2, calculate the corresponding tensions T_A and T_B in the cables. The beam may be treated as a slender bar.

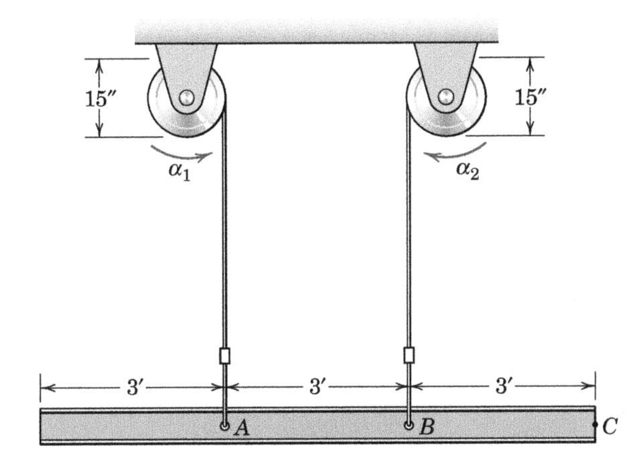

Problem 6/87

Representative Problems

6/88 The system is released from rest with the cable taut, and the homogeneous cylinder does not slip on the rough incline. Determine the angular acceleration of the cylinder and the minimum coefficient μ_s of friction for which the cylinder will not slip.

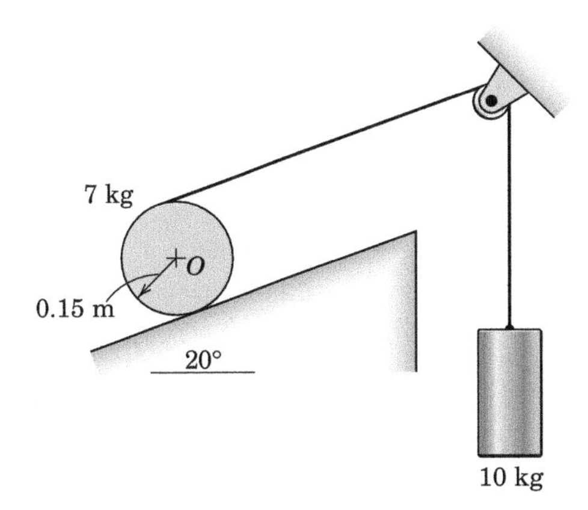

Problem 6/88

6/89 The mass center G of the 20-lb wheel is off center by 0.50 in. If G is in the position shown as the wheel rolls without slipping through the bottom of the circular path of 6-ft radius with an angular velocity ω of 10 rad/sec, compute the force P exerted by the path on the wheel. (Be careful to use the correct mass-center acceleration.)

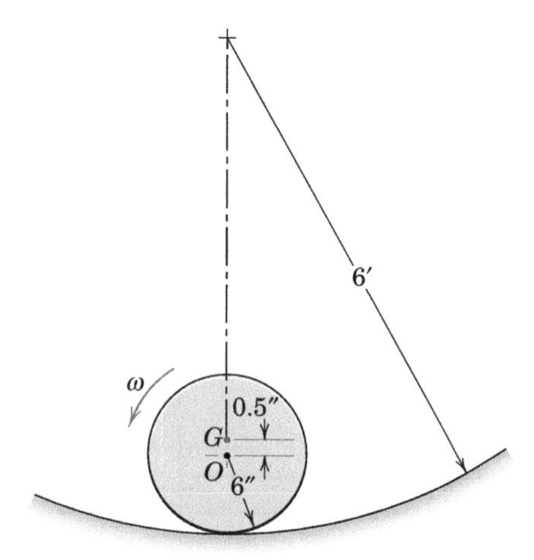

Problem 6/89

6/90 End A of the uniform 5-kg bar is pinned freely to the collar, which has an acceleration $a = 4\ \text{m/s}^2$ along the fixed horizontal shaft. If the bar has a clockwise angular velocity $\omega = 2$ rad/s as it swings past the vertical, determine the components of the force on the bar at A for this instant.

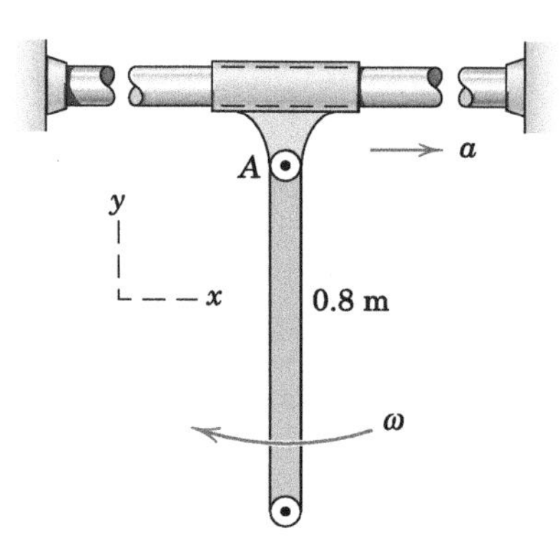

Problem 6/90

6/91 The uniform rectangular panel of mass m is moving to the right when wheel B drops off the horizontal support rail. Determine the resulting angular acceleration and the force T_A in the strap at A immediately after wheel B rolls off the rail. Neglect friction and the mass of the small straps and wheels.

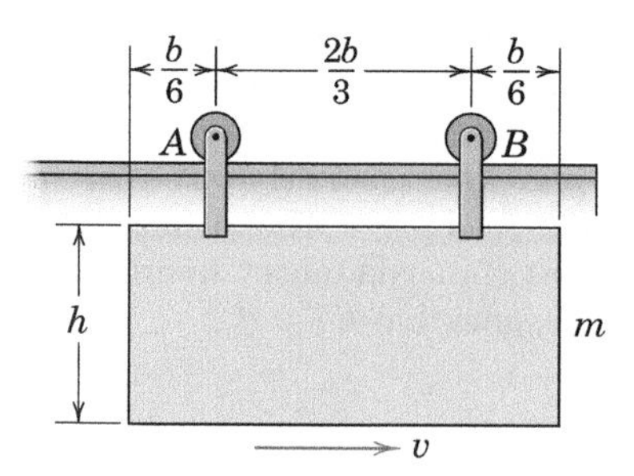

Problem 6/91

6/92 The truck, initially at rest with a solid cylindrical roll of paper in the position shown, moves forward with a constant acceleration a. Find the distance s which the truck goes before the paper rolls off the edge of its horizontal bed. Friction is sufficient to prevent slipping.

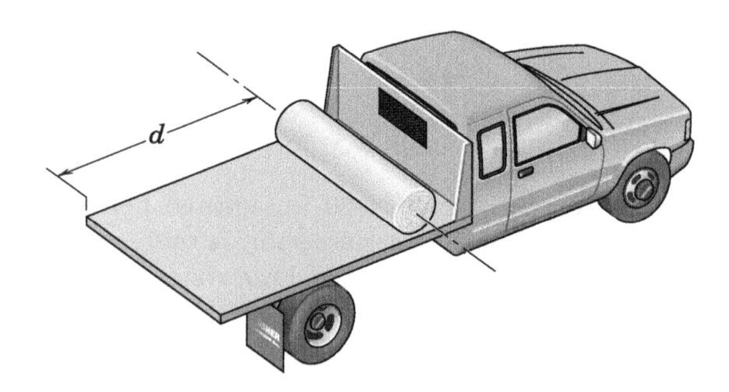

Problem 6/92

6/93 The crank OA rotates in the vertical plane with a constant clockwise angular velocity ω_0 of 4.5 rad/s. For the position where OA is horizontal, calculate the force under the light roller B of the 10-kg slender bar AB.

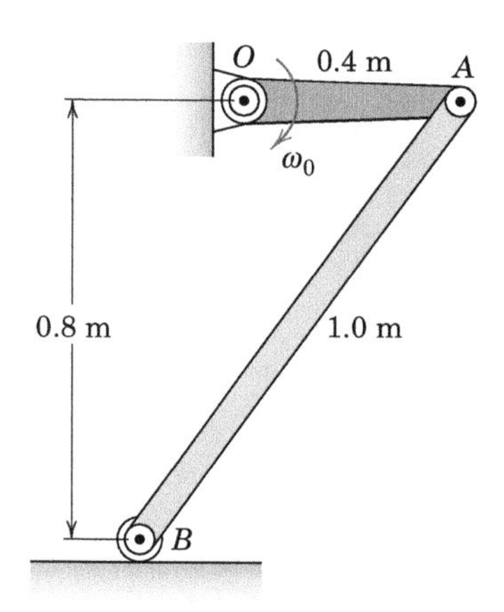

Problem 6/93

6/94 The uniform rectangular 300-lb plate is held in the horizontal position by two cables each of length $L = 3$ ft. If the cable at A suddenly breaks, calculate the tension T_B in the cable at B an instant after the break occurs.

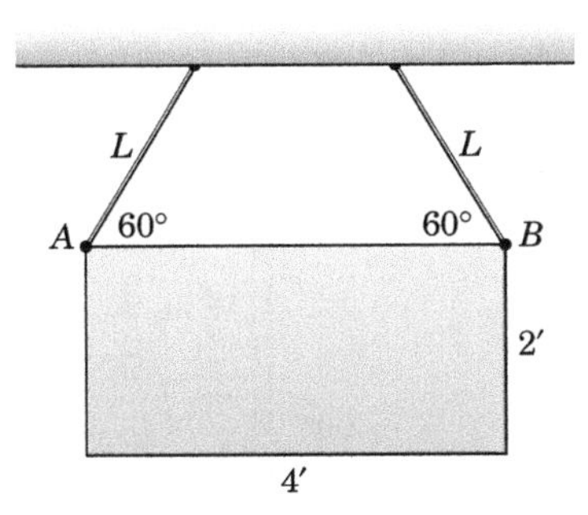

Problem 6/94

6/95 The rectangular plate of Prob. 6/94 is repeated here. The cables at A and B are now attached to a 50-lb trolley which is constrained to move in the horizontal guide. If the cable at A suddenly breaks, calculate the tension T_B in the cable at B an instant after the break occurs and the acceleration a_T of the trolley in the guide. Neglect friction and the mass of the trolley wheels.

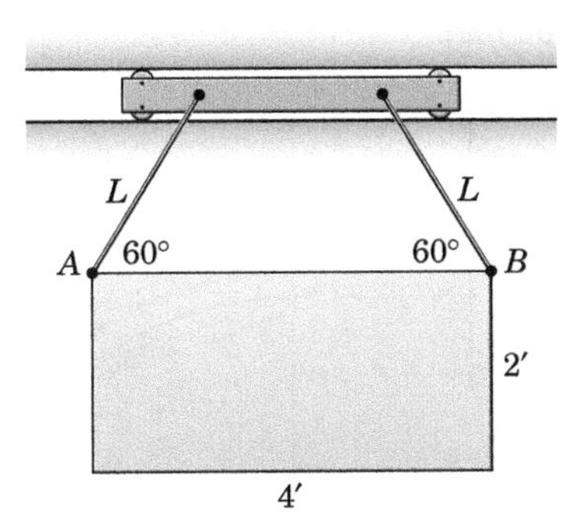

Problem 6/95

6/96 The robotic device of Prob. 6/68 is repeated here. Member AB is rotating about joint A with a counterclockwise angular velocity of 2 rad/s, and this rate is increasing at 4 rad/s^2. Determine the moment M_B exerted by arm AB on arm BC if joint B is held in a locked condition. The mass of arm BC is 4 kg, and the arm may be treated as a uniform slender rod.

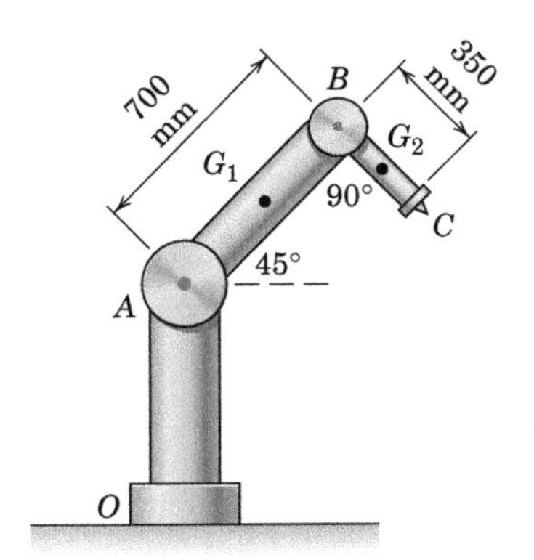

Problem 6/96

6/97 Small ball-bearing rollers mounted on the ends of the slender bar of mass m and length l constrain the motion of the bar in the horizontal x-y slots. If a couple M is applied to the bar initially at rest at $\theta = 45°$, determine the forces exerted on the rollers at A and B as the bar starts to move.

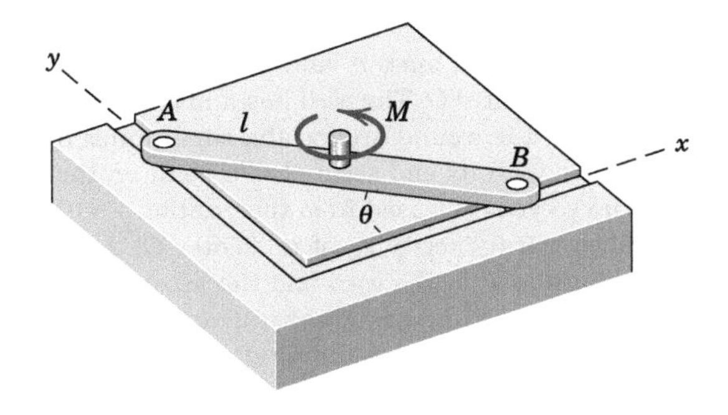

Problem 6/97

6/98 The assembly consisting of a uniform slender bar (mass $m/5$) and a rigidly attached uniform disk (mass $4m/5$) is freely pinned to point O on the collar that in turn slides on the fixed horizontal guide. The assembly is at rest when the collar is given a sudden acceleration a to the left as shown. Determine the initial angular acceleration of the assembly.

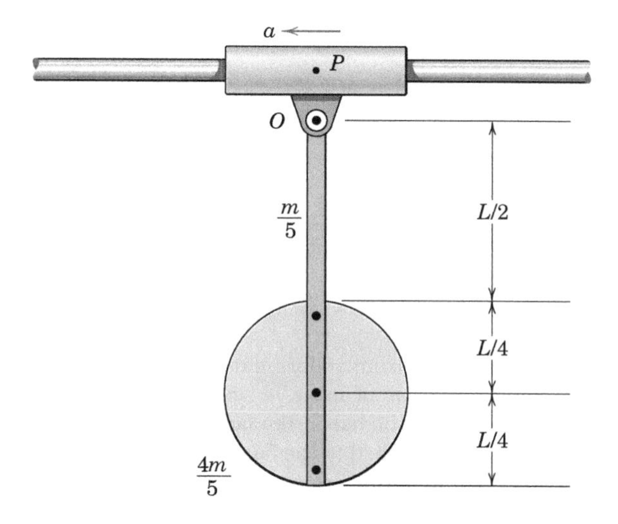

Problem 6/98

6/99 The yo-yo has a mass m and a radius of gyration k about its center O. The cord has a maximum length $y = L$ and is wound around the small inner hub of radius r with its end secured to a point on the hub. If the yo-yo is released from the position $y = 0$ with a downward velocity v_O of its center O, determine the tension T in the cord and the acceleration a of its center during its downward and upward motions. Also find the maximum downward velocity v of its center.

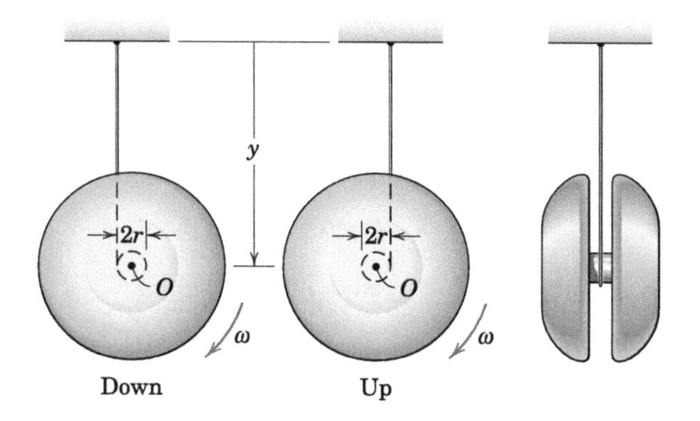

Problem 6/99

6/100 The uniform 12-ft pole is hinged to the truck bed and released from the vertical position as the truck starts from rest with an acceleration of 3 ft/sec^2. If the acceleration remains constant during the motion of the pole, calculate the angular velocity ω of the pole as it reaches the horizontal position.

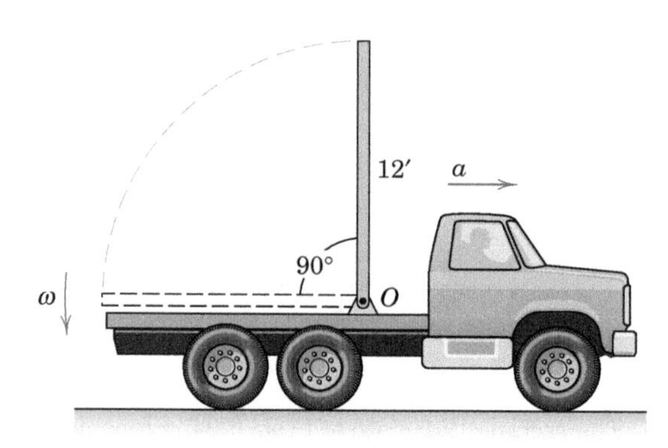

Problem 6/100

6/101 The uniform bar of mass m is constrained by the light rollers which move in the smooth guide, which lies in a vertical plane. If the bar is released from rest, what is the force at each roller an instant after release? Use the values $m = 18$ kg and $r = 150$ mm.

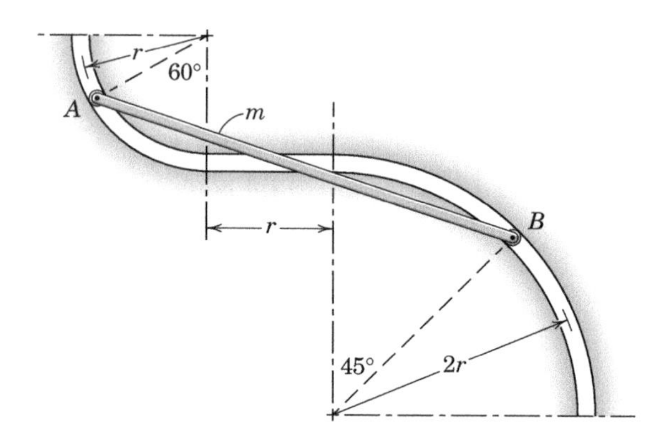

Problem 6/101

6/102 A bowling ball with a circumference of 27 in. weighs 14 lb and has a centroidal radius of gyration of 3.28 in. If the ball is released with a velocity of 20 ft/sec but with no angular velocity as it touches the alley floor, compute the distance traveled by the ball before it begins to roll without slipping. The coefficient of friction between the ball and the floor is 0.20.

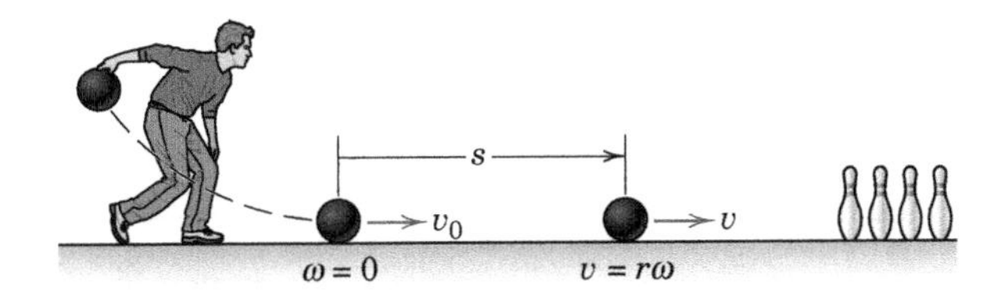

Problem 6/102

6/103 The figure shows the edge view of a uniform concrete slab with a mass of 12 Mg. The slab is being hoisted slowly by the winch D with cable attached to the dolly. At the position $\theta = 60°$, the distance x from the fixed ground position to the dolly is equal to the length $L = 4$ m of the slab. If the hoisting cable should break at this position, determine the initial acceleration a_A of the small dolly, whose mass is negligible, and the initial tension T in the fixed cable. End A of the slab will not slip on the dolly.

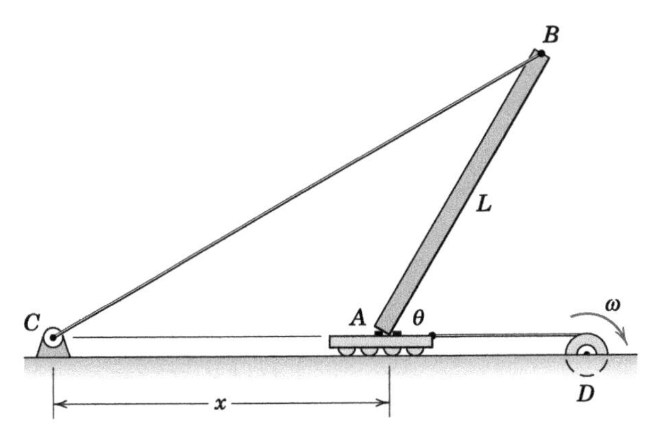

Problem 6/103

6/104 In a study of head injury against the instrument panel of a car during sudden or crash stops where lap belts without shoulder straps or airbags are used, the segmented human model shown in the figure is analyzed. The hip joint O is assumed to remain fixed relative to the car, and the torso above the hip is treated as a rigid body of mass m freely pivoted at O. The center of mass of the torso is at G with the initial position of OG taken as vertical. The radius of gyration of the torso about O is k_O. If the car is brought to a sudden stop with a constant deceleration a, determine the speed v relative to the car with which the model's head strikes the instrument panel. Substitute the values $m = 50$ kg, $\bar{r} = 450$ mm, $r = 800$ mm, $k_O = 550$ mm, $\theta = 45°$, and $a = 10g$ and compute v.

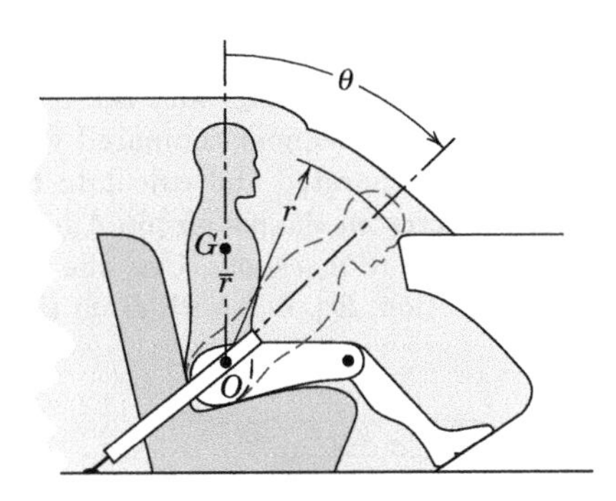

Problem 6/104

6/105 The 25-mm-thick uniform steel plate is being pulled very slowly into position by the cable connected to a winch. If the cable breaks at the instant represented, what are the force under each roller and the acceleration $\mathbf{a}_G$ of the center of mass? Neglect friction and the mass of the rollers, and consider the rollers small. Evaluate your results for $a = 3$ m, $b = 6$ m, and $r = 6$ m. (*Note:* Arc BC is of radius r, the same as arc BD.)

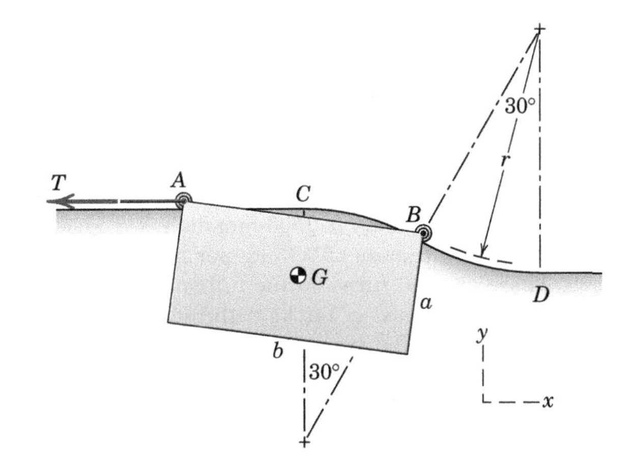

Problem 6/105

6/106 The connecting rod AB of a certain internal-combustion engine weighs 1.2 lb with mass center at G and has a radius of gyration about G of 1.12 in. The piston and piston pin A together weigh 1.80 lb. The engine is running at a constant speed of 3000 rev/min, so that the angular velocity of the crank is $3000(2\pi)/60 = 100\pi$ rad/sec. Neglect the weights of the components and the force exerted by the gas in the cylinder compared with the dynamic forces generated and calculate the magnitude of the force on the piston pin A for the crank angle $\theta = 90°$. (*Suggestion:* Use the alternative moment relation, Eq. 6/3, with B as the moment center.)

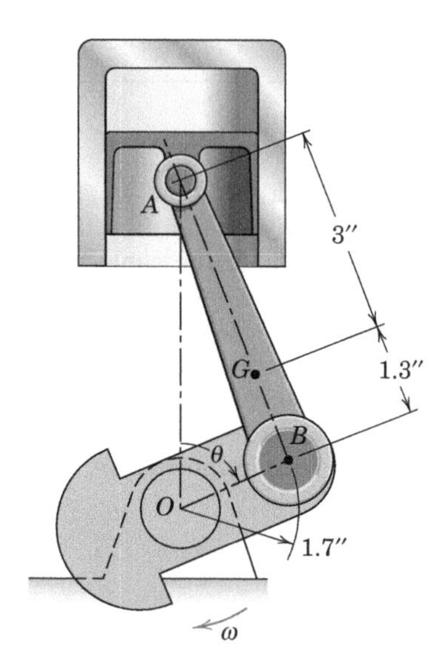

Problem 6/106

6/107 The truck carries a 1500-mm-diameter spool of cable with a mass of 0.75 kg per meter of length. There are 150 turns on the full spool. The empty spool has a mass of 140 kg with radius of gyration of 530 mm. The truck alone has a mass of 2030 kg with mass center at G. If the truck starts from rest with an initial acceleration of $0.2g$, determine (a) the tension T in the cable where it attaches to the wall and (b) the normal reaction under each pair of wheels. Neglect the rotational inertia of the truck wheels.

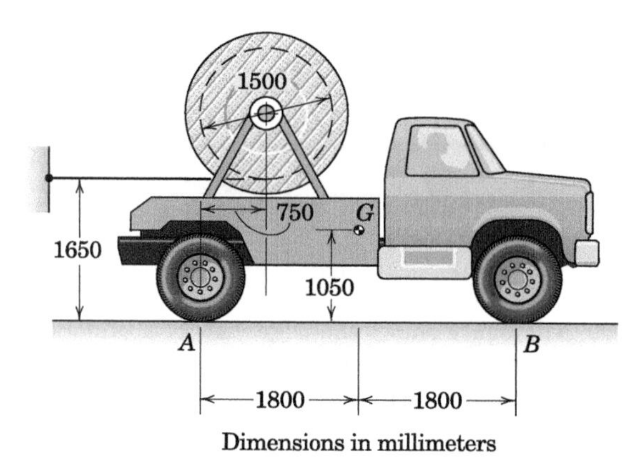

Problem 6/107

▶6/108 The four-bar mechanism lies in a vertical plane and is controlled by crank OA which rotates counterclockwise at a steady rate of 60 rev/min. Determine the torque M which must be applied to the crank at O when the crank angle $\theta = 45°$. The uniform coupler AB has a mass of 7 kg, and the masses of crank OA and the output arm BC may be neglected.

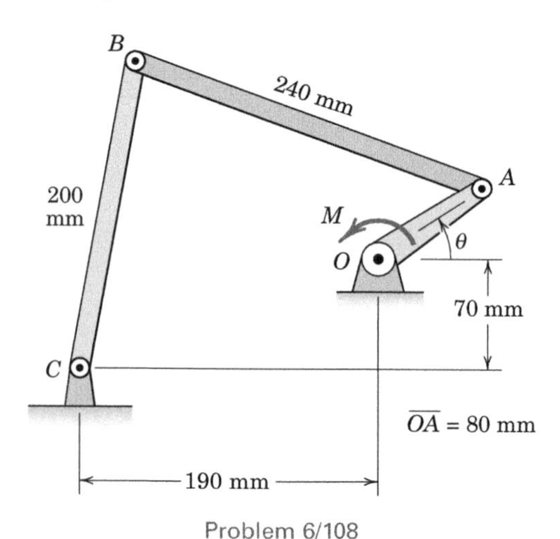

Problem 6/108

▶6/109 Repeat the analysis of Prob. 6/108 with the added information that the mass of crank OA is 1.2 kg and the mass of the output arm BC is 1.8 kg. Each of these bars may be considered uniform for this analysis.

▶6/110 The Ferris wheel at an amusement park has an even number n of gondolas, each freely pivoted at its point of support on the wheel periphery. Each gondola has a loaded mass m, a radius of gyration k about its point of support A, and a mass center a distance h from A. The wheel structure has a moment of inertia I_O about its bearing at O. Determine an expression for the tangential force F which must be transmitted to the wheel periphery at C in order to give the wheel an initial angular acceleration α starting from rest. *Suggestion:* Analyze the gondolas in pairs A and B. Be careful not to assume that the initial angular acceleration of the gondolas is the same as that of the wheel. (*Note:* An American engineer named George Washington Gale Ferris, Jr., created a giant amusement-wheel ride for the World's Columbian Exposition in Chicago in 1893. The wheel was 250 ft in diameter with 36 gondolas, each of which carried up to 60 passengers. Fully loaded, the wheel and gondolas had a mass of 1200 tons. The ride was powered by a 1000-hp steam engine.)

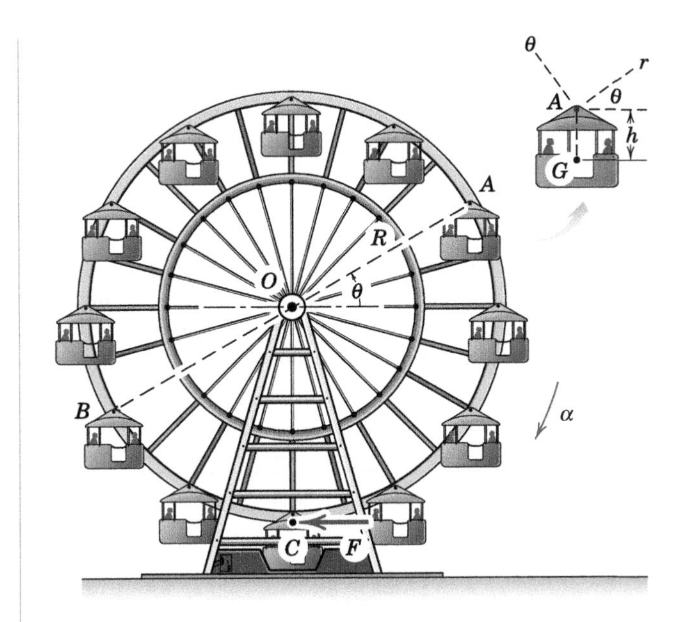

Problem 6/110

PROBLEMS

(In the following problems neglect any energy loss due to kinetic friction unless otherwise instructed.)

Introductory Problems

6/111 The slender rod of mass m and length l has a particle (negligible radius, mass $2m$) attached to its end. If the body is released from rest when in the position shown, determine its angular velocity as it passes the vertical position.

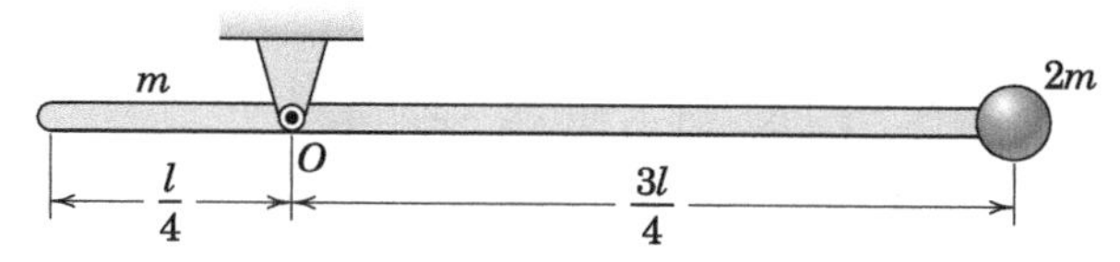

Problem 6/111

6/112 The log is suspended by the two parallel 5-m cables and used as a battering ram. At what angle θ should the log be released from rest in order to strike the object to be smashed with a velocity of 4 m/s?

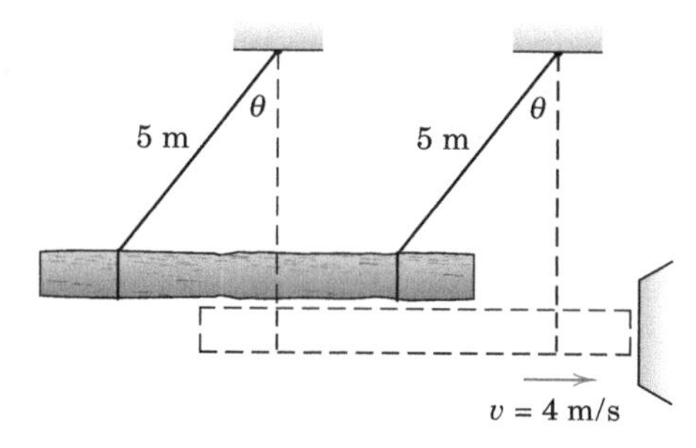

Problem 6/112

6/113 The assembly is constructed of homogeneous slender rod which has a mass ρ per unit length, and it rotates freely about a horizontal axis through the pivot at O. If the assembly is nudged from the starting position shown, determine its angular velocity after it has rotated (*a*) 45°, (*b*) 90°, and (*c*) 180°.

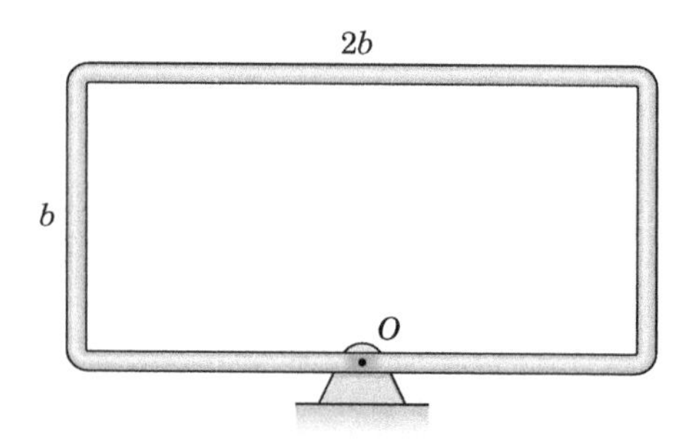

Problem 6/113

6/114 The velocity of the 8-kg cylinder is 0.3 m/s at a certain instant. What is its speed v after dropping an additional 1.5 m? The mass of the grooved drum is 12 kg, its centroidal radius of gyration is $\bar{k} = 210$ mm, and the radius of its groove is $r_i =$ 200 mm. The frictional moment at O is a constant 3 N·m.

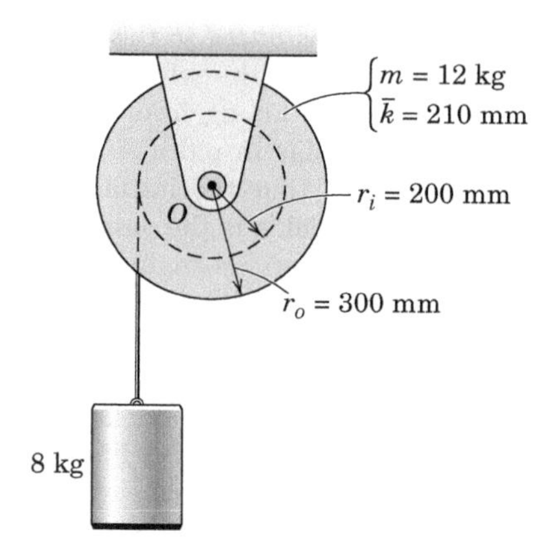

Problem 6/114

6/115 The 32.2-lb wheel is released from rest and rolls on its hubs without slipping. Calculate the speed v of the center O of the wheel after it has moved a distance $x = 10$ ft down the incline. The radius of gyration of the wheel about O is 5 in.

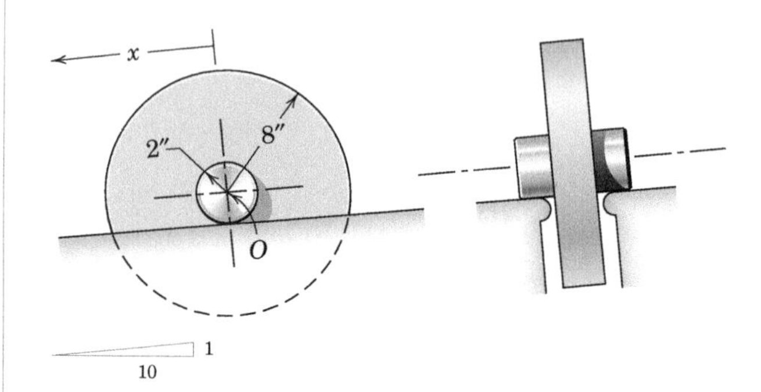

Problem 6/115

6/116 The uniform semicircular bar of radius $r = 75$ mm and mass $m = 3$ kg rotates freely about a horizontal axis through the pivot O. The bar is initially held in position 1 against the action of the torsional spring and then suddenly released. Determine the spring stiffness k_T which will give the bar a counterclockwise angular velocity $\omega =$ 4 rad/s when it reaches position 2, at which the spring is undeformed.

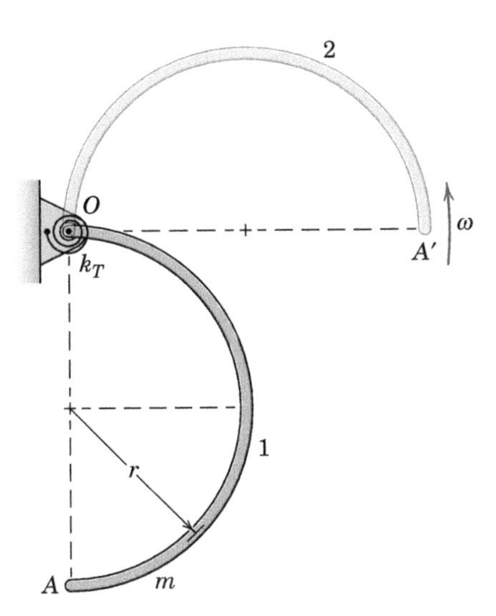

Problem 6/116

6/117 The homogeneous rectangular crate weighs 250 lb and is supported in the horizontal position by the cable at A and the corner hinge at O. If the cable at A is suddenly released, calculate the angular velocity ω of the crate just before it strikes the 30° incline. Does the weight of the crate influence the results, other quantities unchanged?

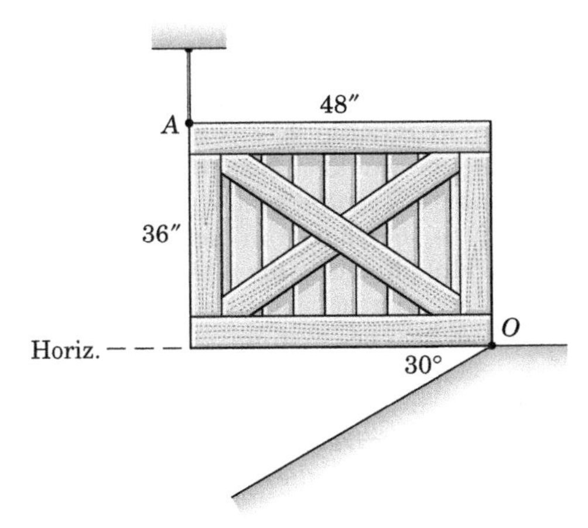

Problem 6/117

6/118 The 24-lb disk is rigidly attached to the 7-lb bar OA, which is pivoted freely about a horizontal axis through point O. If the system is released from rest in the position shown, determine the angular velocity of the bar and the magnitude of the pin reaction at O after the bar has rotated 90°.

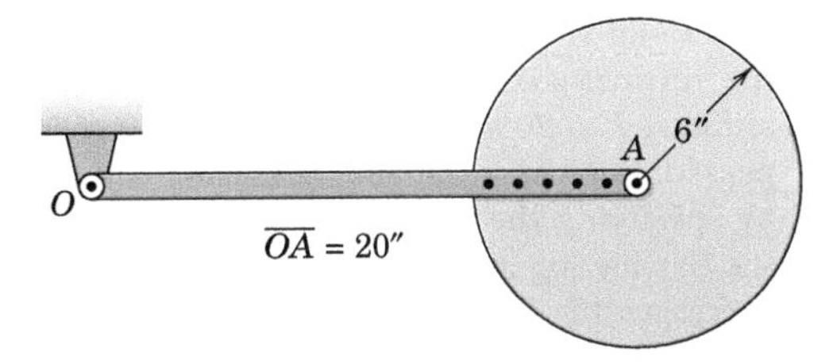

Problem 6/118

6/119 The two wheels of Prob. 6/78, shown again here, represent two extreme conditions of distribution of mass. For case A all of the mass m is assumed to be concentrated in the center of the hoop in the axial bar of negligible diameter. For case B all of the mass m is assumed to be concentrated in the rim. Determine the speed of the center of each hoop after it has traveled a distance x down the incline from rest. The hoops roll without slipping.

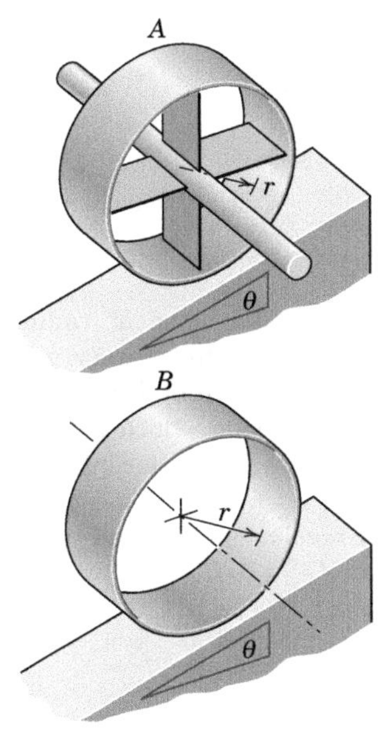

Problem 6/119

6/120 The 15-kg slender bar OA is released from rest in the vertical position and compresses the spring of stiffness $k = 20$ kN/m as the horizontal position is passed. Determine the proper setting of the spring, by specifying the distance h, which will result in the bar having an angular velocity $\omega = 4$ rad/s as it crosses the horizontal position. What is the effect of x on the dynamics of the problem?

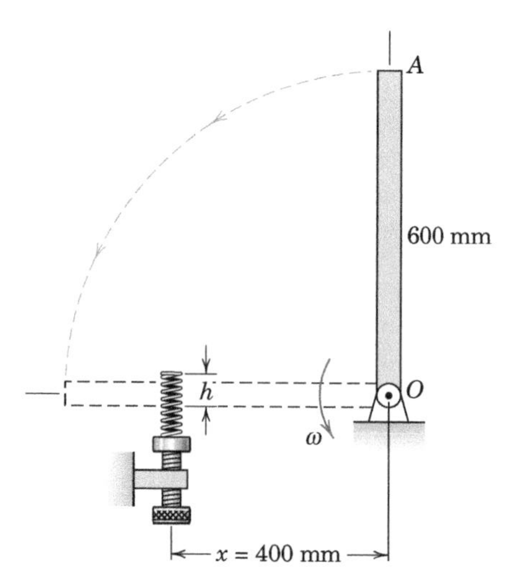

Problem 6/120

6/121 The light circular hoop of radius r contains a quarter-circular sector of mass m and is initially at rest on the horizontal surface. A couple M is applied to the hoop, which rolls without slipping along the horizontal surface. Determine the velocity v of the hoop after it has rolled through one-half of a revolution. Evaluate for $mg = 6$ lb, $r = 9$ in., and $M = 4$ lb-ft.

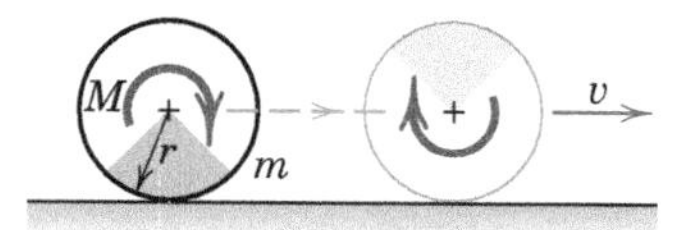

Problem 6/121

6/122 A steady 5-lb force is applied normal to the handle of the hand-operated grinder. The gear inside the housing with its shaft and attached handle together weigh 3.94 lb and have a radius of gyration about their axis of 2.85 in. The grinding wheel with its attached shaft and pinion (inside housing) together weigh 1.22 lb and have a radius of gyration of 2.14 in. If the gear ratio between gear and pinion is 4:1, calculate the speed N of the grinding wheel after 6 complete revolutions of the handle starting from rest.

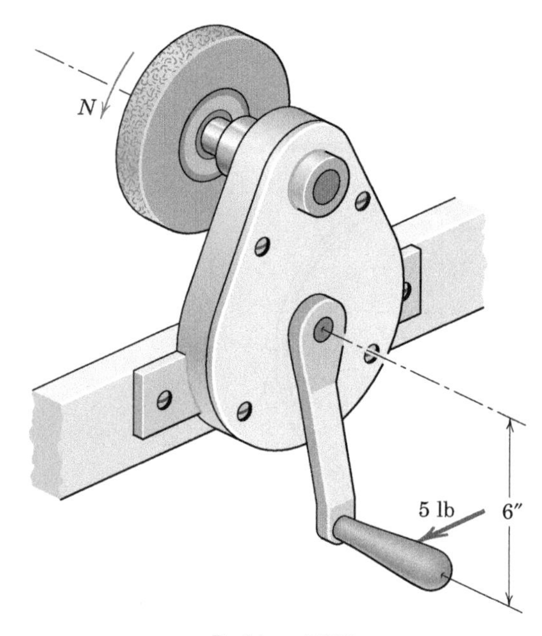

Problem 6/122

Representative Problems

6/123 The figure shows an impact tester used in studying material response to shock loads. The 60-lb pendulum is released from rest and swings through the vertical with negligible resistance. At the bottommost point of the ensuing motion, the pendulum strikes a notched material specimen A. After impact with the specimen, the pendulum swings upward to a height $\bar{h}' = 3.17$ ft. If the impact-energy capacity of the pendulum is 300 ft-lb, determine the change in the angular velocity of the pendulum during the interval from just before to just after impact with the specimen. The center of mass distance $\bar{r}$ from O and the radius of gyration k_O of the pendulum about O are both 35.5 in. (*Note:* Positioning the center of mass directly at the radius of gyration eliminates shock loads on the bearing at O and extends the life of the tester significantly.)

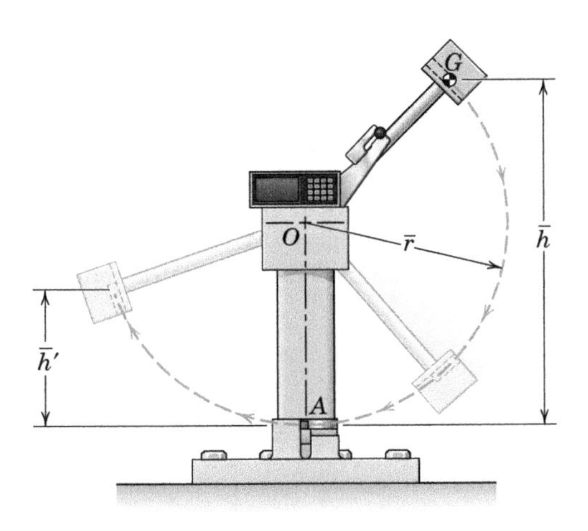

Problem 6/123

6/124 The uniform rectangular plate is released from rest in the position shown. Determine the maximum angular velocity ω during the ensuing motion. Friction at the pivot is negligible.

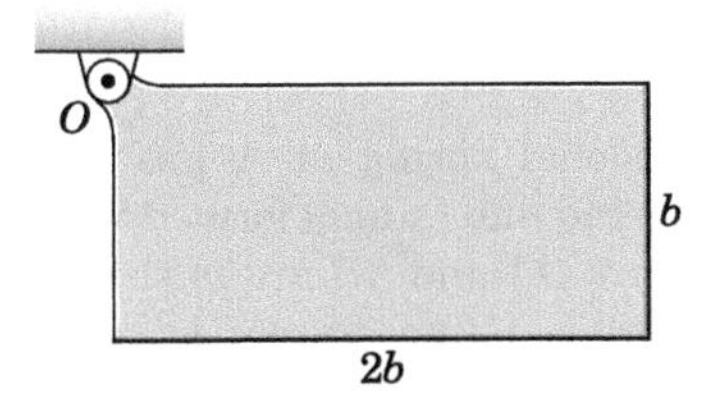

Problem 6/124

6/125 The 50-kg flywheel has a radius of gyration $\bar{k} = 0.4$ m about its shaft axis and is subjected to the torque $M = 2(1 - e^{-0.1\theta})$ N·m, where θ is in radians. If the flywheel is at rest when $\theta = 0$, determine its angular velocity after 5 revolutions.

Problem 6/125

6/126 The 20-kg wheel has an eccentric mass which places the center of mass G a distance $\bar{r} = 70$ mm away from the geometric center O. A constant couple $M = 6$ N·m is applied to the initially stationary wheel, which rolls without slipping along the horizontal surface and enters the curve of radius $R = 600$ mm. Determine the normal force under the wheel just before it exits the curve at C. The wheel has a rolling radius $r = 100$ mm and a radius of gyration $k_O = 65$ mm.

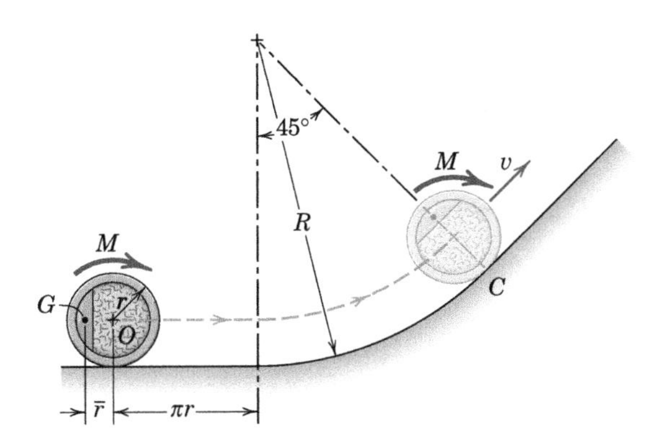

Problem 6/126

6/127 The figure shows the cross section AB of a garage door which is a rectangular 2.5-m by 5-m panel of uniform thickness with a mass of 200 kg. The door is supported by the struts of negligible mass and hinged at O. Two spring-and-cable assemblies, one on each side of the door, control the movement. When the door is in the horizontal open position, each spring is unextended. If the door is given a slight imbalance from the open position and allowed to fall, determine the value of the spring constant k for each spring which will limit the angular velocity of the door to 1.5 rad/s when edge B strikes the floor.

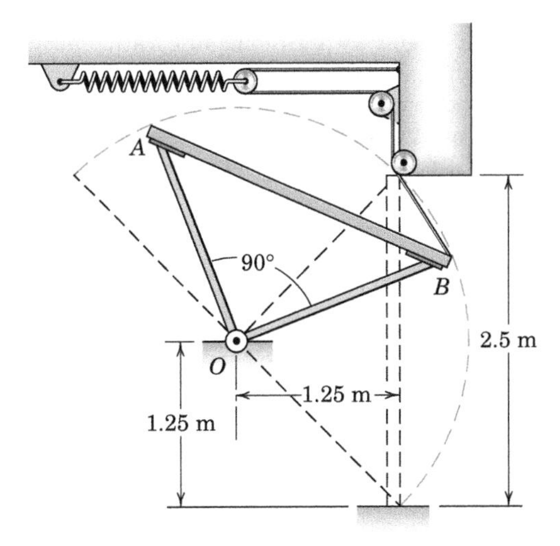

Problem 6/127

6/128 The uniform 40-lb bar with attached 12-lb wheels is released from rest when $\theta = 60°$. If the wheels roll without slipping on the horizontal and vertical surfaces, determine the angular velocity of the bar when $\theta = 45°$. Each wheel has a centroidal radius of gyration of 4.5 inches.

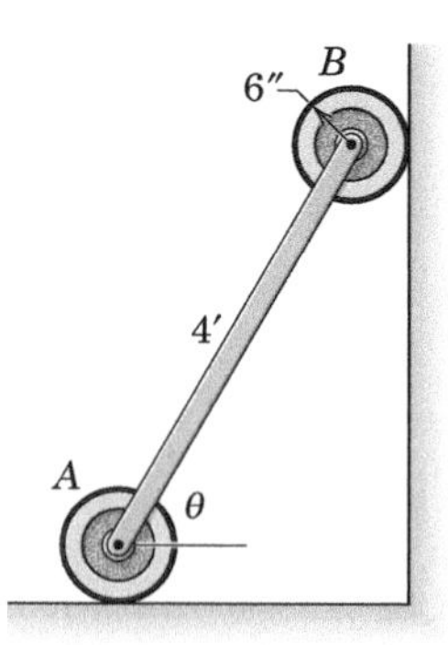

Problem 6/128

6/129 A 1200-kg flywheel with a radius of gyration of 400 mm has its speed reduced from 5000 to 3000 rev/min during a 2-min interval. Calculate the average power supplied by the flywheel. Express your answer both in kilowatts and in horsepower.

6/130 The wheel consists of a 4-kg rim of 250-mm radius with hub and spokes of negligible mass. The wheel is mounted on the 3-kg yoke OA with mass center at G and with a radius of gyration about O of 350 mm. If the assembly is released from rest in the horizontal position shown and if the wheel rolls on the circular surface without slipping, compute the velocity of point A when it reaches A'.

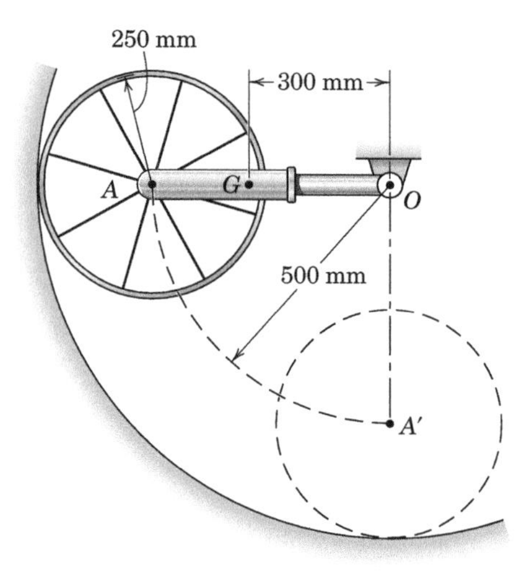

Problem 6/130

6/131 The uniform slender bar ABC weighs 6 lb and is initially at rest with end A bearing against the stop in the horizontal guide. When a constant couple $M = 72$ lb-in. is applied to end C, the bar rotates causing end A to strike the side of the vertical guide with a velocity of 10 ft/sec. Calculate the loss of energy ΔE due to friction in the guides and rollers. The mass of the rollers may be neglected.

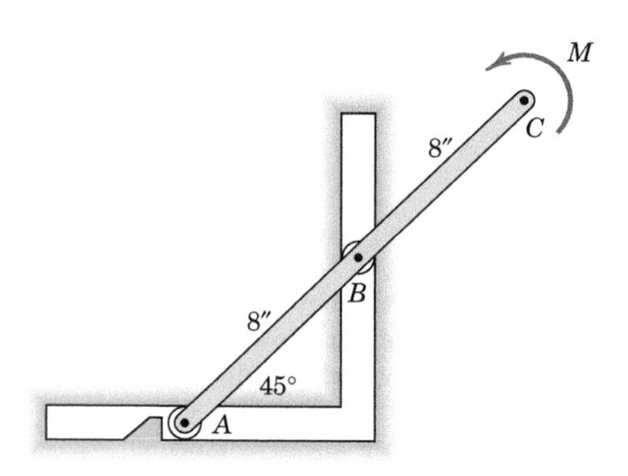

Problem 6/131

6/132 The torsional spring at A has a stiffness $k_T = 10$ N·m/rad and is undeformed when the uniform 10-kg bars OA and AB are in the vertical position and overlap. If the system is released from rest with $\theta = 60°$, determine the angular velocity of wheel B when $\theta = 30°$. The 6-kg wheel at B has a centroidal radius of gyration of 50 mm and is observed to roll without slipping on the horizontal surface.

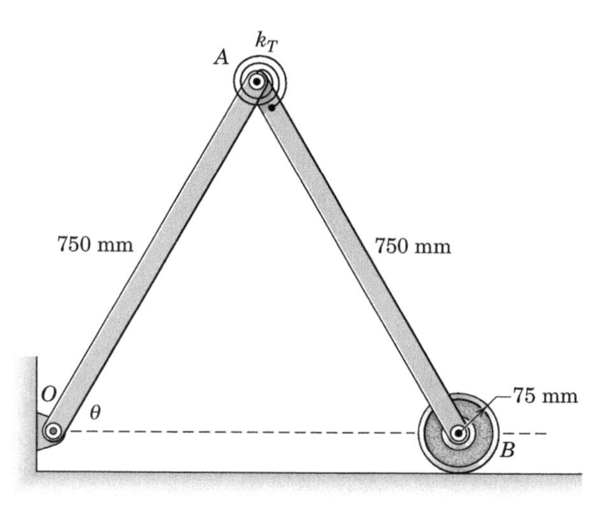

Problem 6/132

6/133 The uniform slender bar of mass m pivots freely about a horizontal axis through O. If the bar is released from rest in the horizontal position shown where the spring is unstretched, it is observed to rotate a maximum of 30° clockwise. The spring constant $k = 200$ N/m and the distance $b = 200$ mm. Determine (a) the mass m of the bar and (b) the angular velocity ω of the bar when the angular displacement is 15° clockwise from the release position.

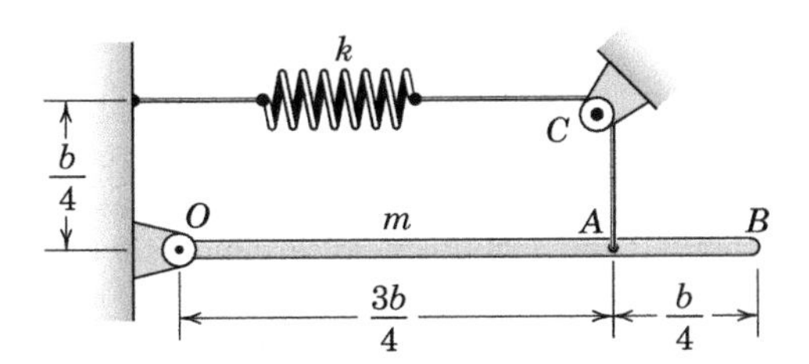

Problem 6/133

6/134 The system is released from rest when the angle $\theta = 90°$. Determine the angular velocity of the uniform slender bar when θ equals 60°. Use the values $m_1 = 1$ kg, $m_2 = 1.25$ kg, and $b = 0.4$ m.

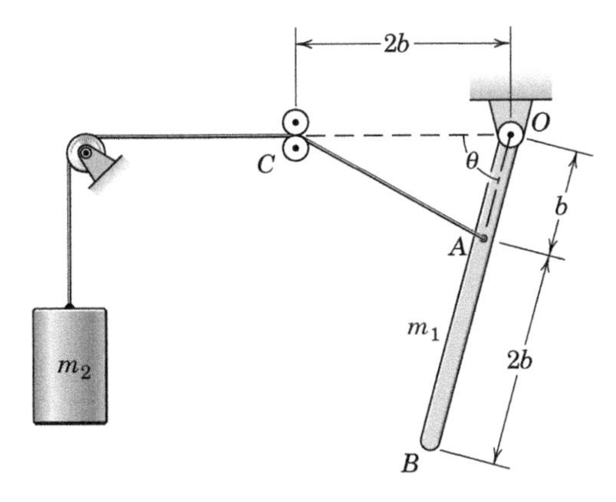

Problem 6/134

6/135 The homogeneous torus and cylindrical ring are released from rest and roll without slipping down the incline. Determine an expression for the velocity difference v_{diff} which develops between the two objects during the ensuing motion as a function of the distance x they have traveled down the incline. Assume that the masses roll straight down the incline and evaluate your expression for the case where $a = 0.2R$. Which object is in the lead, and does the relative size of a ever alter the finishing order?

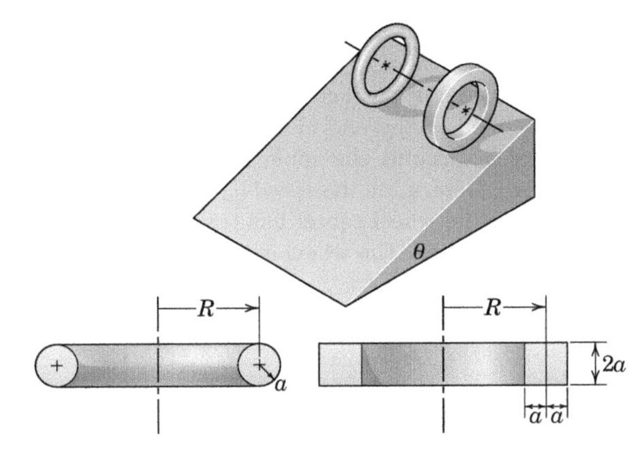

Problem 6/135

6/136 The uniform 12-lb disk pivots freely about a horizontal axis through O. A 4-lb slender bar is fastened to the disk as shown. If the system is nudged from rest while in the position shown, determine its angular velocity ω after it has rotated 180°.

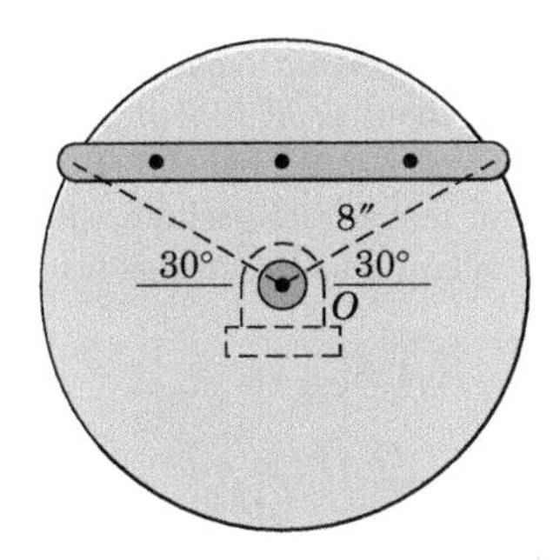

Problem 6/136

6/137 Under active development is the storage of energy in high-speed rotating disks where friction is effectively eliminated by encasing the rotor in an evacuated enclosure and by using magnetic bearings. For a 10-kg rotor with a radius of gyration of 90 mm rotating initially at 80 000 rev/min, calculate the power P which can be extracted from the rotor by applying a constant 2.10-N·m retarding torque (a) when the torque is first applied and (b) at the instant when the torque has been applied for 120 seconds.

6/138 For the pivoted slender rod of length l, determine the distance x for which the angular velocity will be a maximum as the bar passes the vertical position after being released in the horizontal position shown. State the corresponding angular velocity.

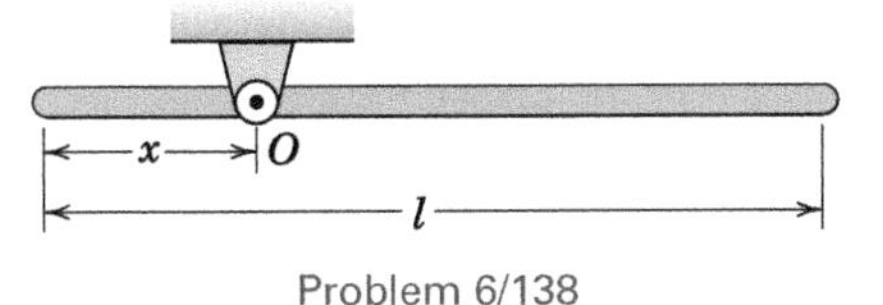

Problem 6/138

6/139 The wheel has mass m and a centroidal radius of gyration $\bar{k}$ and rolls without slipping up the incline under the action of a force P. The force is applied to the end of a cord which is wrapped securely around the inner hub of the wheel as shown. Determine the speed v_O of the wheel center O after the wheel center has traveled a distance d up the incline. The wheel is at rest when the force P is first applied.

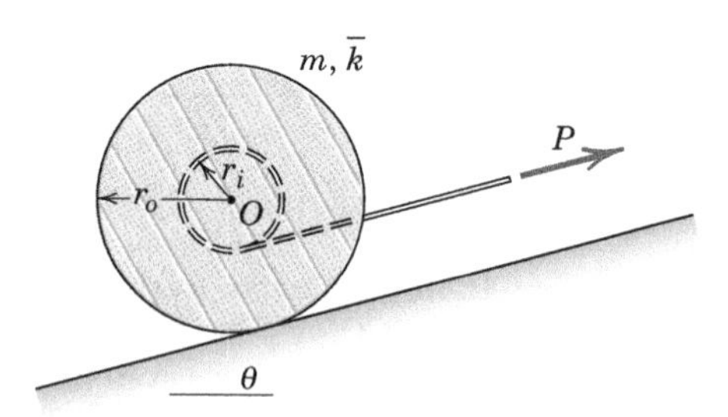

Problem 6/139

6/140 The 8-kg crank OA, with mass center at G and radius of gyration about O of 0.22 m, is connected to the 12-kg uniform slender bar AB. If the linkage is released from rest in the position shown, compute the velocity v of end B as OA swings through the vertical.

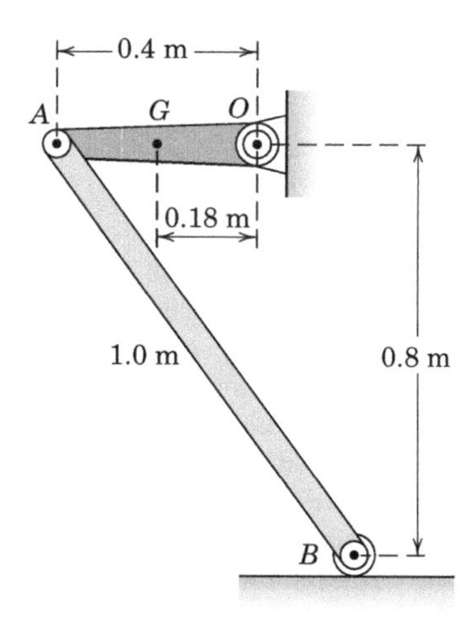

Problem 6/140

6/141 The sheave of 400-mm radius has a mass of 50 kg and a radius of gyration of 300 mm. The sheave and its 100-kg load are suspended by the cable and the spring, which has a stiffness of 1.5 kN/m. If the system is released from rest with the spring initially stretched 100 mm, determine the velocity of O after it has dropped 50 mm.

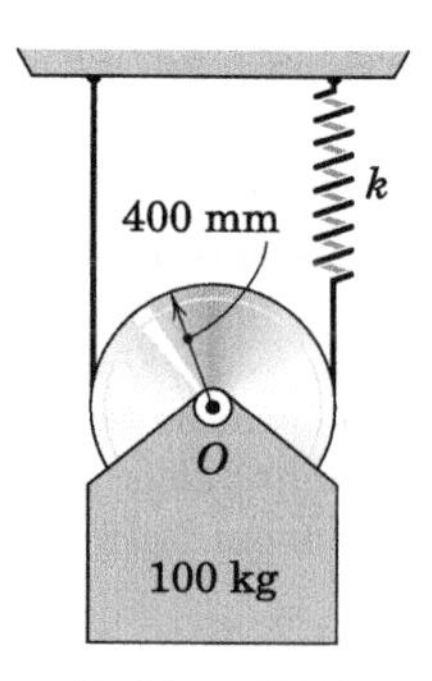

Problem 6/141

6/142 Motive power for the experimental 10-Mg bus comes from the energy stored in a rotating flywheel which it carries. The flywheel has a mass of 1500 kg and a radius of gyration of 500 mm and is brought up to a maximum speed of 4000 rev/min. If the bus starts from rest and acquires a speed of 72 km/h at the top of a hill 20 m above the starting position, compute the reduced speed N of the flywheel. Assume that 10 percent of the energy taken from the flywheel is lost. Neglect the rotational energy of the wheels of the bus. The 10-Mg mass includes the flywheel.

Problem 6/142

6/143 The homogeneous solid semicylinder is released from rest in the position shown. If friction is sufficient to prevent slipping, determine the maximum angular velocity ω reached by the cylinder as it rolls on the horizontal surface.

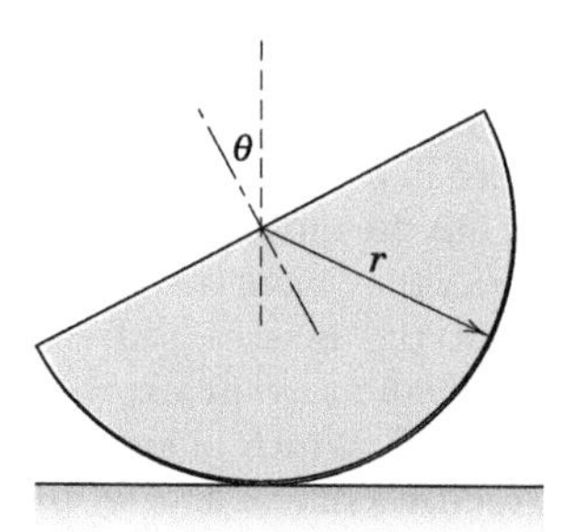

Problem 6/143

6/144 The figure shows the side view of a door to a storage compartment. As the 40-kg uniform door is opened, the light rod slides through the collar at C and compresses the spring of stiffness k. With the door closed ($\theta = 0$), a constant force $P = 225$ N is applied to the end of the door via a cable. If the door has a clockwise angular velocity of 1 rad/s as the position $\theta = 60°$ is passed, determine (*a*) the stiffness of the spring and (*b*) the angular velocity of the door as it passes the position $\theta = 45°$. Neglect all friction and the mass of the pulleys at C and D. The spring is uncompressed when the door is vertical, and $b = 1.25$ m.

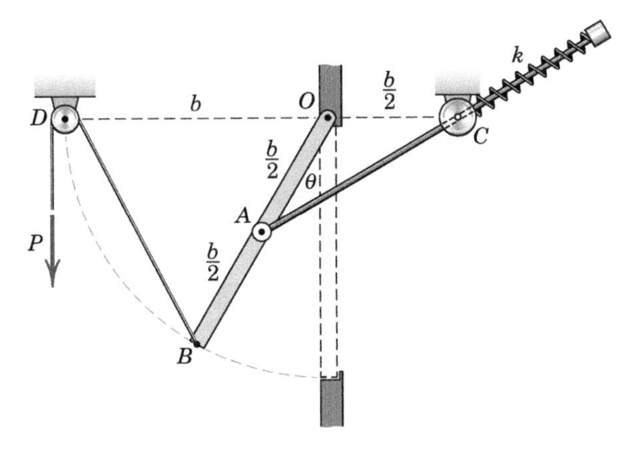

Problem 6/144

***6/145** Reconsider the door of Prob. 6/144. If the door is in the closed vertical position when a constant input force $P = 225$ N is applied through the end of the cable, determine the maximum angle θ_{max} reached by the door before it comes to a stop. Plot the angular velocity of the door over this period and determine the maximum angular velocity of the door along with the corresponding angle θ at which it occurs. The uniform door has a mass $m = 40$ kg, dimension $b = 1.25$ m, and the spring constant is $k = 2280$ N/m.

6/146 A small experimental vehicle has a total mass m of 500 kg including wheels and driver. Each of the four wheels has a mass of 40 kg and a centroidal radius of gyration of 400 mm. Total frictional resistance R to motion is 400 N and is measured by towing the vehicle at a constant speed on a level road with engine disengaged. Determine the power output of the engine for a speed of 72 km/h up the 10-percent grade (*a*) with zero acceleration and (*b*) with an acceleration of 3 m/s^2. (*Hint:* Power equals the time rate of increase of the total energy of the vehicle plus the rate at which frictional work is overcome.)

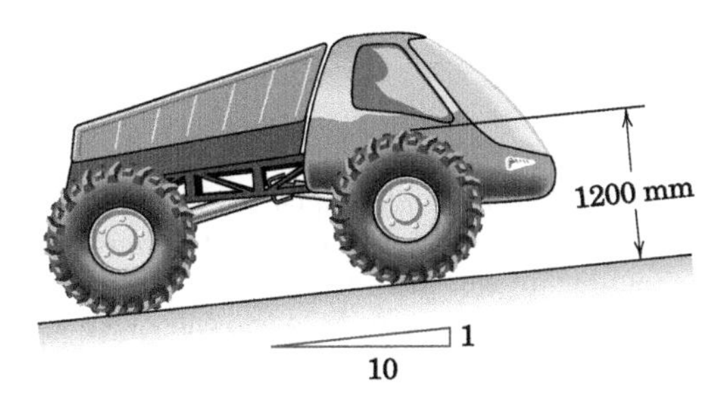

Problem 6/146

6/147 The two slender bars each of mass m and length b are pinned together and move in the vertical plane. If the bars are released from rest in the position shown and move together under the action of a couple M of constant magnitude applied to AB, determine the velocity of A as it strikes O.

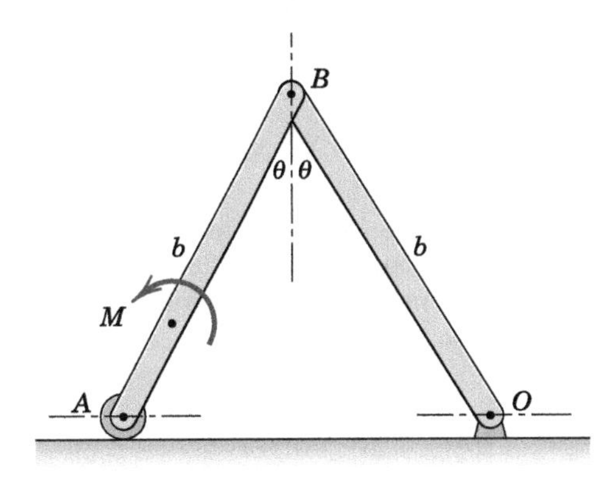

Problem 6/147

6/148 The open square frame is constructed of four identical slender rods, each of length b. If the frame is released from rest in the position shown, determine the speed of corner A (*a*) after A has dropped a distance b and (*b*) after A has dropped a distance $2b$. The small wheels roll without friction in the slots of the vertical surface.

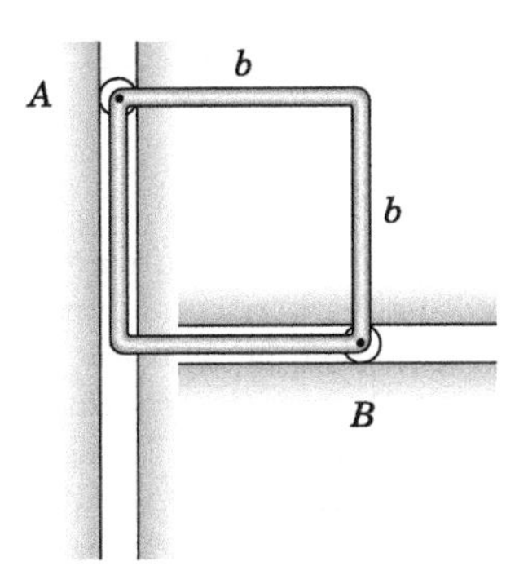

Problem 6/148

PROBLEMS

Introductory Problems

6/149 The load of mass m is supported by the light parallel links and the fixed stop A. Determine the initial angular acceleration α of the links due to the application of the couple M to one end as shown.

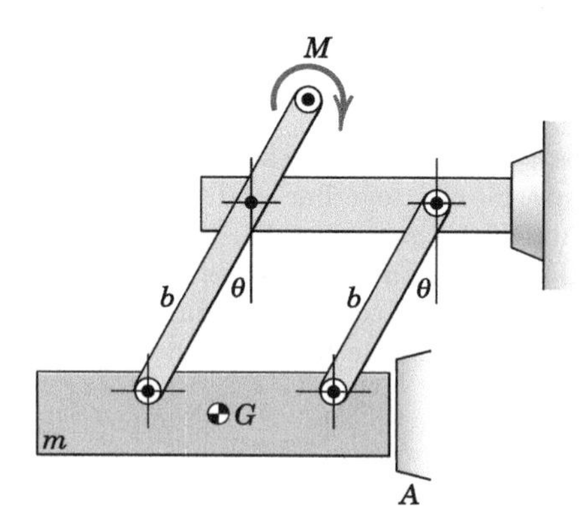

Problem 6/149

6/150 The uniform slender bar of mass m is shown in its equilibrium configuration before the force P is applied. Compute the initial angular acceleration of the bar upon application of P.

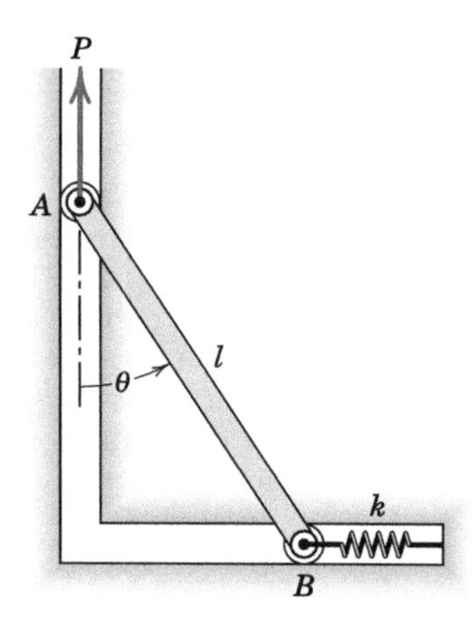

Problem 6/150

6/151 The two uniform slender bars are hinged at O and supported on the horizontal surface by their end rollers of negligible mass. If the bars are released from rest in the position shown, determine their initial angular acceleration α as they collapse in the vertical plane. (*Suggestion:* Make use of the instantaneous center of zero velocity in writing the expression for dT.)

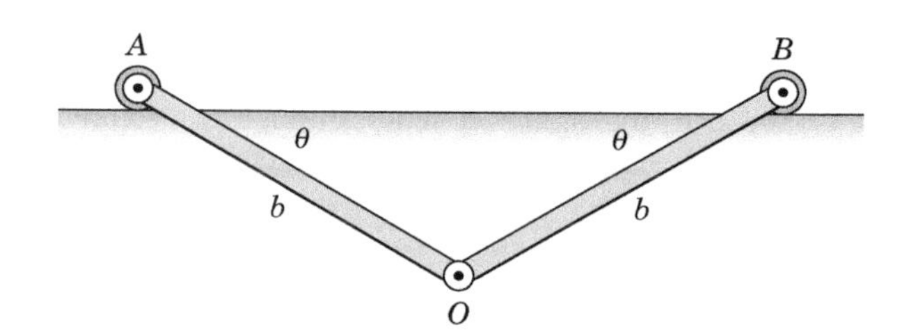

Problem 6/151

6/152 Links A and B each weigh 8 lb, and bar C weighs 12 lb. Calculate the angle θ assumed by the links if the body to which they are pinned is given a steady horizontal acceleration a of 4 ft/sec^2.

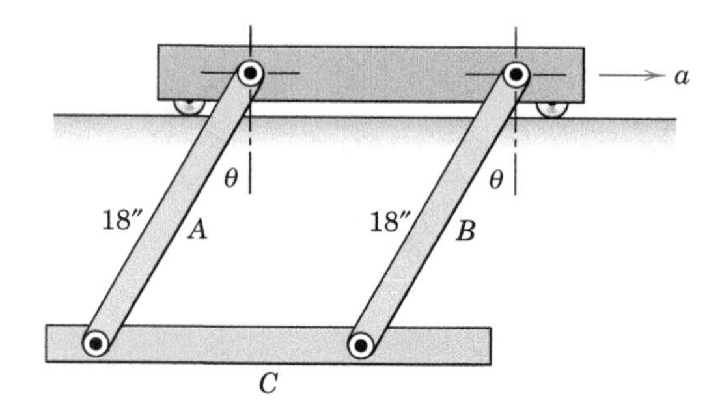

Problem 6/152

6/153 The mechanism shown moves in the vertical plane. The vertical bar AB weighs 10 lb, and each of the two links weighs 6 lb with mass center at G and with a radius of gyration of 10 in. about its bearing (O or C). The spring has a stiffness of 15 lb/ft and an unstretched length of 18 in. If the support at D is suddenly withdrawn, determine the initial angular acceleration α of the links.

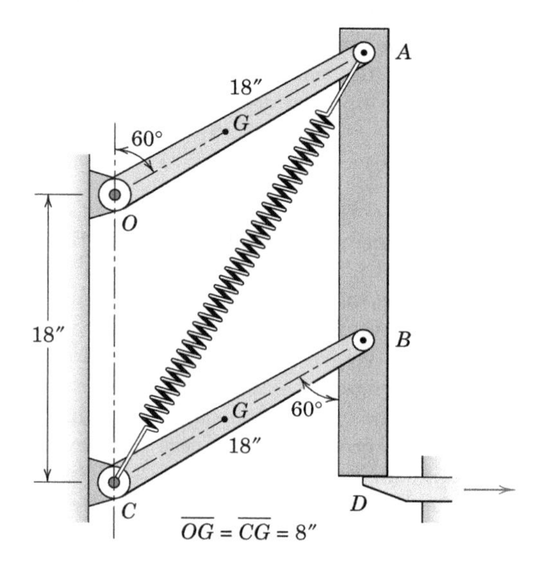

Problem 6/153

Representative Problems

6/154 The load of mass m is given an upward acceleration a from its supported rest position by the application of the forces P. Neglect the mass of the links compared with m and determine the initial acceleration a.

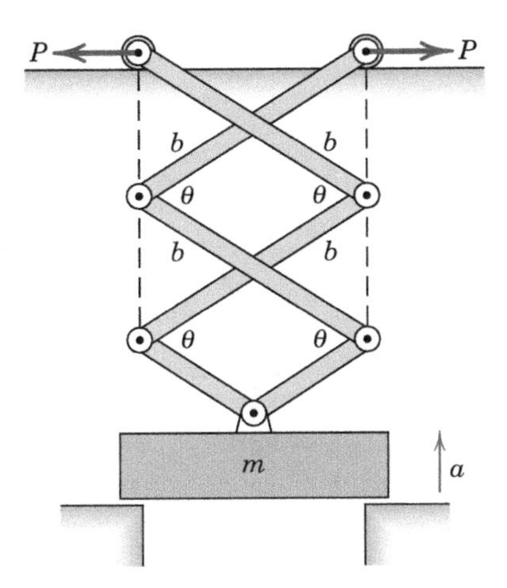

Problem 6/154

6/155 The cargo box of the food-delivery truck for aircraft servicing has a loaded mass m and is elevated by the application of a couple M on the lower end of the link which is hinged to the truck frame. The horizontal slots allow the linkage to unfold as the cargo box is elevated. Determine the upward acceleration of the box in terms of h for a given value of M. Neglect the mass of the links.

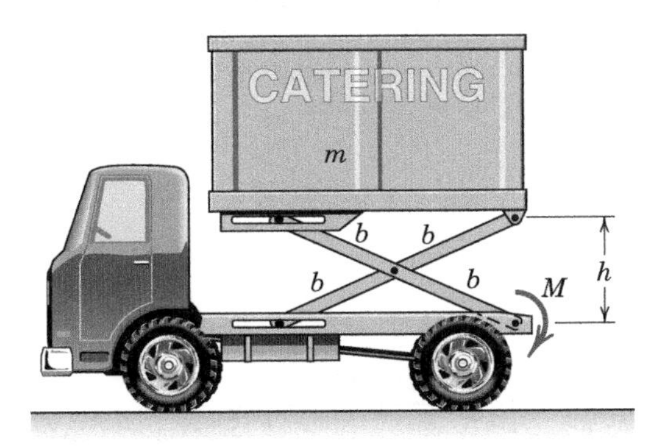

Problem 6/155

6/156 The sliding block is given a horizontal acceleration to the right that is slowly increased to a steady value a. The attached pendulum of mass m and mass center G assumes a steady angular deflection θ. The torsion spring at O exerts a moment $M = k_T\theta$ on the pendulum to oppose the angular deflection. Determine the torsional stiffness k_T that will allow a steady deflection θ.

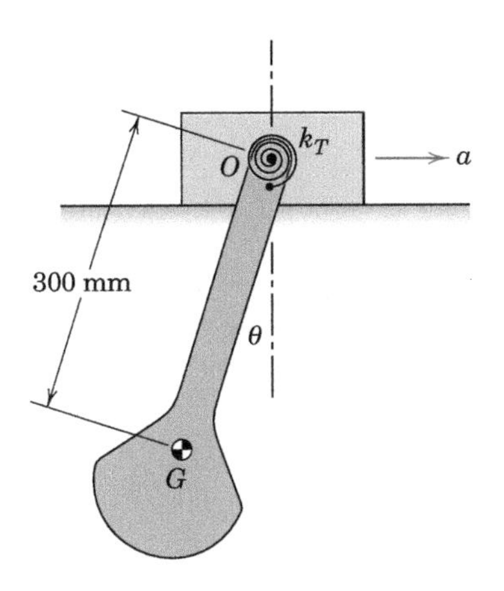

Problem 6/156

6/157 Each of the uniform bars OA and OB has a mass of 2 kg and is freely hinged at O to the vertical shaft, which is given an upward acceleration $a = g/2$. The links which connect the light collar C to the bars have negligible mass, and the collar slides freely on the shaft. The spring has a stiffness $k = 130$ N/m and is uncompressed for the position equivalent to $\theta = 0$. Calculate the angle θ assumed by the bars under conditions of steady acceleration.

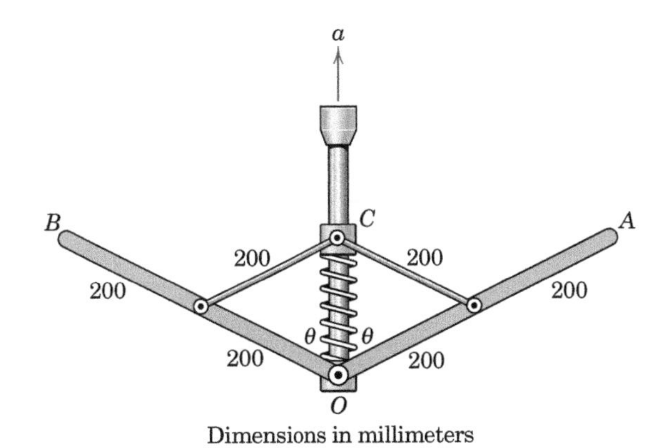

Problem 6/157

6/158 The linkage consists of the two slender bars and moves in the horizontal plane under the influence of force P. Link OC has a mass m and link AC has a mass $2m$. The sliding block at B has negligible mass. Without dismembering the system, determine the initial angular acceleration α of the links as P is applied at A with the links initially at rest. (*Suggestion:* Replace P by its equivalent force-couple system.)

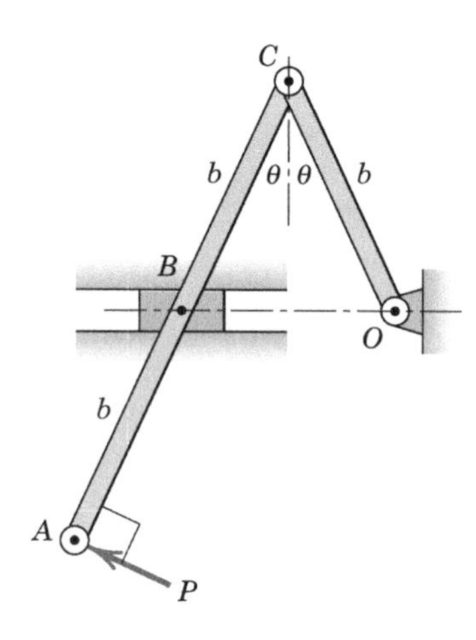

Problem 6/158

6/159 The portable work platform is elevated by means of the two hydraulic cylinders articulated at points C. The pressure in each cylinder produces a force F. The platform, man, and load have a combined mass m, and the mass of the linkage is small and may be neglected. Determine the upward acceleration a of the platform and show that it is independent of both b and θ.

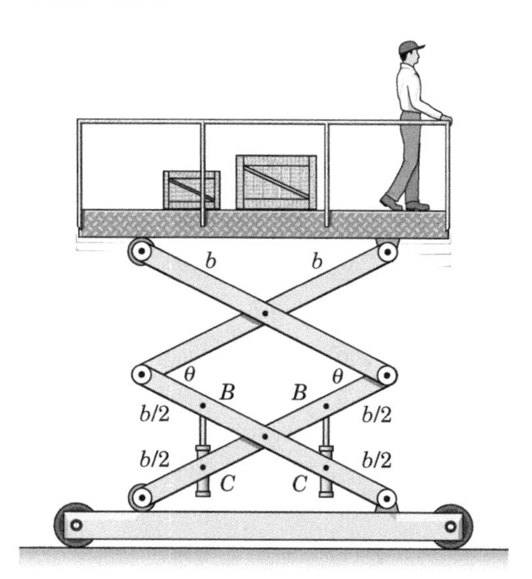

Problem 6/159

6/160 Each of the three identical uniform panels of a segmented industrial door has mass m and is guided in the tracks (one shown dashed). Determine the horizontal acceleration a of the upper panel under the action of the force P. Neglect any friction in the guide rollers.

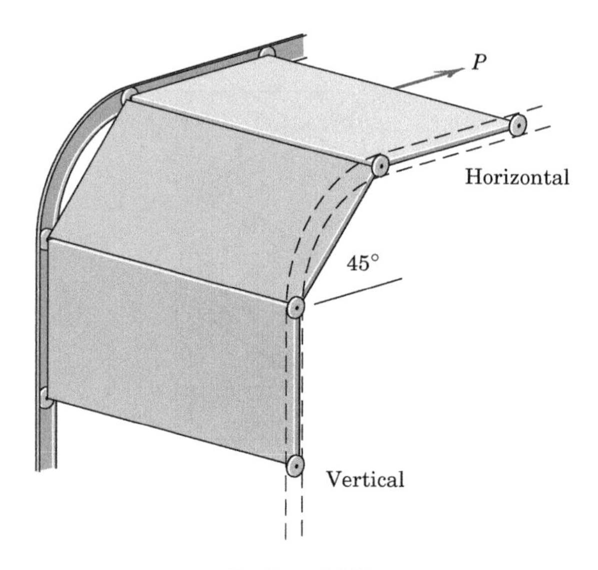

Problem 6/160

6/161 The mechanical tachometer measures the rotational speed N of the shaft by the horizontal motion of the collar B along the rotating shaft. This movement is caused by the centrifugal action of the two 12-oz weights A, which rotate with the shaft. Collar C is fixed to the shaft. Determine the rotational speed N of the shaft for a reading $\beta = 15°$. The stiffness of the spring is 5 lb/in., and it is uncompressed when $\theta = 0$ and $\beta = 0$. Neglect the weights of the links.

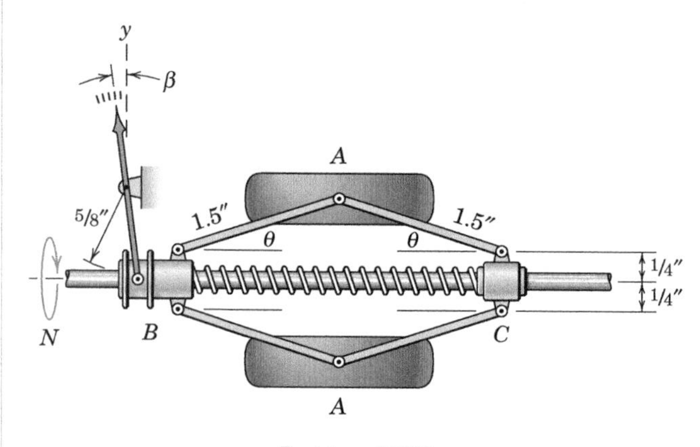

Problem 6/161

6/162 A planetary gear system is shown, where the gear teeth are omitted from the figure. Each of the three identical planet gears A, B, and C has a mass of 0.8 kg, a radius $r = 50$ mm, and a radius of gyration of 30 mm about its center. The spider E has a mass of 1.2 kg and a radius of gyration about O of 60 mm. The ring gear D has a radius $R = 150$ mm and is fixed. If a torque $M = 5$ N·m is applied to the shaft of the spider at O, determine the initial angular acceleration α of the spider.

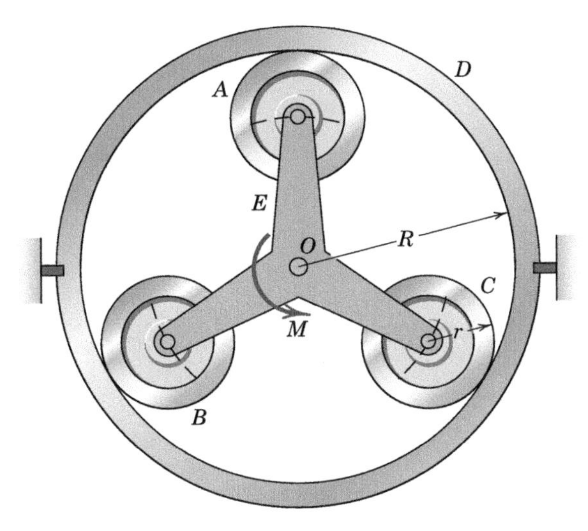

Problem 6/162

6/163 The sector and attached wheels are released from rest in the position shown in the vertical plane. Each wheel is a solid circular disk weighing 12 lb and rolls on the fixed circular path without slipping. The sector weighs 18 lb and is closely approximated by one-fourth of a solid circular disk of 16-in. radius. Determine the initial angular acceleration α of the sector.

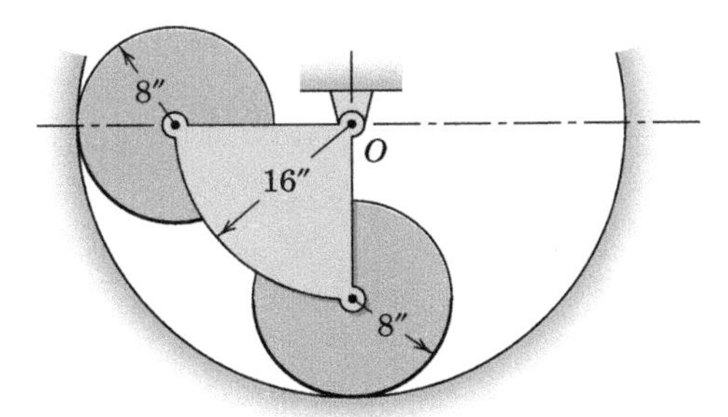

Problem 6/163

6/164 The aerial tower shown is designed to elevate a workman in a vertical direction. An internal mechanism at B maintains the angle between AB and BC at twice the angle θ between BC and the ground. If the combined mass of the man and the cab is 200 kg and if all other masses are neglected, determine the torque M applied to BC at C and the torque M_B in the joint at B required to give the cab an initial vertical acceleration of 1.2 m/s^2 when it is started from rest in the position $\theta = 30°$.

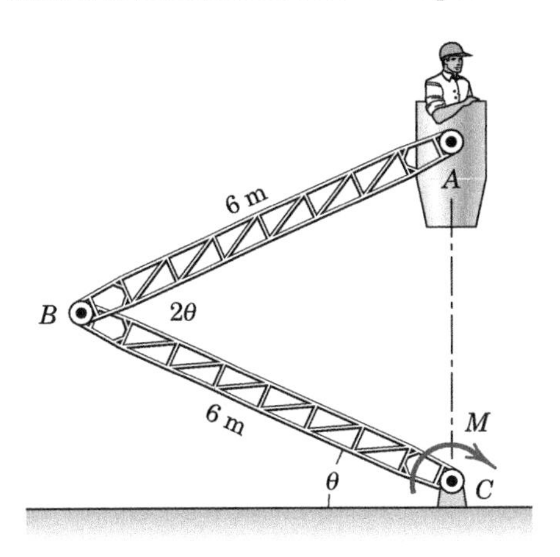

Problem 6/164

6/165 The uniform arm OA has a mass of 4 kg, and the gear D has a mass of 5 kg with a radius of gyration about its center of 64 mm. The large gear B is fixed and cannot rotate. If the arm and small gear are released from rest in the position shown in the vertical plane, calculate the initial angular acceleration α of OA.

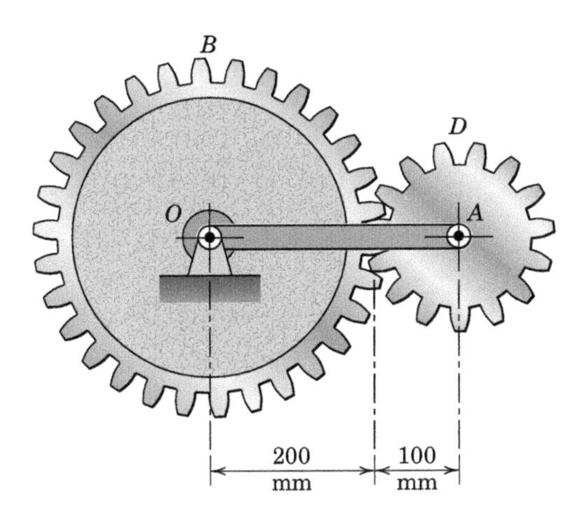

Problem 6/165

6/166 The vehicle is used to transport supplies to and from the bottom of the 25-percent grade. Each pair of wheels, one at A and the other at B, has a mass of 140 kg with a radius of gyration of 150 mm. The drum C has a mass of 40 kg and a radius of gyration of 100 mm. The total mass of the vehicle is 520 kg. The vehicle is released from rest with a restraining force T of 500 N in the control cable which passes around the drum and is secured at D. Determine the initial acceleration a of the vehicle. The wheels roll without slipping.

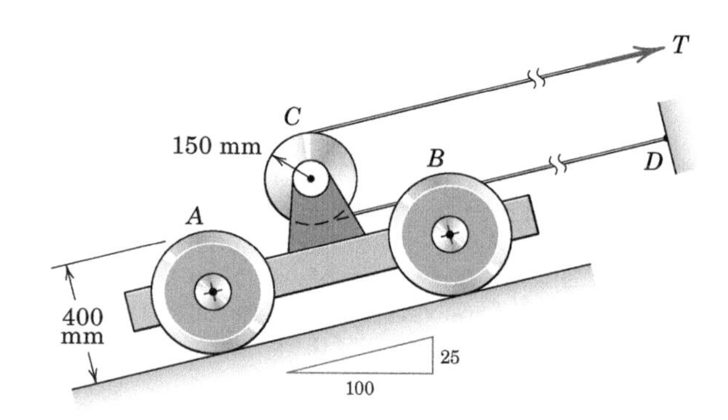

Problem 6/166

PROBLEMS

Introductory Problems

6/167 A person who walks through the revolving door exerts a 90-N horizontal force on one of the four door panels and keeps the 15° angle constant relative to a line which is normal to the panel. If each panel is modeled by a 60-kg uniform rectangular plate which is 1.2 m in length as viewed from above, determine the final angular velocity ω of the door if the person exerts the force for 3 seconds. The door is initially at rest and friction may be neglected.

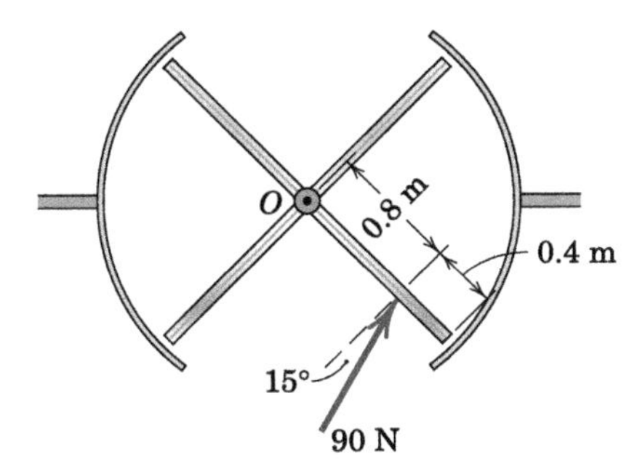

Problem 6/167

6/168 The 75-kg flywheel has a radius of gyration about its shaft axis of $\bar{k} = 0.50$ m and is subjected to the torque $M = 10(1 - e^{-t})$ N·m, where t is in seconds. If the flywheel is at rest at time $t = 0$, determine its angular velocity ω at $t = 3$ s.

Problem 6/168

6/169 Determine the angular momentum of the earth about the center of the sun. Assume a homogeneous earth and a circular earth orbit of radius $149.6(10^6)$ km. Consult Table D/2 of Appendix D for other needed information. Comment on the relative contributions of the terms $\bar{I}\omega$ and $m\bar{v}d$.

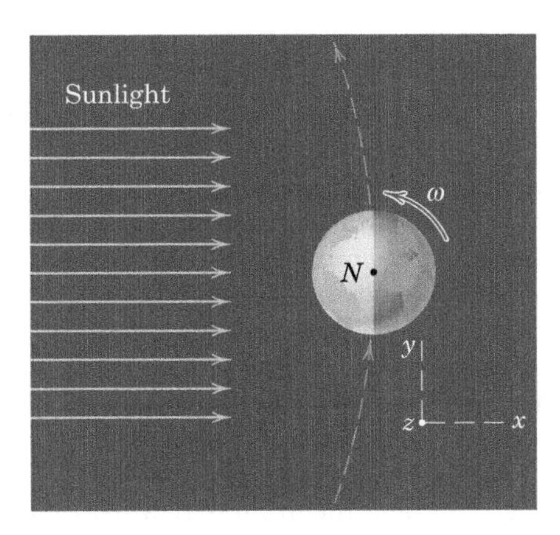

Problem 6/169

6/170 The frame of mass m is welded together from uniform slender rods. The frame is released from rest in the upper position shown and constrained to fall vertically by two light rollers which travel along the smooth slots. The roller at A catches in the support at O without rebounding and serves as a hinge for the frame as it rotates thereafter. Determine the angular velocity of the frame an instant after the roller at A engages the support at O.

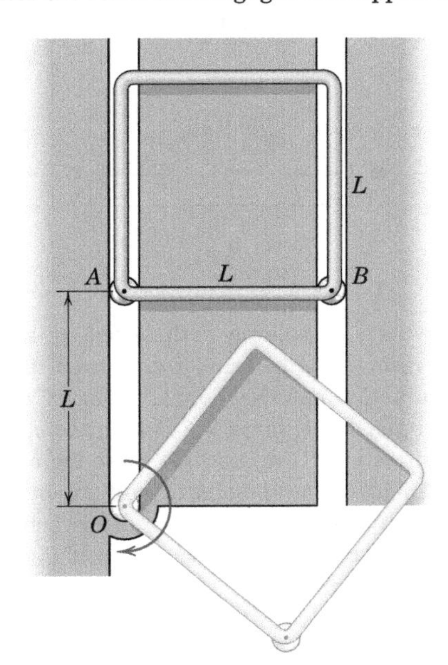

Problem 6/170

6/171 The frictional moment M_f acting on a rotating turbine disk and its shaft is given by $M_f = k\omega^2$ where ω is the angular velocity of the turbine. If the source of power is cut off while the turbine is running with an angular velocity ω_0, determine the time t for the speed of the turbine to drop to one-half of its initial value. The moment of inertia of the turbine disk and shaft is I.

6/172 The cable drum has a mass of 800 kg with radius of gyration of 480 mm about its center O and is mounted in bearings on the 1200-kg carriage. The carriage is initially moving to the left with a speed of 1.5 m/s, and the drum is rotating counterclockwise with an angular velocity of 3 rad/s when a constant horizontal tension $T = 400$ N is applied to the cable at time $t = 0$. Determine the velocity v of the carriage and the angular velocity ω of the drum when $t = 10$ s. Neglect the mass of the carriage wheels.

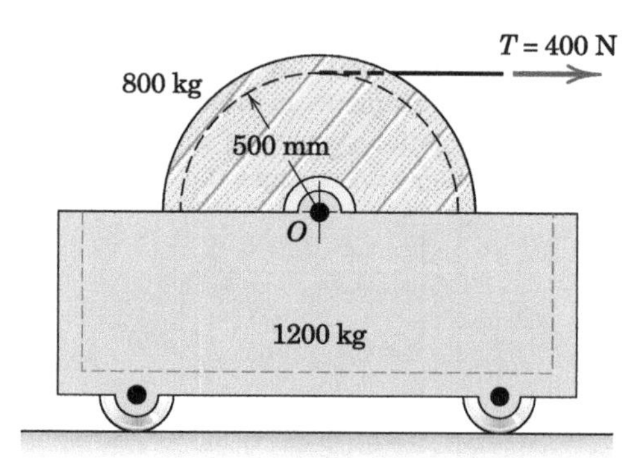

Problem 6/172

6/173 The man is walking with speed $v_1 = 1.2$ m/s to the right when he trips over a small floor discontinuity. Estimate his angular velocity ω just after the impact. His mass is 76 kg with center-of-mass height $h = 0.87$ m, and his mass moment of inertia about the ankle joint O is 66 kg·m^2, where all are properties of the portion of his body above O; i.e., both the mass and moment of inertia do not include the foot.

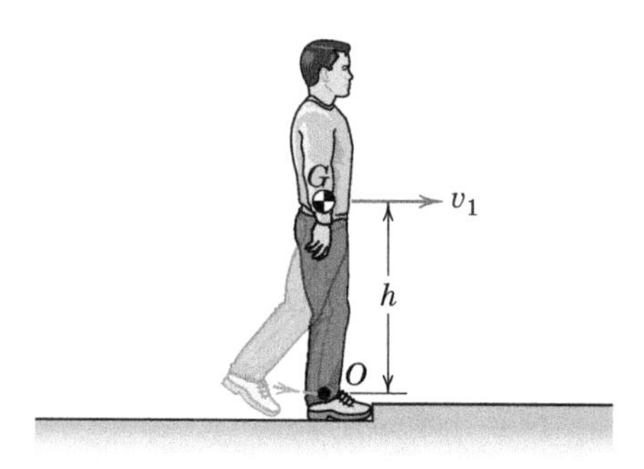

Problem 6/173

6/174 The 15-kg wheel with 150-mm outer radius and 115-mm centroidial radius of gyration is rolling without slipping down the 15° incline at a speed of 2 m/s when a tension $T = 30$ N is applied to a cable wrapped securely around an inner hub with a radius of 100 mm. Determine the time t required for the wheel to come to a stop if the tension is applied first in configuration (a) and then in configuration (b). The wheel is observed to roll without slipping throughout the entire motion for both cases.

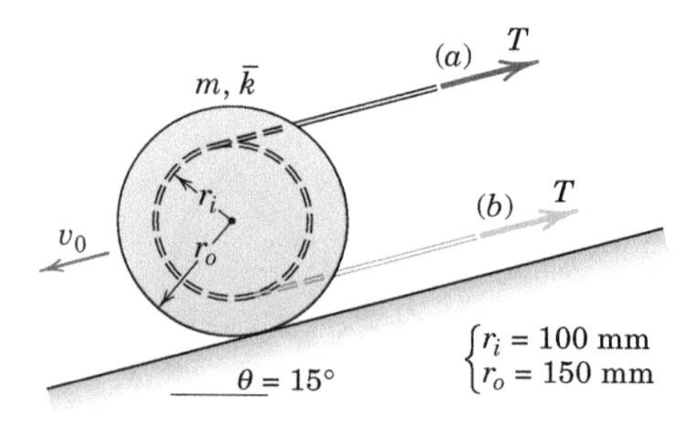

Problem 6/174

6/175 A uniform slender bar of mass M and length L is translating on the smooth horizontal x-y plane with a velocity v_M when a particle of mass m traveling with a velocity v_m as shown strikes and becomes embedded in the bar. Determine the final linear and angular velocities of the bar with its embedded particle.

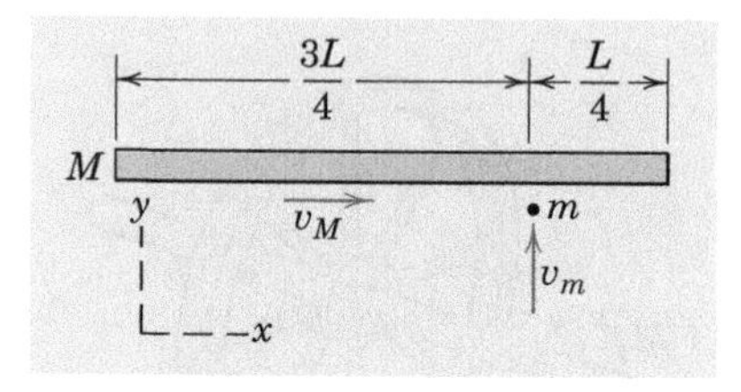

Problem 6/175

6/176 The homogeneous circular cylinder of mass m and radius R carries a slender rod of mass $m/2$ attached to it as shown. If the cylinder rolls on the surface without slipping with a velocity v_O of its center O, determine the angular momenta H_G and H_O of the system about its center of mass G and about O for the instant shown.

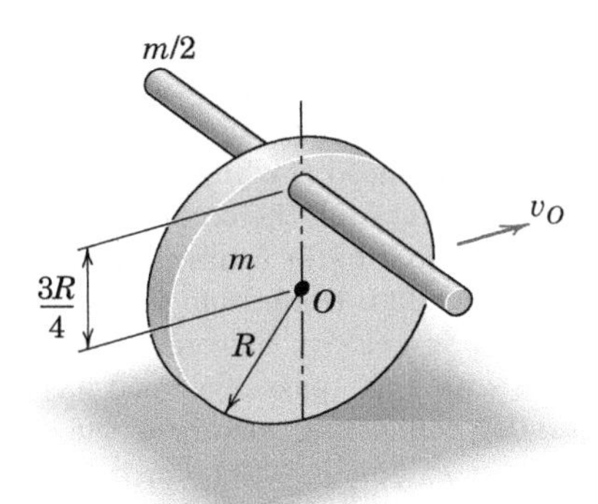

Problem 6/176

6/177 The grooved pulley of mass m is acted on by a constant force F through a cable which is wrapped securely around the exterior of the pulley. The pulley supports a cylinder of mass M which is attached to the end of a cable which is wrapped securely around an inner hub. If the system is stationary when the force F is first applied, determine the upward velocity of the supported mass after 3 seconds. Use the values $m = 40$ kg, $M = 10$ kg, $r_o = 225$ mm, $r_i = 150$ mm, $k_O = 160$ mm, and $F = 75$ N. Assume no mechanical interference for the indicated time frame and neglect friction in the bearing at O. What is the time-averaged value of the force in the cable which supports the 10-kg mass?

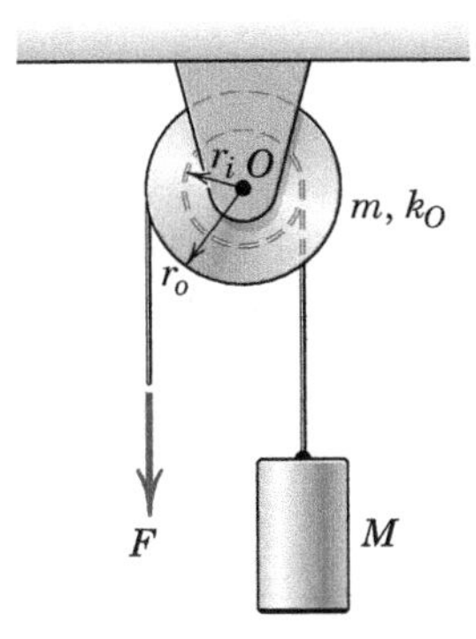

Problem 6/177

6/178 The wad of clay of mass m is initially moving with a horizontal velocity v_1 when it strikes and sticks to the initially stationary uniform slender bar of mass M and length L. Determine the final angular velocity of the combined body and the x-component of the linear impulse applied to the body by the pivot O during the impact.

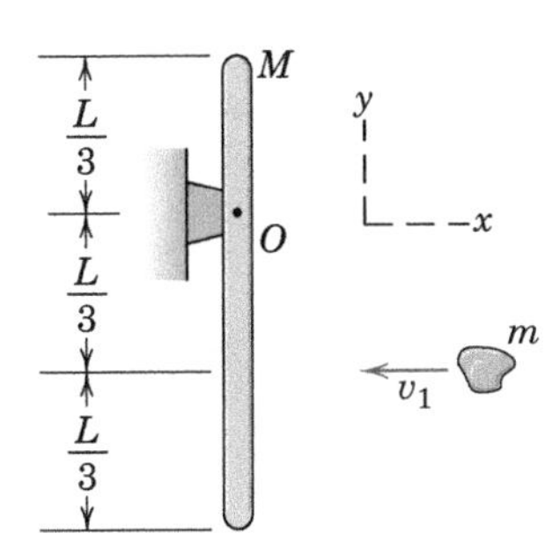

Problem 6/178

Representative Problems

6/179 The 30-lb uniform plate has the indicated velocities at corners A and B. What is the angular momentum of the plate about the mass center G?

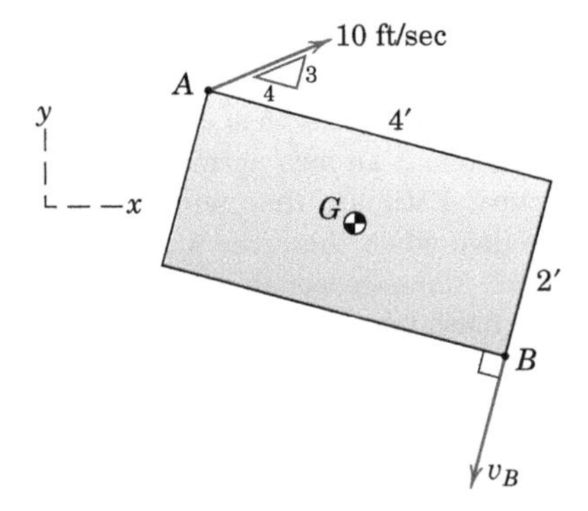

Problem 6/179

6/180 The plate of Prob. 6/179 is repeated here where the coordinates of corner C are established. The plate is falling freely in the x-y vertical plane. What is the linear momentum of the plate and its angular momentum about point O? Additionally, determine the time rate of change of the angular momentum of the plate about point O.

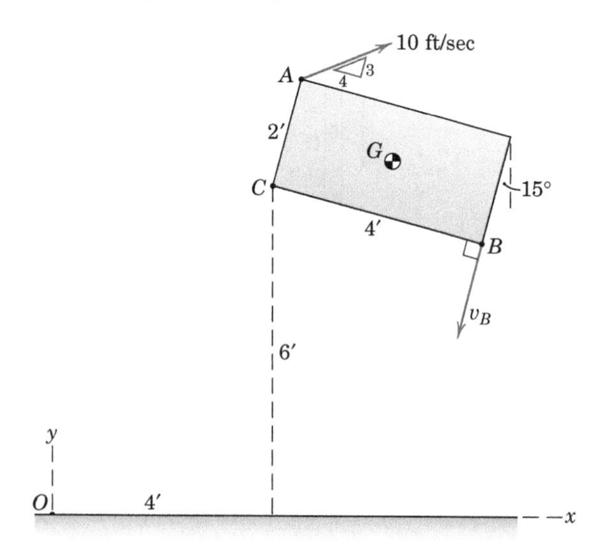

Problem 6/180

6/181 Just after leaving the platform, the diver's fully extended 80-kg body has a rotational speed of 0.3 rev/s about an axis normal to the plane of the trajectory. Estimate the angular velocity N later in the dive when the diver has assumed the tuck position. Make reasonable assumptions concerning the mass moment of inertia of the body in each configuration.

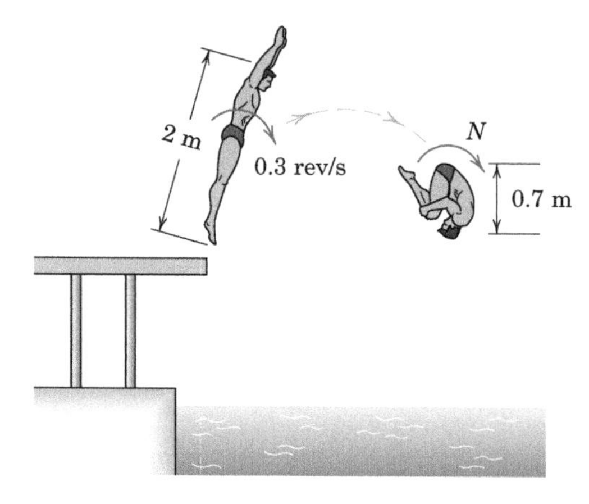

Problem 6/181

6/182 The device shown is a simplified model of an amusement-park ride in which passengers are rotated about the vertical axis of the central post at an angular speed Ω while sitting in a pod which is capable of rotating the occupants 360° about the longitudinal axis of the connecting arm attached to the central collar. Determine the percent increase n in angular velocity between configurations (a) and (b), where the passenger pod has rotated 90° about the connecting arm. For the model, m = 1.2 kg, r = 75 mm, l = 300 mm, and L = 650 mm. The post and connecting arms rotate freely about the z-axis at an initial angular speed Ω = 120 rev/min and have a combined mass moment of inertia about the z-axis of $30(10^{-3})$ kg·m^2.

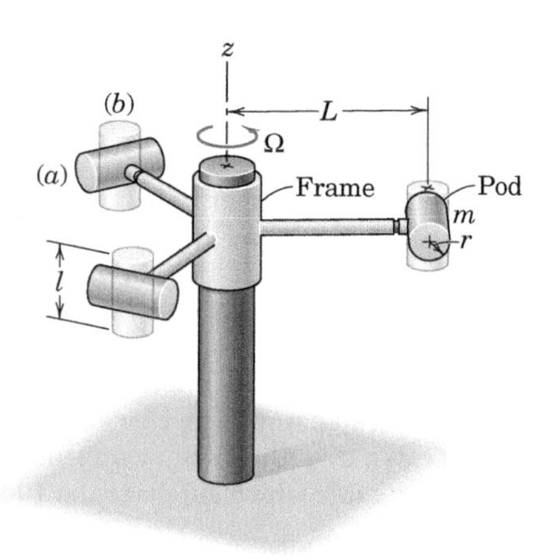

Problem 6/182

6/183 The uniform concrete block, which weighs 171 lb and falls from rest in the horizontal position shown, strikes the fixed corner A and pivots around it with no rebound. Calculate the angular velocity ω of the block immediately after it hits the corner and the percentage loss n of energy due to the impact.

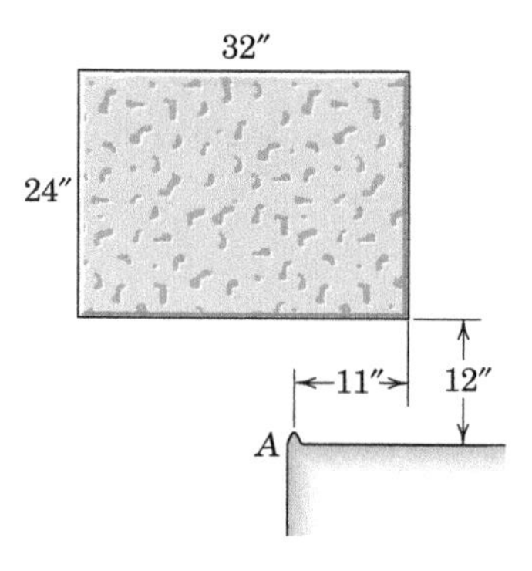

Problem 6/183

6/184 Two small variable-thrust jets are actuated to keep the spacecraft angular velocity about the z-axis constant at $\omega_0 = 1.25$ rad/s as the two telescoping booms are extended from $r_1 = 1.2$ m to $r_2 = 4.5$ m at a constant rate over a 2-min period. Determine the necessary thrust T for each jet as a function of time where $t = 0$ is the time when the telescoping action is begun. The small 10-kg experiment modules at the ends of the booms may be treated as particles, and the mass of the rigid booms is negligible.

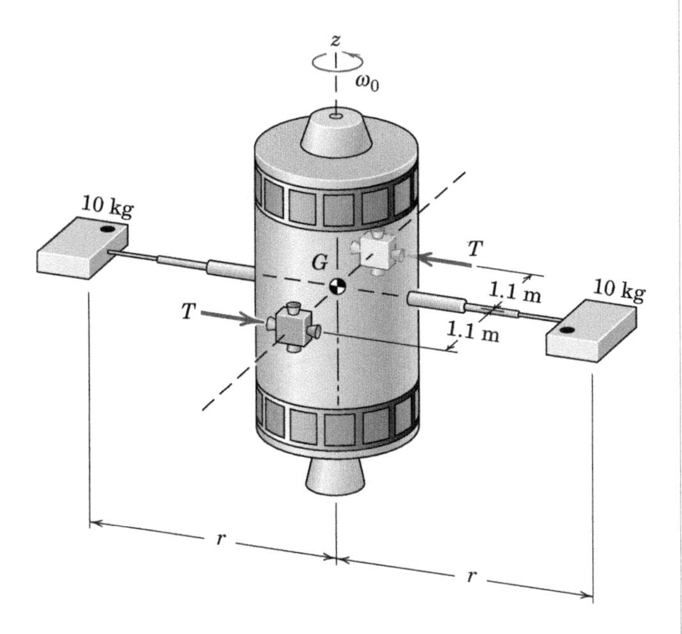

Problem 6/184

6/185 The body composed of slender rods of mass ρ per unit length is lying motionless on the smooth horizontal surface when a linear impluse $\int P\, dt$ is applied as shown. Determine the velocity $\mathbf{v}_B$ of corner B immediately following the application of the impulse if $l = 500$ mm, $\rho = 3$ kg/m, and $\int P\, dt = 8$ N·s.

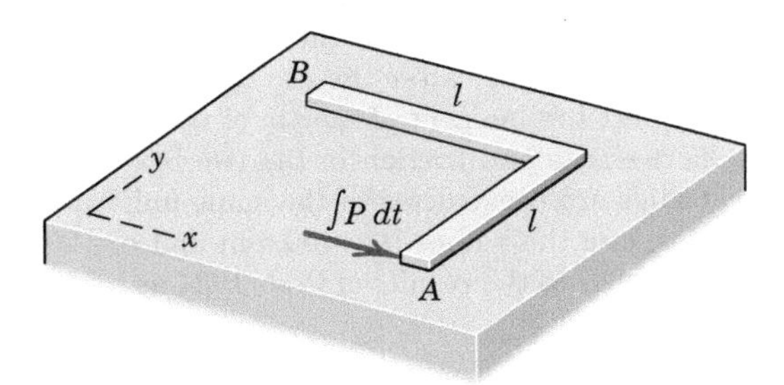

Problem 6/185

6/186 Each of the two 300-mm uniform rods A has a mass of 1.5 kg and is hinged at its end to the rotating base B. The 4-kg base has a radius of gyration of 40 mm and is initially rotating freely about its vertical axis with a speed of 300 rev/min and with the rods latched in the vertical positions. If the latches are released and the rods assume the horizontal positions, calculate the new rotational speed N of the assembly.

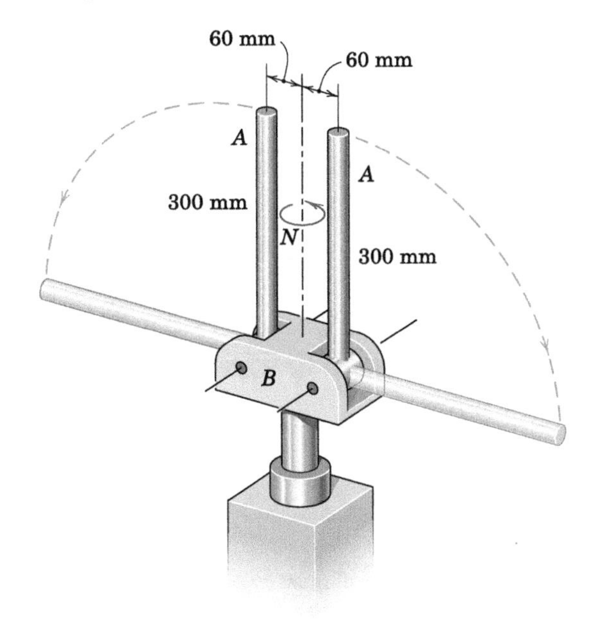

Problem 6/186

6/187 The phenomenon of vehicle "tripping" is investigated here. The sport-utility vehicle is sliding sideways with speed v_1 and no angular velocity when it strikes a small curb. Assume no rebound of the right-side tires and estimate the minimum speed v_1 which will cause the vehicle to roll completely over to its right side. The mass of the SUV is 2300 kg and its mass moment of inertia about a longitudinal axis through the mass center G is 900 kg·m^2.

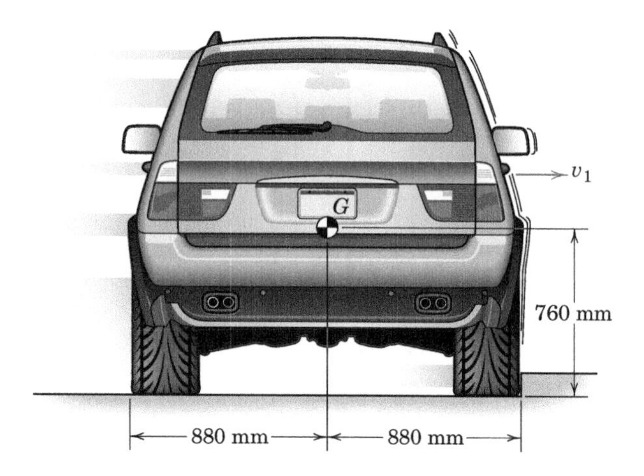

Problem 6/187

6/188 The slender bar of mass m and length l is released from rest in the horizontal position shown. If point A of the bar becomes attached to the pivot at B upon impact, determine the angular velocity ω of the bar immediately after impact in terms of the distance x. Evaluate your expression for $x = 0, l/2$, and l.

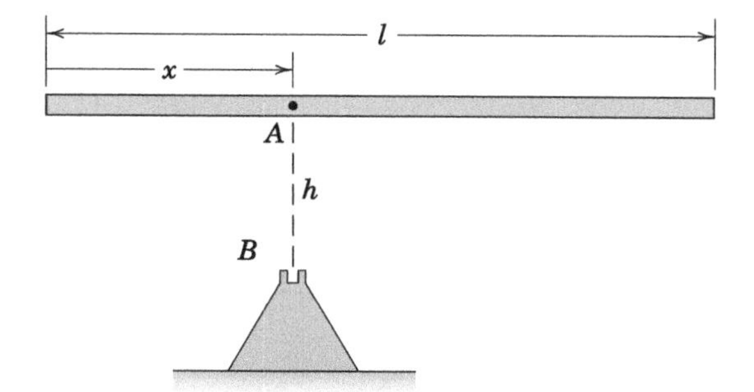

Problem 6/188

6/189 Child A weighs 75 lb and is sitting at rest relative to the merry-go-round which is rotating counterclockwise with an angular velocity $\Omega = 15$ rev/min. Child B, with a weight of 65 lb, runs toward the merry-go-round with a speed $v = 12$ ft/sec and jumps onto the edge. Compute the angular velocity Ω' of the merry-go-round when child B has just jumped aboard the merry-go-round and is standing in position (a). If child B moves to position (b), what is the new angular velocity of the merry-go-round? For simplicity, model the inertia of child A as a 20-in.-diameter uniform disk (horizontal) and model child B as a 14-in.-diameter uniform and upright cylinder. The merry-go-round itself has a centroidal mass moment of inertia $I_O = 110$ lb-ft-sec^2.

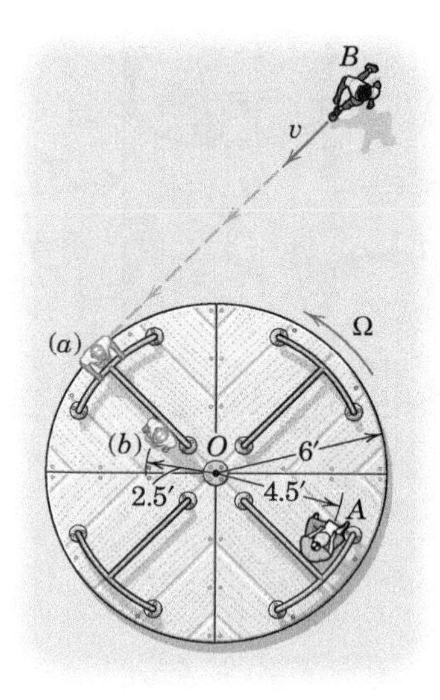

Problem 6/189

6/190 The system is initially rotating freely with angular velocity $\omega_1 = 10$ rad/s when the inner rod A is centered lengthwise within the hollow cylinder B as shown in the figure. Determine the angular velocity of the system (a) if the inner rod A has moved so that a length $b/2$ is protruding from the cylinder, (b) just before the rod leaves the cylinder, and (c) just after the rod leaves the cylinder. Neglect the moment of inertia of the vertical support shafts and friction in the two bearings. Both bodies are constructed of the same uniform material. Use the values $b = 400$ mm and $r = 20$ mm, and refer to the results of Prob. B/34 as needed.

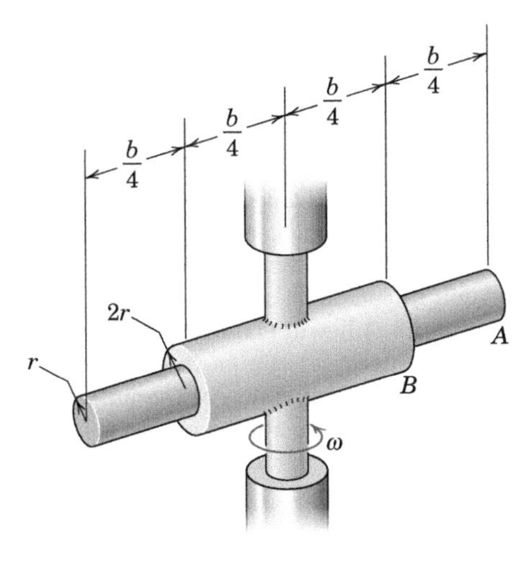

Problem 6/190

6/191 The homogeneous sphere of mass m and radius r is projected along the incline of angle θ with an initial speed v_0 and no angular velocity ($\omega_0 = 0$). If the coefficient of kinetic friction is μ_k, determine the time duration t of the period of slipping. In addition, state the velocity v of the mass center G and the angular velocity ω at the end of the period of slipping.

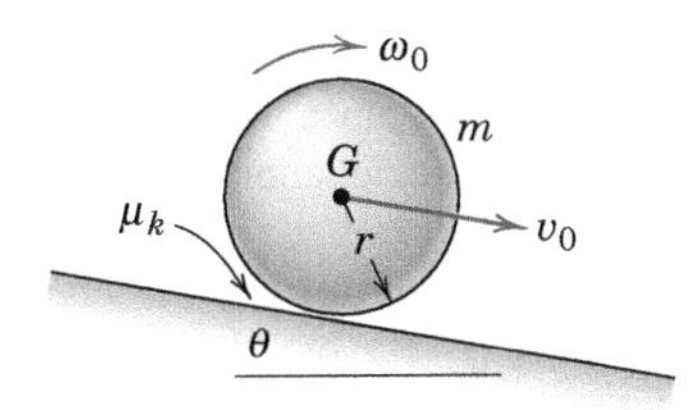

Problem 6/191

6/192 The homogeneous sphere of Prob. 6/191 is placed on the incline with a clockwise angular velocity ω_0 but no linear velocity of its center ($v_0 = 0$). Determine the time duration t of the period of slipping. In addition, state the velocity v of the mass center G and angular velocity ω at the end of the period of slipping.

6/193 The 100-lb platform rolls without slipping along the 10° incline on two pairs of 16-in.-diameter wheels. Each pair of wheels with attached axle weighs 25 lb and has a centroidal radius of gyration of 5.5 in. The platform has an initial speed of 3 ft/sec down the incline when a tension T is applied through a cable attached to the platform. If the platform acquires a speed of 3 ft/sec up the incline after the tension has been applied for 8 seconds, what is the average value of the tension in the cable?

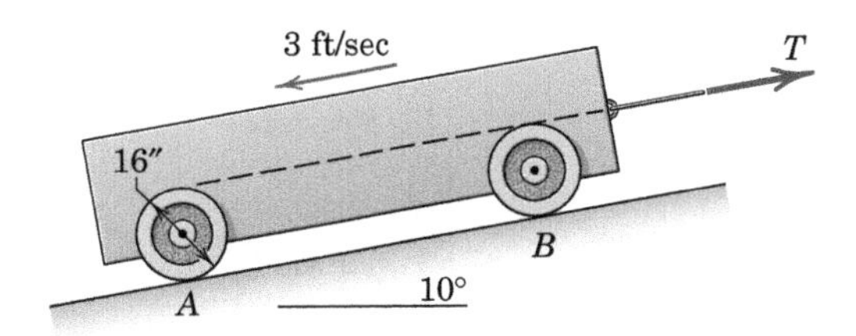

Problem 6/193

6/194 The uniform slender bar of mass m and length l has no angular velocity as end A strikes the ground against the stop with no rebound. If $\alpha = 15°$, what is the minimum magnitude of the initial velocity $\mathbf{v}_1$ for which the bar will rotate about A to the vertical position?

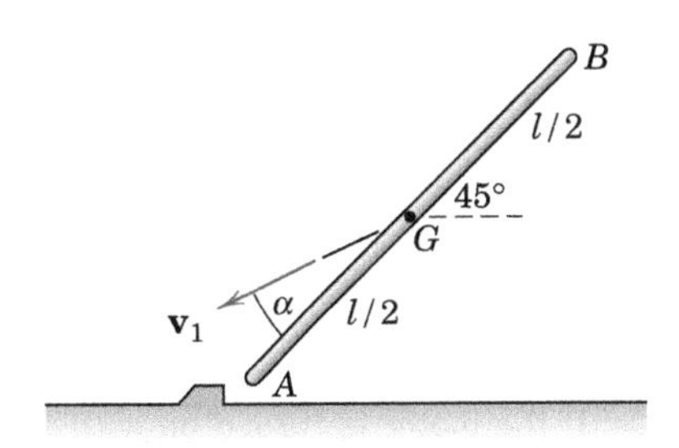

Problem 6/194

6/195 The 165-lb ice skater with arms extended horizontally spins about a vertical axis with a rotational speed of 1 rev/sec. Estimate his rotational speed N if he fully retracts his arms, bringing his hands very close to the centerline of his body. As a reasonable approximation, model the extended arms as uniform slender rods, each of which is 27 in. long and weighs 15 lb. Model the torso as a solid 135-lb cylinder 13 in. in diameter. Treat the man with arms retracted as a solid 165-lb cylinder of 13-in. diameter. Neglect friction at the skate–ice interface.

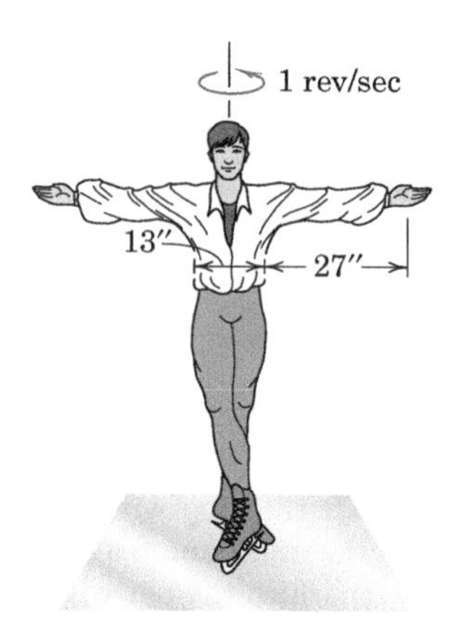

Problem 6/195

6/196 In the rotating assembly shown, arm OA and the attached motor housing B together weigh 10 lb and have a radius of gyration about the z-axis of 7 in. The motor armature and attached 5-in.-radius disk have a combined weight of 15 lb and a radius of gyration of 4 in. about their own axis. The entire assembly is free to rotate about the z-axis. If the motor is turned on with OA initially at rest, determine the angular speed N of OA when the motor has reached a speed of 300 rev/min *relative* to arm OA.

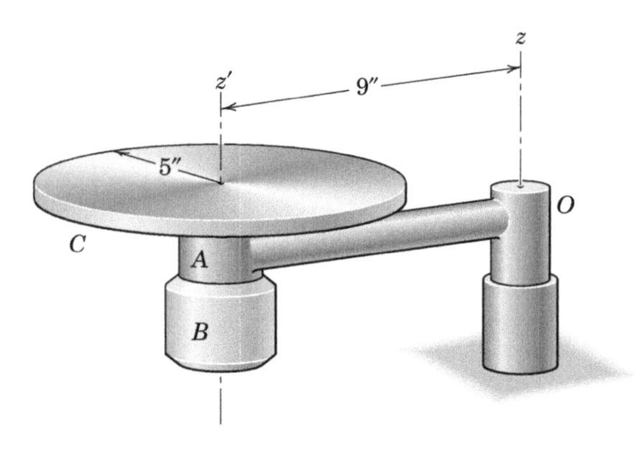

Problem 6/196

6/197 The motor at B supplies a constant torque M which is applied to a 375-mm-diameter internal drum around which is wound the cable shown. This cable then wraps around an 80-kg pulley attached to a 125-kg cart carrying 600 kg of rock. The motor is able to bring the loaded cart to a cruising speed of 1.5 m/s in 3 seconds. What torque M is the motor able to supply, and what is the average value of the tension in each side of the cable which is wrapped around the pulley at O during the speed-up period? The cable does not slip on the pulley and the centroidal radius of gyration of the pulley is 450 mm. What is the power output of the motor when the cart reaches its cruising speed?

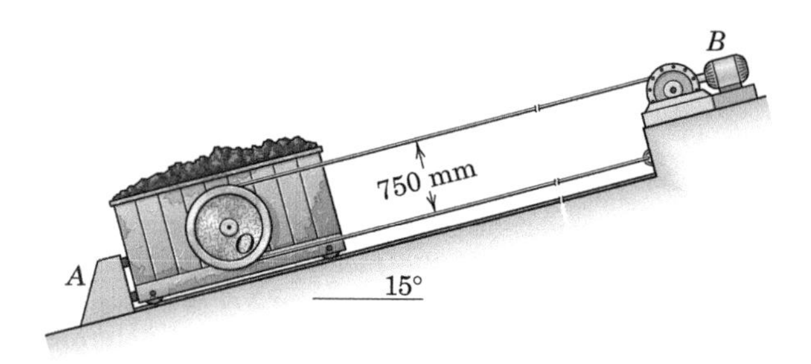

Problem 6/197

6/198 The body of the spacecraft weighs 322 lb on earth and has a radius of gyration about its z-axis of 1.5 ft. Each of the two solar panels may be treated as a uniform flat plate weighing 16.1 lb. If the spacecraft is rotating about its z-axis at the angular rate of 1.0 rad/sec with $\theta = 0$, determine the angular rate ω after the panels are rotated to the position $\theta = \pi/2$ by an internal mechanism. Neglect the small momentum change of the body about the y-axis.

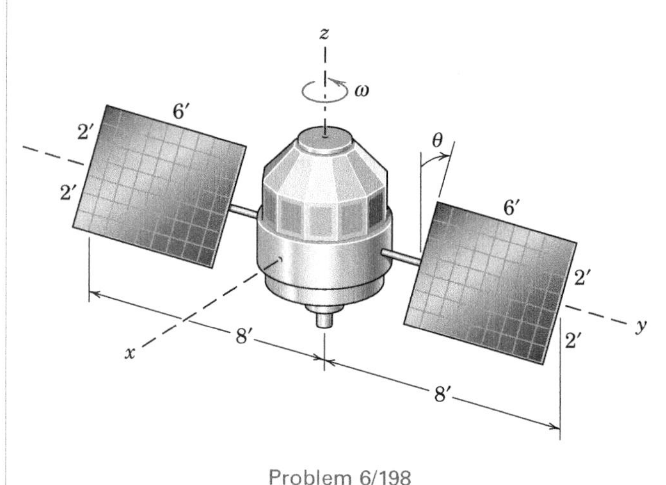

Problem 6/198

6/199 A 55-kg dynamics instructor is demonstrating the principles of angular momentum to her class. She stands on a freely rotating platform with her body aligned with the vertical platform axis. With the platform not rotating, she holds a modified bicycle wheel so that its axis is vertical. She then turns the wheel axis to a horizontal orientation without changing the 600-mm distance from the centerline of her body to the wheel center, and her students observe a platform rotation rate of 30 rev/min. If the rim-weighted wheel has a mass of 10 kg and a centroidal radius of gyration $\bar{k}$ = 300 mm, and is spinning at a fairly constant rate of 250 rev/min, estimate the mass moment of inertia I of the instructor (in the posture shown) about the vertical platform axis.

Problem 6/199

6/200 The 8-lb slotted circular disk has a radius of gyration about its center O of 6 in. and initially is rotating freely about a fixed vertical axis through O with a speed N_1 = 600 rev/min. The 2-lb uniform slender bar A is initially at rest relative to the disk in the centered slot position as shown. A slight disturbance causes the bar to slide to the end of the slot where it comes to rest relative to the disk. Calculate the new angular speed N_2 of the disk, assuming the absence of friction in the shaft bearing at O. Does the presence of any friction in the slot affect the final result?

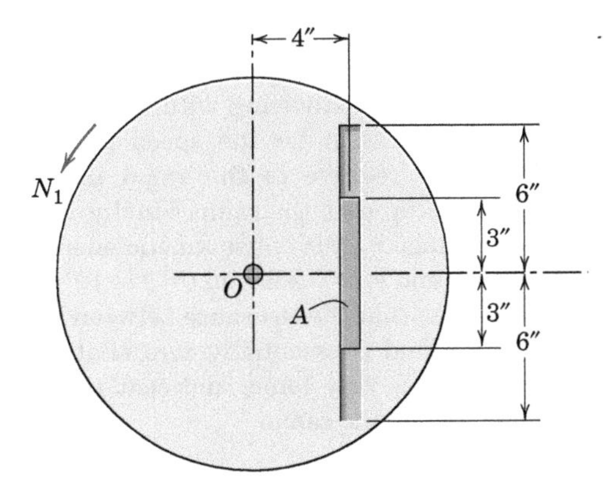

Problem 6/200

6/201 The gear train shown starts from rest and reaches an output speed of ω_C = 240 rev/min in 2.25 s. Rotation of the train is resisted by a constant 150 N·m moment at the output gear C. Determine the required input power to the 86% efficient motor at A just before the final speed is reached. The gears have masses m_A = 6 kg, m_B = 10 kg, and m_C = 24 kg, pitch diameters d_A = 120 mm, d_B = 160 mm, and d_C = 240 mm, and centroidal radii of gyration k_A = 48 mm, k_B = 64 mm, and k_C = 96 mm.

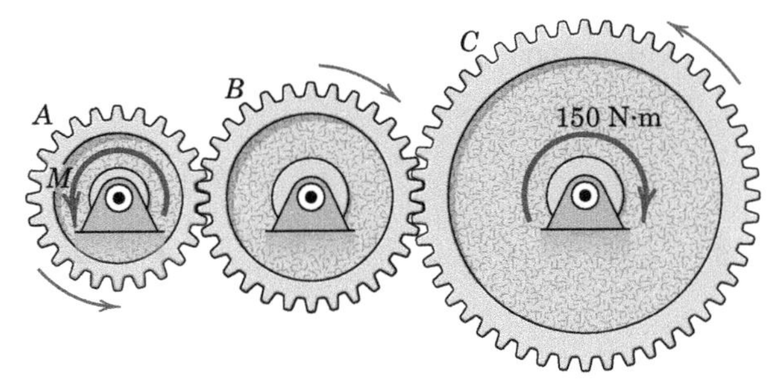

Problem 6/201

6/202 The uniform cylinder is rolling without slip with a velocity v along the horizontal surface when it overtakes a ramp traveling with speed v_0. Determine an expression for the speed v' which the cylinder has relative to the ramp immediately after it rolls up onto the ramp. Finally, determine the percentage n of cylinder kinetic energy lost if (*a*) $\theta = 10°$ and $v_0 = 0.25v$ and (*b*) $\theta = 10°$ and $v_0 = 0.5v$. Assume that the clearance between the ramp and the ground is essentially zero, that the mass of the ramp is very large, and that the cylinder does not slip on the ramp.

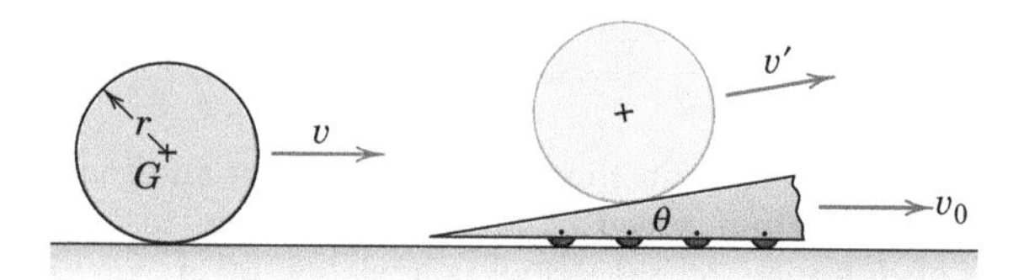

Problem 6/202

6/203 A frozen-juice can rests on the horizontal rack of a freezer door as shown. With what maximum angular velocity Ω can the door be "slammed" shut against its seal and not dislodge the can? Assume that the can rolls without slipping on the corner of the rack, and neglect the dimension d compared with the 500-mm distance.

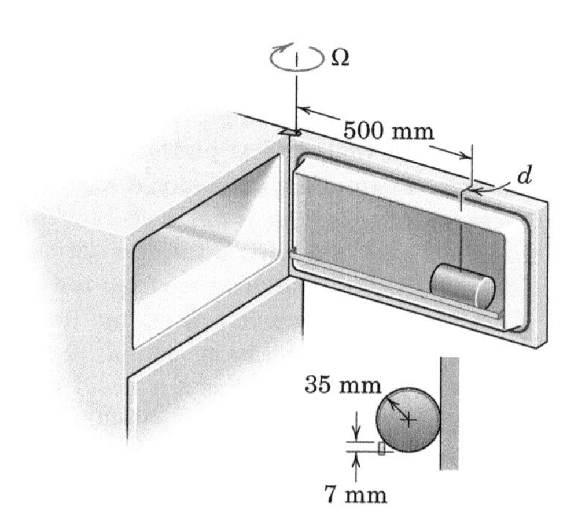

Problem 6/203

▶6/204 The 30-kg wheel has a radius of gyration about its center of 75 mm and is rotating clockwise at the rate of 300 rev/min when it is released onto the incline with no velocity of its center O. While the wheel is slipping, it is observed that the center O remains in a fixed position. Determine the coefficient of kinetic friction μ_k and the time t during which slipping occurs. Also determine the velocity v of the center 4 seconds after the wheel has stopped slipping.

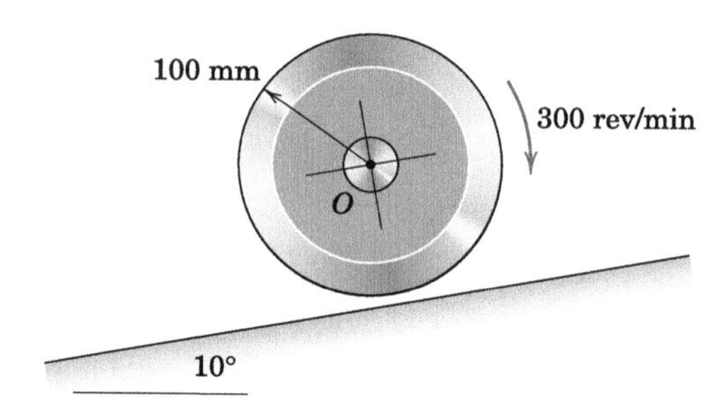

Problem 6/204

PROBLEMS

Introductory Problems

7/1 Place your textbook on your desk, with fixed axes oriented as shown. Rotate the book about the x-axis through a 90° angle and then from this new position rotate it 90° about the y-axis. Sketch the final position of the book. Repeat the process but reverse the order of rotation. From your results, state your conclusion concerning the vector addition of finite rotations. Reconcile your observations with Fig. 7/4.

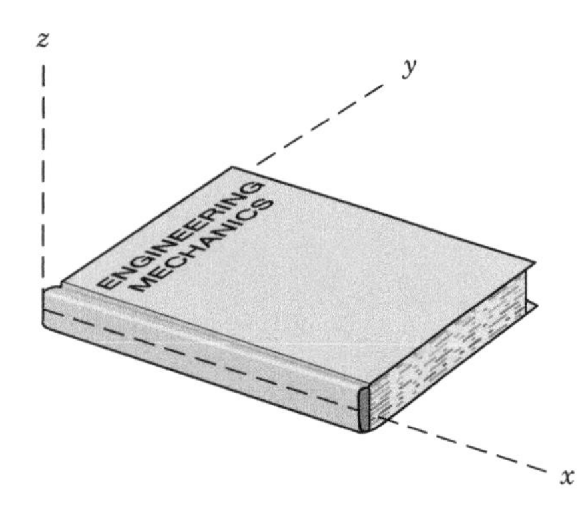

Problem 7/1

7/2 Repeat the experiment of Prob. 7/1 but use a small angle of rotation, say, 5°. Note the near-equal final positions for the two different rotation sequences. What does this observation lead you to conclude for the combination of infinitesimal rotations and for the time derivatives of angular quantities? Reconcile your observations with Fig. 7/5.

7/3 The solid cylinder is rotating about the fixed axis OA with a constant speed $N = 600$ rev/min in the direction shown. If the x- and y-components of the velocity of point P are 12 ft/sec and -6 ft/sec, determine its z-component of velocity and the radial distance R from P to the rotation axis. Also find the magnitude of the acceleration of P.

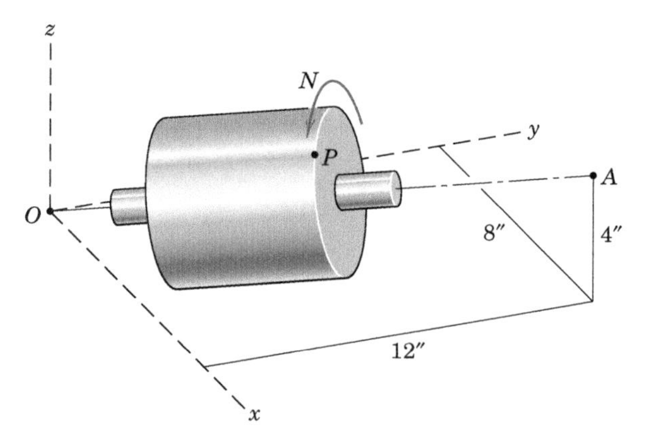

Problem 7/3

7/4 A timing mechanism consists of the rotating distributor arm AB and the fixed contact C. If the arm rotates about the fixed axis OA with a constant angular velocity $\boldsymbol{\omega} = 30(3\mathbf{i} + 2\mathbf{j} + 6\mathbf{k})$ rad/s, and if the coordinates of the contact C expressed in millimeters are (20, 30, 80), determine the magnitude of the acceleration of the tip B of the distributor arm as it passes point C.

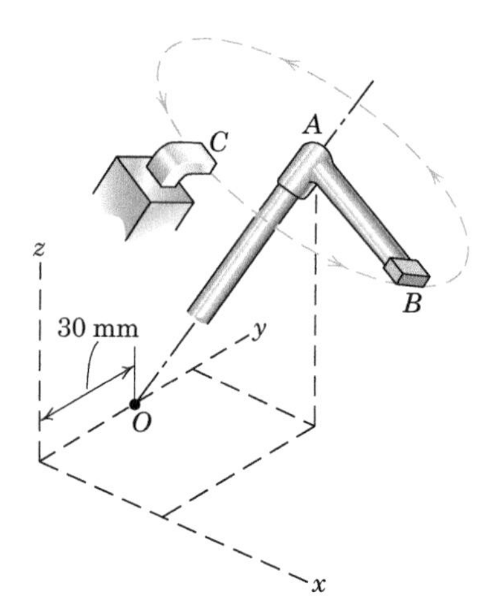

Problem 7/4

7/5 The rotor and shaft are mounted in a clevis which can rotate about the z-axis with an angular velocity Ω. With $\Omega = 0$ and θ constant, the rotor has an angular velocity $\boldsymbol{\omega}_0 = -4\mathbf{j} - 3\mathbf{k}$ rad/s. Find the velocity $\mathbf{v}_A$ of point A on the rim if its position vector at this instant is $\mathbf{r} = 0.5\mathbf{i} + 1.2\mathbf{j} + 1.1\mathbf{k}$ m. What is the rim speed v_B of any point B?

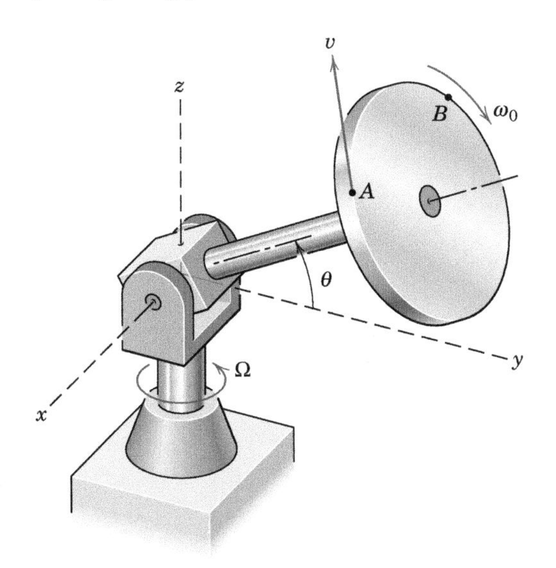

Problem 7/5

7/6 The disk rotates with a spin velocity of 15 rad/s about its horizontal z-axis first in the direction (a) and second in the direction (b). The assembly rotates with the angular velocity $N = 10$ rad/s about the vertical axis. Construct the space and body cones for each case.

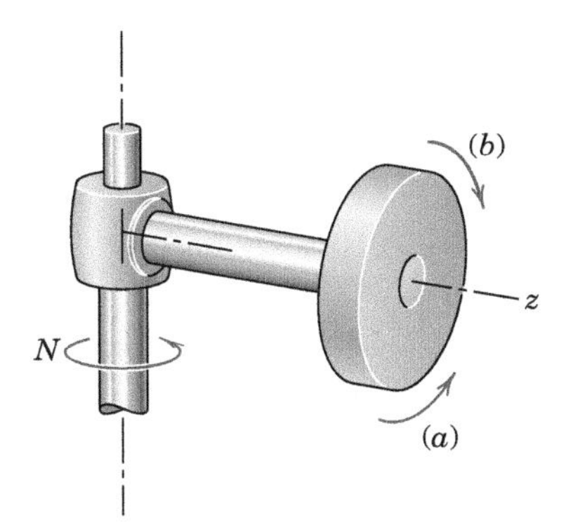

Problem 7/6

7/7 The rotor B spins about its inclined axis OA at the angular speed $N_1 = 200$ rev/min, where $\beta = 30°$. Simultaneously, the assembly rotates about the vertical z-axis at the rate N_2. If the total angular velocity of the rotor has a magnitude of 40 rad/s, determine N_2.

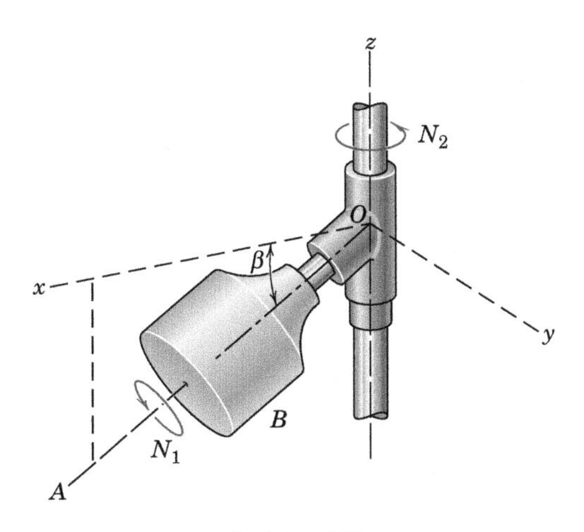

Problem 7/7

7/8 A slender rod bent into the shape shown rotates about the fixed line CD at a constant angular rate ω. Determine the velocity and acceleration of point A.

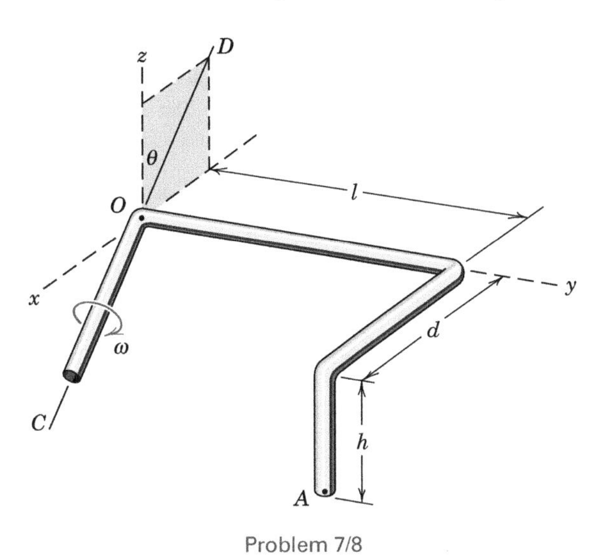

Problem 7/8

7/9 The rod is hinged about the axis O-O of the clevis, which is attached to the end of the vertical shaft. The shaft rotates with a constant angular velocity ω_0 as shown. If θ is decreasing at the constant rate $-\dot{\theta} = p$, write expressions for the angular velocity $\boldsymbol{\omega}$ and angular acceleration $\boldsymbol{\alpha}$ of the rod.

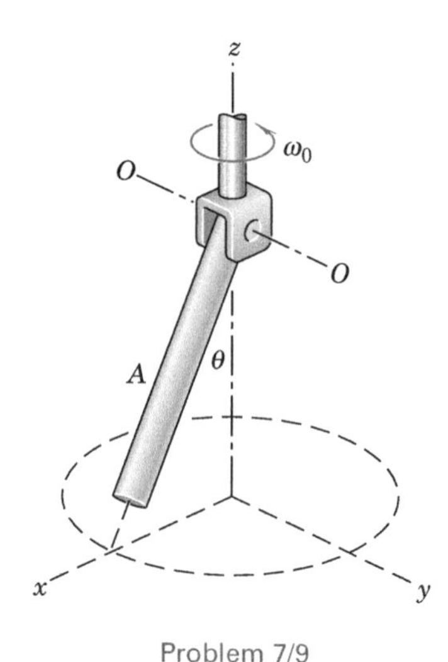

Problem 7/9

7/10 The panel assembly and attached x-y-z axes rotate with a constant angular velocity $\Omega = 0.6$ rad/sec about the vertical z-axis. Simultaneously, the panels rotate about the y-axis as shown with a constant rate $\omega_0 = 2$ rad/sec. Determine the angular acceleration $\boldsymbol{\alpha}$ of panel A and find the acceleration of point P for the instant when $\beta = 90°$.

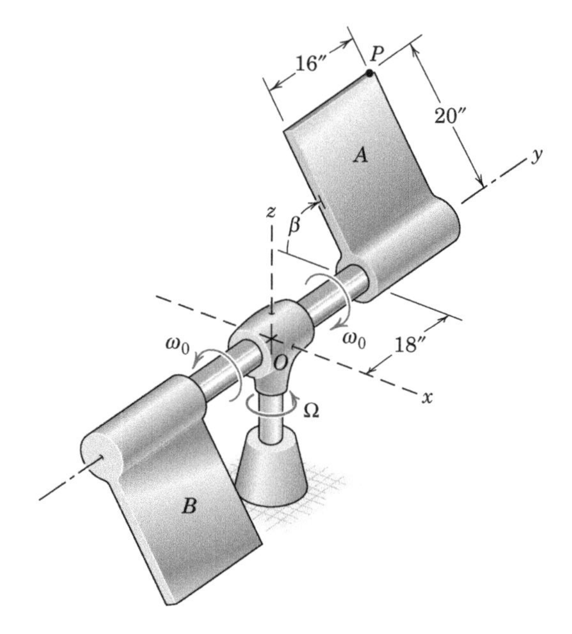

Problem 7/10

Representative Problems

7/11 The motor of Sample Problem 7/2 is shown again here. If the motor pivots about the x-axis at the constant rate $\dot{\gamma} = 3\pi$ rad/sec with no rotation about the Z-axis ($N = 0$), determine the angular acceleration $\boldsymbol{\alpha}$ of the rotor and disk as the position $\gamma = 30°$ is passed. The constant speed of the motor is 120 rev/min. Also find the velocity and acceleration of point A, which is on the top of the disk at this instant.

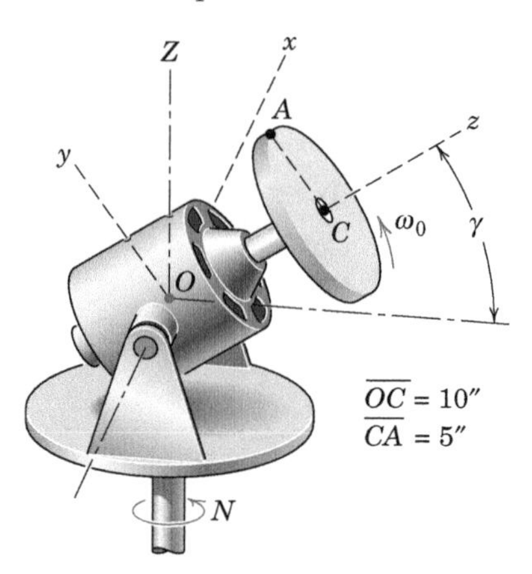

Problem 7/11

7/12 If the motor of Sample Problem 7/2, repeated in Prob. 7/11, reaches a speed of 3000 rev/min in 2 seconds from rest with constant acceleration, determine the total angular acceleration of the rotor and disk $\frac{1}{3}$ second after it is turned on if the turntable is rotating at a constant rate $N = 30$ rev/min. The angle $\gamma = 30°$ is constant.

7/13 The spool A rotates about its axis with an angular velocity of 20 rad/s, first in the sense of ω_a and second in the sense of ω_b. Simultaneously, the assembly rotates about the vertical axis with an angular velocity $\omega_1 = 10$ rad/s. Determine the magnitude ω of the total angular velocity of the spool and construct the body and space cones for the spool for each case.

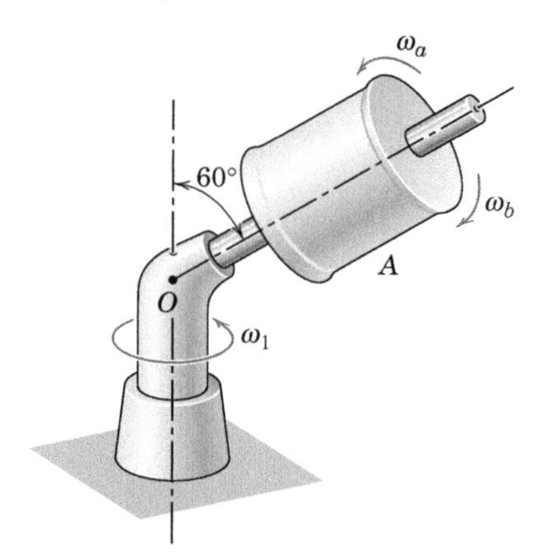

Problem 7/13

7/14 In manipulating the dumbbell, the jaws of the robotic device have an angular velocity $\omega_p = 2$ rad/s about the axis OG with γ fixed at 60°. The entire assembly rotates about the vertical Z-axis at the constant rate $\Omega = 0.8$ rad/s. Determine the angular velocity $\boldsymbol{\omega}$ and angular acceleration $\boldsymbol{\alpha}$ of the dumbbell. Express the results in terms of the given orientation of axes x-y-z, where the y-axis is parallel to the Y-axis.

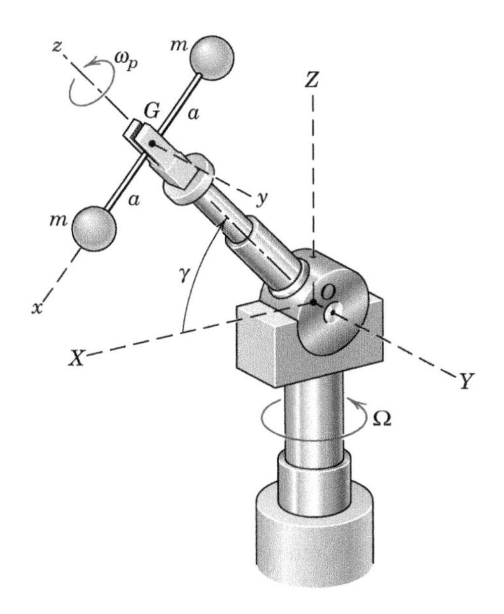

Problem 7/14

7/15 Determine the angular acceleration $\boldsymbol{\alpha}$ of the dumbbell of Prob. 7/14 for the conditions stated, except that Ω is increasing at the rate of 3 rad/s^2 for the instant under consideration.

7/16 The robot shown has five degrees of rotational freedom. The x-y-z axes are attached to the base ring, which rotates about the z-axis at the rate ω_1. The arm O_1O_2 rotates about the x-axis at the rate $\omega_2 = \dot{\theta}$. The control arm O_2A rotates about axis O_1-O_2 at the rate ω_3 and about a perpendicular axis through O_2 which is momentarily parallel to the x-axis at the rate $\omega_4 = \dot{\beta}$. Finally, the jaws rotate about axis O_2-A at the rate ω_5. The magnitudes of all angular rates are constant. For the configuration shown, determine the magnitude ω of the total angular velocity of the jaws for $\theta = 60°$ and $\beta = 45°$ if $\omega_1 = 2$ rad/s, $\dot{\theta} = 1.5$ rad/s, and $\omega_3 = \omega_4 = \omega_5 = 0$. Also express the angular acceleration $\boldsymbol{\alpha}$ of arm O_1O_2 as a vector.

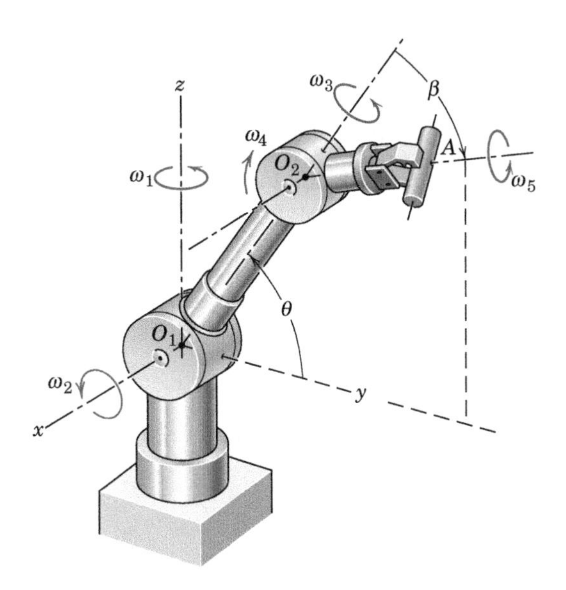

Problem 7/16

7/17 For the robot of Prob. 7/16, determine the angular velocity $\boldsymbol{\omega}$ and angular acceleration $\boldsymbol{\alpha}$ of the jaws A if $\theta = 60°$ and $\beta = 30°$, both constant, and if $\omega_1 = 2$ rad/s, $\omega_2 = \omega_3 = \omega_4 = 0$, and $\omega_5 = 0.8$ rad/s, all constant.

7/18 The wheel rolls without slipping in a circular arc of radius R and makes one complete turn about the vertical y-axis with constant speed in time τ. Determine the vector expression for the angular acceleration $\boldsymbol{\alpha}$ of the wheel and construct the space and body cones.

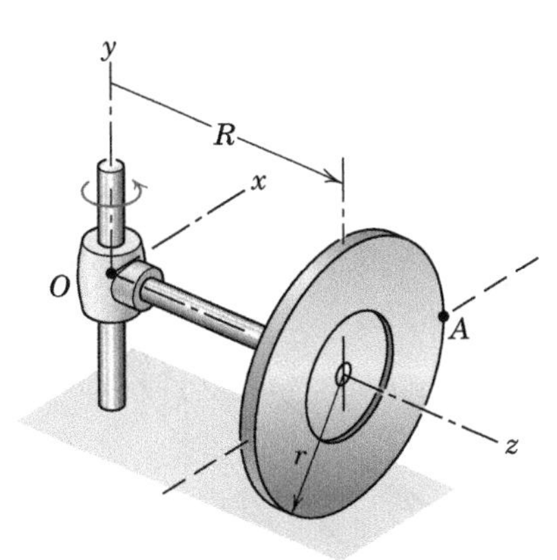

Problem 7/18

7/19 Determine expressions for the velocity **v** and acceleration **a** of point A on the wheel of Prob. 7/18 for the position shown, where A crosses the horizontal line through the center of the wheel.

7/20 The circular disk of 120-mm radius rotates about the z-axis at the constant rate $\omega_z = 20$ rad/s, and the entire assembly rotates about the fixed x-axis at the constant rate $\omega_x = 10$ rad/s. Calculate the magnitudes of the velocity **v** and acceleration **a** of point B for the instant when $\theta = 30°$.

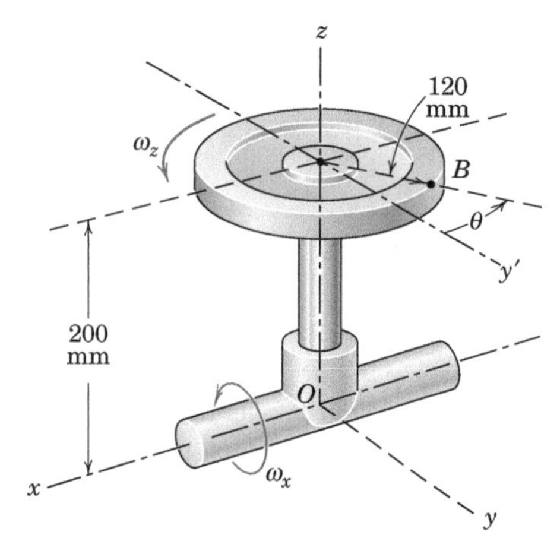

Problem 7/20

7/21 The crane has a boom of length $\overline{OP} = 24$ m and is revolving about the vertical axis at the constant rate of 2 rev/min in the direction shown. Simultaneously, the boom is being lowered at the constant rate $\dot{\beta} = 0.10$ rad/s. Calculate the magnitudes of the velocity and acceleration of the end P of the boom for the instant when it passes the position $\beta = 30°$.

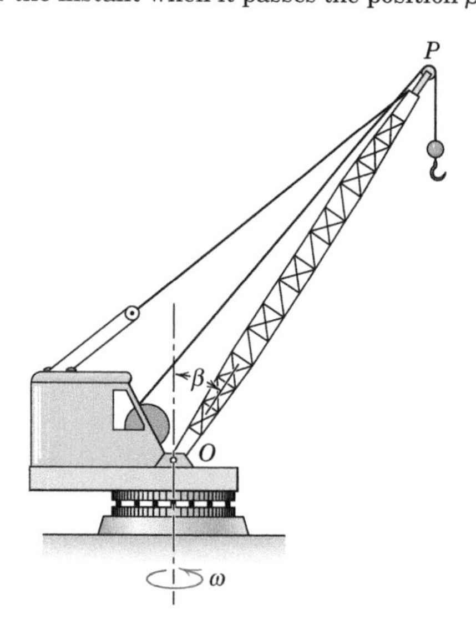

Problem 7/21

7/22 The design of the rotating arm OA of a control mechanism requires that it rotate about the vertical Z-axis at the constant rate $\Omega = \dot{\beta} = \pi$ rad/s. Simultaneously, OA oscillates according to $\theta = \theta_0 \sin 4\Omega t$, where $\theta_0 = \pi/6$ radians and t is in seconds measured from the time when $\beta = 0$. Determine the angular velocity $\boldsymbol{\omega}$ and the angular acceleration $\boldsymbol{\alpha}$ of OA for the instant (a) when $t = 1/2$ s and (b) when $t = 1/8$ s. The x-y reference axes rotate in the X-Y plane with the angular velocity Ω.

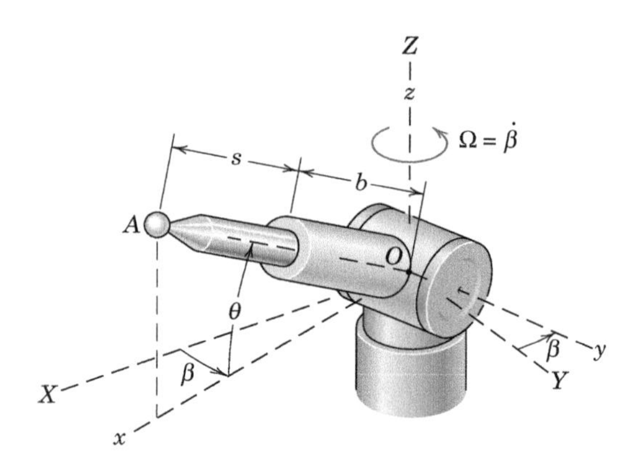

Problem 7/22

7/23 For the rotating and oscillating control arm OA of Prob. 7/22, determine the velocity **v** and acceleration **a** of the ball tip A for the condition when $t = 1/2$ s. Distance $b = 120$ mm, $s = 100$ mm, and $\theta = \theta_0 \sin 4\Omega t$ as defined in Prob. 7/22 with $\Omega = \pi$ rad/s and $\theta_0 = \pi/6$ rad.

7/24 If the angular velocity $\boldsymbol{\omega}_0 = -4\mathbf{j} - 3\mathbf{k}$ rad/s of the rotor in Prob. 7/5 is constant in magnitude, determine the angular acceleration $\boldsymbol{\alpha}$ of the rotor for (a) $\Omega = 0$ and $\dot{\theta} = 2$ rad/s (both constant) and (b) $\theta = \tan^{-1}(\frac{3}{4})$ and $\Omega = 2$ rad/s (both constant). Find the magnitude of the acceleration of point A in each case, where A has the position vector $\mathbf{r} = 0.5\mathbf{i} + 1.2\mathbf{j} + 1.1\mathbf{k}$ m at the instant represented.

7/25 The vertical shaft and attached clevis rotate about the z-axis at the constant rate $\Omega = 4$ rad/s. Simultaneously, the shaft B revolves about its axis OA at the constant rate $\omega_0 = 3$ rad/s, and the angle γ is decreasing at the constant rate of $\pi/4$ rad/s. Determine the angular velocity $\boldsymbol{\omega}$ and the magnitude of the angular acceleration $\boldsymbol{\alpha}$ of shaft B when $\gamma = 30°$. The x-y-z axes are attached to the clevis and rotate with it.

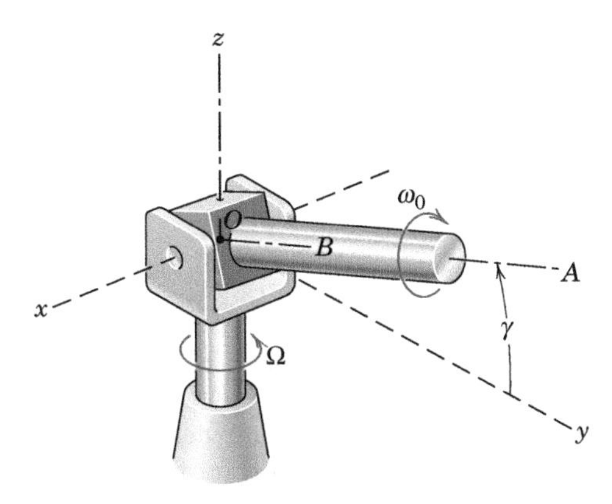

Problem 7/25

▶7/26 The right-circular cone A rolls on the fixed right-circular cone B at a constant rate and makes one complete trip around B every 4 seconds. Compute the magnitude of the angular acceleration $\boldsymbol{\alpha}$ of cone A during its motion.

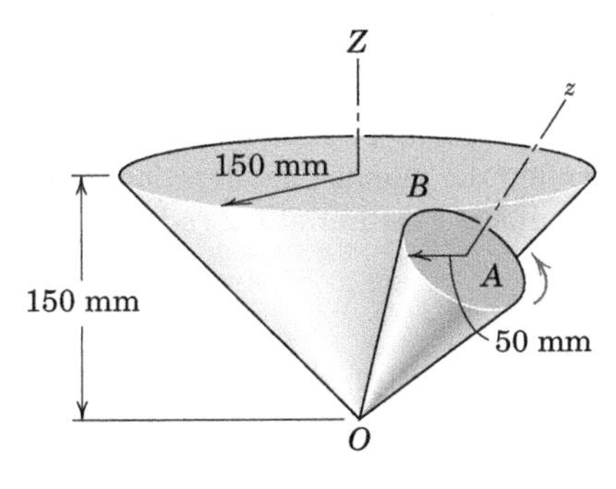

Problem 7/26

▶7/27 The pendulum oscillates about the x-axis according to $\theta = \frac{\pi}{6} \sin 3\pi t$ radians, where t is the time in seconds. Simultaneously, the shaft OA revolves about the vertical z-axis at the constant rate $\omega_z = 2\pi$ rad/sec. Determine the velocity $\mathbf{v}$ and acceleration $\mathbf{a}$ of the center B of the pendulum as well as its angular acceleration $\boldsymbol{\alpha}$ for the instant when $t = 0$.

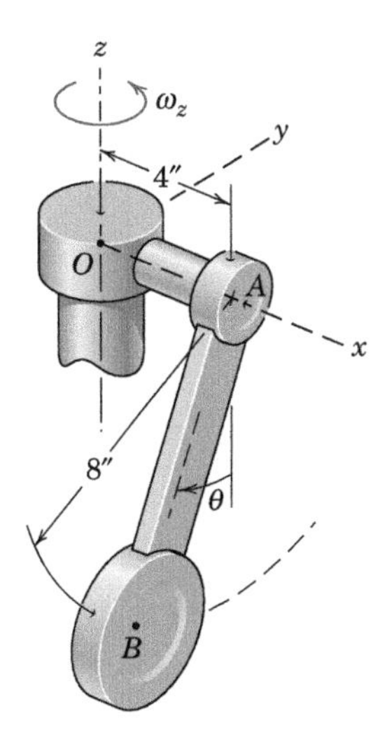

Problem 7/27

▶7/28 The solid right-circular cone of base radius r and height h rolls on a flat surface without slipping. The center B of the circular base moves in a circular path around the z-axis with a constant speed v. Determine the angular velocity $\boldsymbol{\omega}$ and the angular acceleration $\boldsymbol{\alpha}$ of the solid cone.

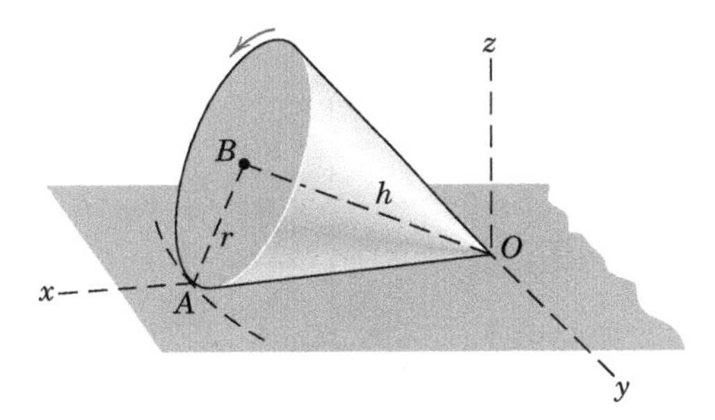

Problem 7/28

PROBLEMS

Introductory Problems

7/29 The solid cylinder has a body cone with a semi-vertex angle of 20°. Momentarily the angular velocity $\boldsymbol{\omega}$ has a magnitude of 30 rad/s and lies in the y-z plane. Determine the rate p at which the cylinder is spinning about its z-axis and write the vector expression for the velocity of B with respect to A.

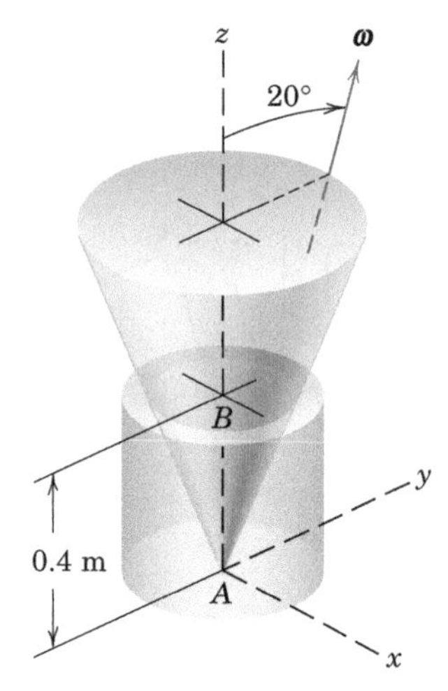

Problem 7/29

7/30 The helicopter is nosing over at the constant rate q rad/s. If the rotor blades revolve at the constant speed p rad/s, write the expression for the angular acceleration $\boldsymbol{\alpha}$ of the rotor. Take the y-axis to be attached to the fuselage and pointing forward perpendicular to the rotor axis.

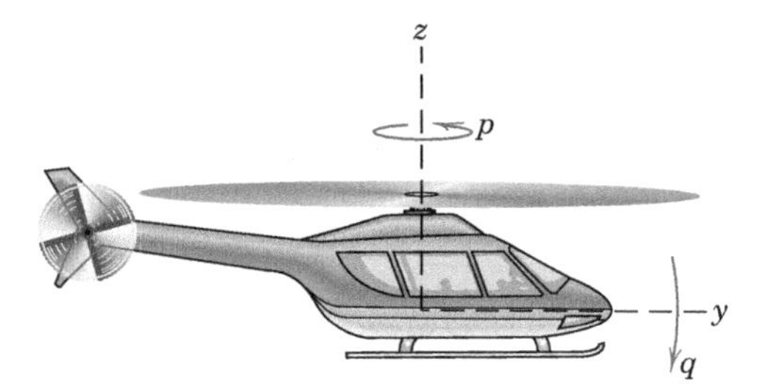

Problem 7/30

7/31 The collar at O and attached shaft OC rotate about the fixed x_0-axis at the constant rate $\Omega = 4$ rad/s. Simultaneously, the circular disk rotates about OC at the constant rate $p = 10$ rad/s. Determine the magnitude of the total angular velocity $\boldsymbol{\omega}$ of the disk and find its angular acceleration $\boldsymbol{\alpha}$.

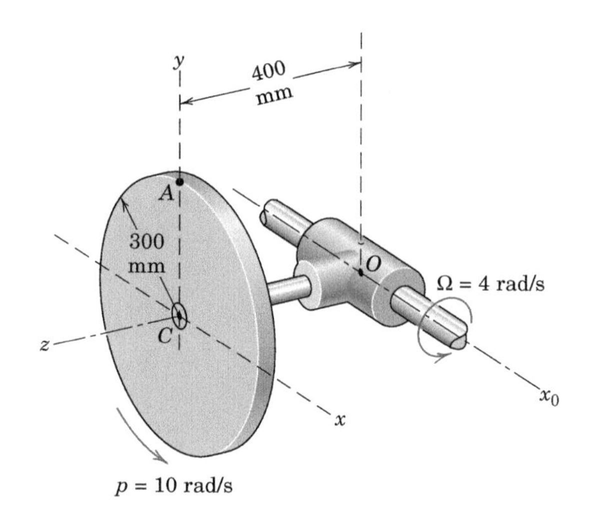

Problem 7/31

7/32 If the angular rate p of the disk in Prob. 7/31 is increasing at the rate of 6 rad/s per second and if Ω remains constant at 4 rad/s, determine the angular acceleration $\boldsymbol{\alpha}$ of the disk at the instant when p reaches 10 rad/s.

7/33 For the conditions of Prob. 7/31, determine the velocity $\mathbf{v}_A$ and acceleration $\mathbf{a}_A$ of point A on the disk as it passes the position shown. Reference axes x-y-z are attached to the collar at O and its shaft OC.

7/34 An unmanned radar-radio controlled aircraft with tilt-rotor propulsion is being designed for reconnaissance purposes. Vertical rise begins with $\theta = 0$ and is followed by horizontal flight as θ approaches 90°. If the rotors turn at a constant speed N of 360 rev/min, determine the angular acceleration $\boldsymbol{\alpha}$ of rotor A for $\theta = 30°$ if $\dot{\theta}$ is constant at 0.2 rad/s.

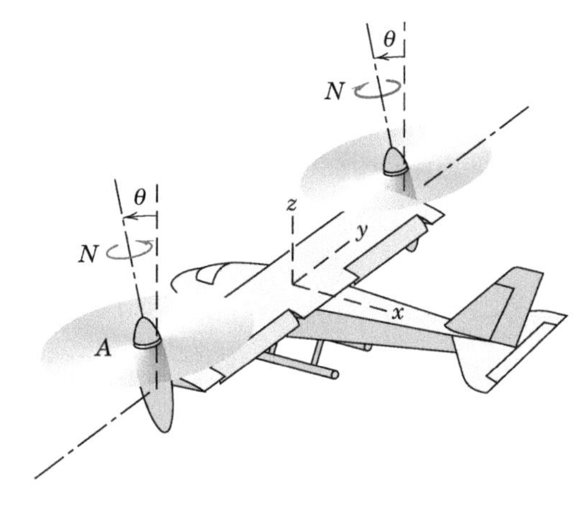

Problem 7/34

7/35 End A of the rigid link is confined to move in the $-x$-direction while end B is confined to move along the z-axis. Determine the component $\boldsymbol{\omega}_n$ normal to AB of the angular velocity of the link as it passes the position shown with $v_A = 3$ ft/sec.

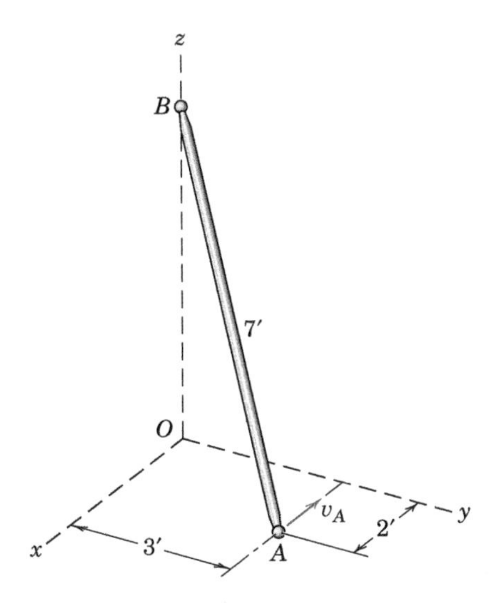

Problem 7/35

Representative Problems

7/36 The small motor M is pivoted about the x-axis through O and gives its shaft OA a constant speed p rad/s in the direction shown relative to its housing. The entire unit is then set into rotation about the vertical Z-axis at the constant angular velocity Ω rad/s. Simultaneously, the motor pivots about the x-axis at the constant rate $\dot{\beta}$ for an interval of motion. Determine the angular acceleration $\boldsymbol{\alpha}$ of the shaft OA in terms of β. Express your result in terms of the unit vectors for the rotating x-y-z axes.

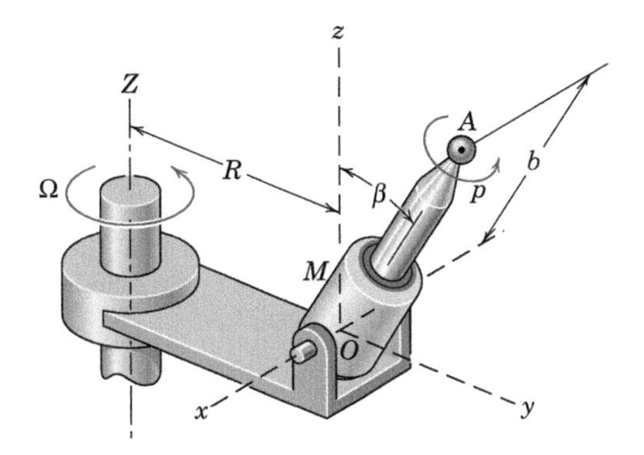

Problem 7/36

7/37 The flight simulator is mounted on six hydraulic actuators connected in pairs to their attachment points on the underside of the simulator. By programming the actions of the actuators, a variety of flight conditions can be simulated with translational and rotational displacements through a limited range of motion. Axes x-y-z are attached to the simulator with origin B at the center of the volume. For the instant represented, B has a velocity and an acceleration in the horizontal y-direction of 3.2 ft/sec and 4 ft/sec^2, respectively. Simultaneously, the angular velocities and their time rates of change are $\omega_x = 1.4$ rad/sec, $\dot{\omega}_x = 2$ rad/sec^2, $\omega_y = 1.2$ rad/sec, $\dot{\omega}_y = 3$ rad/sec^2, $\omega_z = \dot{\omega}_z = 0$. For this instant determine the magnitudes of the velocity and acceleration of point A.

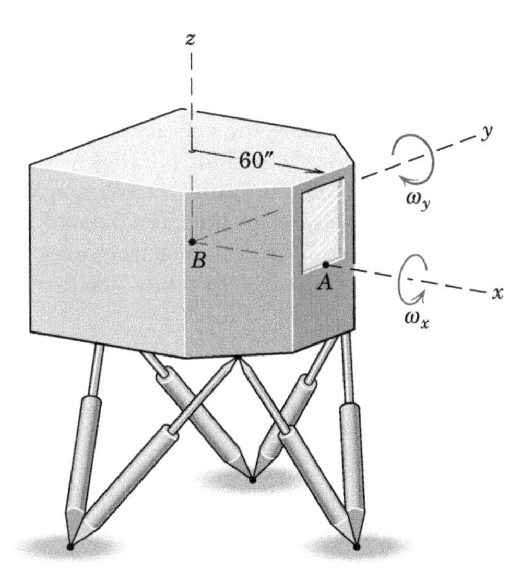

Problem 7/37

7/38 The robot of Prob. 7/16 is shown again here, where the coordinate system x-y-z with origin at O_2 rotates about the X-axis at the rate $\dot{\theta}$. Nonrotating axes X-Y-Z oriented as shown have their origin at O_1. If $\omega_2 = \dot{\theta} = 3$ rad/s constant, $\omega_3 = 1.5$ rad/s constant, $\omega_1 = \omega_5 = 0$, $\overline{O_1O_2} = 1.2$ m, and $\overline{O_2A} = 0.6$ m, determine the velocity of the center A of the jaws for the instant when $\theta = 60°$. The angle β lies in the y-z plane and is constant at 45°.

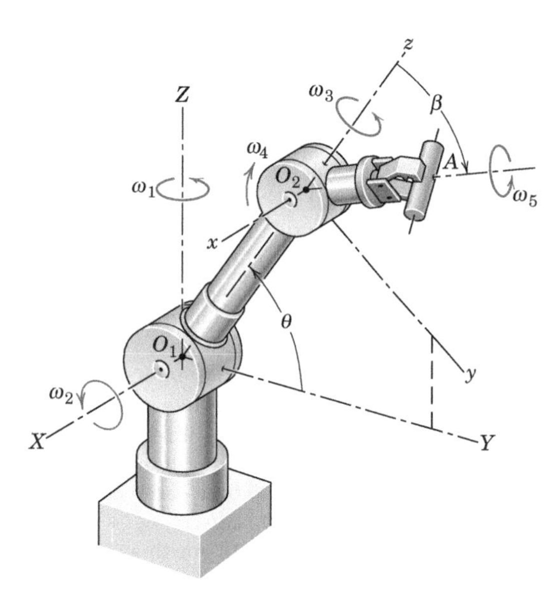

Problem 7/38

7/39 For the instant represented collar B is moving along the fixed shaft in the X-direction with a constant velocity $v_B = 4$ m/s. Also at this instant $X = 0.3$ m and $Y = 0.2$ m. Calculate the velocity of collar A, which moves along the fixed shaft parallel to the Y-axis. Solve, first, by differentiating the relation $X^2 + Y^2 + Z^2 = L^2$ with respect to time and, second, by using the first of Eqs. 7/4 with translating axes attached to B. Each clevis is free to rotate about the axis of the rod.

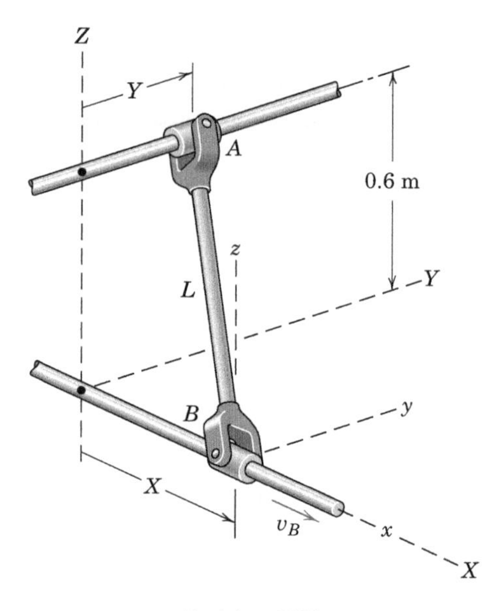

Problem 7/39

7/40 The spacecraft is revolving about its z-axis, which has a fixed space orientation, at the constant rate $p = \frac{1}{10}$ rad/s. Simultaneously, its solar panels are unfolding at the rate $\dot{\beta}$ which is programmed to vary with β as shown in the graph. Determine the angular acceleration $\boldsymbol{\alpha}$ of panel A an instant (a) before and an instant (b) after it reaches the position $\beta = 18°$.

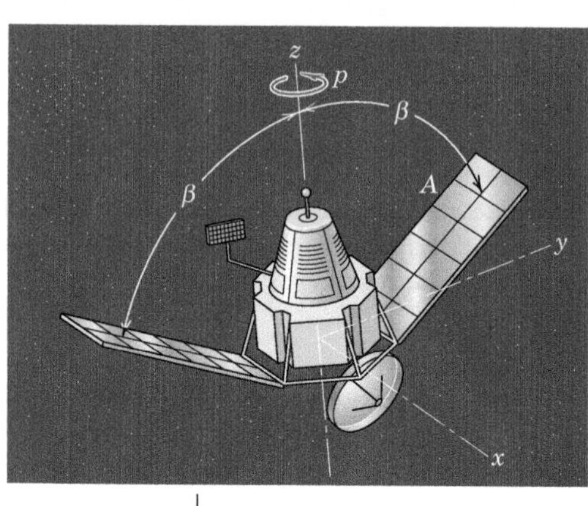

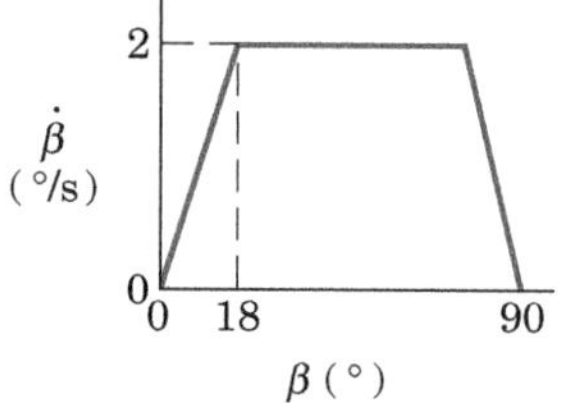

Problem 7/40

7/41 The disk has a constant angular velocity p about its z-axis, and the yoke A has a constant angular velocity ω_2 about its shaft as shown. Simultaneously, the entire assembly revolves about the fixed X-axis with a constant angular velocity ω_1. Determine the expression for the angular acceleration of the disk as the yoke brings it into the vertical plane in the position shown. Solve by picturing the vector changes in the angular-velocity components.

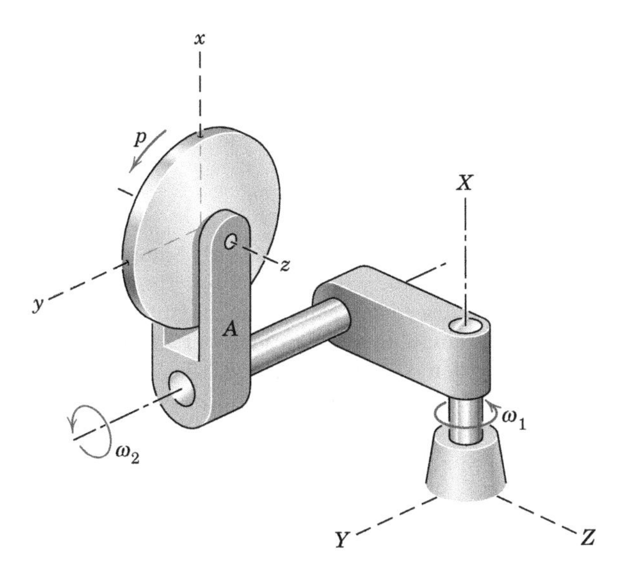

Problem 7/41

7/42 The collar and clevis A are given a constant upward velocity of 8 in./sec for an interval of motion and cause the ball end of the bar to slide in the radial slot in the rotating disk. Determine the angular acceleration of the bar when the bar passes the position for which $z = 3$ in. The disk turns at the constant rate of 2 rad/sec.

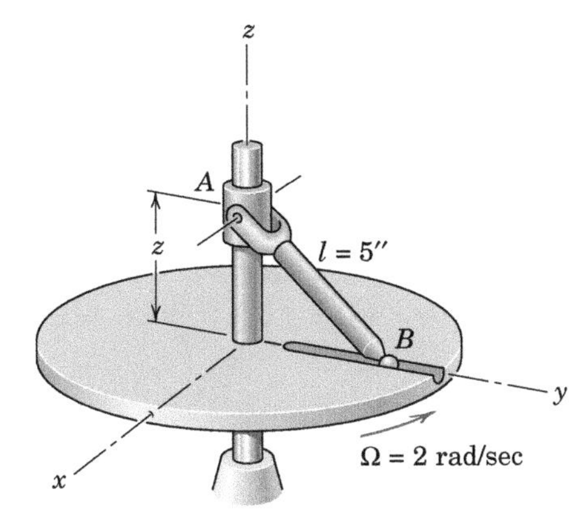

Problem 7/42

7/43 The circular disk of 100-mm radius rotates about its z-axis at the constant speed $p = 240$ rev/min, and arm OCB rotates about the Y-axis at the constant speed $N = 30$ rev/min. Determine the velocity $\mathbf{v}$ and acceleration $\mathbf{a}$ of point A on the disk as it passes the position shown. Use reference axes x-y-z attached to the arm OCB.

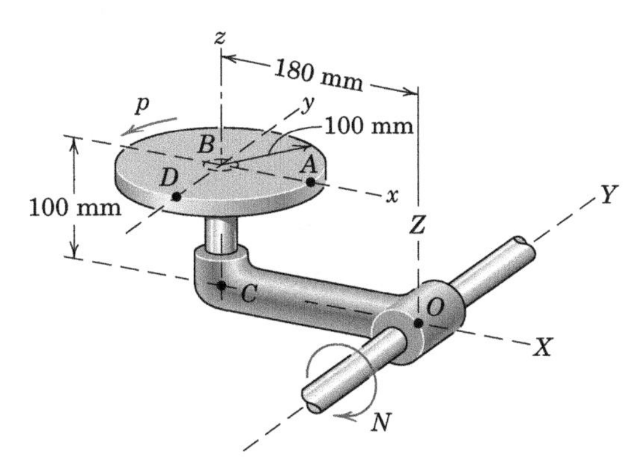

Problem 7/43

7/44 Solve Prob. 7/43 by attaching the reference axes x-y-z to the rotating disk.

7/45 For the conditions described in Prob. 7/36, determine the velocity $\mathbf{v}$ and acceleration $\mathbf{a}$ of the center A of the ball tool in terms of β.

7/46 The circular disk is spinning about its own axis (y-axis) at the constant rate $p = 10\pi$ rad/s. Simultaneously, the frame is rotating about the Z-axis at the constant rate $\Omega = 4\pi$ rad/s. Calculate the angular acceleration $\boldsymbol{\alpha}$ of the disk and the acceleration of point A at the top of the disk. Axes x-y-z are attached to the frame, which has the momentary orientation shown with respect to the fixed axes X-Y-Z.

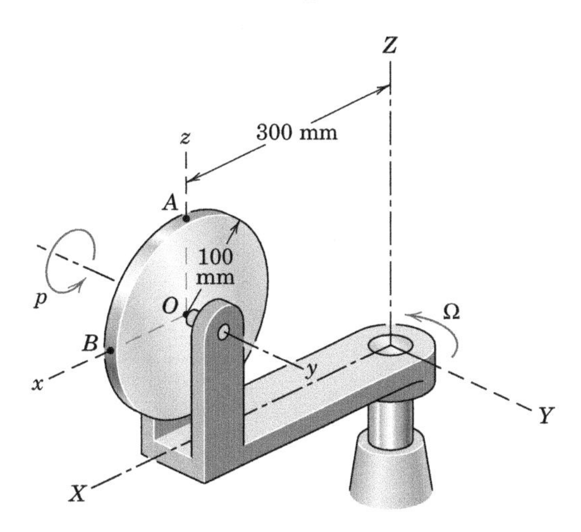

Problem 7/46

7/47 The center O of the spacecraft is moving through space with a constant velocity. During the period of motion prior to stabilization, the spacecraft has a constant rotational rate $\Omega = 0.5$ rad/sec about its z-axis. The x-y-z axes are attached to the body of the craft, and the solar panels rotate about the y-axis at the constant rate $\dot{\theta} = 0.25$ rad/sec with respect to the spacecraft. If $\boldsymbol{\omega}$ is the absolute angular velocity of the solar panels, determine $\dot{\boldsymbol{\omega}}$. Also find the acceleration of point A when $\theta = 30°$.

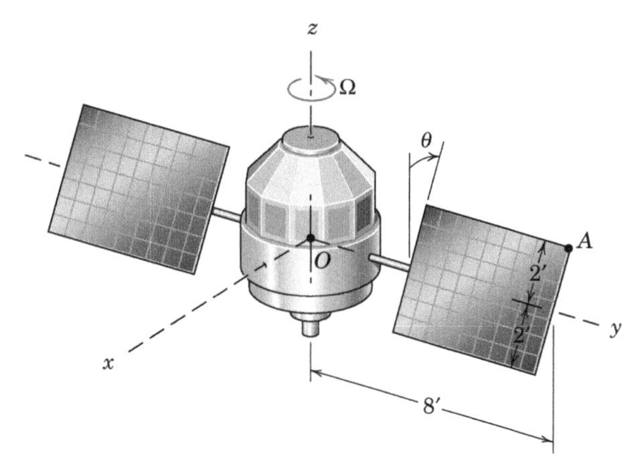

Problem 7/47

7/48 The thin circular disk of mass m and radius r is rotating about its z-axis with a constant angular velocity p, and the yoke in which it is mounted rotates about the x-axis through OB with a constant angular velocity ω_1. Simultaneously, the entire assembly rotates about the fixed Y-axis through O with a constant angular velocity ω_2. Determine the velocity $\mathbf{v}$ and acceleration $\mathbf{a}$ of point A on the rim of the disk as it passes the position shown where the x-y plane of the disk coincides with the X-Y plane. The x-y-z axes are attached to the yoke.

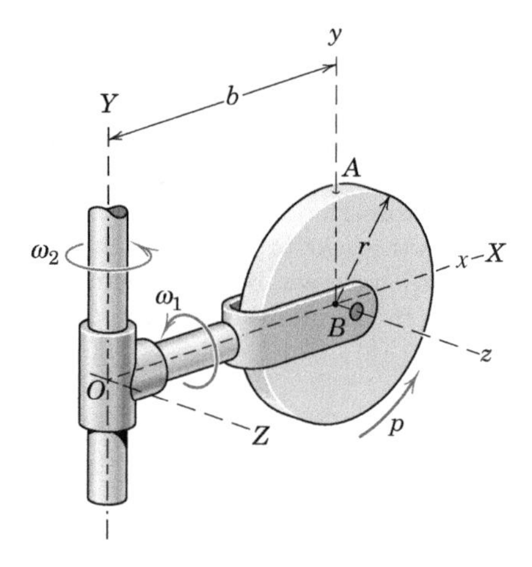

Problem 7/48

▸**7/49** For the conditions specified with Sample Problem 7/2, except that γ is increasing at the steady rate of 3π rad/sec, determine the angular velocity $\boldsymbol{\omega}$ and the angular acceleration $\boldsymbol{\alpha}$ of the rotor when the position $\gamma = 30°$ is passed. (*Suggestion:* Apply Eq. 7/7 to the vector $\boldsymbol{\omega}$ to find $\boldsymbol{\alpha}$. Note that $\boldsymbol{\Omega}$ in Sample Problem 7/2 is no longer the complete angular velocity of the axes.)

▸**7/50** The wheel of radius r is free to rotate about the bent axle CO which turns about the vertical axis at the constant rate p rad/s. If the wheel rolls without slipping on the horizontal circle of radius R, determine the expressions for the angular velocity $\boldsymbol{\omega}$ and angular acceleration $\boldsymbol{\alpha}$ of the wheel. The x-axis is always horizontal.

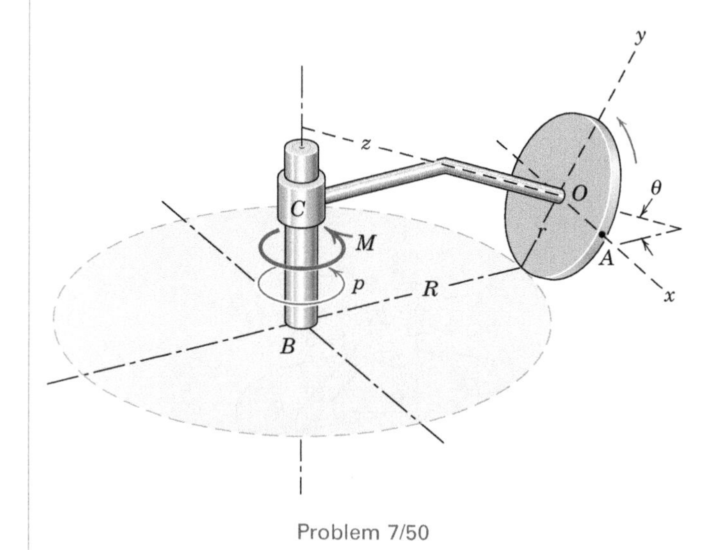

Problem 7/50

▶**7/51** The gyro rotor shown is spinning at the constant rate of 100 rev/min relative to the x-y-z axes in the direction indicated. If the angle γ between the gimbal ring and the horizontal X-Y plane is made to increase at the constant rate of 4 rad/s and if the unit is forced to precess about the vertical at the constant rate $N = 20$ rev/min, calculate the magnitude of the angular acceleration $\boldsymbol{\alpha}$ of the rotor when $\gamma = 30°$. Solve by using Eq. 7/7 applied to the angular velocity of the rotor.

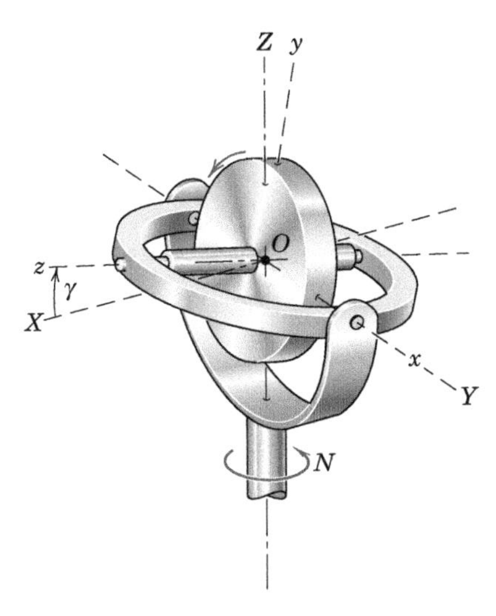

Problem 7/51

▶**7/52** For a short interval of motion, collar A moves along its fixed shaft with a velocity $v_A = 2$ m/s in the Y-direction. Collar B, in turn, slides along its fixed vertical shaft. Link AB is 700 mm in length and can turn within the clevis at A to allow for the angular change between the clevises. For the instant when A passes the position where $y = 200$ mm, determine the velocity of collar B using nonrotating axes attached to B and find the component $\boldsymbol{\omega}_n$, normal to AB, of the angular velocity of the link. Also solve for $\mathbf{v}_B$ by differentiating the appropriate relation $x^2 + y^2 + z^2 = l^2$.

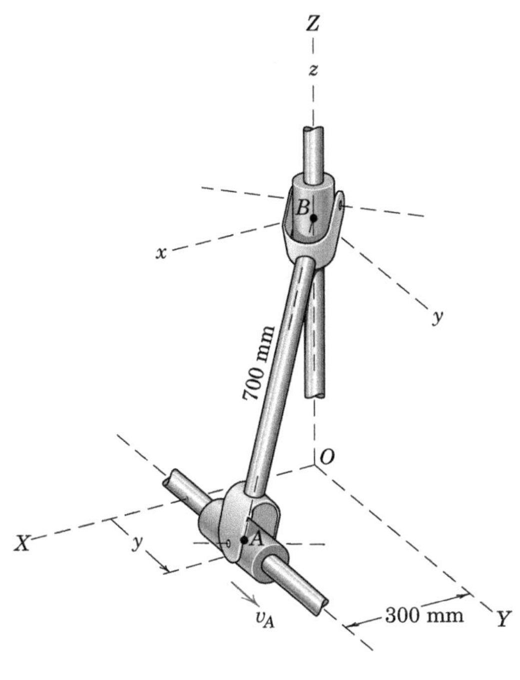

Problem 7/52

PROBLEMS

Introductory Problems

7/53 The three small spheres, each of mass m, are rigidly mounted to the horizontal shaft which rotates with the angular velocity ω as shown. Neglect the radius of each sphere compared with the other dimensions and write expressions for the magnitudes of their linear momentum $\mathbf{G}$ and their angular momentum $\mathbf{H}_O$ about the origin O of the coordinates.

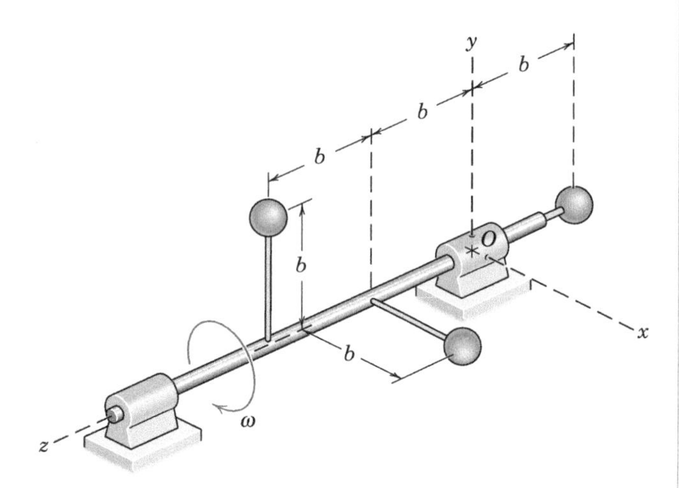

Problem 7/53

7/54 The spheres of Prob. 7/53 are replaced by three rods, each of mass m and length l, mounted at their centers to the shaft, which rotates with the angular velocity ω as shown. The axes of the rods are, respectively, in the x-, y-, and z-directions, and their diameters are negligible compared with the other dimensions. Determine the angular momentum $\mathbf{H}_O$ of the three rods with respect to the coordinate origin O.

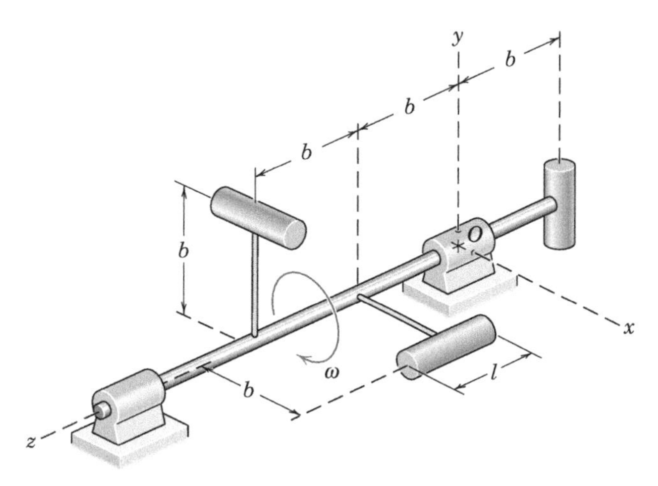

Problem 7/54

7/55 The aircraft landing gear viewed from the front is being retracted immediately after takeoff, and the wheel is spinning at the rate corresponding to the takeoff speed of 200 km/h. The 45-kg wheel has a radius of gyration about its z-axis of 370 mm. Neglect the thickness of the wheel and calculate the angular momentum of the wheel about G and about A for the position where θ is increasing at the rate of 30° per second.

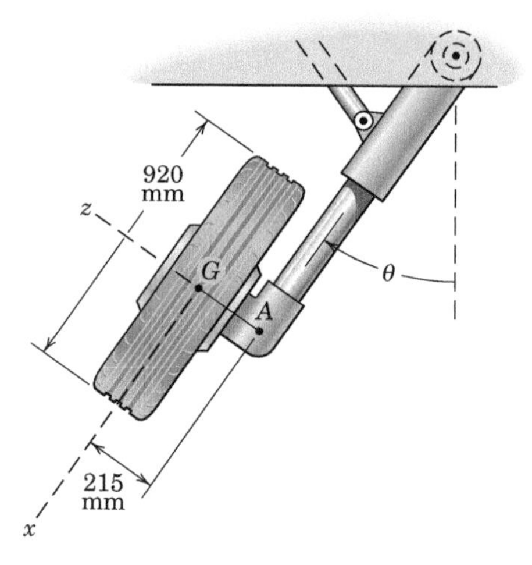

Problem 7/55

7/56 The bent rod has a mass ρ per unit length and rotates about the z-axis with an angular velocity ω. Determine the angular momentum $\mathbf{H}_O$ of the rod about the fixed origin O of the axes, which are attached to the rod. Also find the kinetic energy T of the rod.

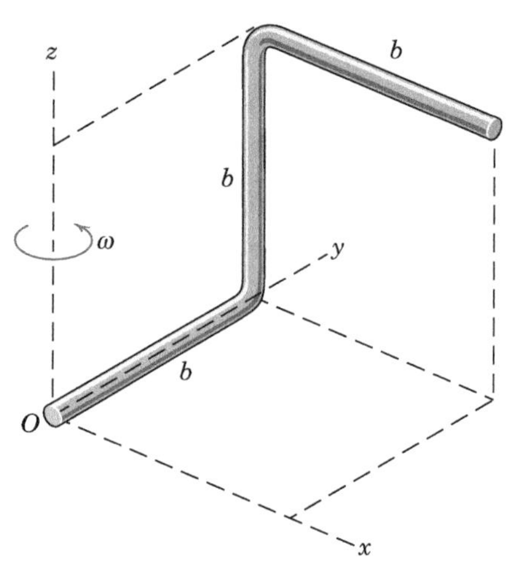

Problem 7/56

7/57 Use the results of Prob. 7/56 and determine the angular momentum $\mathbf{H}_G$ of the bent rod of that problem about its mass center G using the given reference axes.

7/58 The slender rod of mass m and length l rotates about the y-axis as the element of a right-circular cone. If the angular velocity about the y-axis is ω, determine the expression for the angular momentum of the rod with respect to the x-y-z axes for the particular position shown.

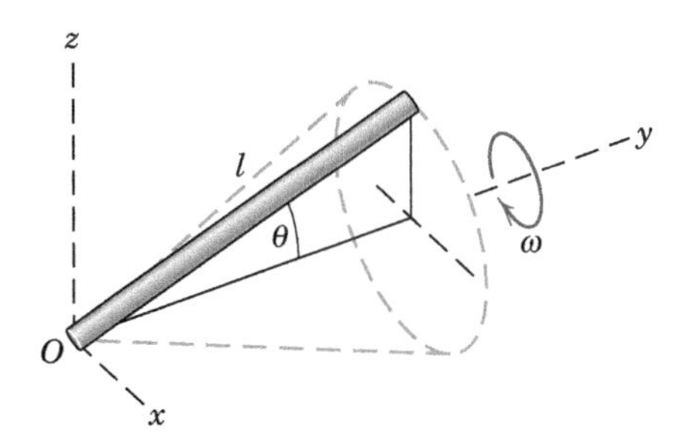

Problem 7/58

Representative Problems

7/59 The solid half-circular cylinder of mass m revolves about the z-axis with an angular velocity ω as shown. Determine its angular momentum $\mathbf{H}$ with respect to the x-y-z axes.

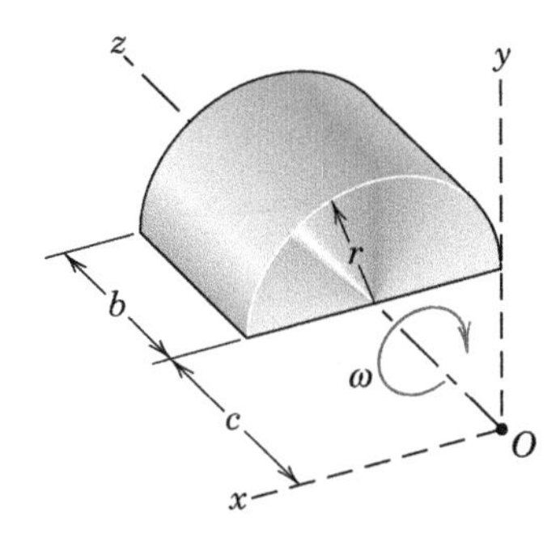

Problem 7/59

7/60 The solid circular cylinder of mass m, radius r, and length b revolves about its geometric axis at an angular rate p rad/s. Simultaneously, the bracket and attached shaft revolve about the x-axis at the rate ω rad/s. Write the expression for the angular momentun $\mathbf{H}_O$ of the cylinder about O with reference axes as shown.

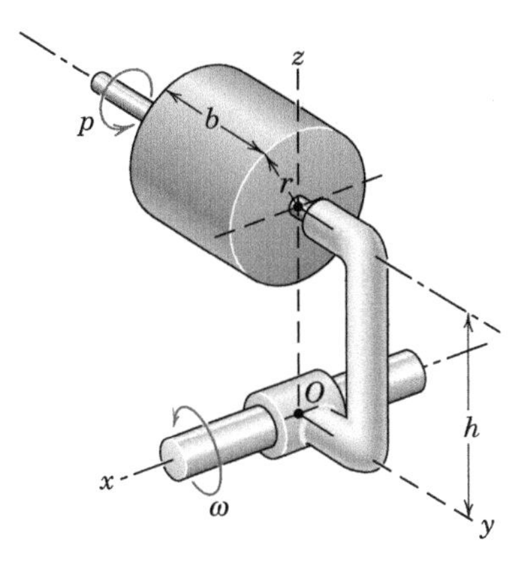

Problem 7/60

7/61 The elements of a reaction-wheel attitude-control system for a spacecraft are shown in the figure. Point G is the center of mass for the system of the spacecraft and wheels, and x, y, z are principal axes for the system. Each wheel has a mass m and a moment of inertia I about its own axis and spins with a relative angular velocity p in the direction indicated. The center of each wheel, which may be treated as a thin disk, is a distance b from G. If the spacecraft has angular velocity components Ω_x, Ω_y, and Ω_z, determine the angular momentum $\mathbf{H}_G$ of the three wheels as a unit.

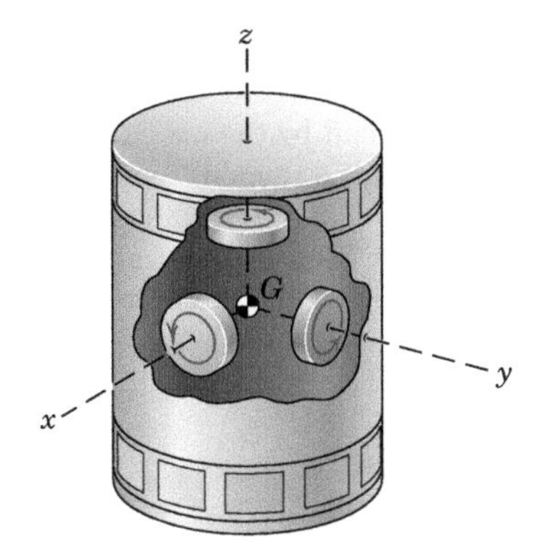

Problem 7/61

7/62 The gyro rotor is spinning at the constant rate p = 100 rev/min relative to the x-y-z axes in the direction indicated. If the angle γ between the gimbal ring and horizontal X-Y plane is made to increase at the rate of 4 rad/sec and if the unit is forced to precess about the vertical at the constant rate N = 20 rev/min, calculate the angular momentum $\mathbf{H}_O$ of the rotor when $\gamma = 30°$. The axial and transverse moments of inertia are $I_{zz} = 5(10^{-3})$ lb-ft-sec^2 and $I_{xx} = I_{yy} = 2.5(10^{-3})$ lb-ft-sec^2.

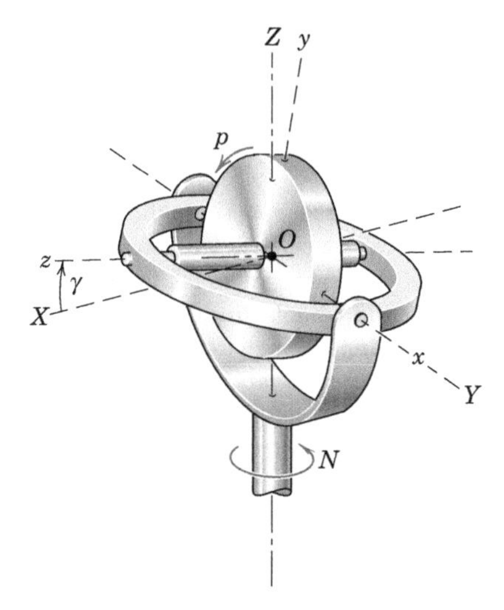

Problem 7/62

7/63 The slender steel rod AB weighs 6.20 lb and is secured to the rotating shaft by the rod OG and its fittings at O and G. The angle β remains constant at 30°, and the entire rigid assembly rotates about the z-axis at the steady rate N = 600 rev/min. Calculate the angular momentum $\mathbf{H}_O$ of AB and its kinetic energy T.

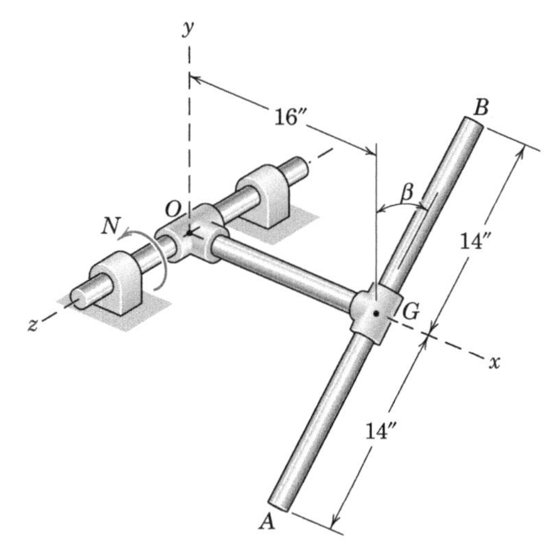

Problem 7/63

7/64 The rectangular plate, with a mass of 3 kg and a uniform small thickness, is welded at the 45° angle to the vertical shaft, which rotates with the angular velocity of 20π rad/s. Determine the angular momentum $\mathbf{H}$ of the plate about O and find the kinetic energy of the plate.

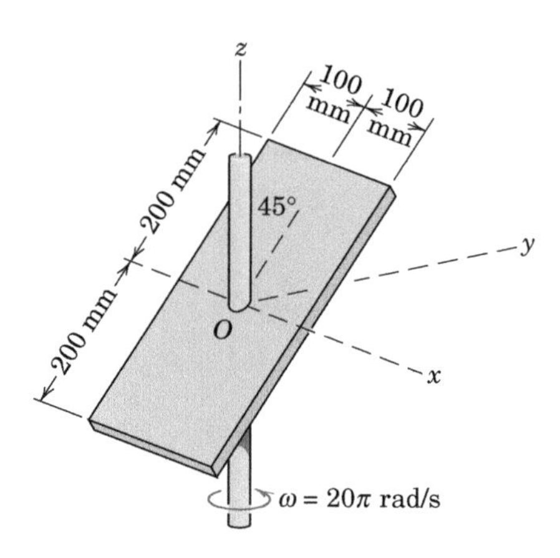

Problem 7/64

7/65 The circular disk of mass m and radius r is mounted on the vertical shaft with an angle α between its plane and the plane of rotation of the shaft. Determine an expression for the angular momentum $\mathbf{H}$ of the disk about O. Find the angle β which the angular momentum $\mathbf{H}$ makes with the shaft if $\alpha = 10°$.

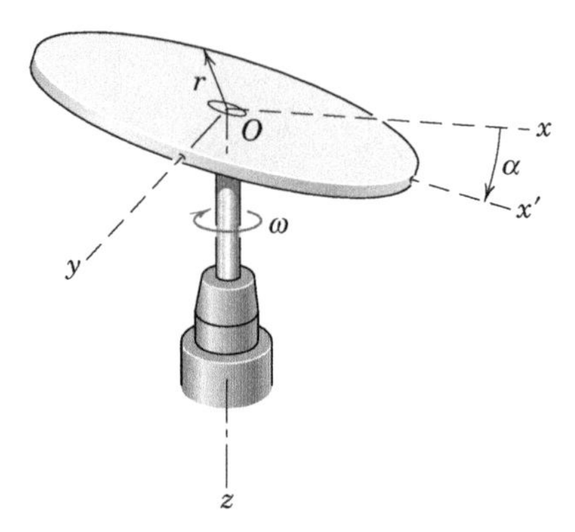

Problem 7/65

7/66 The right-circular cone of height h and base radius r spins about its axis of symmetry with an angular rate p. Simultaneously, the entire cone revolves about the x-axis with angular rate Ω. Determine the angular momentum $\mathbf{H}_O$ of the cone about the origin O of the x-y-z axes and the kinetic energy T for the position shown. The mass of the cone is m.

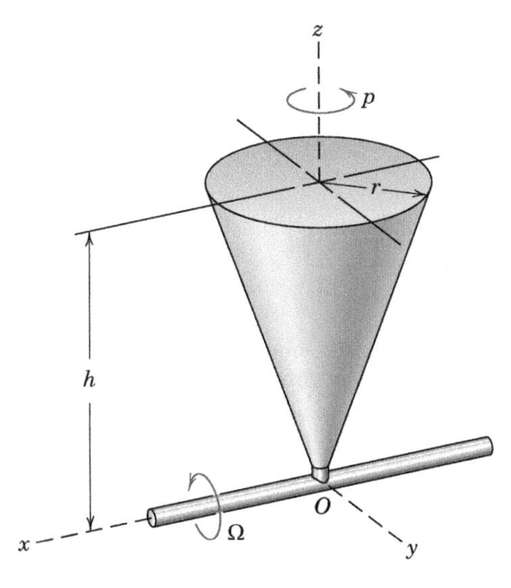

Problem 7/66

7/67 Each of the slender rods of length l and mass m is welded to the circular disk which rotates about the vertical z-axis with an angular velocity ω. Each rod makes an angle β with the vertical and lies in a plane parallel to the y-z plane. Determine an expression for the angular momentum $\mathbf{H}_O$ of the two rods about the origin O of the axes.

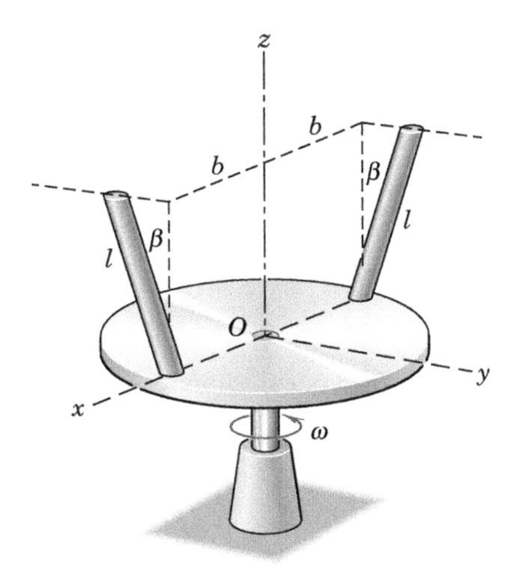

Problem 7/67

7/68 The spacecraft shown has a mass m with mass center G. Its radius of gyration about its z-axis of rotational symmetry is k and that about either the x- or y-axis is k'. In space, the spacecraft spins within its x-y-z reference frame at the rate $p = \dot{\phi}$. Simultaneously, a point C on the z-axis moves in a circle about the z_0-axis with a frequency f (rotations per unit time). The z_0-axis has a constant direction in space. Determine the angular momentum $\mathbf{H}_G$ of the spacecraft relative to the axes designated. Note that the x-axis always lies in the z-z_0 plane and that the y-axis is therefore normal to z_0.

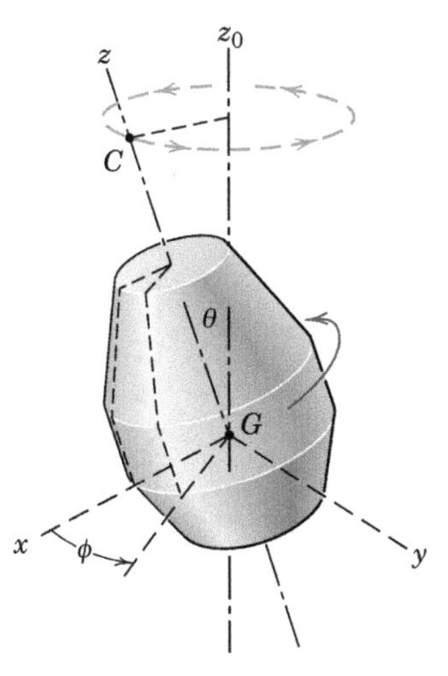

Problem 7/68

7/69 The uniform circular disk of Prob. 7/48 with the three components of angular velocity is shown again here. Determine the kinetic energy T and the angular momentum $\mathbf{H}_O$ with respect to O of the disk for the instant represented, when the x-y plane coincides with the X-Y plane. The mass of the disk is m.

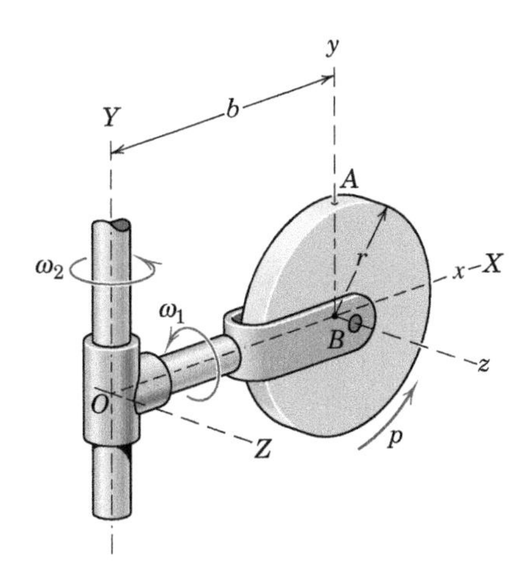

Problem 7/69

7/70 The 4-in.-radius wheel weighs 6 lb and turns about its y'-axis with an angular velocity $p = 40\pi$ rad/sec in the direction shown. Simultaneously, the fork rotates about its x-axis shaft with an angular velocity $\omega = 10\pi$ rad/sec as indicated. Calculate the angular momentum of the wheel about its center O'. Also compute the kinetic energy of the wheel.

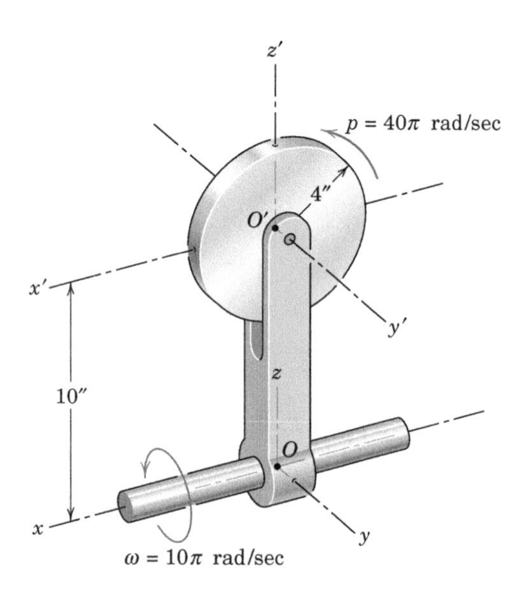

Problem 7/70

7/71 The assembly, consisting of the solid sphere of mass m and the uniform rod of length $2c$ and equal mass m, revolves about the vertical z-axis with an angular velocity ω. The rod of length $2c$ has a diameter which is small compared with its length and is perpendicular to the horizontal rod to which it is welded with the inclination β shown. Determine the combined angular momentum $\mathbf{H}_O$ of the sphere and inclined rod.

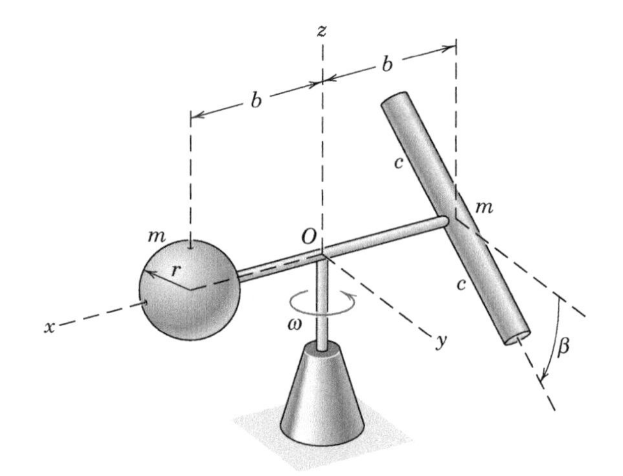

Problem 7/71

7/72 In a test of the solar panels for a spacecraft, the model shown is rotated about the vertical axis at the angular rate ω. If the mass per unit area of panel is ρ, write the expression for the angular momentum $\mathbf{H}_O$ of the assembly about the axes shown in terms of θ. Also determine the maximum, minimum, and intermediate values of the moment of inertia about the axes through O. The combined mass of both panels is m.

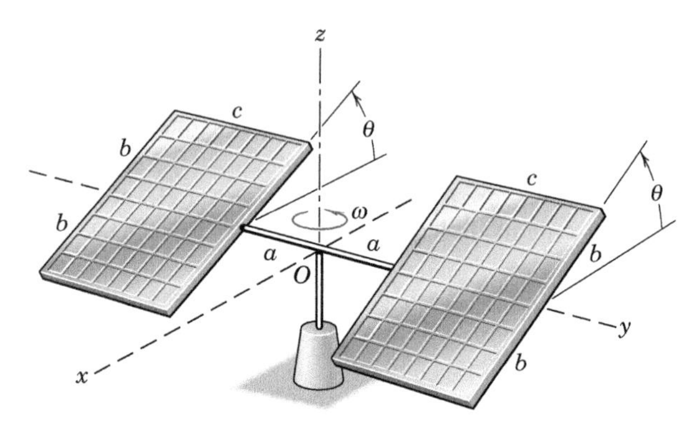

Problem 7/72

PROBLEMS

Introductory Problems

7/73 Each of the two rods of mass m is welded to the face of the disk, which rotates about the vertical axis with a constant angular velocity ω. Determine the bending moment M acting on each rod at its base.

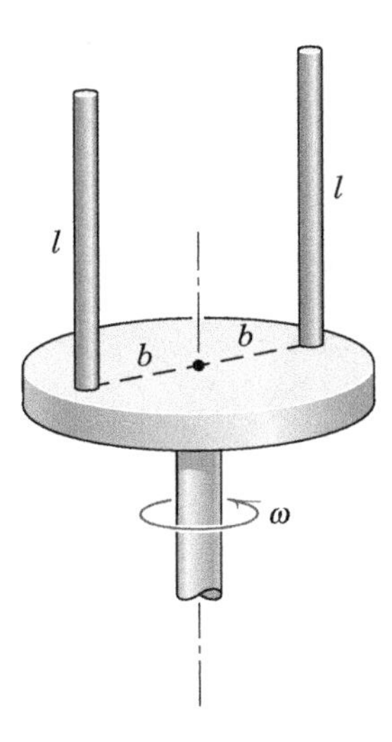

Problem 7/73

7/74 The slender shaft carries two offset particles, each of mass m, and rotates about the z-axis with the constant angular rate ω as indicated. Determine the x- and y-components of the bearing reactions at A and B due to the dynamic imbalance of the shaft for the position shown.

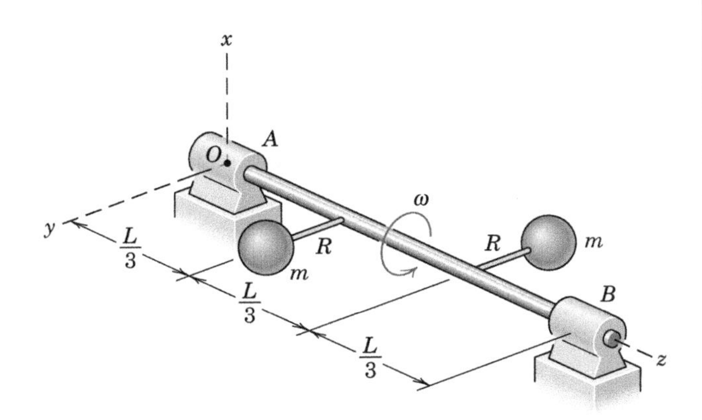

Problem 7/74

7/75 The uniform slender bar of length l and mass m is welded to the shaft, which rotates in bearings A and B with a constant angular velocity ω. Determine the expression for the force supported by the bearing at B as a function of θ. Consider only the force due to the dynamic imbalance and assume that the bearings can support radial forces only.

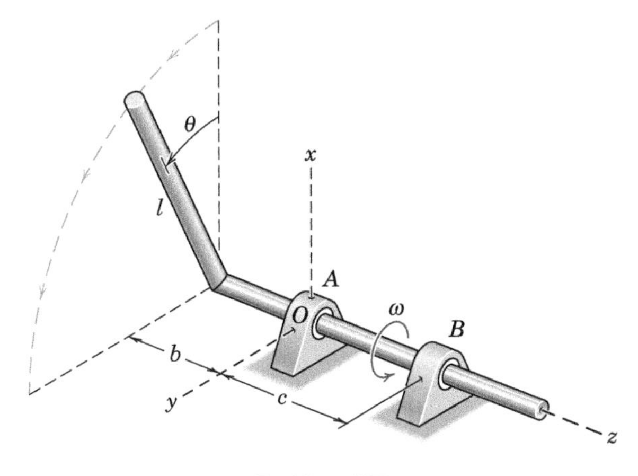

Problem 7/75

7/76 If a torque $\mathbf{M} = M\mathbf{k}$ is applied to the shaft in Prob. 7/75, determine the x- and y-components of the force supported by the bearing B as the bar and shaft start from rest in the position shown. Neglect the mass of the shaft and consider dynamic forces only.

7/77 The paint stirrer shown in the figure is made from a rod of length $7b$ and mass ρ per unit length. Before immersion in the paint, the stirrer is rotating freely at a constant high angular velocity ω about its z-axis. Determine the bending moment $\mathbf{M}$ in the rod at the base O of the chuck.

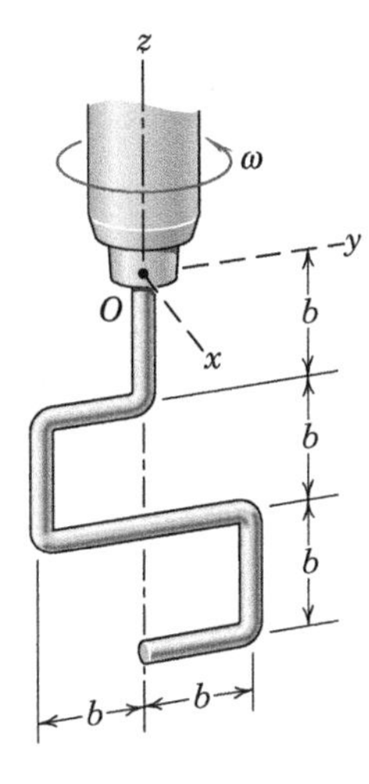

Problem 7/77

7/78 The 6-kg circular disk and attached shaft rotate at a constant speed ω = 10 000 rev/min. If the center of mass of the disk is 0.05 mm off center, determine the magnitudes of the horizontal forces A and B supported by the bearings because of the rotational imbalance.

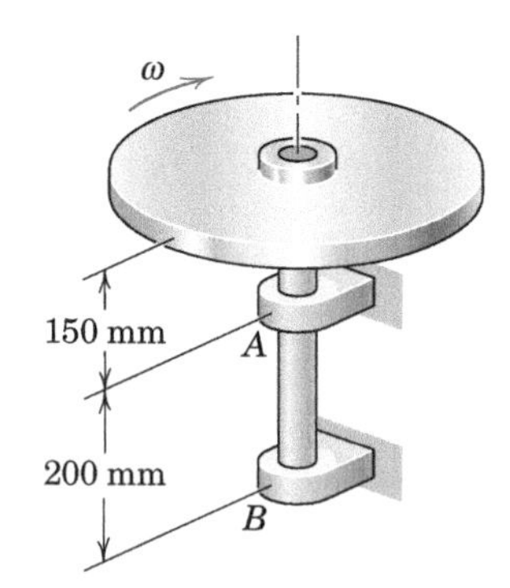

Problem 7/78

Representative Problems

7/79 Determine the bending moment **M** at the tangency point A in the semicircular rod of radius r and mass m as it rotates about the tangent axis with a constant and large angular velocity ω. Neglect the moment mgr produced by the weight of the rod.

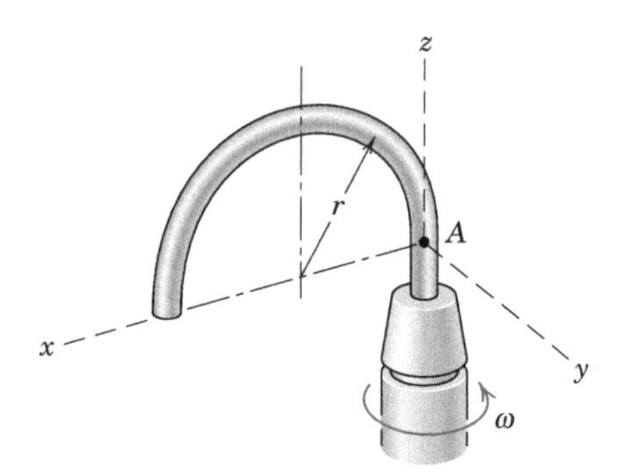

Problem 7/79

7/80 If the semicircular rod of Prob. 7/79 starts from rest under the action of a torque M_O applied through the collar about its z-axis of rotation, determine the initial bending moment **M** in the rod at A.

7/81 The large satellite-tracking antenna has a moment of inertia I about its z-axis of symmetry and a moment of inertia I_O about each of the x- and y-axes. Determine the angular acceleration α of the antenna about the vertical Z-axis caused by a torque M applied about Z by the drive mechanism for a given orientation θ.

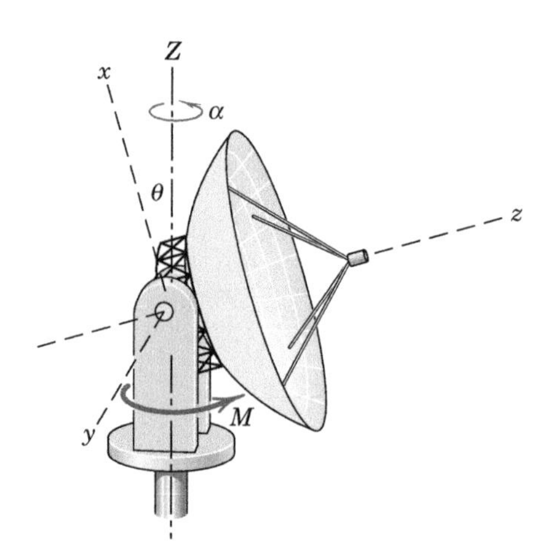

Problem 7/81

7/82 The plate has a mass of 3 kg and is welded to the fixed vertical shaft, which rotates at the constant speed of 20π rad/s. Compute the moment **M** applied *to* the shaft *by* the plate due to dynamic imbalance.

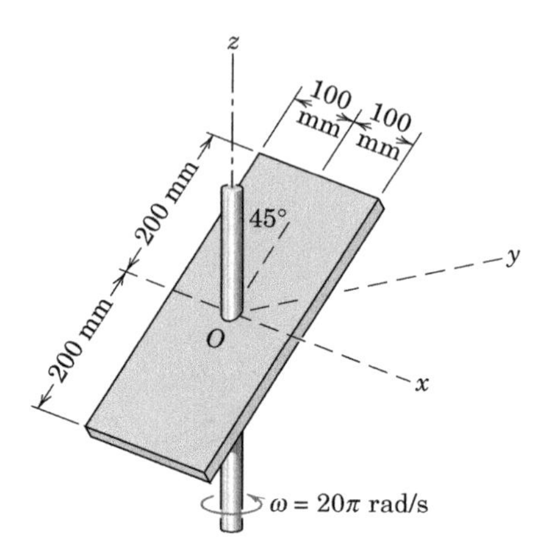

Problem 7/82

7/83 Each of the two semicircular disks has a mass of 1.20 kg and is welded to the shaft supported in bearings A and B as shown. Calculate the forces applied to the shaft by the bearings for a constant angular speed $N = 1200$ rev/min. Neglect the forces of static equilibrium.

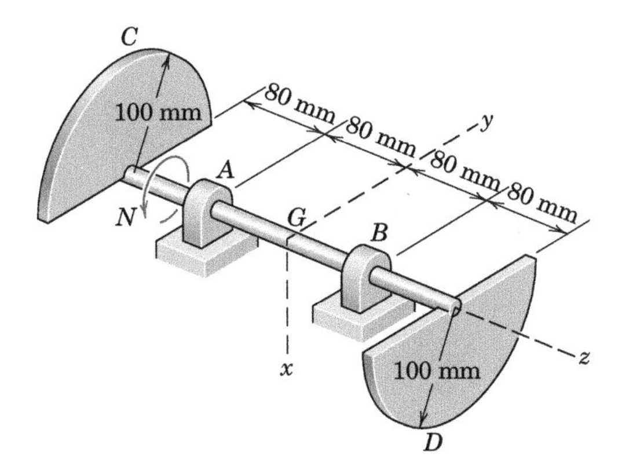

Problem 7/83

7/84 Solve Prob. 7/83 for the case where the assembly starts from rest with an initial angular acceleration $\alpha = 900$ rad/s^2 as a result of a starting torque (couple) M applied to the shaft in the same sense as N. Neglect the moment of inertia of the shaft about its z-axis and calculate M.

7/85 The uniform slender bar of mass ρ per unit length is freely pivoted about the y-axis at the clevis, which rotates about the fixed vertical z-axis with a constant angular velocity ω. Determine the steady-state angle θ assumed by the bar. Length b is greater than length c.

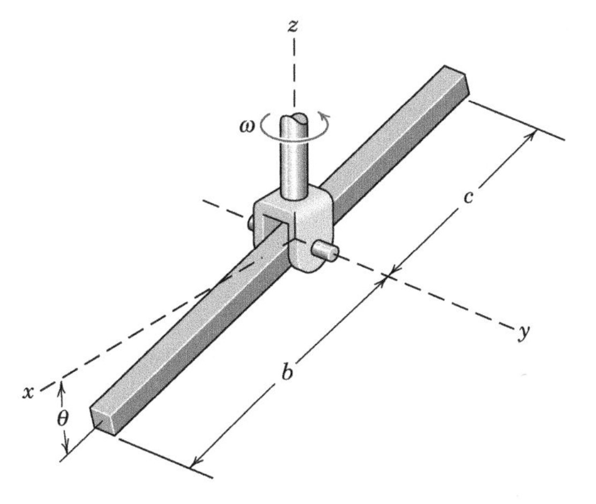

Problem 7/85

7/86 The circular disk of mass m and radius r is mounted on the vertical shaft with a small angle α between its plane and the plane of rotation of the shaft. Determine the expression for the bending moment **M** acting *on* the shaft due to the wobble of the disk at a shaft speed of ω rad/s.

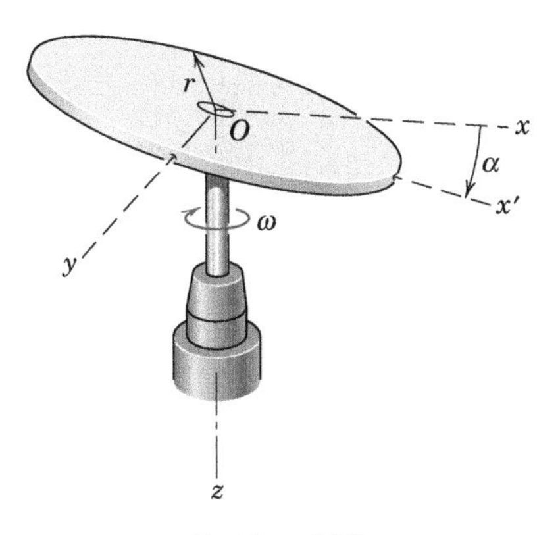

Problem 7/86

7/87 The thin circular disk of mass m and radius R is hinged about its horizontal tangent axis to the end of a shaft rotating about its vertical axis with an angular velocity ω. Determine the steady-state angle β assumed by the plane of the disk with the vertical axis. Observe any limitation on ω to ensure that $\beta > 0$.

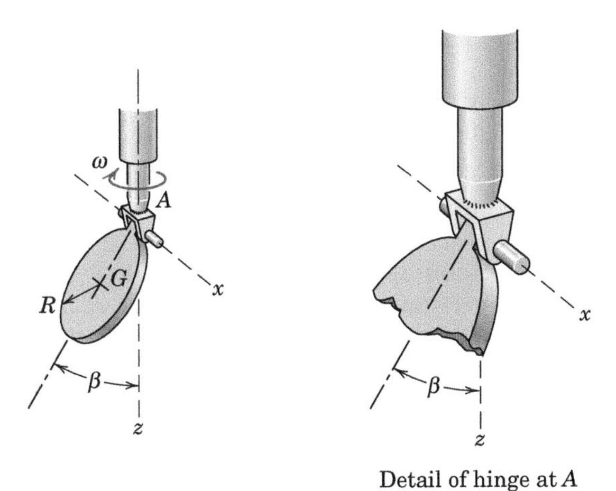

Problem 7/87

7/88 Determine the normal forces under the two disks of Sample Problem 7/7 for the position where the plane of the curved bar is vertical. Take the curved bar to be at the top of disk A and at the bottom of disk B.

7/89 The uniform square plate of mass m is welded at O to the end of the shaft, which rotates about the vertical z-axis with a constant angular velocity ω. Determine the moment applied to the plate by the weld due only to the rotation.

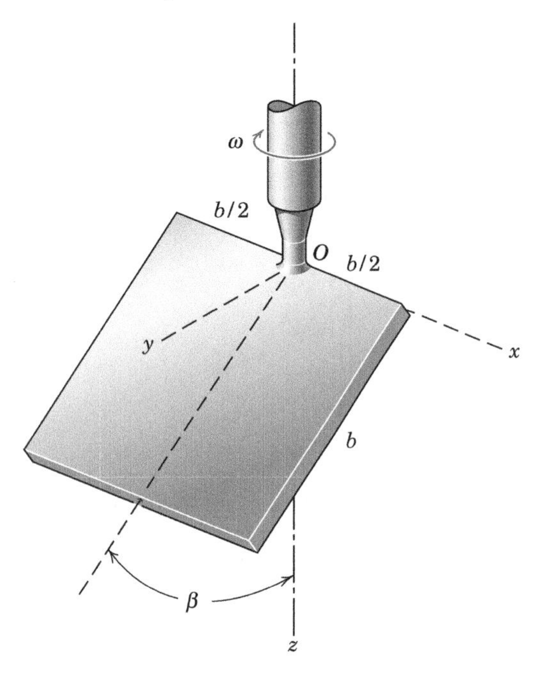

Problem 7/89

7/90 For the plate of mass m in Prob. 7/89, determine the y- and z-components of the moment applied to the plate by the weld at O necessary to give the plate an angular acceleration $\alpha = \dot{\omega}$ starting from rest. Neglect the moment due to the weight.

7/91 The uniform slender rod of length l is welded to the bracket at A on the underside of the disk B. The disk rotates about a vertical axis with a constant angular velocity ω. Determine the value of ω which will result in a zero moment supported by the weld at A for the position $\theta = 60°$ with $b = l/4$.

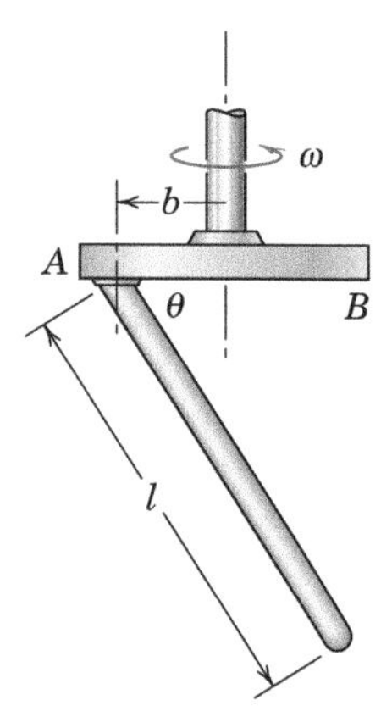

Problem 7/91

7/92 The half-cylindrical shell of radius r, length $2b$, and mass m revolves about the vertical z-axis with a constant angular velocity ω as indicated. Determine the magnitude M of the bending moment in the shaft at A due to both the weight and the rotational motion of the shell.

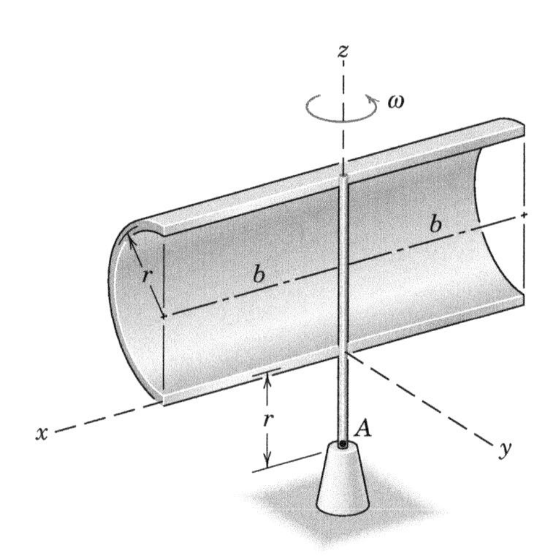

Problem 7/92

▶7/93 The homogeneous thin triangular plate of mass m is welded to the horizontal shaft, which rotates freely in the bearings at A and B. If the plate is released from rest in the horizontal position shown, determine the magnitude of the bearing reaction at A for the instant just after release.

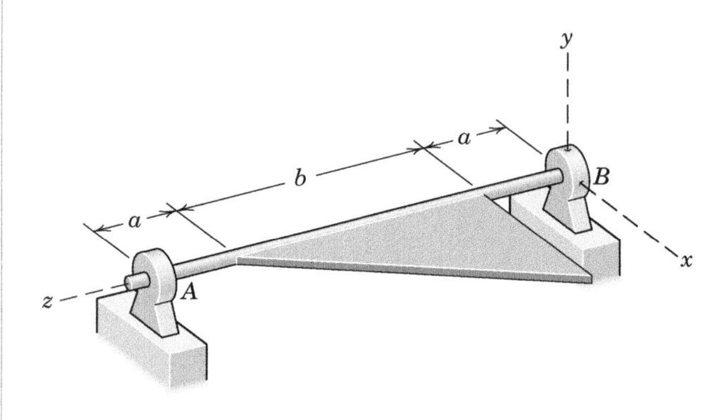

Problem 7/93

▶7/94 If the homogeneous triangular plate of Prob. 7/93 is released from rest in the position shown, determine the magnitude of the bearing reaction at A after the plate has rotated 90°.

PROBLEMS

Introductory Problems

7/95 A dynamics instructor demonstrates gyroscopic principles to his students. He suspends a rapidly spinning wheel with a string attached to one end of its horizontal axle. Describe the precession motion of the wheel.

Problem 7/95

7/96 The student has volunteered to assist in a classroom demonstration involving a momentum wheel which is rapidly spinning with angular speed p as shown. The instructor has asked her to hold the axle of the wheel in the horizontal position shown and then attempt to tilt the axis upward in a vertical plane. What motion tendency of the wheel assembly will the student sense?

Problem 7/96

7/97 A car makes a turn to the right on a level road. Determine whether the normal reaction under the right rear wheel is increased or decreased as a result of the gyroscopic effect of the precessing wheels.

7/98 The 50-kg wheel is a solid circular disk which rolls on the horizontal plane in a circle of 600-mm radius. The wheel shaft is pivoted about the axis O-O and is driven by the vertical shaft at the constant rate $N = 48$ rev/min about the Z-axis. Determine the normal force R between the wheel and the horizontal surface. Neglect the weight of the horizontal shaft.

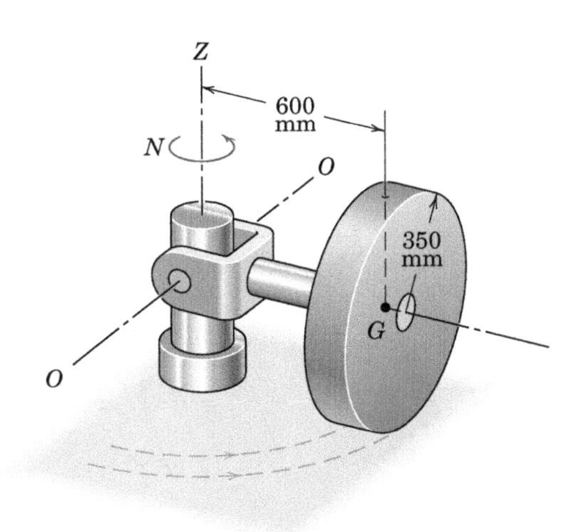

Problem 7/98

7/99 The special-purpose fan is mounted as shown. The motor armature, shaft, and blades have a combined mass of 2.2 kg with radius of gyration of 60 mm. The axial position b of the 0.8-kg block A can be adjusted. With the fan turned off, the unit is balanced about the x-axis when $b = 180$ mm. The motor and fan operate at 1725 rev/min in the direction shown. Determine the value of b which will produce a steady precession of 0.2 rad/s about the positive y-axis.

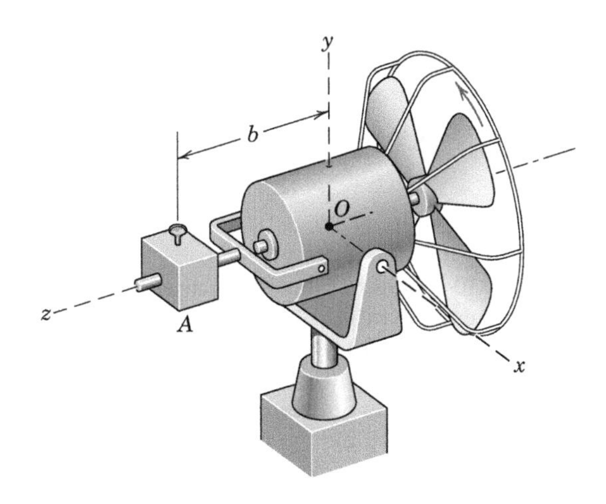

Problem 7/99

7/100 An airplane has just cleared the runway with a takeoff speed v. Each of its freely spinning wheels has a mass m, with a radius of gyration k about its axle. As seen from the front of the airplane, the wheel precesses at the angular rate Ω as the landing strut is folded into the wing about its pivot O. As a result of the gyroscopic action, the supporting member A exerts a torsional moment M on B to prevent the tubular member from rotating in the sleeve at B. Determine M and identify whether it is in the sense of M_1 or M_2.

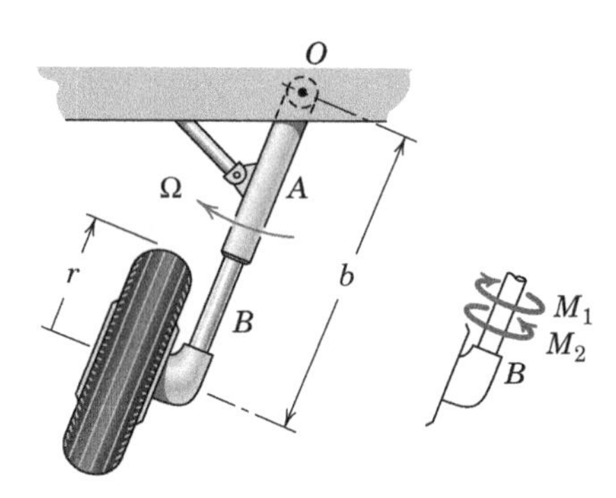

Problem 7/100

7/101 An experimental antipollution bus is powered by the kinetic energy stored in a large flywheel which spins at a high speed p in the direction indicated. As the bus encounters a short upward ramp, the front wheels rise, thus causing the flywheel to precess. What changes occur to the forces between the tires and the road during this sudden change?

Problem 7/101

7/102 The 210-kg rotor of a turbojet aircraft engine has a radius of gyration of 220 mm and rotates counterclockwise at 18 000 rev/min as viewed from the front. If the aircraft is traveling at 1200 km/h and starts to execute an inside vertical loop of 3800-m radius, compute the gyroscopic moment M transmitted to the airframe. What correction to the controls does the pilot have to make in order to remain in the vertical plane?

Representative Problems

7/103 A small air compressor for an aircraft cabin consists of the 3.50-kg turbine A which drives the 2.40-kg blower B at a speed of 20 000 rev/min. The shaft of the assembly is mounted transversely to the direction of flight and is viewed from the rear of the aircraft in the figure. The radii of gyration of A and B are 79.0 and 71.0 mm, respectively. Calculate the radial forces exerted on the shaft by the bearings at C and D if the aircraft executes a clockwise roll (rotation about the longitudinal flight axis) of 2 rad/s viewed from the rear of the aircraft. Neglect the small moments caused by the weights of the rotors. Draw a free-body diagram of the shaft as viewed from above and indicate the shape of its deflected centerline.

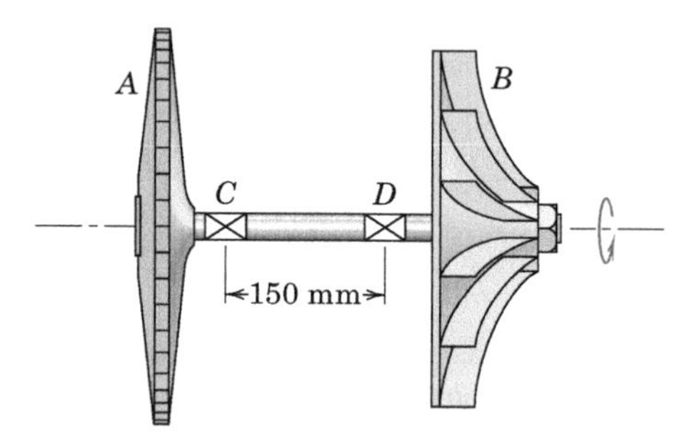

Problem 7/103

7/104 The two solid cones with the same base and equal altitudes are spinning in space about their common axis at the rate p. For what ratio h/r will precession of their spin axis be impossible?

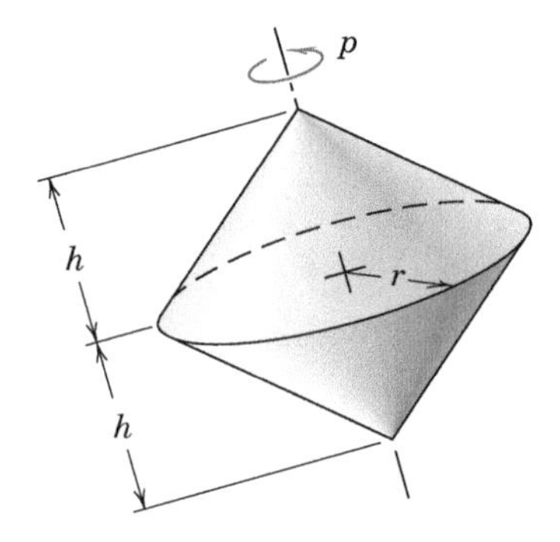

Problem 7/104

7/105 The blades and hub of the helicopter rotor weigh 140 lb and have a radius of gyration of 10 ft about the z-axis of rotation. With the rotor turning at 500 rev/min during a short interval following vertical liftoff, the helicopter tilts forward at the rate $\dot{\theta} = 10$ deg/sec in order to acquire forward velocity. Determine the gyroscopic moment M transmitted to the body of the helicopter by its rotor and indicate whether the helicopter tends to deflect clockwise or counterclockwise, as viewed by a passenger facing forward.

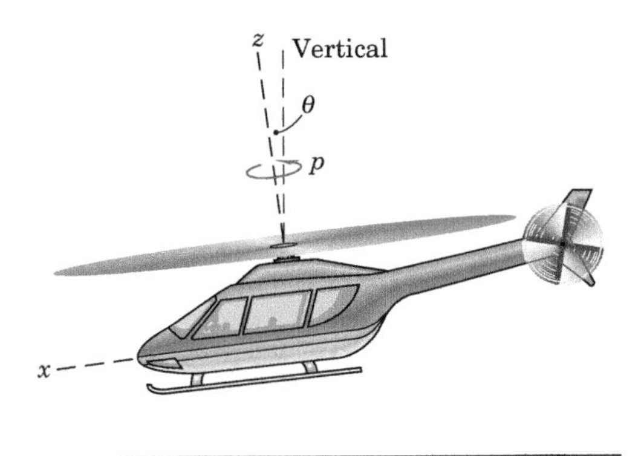

Problem 7/105

7/106 The 4-oz top with radius of gyration about its spin axis of 0.62 in. is spinning at the rate $p = 3600$ rev/min in the sense shown, with its spin axis making an angle $\theta = 20°$ with the vertical. The distance from its tip O to its mass center G is $\bar{r} = 2.5$ in. Determine the precession $\mathbf{\Omega}$ of the top and explain why θ gradually decreases as long as the spin rate remains large. An enlarged view of the contact of the tip is shown.

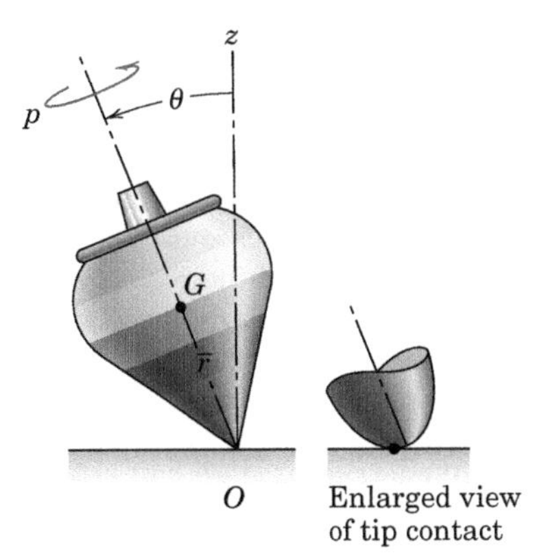

Problem 7/106

7/107 The figure shows a gyro mounted with a vertical axis and used to stabilize a hospital ship against rolling. The motor A turns the pinion which precesses the gyro by rotating the large precession gear B and attached rotor assembly about a horizontal transverse axis in the ship. The rotor turns inside the housing at a clockwise speed of 960 rev/min as viewed from the top and has a mass of 80 Mg with radius of gyration of 1.45 m. Calculate the moment exerted on the hull structure by the gyro if the motor turns the precession gear B at the rate of 0.320 rad/s. In which of the two directions, (a) or (b), should the motor turn in order to counteract a roll of the ship to port?

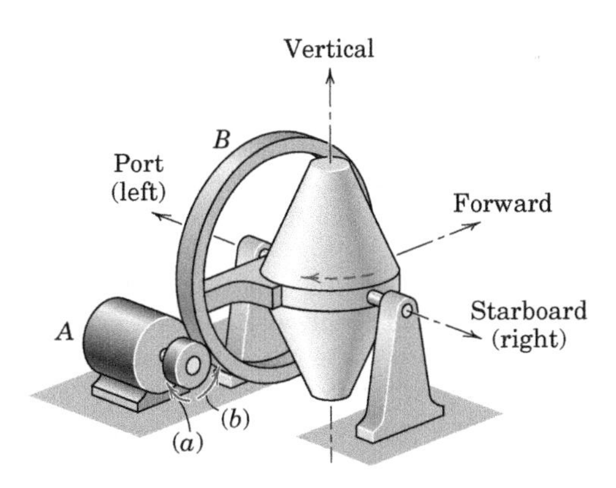

Problem 7/107

7/108 Each of the identical wheels has a mass of 4 kg and a radius of gyration $k_z = 120$ mm and is mounted on a horizontal shaft AB secured to the vertical shaft at O. In case (a), the horizontal shaft is fixed to a collar at O which is free to rotate about the vertical y-axis. In case (b), the shaft is secured by a yoke hinged about the x-axis to the collar. If the wheel has a large angular velocity $p = 3600$ rev/min about its z-axis in the position shown, determine any precession which occurs and the bending moment M_A in the shaft at A for each case. Neglect the small mass of the shaft and fitting at O.

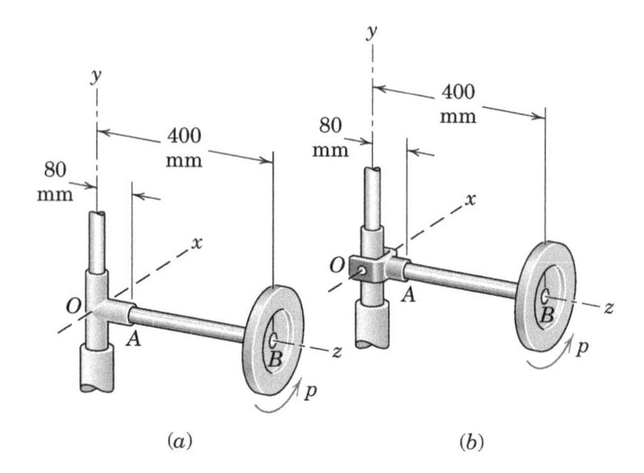

Problem 7/108

7/109 If the wheel in case (a) of Prob. 7/108 is forced to precess about the vertical by a mechanical drive at the steady rate $\mathbf{\Omega} = 2\mathbf{j}$ rad/s, determine the bending moment in the horizontal shaft at A. In the absence of friction, what torque M_O is applied to the collar at O to sustain this motion?

7/110 The figure shows the side view of the wheel carriage (truck) of a railway passenger car where the vertical load is transmitted to the frame in which the journal wheel bearings are located. The lower view shows only one pair of wheels and their axle which rotates with the wheels. Each of the 33-in.-diameter wheels weighs 560 lb, and the axle weighs 300 lb with a diameter of 5 in. Both wheels and axle are made of steel with a specific weight of 489 lb/ft^3. If the train is traveling at 80 mi/hr while rounding an 8° curve to the right (radius of curvature 717 ft), calculate the change ΔR in the vertical force supported by each wheel due only to the gyroscopic action. As a close approximation, treat each wheel as a uniform circular disk and the axle as a uniform solid cylinder. Also assume that both rails are in the same horizontal plane.

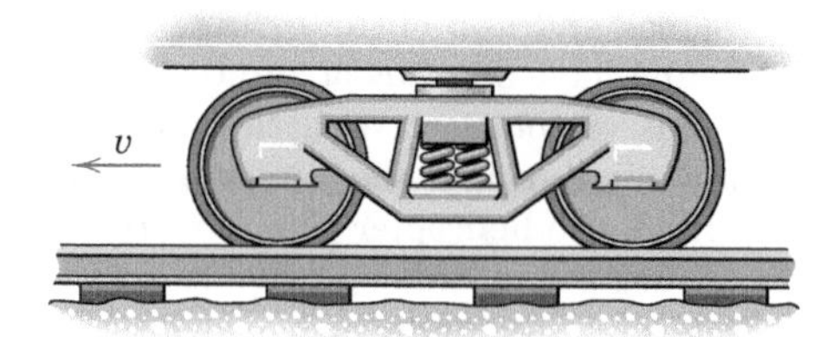

Side view of carriage

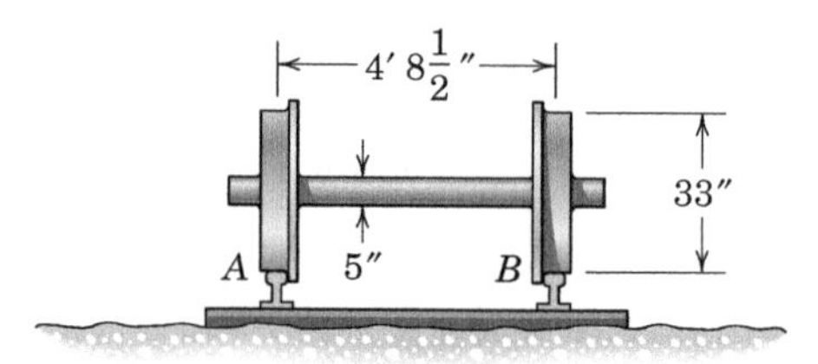

View of wheels and axle

Problem 7/110

7/111 The primary structure of a proposed space station consists of five spherical shells connected by tubular spokes. The moment of inertia of the structure about its geometric axis A-A is twice as much as that about any axis through O normal to A-A. The station is designed to rotate about its geometric axis at the constant rate of 3 rev/min. If the spin axis A-A precesses about the Z-axis of fixed orientation and makes a very small angle with it, calculate the rate $\dot{\psi}$ at which the station wobbles. The mass center O has negligible acceleration.

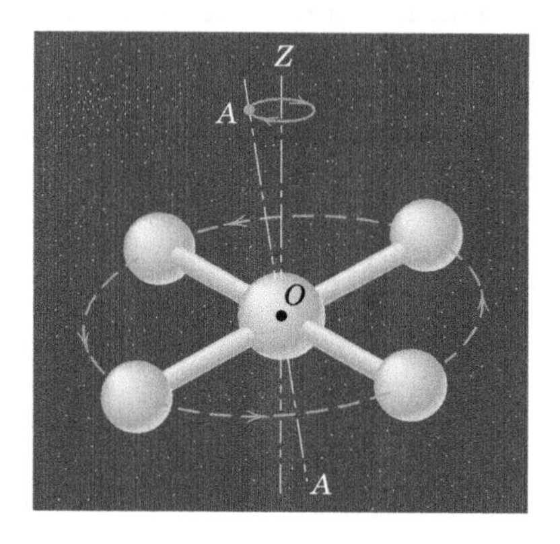

Problem 7/111

7/112 The uniform 640-mm rod has a mass of 3 kg and is welded centrally to the uniform 160-mm-radius circular disk which has a mass of 8 kg. The unit is given a spin velocity $p = 60$ rad/s in the direction shown. The axis of the rod is seen to wobble through a total angle of 30°. Calculate the angular velocity $\dot{\psi}$ of precession and determine whether it is $\dot{\psi}_1$ or $\dot{\psi}_2$.

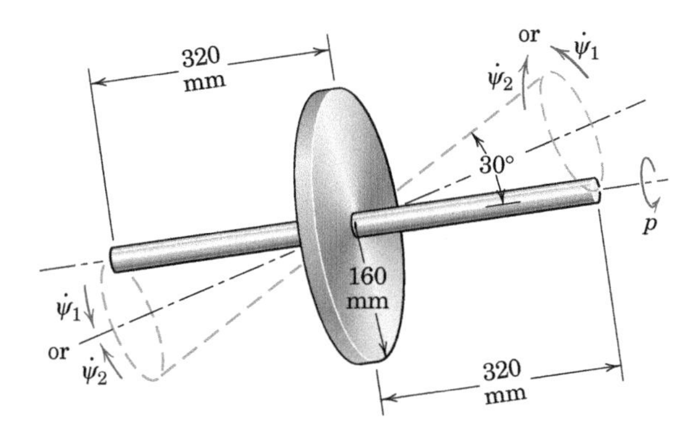

Problem 7/112

7/113 The electric motor has a total weight of 20 lb and is supported by the mounting brackets A and B attached to the rotating disk. The armature of the motor has a weight of 5 lb and a radius of gyration of 1.5 in. and turns counterclockwise at a speed of 1725 rev/min as viewed from A to B. The turntable revolves about its vertical axis at the constant rate of 48 rev/min in the direction shown. Determine the vertical components of the forces supported by the mounting brackets at A and B.

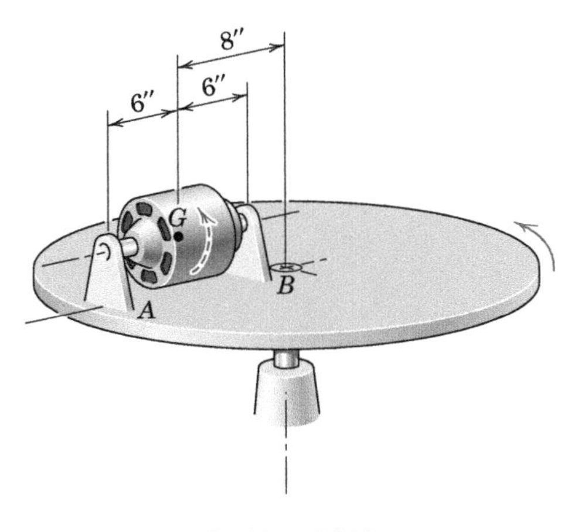

Problem 7/113

7/114 The spacecraft shown is symmetrical about its z-axis and has a radius of gyration of 720 mm about this axis. The radii of gyration about the x- and y-axes through the mass center are both equal to 540 mm. When moving in space, the z-axis is observed to generate a cone with a total vertex angle of 4° as it precesses about the axis of total angular momentum. If the spacecraft has a spin velocity $\dot{\phi}$ about its z-axis of 1.5 rad/s, compute the period τ of each full precession. Is the spin vector in the positive or negative z-direction?

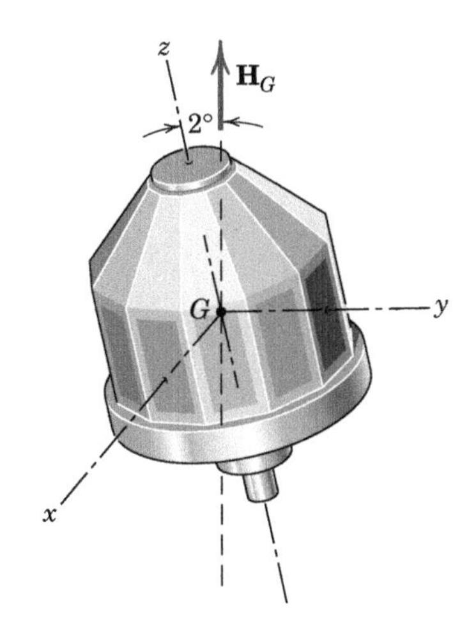

Problem 7/114

7/115 The 8-lb rotor with radius of gyration of 3 in. rotates on ball bearings at a speed of 3000 rev/min about its shaft OG. The shaft is free to pivot about the X-axis, as well as to rotate about the Z-axis. Calculate the vector $\mathbf{\Omega}$ for precession about the Z-axis. Neglect the mass of shaft OG and compute the gyroscopic couple $\mathbf{M}$ exerted by the shaft on the rotor at G.

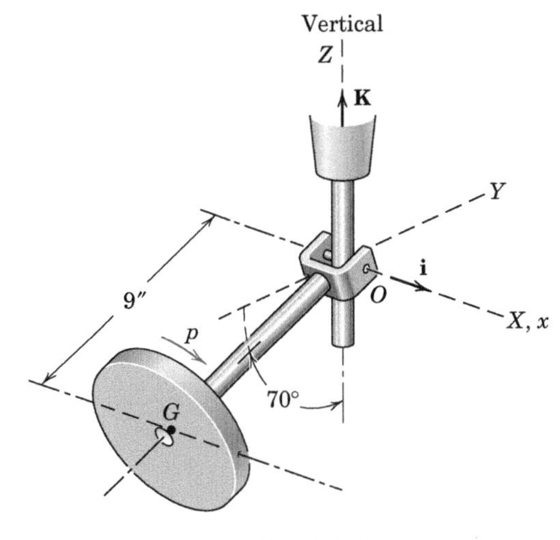

Problem 7/115

7/116 The housing of the electric motor is freely pivoted about the horizontal x-axis, which passes through the mass center G of the rotor. If the motor is turning at the constant rate $\dot{\phi} = p$, determine the angular acceleration $\ddot{\psi}$ which will result from the application of the moment M about the vertical shaft if $\dot{\gamma} = \dot{\psi} = 0$. The mass of the frame and housing is considered negligible compared with the mass m of the rotor. The radius of gyration of the rotor about the z-axis is k_z and that about the x-axis is k_x.

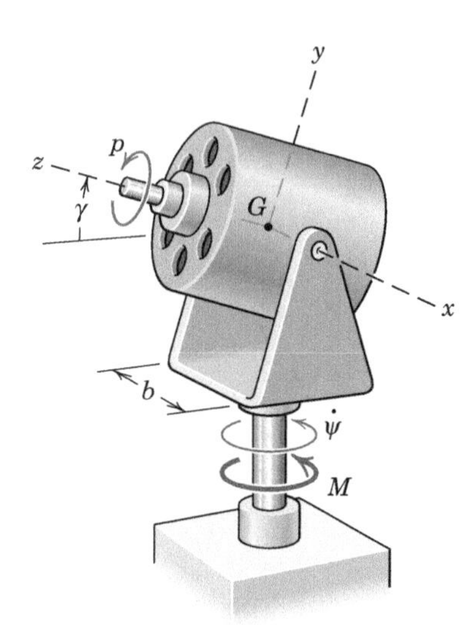

Problem 7/116

7/117 The thin ring is projected into the air with a spin velocity of 300 rev/min. If its geometric axis is observed to have a very slight precessional wobble, determine the frequency f of the wobble.

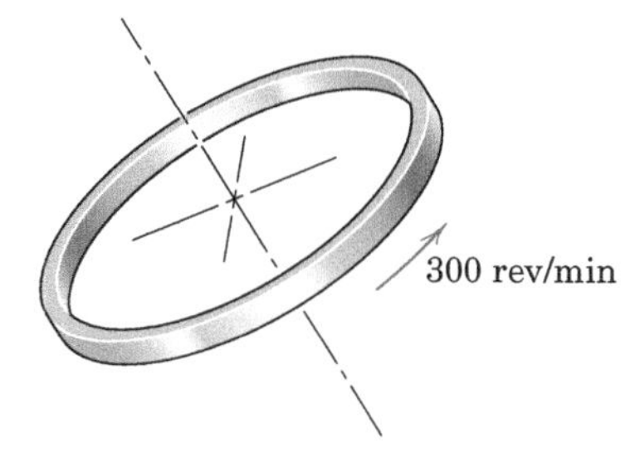

Problem 7/117

7/118 A boy throws a thin circular disk (like a Frisbee) with a spin rate of 300 rev/min. The plane of the disk is seen to wobble through a total angle of 10°. Calculate the period τ of the wobble and indicate whether the precession is direct or retrograde.

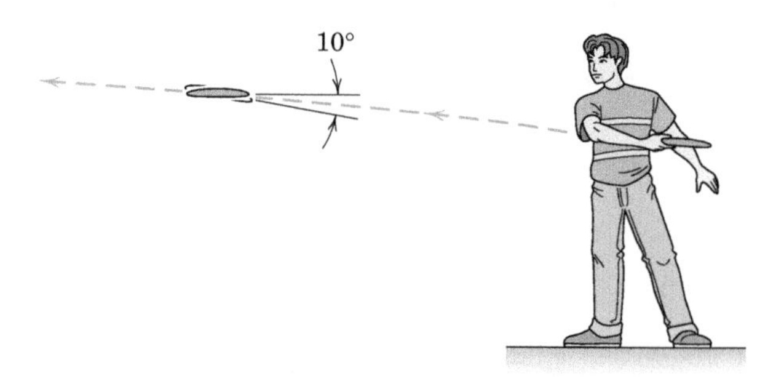

Problem 7/118

7/119 The figure shows a football in three common in-flight configurations. Case (a) is a perfectly thrown spiral pass with a spin rate of 120 rev/min. Case (b) is a wobbly spiral pass again with a spin rate of 120 rev/min about its own axis, but with the axis wobbling through a total angle of 20°. Case (c) is an end-over-end place kick with a rotational rate of 120 rev/min. For each case, specify the values of p, θ, β, and $\dot{\psi}$ as defined in this article. The moment of inertia about the long axis of the ball is 0.3 of that about the transverse axis of symmetry.

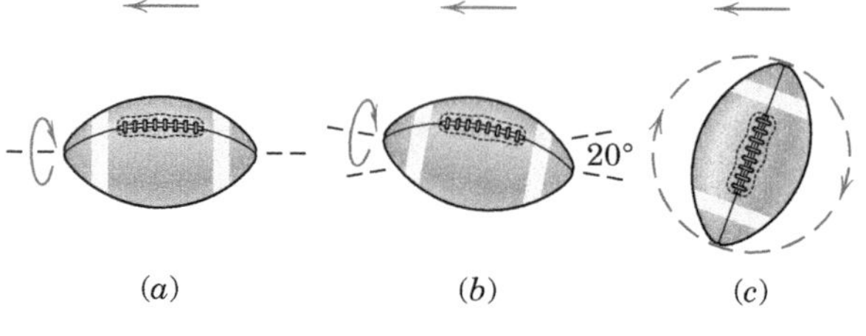

Problem 7/119

7/120 The rectangular bar is spinning in space about its longitudinal axis at the rate $p = 200$ rev/min. If its axis wobbles through a total angle of 20° as shown, calculate the period τ of the wobble.

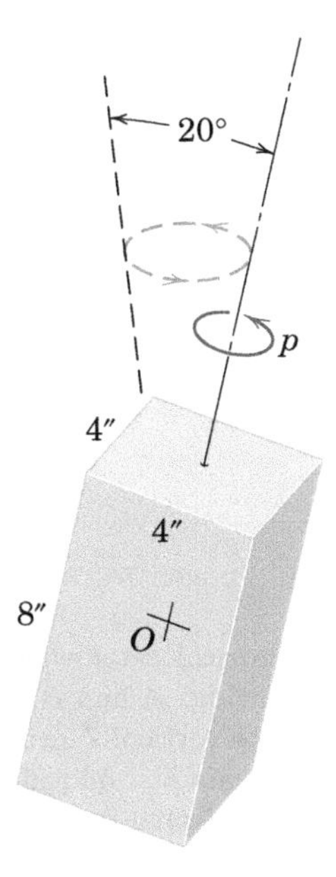

Problem 7/120

7/121 The 5-kg disk and hub A have a radius of gyration of 85 mm about the z_0-axis and spin at the rate $p = 1250$ rev/min. Simultaneously, the assembly rotates about the vertical z-axis at the rate $\Omega = 400$ rev/min. Calculate the gyroscopic moment **M** exerted *on* the shaft at C *by* the disk and the bending moment M_O in the shaft at O. Neglect the mass of the shaft but otherwise account for all forces acting on it.

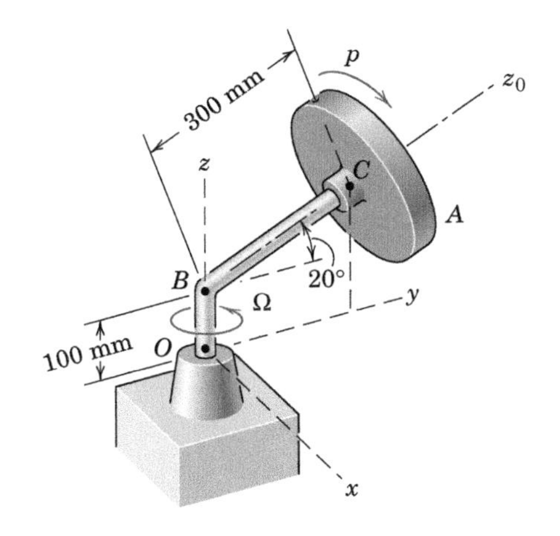

Problem 7/121

7/122 The uniform slender bar of mass m and length l is centrally mounted on the shaft A-A, about which it rotates with a constant speed $\dot{\phi} = p$. Simultaneously, the yoke is forced to rotate about the x-axis with a constant speed ω_0. As a function of ϕ, determine the magnitude of the torque M required to maintain the constant speed ω_0. (*Hint:* Apply Eq. 7/19 to obtain the x-component of M.)

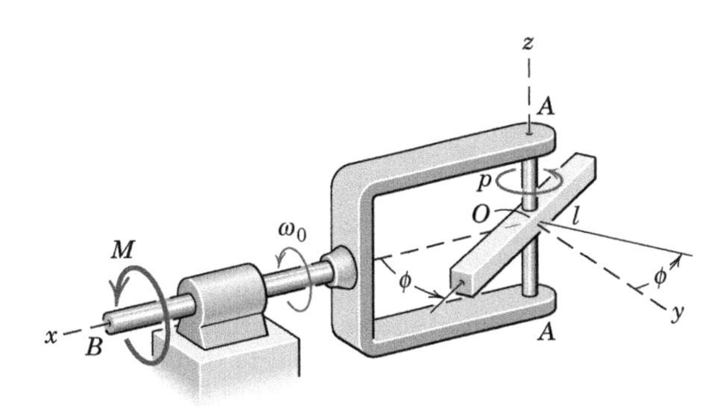

Problem 7/122

►7/123 The solid circular disk of mass m and small thickness is spinning freely on its shaft at the rate p. If the assembly is released in the vertical position at $\theta = 0$ with $\dot{\theta} = 0$, determine the horizontal components of the forces A and B exerted by the respective bearings on the horizontal shaft as the position $\theta = \pi/2$ is passed. Neglect the mass of the two shafts compared with m and neglect all friction. Solve by using the appropriate moment equations.

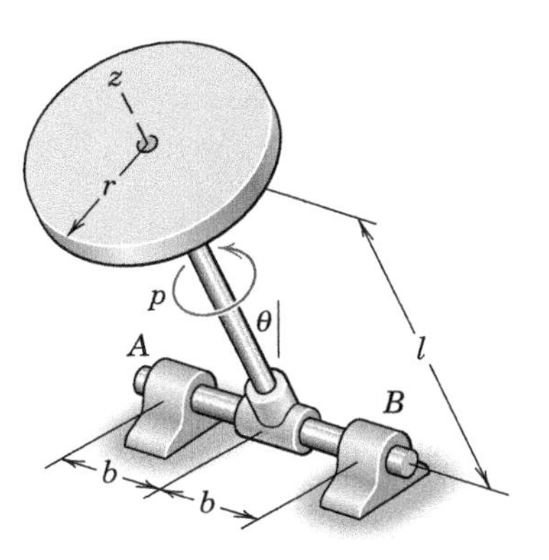

Problem 7/123

▸7/124 The earth-scanning satellite is in a circular orbit of period τ. The angular velocity of the satellite about its y- or pitch-axis is $\omega = 2\pi/\tau$, and the angular rates about the x- and z-axes are zero. Thus, the x-axis of the satellite always points to the center of the earth. The satellite has a reaction-wheel attitude-control system consisting of the three wheels shown, each of which may be variably torqued by its individual motor. The angular rate Ω_z of the z-wheel relative to the satellite is Ω_0 at time $t = 0$, and the x- and y-wheels are at rest relative to the satellite at $t = 0$. Determine the axial torques M_x, M_y, and M_z which must be exerted by the motors on the shafts of their respective wheels in order that the angular velocity $\boldsymbol{\omega}$ of the satellite will remain constant. The moment of inertia of each reaction wheel about its axis is I. The x and z reaction-wheel speeds are harmonic functions of the time with a period equal to that of the orbit. Plot the variations of the torques and the relative wheel speeds Ω_x, Ω_y, and Ω_z as functions of the time during one orbit period. (*Hint:* The torque to accelerate the x-wheel equals the reaction of the gyroscopic moment on the z-wheel, and vice versa.)

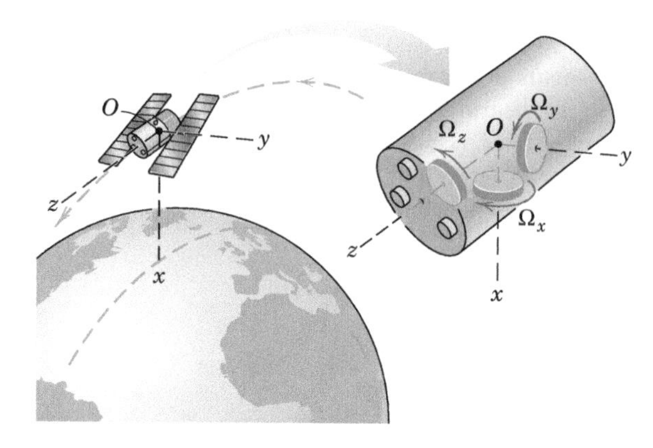

Problem 7/124

▸7/125 The two solid homogeneous right-circular cones, each of mass m, are fastened together at their vertices to form a rigid unit and are spinning about their axis of radial symmetry at the rate $p = 200$ rev/min. (*a*) Determine the ratio h/r for which the rotation axis will not precess. (*b*) Sketch the space and body cones for the case where h/r is less than the critical ratio. (*c*) Sketch the space and body cones when $h = r$ and the precessional velocity is $\dot{\psi} = 18$ rad/s.

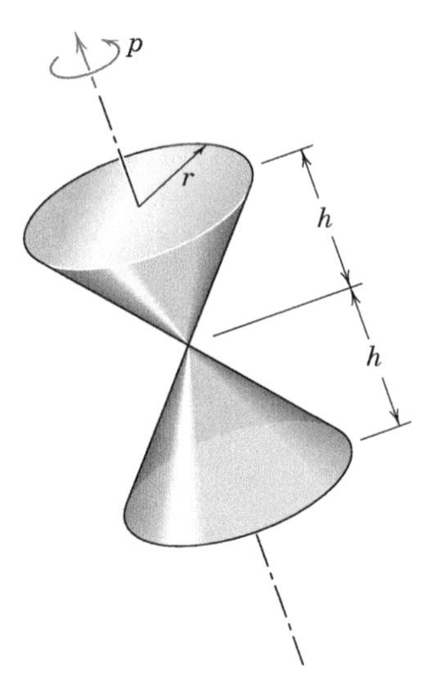

Problem 7/125

▸7/126 The solid cylindrical rotor weighs 64.4 lb and is mounted in bearings A and B of the frame which rotates about the vertical Z-axis. If the rotor spins at the constant rate $p = 50$ rad/sec relative to the frame and if the frame itself rotates at the constant rate $\Omega = 30$ rad/sec, compute the bending moment $\mathbf{M}$ in the shaft at C which the lower portion of the shaft exerts on the upper portion. Also compute the kinetic energy T of the rotor. Neglect the mass of the frame.

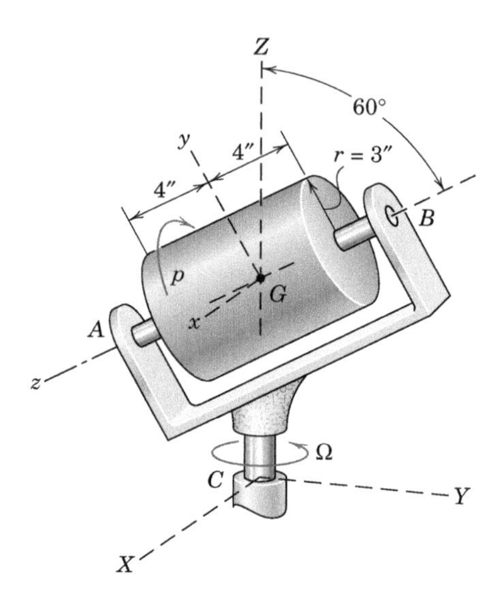

Problem 7/126

PROBLEMS

(Unless otherwise indicated, all motion variables are referred to the equilibrium position.)

Undamped, Free Vibrations

8/1 When a 3-kg collar is placed upon the pan which is attached to the spring of unknown constant, the additional static deflection of the pan is observed to be 42 mm. Determine the spring constant k in N/m, lb/in., and lb/ft.

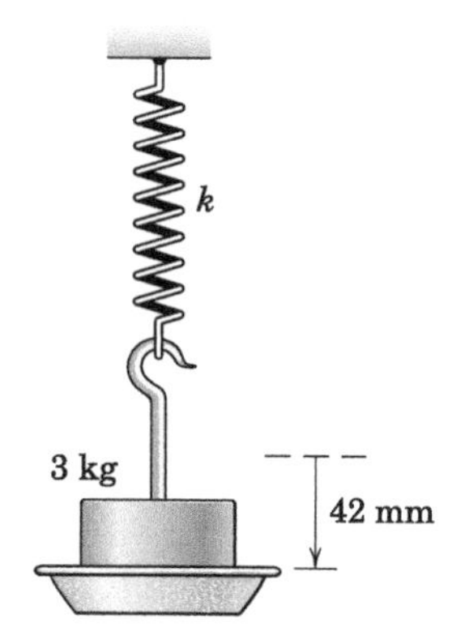

Problem 8/1

8/2 Determine the natural frequency of the spring-mass system in both radians per second and cycles per second (Hz).

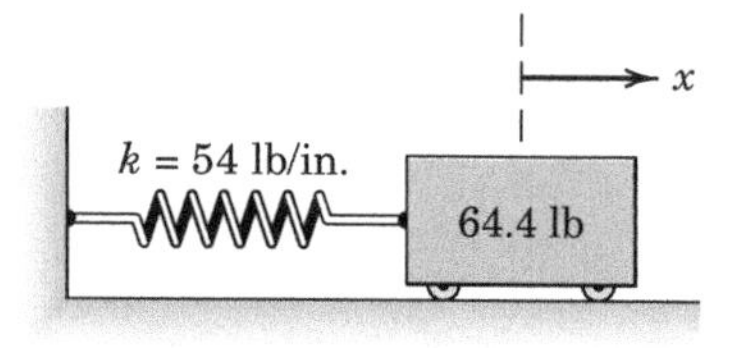

Problem 8/2

8/3 For the system of Prob. 8/2, determine the position x of the mass as a function of time if the mass is released from rest at time $t = 0$ from a position 2 inches to the left of the equilibrium position. Determine the maximum velocity and maximum acceleration of the mass over one cycle of motion.

8/4 For the system of Prob. 8/2, determine the position x as a function of time if the mass is released at time $t = 0$ from a position 2 inches to the right of the equilibrium position with an initial velocity of 9 in./sec to the left. Determine the amplitude C and period τ of the motion.

8/5 For the spring-mass system shown, determine the static deflection δ_{st}, the system period τ, and the maximum velocity v_{max} which result if the cylinder is displaced 100 mm downward from its equilibrium position and released from rest.

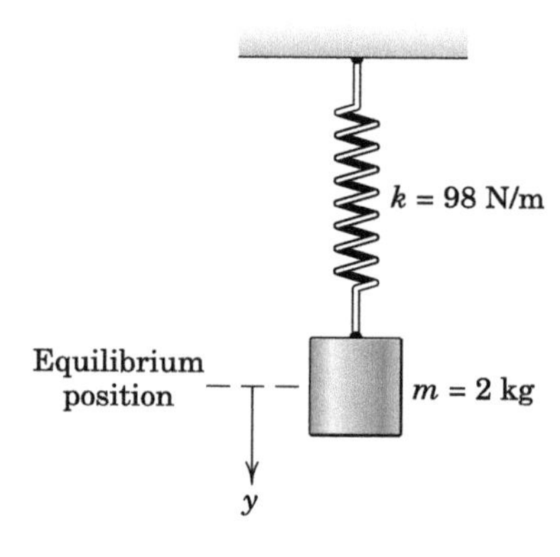

Problem 8/5

8/6 The cylinder of the system of Prob. 8/5 is displaced 100 mm downward from its equilibrium position and released at time $t = 0$. Determine the position y, velocity v, and acceleration a when $t = 3$ s. What is the maximum acceleration?

8/7 Determine the natural frequency in cycles per second for the system shown. Neglect the mass and friction of the pulleys. Assume that the block of mass m remains horizontal.

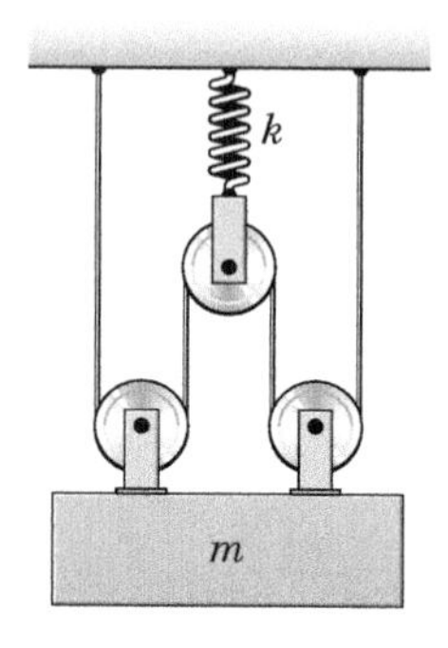

Problem 8/7

8/8 The vertical plunger has a mass of 2.5 kg and is supported by the two springs, which are always in compression. Calculate the natural frequency f_n of vibration of the plunger if it is deflected from the equilibrium position and released from rest. Friction in the guide is negligible.

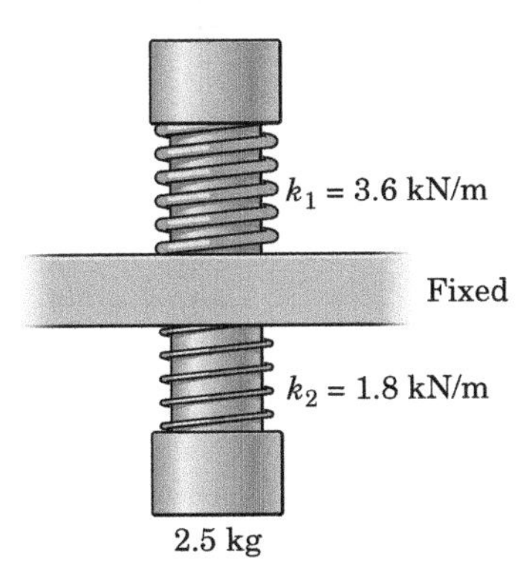

Problem 8/8

8/9 Determine the period τ for the system shown. The cable is always taut, and the mass and friction of the pulley are to be neglected.

Problem 8/9

8/10 In the equilibrium position, the 30-kg cylinder causes a static deflection of 50 mm in the coiled spring. If the cylinder is depressed an additional 25 mm and released from rest, calculate the resulting natural frequency f_n of vertical vibration of the cylinder in cycles per second (Hz).

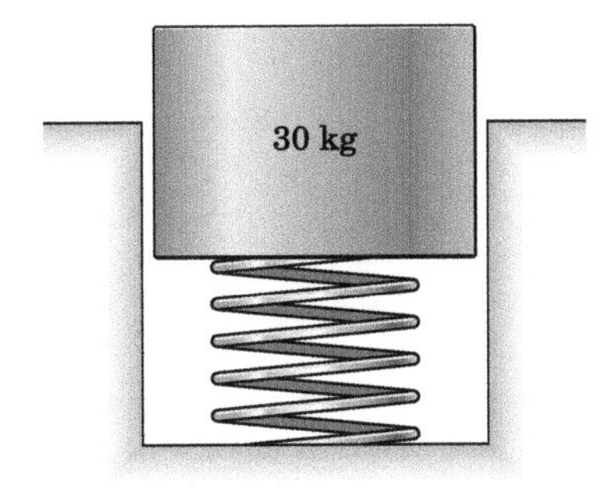

Problem 8/10

8/11 For the cylinder of Prob. 8/10, determine the vertical displacement x, measured positive down in millimeters from the equilibrium position, in terms of the time t in seconds measured from the instant of release from the position of 25 mm added deflection.

8/12 Determine the natural frequency in radians per second for the system shown. Neglect the mass and friction of the pulleys.

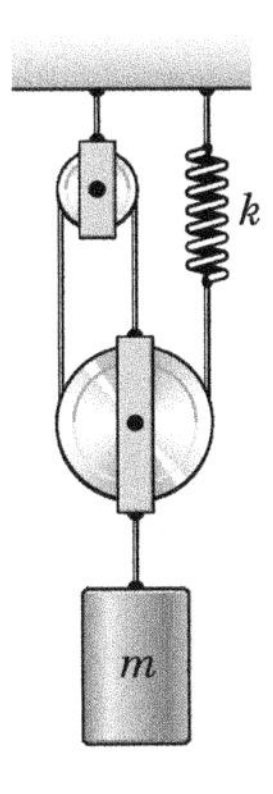

Problem 8/12

8/13 An old car being moved by a magnetic crane pickup is dropped from a short distance above the ground. Neglect any damping effects of its worn-out shock absorbers and calculate the natural frequency f_n in cycles per second (Hz) of the vertical vibration which occurs after impact with the ground. Each of the four springs on the 1000-kg car has a constant of 17.5 kN/m. Because the center of mass is located midway between the axles and the car is level when dropped, there is no rotational motion. State any assumptions.

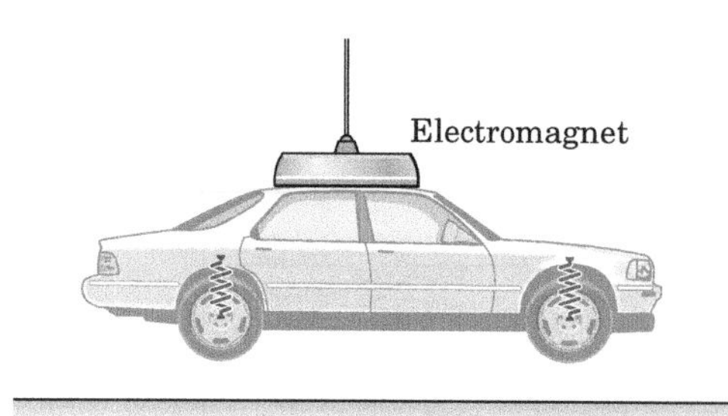

Problem 8/13

8/14 The 4-oz slider oscillates in the fixed slot under the action of the three springs, each of stiffness $k = 0.5$ lb/in. If the initial conditions at time $t = 0$ are $x_0 = 0.1$ in. and $\dot{x}_0 = 0.5$ in./sec, determine the position and velocity of the slider at time $t = 2$ sec. What is the system period?

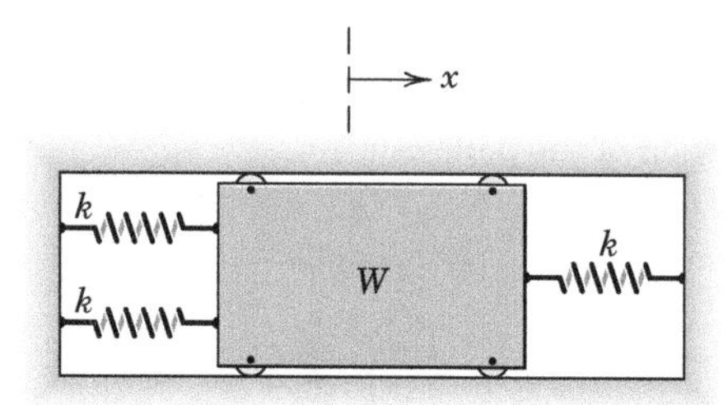

Problem 8/14

8/15 During the design of the spring-support system for the 4000-kg weighing platform, it is decided that the frequency of free vertical vibration in the unloaded condition shall not exceed 3 cycles per second. (*a*) Determine the maximum acceptable spring constant k for each of the three identical springs. (*b*) For this spring constant, what would be the natural frequency f_n of vertical vibration of the platform loaded by the 40-Mg truck?

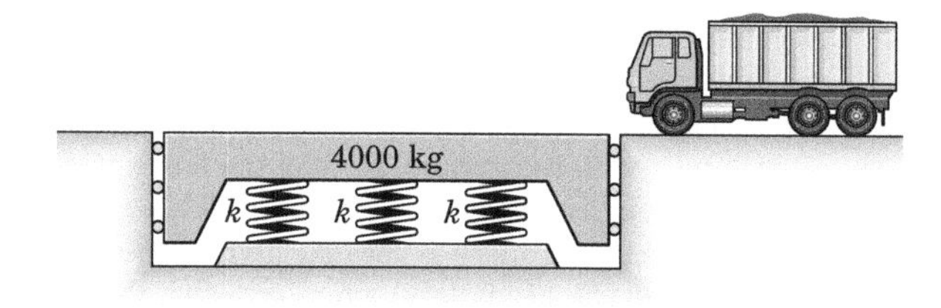

Problem 8/15

8/16 Calculate the natural frequency f_n of vibration if the mass is deflected from its equilibrium position and released from rest. Each pair of springs is connected by an inextensible cable. Evaluate your results for $m = 15$ kg, $k_1 = 225$ N/m, and $k_2 = 150$ N/m.

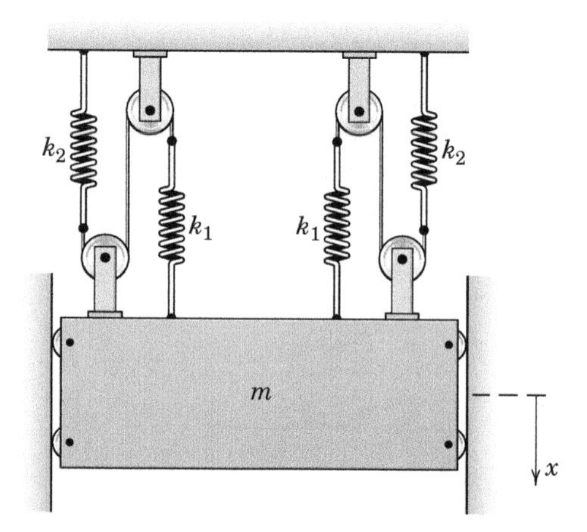

Problem 8/16

8/17 Replace the springs in each of the two cases shown by a single spring of stiffness k (equivalent spring stiffness) which will cause each mass to vibrate with its original frequency.

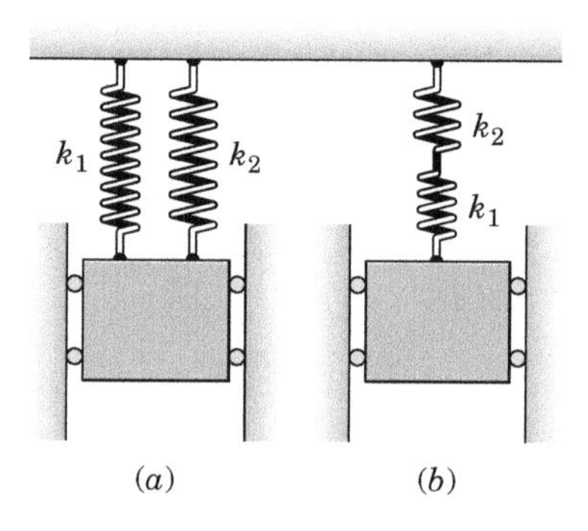

Problem 8/17

8/18 With the assumption of no slipping, determine the mass m of the block which must be placed on the top of the 6-kg cart in order that the system period be 0.75 s. What is the minimum coefficient μ_s of static friction for which the block will not slip relative to the cart if the cart is displaced 50 mm from the equilibrium position and released?

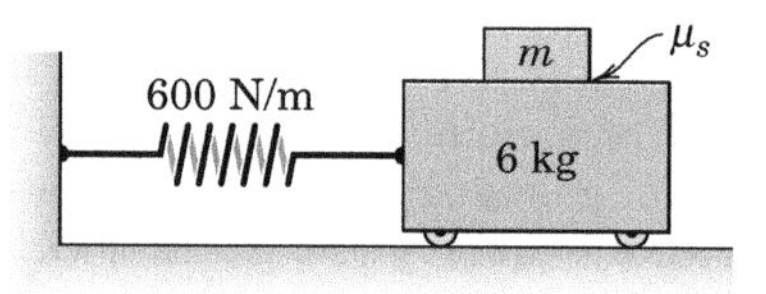

Problem 8/18

8/19 If both springs are unstretched when the mass is in the central position shown, determine the static deflection δ_{st} of the mass. What is the period of oscillatory motion about the position of static equilibrium?

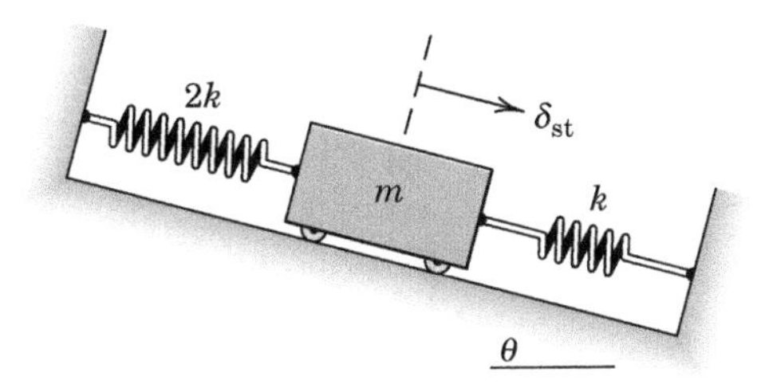

Problem 8/19

8/20 An energy-absorbing car bumper with its springs initially undeformed has an equivalent spring constant of 3000 lb/in. If the 2500-lb car approaches a massive wall with a speed of 5 mi/hr, determine (*a*) the velocity v of the car as a function of time during contact with the wall, where $t = 0$ is the beginning of the impact, and (*b*) the maximum deflection x_{max} of the bumper.

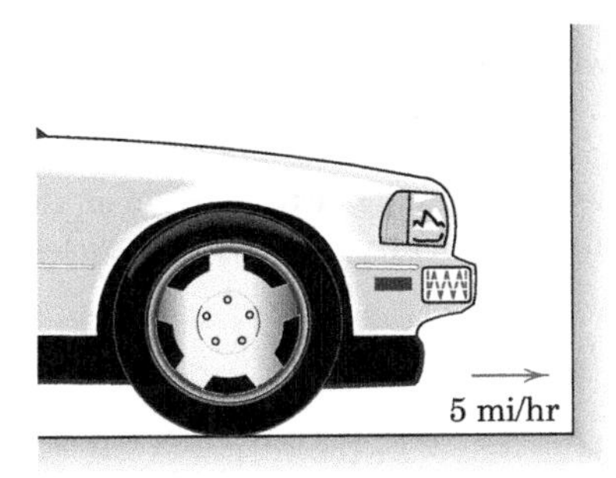

Problem 8/20

8/21 A 90-kg man stands at the end of a diving board and causes a vertical oscillation which is observed to have a period of 0.6 s. What is the static deflection δ_{st} at the end of the board? Neglect the mass of the board.

Problem 8/21

8/22 Shown in the figure is a model of a one-story building. The bar of mass m is supported by two light elastic upright columns whose upper and lower ends are fixed against rotation. For each column, if a force P and corresponding moment M were applied as shown in the right-hand part of the figure, the deflection δ would be given by $\delta = PL^3/12EI$, where L is the effective column length, E is Young's modulus, and I is the area moment of inertia of the column cross section with respect to its neutral axis. Determine the natural frequency of horizontal oscillation of the bar when the columns bend as shown in the figure.

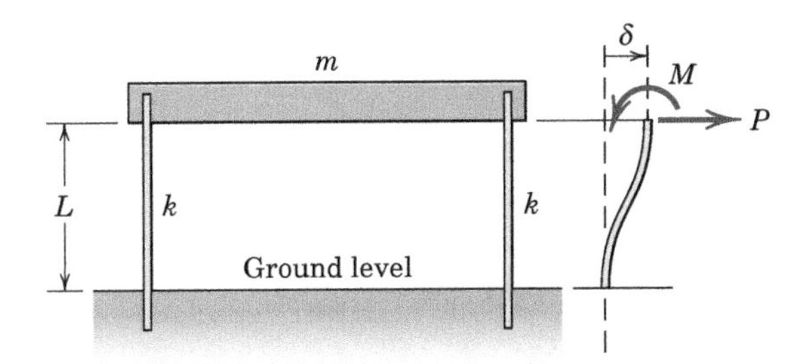

Problem 8/22

8/23 Calculate the natural circular frequency ω_n of the system shown in the figure. The mass and friction of the pulleys are negligible.

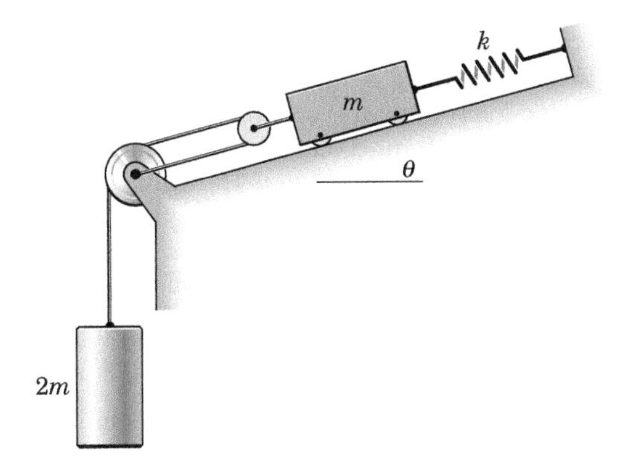

Problem 8/23

▶8/24 The slider of mass m is confined to the horizontal slot shown. The two springs each of constant k are linear. Derive the nonlinear equation of motion for small values of y, retaining terms of order y^3 and larger. Both springs are unstretched when $y = 0$. Neglect friction.

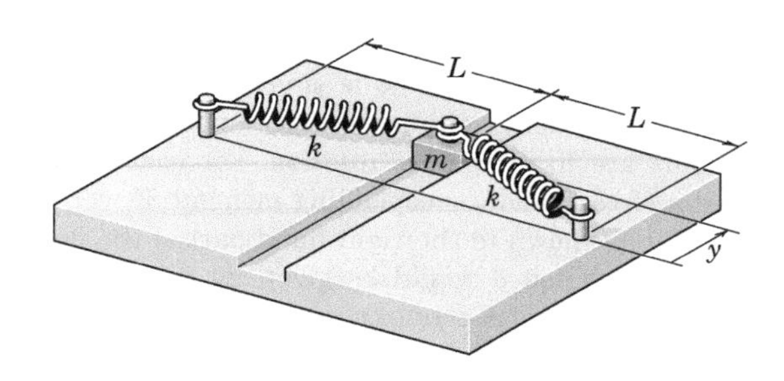

Problem 8/24

Damped, Free Vibrations

8/25 Determine the value of the damping ratio ζ for the simple spring-mass-dashpot system shown.

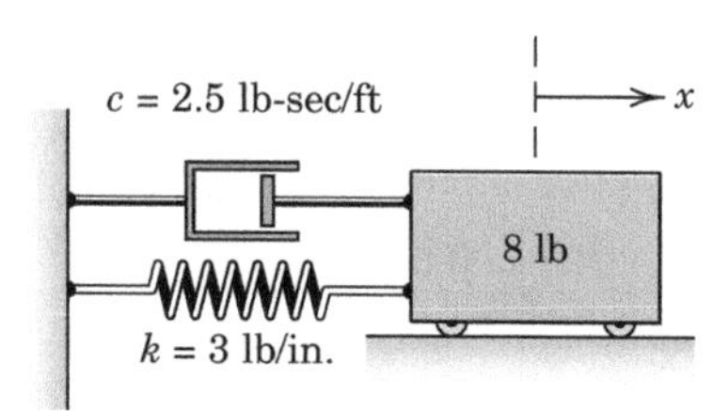

Problem 8/25

8/26 The period τ_d of damped linear oscillation for a certain 1-kg mass is 0.3 s. If the stiffness of the supporting linear spring is 800 N/m, calculate the damping coefficient c.

8/27 Viscous damping is added to an initially undamped spring-mass system. For what value of the damping ratio ζ will the damped natural frequency ω_d be equal to 90 percent of the natural frequency of the original undamped system?

8/28 The addition of damping to an undamped spring-mass system causes its period to increase by 25 percent. Determine the damping ratio ζ.

8/29 Determine the value of the viscous damping coefficient c for which the system shown is critically damped.

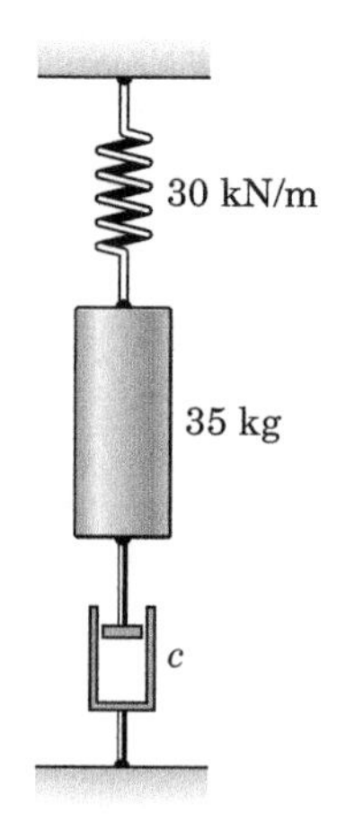

Problem 8/29

8/30 The 8-lb body of Prob. 8/25 is released from rest a distance x_0 to the right of the equilibrium position. Determine the displacement x as a function of time t, where $t = 0$ is the time of release.

8/31 The figure represents the measured displacement-time relationship for a vibration with small damping where it is impractical to achieve accurate results by measuring the nearly equal amplitudes of two successive cycles. Modify the expression for the viscous damping factor ζ based on the measured amplitudes x_0 and x_N which are N cycles apart.

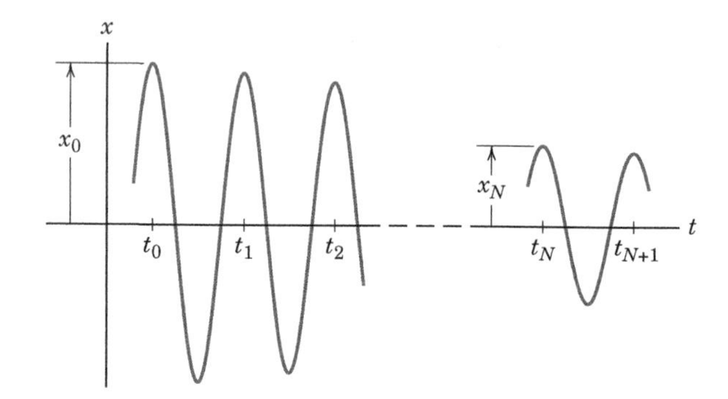

Problem 8/31

8/32 A linear harmonic oscillator having a mass of 1.10 kg is set into motion with viscous damping. If the frequency is 10 Hz and if two successive amplitudes a full cycle apart are measured to be 4.65 mm and 4.30 mm as shown, compute the viscous damping coefficient c.

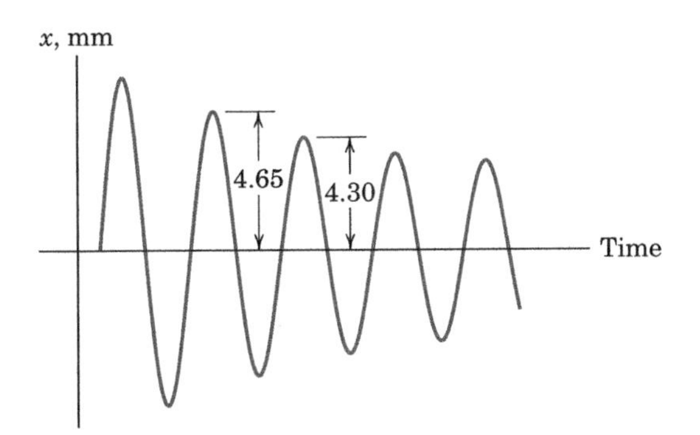

Problem 8/32

8/33 Determine the damping ratio ζ for the system shown. The system parameters are $m = 4$ kg, $k = 500$ N/m, and $c = 100$ N·s/m. Neglect the mass and friction of all pulleys, and assume that the cord remains taut throughout a motion cycle.

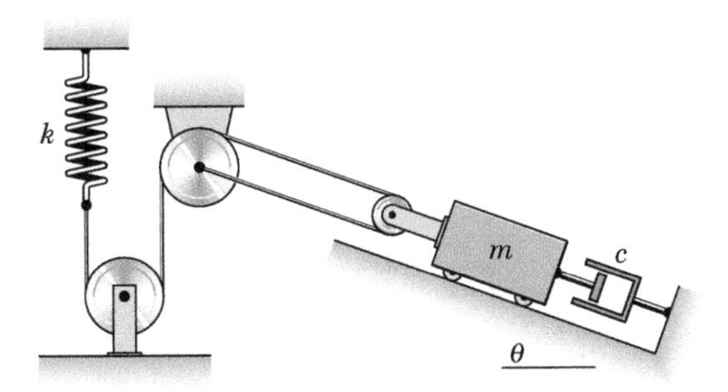

Problem 8/33

8/34 Further design refinement for the weighing platform of Prob. 8/15 is shown here where two viscous dampers are to be added to limit the ratio of successive positive amplitudes of vertical vibration in the unloaded condition to 4. Determine the necessary viscous damping coefficient c for each of the dampers.

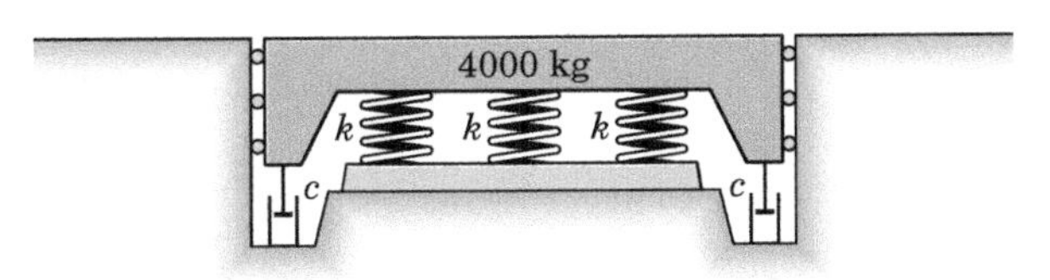

Problem 8/34

8/35 Derive the differential equation of motion for the system shown in terms of the variable x_1. Neglect friction and the mass of the linkage.

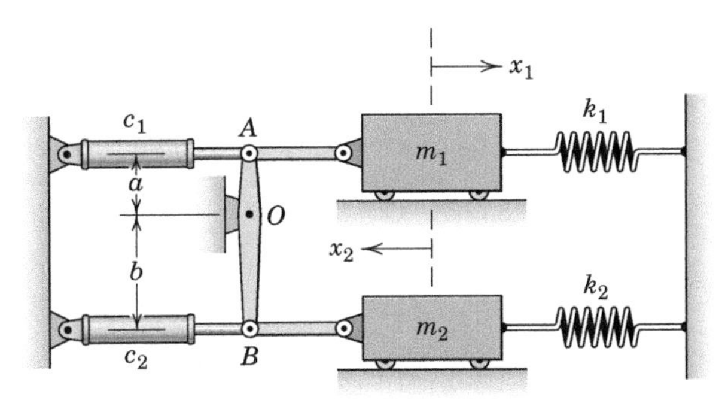

Problem 8/35

8/36 The system shown is released from rest from an initial position x_0. Determine the overshoot displacement x_1. Assume translational motion in the x-direction.

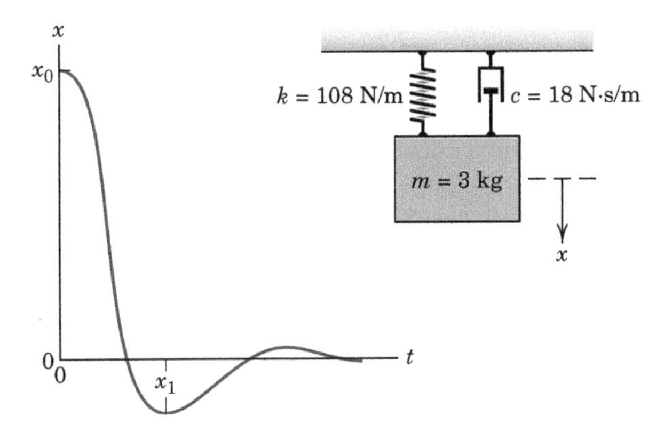

Problem 8/36

8/37 Determine the equation of motion for the system in terms of the variable x. The cables remain taut at all times, and the pulleys turn independently. Neglect friction and the mass of the pulleys. Additionally, determine expressions for the natural circular frequency ω_n and the damping ratio ζ.

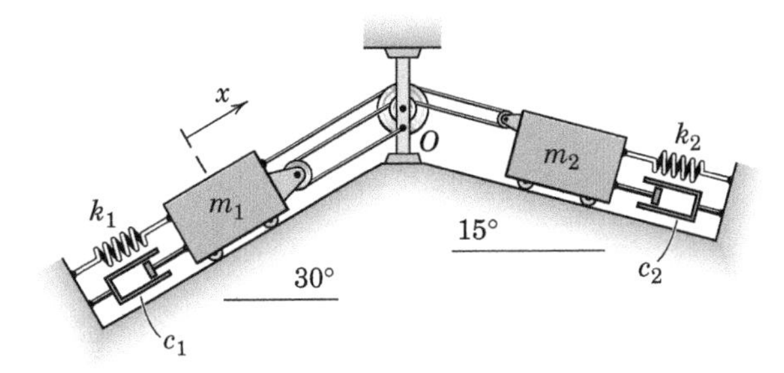

Problem 8/37

8/38 The mass of a given critically damped system is released at time $t = 0$ from the position $x_0 > 0$ with a negative initial velocity. Determine the critical value $(\dot{x}_0)_c$ of the initial velocity below which the mass will pass through the equilibrium position.

8/39 The mass of the system shown is released from rest at $x_0 = 6$ in. when $t = 0$. Determine the displacement x at $t = 0.5$ sec if (*a*) $c = 12$ lb-sec/ft and (*b*) $c = 18$ lb-sec/ft.

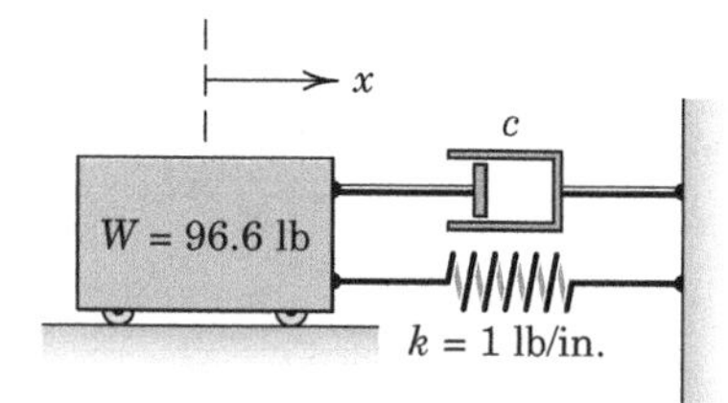

Problem 8/39

8/40 Derive the equation of motion for the system shown in terms of the displacement x. The masses are coupled through the light connecting rod ABC which pivots about the smooth bearing at point O. Neglect all friction, consider the rollers on m_2 and m_3 to be light, and assume small oscillations about the equilibrium position. State the system natural circular frequency ω_n and the viscous damping ratio ζ for $m_1 = 15$ kg, $m_2 = 12$ kg, $m_3 = 8$ kg, $k_1 = 400$ N/m, $k_2 = 650$ N/m, $k_3 = 225$ N/m, $c_1 = 44$ N·s/m, $c_2 = 36$ N·s/m, $c_3 = 52$ N·s/m, $a = 1.2$ m, $b = 1.8$ m, and $c = 0.9$ m.

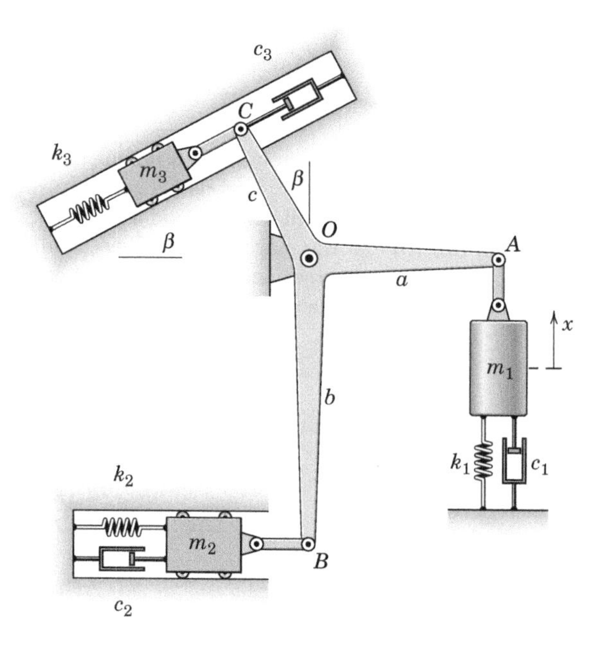

Problem 8/40

8/41 The owner of a 3400-lb pickup truck tests the action of his rear-wheel shock absorbers by applying a steady 100-lb force to the rear bumper and measuring a static deflection of 3 in. Upon sudden release of the force, the bumper rises and then falls to a maximum of $\frac{1}{2}$ in. below the unloaded equilibrium position of the bumper on the first rebound. Treat the action as a one-dimensional problem with an equivalent mass of half the truck mass. Find the viscous damping factor ζ for the rear end and the viscous damping coefficient c for each shock absorber assuming its action to be vertical.

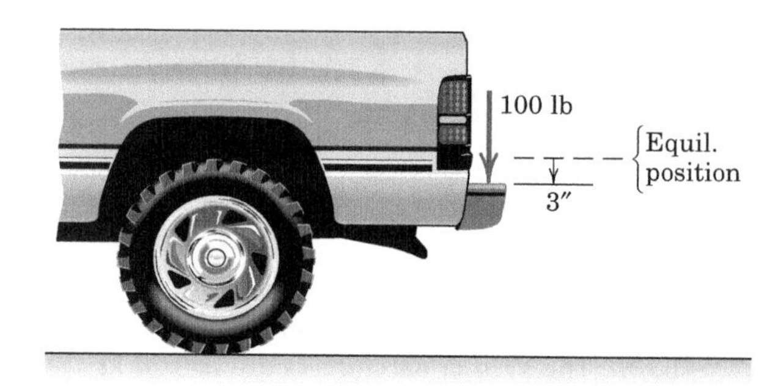

Problem 8/41

8/42 The 2-kg mass is released from rest at a distance x_0 to the right of the equilibrium position. Determine the displacement x as a function of time.

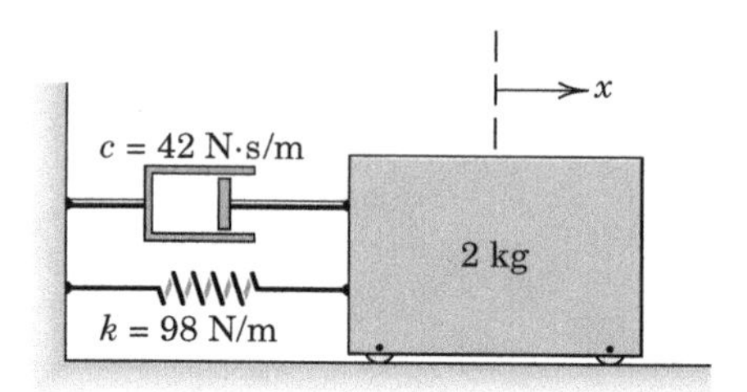

Problem 8/42

8/43 Develop the equation of motion in terms of the variable x for the system shown. Determine an expression for the damping ratio ζ in terms of the given system properties. Neglect the mass of the crank AB and assume small oscillations about the equilibrium position shown.

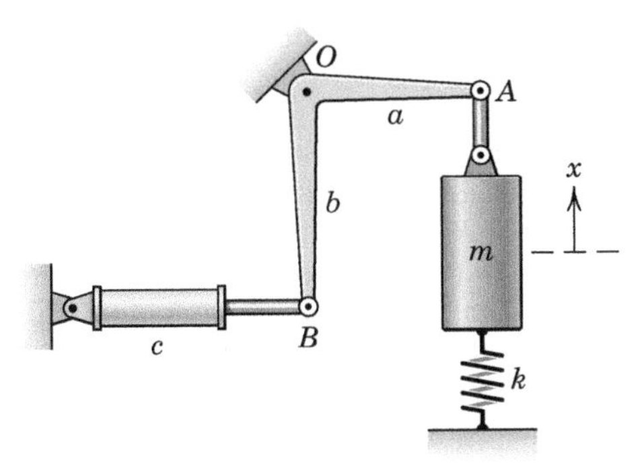

Problem 8/43

▶8/44 Investigate the case of Coulomb damping for the block shown, where the coefficient of kinetic friction is μ_k and each spring has a stiffness $k/2$. The block is displaced a distance x_0 from the neutral position and released. Determine and solve the differential equation of motion. Plot the resulting vibration and indicate the rate r of decay of the amplitude with time.

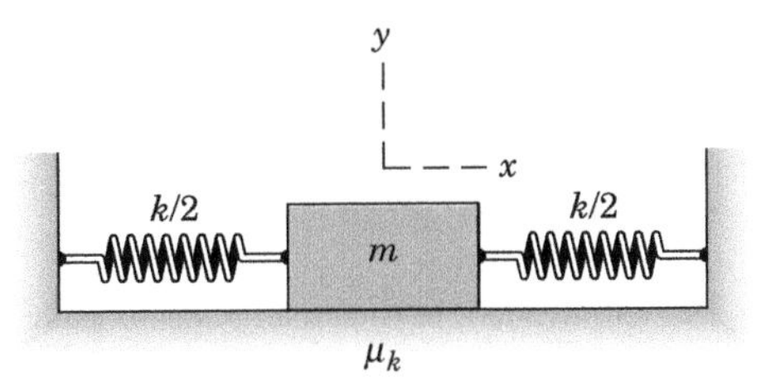

Problem 8/44

PROBLEMS

(Unless otherwise instructed, assume that the damping is light to moderate so that the amplitude of the forced response is a maximum at $\omega/\omega_n \cong 1$.)

Introductory Problems

8/45 A viscously damped spring-mass system is excited by a harmonic force of constant amplitude F_0 but varying frequency ω. If the amplitude of the steady-state motion is observed to decrease by a factor of 8 as the frequency ratio ω/ω_n is varied from 1 to 2, determine the damping ratio ζ of the system.

8/46 Determine the amplitude X of the steady-state motion of the 10-kg mass if (*a*) $c = 500$ N·s/m and (*b*) $c = 0$.

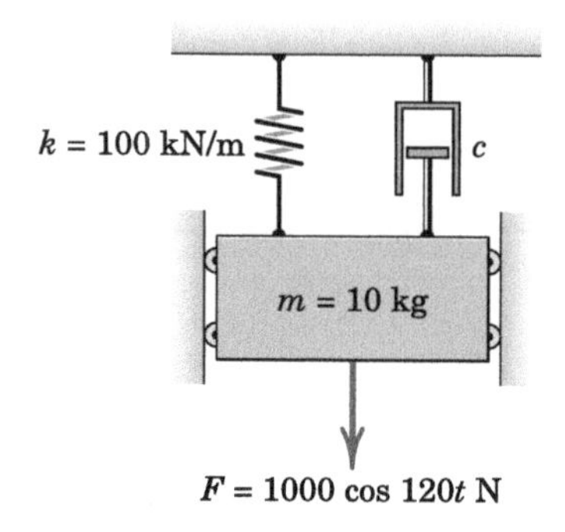

Problem 8/46

8/47 The 30-kg cart is acted upon by the harmonic force shown in the figure. If $c = 0$, determine the range of the driving frequency ω for which the magnitude of the steady-state response is less than 75 mm.

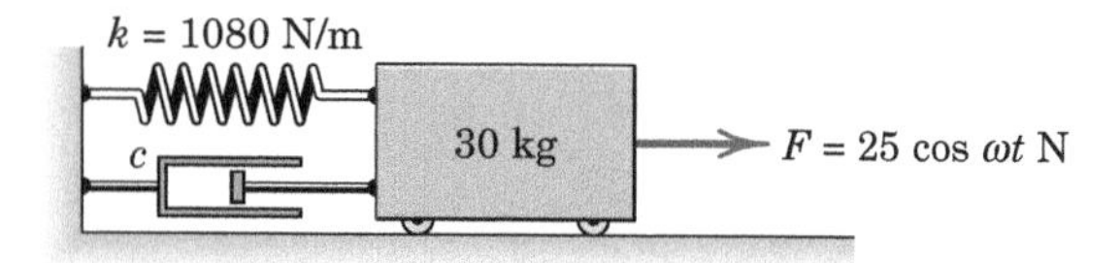

Problem 8/47

8/48 If the viscous damping coefficient of the damper in the system of Prob. 8/47 is $c = 36$ N·s/m, determine the range of the driving frequency ω for which the magnitude of the steady-state response is less than 75 mm.

8/49 If the driving frequency for the system of Prob. 8/47 is $\omega = 6$ rad/s, determine the required value of the damping coefficient c if the steady-state amplitude is not to exceed 75 mm.

8/50 A spring-mounted machine with a mass of 24 kg is observed to vibrate harmonically in the vertical direction with an amplitude of 0.30 mm under the action of a vertical force which varies harmonically between F_0 and $-F_0$ with a frequency of 4 Hz. Damping is negligible. If a static force of magnitude F_0 causes a deflection of 0.60 mm, calculate the equivalent spring constant k for the springs which support the machine.

8/51 The block of weight $W = 100$ lb is suspended by two springs each of stiffness $k = 200$ lb/ft and is acted upon by the force $F = 75 \cos 15t$ lb where t is the time in seconds. Determine the amplitude X of the steady-state motion if the viscous damping coefficient c is (*a*) 0 and (*b*) 60 lb-sec/ft. Compare these amplitudes to the static spring deflection δ_{st}.

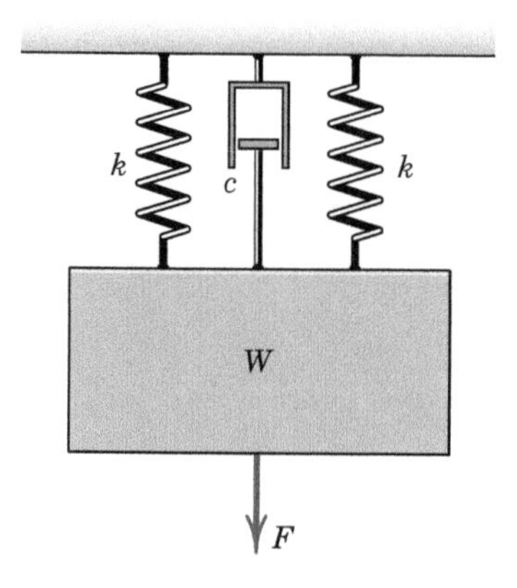

Problem 8/51

8/52 An external force $F = F_0 \sin \omega t$ is applied to the cylinder as shown. What value ω_c of the driving frequency would cause excessively large oscillations of the system?

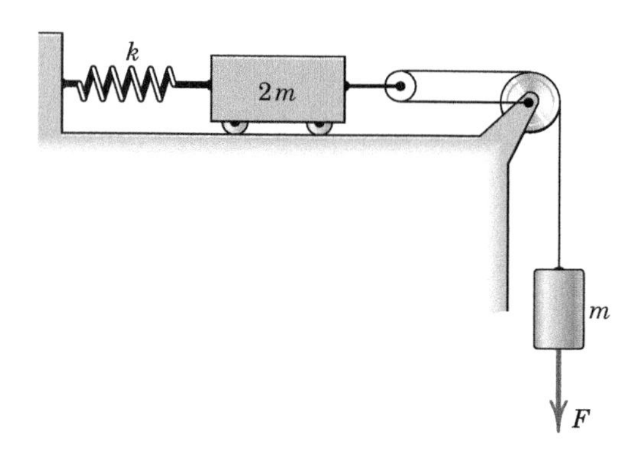

Problem 8/52

8/53 A viscously damped spring-mass system is forced harmonically at the undamped natural frequency ($\omega/\omega_n = 1$). If the damping ratio ζ is doubled from 0.1 to 0.2, compute the percentage reduction R_1 in the steady-state amplitude. Compare with the result R_2 of a similar calculation for the condition $\omega/\omega_n = 2$. Verify your results by inspecting Fig. 8/11.

8/54 The 4-lb body is attached to two springs, each of which has a stiffness of 6 lb/in. The body is mounted on a shake table which vibrates harmonically in the horizontal direction with an amplitude of 0.5 in. and a frequency f which can be varied. Power to the shake table is turned off when electrical contact is made at A or B. Determine the maximum value of the frequency f at which the shake table may be operated without turning the power off as it starts from rest and increases its frequency gradually. Damping may be neglected. The equilibrium position is centered between the fixed contacts.

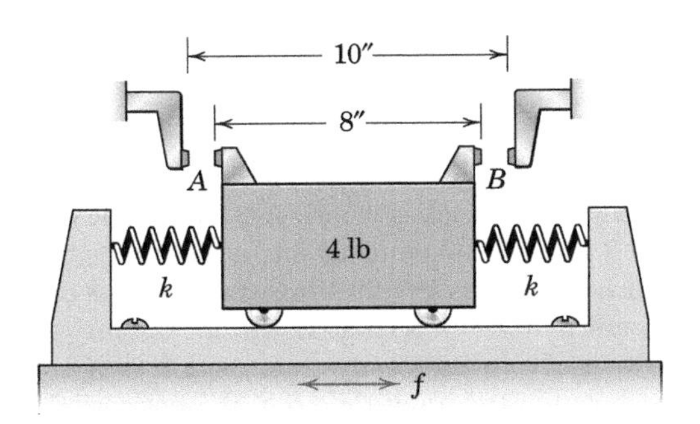

Problem 8/54

Representative Problems

8/55 It was noted in the text that the maxima of the curves for the magnification factor M are not located at $\omega/\omega_n = 1$. Determine an expression in terms of the damping ratio ζ for the frequency ratio at which the maxima occur.

8/56 The motion of the outer frame B is given by $x_B = b \sin \omega t$. For what range of the driving frequency ω is the amplitude of the motion of the mass m relative to the frame less than $2b$?

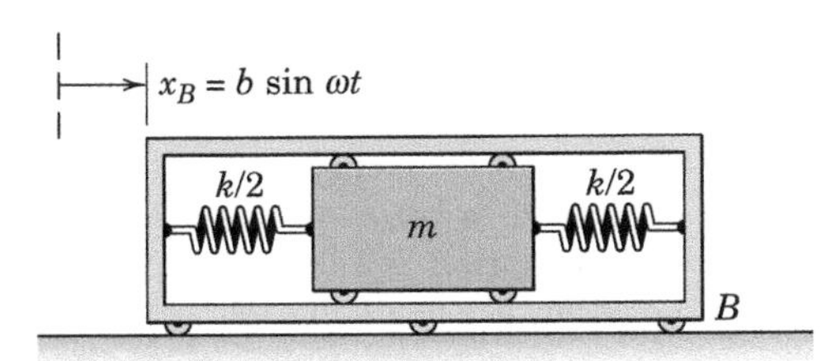

Problem 8/56

8/57 The 20-kg variable-speed motorized unit is restrained in the horizontal direction by two springs, each of which has a stiffness of 2.1 kN/m. Each of the two dashpots has a viscous damping coefficient $c = 58$ N·s/m. In what ranges of speeds N can the motor be run for which the magnification factor M will not exceed 2?

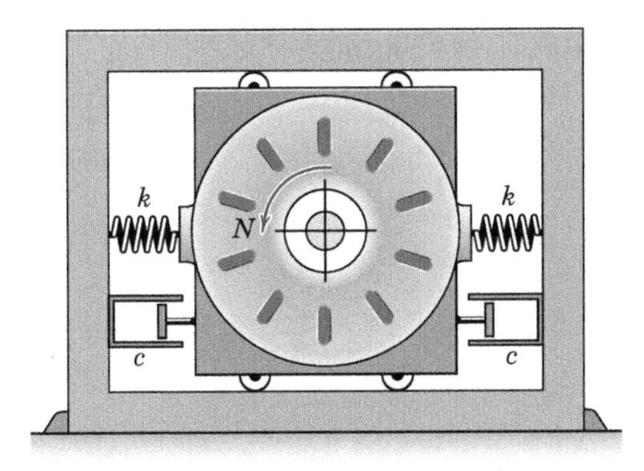

Problem 8/57

8/58 A single-cylinder four-stroke gasoline engine with a mass of 90 kg is mounted on four stiff spring pads, each with a stiffness of $30(10^3)$ kN/m, and is designed to run at 3600 rev/min. The mounting system is equipped with viscous dampers which have a large enough combined viscous damping coefficient c so that the system is critically damped when it is given a vertical displacement and then released while not running. When the engine is running, it fires every other revolution, causing a periodic vertical displacement modeled by 1.2 cos ωt mm with t in seconds. Determine the magnification factor M and the overall damping coefficient c.

8/59 When the person stands in the center of the floor system shown, he causes a static deflection δ_{st} of the floor under his feet. If he walks (or runs quickly!) in the same area, how many steps per second would cause the floor to vibrate with the greatest vertical amplitude?

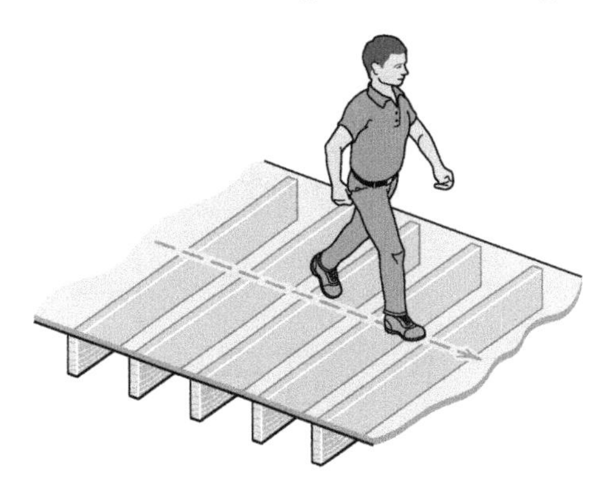

Problem 8/59

8/60 The instrument shown has a mass of 43 kg and is spring-mounted to the horizontal base. If the amplitude of vertical vibration of the base is 0.10 mm, calculate the range of frequencies f_n of the base vibration which must be prohibited if the amplitude of vertical vibration of the instrument is not to exceed 0.15 mm. Each of the four identical springs has a stiffness of 7.2 kN/m.

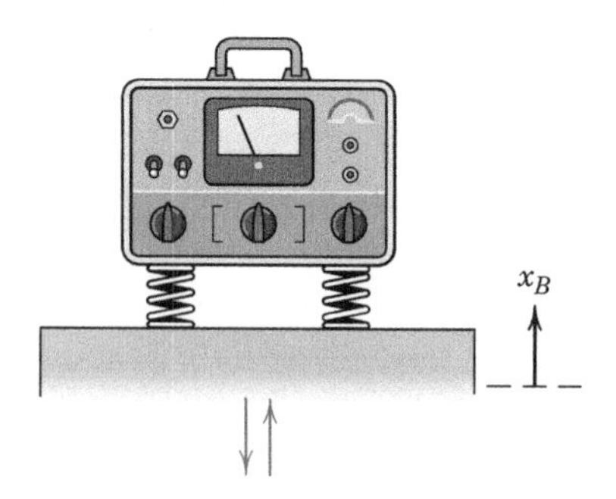

Problem 8/60

8/61 Derive the equation of motion for the *inertial* displacement x_i of the mass of Fig. 8/14. Comment on, but do not carry out, the solution to the equation of motion.

8/62 Attachment B is given a horizontal motion $x_B = b \cos \omega t$. Derive the equation of motion for the mass m and state the critical frequency ω_c for which the oscillations of the mass become excessively large.

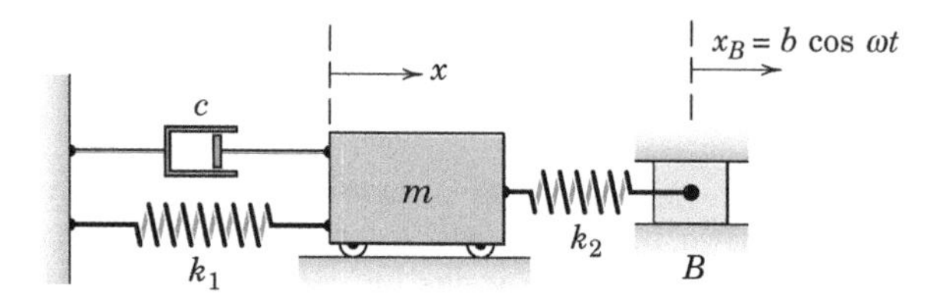

Problem 8/62

8/63 Attachment B is given a horizontal motion $x_B = b \cos \omega t$. Derive the equation of motion for the mass m and state the critical frequency ω_c for which the oscillations of the mass become excessively large. What is the damping ratio ζ for the system?

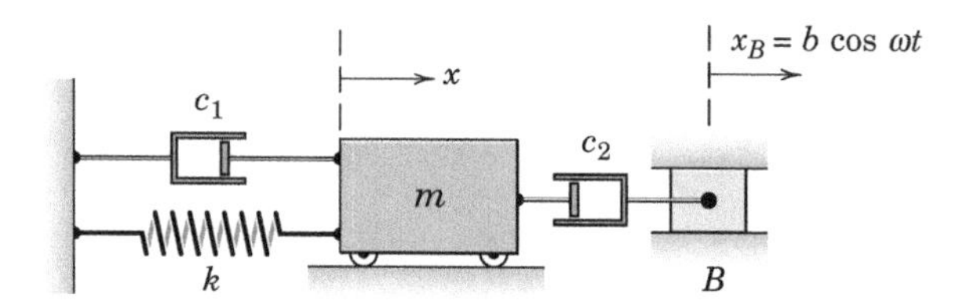

Problem 8/63

8/64 A device to produce vibrations consists of the two counter-rotating wheels, each carrying an eccentric mass $m_0 = 1$ kg with a center of mass at a distance $e = 12$ mm from its axis of rotation. The wheels are synchronized so that the vertical positions of the unbalanced masses are always identical. The total mass of the device is 10 kg. Determine the two possible values of the equivalent spring constant k for the mounting which will permit the amplitude of the periodic force transmitted to the fixed mounting to be 1500 N due to the imbalance of the rotors at a speed of 1800 rev/min. Neglect damping.

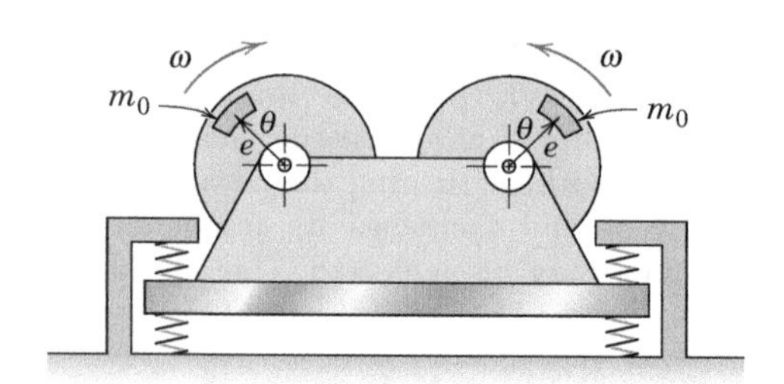

Problem 8/64

8/65 The seismic instrument shown is attached to a structure which has a horizontal harmonic vibration at 3 Hz. The instrument has a mass $m = 0.5$ kg, a spring stiffness $k = 20$ N/m, and a viscous damping coefficient $c = 3$ N·s/m. If the maximum recorded value of x in its steady-state motion is $X = 2$ mm, determine the amplitude b of the horizontal movement x_B of the structure.

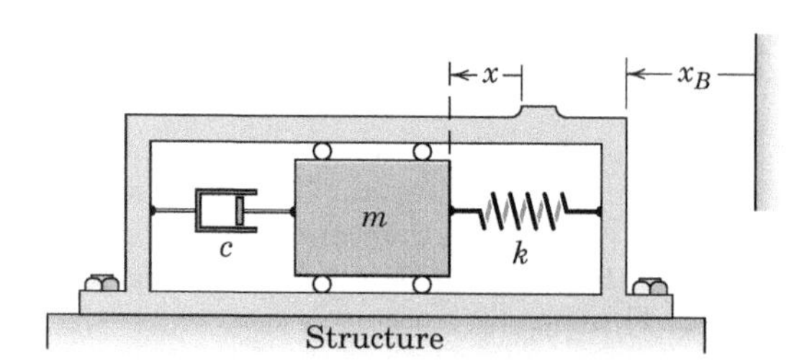

Problem 8/65

8/66 The equilibrium position of the mass m occurs where $y = 0$ and $y_B = 0$. When the attachment B is given a steady vertical motion $y_B = b \sin \omega t$, the mass m will acquire a steady vertical oscillation. Derive the differential equation of motion for m and specify the circular frequency ω_c for which the oscillations of m tend to become excessively large. The stiffness of the spring is k, and the mass and friction of the pulley are negligible.

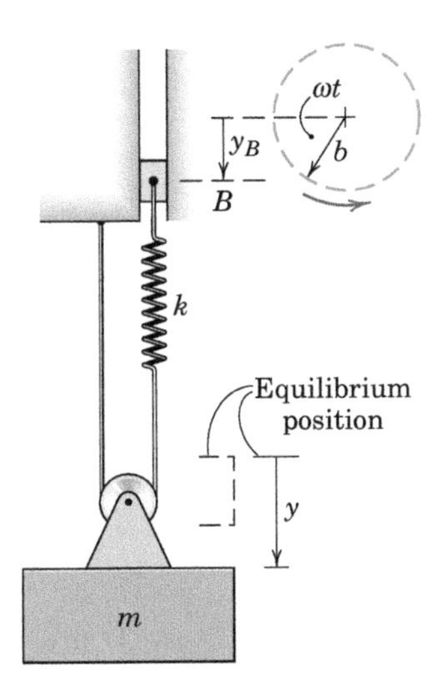

Problem 8/66

8/67 Derive and solve the equation of motion for the mass m in terms of the variable x for the system shown. Neglect the mass of the lever AOC and assume small oscillations.

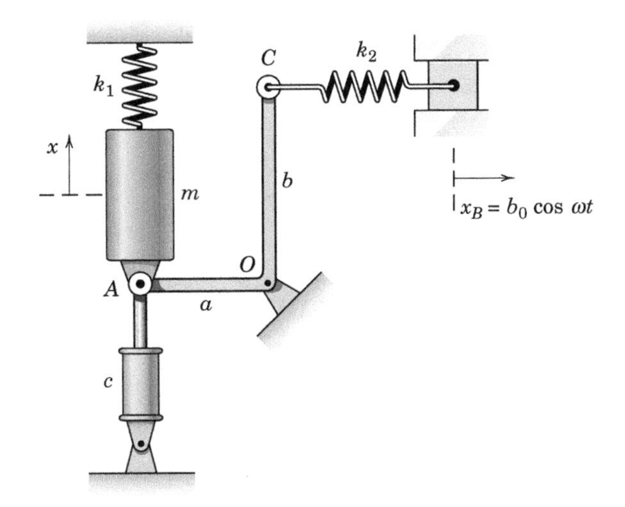

Problem 8/67

8/68 The seismic instrument is mounted on a structure which has a vertical vibration with a frequency of 5 Hz and a double amplitude of 18 mm. The sensing element has a mass $m = 2$ kg, and the spring stiffness is $k = 1.5$ kN/m. The motion of the mass relative to the instrument base is recorded on a revolving drum and shows a double amplitude of 24 mm during the steady-state condition. Calculate the viscous damping constant c.

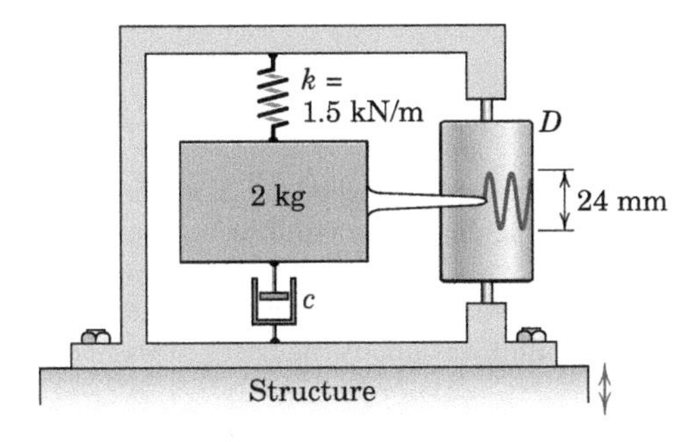

Problem 8/68

►8/69 Derive the expression for the energy loss E over a complete steady-state cycle due to the frictional dissipation of energy in a viscously damped linear oscillator. The forcing function is $F_0 \sin \omega t$, and the displacement-time relation for steady-state motion is $x_P = X \sin (\omega t - \phi)$ where the amplitude X is given by Eq. 8/20. (*Hint:* The frictional energy loss during a displacement dx is $c\dot{x}\,dx$, where c is the viscous damping coefficient. Integrate this expression over a complete cycle.)

►8/70 Determine the amplitude of vertical vibration of the car as it travels at a velocity $v = 40$ km/h over the wavy road whose contour may be expressed as a sine or cosine function with a double amplitude $2b = 50$ mm. The mass of the car is 1800 kg and the stiffness of each of the four car springs is 35 kN/m. Assume that all four wheels are in continuous contact with the road, and neglect damping. Note that the wheelbase of the car and the spatial period of the road are the same at $L = 3$ m, so that it may be assumed that the car translates but does not rotate. At what critical speed v_c is the vertical vibration of the car at its maximum?

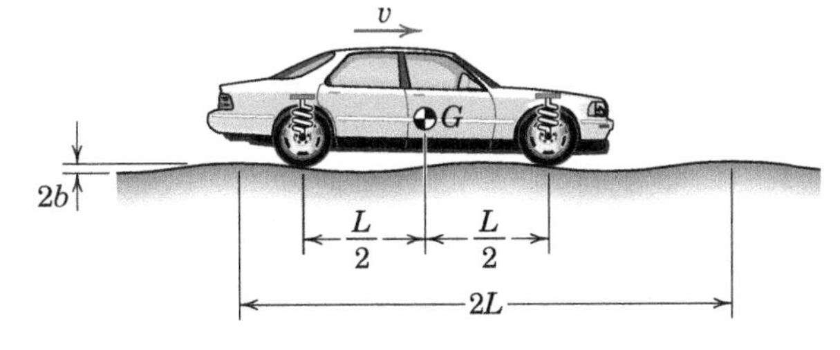

Problem 8/70

PROBLEMS

Introductory Problems

8/71 The light rod and attached small spheres of mass m each are shown in the equilibrium position, where all four springs are equally precompressed. Determine the natural frequency ω_n and period τ for small oscillations about the frictionless pivot O.

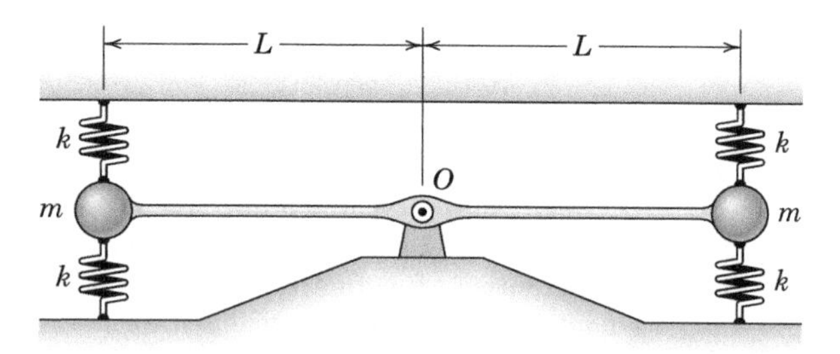

Problem 8/71

8/72 A uniform rectangular plate pivots about a horizontal axis through one of its corners as shown. Determine the natural frequency ω_n of small oscillations.

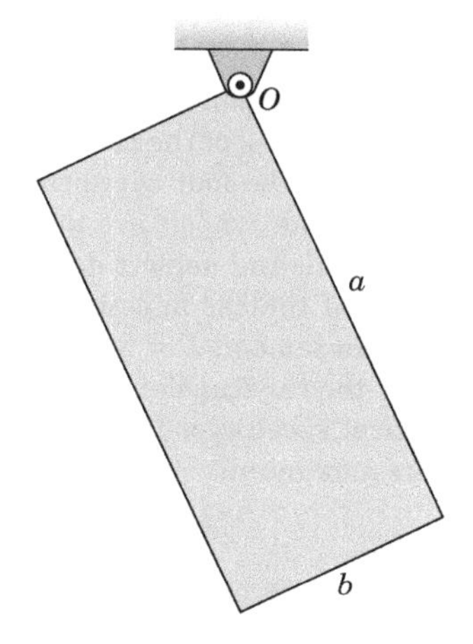

Problem 8/72

8/73 The thin square plate is suspended from a socket (not shown) which fits the small ball attachment at O. If the plate is made to swing about axis A-A, determine the period for small oscillations. Neglect the small offset, mass, and friction of the ball.

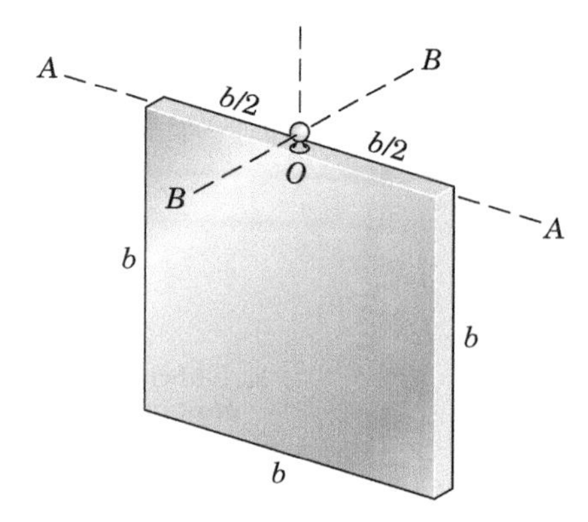

Problem 8/73

8/74 If the square plate of Prob. 8/73 is made to oscillate about axis B-B, determine the period of small oscillations.

8/75 The 20-lb spoked wheel has a centroidal radius of gyration $k = 6$ in. A torsional spring of constant $k_T = 160$ lb-ft/rad resists rotation about the smooth bearing. If an external torque of form $M = M_0 \cos \omega t$ is applied to the wheel, what is the magnitude of its steady-state angular displacement? The moment magnitude is $M_0 = 8$ lb-ft and the driving frequency is $\omega = 25$ rad/sec.

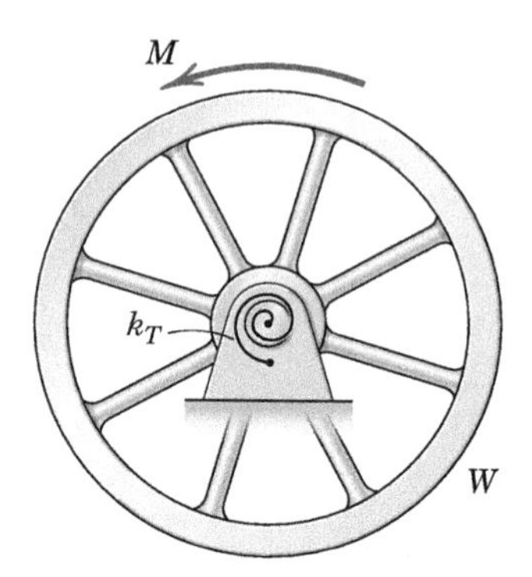

Problem 8/75

8/76 The uniform rod of length l and mass m is suspended at its midpoint by a wire of length L. The resistance of the wire to torsion is proportional to its angle of twist θ and equals $(JG/L)\theta$ where J is the polar moment of inertia of the wire cross section and G is the shear modulus of elasticity. Derive the expression for the period τ of oscillation of the bar when it is set into rotation about the axis of the wire.

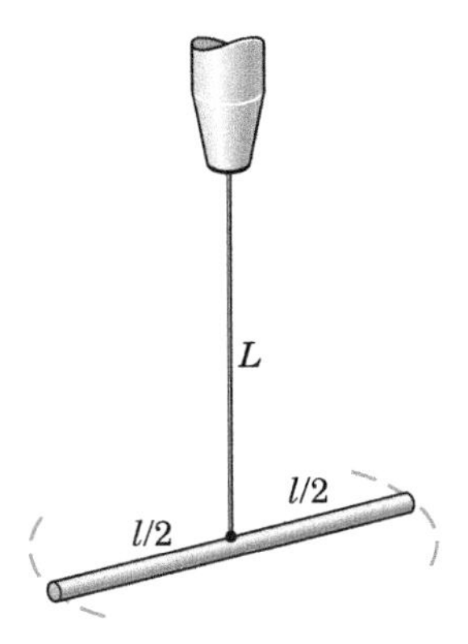

Problem 8/76

8/77 The uniform sector has mass m and is freely hinged about a horizontal axis through point O. Determine the equation of motion of the sector for large-amplitude vibrations about the equilibrium position. State the period τ for small oscillations about the equilibrium position if $r = 325$ mm and $\beta = 45°$.

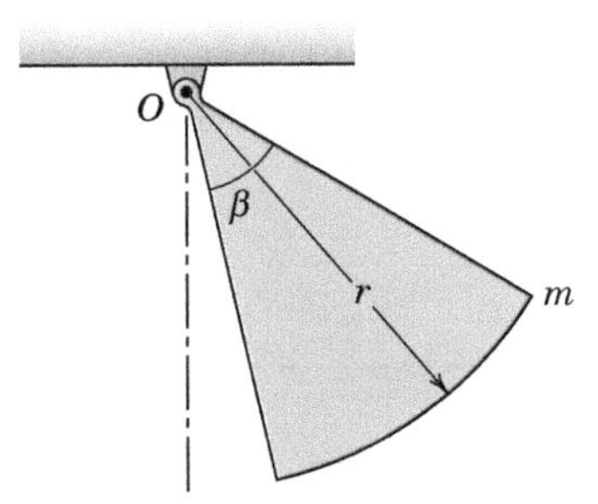

Problem 8/77

8/78 The assembly of mass m is formed from uniform and slender welded rods and is freely hinged about a horizontal axis through O. Determine the equation of motion of the assembly for large-amplitude vibrations about the equilibrium position. State the period τ for small oscillations about the equilibrium position if $r = 325$ mm and $\beta = 45°$.

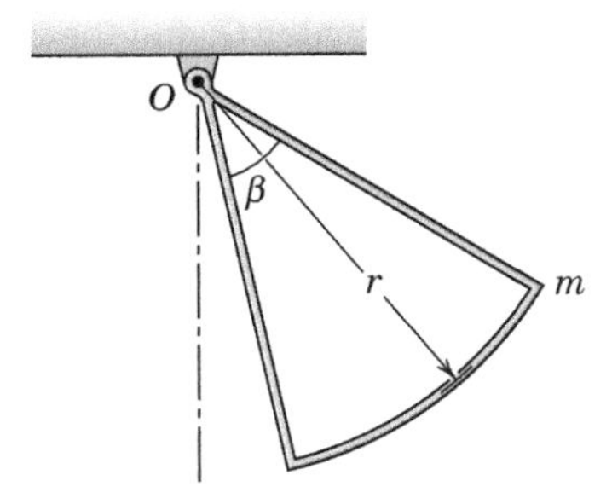

Problem 8/78

8/79 The thin-walled cylindrical shell of radius r and height h is welded to the small shaft at its upper end as shown. Determine the natural circular frequency ω_n for small oscillations of the shell about the y-axis.

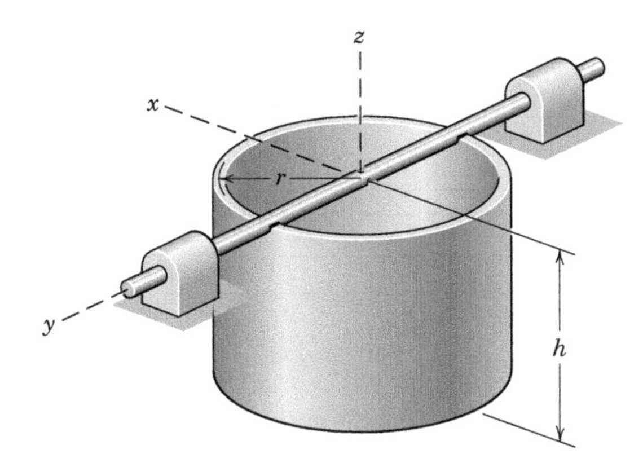

Problem 8/79

8/80 Determine the system equation of motion in terms of the variable θ shown in the figure. Assume small angular motion of bar OA, and neglect the mass of link CD.

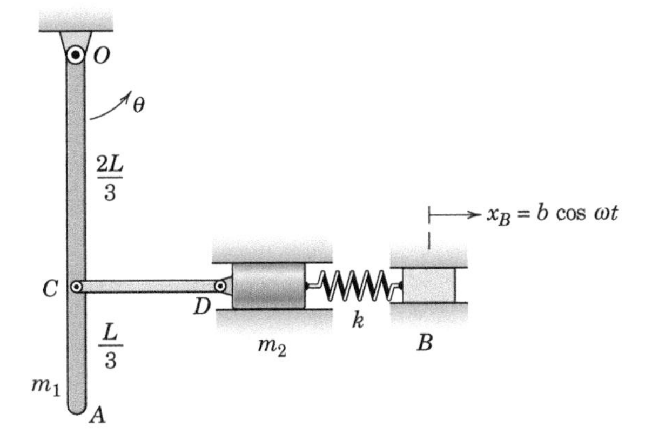

Problem 8/80

8/81 The uniform rod of mass m is freely pivoted about a horizontal axis through point O. Assume small oscillations and determine an expression for the damping ratio ζ. For what value c_{cr} of the damping coefficient c will the system be critically damped?

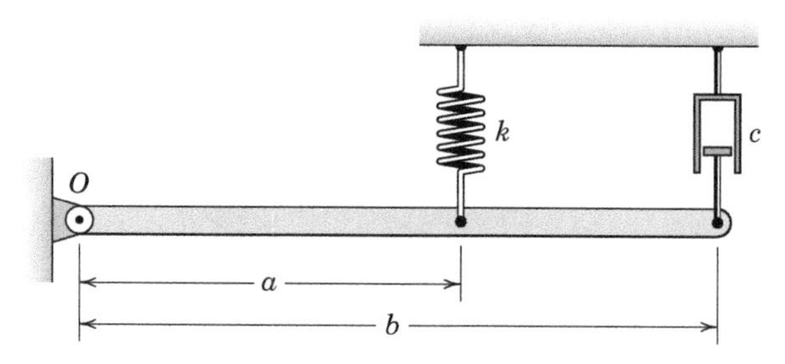

Problem 8/81

8/82 The mass of the uniform slender rod is 3 kg. Determine the position x for the 1.2-kg slider such that the system period is 1 s. Assume small oscillations about the horizontal equilibrium position shown.

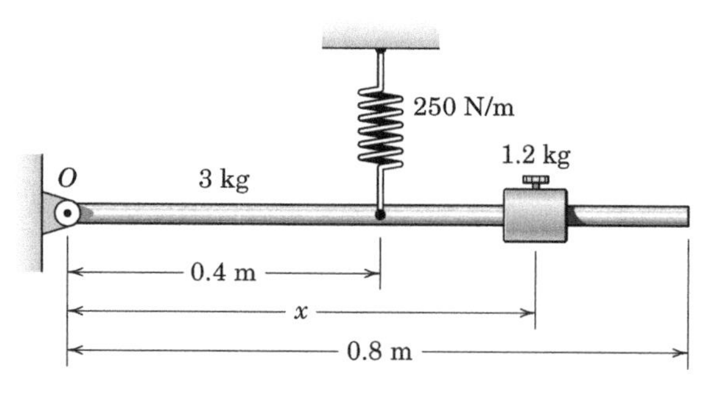

Problem 8/82

Representative Problems

8/83 The triangular frame of mass m is formed from uniform slender rod and is suspended from a socket (not shown) which fits the small ball attachment at O. If the frame is made to swing about axis A-A, determine the natural circular frequency ω_n for small oscillations. Neglect the small offset, mass, and friction of the ball. Evaluate for $l = 200$ mm.

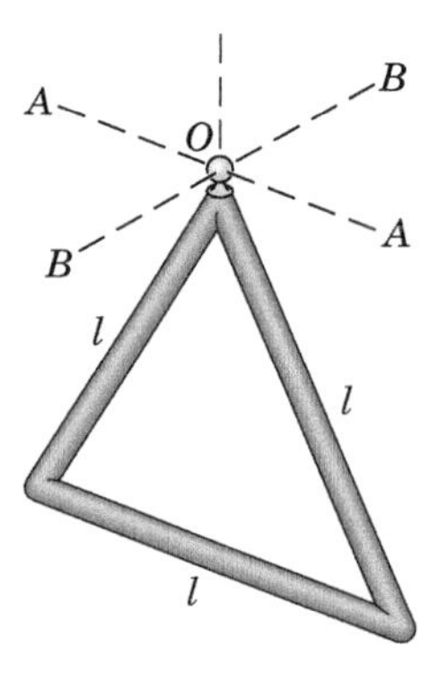

Problem 8/83

8/84 If the triangular frame of Prob. 8/83 is made to oscillate about axis B-B, determine the natural circular frequency ω_n for small oscillations. Evaluate for $l = 200$ mm and compare your answer with that of Prob. 8/83.

8/85 The uniform rod of mass m is freely pivoted about point O. Assume small oscillations and determine an expression for the damping ratio ζ. For what value c_{cr} of the damping coefficient c will the system be critically damped?

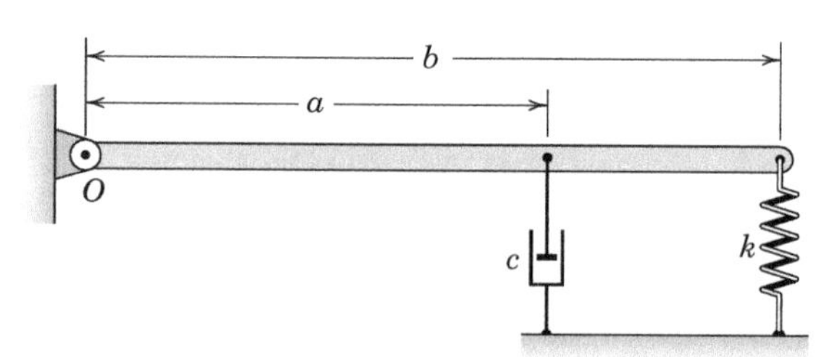

Problem 8/85

8/86 The mechanism shown oscillates in the vertical plane about the pivot O. The springs of equal stiffness k are both compressed in the equilibrium position $\theta = 0$. Determine an expression for the period τ of small oscillations about O. The mechanism has a mass m with mass center G, and the radius of gyration of the assembly about O is k_O.

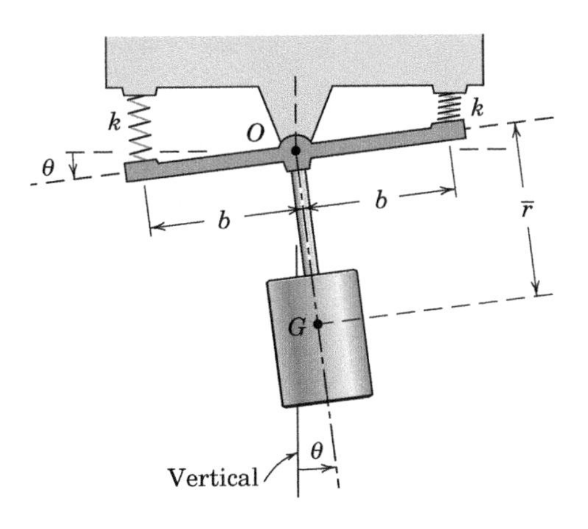

Problem 8/86

8/87 When the motor is slowly brought up to speed, a rather large vibratory oscillation of the entire motor about *O-O* occurs at a speed of 360 rev/min, which shows that this speed corresponds to the natural frequency of free oscillation of the motor. If the motor has a mass of 43 kg and radius of gyration of 100 mm about *O-O*, determine the stiffness k of each of the four identical spring mounts.

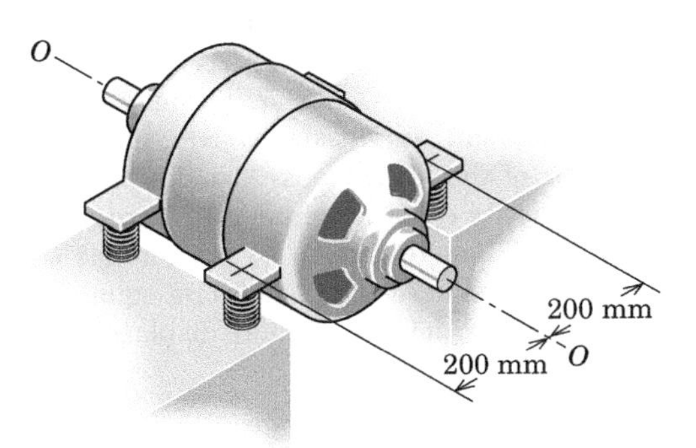

Problem 8/87

8/88 The system of Prob. 8/35 is repeated here with the added information that link *AOB* now has mass m_3 and radius of gyration k_O about point O. Ignore friction and derive the differential equation of motion for the system shown in terms of the variable x_1.

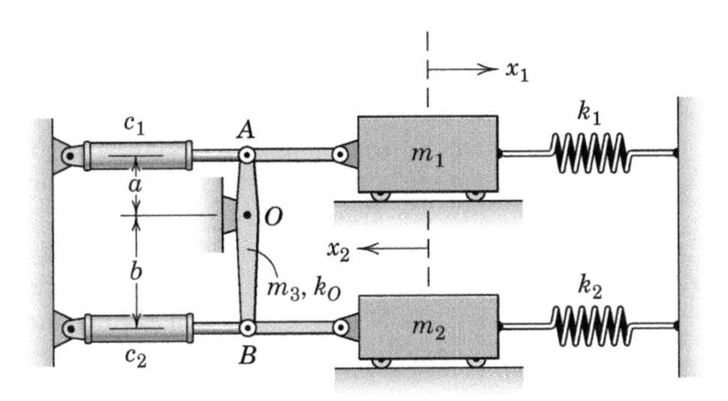

Problem 8/88

8/89 Determine the value m_{eff} of the mass of system (b) so that the frequency of system (b) is equal to that of system (a). Note that the two springs are identical and that the wheel of system (a) is a solid homogeneous cylinder of mass m_2. The cord does not slip on the cylinder.

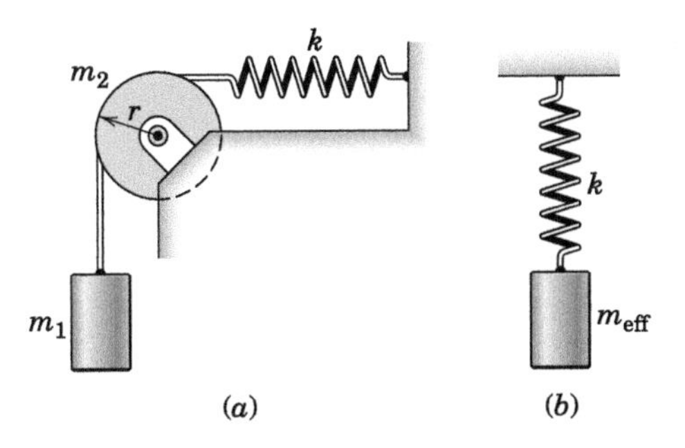

Problem 8/89

8/90 The system of Prob. 8/43 is repeated here. If the crank AB now has mass m_2 and a radius of gyration k_O about point O, determine expressions for the undamped natural frequency ω_n and the damping ratio ζ in terms of the given system properties. Assume small oscillations. The damping coefficient for the damper is c.

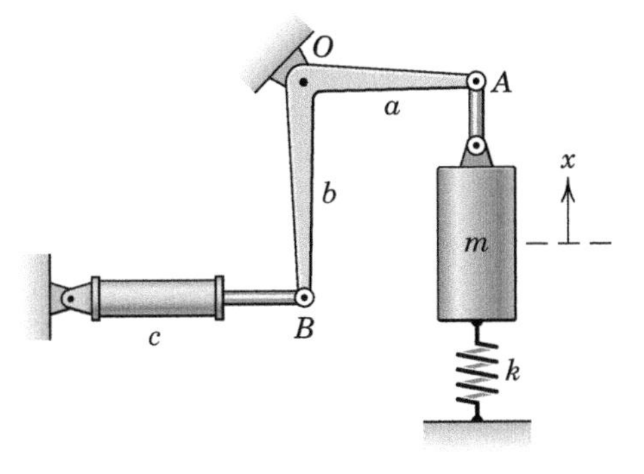

Problem 8/90

8/91 The two masses are connected by an inextensible cable which passes securely over the periphery of the cylindrical pulley of mass m_3, radius r, and radius of gyration k_O. Determine the equation of motion for the system in terms of the variable x. State the critical driving frequency ω_c of the block B which will result in excessively large oscillations of the assembly. Evaluate for $m_1 = 25$ kg, $m_2 = 10$ kg, $m_3 = 15$ kg, $k_1 = 450$ N/m, $k_2 = 300$ N/m, $r = 300$ mm, and $k_O = 200$ mm.

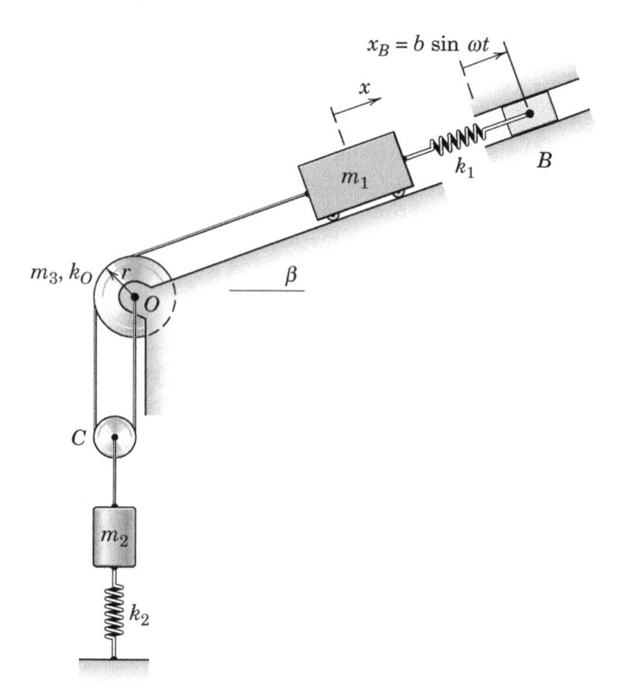

Problem 8/91

8/92 The lower spring of Prob. 8/91 is replaced by a damper. Determine the equation of motion for the system in terms of the variable x. State the value of the viscous damping coefficient c which will give a damping ratio $\zeta = 0.2$. Evaluate for $m_1 = 25$ kg, $m_2 = 10$ kg, $m_3 = 15$ kg, $k = 450$ N/m, $r = 300$ mm, and $k_O = 200$ mm.

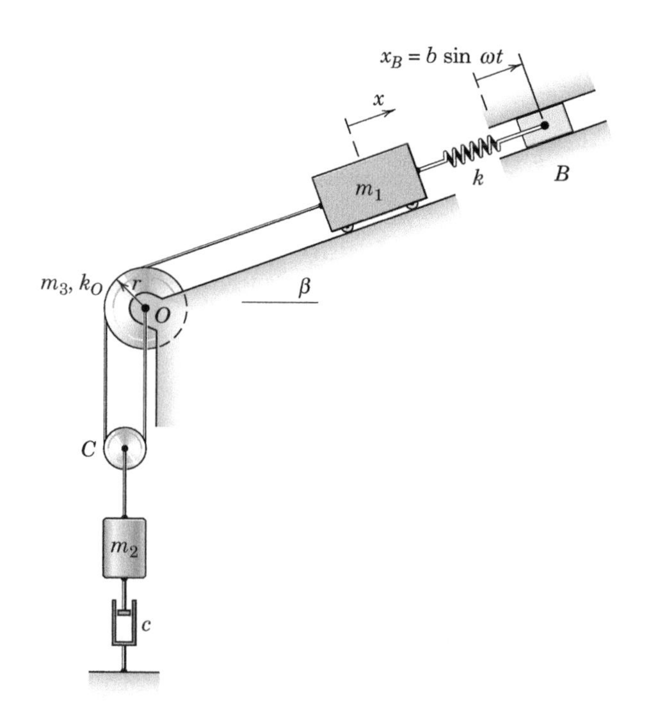

Problem 8/92

8/93 The uniform solid cylinder of mass m and radius r rolls without slipping during its oscillation on the circular surface of radius R. If the motion is confined to small amplitudes $\theta = \theta_0$, determine the period τ of the oscillations. Also determine the angular velocity ω of the cylinder as it crosses the vertical centerline. (*Caution:* Do not confuse ω with $\dot{\theta}$ or with ω_n as used in the defining equations. Note also that θ is not the angular displacement of the cylinder.)

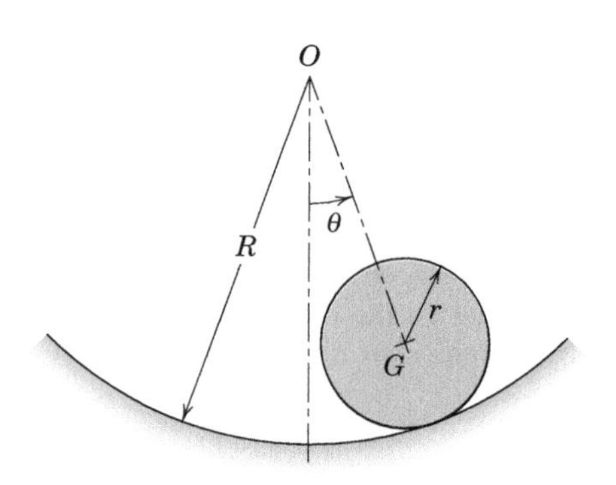

Problem 8/93

8/94 The cart B is given the harmonic displacement $x_B = b \sin \omega t$. Determine the steady-state amplitude Θ of the periodic oscillation of the uniform slender bar which is pinned to the cart at P. Assume small angles and neglect friction at the pivot. The torsional spring is undeformed when $\theta = 0$.

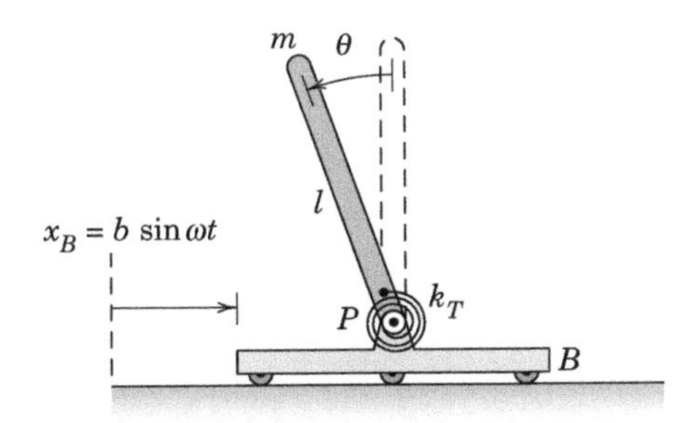

Problem 8/94

▶**8/95** The assembly of Prob. 8/40 is repeated here with the additional information that body ABC now has mass m_4 and a radius of gyration k_O about its pivot at O, about which it is balanced. If a harmonic torque $M = M_0 \cos \omega t$ is applied to body ABC, determine the equation of motion for the system in terms of the variable x. State the critical frequency ω_c of the harmonic torque which will result in an excessively large system response. Evaluate ω_c for $m_1 = 15$ kg, $m_2 = 12$ kg, $m_3 = 8$ kg, $m_4 = 6$ kg, $k_1 = 400$ N/m, $k_2 = 650$ N/m, $k_3 = 225$ N/m, $c_1 = 44$ N·s/m, $c_2 = 36$ N·s/m, $c_3 = 52$ N·s/m, $a = 1.2$ m, $b = 1.8$ m, $c = 0.9$ m, and $k_O = 0.75$ m. What is the damping ratio ζ for these conditions?

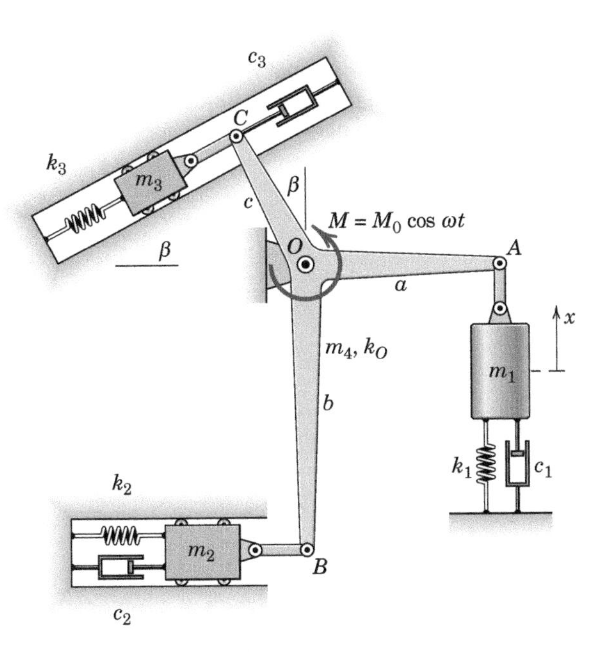

Problem 8/90

▶**8/96** The elements of the "swing-axle" type of independent rear suspension for automobiles are depicted in the figure. The differential D is rigidly attached to the car frame. The half-axles are pivoted at their inboard ends (point O for the half-axle shown) and are rigidly attached to the wheels. Suspension elements not shown constrain the wheel motion to the plane of the figure. The weight of the wheel–tire assembly is $W = 100$ lb, and its mass moment of inertia about a diametral axis passing through its mass center G is 1 lb-ft-sec^2. The weight of the half-axle is negligible. The spring rate and shock-absorber damping coefficient are $k = 50$ lb/in. and $c = 200$ lb-sec/ft, respectively. If a static tire imbalance is present, as represented by the additional concentrated weight $w = 0.5$ lb as shown, determine the angular velocity ω which results in the suspension system being driven at its undamped natural frequency. What would be the corresponding vehicle speed v? Determine the damping ratio ζ. Assume small angular deflections and neglect gyroscopic effects and any car frame vibration. In order to avoid the complications associated with the varying normal force exerted by the road on the tire, treat the vehicle as being on a lift with the wheels hanging free.

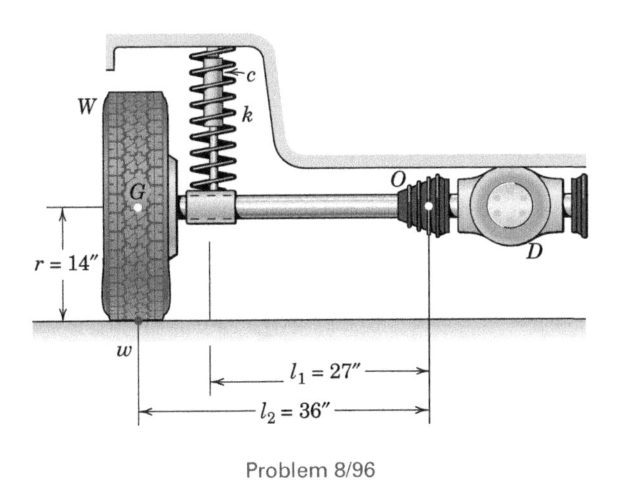

Problem 8/96

PROBLEMS

(Solve the following problems by the energy method of Art. 8/5.)

Introductory Problems

8/97 The 1.5-kg bar OA is suspended vertically from the bearing O and is constrained by the two springs each of stiffness $k = 120$ N/m and both equally precompressed with the bar in the vertical equilibrium position. Treat the bar as a uniform slender rod and compute the natural frequency f_n of small oscillations about O.

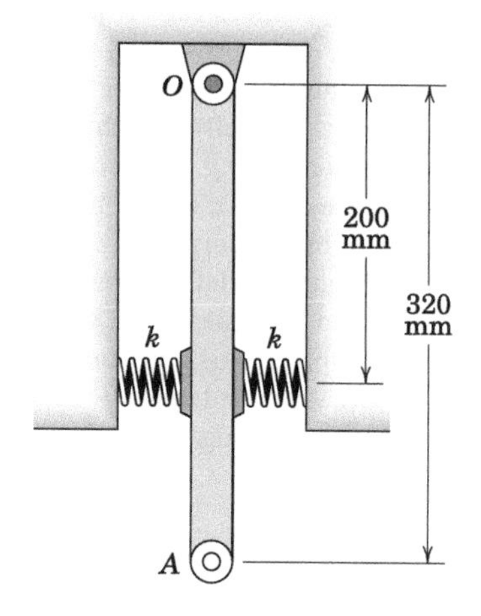

Problem 8/97

8/98 The light rod and attached sphere of mass m are at rest in the horizontal position shown. Determine the period τ for small oscillations in the vertical plane about the pivot O.

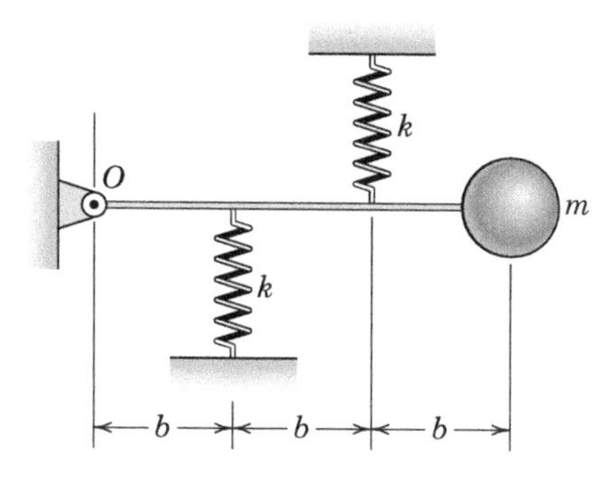

Problem 8/98

8/99 A uniform rod of mass m and length l is welded at one end to the rim of a light circular hoop of radius l. The other end lies at the center of the hoop. Determine the period τ for small oscillations about the vertical position of the bar if the hoop rolls on the horizontal surface without slipping.

Problem 8/99

8/100 The spoked wheel of radius r, mass m, and centroidal radius of gyration $\bar{k}$ rolls without slipping on the incline. Determine the natural frequency of oscillation and explore the limiting cases of $\bar{k} = 0$ and $\bar{k} = r$.

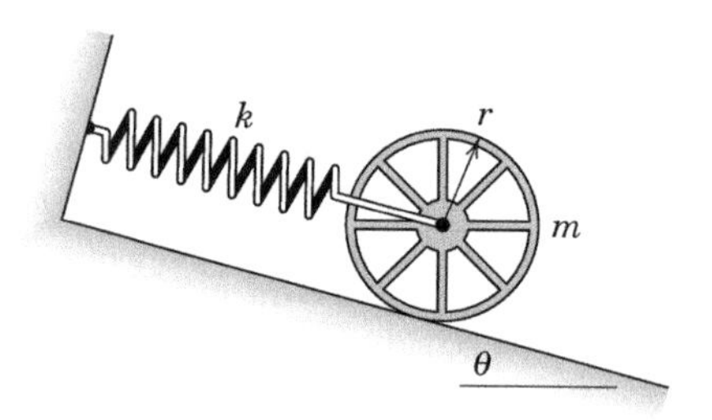

Problem 8/100

8/101 Determine the period τ for the uniform circular hoop of radius r as it oscillates with small amplitude about the horizontal knife edge.

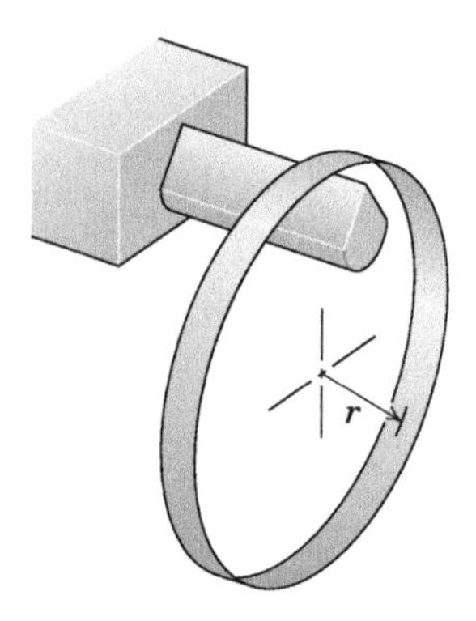

Problem 8/101

8/102 The length of the spring is adjusted so that the equilibrium position of the arm is horizontal as shown. Neglect the mass of the spring and the arm and calculate the natural frequency f_n for small oscillations.

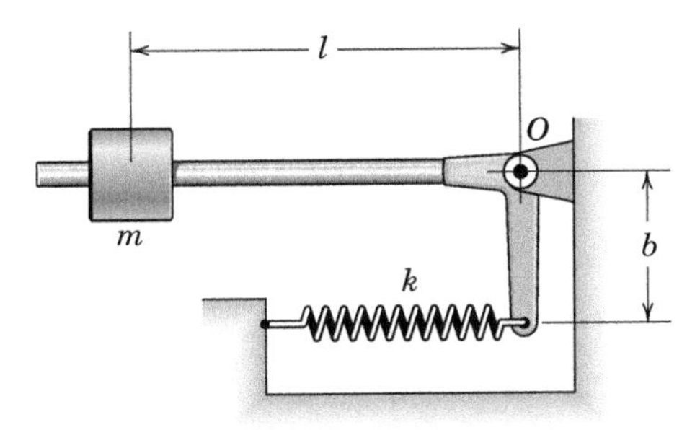

Problem 8/102

8/103 The body consists of two slender uniform rods which have a mass ρ per unit length. The rods are welded together and pivot about a horizontal axis through O against the action of a torsional spring of stiffness k_T. By the method of this article, determine the natural circular frequency ω_n for small oscillations about the equilibrium position. The spring is undeformed when $\theta = 0$, and friction in the pivot at O is negligible.

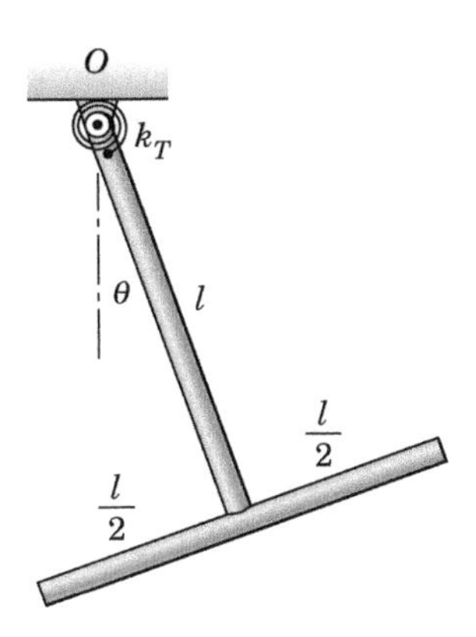

Problem 8/103

8/104 By the method of this article, determine the period of vertical oscillation. Each spring has a stiffness of 6 lb/in., and the mass of the pulleys may be neglected.

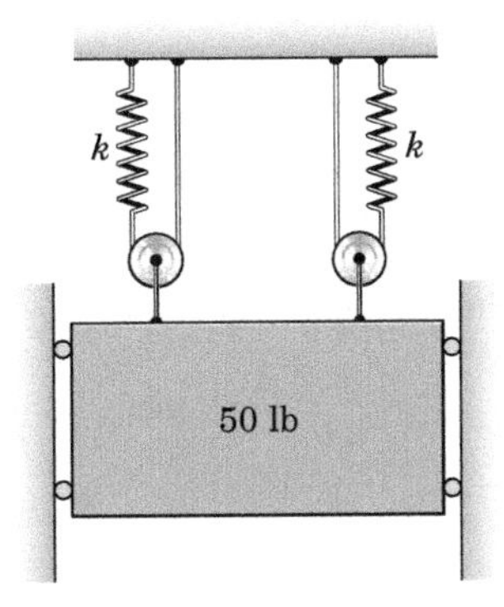

Problem 8/104

Representative Problems

8/105 The homogeneous circular cylinder of Prob. 8/93, repeated here, rolls without slipping on the track of radius R. Determine the period τ for small oscillations.

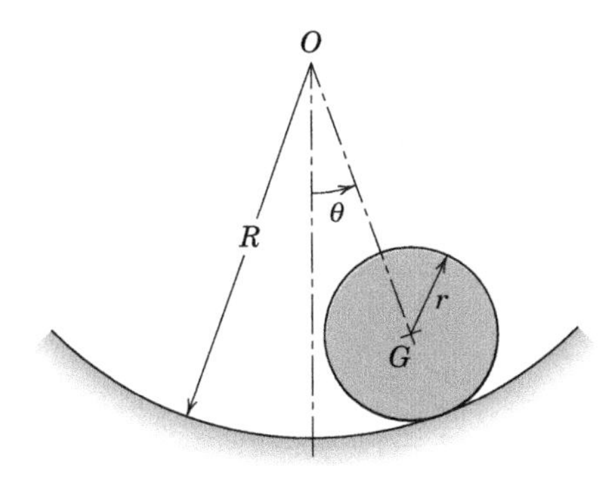

Problem 8/105

8/106 The disk has mass moment of inertia I_O about O and is acted upon by a torsional spring of constant k_T. The position of the small sliders, each of which has mass m, is adjustable. Determine the value of x for which the system has a given period τ.

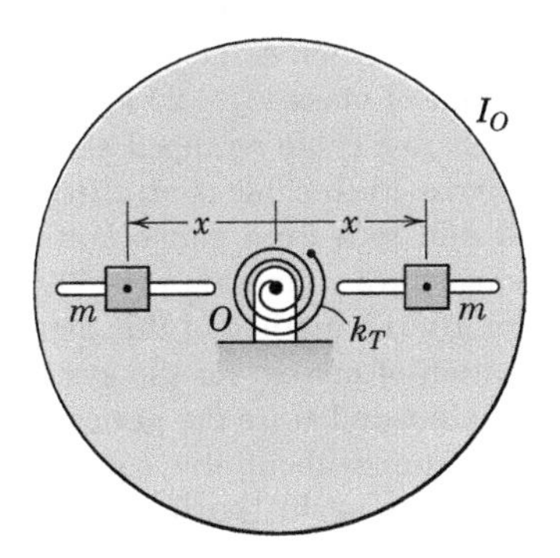

Problem 8/106

8/107 The uniform slender rod of length l and mass m_2 is secured to the uniform disk of radius $l/5$ and mass m_1. If the system is shown in its equilibrium position, determine the natural frequency ω_n and the maximum angular velocity ω for small oscillations of amplitude θ_0 about the pivot O.

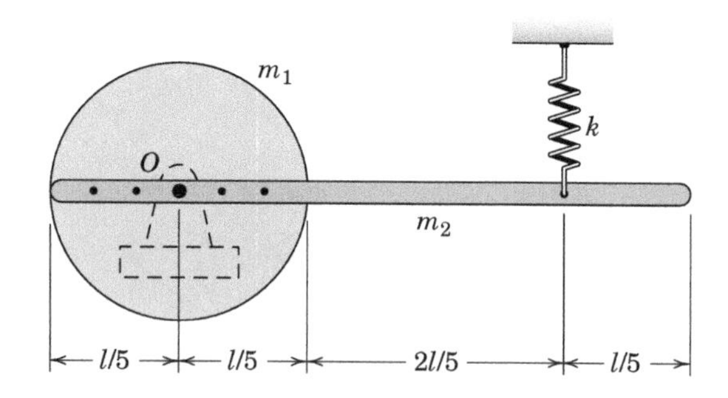

Problem 8/107

8/108 The assembly shown consists of two sheaves of mass $m_1 = 35$ kg and $m_2 = 15$ kg, outer groove radii $r_1 = 525$ mm and $r_2 = 250$ mm, and centroidal radii of gyration $(k_O)_1 = 350$ mm and $(k_O)_2 = 150$ mm. The sheaves are fitted to a central shaft at O with bearings which allow them to rotate independently of each other. Attached to the central shaft is a carriage of mass $m_3 = 25$ kg. Each sheave has an inextensible cable wrapped securely within its outer groove. Each cable is attached to a spring at one end and to a fixed support at the other end. The springs have stiffnesses $k_1 = 800$ N/m and $k_2 = 650$ N/m. By the method of this article, determine the equation of motion for the system in terms of the variable x and state the period τ for small vertical oscillations about the equilibrium position. Neglect friction in the bearings at O.

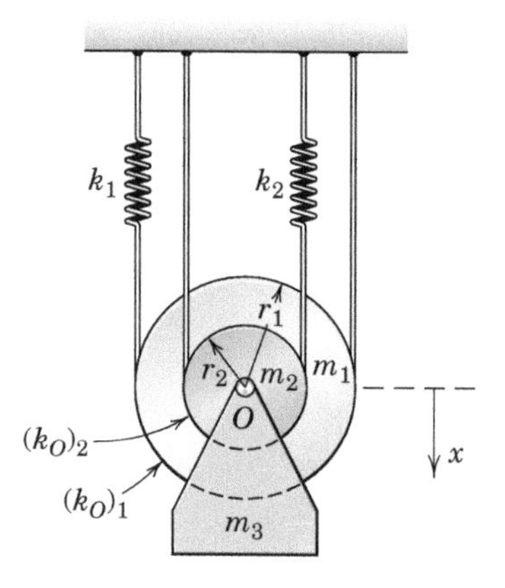

Problem 8/108

8/109 Derive the natural frequency f_n of the system composed of two homogeneous circular cylinders, each of mass M, and the connecting link AB of mass m. Assume small oscillations.

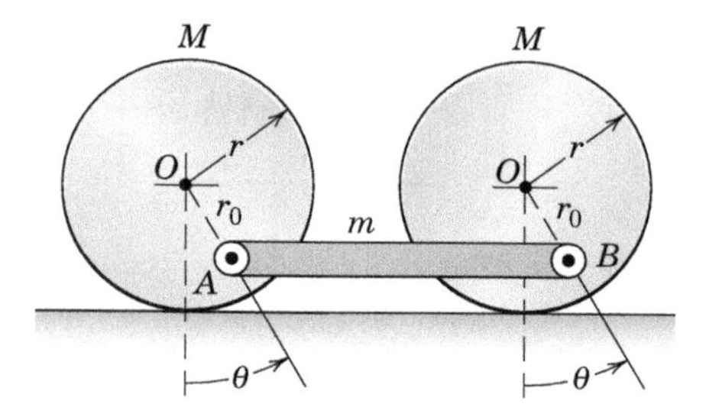

Problem 8/109

8/110 The rotational axis of the turntable is inclined at an angle α from the vertical. The turntable shaft pivots freely in bearings which are not shown. If a small block of mass m is placed a distance r from point O, determine the natural frequency ω_n for small rotational oscillations through the angle θ. The mass moment of inertia of the turntable about the axis of its shaft is I.

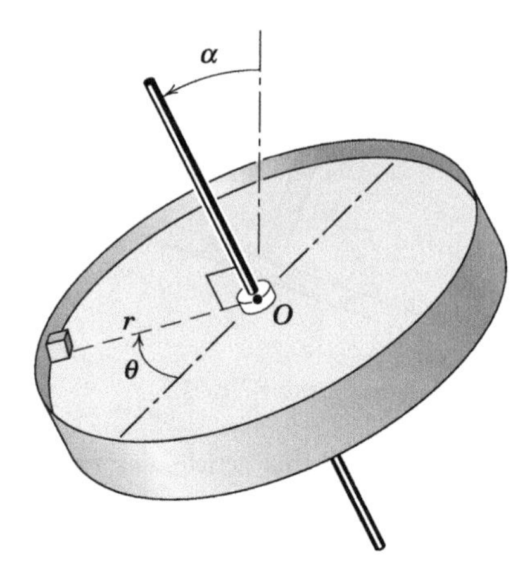

Problem 8/110

8/111 The assembly of Prob. 8/95 is repeated here without the applied harmonic torque. By the method of this article, determine the equation of motion of the system in terms of x for small oscillations about the equilibrium configuration if $c_1 = c_2 = c_3 = 0$.

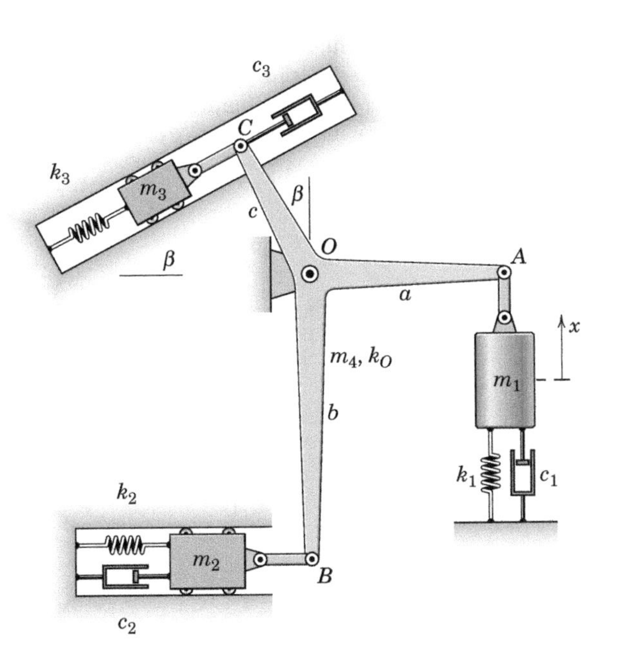

Problem 8/111

8/112 The ends of the uniform bar of mass m slide freely in the vertical and horizontal slots as shown. If the bar is in static equilibrium when $\theta = 0$, determine the natural frequency ω_n of small oscillations. What condition must be imposed on the spring constant k in order that oscillations take place?

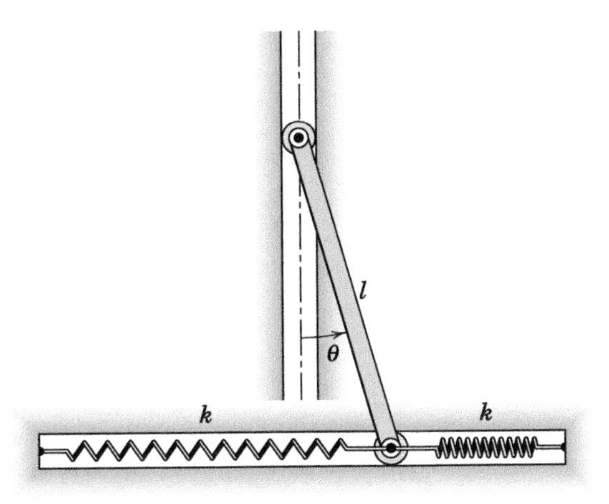

Problem 8/112

8/113 The 12-kg block is supported by the two 5-kg links with two torsion springs, each of constant $k_T = 500$ N·m/rad, arranged as shown. The springs are sufficiently stiff so that stable equilibrium is established in the position shown. Determine the natural frequency f_n for small oscillations about this equilibrium position.

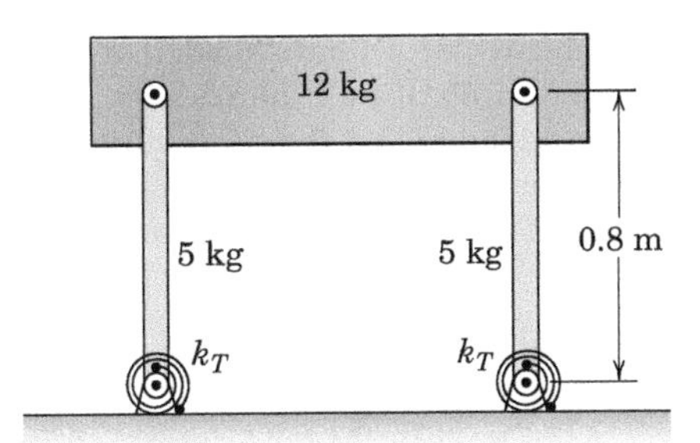

Problem 8/113

8/114 The front-end suspension of an automobile is shown. Each of the coil springs has a stiffness of 270 lb/in. If the weight of the front-end frame and equivalent portion of the body attached to the front end is 1800 lb, determine the natural frequency f_n of vertical oscillation of the frame and body in the absence of shock absorbers. (*Hint:* To relate the spring deflection to the deflection of the frame and body, consider the frame fixed and let the ground and wheels move vertically.)

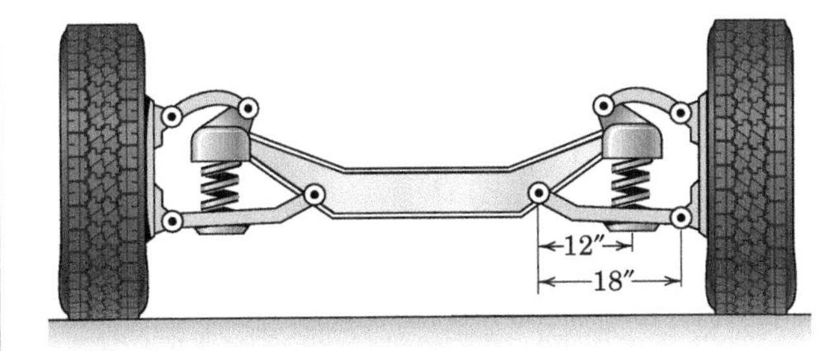

Problem 8/114

8/115 The system shown features a nonlinear spring whose resisting force F increases with deflection from the neutral position according to the graph shown. Determine the equation of motion for the system by the method of this article.

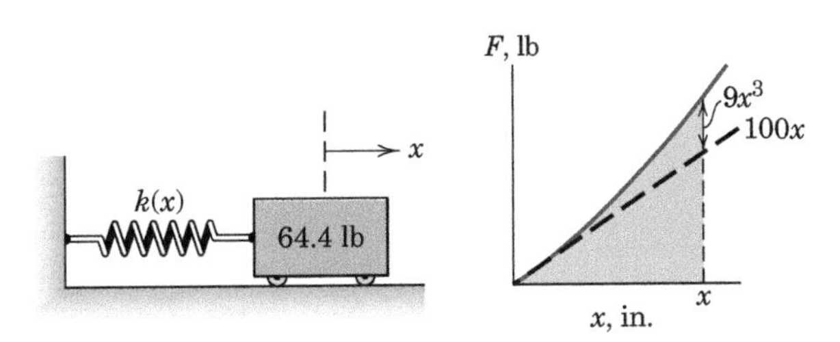

Problem 8/115

8/116 The semicircular cylindrical shell of radius r with small but uniform wall thickness is set into small rocking oscillation on the horizontal surface. If no slipping occurs, determine the expression for the period τ of each complete oscillation.

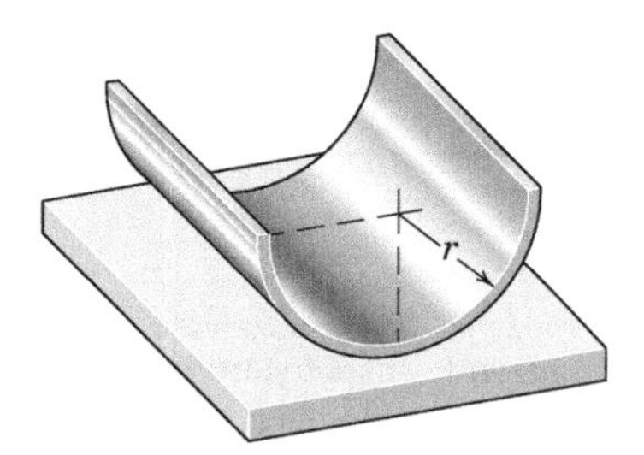

Problem 8/116

►**8/117** A hole of radius $R/4$ is drilled through a cylinder of radius R to form a body of mass m as shown. If the body rolls on the horizontal surface without slipping, determine the period τ for small oscillations.

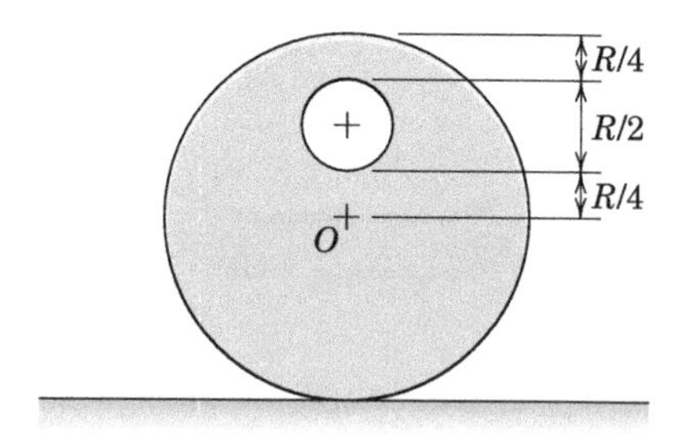

Problem 8/117

►**8/118** The quarter-circular sector of mass m and radius r is set into small rocking oscillation on the horizontal surface. If no slipping occurs, determine the expression for the period τ of each complete oscillation.

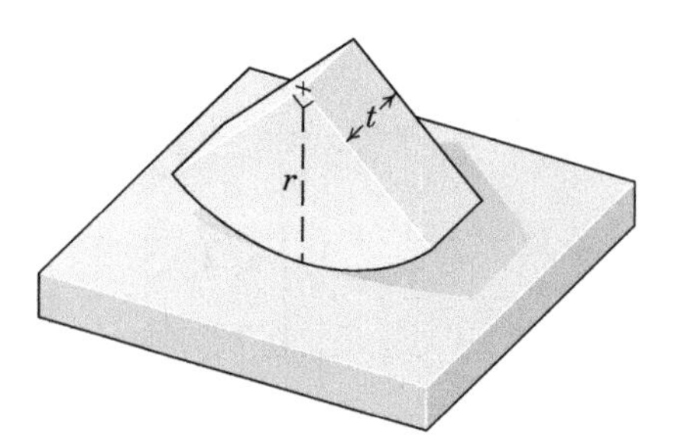

Problem 8/118

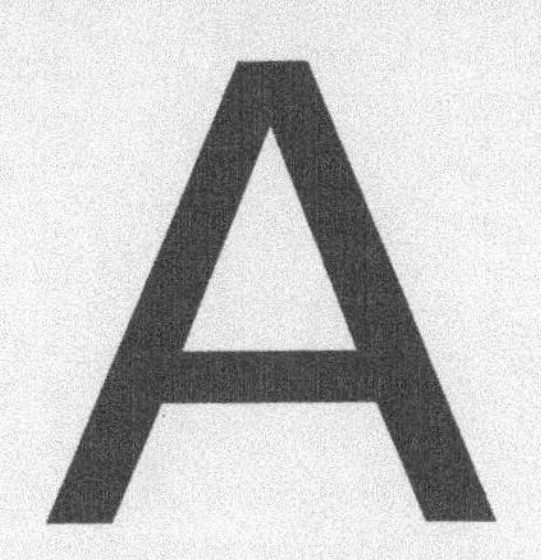

Area Moments of Inertia

See Appendix A of *Vol. 1 Statics* for a treatment of the theory and calculation of area moments of inertia. Because this quantity plays an important role in the design of structures, especially those dealt with in statics, we present only a brief definition in this *Dynamics* volume so that the student can appreciate the basic differences between area and mass moments of inertia.

The moments of inertia of a plane area A about the x- and y-axes in its plane and about the z-axis normal to its plane, Fig. A/1, are defined by

$$I_x = \int y^2\,dA \qquad I_y = \int x^2\,dA \qquad I_z = \int r^2\,dA$$

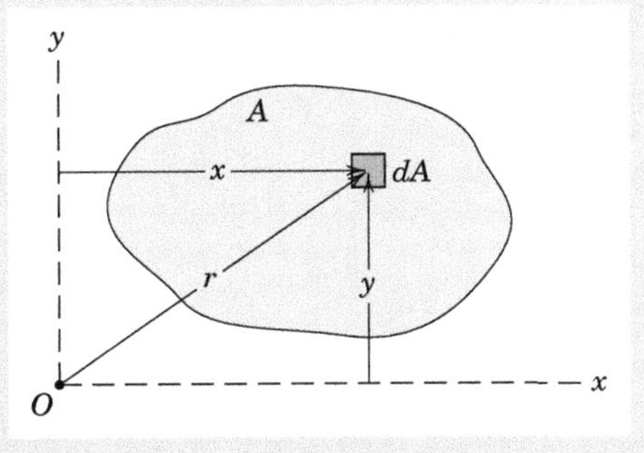

Figure A/1

where dA is the differential element of area and $r^2 = x^2 + y^2$. Clearly, the polar moment of inertia I_z equals the sum $I_x + I_y$ of the rectangular moments of inertia. For thin flat plates, the area moment of inertia is useful in the calculation of the mass moment of inertia, as explained in Appendix B.

The area moment of inertia is a measure of the distribution of area about the axis in question and, for that axis, is a constant property of the area. The dimensions of area moment of inertia are (distance)^4 expressed in m^4 or mm^4 in SI units and ft^4 or in.^{4} in U.S. customary units. In contrast, mass moment of inertia is a measure of the distribution of mass about the axis in question, and its dimensions are (mass)(distance)^2, which are expressed in kg·m^2 in SI units and in lb-ft-sec^2 or lb-in.-sec^2 in U.S. customary units.

B MASS MOMENTS OF INERTIA

APPENDIX OUTLINE

B/1 MASS MOMENTS OF INERTIA ABOUT AN AXIS

The equation of rotational motion about an axis normal to the plane of motion for a rigid body in plane motion contains an integral which depends on the distribution of mass with respect to the moment axis. This integral occurs whenever a rigid body has an angular acceleration about its axis of rotation. Thus, to study the dynamics of rotation, you should be thoroughly familiar with the calculation of mass moments of inertia for rigid bodies.

Consider a body of mass m, Fig. B/1, rotating about an axis O-O with an angular acceleration α. All particles of the body move in parallel planes which are normal to the rotation axis O-O. We may choose any one of the planes as the plane of motion, although the one containing the center of mass is usually the one so designated. An element of mass dm has a component of acceleration tangent to its circular path equal to $r\alpha$, and by Newton's second law of motion the resultant tangential force on this element equals $r\alpha\, dm$. The moment of this force about the axis O-O is $r^2\alpha\, dm$, and the sum of the moments of these forces for all elements is $\int r^2\alpha\, dm$.

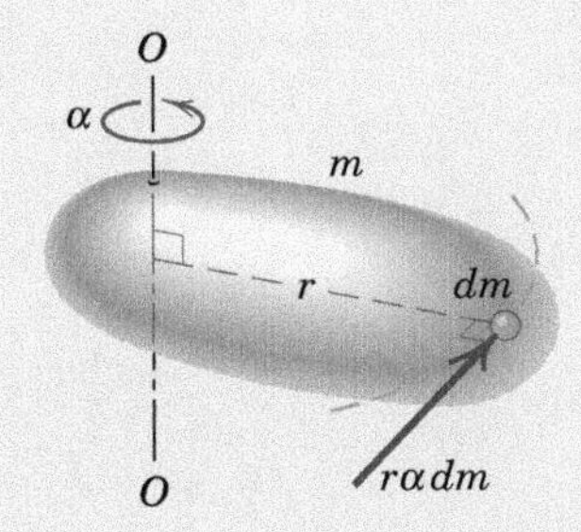

Figure B/1

For a rigid body, α is the same for all radial lines in the body and we may take it outside the integral sign. The remaining integral is called the mass moment of inertia I of the body about the axis O-O and is

$$I = \int r^2\, dm \qquad \textbf{(B/1)}$$

This integral represents an important property of a body and is involved in the analysis of any body which has rotational acceleration about a given axis. Just as the mass m of a body is a measure of the resistance to translational acceleration, the moment of inertia I is a measure of resistance to rotational acceleration of the body.

The moment-of-inertia integral may be expressed alternatively as

$$I = \Sigma r_i^2 m_i \qquad \textbf{(B/1}\boldsymbol{a}\textbf{)}$$

where r_i is the radial distance from the inertia axis to the representative particle of mass m_i and where the summation is taken over all particles of the body.

If the density ρ is constant throughout the body, the moment of inertia becomes

$$I = \rho \int r^2 \, dV$$

where dV is the element of volume. In this case, the integral by itself defines a purely geometrical property of the body. When the density is not constant but is expressed as a function of the coordinates of the body, it must be left within the integral sign and its effect accounted for in the integration process.

In general, the coordinates which best fit the boundaries of the body should be used in the integration. It is particularly important that we make a good choice of the element of volume dV. To simplify the integration, an element of lowest possible order should be chosen, and the correct expression for the moment of inertia of the element about the axis involved should be used. For example, in finding the moment of inertia of a solid right-circular cone about its central axis, we may choose an element in the form of a circular slice of infinitesimal thickness, Fig. B/2*a*. The differential moment of inertia for this element is the expression for the moment of inertia of a circular cylinder of infinitesimal altitude about its central axis. (This expression will be obtained in Sample Problem B/1.)

Alternatively, we could choose an element in the form of a cylindrical shell of infinitesimal thickness as shown in Fig. B/2*b*. Because all of the mass of the element is at the same distance r from the inertia axis, the differential moment of inertia for this element is merely $r^2 \, dm$ where dm is the differential mass of the elemental shell.

From the definition of mass moment of inertia, its dimensions are (mass)(distance)^2 and are expressed in the units kg·m^2 in SI units and lb-ft-sec^2 in U.S. customary units.

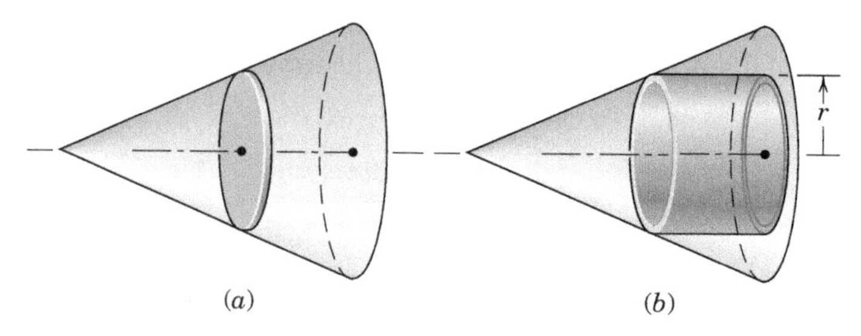

Figure B/2

Radius of Gyration

The radius of gyration k of a mass m about an axis for which the moment of inertia is I is defined as

$$k = \sqrt{\frac{1}{m}} \qquad \text{or} \qquad I = k^2 m \tag{B/2}$$

Thus, k is a measure of the distribution of mass of a given body about the axis in question, and its definition is analogous to the definition of the radius of gyration for area moments of inertia. If all the mass m of a body could be concentrated at a distance k from the axis, the moment of inertia would be unchanged.

The moment of inertia of a body about a particular axis is frequently indicated by specifying the mass of the body and the radius of gyration of the body about the axis. The moment of inertia is then calculated from Eq. B/2.

Transfer of Axes

If the moment of inertia of a body is known about an axis passing through the mass center, it may be determined easily about any parallel axis. To prove this statement, consider the two parallel axes in Fig. B/3, one being an axis through the mass center G and the other a parallel axis through some other point C. The radial distances from the two axes to any element of mass dm are r_0 and r, and the separation of the axes is d. Substituting the law of cosines $r^2 = r_0{}^2 + d^2 + 2r_0 d \cos\theta$ into the definition for the moment of inertia about the axis through C gives

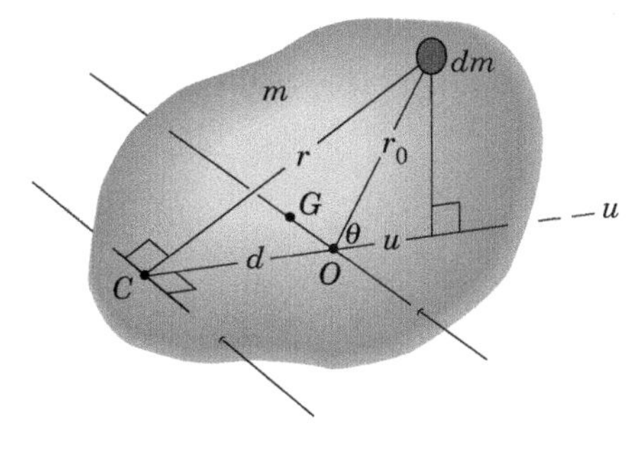

Figure B/3

$$\begin{aligned} I &= \int r^2\, dm = \int (r_0{}^2 + d^2 + 2r_0 d \cos\theta)\, dm \\ &= \int r_0{}^2\, dm + d^2 \int dm + 2d \int u\, dm \end{aligned}$$

The first integral is the moment of inertia $\bar{I}$ about the mass-center axis, the second term is md^2, and the third integral equals zero, since the u-coordinate of the mass center with respect to the axis through G is zero. Thus, the parallel-axis theorem is

$$I = \bar{I} + md^2 \tag{B/3}$$

Remember that the transfer cannot be made unless one axis passes through the center of mass and unless the axes are parallel.

When the expressions for the radii of gyration are substituted in Eq. B/3, there results

$$k^2 = \bar{k}^2 + d^2 \tag{B/3a}$$

Equation B/3a is the parallel-axis theorem for obtaining the radius of gyration k about an axis which is a distance d from a parallel axis through the mass center, for which the radius of gyration is $\bar{k}$.

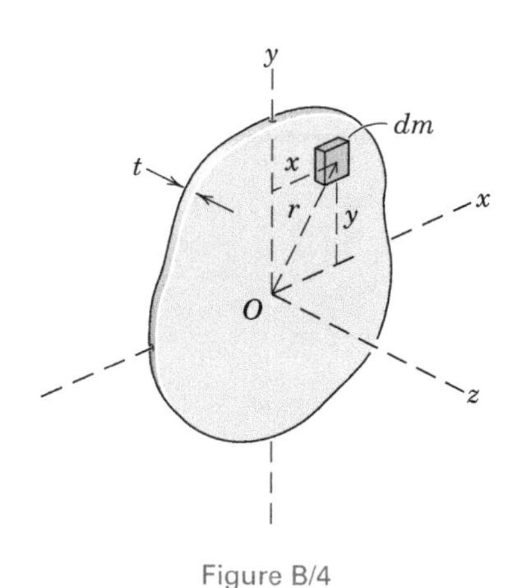

Figure B/4

For plane-motion problems where rotation occurs about an axis normal to the plane of motion, a single subscript for I is sufficient to designate the inertia axis. Thus, if the plate of Fig. B/4 has plane motion in the x-y plane, the moment of inertia of the plate about the z-axis through O is designated I_O. For three-dimensional motion, however, where components of rotation may occur about more than one axis, we use a double subscript to preserve notational symmetry with product-of-inertia terms, which are described in Art. B/2. Thus, the moments of inertia about the x-, y-, and z-axes are labeled I_{xx}, I_{yy}, and I_{zz}, respectively, and from Fig. B/5 we see that they become

$$\begin{aligned} I_{xx} &= \int r_x^{\,2}\,dm = \int (y^2 + z^2)\,dm \\ I_{yy} &= \int r_y^{\,2}\,dm = \int (z^2 + x^2)\,dm \\ I_{zz} &= \int r_z^{\,2}\,dm = \int (x^2 + y^2)\,dm \end{aligned} \tag{B/4}$$

These integrals are cited in Eqs. 7/10 of Art. 7/7 on angular momentum in three-dimensional rotation.

The defining expressions for mass moments of inertia and area moments of inertia are similar. An exact relationship between the two moment-of-inertia expressions exists in the case of flat plates. Consider the flat plate of uniform thickness in Fig. B/4. If the constant thickness is t and the density is ρ, the mass moment of inertia I_{zz} of the plate about the z-axis normal to the plate is

$$I_{zz} = \int r^2\,dm = \rho t \int r^2\,dA = \rho t I_z \tag{B/5}$$

Thus, the mass moment of inertia about the z-axis equals the mass per unit area ρt times the polar moment of inertia I_z of the plate area about the z-axis. If t is small compared with the dimensions of the plate in its

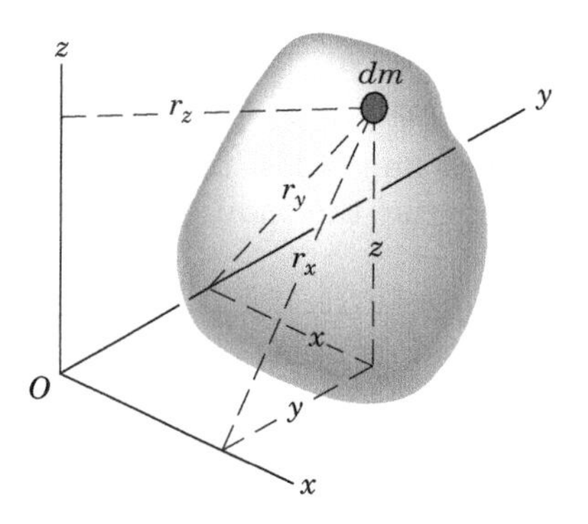

Figure B/5

plane, the mass moments of inertia I_{xx} and I_{yy} of the plate about the x- and y-axes are closely approximated by

$$I_{xx} = \int y^2 \, dm = \rho t \int y^2 \, dA = \rho t I_x$$
$$I_{yy} = \int x^2 \, dm = \rho t \int x^2 \, dA = \rho t I_y \tag{B/6}$$

Thus, the mass moments of inertia equal the mass per unit area ρt times the corresponding area moments of inertia. The double subscripts for mass moments of inertia distinguish these quantities from area moments of inertia.

Inasmuch as $I_z = I_x + I_y$ for area moments of inertia, we have

$$I_{zz} = I_{xx} + I_{yy} \tag{B/7}$$

which holds *only* for a thin flat plate. This restriction is observed from Eqs. B/6, which do not hold true unless the thickness t or the z-coordinate of the element is negligible compared with the distance of the element from the corresponding x- or y-axis. Equation B/7 is very useful when dealing with a differential mass element taken as a flat slice of differential thickness, say, dz. In this case, Eq. B/7 holds exactly and becomes

$$dI_{zz} = dI_{xx} + dI_{yy} \tag{B/7a}$$

for axes x and y in the plane of the plate.

Composite Bodies

As in the case of area moments of inertia, the mass moment of inertia of a composite body is the sum of the moments of inertia of the individual parts about the same axis. It is often convenient to treat a composite body as defined by positive volumes and negative volumes. The moment of inertia of a negative element, such as the material removed to form a hole, must be considered a negative quantity.

A summary of some of the more useful formulas for mass moments of inertia of various masses of common shape is given in Table D/4, Appendix D.

The problems which follow the sample problems are divided into the categories *Integration Exercises* and *Composite and Parallel-Axis Exercises*. The parallel-axis theorem will also be useful in some of the problems in the first category.

Sample Problem B/1

Determine the moment of inertia and radius of gyration of a homogeneous right-circular cylinder of mass m and radius r about its central axis O-O.

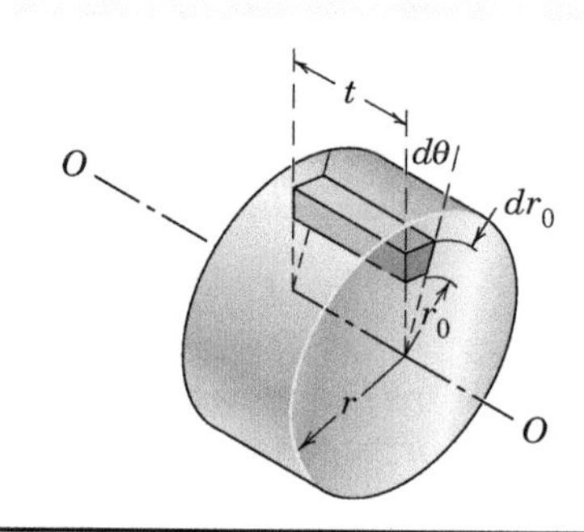

Solution. An element of mass in cylindrical coordinates is $dm = \rho\, dV = \rho t r_0\, dr_0\, d\theta$, where ρ is the density of the cylinder. The moment of inertia about the axis of the cylinder is ①

$$I = \int r_0^{\,2}\, dm = \rho t \int_0^{2\pi} \int_0^r r_0^{\,3}\, dr_0\, d\theta = \rho t \frac{\pi r^4}{2} = \frac{1}{2} m r^2 \qquad \textit{Ans.}$$ ②

The radius of gyration is

$$k = \sqrt{\frac{I}{m}} = \frac{r}{\sqrt{2}} \qquad \textit{Ans.}$$

Helpful Hints

① If we had started with a cylindrical shell of radius r_0 and axial length t as our mass element dm, then $dI = r_0^{\,2}\, dm$ directly. You should evaluate the integral.

② The result $I = \frac{1}{2}mr^2$ applies *only* to a solid homogeneous circular cylinder and cannot be used for any other wheel of circular periphery.

Sample Problem B/2

Determine the moment of inertia and radius of gyration of a homogeneous solid sphere of mass m and radius r about a diameter.

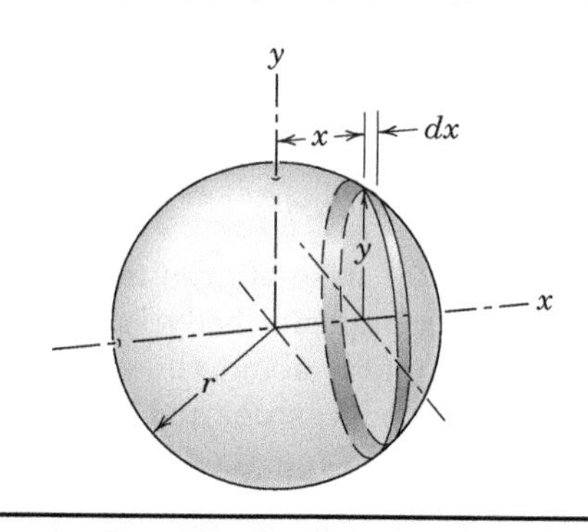

Solution. A circular slice of radius y and thickness dx is chosen as the volume element. From the results of Sample Problem B/1, the moment of inertia about the x-axis of the elemental cylinder is

$$dI_{xx} = \tfrac{1}{2}(dm)y^2 = \tfrac{1}{2}(\pi\rho y^2\, dx)y^2 = \frac{\pi\rho}{2}(r^2 - x^2)^2\, dx$$ ①

where ρ is the constant density of the sphere. The total moment of inertia about the x-axis is

$$I_{xx} = \frac{\pi\rho}{2}\int_{-r}^{r} (r^2 - x^2)^2\, dx = \tfrac{8}{15}\pi\rho r^5 = \tfrac{2}{5}mr^2 \qquad \textit{Ans.}$$

The radius of gyration about the x-axis is

$$k_x = \sqrt{\frac{I_{xx}}{m}} = \sqrt{\frac{2}{5}}\,r \qquad \textit{Ans.}$$

Helpful Hint

① Here is an example where we utilize a previous result to express the moment of inertia of the chosen element, which in this case is a right-circular cylinder of differential axial length dx. It would be foolish to start with a third-order element, such as $\rho\, dx\, dy\, dz$, when we can easily solve the problem with a first-order element.

Sample Problem B/3

Determine the moments of inertia of the homogeneous rectangular parallelepiped of mass m about the centroidal x_0- and z-axes and about the x-axis through one end.

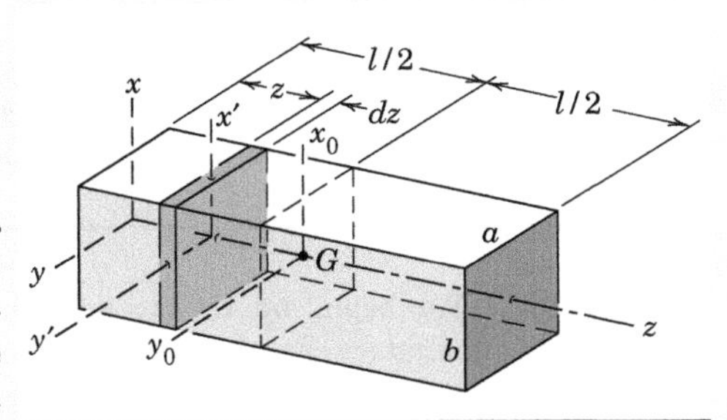

Solution. A transverse slice of thickness dz is selected as the element of volume. The moment of inertia of this slice of infinitesimal thickness equals the moment of inertia of the area of the section times the mass per unit area $\rho\, dz$. Thus, the moment of inertia of the transverse slice about the y'-axis is

$$dI_{y'y'} = (\rho\, dz)(\tfrac{1}{12}ab^3)$$

and that about the x'-axis is

① 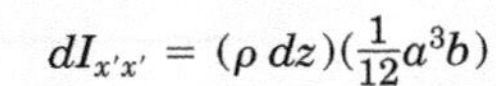

$$dI_{x'x'} = (\rho\, dz)(\tfrac{1}{12}a^3b)$$

As long as the element is a plate of differential thickness, the principle given by Eq. B/7*a* may be applied to give

$$dI_{zz} = dI_{x'x'} + dI_{y'y'} = (\rho\, dz)\frac{ab}{12}(a^2 + b^2)$$

These expressions may now be integrated to obtain the desired results.

The moment of inertia about the z-axis is

$$I_{zz} = \int dI_{zz} = \frac{\rho ab}{12}(a^2 + b^2)\int_0^l dz = \tfrac{1}{12}m(a^2 + b^2) \qquad \textit{Ans.}$$

where m is the mass of the block. By interchange of symbols, the moment of inertia about the x_0-axis is

$$I_{x_0x_0} = \tfrac{1}{12}m(a^2 + l^2) \qquad \textit{Ans.}$$

The moment of inertia about the x-axis may be found by the parallel-axis theorem, Eq. B/3. Thus,

$$I_{xx} = I_{x_0x_0} + m\left(\frac{l}{2}\right)^2 = \tfrac{1}{12}m(a^2 + 4l^2) \qquad \textit{Ans.}$$

This last result may be obtained by expressing the moment of inertia of the elemental slice about the x-axis and integrating the expression over the length of the bar. Again, by the parallel-axis theorem

$$dI_{xx} = dI_{x'x'} + z^2\, dm = (\rho\, dz)(\tfrac{1}{12}a^3b) + z^2\rho ab\, dz = \rho ab\left(\frac{a^2}{12} + z^2\right)dz$$

Integrating gives the result obtained previously:

$$I_{xx} = \rho ab\int_0^l \left(\frac{a^2}{12} + z^2\right)dz = \frac{\rho abl}{3}\left(l^2 + \frac{a^2}{4}\right) = \tfrac{1}{12}m(a^2 + 4l^2)$$

The expression for I_{xx} may be simplified for a long prismatic bar or slender rod whose transverse dimensions are small compared with the length. In this case, a^2 may be neglected compared with $4l^2$, and the moment of inertia of such a slender bar about an axis through one end normal to the bar becomes $I = \frac{1}{3}ml^2$. By the same approximation, the moment of inertia about a centroidal axis normal to the bar is $I = \frac{1}{12}ml^2$.

Helpful Hint

① Refer to Eqs. B/6 and recall the expression for the area moment of inertia of a rectangle about an axis through its center parallel to its base.

Sample Problem B/4

The upper edge of the thin homogeneous plate of mass m is parabolic with a vertical slope at the origin O. Determine its mass moments of inertia about the x-, y-, and z-axes.

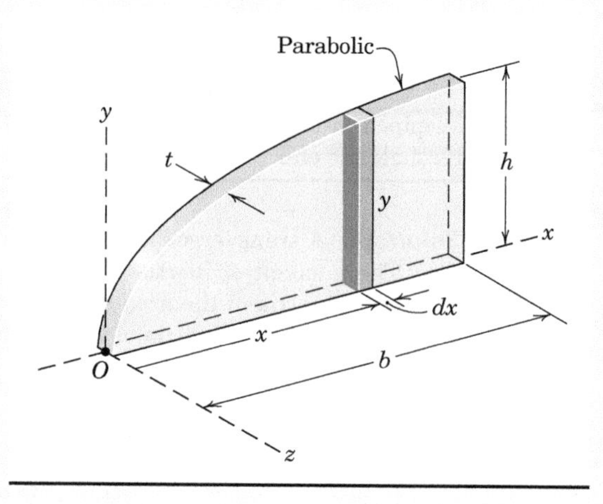

Solution. We begin by clearly establishing the function associated with ① the upper boundary. From $y = k\sqrt{x}$ evaluated at $(x, y) = (b, h)$, we find that $k = h/\sqrt{b}$ so that $y = \frac{h}{\sqrt{b}}\sqrt{x}$. We choose a transverse slice of thickness dx for the integrations leading to I_{xx} and I_{yy}. The mass of this slice is

$$dm = \rho ty\,dx$$

and the total mass of the plate is

② $$m = \int dm = \int \rho ty\,dx = \int_0^b \rho t \frac{h}{\sqrt{b}}\sqrt{x}\,dx = \tfrac{2}{3}\rho thb$$

The moment of inertia of the slice about the x-axis is

③ $$dI_{xx} = \tfrac{1}{3}\,dm\,y^2 = \tfrac{1}{3}(\rho ty\,dx)y^2 = \tfrac{1}{3}\rho ty^3\,dx$$

For the entire plate, we have

$$I_{xx} = \int dI_{xx} = \int_0^b \tfrac{1}{3}\rho t\left(\frac{h}{\sqrt{b}}\sqrt{x}\right)^3 dx = \tfrac{2}{15}\rho th^3 b$$

In terms of the mass m:

④ $$I_{xx} = \tfrac{2}{15}\rho th^3 b\left(\frac{m}{\frac{2}{3}\rho thb}\right) = \tfrac{1}{5}mh^2 \qquad \textit{Ans.}$$

The moment of inertia of the element about the y-axis is

$$dI_{yy} = dm\,y^2 = (\rho ty\,dx)y^2 = \rho ty^3\,dx$$

For the entire plate,

⑤ $$I_{yy} = \int dI_{yy} = \int_0^b \rho t\left(\frac{h}{\sqrt{b}}\sqrt{x}\right)^3 dx = \tfrac{2}{7}\rho thb^3\left(\frac{m}{\frac{2}{3}\rho thb}\right) = \tfrac{3}{7}mb^2 \qquad \textit{Ans.}$$

For thin plates which lie in the x-y plane,

$$I_{zz} = I_{xx} + I_{yy} = \tfrac{1}{5}mh^2 + \tfrac{3}{7}mb^2$$

$$I_{zz} = m\left(\frac{h^2}{5} + \frac{3b^2}{7}\right) \qquad \textit{Ans.}$$

Helpful Hints

① If we have $y = kx^2$, saying that "y gets large faster than x" helps establish that the parabola opens upward. Here, we have $y^2 = k^2x$, which says that "x gets large faster than y", helping establish that the parabola opens rightward.

② For a full b by h rectangular plate of thickness t, the mass would be ρthb (density times volume). So the factor of $\frac{2}{3}$ for the parabolic plate makes sense.

③ Recall that for a slender rod of mass m and length l, the moment of inertia about an axis perpendicular to the rod and passing through one end is $\frac{1}{3}ml^2$.

④ Note that I_{xx} is independent of the width b.

⑤ Note that I_{yy} is independent of the height h.

Sample Problem B/5

The radius of the homogeneous solid of revolution is proportional to the square of its x-coordinate. If the mass of the body is m, determine its mass moments of inertia about the x- and y-axes.

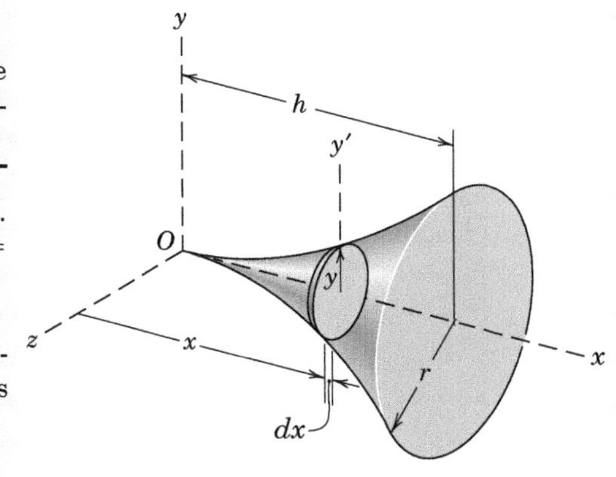

Solution. We begin by writing the boundary in the x-y plane as $y = kx^2$. The constant k is determined by evaluating this equation at the point $(x, y) = (h, r)$: $r = kh^2$, which gives $k = r/h^2$, so that $y = \frac{r}{h^2}x^2$.

As is usually convenient for bodies with axial symmetry, we choose a disk-shaped slice as our differential element, as shown in the given figure. The mass of this element is

① $$dm = \rho\pi y^2\,dx$$

where ρ represents the density of the body. The moment of inertia of the element about the x-axis is

② $$dI_{xx} = \frac{1}{2}\,dm\,y^2 = \frac{1}{2}(\rho\pi y^2\,dx)\,y^2 = \frac{1}{2}\rho\pi y^4\,dx$$

The mass of the entire body is

③ $$m = \int dm = \int_0^h \rho\pi y^2\,dx = \int_0^h \rho\pi\left(\frac{r}{h^2}x^2\right)^2 dx = \rho\pi\frac{r^2}{h^4}\frac{x^5}{5}\bigg|_0^h = \frac{1}{5}\rho\pi r^2 h$$

and the moment of inertia of the entire body is

$$I_{xx} = \int dI_{xx} = \int_0^h \frac{1}{2}\rho\pi y^4\,dx = \int_0^h \frac{1}{2}\rho\pi\left(\frac{r}{h^2}x^2\right)^4 dx = \frac{1}{18}\rho\pi r^4 h$$

All that remains is to express I_{xx} more conventionally in terms of its mass. We do so by writing

④ ⑤ $$I_{xx} = \frac{1}{18}\rho\pi r^4 h\left(\frac{m}{\frac{1}{5}\rho\pi r^2 h}\right) = \frac{5}{18}mr^2 \qquad \textit{Ans.}$$

By the parallel-axis theorem, the moment of inertia of the disk-shaped element about the y-axis is

$$\begin{aligned} dI_{yy} &= dI_{y'y'} + x^2\,dm = \frac{1}{4}\,dm\,y^2 + x^2\,dm \\ &= dm\left(\frac{1}{4}\left(\frac{r}{h^2}x^2\right)^2 + x^2\right) = \rho\pi y^2\,dx\left(\frac{1}{4}\frac{r^2}{h^4}x^4 + x^2\right) \\ &= \rho\pi\left(\frac{r}{h^2}x^2\right)^2\left(\frac{1}{4}\frac{r^2}{h^4}x^4 + x^2\right)dx = \rho\pi\frac{r^2}{h^4}\left(\frac{1}{4}\frac{r^2}{h^4}x^8 + x^6\right)dx \end{aligned}$$

For the entire body, we have

$$\begin{aligned} I_{yy} &= \int dI_{yy} = \int_0^h \rho\pi\frac{r^2}{h^4}\left(\frac{1}{4}\frac{r^2}{h^4}x^8 + x^6\right)dx = \rho\pi\frac{r^2}{h^4}\left(\frac{1}{4}\frac{r^2}{h^4}\frac{x^9}{9} + \frac{x^7}{7}\right)\bigg|_0^h \\ &= \rho\pi r^2 h\left(\frac{r^2}{36} + \frac{h^2}{7}\right) \end{aligned}$$

Finally, we multiply by the same unit expression as above to obtain a result in terms of the body mass m.

$$I_{yy} = \rho\pi r^2 h\left(\frac{r^2}{36} + \frac{h^2}{7}\right)\left(\frac{m}{\frac{1}{5}\rho\pi r^2 h}\right) = 5m\left(\frac{r^2}{36} + \frac{h^2}{7}\right) \qquad \textit{Ans.}$$

Helpful Hints

① The volume of a disk is the area of its face times its thickness. Then density times volume gives mass.

② From Sample Problem B/1, the mass moment of inertia of a uniform cylinder (or disk) about its longitudinal axis is $\frac{1}{2}mr^2$.

③ Remember to regard an integral operation as an infinite summation.

④ The parenthetical expression here is unity, because its numerator and denominator are equal.

⑤ We note that I_{xx} is independent of h. So the body could be compressed to $h \cong 0$ or elongated to a large value of h with no resulting change in I_{xx}. This is true because no particle of the body would be changing its distance from the x-axis.

PROBLEMS

Integration Exercises

B/1 Determine the mass moment of inertia of the bent uniform slender rod about the x- and y-axes shown, and about the z-axis. The rod has a mass ρ per unit length.

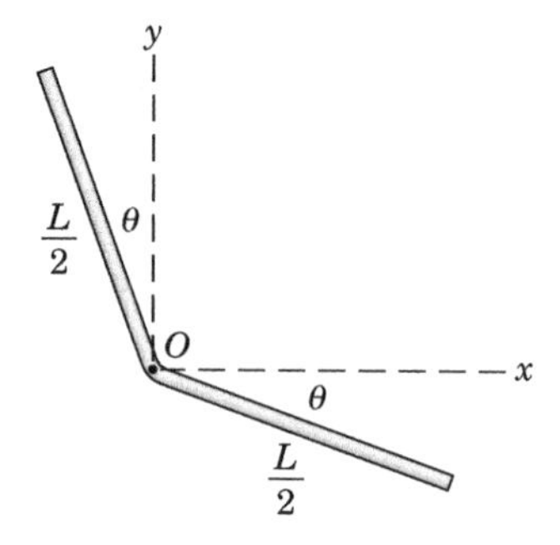

Problem B/1

B/2 Determine the mass moment of inertia of the uniform thin triangular plate of mass m about the x-axis. Also determine the radius of gyration about the x-axis. By analogy state I_{yy} and k_y. Then determine I_{zz} and k_z.

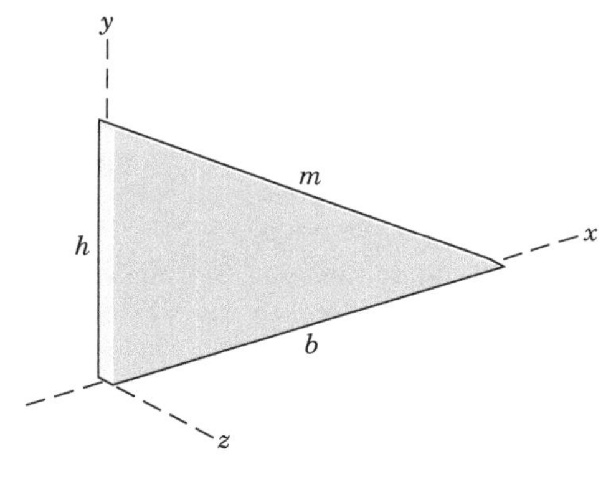

Problem B/2

B/3 Determine the mass moment of inertia about the y-axis for the uniform thin equilateral triangular plate of mass m. Also determine its radius of gyration about the y-axis.

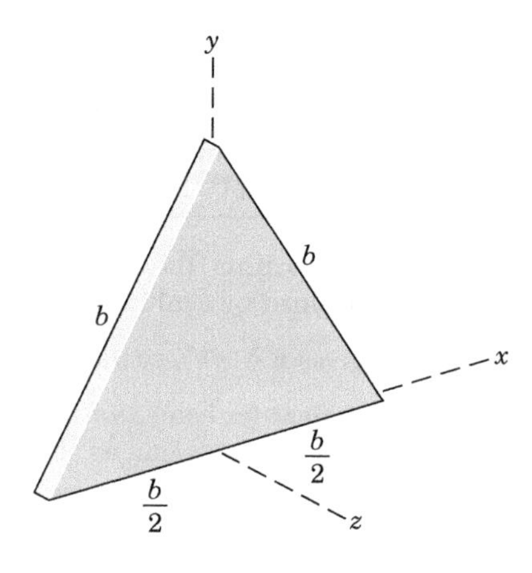

Problem B/3

B/4 Calculate the mass moment of inertia of the homogeneous right-circular cone of mass m, base radius r, and altitude h about the cone axis x and about the y-axis through its vertex.

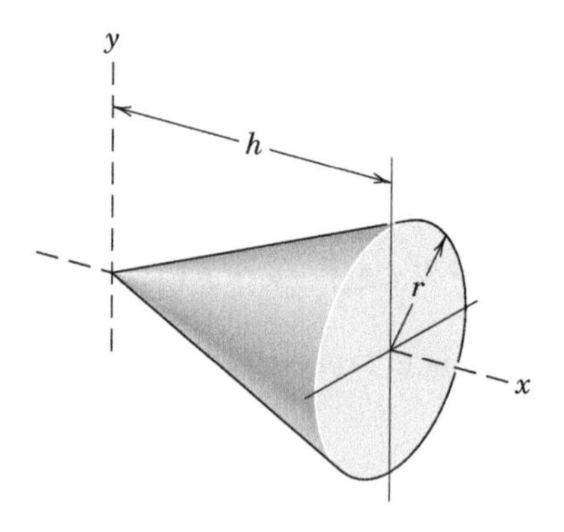

Problem B/4

B/5 Determine the mass moment of inertia of the uniform thin parabolic plate of mass m about the x-axis. State the corresponding radius of gyration.

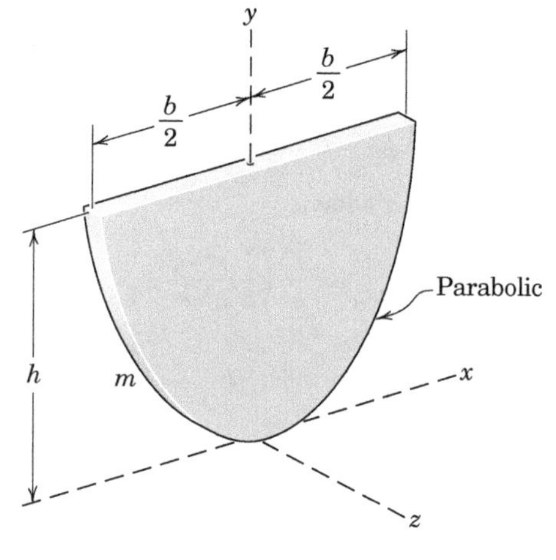

Problem B/5

B/6 Determine the mass moment of inertia about the y-axis for the parabolic plate of the previous problem. State the radius of gyration about the y-axis.

B/7 For the thin homogeneous plate of uniform thickness t and mass m, determine the mass moments of inertia about the axes x'-, y'-, and z'-through the end of the pate at A. Refer to the results of Sample Problem B/4 and Table D/3 in Appendix D as needed.

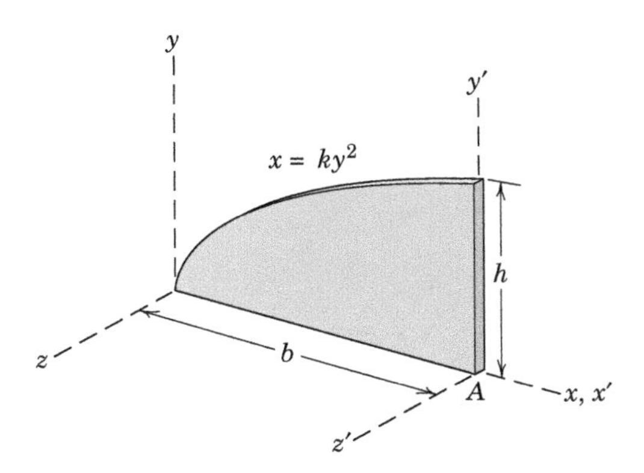

Problem B/7

B/8 Determine the mass moment of inertia about the x-axis of the thin elliptical plate of mass m.

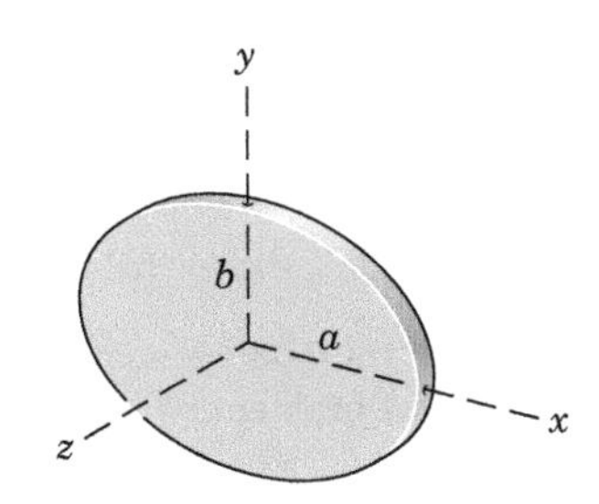

Problem B/8

B/9 Determine the mass moment of inertia of the homogeneous solid of revolution of mass m about the x-axis.

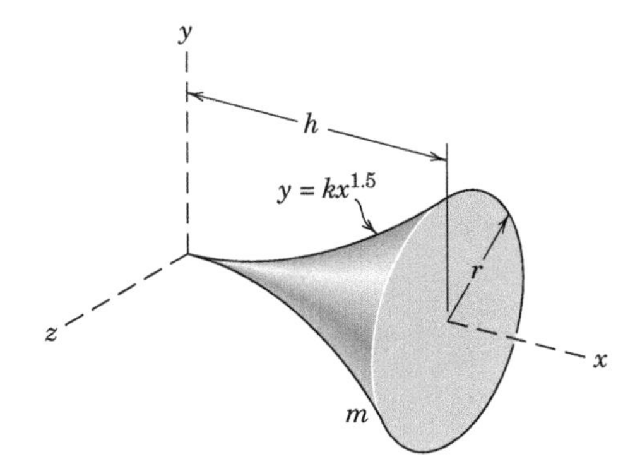

Problem B/9

B/10 Determine the mass moment of inertia of the homogeneous solid of revolution of the previous problem about the y- and z-axes.

B/11 Determine the radius of gyration about the y-axis for the steel part shown in section. The part is formed by revolving one of the trapezoidal areas 360° around the y-axis.

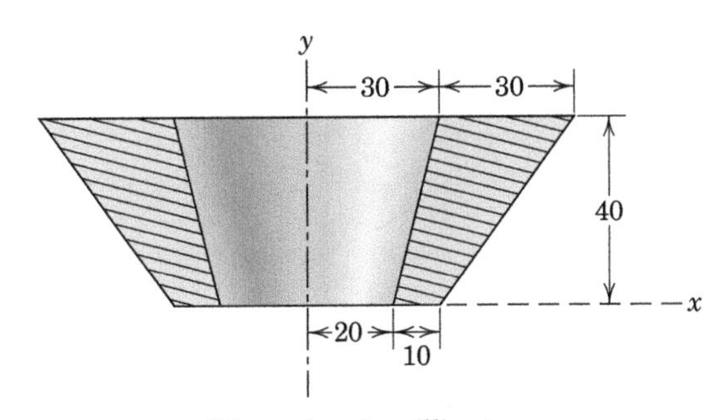

Problem B/11

B/12 For the steel part of Prob. B/11, determine the radius of gyration about the x-axis.

B/13 Determine the radius of gyration about the z-axis of the paraboloid of revolution shown. The mass of the homogenous body is m.

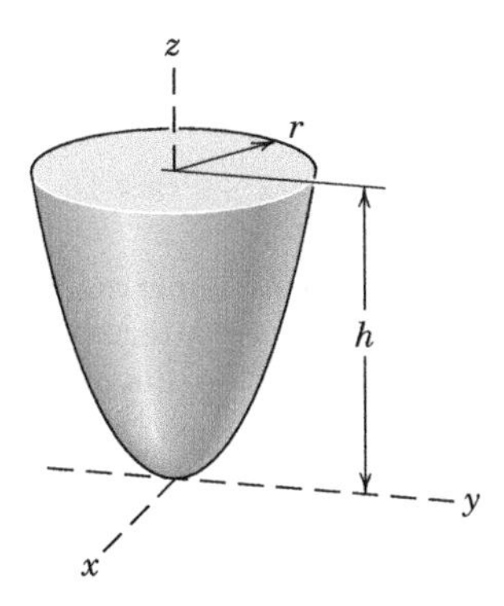

Problem B/13

B/14 Determine the moment of inertia about the y-axis for the paraboloid of revolution of Prob. B/13.

B/15 Develop an expression for the mass moment of inertia of the homogeneous solid of revolution of mass m about the y-axis.

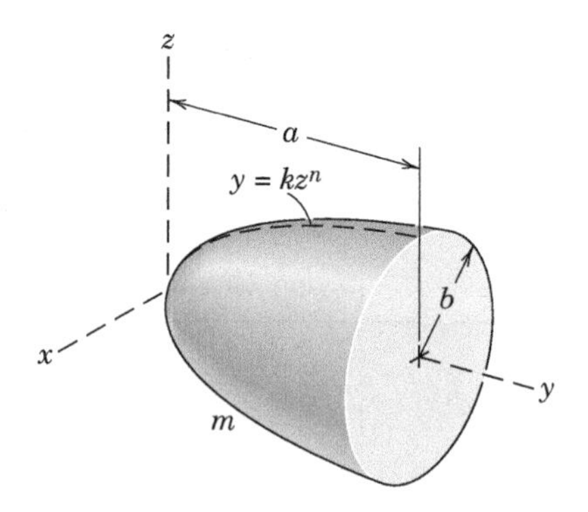

Problem B/15

B/16 Determine the mass moment of inertia about the x-axis of the solid spherical segment of mass m.

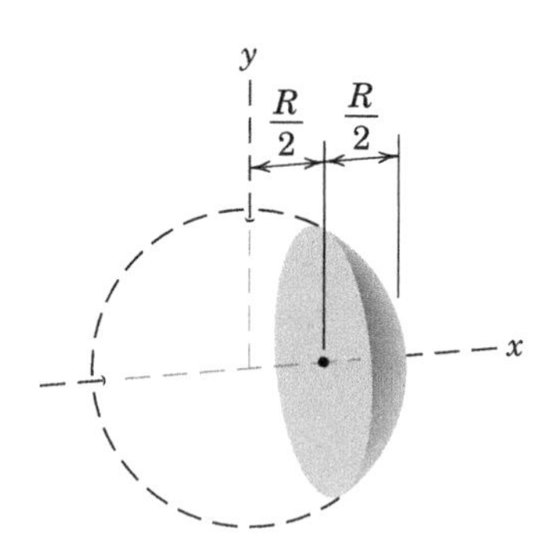

Problem B/16

B/17 Determine the moment of inertia about the generating axis of a complete ring (torus) of mass m having a circular section with the dimensions shown in the sectional view.

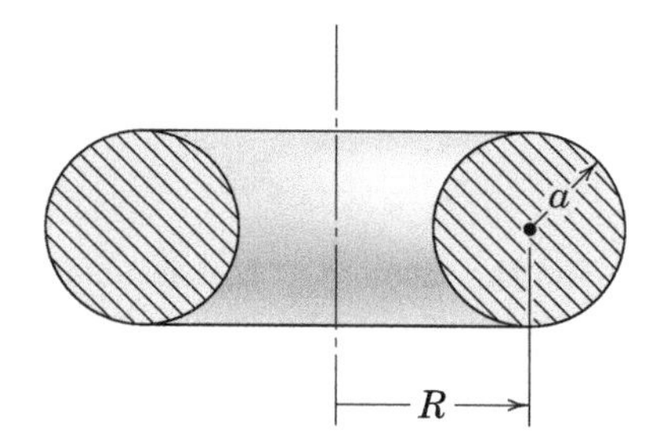

Problem B/17

B/18 The plane area shown in the top portion of the figure is rotated 180° about the x-axis to form the body of revolution of mass m shown in the lower portion of the figure. Determine the mass moment of inertia of the body about the x-axis.

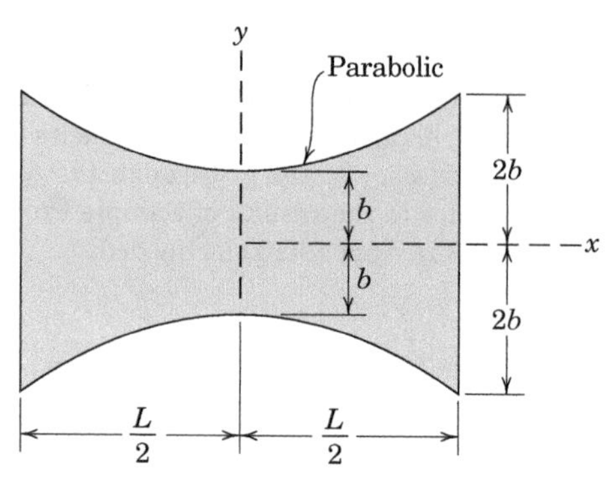

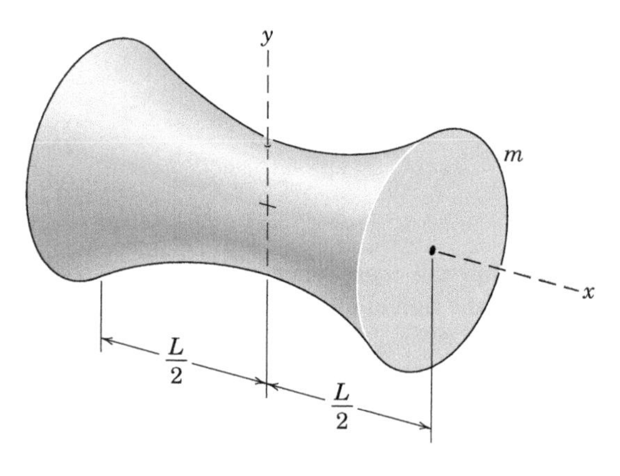

Problem B/18

B/19 Determine I_{yy} for the homogeneous body of revolution of the previous problem.

B/20 Determine the mass moment of inertia of the homogeneous square pyramid about the z-axis. The pyramid has mass m.

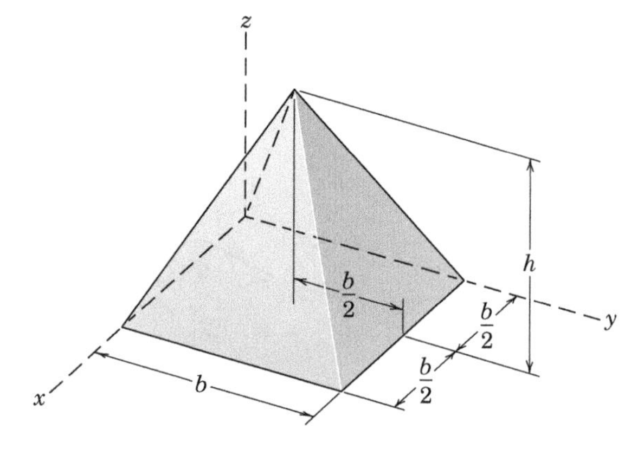

Problem B/20

B/21 For the square pyramid of Prob. B/20, determine the mass moment of inertia about the x-axis.

B/22 The thickness of the homogeneous triangular plate of mass m varies linearly with the distance from the vertex toward the base. The thickness a at the base is small compared with the other dimensions. Determine the moment of inertia of the plate about the y-axis along the centerline of the base.

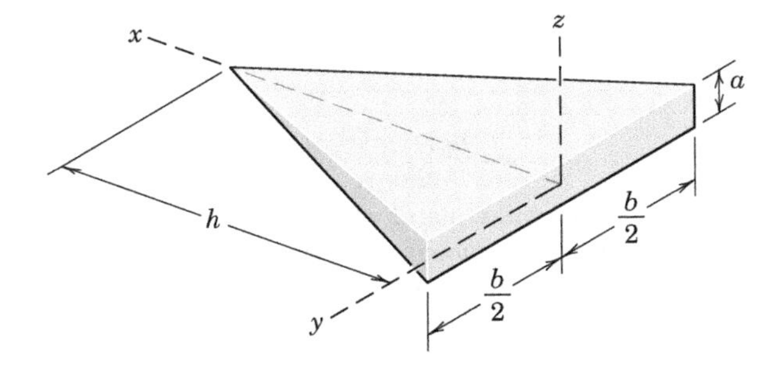

Problem B/22

B/23 Determine the moment of inertia, about the generating axis, of the hollow circular tube of mass m obtained by revolving the thin ring shown in the sectional view completely around the generating axis.

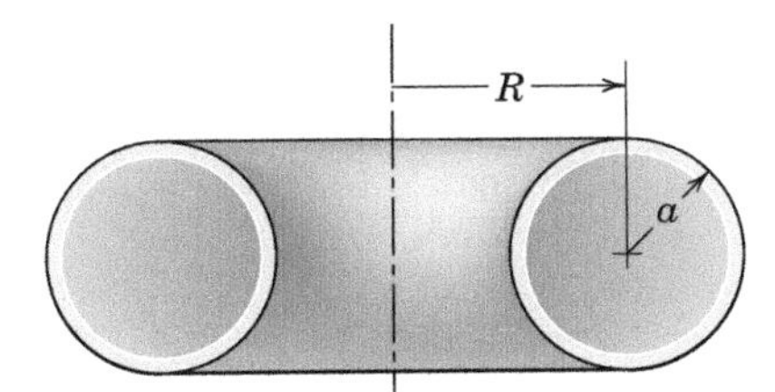

Problem B/23

B/24 Determine the moments of inertia of the hemispherical shell with respect to the x- and z-axes. The mass of the shell is m, and its thickness is negligible compared with the radius r.

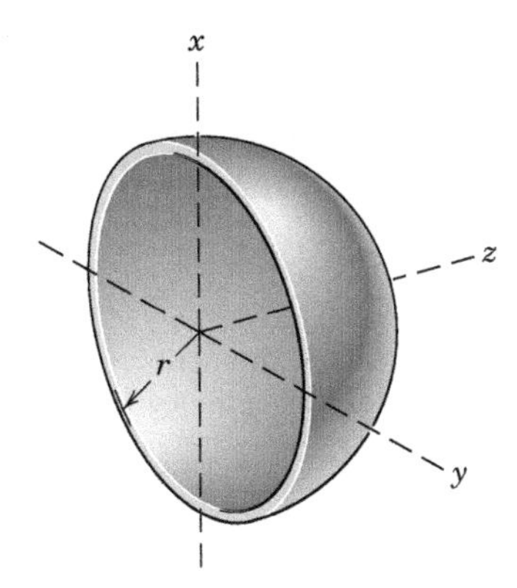

Problem B/24

B/25 The partial solid of revolution is formed by revolving the shaded area in the x-z plane 90° about the z-axis. If the mass of the solid is m, determine its mass moment of inertia about the z-axis.

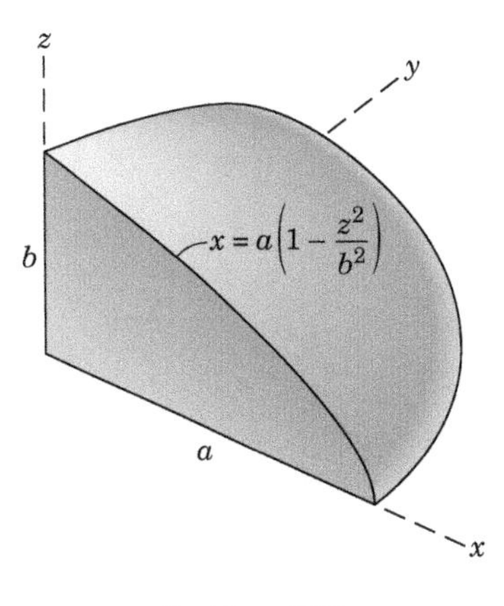

Problem B/25

B/26 For the partial solid of revolution in Prob. B/25, determine the mass moment of inertia about the x-axis.

▶B/27 A shell of mass m is obtained by revolving the quarter-circular section about the z-axis. If the thickness of the shell is small compared with a and if $r = a/3$, determine the radius of gyration of the shell about the z-axis.

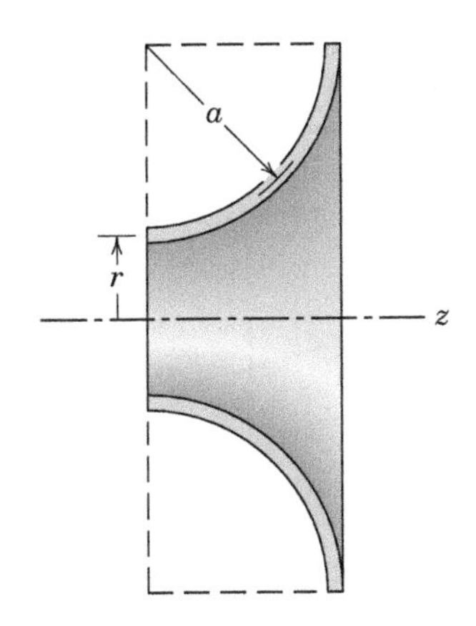

Problem B/27

►B/28 Determine the mass moment of inertia and corresponding radius of gyration of the thin homogeneous parabolic shell about the y-axis. The shell has dimensions $r = 70$ mm and $h = 200$ mm, and is made of metal plate having a mass per unit area of 32 kg/m^2.

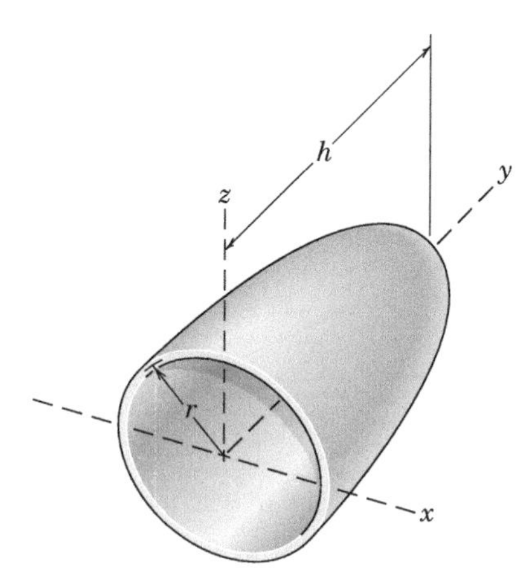

Problem B/28

►B/29 For the parabolic shell of Prob. B/28, determine the mass moment of inertia and corresponding radius of gyration about the z-axis.

Composite and Parallel-Axis Exercises

B/30 Every "slender" rod has a finite radius r. Refer to Table D/4 and derive an expression for the percentage error e which results if one neglects the radius r of a homogeneous solid cylindrical rod of length l when calculating its moment of inertia I_{zz}. Evaluate your expression for the ratios $r/l = 0.01$, 0.1, and 0.5.

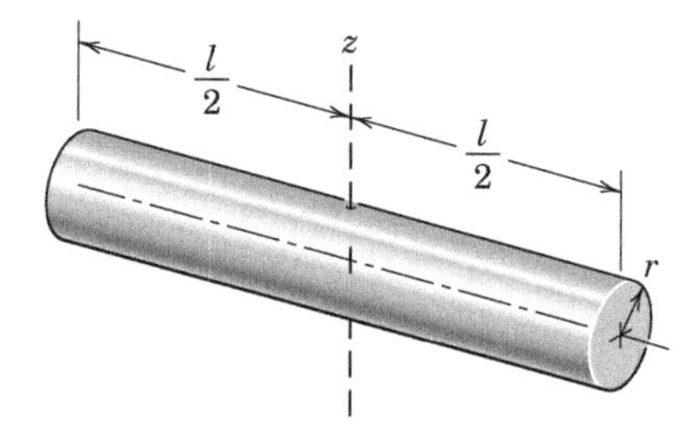

Problem B/30

B/31 The two small spheres of mass m each are connected by the light rigid rod which lies in the x-z plane. Determine the mass moments of inertia of the assembly about the x-, y-, and z-axes.

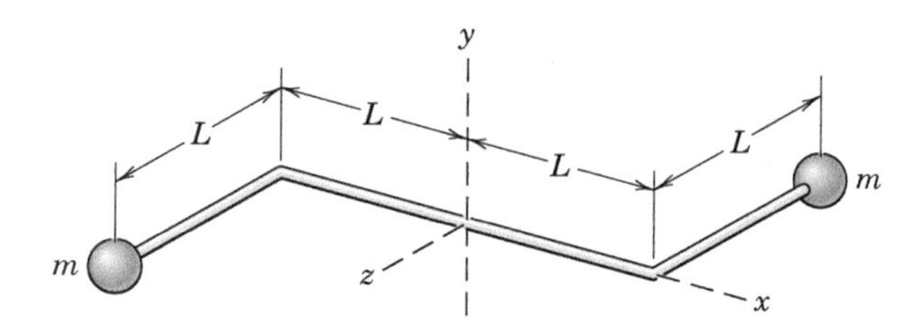

Problem B/31

B/32 The rectangular metal plate has a mass of 15 kg. Compute its moment of inertia about the y-axis. What is the magnitude of the percentage error e introduced by using the approximate relation $\frac{1}{3}ml^2$ for I_{xx}?

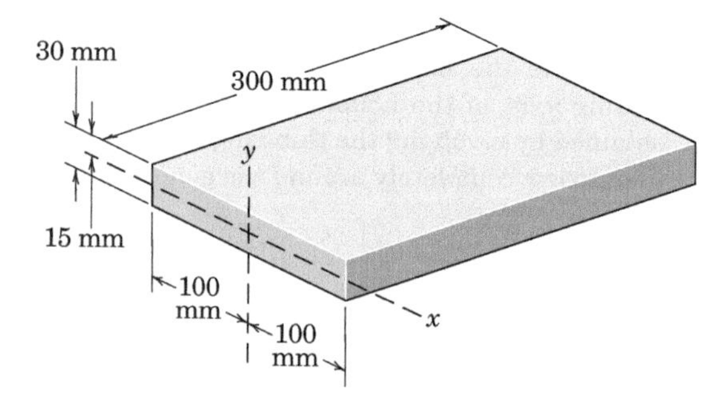

Problem B/32

B/33 Determine I_{xx} for the cylinder with a centered circular hole. The mass of the body is m.

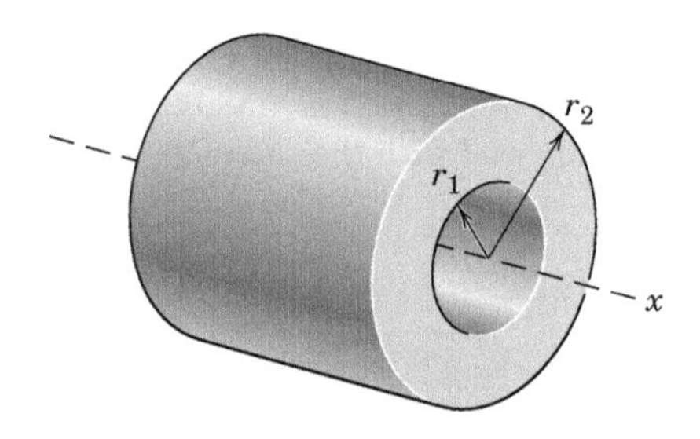

Problem B/33

B/34 Determine the mass moment of inertia about the z-axis for the right-circular cylinder with a central longitudinal hole.

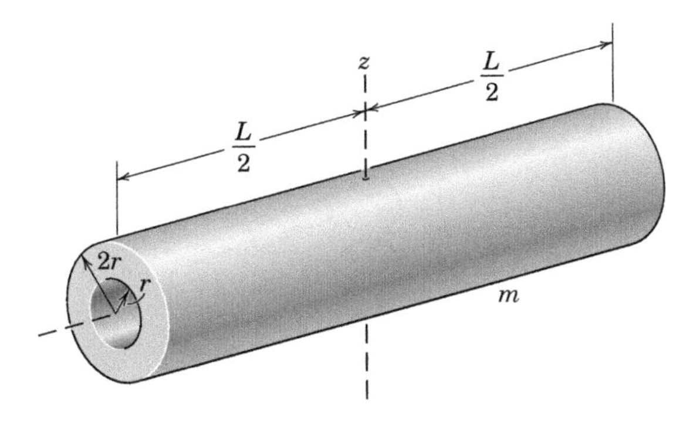

Problem B/34

B/35 Determine the moment of inertia of the half-ring of mass m about its diametral axis a-a and about axis b-b through the midpoint of the arc normal to the plane of the ring. The radius of the circular cross section is small compared with r.

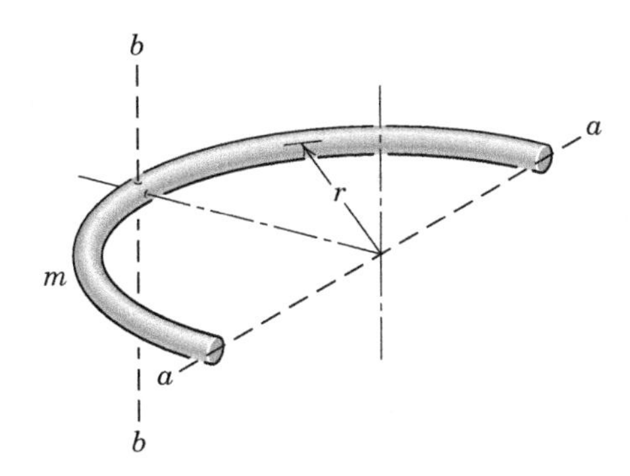

Problem B/35

B/36 A 6-in. steel cube is cut along its diagonal plane. Calculate the moment of inertia of the resulting prism about the edge x-x.

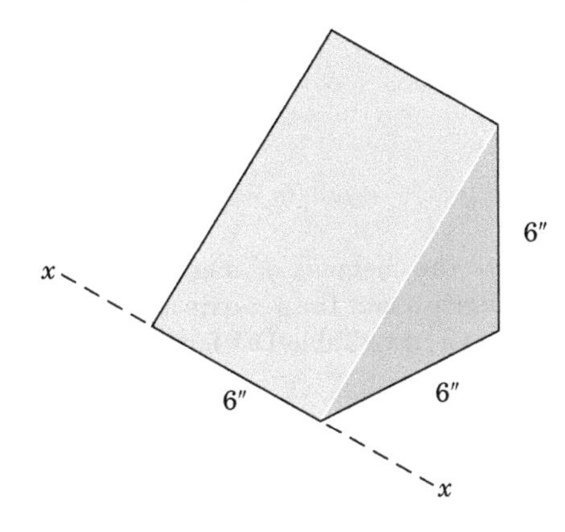

Problem B/36

B/37 The uniform coiled spring weighs 4 lb. Approximate its moments of inertia about the x-, y-, and z-axes from the analogy to the properties of a cylindrical shell.

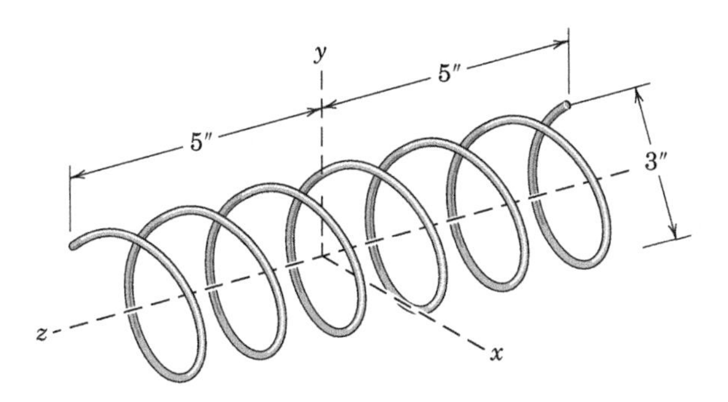

Problem B/37

B/38 Determine the length L of each of the slender rods of mass $m/2$ which must be centrally attached to the faces of the thin homogeneous disk of mass m in order to make the mass moments of inertia of the unit about the x- and z-axes equal.

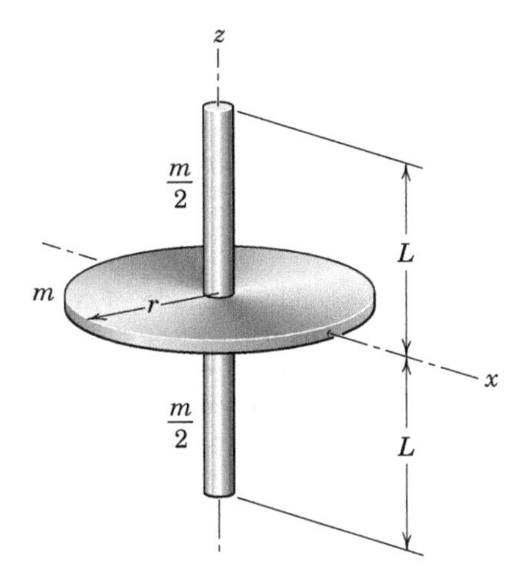

Problem B/38

B/39 A badminton racket is constructed of uniform slender rods bent into the shape shown. Neglect the strings and the built-up wooden grip and estimate the mass moment of inertia about the y-axis through O, which is the location of the player's hand. The mass per unit length of the rod material is ρ.

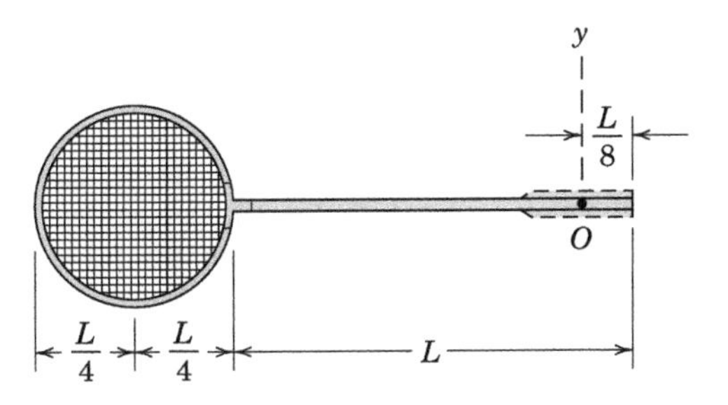

Problem B/39

B/40 Calculate the moment of inertia of the steel control wheel, shown in section, about its central axis. There are eight spokes, each of which has a constant cross-sectional area of 200 mm^2. What percent n of the total moment of inertia is contributed by the outer rim?

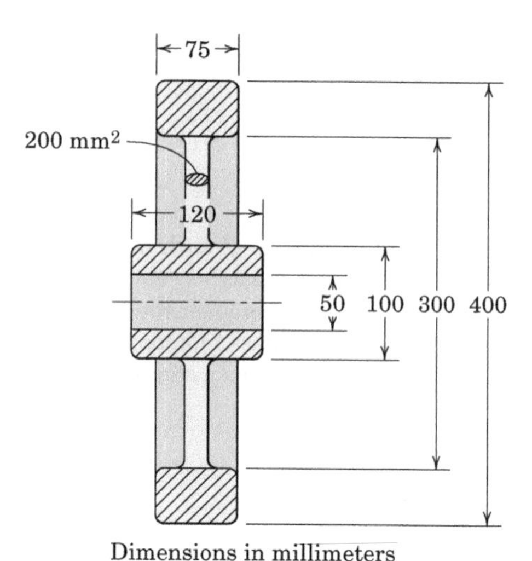

Problem B/40

B/41 The welded assembly is made of a uniform rod which weighs 0.370 lb per foot of length and the semicircular plate which weighs 8 lb per square foot. Determine the mass moment of inertia of the assembly about the three coordinate axes shown.

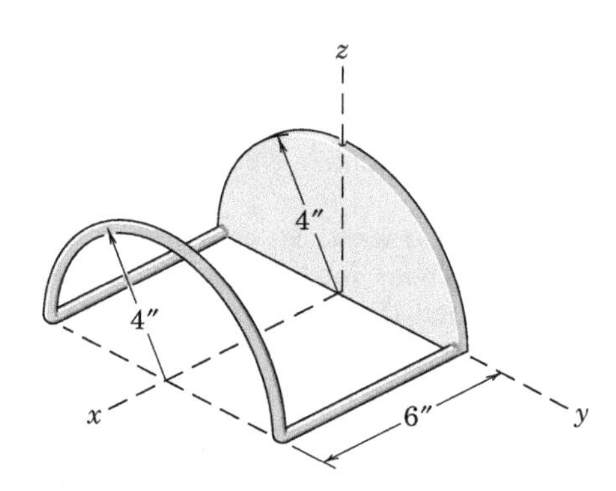

Problem B/41

B/42 The uniform rod of length $4b$ and mass m is bent into the shape shown. The diameter of the rod is small compared with its length. Determine the moments of inertia of the rod about the three coordinate axes.

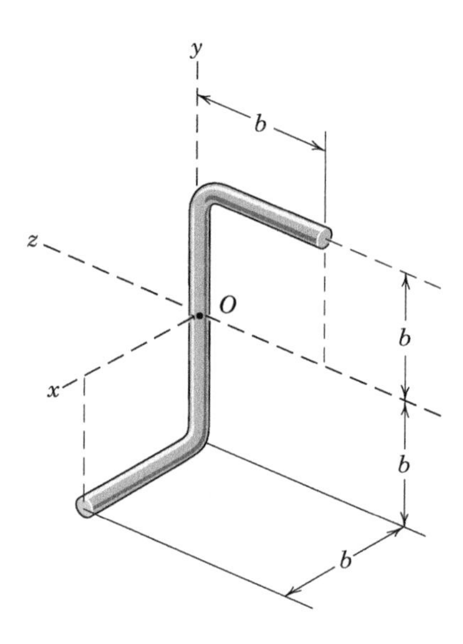

Problem B/42

B/43 The welded assembly shown in made from a steel rod which weighs 0.455 lb per foot of length. Determine the mass moment of inertia of the assembly (a) about the y-axis and (b) about the z-axis.

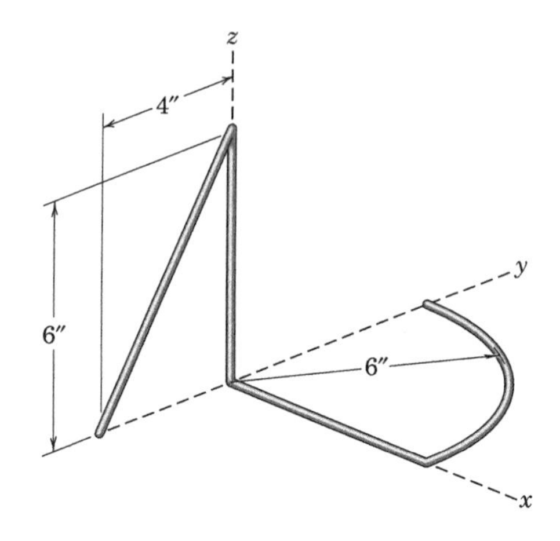

Problem B/43

B/44 Calculate the moment of inertia of the solid steel semicylinder about the x-x axis and about the parallel x_0-x_0 axis. (See Table D/1 for the density of steel.)

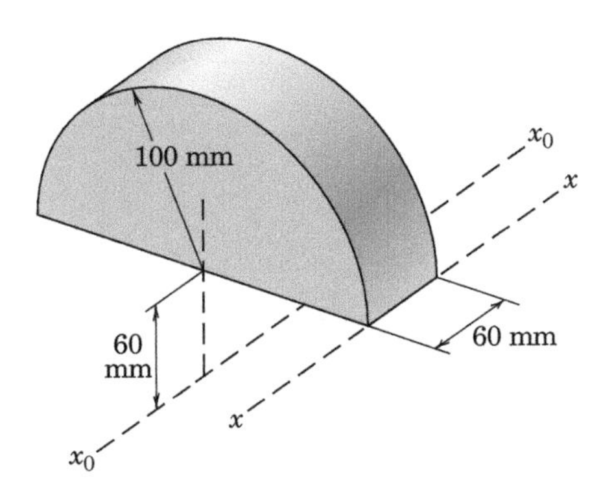

Problem B/44

B/45 The body is constructed of a uniform square plate, a uniform straight rod, a uniform quarter-circular rod, and a particle (negligible dimensions). If each part has the indicated mass, determine the mass moments of inertia of the body about the x-, y-, and z-axes.

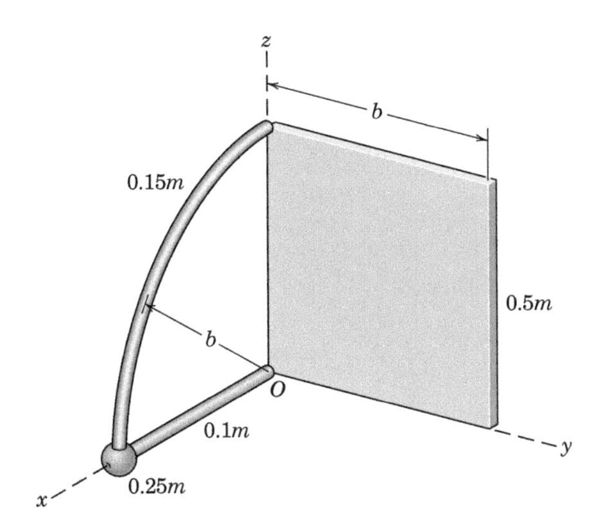

Problem B/45

B/46 The clock pendulum consists of the slender rod of length l and mass m and the bob of mass $7m$. Neglect the effects of the radius of the bob and determine I_O in terms of the bob position x. Calculate the ratio R of I_O evaluated for $x = \frac{3}{4}l$ to I_O evaluated for $x = l$.

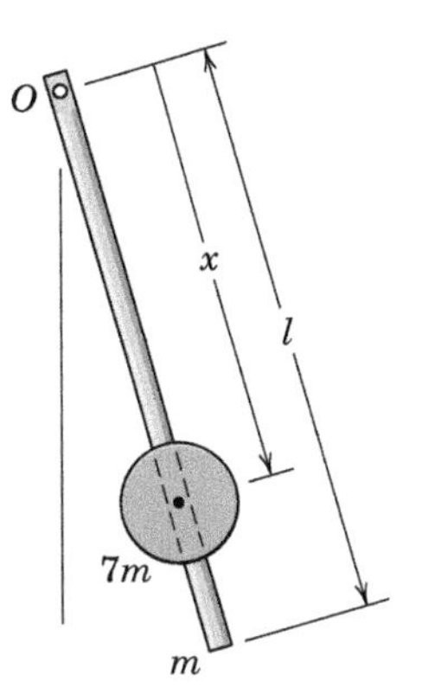

Problem B/46

B/47 Determine the mass moments of inertia of the bracket about the x- and x'-axes. The bracket is made from thin plate of uniform thickness and has a mass of 0.35 kg per square meter of area.

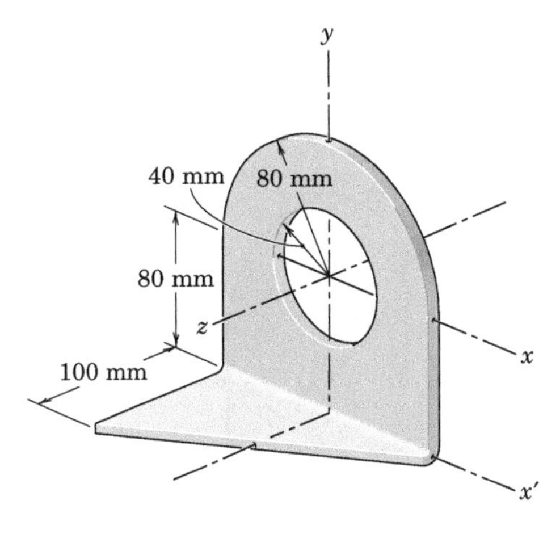

Problem B/47

B/48 A square plate with a quarter-circular sector removed has a net mass m. Determine its moment of inertia about axis A-A normal to the plane of the plate.

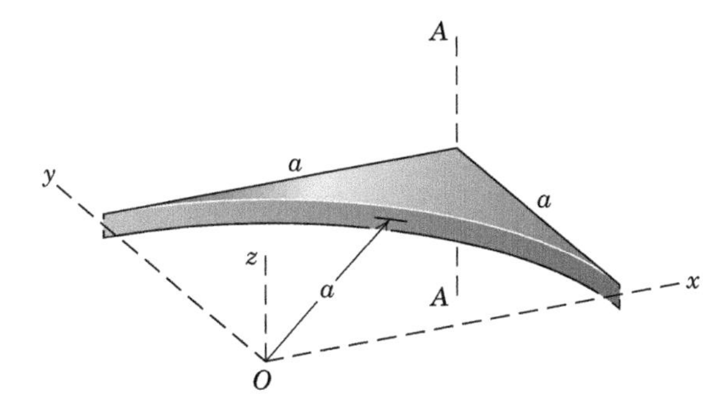

Problem B/48

B/49 The welded assembly consists of a cylindrical shell with a closed semicircular end. The shell is made from sheet metal with a mass of 24 kg/m^2, and the end is made from metal plate with a mass of 36 kg/m^2. Determine the mass moments of inertia of the assembly about the coordinate axes shown.

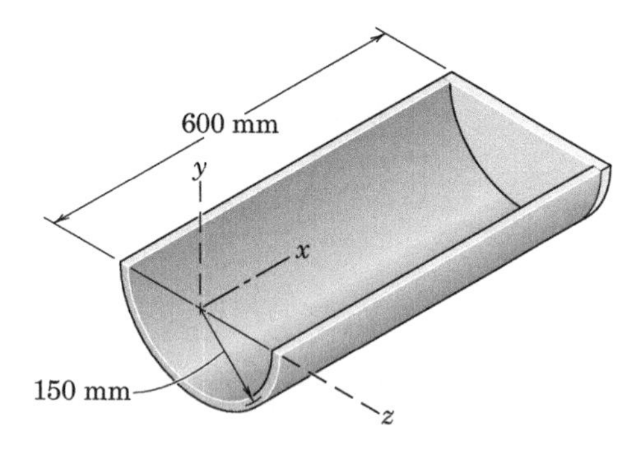

Problem B/49

B/50 Determine the mass moment of inertia of the steel bracket about the z-axis which passes through the midline of the base.

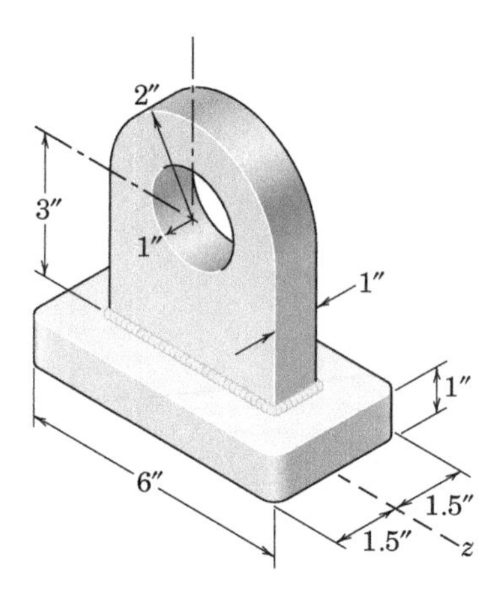

Problem B/50

B/51 The slender metal rods are welded together in the configuration shown. Each 6-in. segment weighs 0.30 lb. Compute the moment of inertia of the assembly about the y-axis.

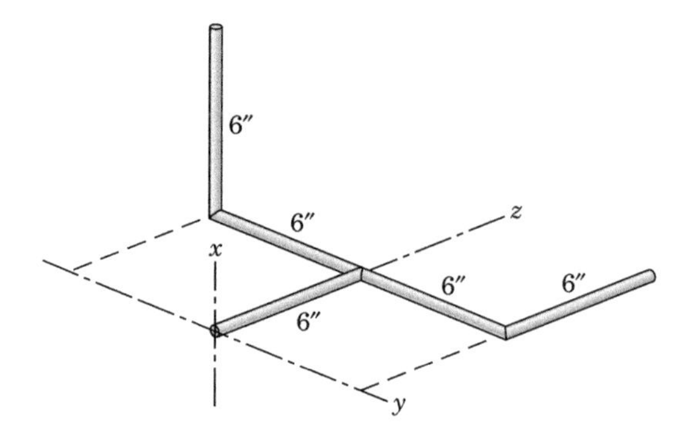

Problem B/51

B/52 The welded assembly is formed from thin sheet metal having a mass of 19 kg/m^2. Determine the mass moments of inertia for the assembly about the x- and y-axes.

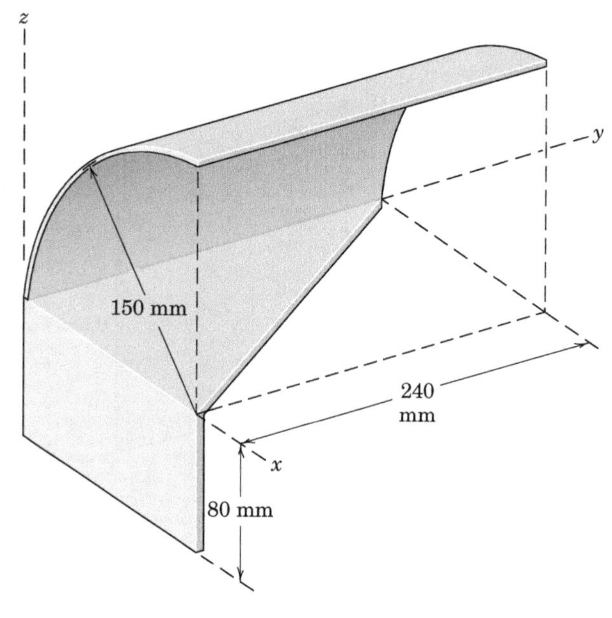

Problem B/52

B/53 Determine I_{xx} for the cone frustum, which has base radii r_1 and r_2 and mass m.

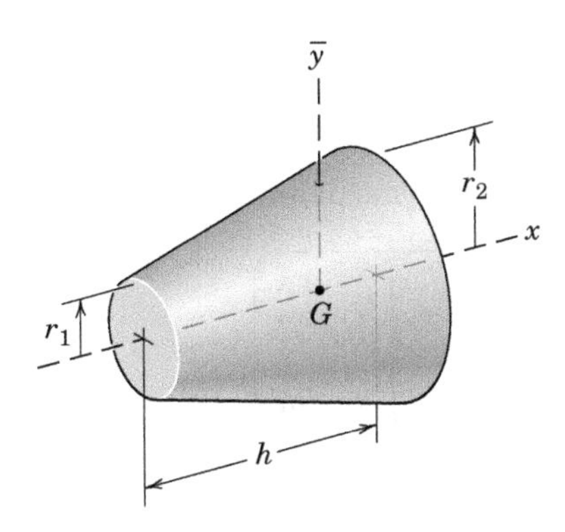

Problem B/53

***B/54** A preliminary design model to ensure rotational stability for a spacecraft consists of the cylindrical shell and the two square panels as shown. The shell and panels have the same thickness and density. It can be shown that rotational stability about the z-axis can be maintained if I_{zz} is less than I_{xx} and I_{yy}. For a given value of r, determine the limitation for L.

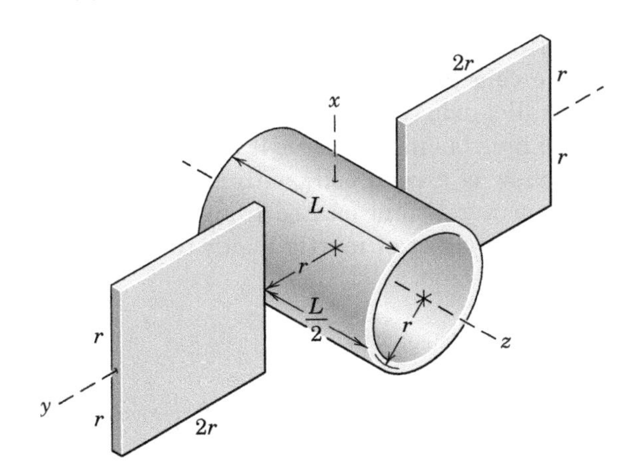

Problem B/54

►B/55 Determine the radius of gyration of the aluminum part about the z-axis. The hole in the upper surface is drilled completely through the part.

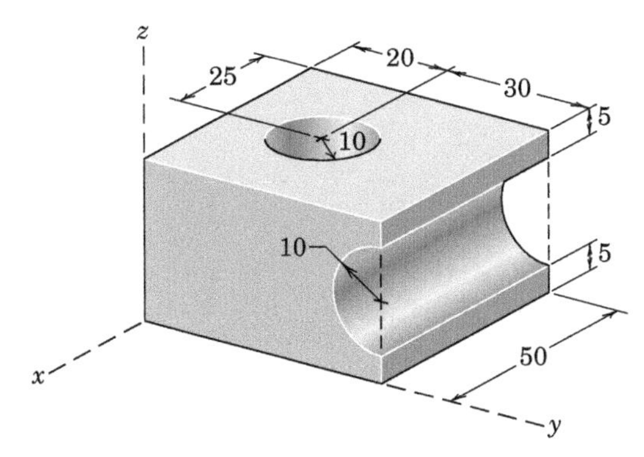

Problem B/55

►B/56 Compute the moment of inertia of the mallet about the O-O axis. The mass of the head is 0.8 kg, and the mass of the handle is 0.5 kg.

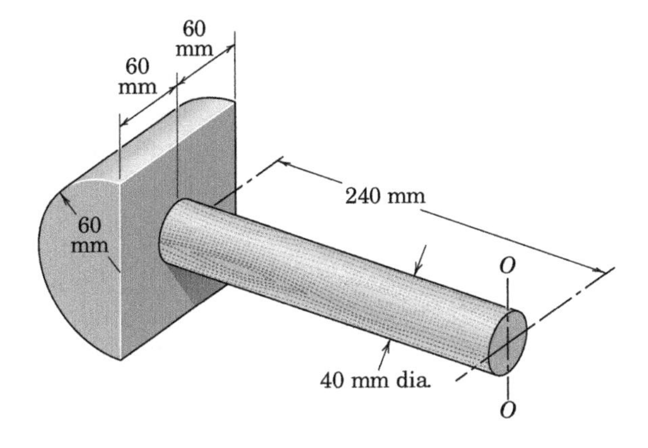

Problem B/56

B/2 Products of Inertia

For problems in the rotation of three-dimensional rigid bodies, the expression for angular momentum contains, in addition to the moment-of-inertia terms, *product-of-inertia* terms defined as

$$\begin{aligned} I_{xy} &= I_{yx} = \int xy\, dm \\ I_{xz} &= I_{zx} = \int xz\, dm \\ I_{yz} &= I_{zy} = \int yz\, dm \end{aligned} \qquad \textbf{(B/8)}$$

These expressions were cited in Eqs. 7/10 in the expansion of the expression for angular momentum, Eq. 7/9.

The calculation of products of inertia involves the same basic procedure which we have followed in calculating moments of inertia and in evaluating other volume integrals as far as the choice of element and the limits of integration are concerned. The only special precaution we need to observe is to be doubly watchful of the algebraic signs in the expressions. Whereas moments of inertia are always positive, products of inertia may be either positive or negative. The units of products of inertia are the same as those of moments of inertia.

We have seen that the calculation of moments of inertia is often simplified by using the parallel-axis theorem. A similar theorem exists for transferring products of inertia, and we prove it easily as follows. In Fig. B/6 is shown the x-y view of a rigid body with parallel axes x_0-y_0 passing through the mass center G and located from the x-y axes by the distances d_x and d_y. The product of inertia about the x-y axes by definition is

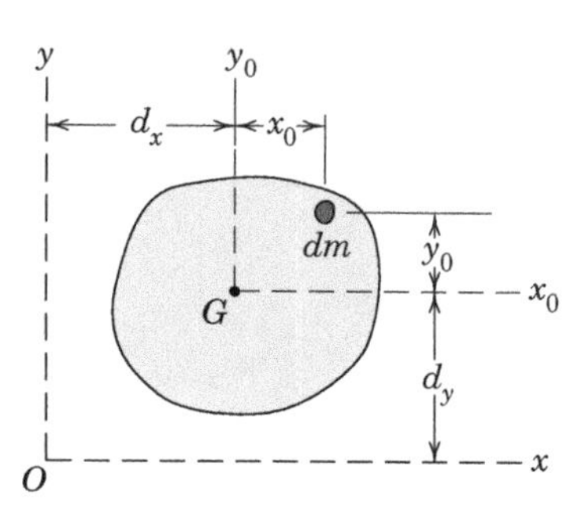

Figure B/6

$$\begin{aligned} I_{xy} &= \int xy\, dm = \int (x_0 + d_x)(y_0 + d_y)\, dm \\ &= \int x_0 y_0\, dm + d_x d_y \int dm + d_x \int y_0\, dm + d_y \int x_0\, dm \\ &= I_{x_0 y_0} + m d_x d_y \end{aligned}$$

The last two integrals vanish since the first moments of mass about the mass center are necessarily zero.

Similar relations exist for the remaining two product-of-inertia terms. Dropping the zero subscripts and using the bar to designate the mass-center quantity, we obtain

$$\begin{aligned} I_{xy} &= \bar{I}_{xy} + m d_x d_y \\ I_{xz} &= \bar{I}_{xz} + m d_x d_z \\ I_{yz} &= \bar{I}_{yz} + m d_y d_z \end{aligned} \qquad \textbf{(B/9)}$$

These transfer-of-axis relations are valid *only* for transfer to or from *parallel axes* through the *mass center*.

With the aid of the product-of-inertia terms, we can calculate the moment of inertia of a rigid body about any prescribed axis through the coordinate origin. For the rigid body of Fig. B/7, suppose we must determine the moment of inertia about axis $O\text{-}M$. The direction cosines of $O\text{-}M$ are l, m, n, and a unit vector $\boldsymbol{\lambda}$ along $O\text{-}M$ may be written $\boldsymbol{\lambda} = l\mathbf{i} + m\mathbf{j} + n\mathbf{k}$. The moment of inertia about $O\text{-}M$ is

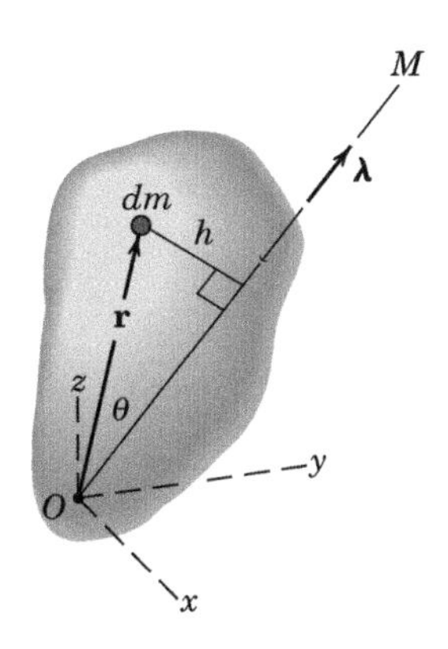

Figure B/7

$$I_M = \int h^2\, dm = \int (\mathbf{r} \times \boldsymbol{\lambda}) \cdot (\mathbf{r} \times \boldsymbol{\lambda})\, dm$$

where $|\mathbf{r} \times \boldsymbol{\lambda}| = r \sin\theta = h$. The cross product is

$$(\mathbf{r} \times \boldsymbol{\lambda}) = (yn - zm)\mathbf{i} + (zl - xn)\mathbf{j} + (xm - yl)\mathbf{k}$$

and, after we collect terms, the dot-product expansion gives

$$\begin{aligned}(\mathbf{r} \times \boldsymbol{\lambda}) \cdot (\mathbf{r} \times \boldsymbol{\lambda}) = h^2 &= (y^2 + z^2)l^2 + (x^2 + z^2)m^2 + (x^2 + y^2)n^2 \\ &\quad - 2xylm - 2xzln - 2yzmn\end{aligned}$$

Thus, with the substitution of the expressions of Eqs. B/4 and B/8, we have

$$\boxed{I_M = I_{xx}l^2 + I_{yy}m^2 + I_{zz}n^2 - 2I_{xy}lm - 2I_{xz}ln - 2I_{yz}mn} \qquad \textbf{(B/10)}$$

This expression gives the moment of inertia about any axis $O\text{-}M$ in terms of the direction cosines of the axis and the moments and products of inertia about the coordinate axes.

Principal Axes of Inertia

As noted in Art. 7/7, the array

$$\begin{bmatrix} I_{xx} & -I_{xy} & -I_{xz} \\ -I_{yx} & I_{yy} & -I_{yz} \\ -I_{zx} & -I_{zy} & I_{zz} \end{bmatrix}$$

whose elements appear in the expansion of the angular-momentum expression, Eq. 7/11, for a rigid body with attached axes, is called the *inertia matrix* or *inertia tensor*. If we examine the moment- and product-of-inertia terms for all possible orientations of the axes with respect to the body for a given origin, we will find in the general case an orientation of the x-y-z axes for which the product-of-inertia terms vanish and the array takes the diagonalized form

$$\begin{bmatrix} I_{xx} & 0 & 0 \\ 0 & I_{yy} & 0 \\ 0 & 0 & I_{zz} \end{bmatrix}$$

Such axes x-y-z are called the *principal axes of inertia*, and I_{xx}, I_{yy}, and I_{zz} are called the *principal moments of inertia* and represent the maximum, minimum, and intermediate values of the moments of inertia for the particular origin chosen.

It may be shown* that for any given orientation of axes x-y-z the solution of the determinant equation

$$\begin{vmatrix} I_{xx} - I & -I_{xy} & -I_{xz} \\ -I_{yx} & I_{yy} - I & -I_{yz} \\ -I_{zx} & -I_{zy} & I_{zz} - I \end{vmatrix} = 0 \qquad \textbf{(B/11)}$$

for I yields three roots I_1, I_2, and I_3 of the resulting cubic equation which are the three principal moments of inertia. Also, the direction cosines l, m, and n of a principal inertia axis are given by

$$\begin{aligned} (I_{xx} - I)l - I_{xy}m - I_{xz}n &= 0 \\ -I_{yx}l + (I_{yy} - I)m - I_{yz}n &= 0 \\ -I_{zx}l - I_{zy}m + (I_{zz} - I)n &= 0 \end{aligned} \qquad \textbf{(B/12)}$$

These equations along with $l^2 + m^2 + n^2 = 1$ will enable a solution for the direction cosines to be made for each of the three I's.

To assist with the visualization of these conclusions, consider the rectangular block, Fig. B/8, having an arbitrary orientation with respect to the x-y-z axes. For simplicity, the mass center G is located at the origin of the coordinates. If the moments and products of inertia for the block about the x-y-z axes are known, then solution of Eq. B/11 would give the three roots, I_1, I_2, and I_3, which are the principal moments of inertia. Solution of Eq. B/12 using each of the three I's, in turn, along with $l^2 + m^2 + n^2 = 1$, would give the direction cosines l, m, and n for each of the respective principal axes, which are always mutually perpendicular. From the proportions of the block as drawn, we see that I_1 is the maximum moment of inertia, I_2 is the intermediate value, and I_3 is the minimum value.

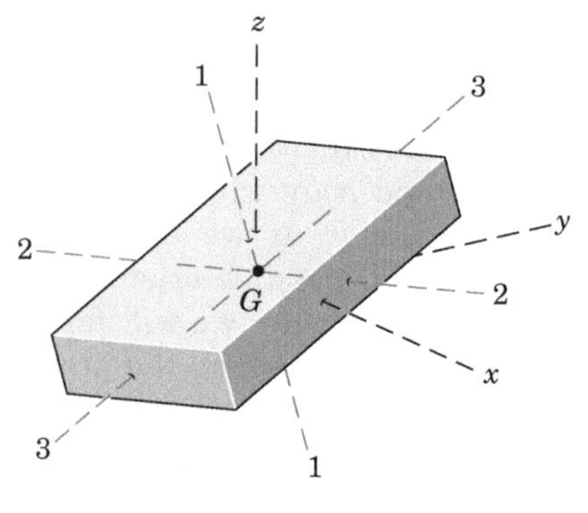

Figure B/8

*See, for example, the first author's *Dynamics, SI Version*, 1975, John Wiley & Sons, Art. 41.

Sample Problem B/6

The bent plate has a uniform thickness t which is negligible compared with its other dimensions. The density of the plate material is ρ. Determine the products of inertia of the plate with respect to the axes as chosen.

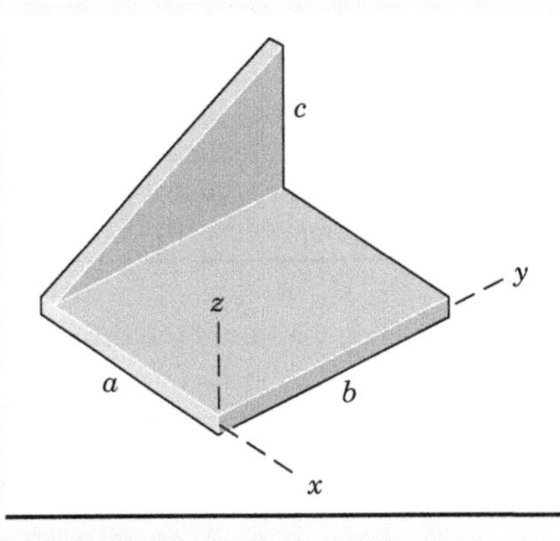

Solution. Each of the two parts is analyzed separately.

Rectangular part. In the separate view of this part, we introduce parallel ① axes x_0-y_0 through the mass center G and use the transfer-of-axis theorem. By symmetry, we see that $\bar{I}_{xy} = I_{x_0y_0} = 0$ so that

$$[I_{xy} = \bar{I}_{xy} + md_xd_y] \qquad I_{xy} = 0 + \rho tab\left(-\frac{a}{2}\right)\left(\frac{b}{2}\right) = -\frac{1}{4}\rho ta^2b^2$$

Because the z-coordinate of all elements of the plate is zero, it follows that $I_{xz} = I_{yz} = 0$.

Triangular part. In the separate view of this part, we locate the mass center G and construct x_0-, y_0-, and z_0-axes through G. Since the x_0-coordinate of all elements is zero, it follows that $\bar{I}_{xy} = I_{x_0y_0} = 0$ and $\bar{I}_{xz} = I_{x_0z_0} = 0$. The transfer-of-axis theorems then give us

$$[I_{xy} = \bar{I}_{xy} + md_xd_y] \qquad I_{xy} = 0 + \rho t\frac{b}{2}c(-a)\left(\frac{2b}{3}\right) = -\frac{1}{3}\rho tab^2c$$

$$[I_{xz} = \bar{I}_{xz} + md_xd_z] \qquad I_{xz} = 0 + \rho t\frac{b}{2}c(-a)\left(\frac{c}{3}\right) = -\frac{1}{6}\rho tabc^2$$

We obtain I_{yz} by direct integration, noting that the distance a of the plane of the triangle from the y-z plane in no way affects the y- and z-coordinates. With the mass element $dm = \rho t\,dy\,dz$, we have

② $$\left[I_{yz} = \int yz\,dm\right] \qquad I_{yz} = \rho t\int_0^b\int_0^{cy/b} yz\,dz\,dy = \rho t\int_0^b y\left[\frac{z^2}{2}\right]_0^{cy/b} dy$$

$$= \frac{\rho tc^2}{2b^2}\int_0^b y^3\,dy = \frac{1}{8}\rho tb^2c^2$$

Adding the expressions for the two parts gives

$$I_{xy} = -\tfrac{1}{4}\rho ta^2b^2 - \tfrac{1}{3}\rho tab^2c = -\tfrac{1}{12}\rho tab^2(3a + 4c) \qquad \textit{Ans.}$$

$$I_{xz} = 0 \quad -\tfrac{1}{6}\rho tabc^2 = -\tfrac{1}{6}\rho tabc^2 \qquad \textit{Ans.}$$

$$I_{yz} = 0 \quad +\tfrac{1}{8}\rho tb^2c^2 = +\tfrac{1}{8}\rho tb^2c^2 \qquad \textit{Ans.}$$

Helpful Hints

① We must be careful to preserve the same sense of the coordinates. Thus, plus x_0 and y_0 must agree with plus x and y.

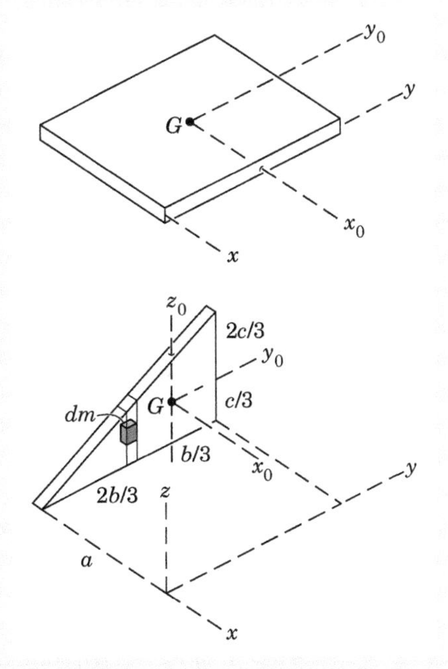

② We choose to integrate with respect to z first, where the upper limit is the variable height $z = cy/b$. If we were to integrate first with respect to y, the limits of the first integral would be from the variable $y = bz/c$ to b.

Sample Problem B/7

The angle bracket is made from aluminum plate with a mass of 13.45 kg per square meter. Calculate the principal moments of inertia about the origin O and the direction cosines of the principal axes of inertia. The thickness of the plate is small compared with the other dimensions.

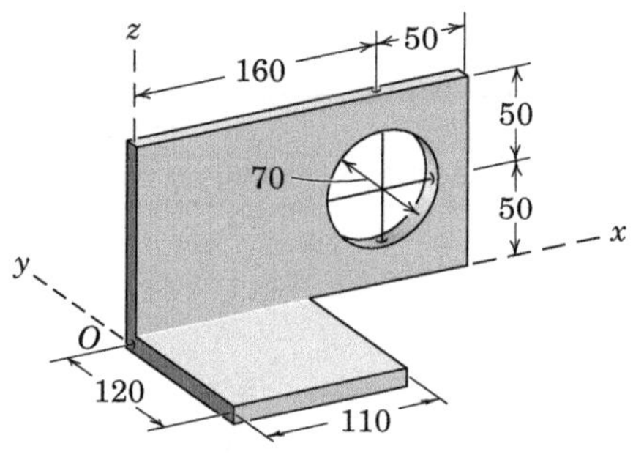

Dimensions in millimeters

Solution. The masses of the three parts are

$$m_1 = 13.45(0.21)(0.1) = 0.282 \text{ kg}$$

① $$m_2 = -13.45\pi(0.035)^2 = -0.0518 \text{ kg}$$

$$m_3 = 13.45(0.12)(0.11) = 0.1775 \text{ kg}$$

Part 1

$$I_{xx} = \frac{1}{3}mb^2 = \frac{1}{3}(0.282)(0.1)^2 = 9.42(10^{-4}) \text{ kg}\cdot\text{m}^2$$

② $$I_{yy} = \frac{1}{3}m(a^2 + b^2) = \frac{1}{3}(0.282)[(0.21)^2 + (0.1)^2] = 50.9(10^{-4}) \text{ kg}\cdot\text{m}^2$$

$$I_{zz} = \frac{1}{3}ma^2 = \frac{1}{3}(0.282)(0.21)^2 = 41.5(10^{-4}) \text{ kg}\cdot\text{m}^2$$

$$I_{xy} = 0 \qquad I_{yz} = 0$$

$$I_{xz} = \bar{I}_{xz} + md_xd_z$$

$$= 0 + m\frac{a}{2}\frac{b}{2} = 0.282(0.105)(0.05) = 14.83(10^{-4}) \text{ kg}\cdot\text{m}^2$$

Part 2

$$I_{xx} = \frac{1}{4}mr^2 + md_z^{\,2} = -0.0518\left[\frac{(0.035)^2}{4} + (0.050)^2\right]$$

$$= -1.453(10^{-4}) \text{ kg}\cdot\text{m}^2$$

$$I_{yy} = \frac{1}{2}mr^2 + m(d_x^{\,2} + d_z^{\,2})$$

$$= -0.0518\left[\frac{(0.035)^2}{2} + (0.16)^2 + (0.05)^2\right]$$

$$= -14.86(10^{-4}) \text{ kg}\cdot\text{m}^2$$

$$I_{zz} = \frac{1}{4}mr^2 + md_x^{\,2} = -0.0518\left[\frac{(0.035)^2}{4} + (0.16)^2\right]$$

$$= -13.41(10^{-4}) \text{ kg}\cdot\text{m}^2$$

$$I_{xy} = 0 \qquad I_{yz} = 0$$

$$I_{xz} = \bar{I}_{xz} + md_xd_z = 0 - 0.0518(0.16)(0.05) = -4.14(10^{-4}) \text{ kg}\cdot\text{m}^2$$

Helpful Hints

① Note that the mass of the hole is treated as a negative number.

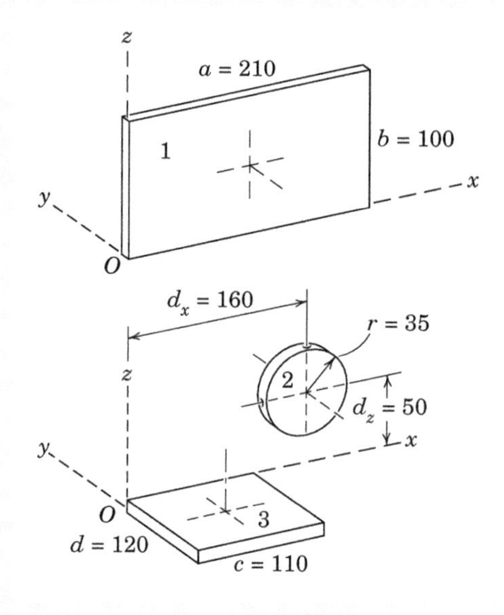

② You can easily derive this formula. Also check Table D/4.

Sample Problem B/7 (Continued)

Part 3

$$I_{xx} = \frac{1}{3}md^2 = \frac{1}{3}(0.1775)(0.12)^2 = 8.52(10^{-4})\ \text{kg}\cdot\text{m}^2$$

$$I_{yy} = \frac{1}{3}mc^2 = \frac{1}{3}(0.1775)(0.11)^2 = 7.16(10^{-4})\ \text{kg}\cdot\text{m}^2$$

$$I_{zz} = \frac{1}{3}m(c^2 + d^2) = \frac{1}{3}(0.1775)[(0.11)^2 + (0.12)^2]$$

$$= 15.68(10^{-4})\ \text{kg}\cdot\text{m}^2$$

$$I_{xy} = \bar{I}_{xy} + md_xd_y$$

$$= 0 + m\frac{c}{2}\left(\frac{-d}{2}\right) = 0.1775(0.055)(-0.06) = -5.86(10^{-4})\ \text{kg}\cdot\text{m}^2$$

$$I_{yz} = 0 \qquad I_{xz} = 0$$

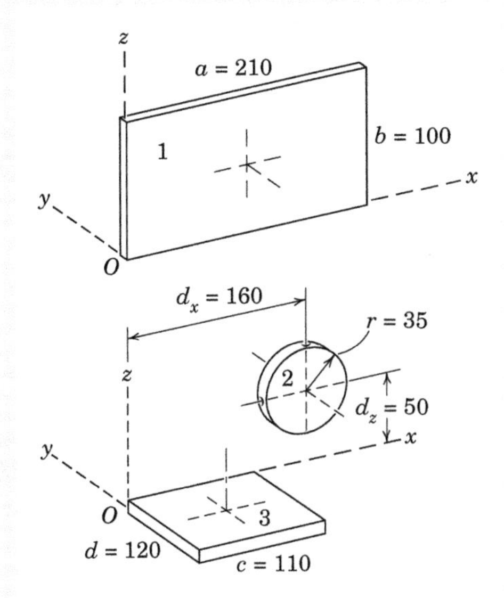

Totals

$$I_{xx} = 16.48(10^{-4})\ \text{kg}\cdot\text{m}^2 \qquad I_{xy} = -5.86(10^{-4})\ \text{kg}\cdot\text{m}^2$$

$$I_{yy} = 43.2(10^{-4})\ \text{kg}\cdot\text{m}^2 \qquad I_{yz} = 0$$

$$I_{zz} = 43.8(10^{-4})\ \text{kg}\cdot\text{m}^2 \qquad I_{xz} = 10.69(10^{-4})\ \text{kg}\cdot\text{m}^2$$

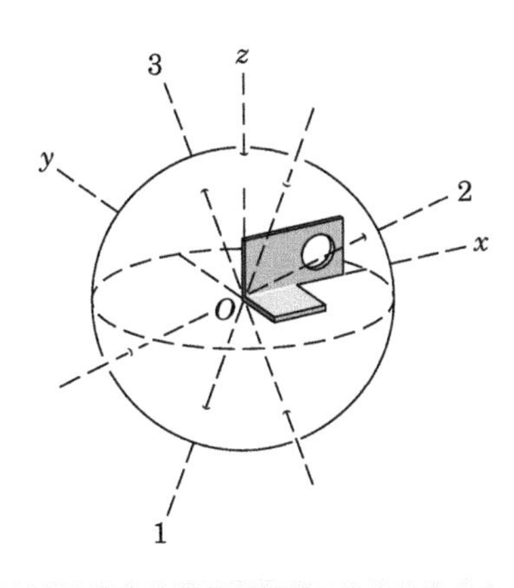

Substitution into Eq. B/11, expansion of the determinant, and simplification yield

$$I^3 - 103.5(10^{-4})I^2 + 3180(10^{-8})I - 24\,800(10^{-12}) = 0$$

③ Solution of this cubic equation yields the following roots, which are the principal moments of inertia

$$I_1 = 48.3(10^{-4})\ \text{kg}\cdot\text{m}^2$$

$$I_2 = 11.82(10^{-4})\ \text{kg}\cdot\text{m}^2 \qquad \textit{Ans.}$$

$$I_3 = 43.4(10^{-4})\ \text{kg}\cdot\text{m}^2$$

The direction cosines of each principal axis are obtained by substituting each root, in turn, into Eq. B/12 and using $l^2 + m^2 + n^2 = 1$. The results are

$$l_1 = 0.357 \qquad l_2 = 0.934 \qquad l_3 = 0.01830$$

$$m_1 = 0.410 \qquad m_2 = -0.1742 \qquad m_3 = 0.895 \qquad \textit{Ans.}$$

$$n_1 = -0.839 \qquad n_2 = 0.312 \qquad n_3 = 0.445$$

The bottom figure shows a pictorial view of the bracket and the orientation of its principal axes of inertia.

③ A computer program for the solution of a cubic equation may be used, or an algebraic solution using the formula cited in item 4 of Art. C/4, Appendix C, may be employed.

PROBLEMS

Introductory Problems

B/57 Determine the products of inertia about the coordinate axes for the unit which consists of four small particles, each of mass m, connected by the light but rigid slender rods.

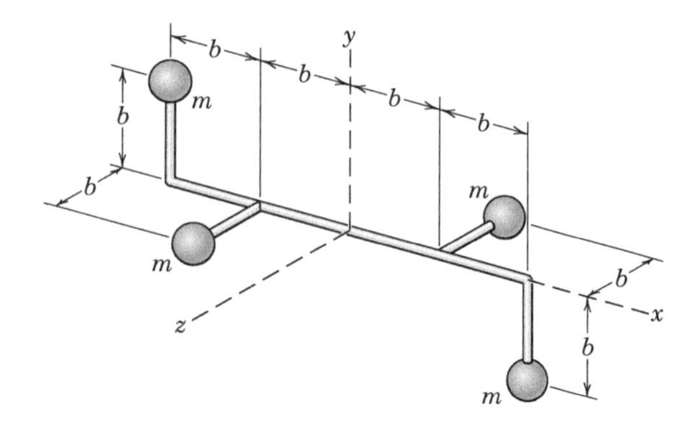

Problem B/57

B/58 Determine the products of inertia about the coordinate axes for the unit which consists of three small spheres, each of mass m, connected by the light but rigid slender rods.

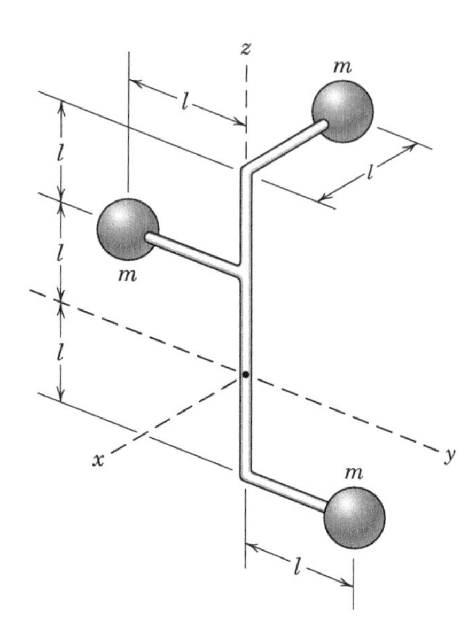

Problem B/58

B/59 Determine the product of inertia I_{xy} for the slender rod of mass m.

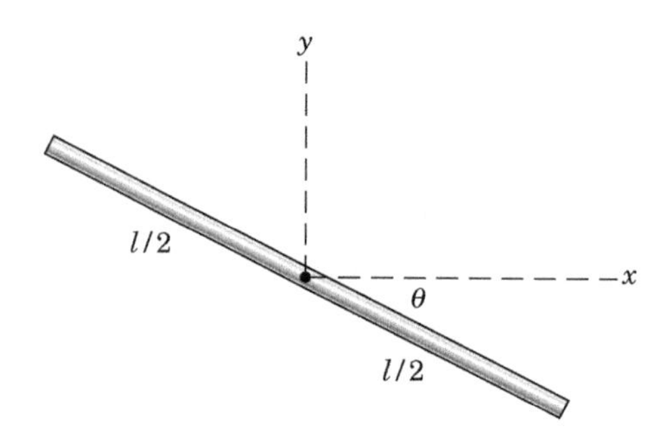

Problem B/59

B/60 The rod of Prob. B/1 is repeated here. Determine the product of inertia for the rod about the x-y axes. The rod has a mass ρ per unit length.

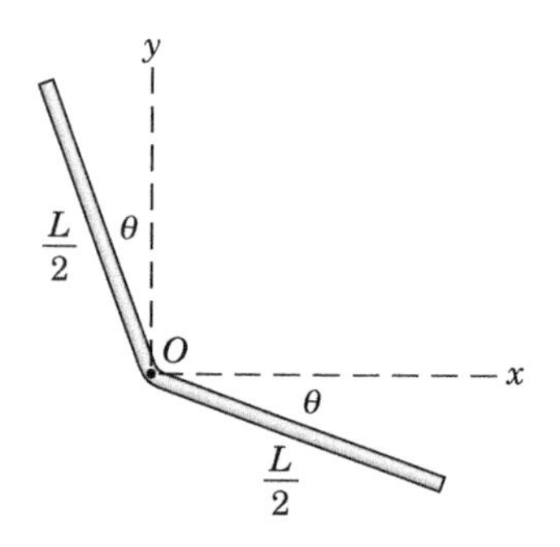

Problem B/60

B/61 Determine the products of inertia of the uniform slender rod of mass m about the coordinate axes shown.

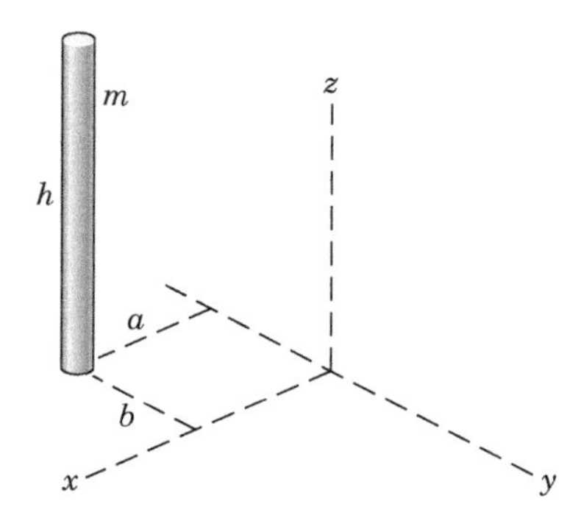

Problem B/61

B/62 Determine the products of inertia about the coordinate axes for the thin square plate with two circular holes. The mass of the plate material per unit area is ρ.

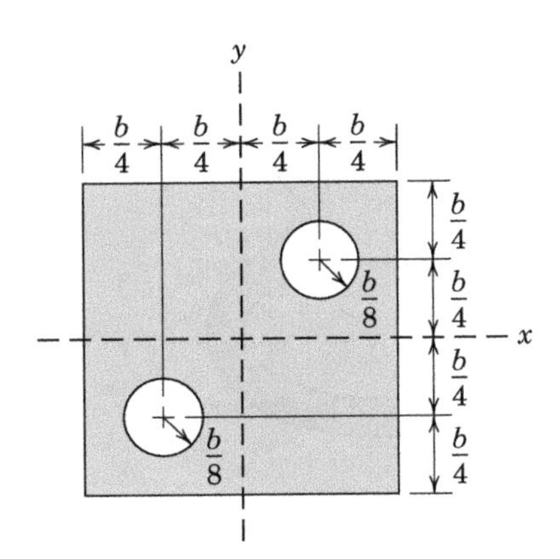

Problem B/62

B/63 Determine the products of inertia of the solid homogeneous half-cylinder of mass m for the axes shown.

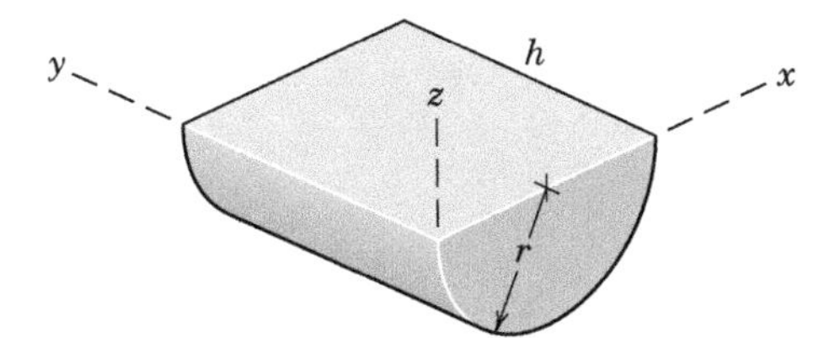

Problem B/63

B/64 Determine the products of inertia about the coordinate axes for the thin plate of mass m which has the shape of a circular sector of radius a and angle β as shown.

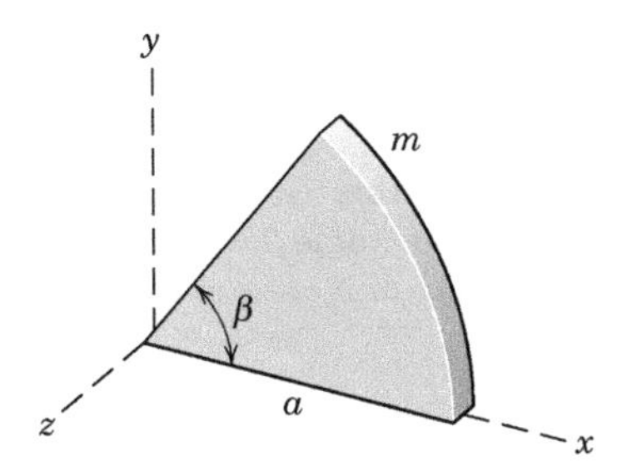

Problem B/64

B/65 The homogeneous plate of Prob. B/7 is repeated here. Determine the product of inertia for the plate about the x-y axes. The plate has mass m and uniform thickness t.

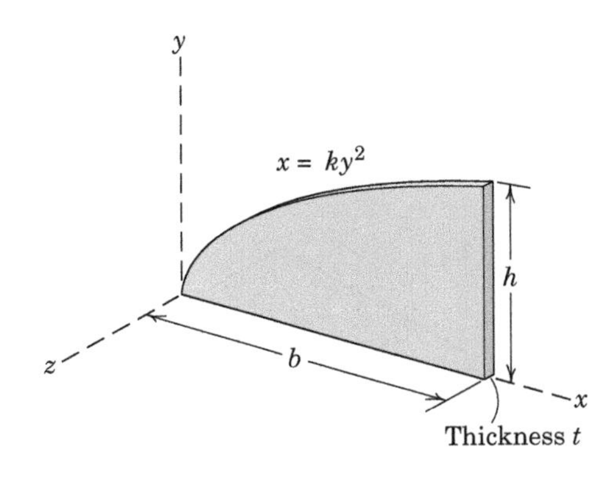

Problem B/65

B/66 Determine by direct integration the product of inertia of the thin homogenous triangular plate of mass m about the x-y axes. Then, use the parallel-axis theorem to determine the product of inertia for the plate about the x'-y' axes and the y''-y'' axes. What is the product of inertia about the centroid of the plate?

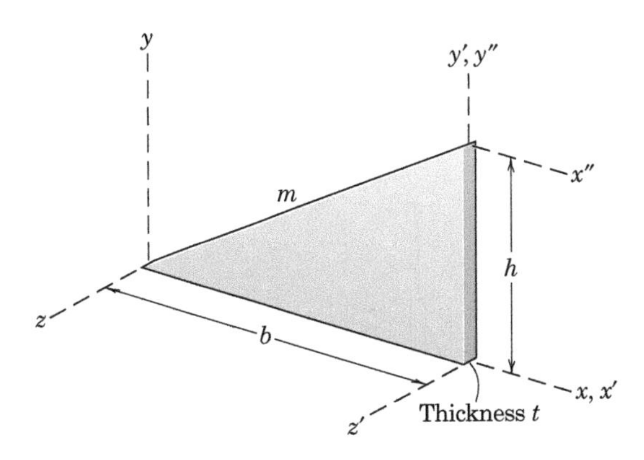

Problem B/66

Representative Problems

B/67 The aluminium casting consists of a 6-in. cube with a 4-in. cubical recess. Calculate the products of inertia of the casting about the axes shown.

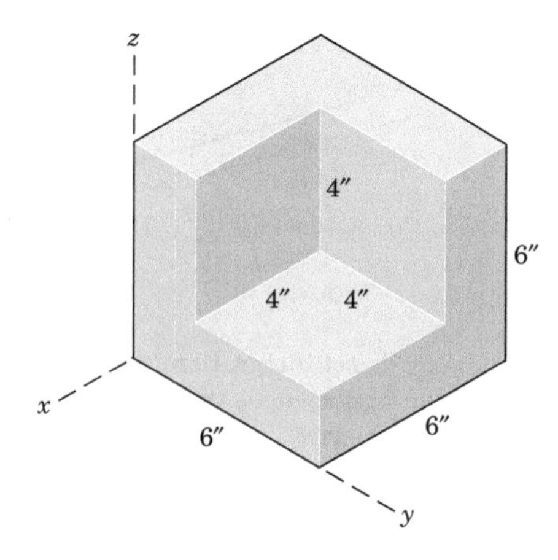

Problem B/67

B/68 Determine the products of inertia about the coordinate axes for the assembly which consists of uniform slender rods. Each rod has a mass ρ per unit length.

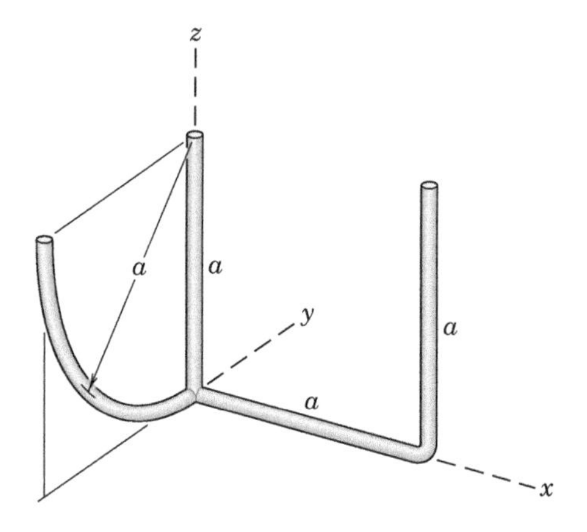

Problem B/68

B/69 The S-shaped piece is formed from a rod of diameter d and bent into the two semicircular shapes. Determine the products of inertia for the rod, for which d is small compared with r.

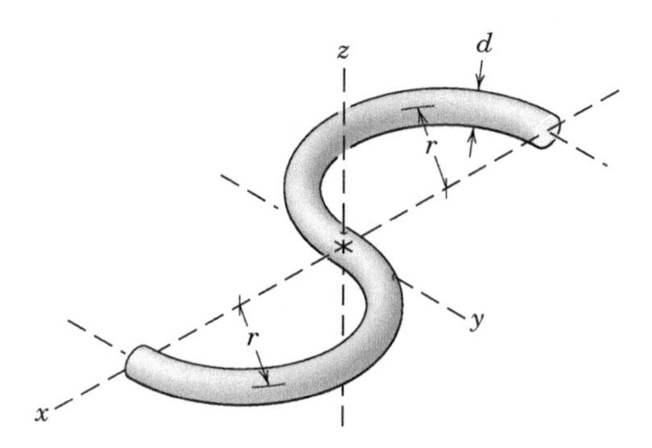

Problem B/69

B/70 Prove that the moment of inertia of the rigid assembly of three identical balls, each of mass m and radius r, has the same value for all axes through O. Neglect the mass of the connecting rods.

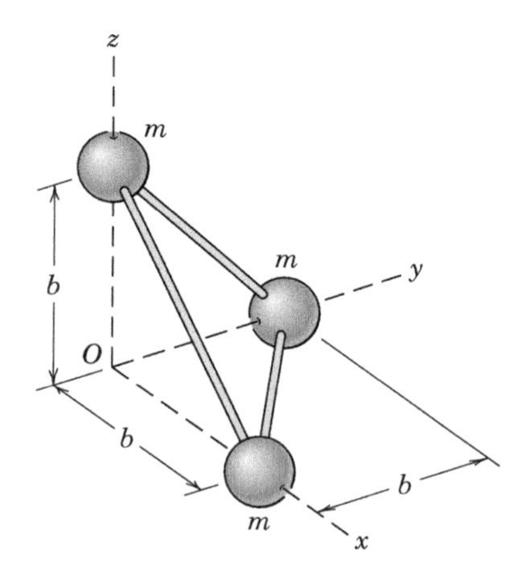

Problem B/70

►**B/71** The plane of the thin circular disk of mass m and radius r makes an angle β with the x-z plane. Determine the product of inertia of the disk with respect to the y-z plane.

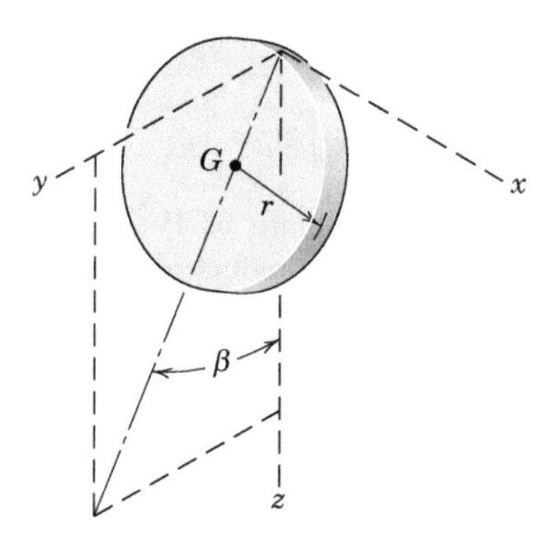

Problem B/71

*Computer-Oriented Problems

***B/72** The L-shaped piece is cut from steel plate having a mass per unit area of 160 kg/m^2. Determine and plot the moment of inertia of the piece about axis A-A as a function of θ from $\theta = 0$ to $\theta = 90°$ and find its minimum value.

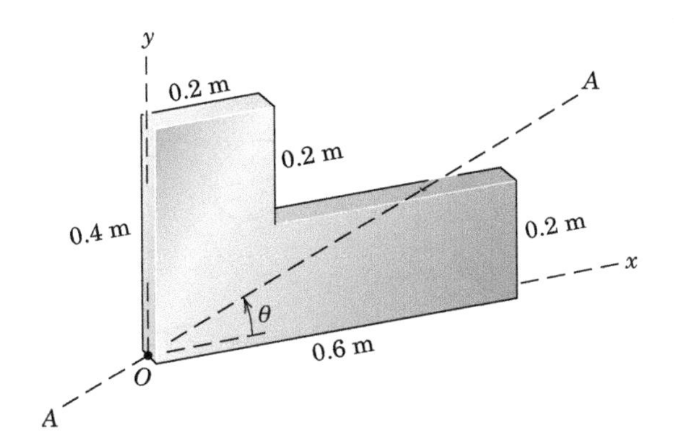

Problem B/72

***B/73** Determine the moment of inertia I about axis O-M for the uniform slender rod bent into the shape shown. Plot I versus θ from $\theta = 0$ to $\theta = 90°$ and determine the minimum value of I and the angle α which its axis makes with the x-direction. (*Note:* Because the analysis does not involve the z-coordinate, the expressions developed for area moments of inertia, Eqs. A/9, A/10, and A/11 in Appendix A of *Vol. 1 Statics*, may be utilized for this problem in place of the three-dimensional relations of Appendix B.) The rod has a mass ρ per unit length.

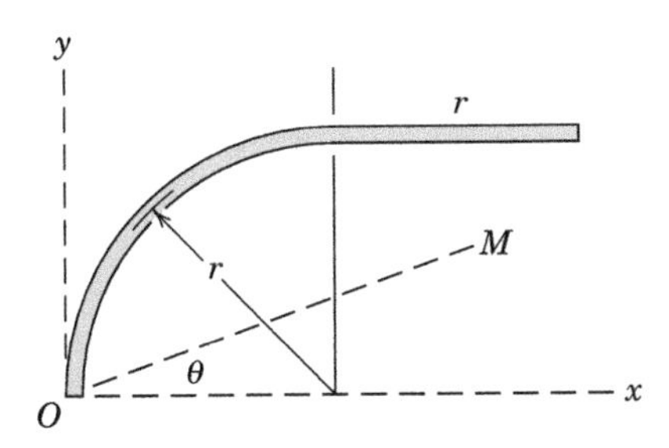

Problem B/73

***B/74** The assembly of three small spheres connected by light rigid bars of Prob. B/58 is repeated here. Determine the principal moments of inertia and the direction cosines associated with the axis of maximum moment of inertia.

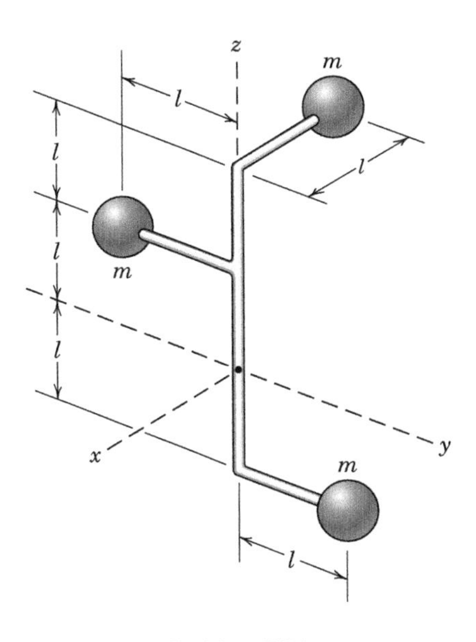

Problem B/74

***B/75** Determine the inertia tensor for the homogeneous thin plate about the x-, y-, and z-axes. The plate has a mass m and uniform thickness t. What is the minimum angle, measured from the x-axis, which will rotate the plate into principal directions?

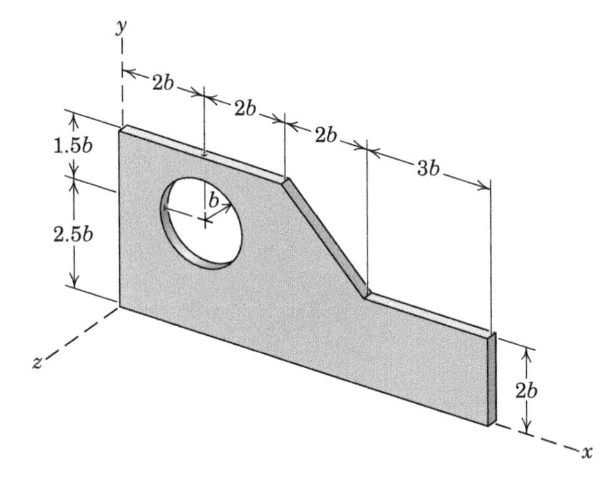

Problem B/75

*B/76 The thin plate has a mass ρ per unit area and is formed into the shape shown. Determine the principal moments of inertia of the plate about axes through O.

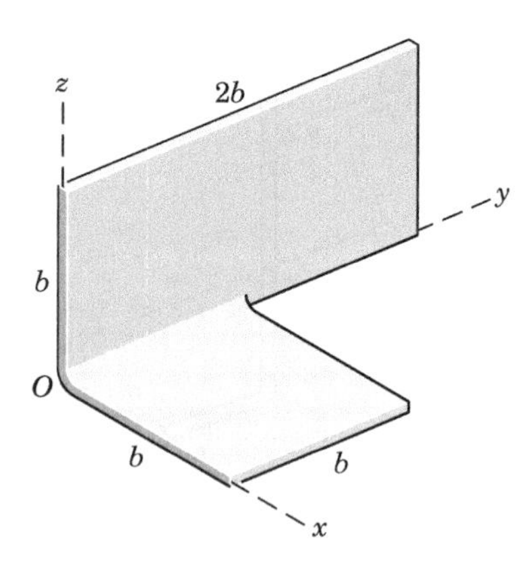

Problem B/76

*B/77 The slender rod has a mass ρ per unit length and is formed into the shape shown. Determine the principal moments of inertia about axes through O and calculate the direction cosines of the axis of minimum moment of inertia.

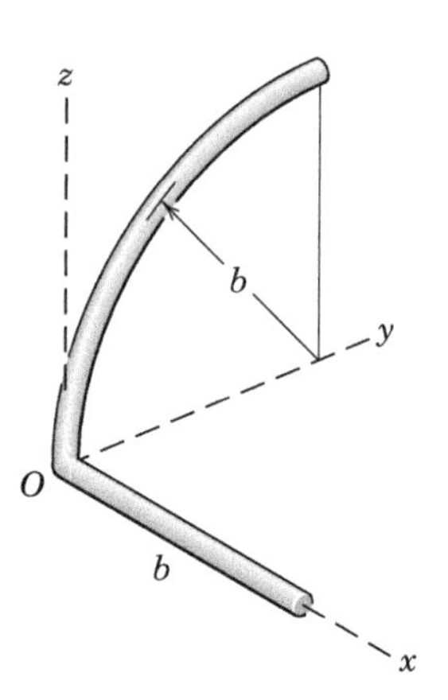

Problem B/77

*B/78 The welded assembly is formed from uniform sheet metal with a mass of 32 kg/m^2. Determine the principal mass moments of inertia for the assembly and the corresponding direction cosines for each principal axis.

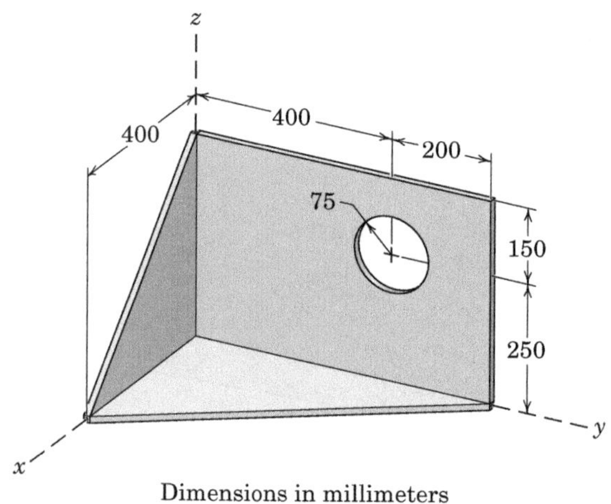

Dimensions in millimeters

Problem B/78

C SELECTED TOPICS OF MATHEMATICS

C/1 INTRODUCTION

Appendix C contains an abbreviated summary and reminder of selected topics in basic mathematics which find frequent use in mechanics. The relationships are cited without proof. The student of mechanics will have frequent occasion to use many of these relations, and he or she will be handicapped if they are not well in hand. Other topics not listed will also be needed from time to time.

As the reader reviews and applies mathematics, he or she should bear in mind that mechanics is an applied science descriptive of real bodies and actual motions. Therefore, the geometric and physical interpretation of the applicable mathematics should be kept clearly in mind during the development of theory and the formulation and solution of problems.

C/2 PLANE GEOMETRY

1. When two intersecting lines are, respectively, perpendicular to two other lines, the angles formed by the two pairs are equal.

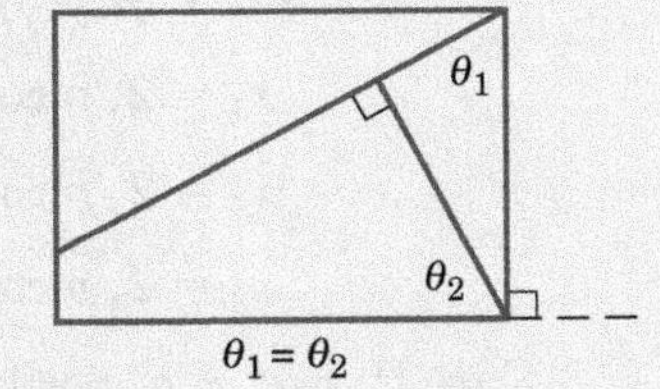

2. Similar triangles

$$\frac{x}{b} = \frac{h - y}{h}$$

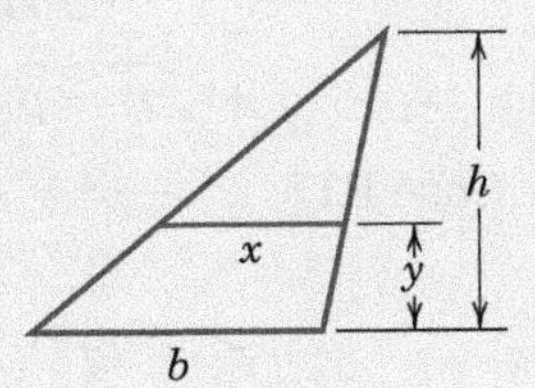

3. Any triangle

Area $= \frac{1}{2}bh$

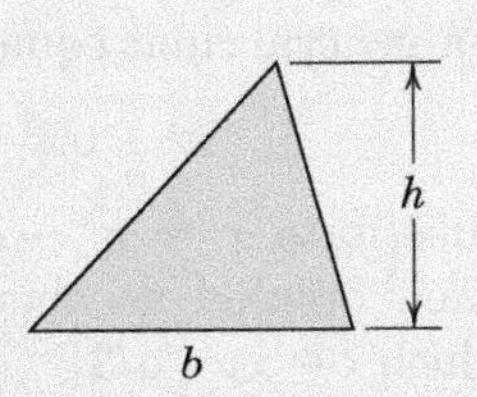

4. Circle

Circumference $= 2\pi r$
Area $= \pi r^2$
Arc length $s = r\theta$
Sector area $= \frac{1}{2}r^2\theta$

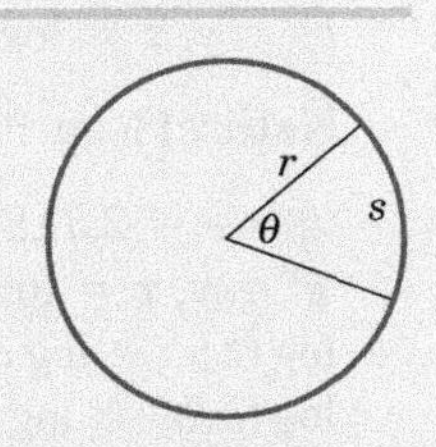

5. Every triangle inscribed within a semicircle is a right triangle.

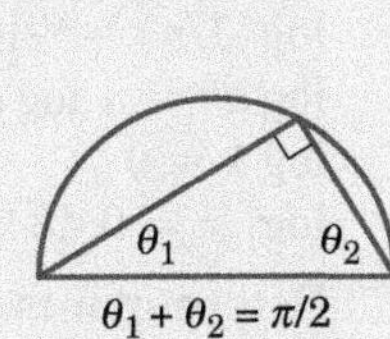

6. Angles of a triangle

$\theta_1 + \theta_2 + \theta_3 = 180°$
$\theta_4 = \theta_1 + \theta_2$

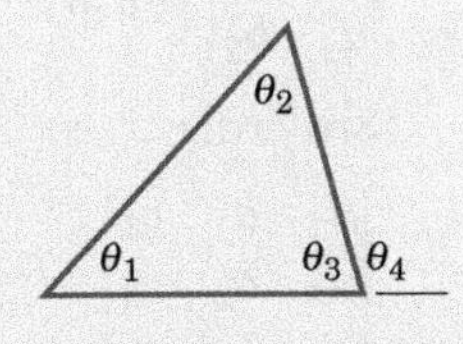

C/3 Solid Geometry

1. Sphere

   Volume $= \frac{4}{3}\pi r^3$

   Surface area $= 4\pi r^2$

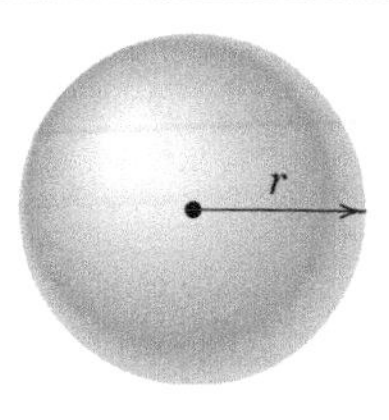

2. Spherical wedge

   Volume $= \frac{2}{3}r^3\theta$

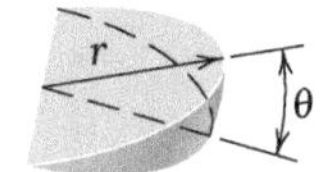

3. Right-circular cone

   Volume $= \frac{1}{3}\pi r^2 h$

   Lateral area $= \pi r L$

   $L = \sqrt{r^2 + h^2}$

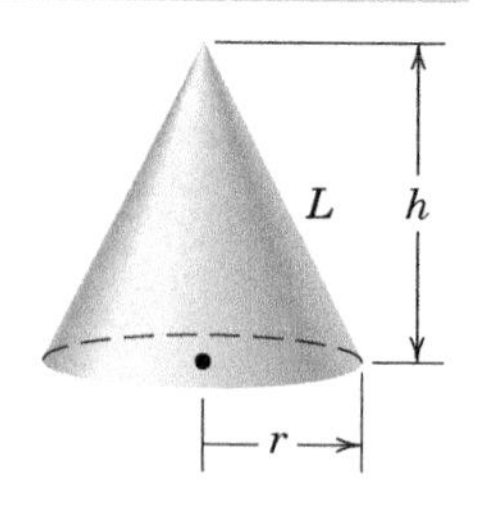

4. Any pyramid or cone

   Volume $= \frac{1}{3}Bh$

   where B = area of base

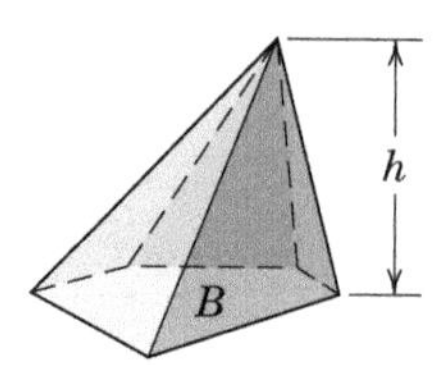

C/4 Algebra

1. Quadratic equation

   $$ax^2 + bx + c = 0$$

   $$x = \frac{-b \pm \sqrt{b^2 - 4ac}}{2a}, \; b^2 \geq 4ac \text{ for real roots}$$

2. Logarithms

   $$b^x = y, \; x = \log_b y$$

   Natural logarithms

   $$b = e = 2.718\,282$$
   $$e^x = y, \; x = \log_e y = \ln y$$
   $$\log(ab) = \log a + \log b$$
   $$\log(a/b) = \log a - \log b$$
   $$\log(1/n) = -\log n$$
   $$\log a^n = n \log a$$
   $$\log 1 = 0$$
   $$\log_{10} x = 0.4343 \ln x$$

3. Determinants

   2nd order

   $$\begin{vmatrix} a_1 & b_1 \\ a_2 & b_2 \end{vmatrix} = a_1 b_2 - a_2 b_1$$

   3rd order

   $$\begin{vmatrix} a_1 & b_1 & c_1 \\ a_2 & b_2 & c_2 \\ a_3 & b_3 & c_3 \end{vmatrix} = \begin{array}{l} +a_1 b_2 c_3 + a_2 b_3 c_1 + a_3 b_1 c_2 \\ -a_3 b_2 c_1 - a_2 b_1 c_3 - a_1 b_3 c_2 \end{array}$$

4. Cubic equation

   $$x^3 = Ax + B$$

   Let $p = A/3$, $q = B/2$.

   Case I: $q^2 - p^3$ negative (three roots real and distinct)

   $$\cos u = q/(p\sqrt{p}), \; 0 < u < 180°$$
   $$x_1 = 2\sqrt{p}\cos(u/3)$$
   $$x_2 = 2\sqrt{p}\cos(u/3 + 120°)$$
   $$x_3 = 2\sqrt{p}\cos(u/3 + 240°)$$

   Case II: $q^2 - p^3$ positive (one root real, two roots imaginary)

   $$x_1 = (q + \sqrt{q^2 - p^3})^{1/3} + (q - \sqrt{q^2 - p^3})^{1/3}$$

   Case III: $q^2 - p^3 = 0$ (three roots real, two roots equal)

   $$x_1 = 2q^{1/3}, \; x_2 = x_3 = -q^{1/3}$$

   For general cubic equation

   $$x^3 + ax^2 + bx + c = 0$$

   Substitute $x = x_0 - a/3$ and get $x_0{}^3 = Ax_0 + B$. Then proceed as above to find values of x_0 from which $x = x_0 - a/3$.

C/5 Analytic Geometry

1. Straight line

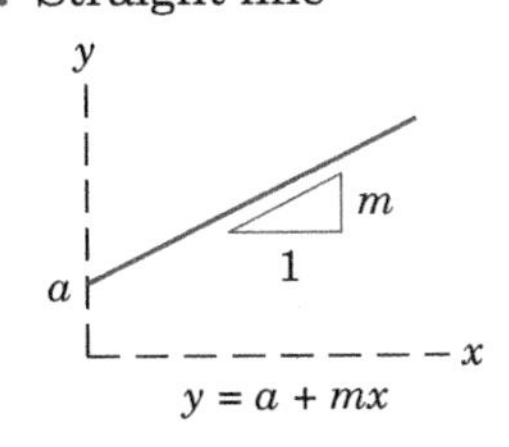

$y = a + mx$

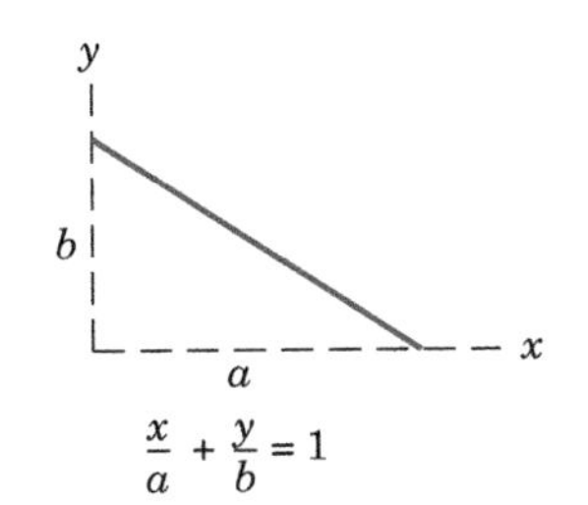

$\frac{x}{a} + \frac{y}{b} = 1$

2. Circle

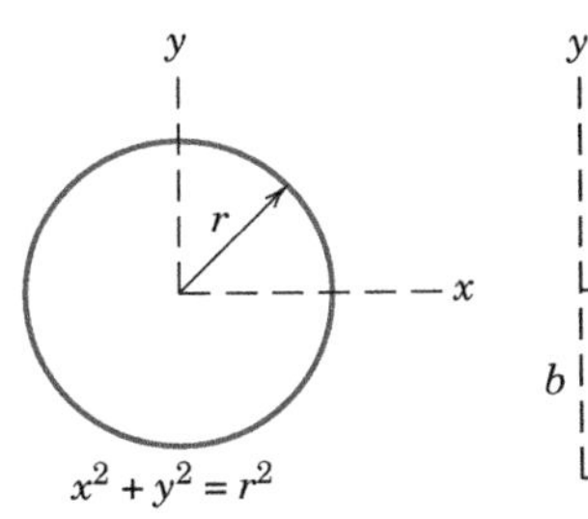

$x^2 + y^2 = r^2$

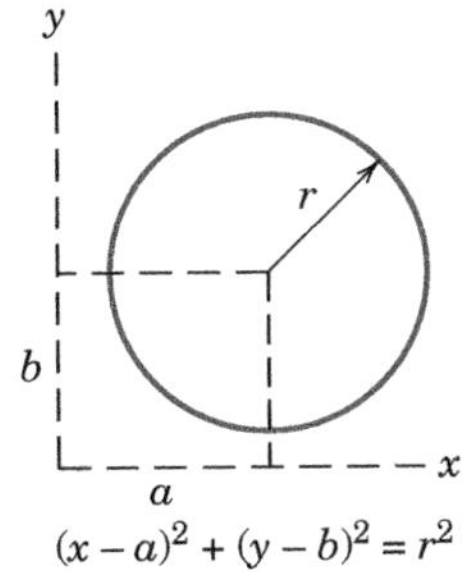

$(x - a)^2 + (y - b)^2 = r^2$

3. Parabola

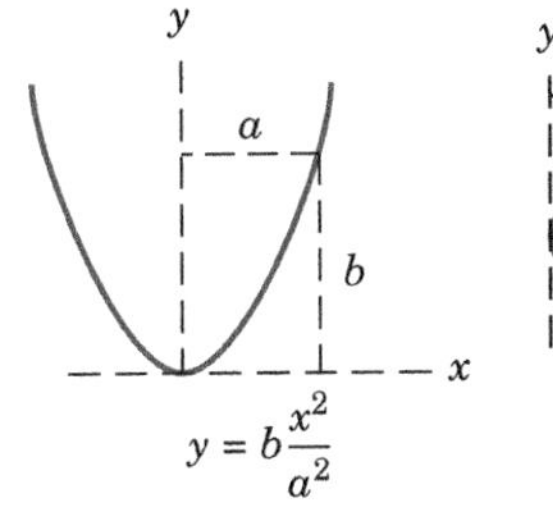

$y = b\frac{x^2}{a^2}$

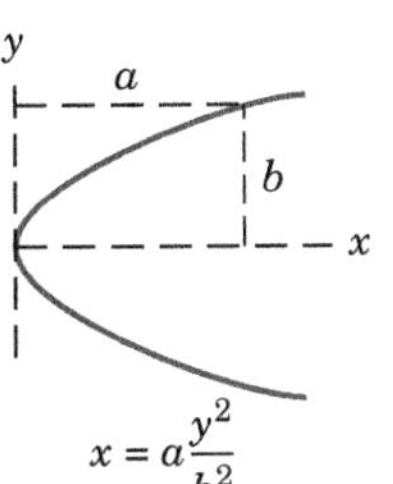

$x = a\frac{y^2}{b^2}$

4. Ellipse

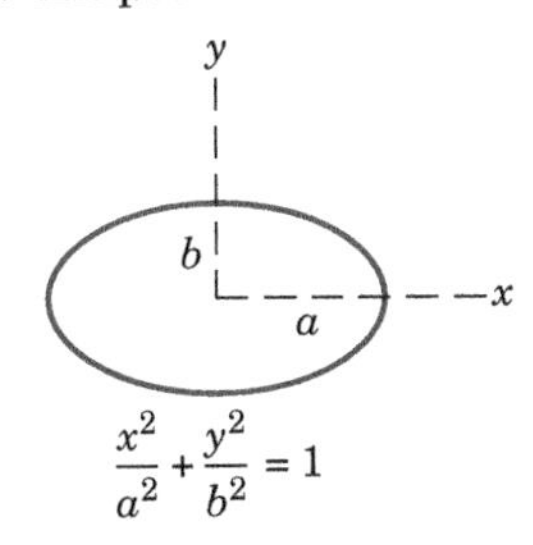

$\frac{x^2}{a^2} + \frac{y^2}{b^2} = 1$

5. Hyperbola

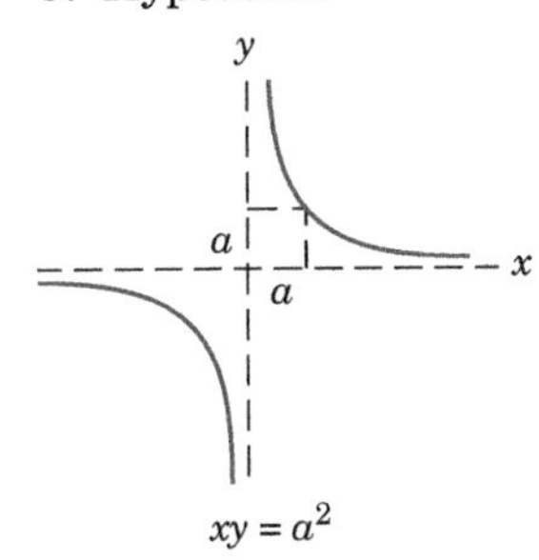

$xy = a^2$

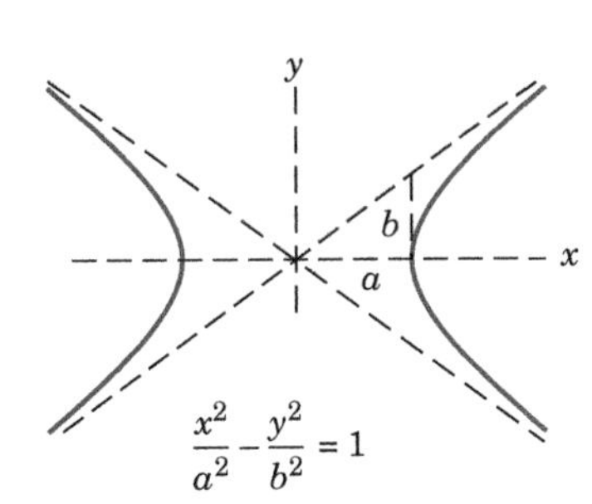

$\frac{x^2}{a^2} - \frac{y^2}{b^2} = 1$

C/6 Trigonometry

1. Definitions

$\sin\theta = a/c \quad \csc\theta = c/a$

$\cos\theta = b/c \quad \sec\theta = c/b$

$\tan\theta = a/b \quad \cot\theta = b/a$

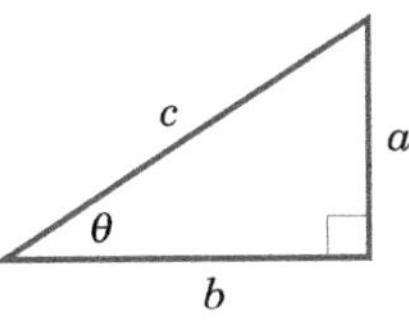

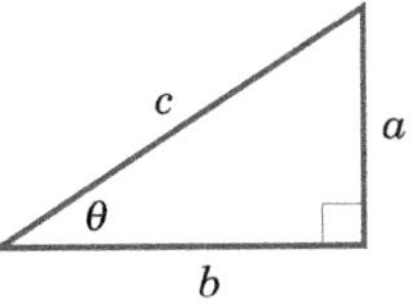

2. Signs in the four quadrants

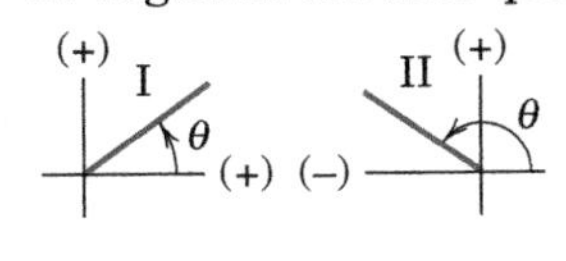

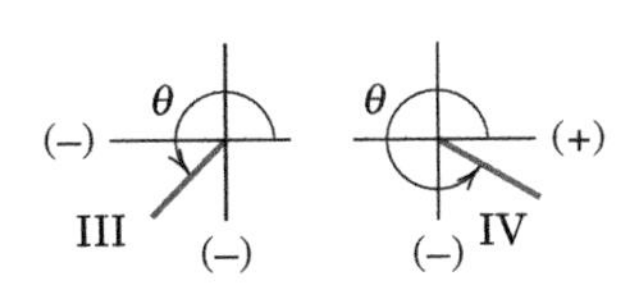

	I	II	III	IV
$\sin\theta$	+	+	−	−
$\cos\theta$	+	−	−	+
$\tan\theta$	+	−	+	−
$\csc\theta$	+	+	−	−
$\sec\theta$	+	−	−	+
$\cot\theta$	+	−	+	−

3. Miscellaneous relations

$\sin^2\theta + \cos^2\theta = 1$
$1 + \tan^2\theta = \sec^2\theta$
$1 + \cot^2\theta = \csc^2\theta$

$\sin\frac{\theta}{2} = \sqrt{\frac{1}{2}(1 - \cos\theta)}$

$\cos\frac{\theta}{2} = \sqrt{\frac{1}{2}(1 - \cos\theta)}$

$\sin 2\theta = 2\sin\theta\cos\theta$
$\cos 2\theta = \cos^2\theta - \sin^2\theta$
$\sin(a \pm b) = \sin a \cos b \pm \cos a \sin b$
$\cos(a \pm b) = \cos a \cos b \mp \sin a \sin b$

4. Law of sines

$\frac{a}{b} = \frac{\sin A}{\sin B}$

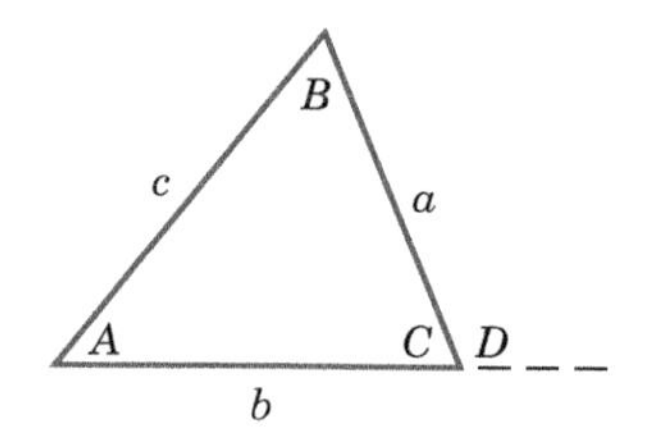

5. Law of cosines

$c^2 = a^2 + b^2 - 2ab\cos C$
$c^2 = a^2 + b^2 + 2ab\cos D$

C/7 Vector Operations

1. *Notation.* Vector quantities are printed in boldface type, and scalar quantities appear in lightface italic type. Thus, the vector quantity **V** has a scalar magnitude V. In longhand work vector quantities should always be consistently indicated by a symbol such as $\underline{V}$ or $\vec{V}$ to distinguish them from scalar quantities.

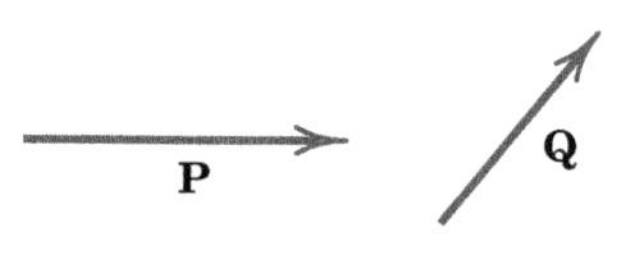

2. *Addition*

Triangle addition $\mathbf{P} + \mathbf{Q} = \mathbf{R}$

Parallelogram addition $\mathbf{P} + \mathbf{Q} = \mathbf{R}$

Commutative law $\mathbf{P} + \mathbf{Q} = \mathbf{Q} + \mathbf{P}$

Associative law $\mathbf{P} + (\mathbf{Q} + \mathbf{R}) = (\mathbf{P} + \mathbf{Q}) + \mathbf{R}$

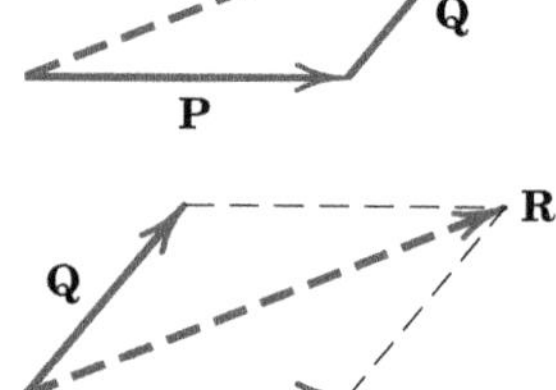

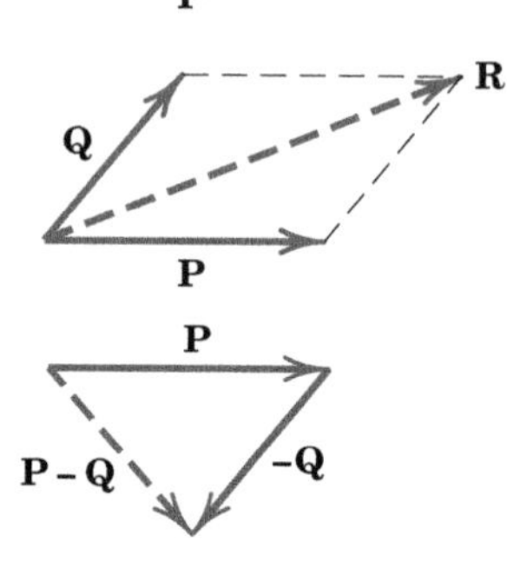

3. *Subtraction*

$$\mathbf{P} - \mathbf{Q} = \mathbf{P} + (-\mathbf{Q})$$

4. *Unit vectors* **i**, **j**, **k**

$$\mathbf{V} = V_x\mathbf{i} + V_y\mathbf{j} + V_z\mathbf{k}$$

where $|\mathbf{V}| = V = \sqrt{V_x^{\,2} + V_y^{\,2} + V_z^{\,2}}$

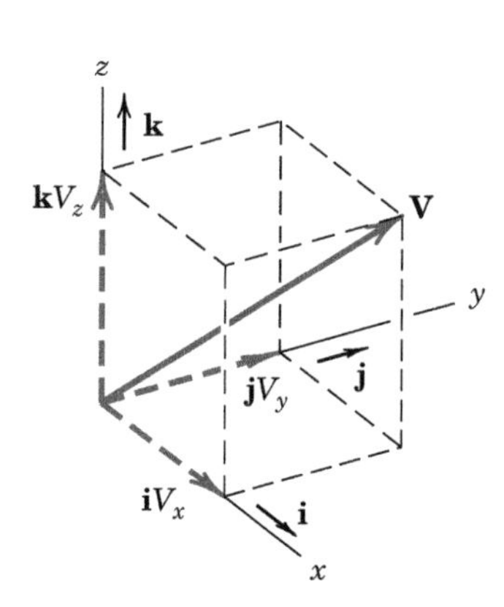

5. *Direction cosines* l, m, n are the cosines of the angles between **V** and the x-, y-, z-axes. Thus,

$$l = V_x/V \qquad m = V_y/V \qquad n = V_z/V$$

so that $\mathbf{V} = V(l\mathbf{i} + m\mathbf{j} + n\mathbf{k})$

and $l^2 + m^2 + n^2 = 1$

6. ***Dot or scalar product***

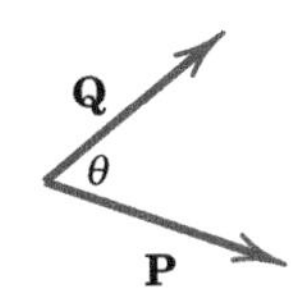

$$\mathbf{P}\cdot\mathbf{Q} = PQ \cos\theta$$

This product may be viewed as the magnitude of **P** multiplied by the component $Q\cos\theta$ of **Q** in the direction of **P**, or as the magnitude of **Q** multiplied by the component $P\cos\theta$ of **P** in the direction of **Q**.

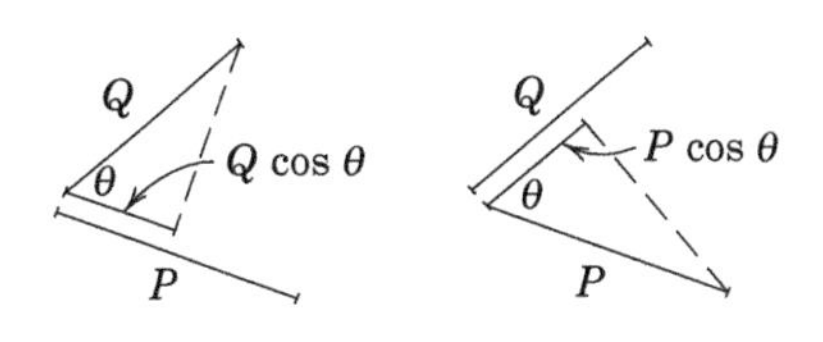

Commutative law $\quad \mathbf{P}\cdot\mathbf{Q} = \mathbf{Q}\cdot\mathbf{P}$

From the definition of the dot product

$$\mathbf{i}\cdot\mathbf{i} = \mathbf{j}\cdot\mathbf{j} = \mathbf{k}\cdot\mathbf{k} = 1$$
$$\mathbf{i}\cdot\mathbf{j} = \mathbf{j}\cdot\mathbf{i} = \mathbf{i}\cdot\mathbf{k} = \mathbf{k}\cdot\mathbf{i} = \mathbf{j}\cdot\mathbf{k} = \mathbf{k}\cdot\mathbf{j} = 0$$
$$\mathbf{P}\cdot\mathbf{Q} = (P_x\mathbf{i} + P_y\mathbf{j} + P_z\mathbf{k})\cdot(Q_x\mathbf{i} + Q_y\mathbf{j} + Q_z\mathbf{k})$$
$$= P_xQ_x + P_yQ_y + P_zQ_z$$
$$\mathbf{P}\cdot\mathbf{P} = P_x^{\,2} + P_y^{\,2} + P_z^{\,2}$$

It follows from the definition of the dot product that two vectors **P** and **Q** are perpendicular when their dot product vanishes, $\mathbf{P}\cdot\mathbf{Q} = 0$.

The angle θ between two vectors $\mathbf{P}_1$ and $\mathbf{P}_2$ may be found from their dot product expression $\mathbf{P}_1\cdot\mathbf{P}_2 = P_1P_2\cos\theta$, which gives

$$\cos\theta = \frac{\mathbf{P}_1\cdot\mathbf{P}_2}{P_1P_2} = \frac{P_{1_x}P_{2_x} + P_{1_y}P_{2_y} + P_{1_z}P_{2_z}}{P_1P_2} = l_1l_2 + m_1m_2 + n_1n_2$$

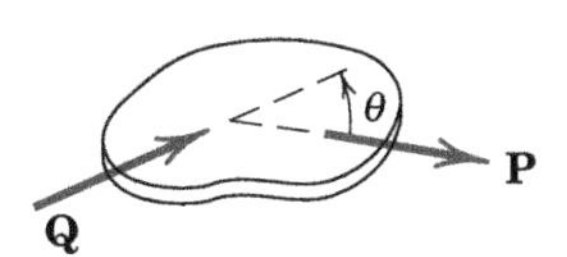

where l, m, n stand for the respective direction cosines of the vectors. It is also observed that two vectors are perpendicular to each other when their direction cosines obey the relation $l_1l_2 + m_1m_2 + n_1n_2 = 0$.

Distributive law $\quad \mathbf{P}\cdot(\mathbf{Q} + \mathbf{R}) = \mathbf{P}\cdot\mathbf{Q} + \mathbf{P}\cdot\mathbf{R}$

7. ***Cross or vector product.*** The cross product $\mathbf{P} \times \mathbf{Q}$ of the two vectors **P** and **Q** is defined as a vector with a magnitude

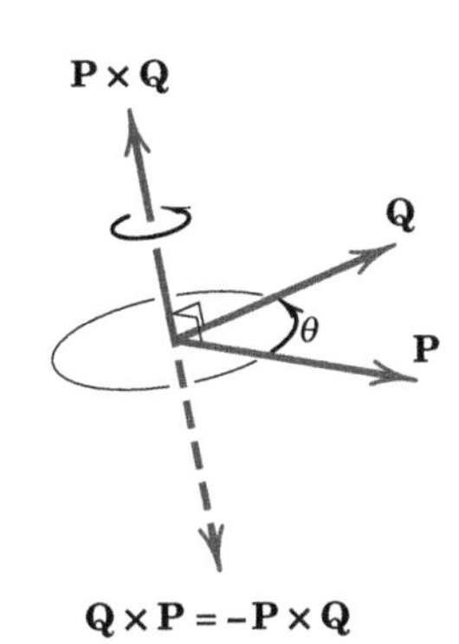

$$|\mathbf{P} \times \mathbf{Q}| = PQ\sin\theta$$

and a direction specified by the right-hand rule as shown. Reversing the vector order and using the right-hand rule give $\mathbf{Q} \times \mathbf{P} = -\mathbf{P} \times \mathbf{Q}$.

Distributive law $\quad \mathbf{P} \times (\mathbf{Q} + \mathbf{R}) = \mathbf{P} \times \mathbf{Q} + \mathbf{P} \times \mathbf{R}$

From the definition of the cross product, using a *right-handed coordinate system*, we get

$$\mathbf{i} \times \mathbf{j} = \mathbf{k} \qquad \mathbf{j} \times \mathbf{k} = \mathbf{i} \qquad \mathbf{k} \times \mathbf{i} = \mathbf{j}$$
$$\mathbf{j} \times \mathbf{i} = -\mathbf{k} \qquad \mathbf{k} \times \mathbf{j} = -\mathbf{i} \qquad \mathbf{i} \times \mathbf{k} = -\mathbf{j}$$
$$\mathbf{i} \times \mathbf{i} = \mathbf{j} \times \mathbf{j} = \mathbf{k} \times \mathbf{k} = \mathbf{0}$$

With the aid of these identities and the distributive law, the vector product may be written

$$\mathbf{P} \times \mathbf{Q} = (P_x\mathbf{i} + P_y\mathbf{j} + P_z\mathbf{k}) \times (Q_x\mathbf{i} + Q_y\mathbf{j} + Q_z\mathbf{k})$$
$$= (P_yQ_z - P_zQ_y)\mathbf{i} + (P_zQ_x - P_xQ_z)\mathbf{j} + (P_xQ_y - P_yQ_x)\mathbf{k}$$

The cross product may also be expressed by the determinant

$$\mathbf{P} \times \mathbf{Q} = \begin{vmatrix} \mathbf{i} & \mathbf{j} & \mathbf{k} \\ P_x & P_y & P_z \\ Q_x & Q_y & Q_z \end{vmatrix}$$

8. ***Additional relations***

Triple scalar product $(\mathbf{P} \times \mathbf{Q})\cdot\mathbf{R} = \mathbf{R}\cdot(\mathbf{P} \times \mathbf{Q})$. The dot and cross may be interchanged as long as the order of the vectors is maintained. Parentheses are unnecessary since $\mathbf{P} \times (\mathbf{Q}\cdot\mathbf{R})$ is meaningless because a vector $\mathbf{P}$ cannot be crossed into a scalar $\mathbf{Q}\cdot\mathbf{R}$. Thus, the expression may be written

$$\mathbf{P} \times \mathbf{Q}\cdot\mathbf{R} = \mathbf{P}\cdot\mathbf{Q} \times \mathbf{R}$$

The triple scalar product has the determinant expansion

$$\mathbf{P} \times \mathbf{Q}\cdot\mathbf{R} = \begin{vmatrix} P_x & P_y & P_z \\ Q_x & Q_y & Q_z \\ R_x & R_y & R_z \end{vmatrix}$$

Triple vector product $(\mathbf{P} \times \mathbf{Q}) \times \mathbf{R} = -\mathbf{R} \times (\mathbf{P} \times \mathbf{Q}) = \mathbf{R} \times (\mathbf{Q} \times \mathbf{P})$. Here we note that the parentheses must be used since an expression $\mathbf{P} \times \mathbf{Q} \times \mathbf{R}$ would be ambiguous because it would not identify the vector to be crossed. It may be shown that the triple vector product is equivalent to

$$(\mathbf{P} \times \mathbf{Q}) \times \mathbf{R} = \mathbf{R}\cdot\mathbf{P}\mathbf{Q} - \mathbf{R}\cdot\mathbf{Q}\mathbf{P}$$

or

$$\mathbf{P} \times (\mathbf{Q} \times \mathbf{R}) = \mathbf{P}\cdot\mathbf{R}\mathbf{Q} - \mathbf{P}\cdot\mathbf{Q}\mathbf{R}$$

The first term in the first expression, for example, is the dot product $\mathbf{R}\cdot\mathbf{P}$, a scalar, multiplied by the vector $\mathbf{Q}$.

9. ***Derivatives of vectors*** obey the same rules as they do for scalars.

$$\frac{d\mathbf{P}}{dt} = \dot{\mathbf{P}} = \dot{P}_x\mathbf{i} + \dot{P}_y\mathbf{j} + \dot{P}_z\mathbf{k}$$
$$\frac{d(\mathbf{P}u)}{dt} = \mathbf{P}\dot{u} + \dot{\mathbf{P}}u$$
$$\frac{d(\mathbf{P}\cdot\mathbf{Q})}{dt} = \mathbf{P}\cdot\dot{\mathbf{Q}} + \dot{\mathbf{P}}\cdot\mathbf{Q}$$
$$\frac{d(\mathbf{P} \times \mathbf{Q})}{dt} = \mathbf{P} \times \dot{\mathbf{Q}} + \dot{\mathbf{P}} \times \mathbf{Q}$$

10. ***Integration of vectors.*** If $\mathbf{V}$ is a function of x, y, and z and an element of volume is $d\tau = dx\, dy\, dz$, the integral of $\mathbf{V}$ over the volume may be written as the vector sum of the three integrals of its components. Thus,

$$\int \mathbf{V}\, d\tau = \mathbf{i}\int V_x\, d\tau + \mathbf{j}\int V_y\, d\tau + \mathbf{k}\int V_z\, d\tau$$

C/8 Series

(The expression in brackets following a series indicates the range of convergence.)

$$(1 \pm x)^n = 1 \pm nx + \frac{n(n-1)}{2!}x^2 \pm \frac{n(n-1)(n-2)}{3!}x^3 + \cdots \quad [x^2 < 1]$$

$$\sin x = x - \frac{x^3}{3!} + \frac{x^5}{5!} - \frac{x^7}{7!} + \cdots \quad [x^2 < \infty]$$

$$\cos x = 1 - \frac{x^2}{2!} + \frac{x^4}{4!} - \frac{x^6}{6!} + \cdots \quad [x^2 < \infty]$$

$$\sinh x = \frac{e^x - e^{-x}}{2} = x + \frac{x^3}{3!} + \frac{x^5}{5!} + \frac{x^7}{7!} + \cdots \quad [x^2 < \infty]$$

$$\cosh x = \frac{e^x + e^{-x}}{2} = 1 + \frac{x^2}{2!} + \frac{x^4}{4!} + \frac{x^6}{6!} + \cdots \quad [x^2 < \infty]$$

$$f(x) = \frac{a_0}{2} + \sum_{n=1}^{\infty} a_n \cos\frac{n\pi x}{l} + \sum_{n=1}^{\infty} b_n \sin\frac{n\pi x}{l}$$

where $a_n = \frac{1}{l}\int_{-l}^{l} f(x)\cos\frac{n\pi x}{l}\,dx, \qquad b_n = \frac{1}{l}\int_{-l}^{l} f(x)\sin\frac{n\pi x}{l}\,dx$

[Fourier expansion for $-l < x < l$]

C/9 Derivatives

$$\frac{dx^n}{dx} = nx^{n-1}, \qquad \frac{d(uv)}{dx} = u\frac{dv}{dx} + v\frac{du}{dx}, \qquad \frac{d\left(\frac{u}{v}\right)}{dx} = \frac{v\frac{du}{dx} - u\frac{dv}{dx}}{v^2}$$

$$\lim_{\Delta x \to 0} \sin \Delta x = \sin dx = \tan dx = dx$$

$$\lim_{\Delta x \to 0} \cos \Delta x = \cos dx = 1$$

$$\frac{d \sin x}{dx} = \cos x, \qquad \frac{d \cos x}{dx} = -\sin x, \qquad \frac{d \tan x}{dx} = \sec^2 x$$

$$\frac{d \sinh x}{dx} = \cosh x, \qquad \frac{d \cosh x}{dx} = \sinh x, \qquad \frac{d \tanh x}{dx} = \operatorname{sech}^2 x$$

C/10 INTEGRALS

$$\int x^n \, dx = \frac{x^{n+1}}{n+1}$$

$$\int \frac{dx}{x} = \ln x$$

$$\int \sqrt{a+bx} \, dx = \frac{2}{3b}\sqrt{(a+bx)^3}$$

$$\int x\sqrt{a+bx} \, dx = \frac{2}{15b^2}(3bx - 2a)\sqrt{(a+bx)^3}$$

$$\int x^2\sqrt{a+bx} \, dx = \frac{2}{105b^3}(8a^2 - 12abx + 15b^2x^2)\sqrt{(a+bx)^3}$$

$$\int \frac{dx}{\sqrt{a+bx}} = \frac{2\sqrt{a+bx}}{b}$$

$$\int \frac{\sqrt{a+x}}{\sqrt{b-x}} \, dx = -\sqrt{a+x}\,\sqrt{b-x} + (a+b)\sin^{-1}\sqrt{\frac{a+x}{a+b}}$$

$$\int \frac{x\,dx}{a+bx} = \frac{1}{b^2}[a + bx - a\ln(a+bx)]$$

$$\int \frac{x\,dx}{(a+bx)^n} = \frac{(a+bx)^{1-n}}{b^2}\left(\frac{a+bx}{2-n} - \frac{a}{1-n}\right)$$

$$\int \frac{dx}{a+bx^2} = \frac{1}{\sqrt{ab}}\tan^{-1}\frac{x\sqrt{ab}}{a} \quad \text{or} \quad \frac{1}{\sqrt{-ab}}\tanh^{-1}\frac{x\sqrt{-ab}}{a}$$

$$\int \frac{x\,dx}{a+bx^2} = \frac{1}{2b}\ln(a+bx^2)$$

$$\int \sqrt{x^2 \pm a^2}\,dx = \tfrac{1}{2}[x\sqrt{x^2 \pm a^2} \pm a^2 \ln(x + \sqrt{x^2 \pm a^2})]$$

$$\int \sqrt{a^2 - x^2}\,dx = \tfrac{1}{2}\left(x\sqrt{a^2 - x^2} + a^2\sin^{-1}\frac{x}{a}\right)$$

$$\int x\sqrt{a^2 - x^2}\,dx = -\tfrac{1}{3}\sqrt{(a^2 - x^2)^3}$$

$$\int x^2\sqrt{a^2 - x^2}\,dx = -\frac{x}{4}\sqrt{(a^2 - x^2)^3} + \frac{a^2}{8}\left(x\sqrt{a^2 - x^2} + a^2\sin^{-1}\frac{x}{a}\right)$$

$$\int x^3\sqrt{a^2 - x^2}\,dx = -\tfrac{1}{5}(x^2 + \tfrac{2}{3}a^2)\sqrt{(a^2 - x^2)^3}$$

$$\int \frac{dx}{\sqrt{a+bx+cx^2}} = \frac{1}{\sqrt{c}}\ln\left(\sqrt{a+bx+cx^2} + x\sqrt{c} + \frac{b}{2\sqrt{c}}\right) \quad \text{or} \quad \frac{-1}{\sqrt{-c}}\sin^{-1}\left(\frac{b+2cx}{\sqrt{b^2 - 4ac}}\right)$$

$$\int \frac{dx}{\sqrt{x^2 \pm a^2}} = \ln(x + \sqrt{x^2 \pm a^2})$$

$$\int \frac{dx}{\sqrt{a^2 - x^2}} = \sin^{-1} \frac{x}{a}$$

$$\int \frac{x\,dx}{\sqrt{x^2 - a^2}} = \sqrt{x^2 - a^2}$$

$$\int \frac{x\,dx}{\sqrt{a^2 \pm x^2}} = \pm\sqrt{a^2 \pm x^2}$$

$$\int x\sqrt{x^2 \pm a^2}\,dx = \frac{1}{3}\sqrt{(x^2 \pm a^2)^3}$$

$$\int x^2\sqrt{x^2 \pm a^2}\,dx = \frac{x}{4}\sqrt{(x^2 \pm a^2)^3} \mp \frac{a^2}{8}x\sqrt{x^2 \pm a^2} - \frac{a^4}{8}\ln\left(x + \sqrt{x^2 \pm a^2}\right)$$

$$\int \sin x\,dx = -\cos x$$

$$\int \cos x\,dx = \sin x$$

$$\int \sec x\,dx = \frac{1}{2}\ln\frac{1 + \sin x}{1 - \sin x}$$

$$\int \sin^2 x\,dx = \frac{x}{2} - \frac{\sin 2x}{4}$$

$$\int \cos^2 x\,dx = \frac{x}{2} + \frac{\sin 2x}{4}$$

$$\int \sin x \cos x\,dx = \frac{\sin^2 x}{2}$$

$$\int \sinh x\,dx = \cosh x$$

$$\int \cosh x\,dx = \sinh x$$

$$\int \tanh x\,dx = \ln \cosh x$$

$$\int \ln x\,dx = x \ln x - x$$

$$\int e^{ax}\,dx = \frac{e^{ax}}{a}$$

$$\int xe^{ax}\,dx = \frac{e^{ax}}{a^2}(ax - 1)$$

$$\int e^{ax} \sin px\,dx = \frac{e^{ax}(a \sin px - p \cos px)}{a^2 + p^2}$$

$$\int e^{ax} \cos px\, dx = \frac{e^{ax}\,(a \cos px + p \sin px)}{a^2 + p^2}$$

$$\int e^{ax} \sin^2 x\, dx = \frac{e^{ax}}{4 + a^2}\left(a \sin^2 x - \sin 2x + \frac{2}{a}\right)$$

$$\int e^{ax} \cos^2 x\, dx = \frac{e^{ax}}{4 + a^2}\left(a \cos^2 x + \sin 2x + \frac{2}{a}\right)$$

$$\int e^{ax} \sin x \cos x\, dx = \frac{e^{ax}}{4 + a^2}\left(\frac{a}{2} \sin 2x - \cos 2x\right)$$

$$\int \sin^3 x\, dx = -\frac{\cos x}{3}\,(2 + \sin^2 x)$$

$$\int \cos^3 x\, dx = \frac{\sin x}{3}\,(2 + \cos^2 x)$$

$$\int \cos^5 x\, dx = \sin x - \tfrac{2}{3} \sin^3 x + \tfrac{1}{5} \sin^5 x$$

$$\int x \sin x\, dx = \sin x - x \cos x$$

$$\int x \cos x\, dx = \cos x + x \sin x$$

$$\int x^2 \sin x\, dx = 2x \sin x - (x^2 - 2) \cos x$$

$$\int x^2 \cos x\, dx = 2x \cos x + (x^2 - 2) \sin x$$

Radius of curvature

$$\rho_{xy} = \frac{\left[1 + \left(\frac{dy}{dx}\right)^2\right]^{3/2}}{\frac{d^2y}{dx^2}}$$

$$\rho_{r\theta} = \frac{\left[r^2 + \left(\frac{dr}{d\theta}\right)^2\right]^{3/2}}{r^2 + 2\left(\frac{dr}{d\theta}\right)^2 - r\frac{d^2r}{d\theta^2}}$$

C/11 Newton's Method for Solving Intractable Equations

Frequently, the application of the fundamental principles of mechanics leads to an algebraic or transcendental equation which is not solvable (or easily solvable) in closed form. In such cases, an iterative technique, such as Newton's method, can be a powerful tool for obtaining a good estimate to the root or roots of the equation.

Let us place the equation to be solved in the form $f(x) = 0$. Part a of the accompanying figure depicts an arbitrary function $f(x)$ for values of x in the vicinity of the desired root x_r. Note that x_r is merely the value

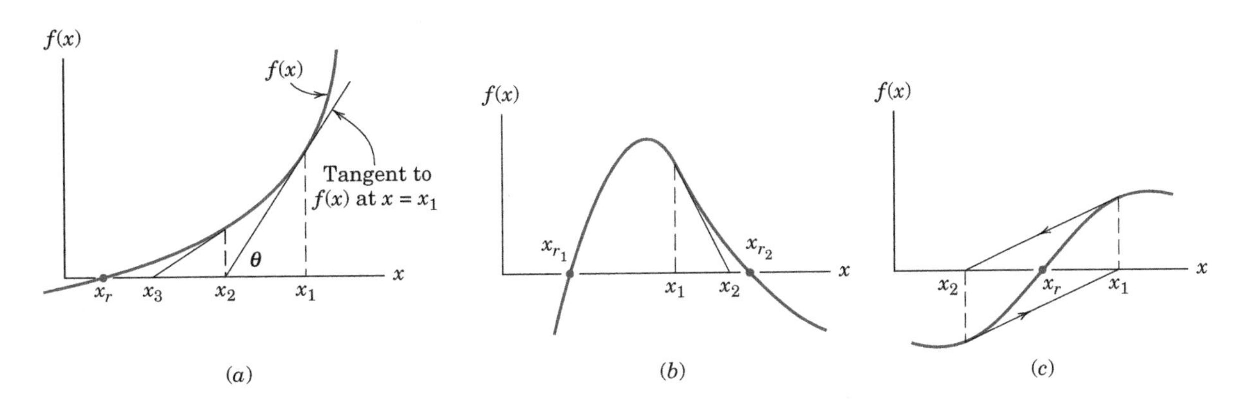

of x at which the function crosses the x-axis. Suppose that we have available (perhaps via a hand-drawn plot) a rough estimate x_1 of this root. Provided that x_1 does not closely correspond to a maximum or minimum value of the function $f(x)$, we may obtain a better estimate of the root x_r by extending the tangent to $f(x)$ at x_1 so that it intersects the x-axis at x_2. From the geometry of the figure, we may write

$$\tan \theta = f'(x_1) = \frac{f(x_1)}{x_1 - x_2}$$

where $f'(x_1)$ denotes the derivative of $f(x)$ with respect to x evaluated at $x = x_1$. Solving the above equation for x_2 results in

$$x_2 = x_1 - \frac{f(x_1)}{f'(x_1)}$$

The term $-f(x_1)/f'(x_1)$ is the correction to the initial root estimate x_1. Once x_2 is calculated, we may repeat the process to obtain x_3, and so forth.

Thus, we generalize the above equation to

$$x_{k+1} = x_k - \frac{f(x_k)}{f'(x_k)}$$

where

x_{k+1} = the $(k + 1)$th estimate of the desired root x_r

x_k = the kth estimate of the desired root x_r

$f(x_k)$ = the function $f(x)$ evaluated at $x = x_k$

$f'(x_k)$ = the function derivative evaluated at $x = x_k$

This equation is repeatedly applied until $f(x_{k+1})$ is sufficiently close to zero and $x_{k+1} \cong x_k$. The student should verify that the equation is valid for all possible sign combinations of x_k, $f(x_k)$, and $f'(x_k)$.

Several cautionary notes are in order:

1. Clearly, $f'(x_k)$ must not be zero or close to zero. This would mean, as restricted above, that x_k exactly or approximately corresponds to a minimum or maximum of $f(x)$. If the slope $f'(x_k)$ is zero, then the tangent to the curve never intersects the x-axis. If the slope $f'(x_k)$ is small, then the correction to x_k may be so large that x_{k+1} is a worse root estimate than x_k. For this reason, experienced engineers usually limit the size of the correction term; that is, if the absolute value of $f(x_k)/f'(x_k)$ is larger than a preselected maximum value, that maximum value is used.
2. If there are several roots of the equation $f(x) = 0$, we must be in the vicinity of the desired root x_r in order that the algorithm actually converges to that root. Part *b* of the figure depicts the condition when the initial estimate x_1 will result in convergence to x_{r_2} rather than x_{r_1}.
3. Oscillation from one side of the root to the other can occur if, for example, the function is antisymmetric about a root which is an inflection point. The use of one-half of the correction will usually prevent this behavior, which is depicted in part *c* of the accompanying figure.

Example: Beginning with an initial estimate of $x_1 = 5$, estimate the single root of the equation $e^x - 10 \cos x - 100 = 0$.

The table below summarizes the application of Newton's method to the given equation. The iterative process was terminated when the absolute value of the correction $-f(x_k)/f'(x_k)$ became less than 10^{-6}.

k	x_k	$f(x_k)$	$f'(x_k)$	$x_{k+1} - x_k = -\dfrac{f(x_k)}{f'(x_k)}$
1	5.000 000	45.576 537	138.823 916	−0.328 305
2	4.671 695	7.285 610	96.887 065	−0.075 197
3	4.596 498	0.292 886	89.203 650	−0.003 283
4	4.593 215	0.000 527	88.882 536	−0.000 006
5	4.593 209	$-2(10^{-8})$	88.881 956	$2.25(10^{-10})$

C/12 Selected Techniques for Numerical Integration

*1. **Area determination.*** Consider the problem of determining the shaded area under the curve $y = f(x)$ from $x = a$ to $x = b$, as depicted in part *a* of the figure, and suppose that analytical integration is not feasible. The function may be known in tabular form from experimental measurements, or it may be known in analytical form. The function is taken to be continuous within the interval $a < x < b$. We may divide the area into n vertical strips, each of width $\Delta x = (b - a)/n$, and then add the areas of all strips to obtain $A = \int y\, dx$. A representative strip of area A_i is shown with darker shading in the figure. Three useful numerical approximations are cited. In each case the greater the number of strips, the more accurate becomes the approximation geometrically. As a general rule, one can begin with a relatively small number of strips and increase the number until the resulting changes in the area approximation no longer improve the accuracy obtained.

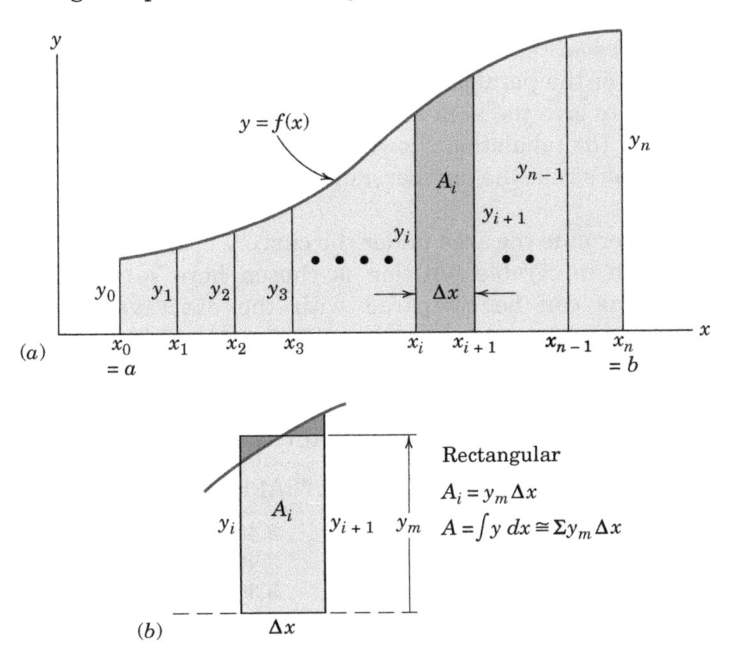

I. *Rectangular* [Figure (*b*)] The areas of the strips are taken to be rectangles, as shown by the representative strip whose height y_m is chosen visually so that the small cross-hatched areas are as nearly equal as possible. Thus, we form the sum Σy_m of the effective heights and multiply by Δx. For a function known in analytical form, a value for y_m equal to that of the function at the midpoint $x_i + \Delta x/2$ may be calculated and used in the summation.

II. *Trapezoidal* [Figure (*c*)] The areas of the strips are taken to be trapezoids, as shown by the representative strip. The area A_i is the average height $(y_i + y_{i+1})/2$ times Δx. Adding the areas gives the area approximation as tabulated. For the example with the curvature shown, clearly the approximation will be on the low side. For the reverse curvature, the approximation will be on the high side.

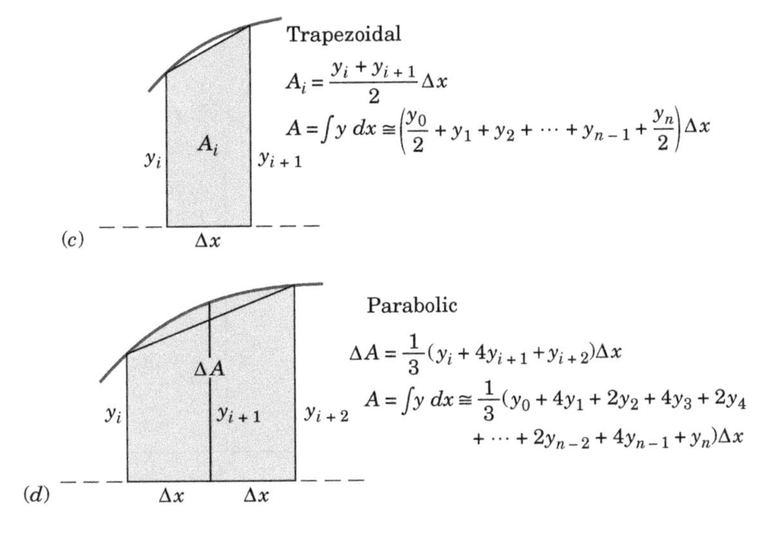

III. *Parabolic* [Figure (*d*)] The area between the chord and the curve (neglected in the trapezoidal solution) may be accounted for by approximating the function by a parabola passing through the points defined by three successive values of y. This area may be calculated from the geometry of the parabola and added to the trapezoidal area of the pair of strips to give the area ΔA of the pair as cited. Adding all of the ΔA's produces the tabulation shown, which is known as Simpson's rule. To use Simpson's rule, the number n of strips must be even.

Example: Determine the area under the curve $y = x\sqrt{1 + x^2}$ from $x = 0$ to $x = 2$. (An integrable function is chosen here so that the three approximations can be compared with the exact value, which is $A = \int_0^2 x\sqrt{1 + x^2}\,dx = \frac{1}{3}(1 + x^2)^{3/2}\big|_0^2 = \frac{1}{3}(5\sqrt{5} - 1) = 3.393\ 447$).

NUMBER OF SUBINTERVALS	AREA APPROXIMATIONS		
	RECTANGULAR	TRAPEZOIDAL	PARABOLIC
4	3.361 704	3.456 731	3.392 214
10	3.388 399	3.403 536	3.393 420
50	3.393 245	3.393 850	3.393 447
100	3.393 396	3.393 547	3.393 447
1000	3.393 446	3.393 448	3.393 447
2500	3.393 447	3.393 447	3.393 447

Note that the worst approximation error is less than 2 percent, even with only four strips.

2. *Integration of first-order ordinary differential equations.* The application of the fundamental principles of mechanics frequently results in differential relationships. Let us consider the first-order form $dy/dt = f(t)$, where the function $f(t)$ may not be readily integrable or may be known only in tabular form. We may numerically integrate by means of a simple slope-projection technique, known as Euler integration, which is illustrated in the figure.

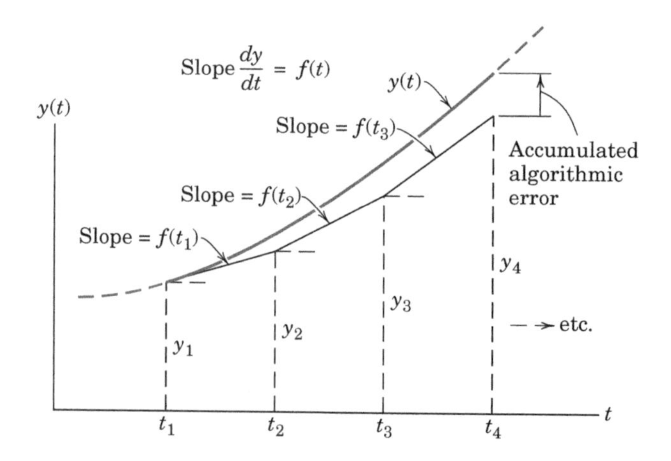

Beginning at t_1, at which the value y_1 is known, we project the slope over a horizontal subinterval or step $(t_2 - t_1)$ and see that $y_2 = y_1 + f(t_1)(t_2 - t_1)$. At t_2, the process may be repeated beginning at y_2, and so forth until the desired value of t is reached. Hence, the general expression is

$$y_{k+1} = y_k + f(t_k)(t_{k+1} - t_k)$$

If y versus t were linear, i.e., if $f(t)$ were constant, the method would be exact, and there would be no need for a numerical approach in that case. Changes in the slope over the subinterval introduce error. For the case shown in the figure, the estimate y_2 is clearly less than the true value of the function $y(t)$ at t_2. More accurate integration techniques (such as Runge-Kutta methods) take into account changes in the slope over the subinterval and thus provide better results.

As with the area-determination techniques, experience is helpful in the selection of a subinterval or step size when dealing with analytical functions. As a rough rule, one begins with a relatively large step size and then steadily decreases the step size until the corresponding changes in the integrated result are much smaller than the desired accuracy. A step size which is too small, however, can result in increased error due to a very large number of computer operations. This type of error is generally known as "round-off error," while the error which results from a large step size is known as algorithm error.

Example: For the differential equation $dy/dt = 5t$ with the initial condition $y = 2$ when $t = 0$, determine the value of y for $t = 4$.

Application of the Euler integration technique yields the following results:

NUMBER OF SUBINTERVALS	STEP SIZE	y at $t = 4$	PERCENT ERROR
10	0.4	38	9.5
100	0.04	41.6	0.95
500	0.008	41.92	0.19
1000	0.004	41.96	0.10

This simple example may be integrated analytically. The result is $y = 42$ (exactly).

USEFUL TABLES

TABLE D/1 PHYSICAL PROPERTIES

***Density* (kg/m^3) and *specific weight* (lb/ft^3)**

	kg/m^3	lb/ft^3		kg/m^3	lb/ft^3
Air*	1.2062	0.07530	Lead	11 370	710
Aluminum	2 690	168	Mercury	13 570	847
Concrete (av.)	2 400	150	Oil (av.)	900	56
Copper	8 910	556	Steel	7 830	489
Earth (wet, av.)	1 760	110	Titanium	4 510	281
(dry, av.)	1 280	80	Water (fresh)	1 000	62.4
Glass	2 590	162	(salt)	1 030	64
Gold	19 300	1205	Wood (soft pine)	480	30
Ice	900	56	(hard oak)	800	50
Iron (cast)	7 210	450			

* At 20°C (68°F) and atmospheric pressure

Coefficients of friction

(The coefficients in the following table represent typical values under normal working conditions. Actual coefficients for a given situation will depend on the exact nature of the contacting surfaces. A variation of 25 to 100 percent or more from these values could be expected in an actual application, depending on prevailing conditions of cleanliness, surface finish, pressure, lubrication, and velocity.)

	TYPICAL VALUES OF COEFFICIENT OF FRICTION	
CONTACTING SURFACE	STATIC, μ_s	KINETIC, μ_k
Steel on steel (dry)	0.6	0.4
Steel on steel (greasy)	0.1	0.05
Teflon on steel	0.04	0.04
Steel on babbitt (dry)	0.4	0.3
Steel on babbitt (greasy)	0.1	0.07
Brass on steel (dry)	0.5	0.4
Brake lining on cast iron	0.4	0.3
Rubber tires on smooth pavement (dry)	0.9	0.8
Wire rope on iron pulley (dry)	0.2	0.15
Hemp rope on metal	0.3	0.2
Metal on ice		0.02

TABLE D/2 SOLAR SYSTEM CONSTANTS

Universal gravitational constant	$G = 6.673(10^{-11})\ m^3/(kg \cdot s^2)$
	$= 3.439(10^{-8})\ ft^4/(lbf\text{-}s^4)$
Mass of Earth	$m_e = 5.976(10^{24})\ kg$
	$= 4.095(10^{23})\ lbf\text{-}s^2/ft$
Period of Earth's rotation (1 sidereal day)	= 23 h 56 min 4 s
	= 23.9344 h
Angular velocity of Earth	$\omega = 0.7292(10^{-4})$ rad/s
Mean angular velocity of Earth–Sun line	$\omega' = 0.1991(10^{-6})$ rad/s
Mean velocity of Earth's center about Sun	= 107 200 km/h
	= 66,610 mi/h

BODY	MEAN DISTANCE TO SUN km (mi)	ECCENTRICITY OF ORBIT e	PERIOD OF ORBIT solar days	MEAN DIAMETER km (mi)	MASS RELATIVE TO EARTH	SURFACE GRAVITATIONAL ACCELERATION m/s^2 (ft/s^2)	ESCAPE VELOCITY km/s (mi/s)
Sun	—	—	—	1 392 000 (865 000)	333 000	274 (898)	616 (383)
Moon	384 398[1] (238 854)[1]	0.055	27.32	3 476 (2 160)	0.0123	1.62 (5.32)	2.37 (1.47)
Mercury	57.3×10^6 (35.6×10^6)	0.206	87.97	5 000 (3 100)	0.054	3.47 (11.4)	4.17 (2.59)
Venus	108×10^6 (67.2×10^6)	0.0068	224.70	12 400 (7 700)	0.815	8.44 (27.7)	10.24 (6.36)
Earth	149.6×10^6 (92.96×10^6)	0.0167	365.26	12 742[2] (7 918)[2]	1.000	9.821[3] (32.22)[3]	11.18 (6.95)
Mars	227.9×10^6 (141.6×10^6)	0.093	686.98	6 788 (4 218)	0.107	3.73 (12.3)	5.03 (3.13)
Jupiter[4]	778×10^6 (483×10^6)	0.0489	4333	139 822 (86 884)	317.8	24.79 (81.3)	59.5 (36.8)

[1] Mean distance to Earth (center-to-center)

[2] Diameter of sphere of equal volume, based on a spheroidal Earth with a polar diameter of 12 714 km (7900 mi) and an equatorial diameter of 12 756 km (7926 mi)

[3] For nonrotating spherical Earth, equivalent to absolute value at sea level and latitude 37.5°

[4] Note that Jupiter is not a solid body.

TABLE D/3 PROPERTIES OF PLANE FIGURES

FIGURE	CENTROID	AREA MOMENTS OF INERTIA
Arc Segment 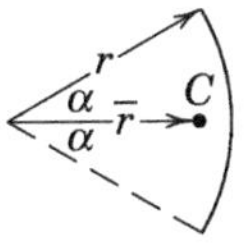	$\bar{r} = \dfrac{r \sin \alpha}{\alpha}$	—
Quarter and Semicircular Arcs 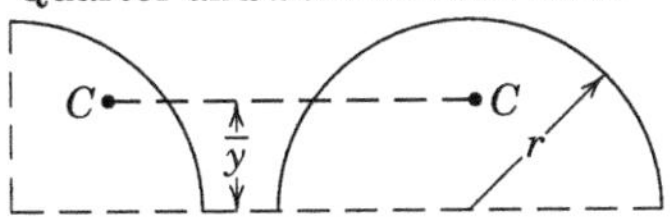	$\bar{y} = \dfrac{2r}{\pi}$	—
Circular Area 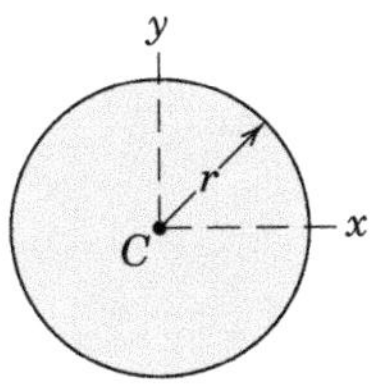	—	$I_x = I_y = \dfrac{\pi r^4}{4}$ $I_z = \dfrac{\pi r^4}{2}$
Semicircular Area 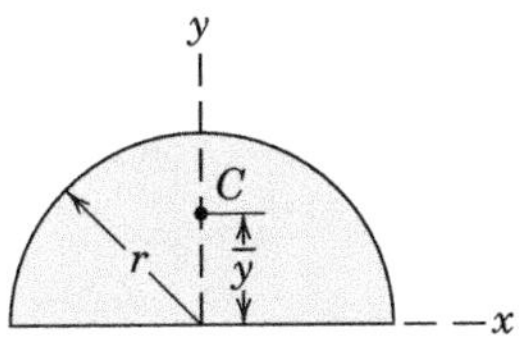	$\bar{y} = \dfrac{4r}{3\pi}$	$I_x = I_y = \dfrac{\pi r^4}{8}$ $\bar{I}_x = \left(\dfrac{\pi}{8} - \dfrac{8}{9\pi}\right) r^4$ $I_z = \dfrac{\pi r^4}{4}$
Quarter-Circular Area 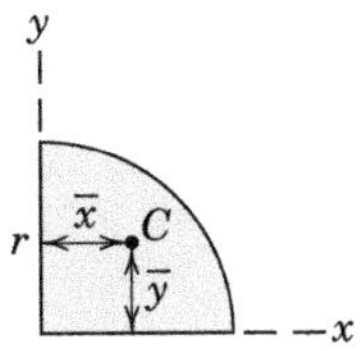	$\bar{x} = \bar{y} = \dfrac{4r}{3\pi}$	$I_x = I_y = \dfrac{\pi r^4}{16}$ $\bar{I}_x = \bar{I}_y = \left(\dfrac{\pi}{16} - \dfrac{4}{9\pi}\right) r^4$ $I_z = \dfrac{\pi r^4}{8}$
Area of Circular Sector 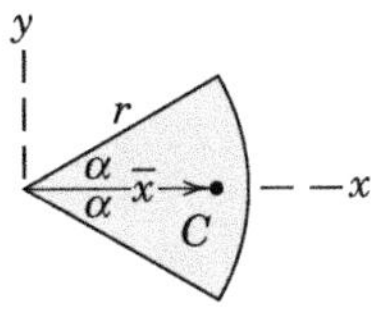	$\bar{x} = \dfrac{2}{3} \dfrac{r \sin \alpha}{\alpha}$	$I_x = \dfrac{r^4}{4}(\alpha - \dfrac{1}{2} \sin 2\alpha)$ $I_y = \dfrac{r^4}{4}(\alpha + \dfrac{1}{2} \sin 2\alpha)$ $I_z = \dfrac{1}{2} r^4 \alpha$

TABLE D/3 PROPERTIES OF PLANE FIGURES *Continued*

FIGURE	CENTROID	AREA MOMENTS OF INERTIA
Rectangular Area 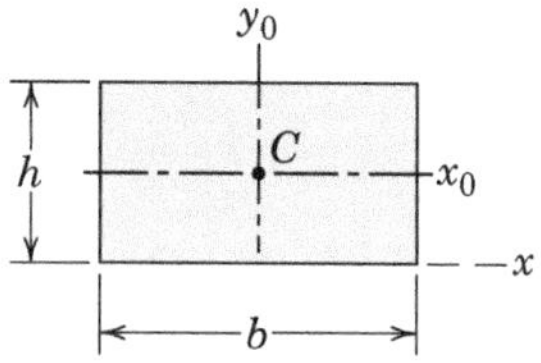 y_0, C, h, x_0, x, b	—	$I_x = \frac{bh^3}{3}$ $\bar{I}_x = \frac{bh^3}{12}$ $\bar{I}_z = \frac{bh}{12}(b^2 + h^2)$
Triangular Area 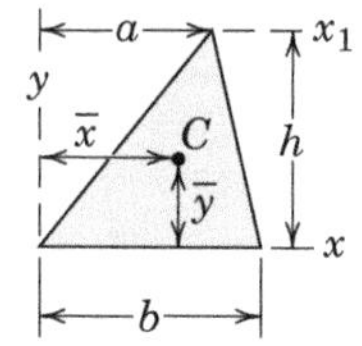 a, x_1, y, $\bar{x}$, C, h, $\bar{y}$, x, b	$\bar{x} = \frac{a+b}{3}$ $\bar{y} = \frac{h}{3}$	$I_x = \frac{bh^3}{12}$ $\bar{I}_x = \frac{bh^3}{36}$ $I_{x_1} = \frac{bh^3}{4}$
Area of Elliptical Quadrant 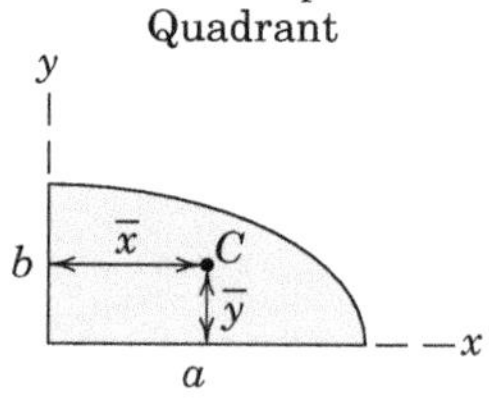 y, b, $\bar{x}$, C, $\bar{y}$, x, a	$\bar{x} = \frac{4a}{3\pi}$ $\bar{y} = \frac{4b}{3\pi}$	$I_x = \frac{\pi ab^3}{16}$, $\bar{I}_x = \left(\frac{\pi}{16} - \frac{4}{9\pi}\right)ab^3$ $I_y = \frac{\pi a^3 b}{16}$, $\bar{I}_y = \left(\frac{\pi}{16} - \frac{4}{9\pi}\right)a^3 b$ $I_z = \frac{\pi ab}{16}(a^2 + b^2)$
Subparabolic Area $y = kx^2 = \frac{b}{a^2}x^2$ Area $A = \frac{ab}{3}$ 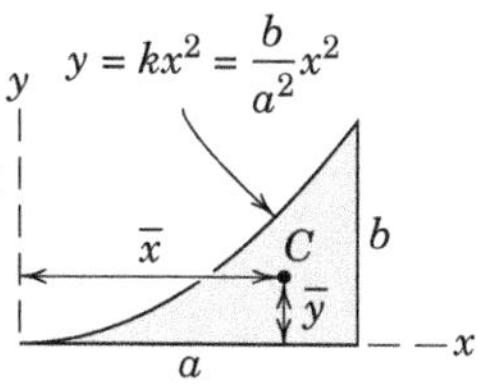 y, $\bar{x}$, C, b, $\bar{y}$, x, a	$\bar{x} = \frac{3a}{4}$ $\bar{y} = \frac{3b}{10}$	$I_x = \frac{ab^3}{21}$ $I_y = \frac{a^3 b}{5}$ $I_z = ab\left(\frac{a^2}{5} + \frac{b^2}{21}\right)$
Parabolic Area $y = kx^2 = \frac{b}{a^2}x^2$ Area $A = \frac{2ab}{3}$ 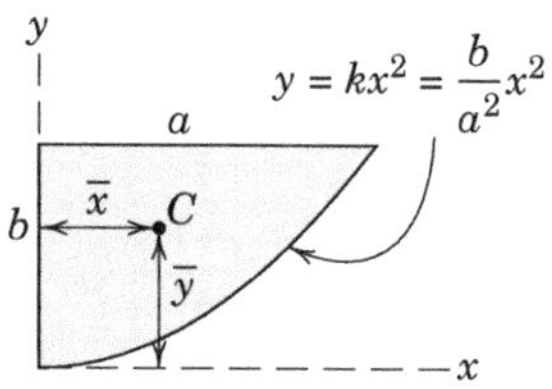 y, a, b, $\bar{x}$, C, $\bar{y}$, x	$\bar{x} = \frac{3a}{8}$ $\bar{y} = \frac{3b}{5}$	$I_x = \frac{2ab^3}{7}$ $I_y = \frac{2a^3 b}{15}$ $I_z = 2ab\left(\frac{a^2}{15} + \frac{b^2}{7}\right)$

TABLE D/4 PROPERTIES OF HOMOGENEOUS SOLIDS
(m = mass of body shown)

BODY	MASS CENTER	MASS MOMENTS OF INERTIA
Circular Cylindrical Shell	—	$I_{xx} = \frac{1}{2}mr^2 + \frac{1}{12}ml^2$ $I_{x_1x_1} = \frac{1}{2}mr^2 + \frac{1}{3}ml^2$ $I_{zz} = mr^2$
Half Cylindrical Shell	$\bar{x} = \frac{2r}{\pi}$	$I_{xx} = I_{yy}$ $= \frac{1}{2}mr^2 + \frac{1}{12}ml^2$ $I_{x_1x_1} = I_{y_1y_1}$ $= \frac{1}{2}mr^2 + \frac{1}{3}ml^2$ $I_{zz} = mr^2$ $\bar{I}_{zz} = \left(1 - \frac{4}{\pi^2}\right)mr^2$
Circular Cylinder	—	$I_{xx} = \frac{1}{4}mr^2 + \frac{1}{12}ml^2$ $I_{x_1x_1} = \frac{1}{4}mr^2 + \frac{1}{3}ml^2$ $I_{zz} = \frac{1}{2}mr^2$
Semicylinder	$\bar{x} = \frac{4r}{3\pi}$	$I_{xx} = I_{yy}$ $= \frac{1}{4}mr^2 + \frac{1}{12}ml^2$ $I_{x_1x_1} = I_{y_1y_1}$ $= \frac{1}{4}mr^2 + \frac{1}{3}ml^2$ $I_{zz} = \frac{1}{2}mr^2$ $\bar{I}_{zz} = \left(\frac{1}{2} - \frac{16}{9\pi^2}\right)mr^2$
Rectangular Parallelepiped	—	$I_{xx} = \frac{1}{12}m(a^2 + l^2)$ $I_{yy} = \frac{1}{12}m(b^2 + l^2)$ $I_{zz} = \frac{1}{12}m(a^2 + b^2)$ $I_{y_1y_1} = \frac{1}{12}mb^2 + \frac{1}{3}ml^2$ $I_{y_2y_2} = \frac{1}{3}m(b^2 + l^2)$

TABLE D/4 PROPERTIES OF HOMOGENEOUS SOLIDS ***Continued***

(m = mass of body shown)

BODY	MASS CENTER	MASS MOMENTS OF INERTIA
Spherical Shell	—	$I_{zz} = \frac{2}{3}mr^2$
Hemispherical Shell	$\bar{x} = \frac{r}{2}$	$I_{xx} = I_{yy} = I_{zz} = \frac{2}{3}mr^2$ $\bar{I}_{yy} = \bar{I}_{zz} = \frac{5}{12}mr^2$
Sphere	—	$I_{zz} = \frac{2}{5}mr^2$
Hemisphere	$\bar{x} = \frac{3r}{8}$	$I_{xx} = I_{yy} = I_{zz} = \frac{2}{5}mr^2$ $\bar{I}_{yy} = \bar{I}_{zz} = \frac{83}{320}mr^2$
Uniform Slender Rod	—	$I_{yy} = \frac{1}{12}ml^2$ $I_{y_1y_1} = \frac{1}{3}ml^2$

TABLE D/4 PROPERTIES OF HOMOGENEOUS SOLIDS *Continued*
(m = mass of body shown)

BODY	MASS CENTER	MASS MOMENTS OF INERTIA
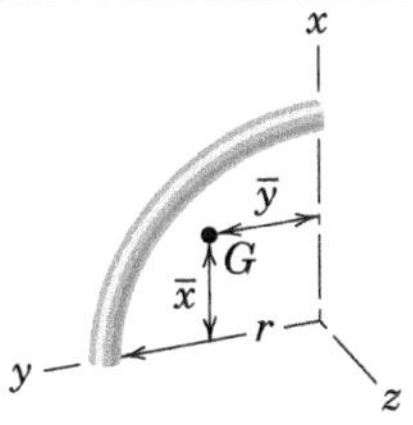 Quarter-Circular Rod	$\bar{x} = \bar{y}$ $= \frac{2r}{\pi}$	$I_{xx} = I_{yy} = \frac{1}{2}mr^2$ $I_{zz} = mr^2$
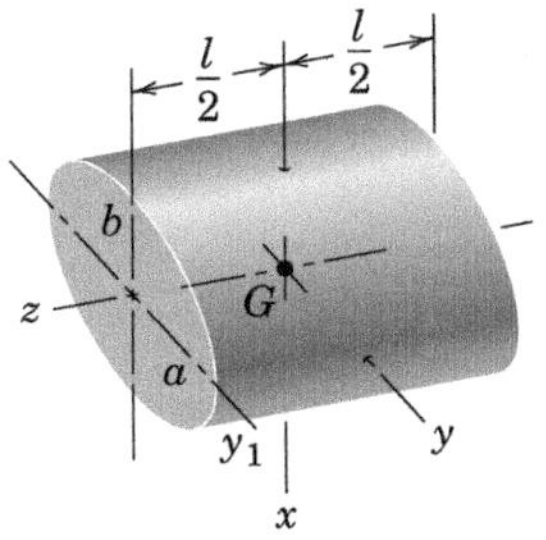 Elliptical Cylinder	—	$I_{xx} = \frac{1}{4}ma^2 + \frac{1}{12}ml^2$ $I_{yy} = \frac{1}{4}mb^2 + \frac{1}{12}ml^2$ $I_{zz} = \frac{1}{4}m(a^2 + b^2)$ $I_{y_1y_1} = \frac{1}{4}mb^2 + \frac{1}{3}ml^2$
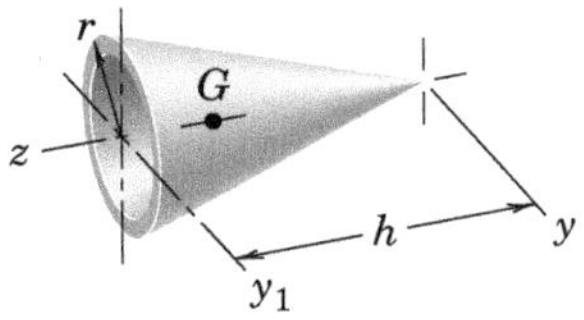 Conical Shell	$\bar{z} = \frac{2h}{3}$	$I_{yy} = \frac{1}{4}mr^2 + \frac{1}{2}mh^2$ $I_{y_1y_1} = \frac{1}{4}mr^2 + \frac{1}{6}mh^2$ $I_{zz} = \frac{1}{2}mr^2$ $\bar{I}_{yy} = \frac{1}{4}mr^2 + \frac{1}{18}mh^2$
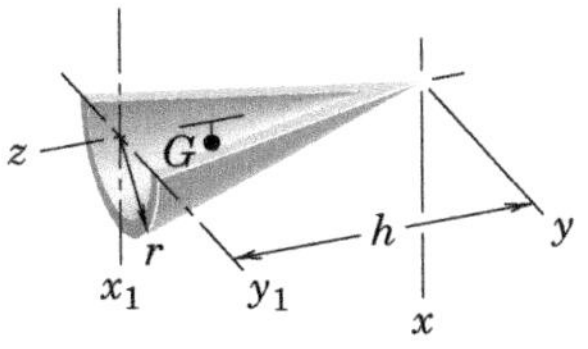 Half Conical Shell	$\bar{x} = \frac{4r}{3\pi}$ $\bar{z} = \frac{2h}{3}$	$I_{xx} = I_{yy}$ $= \frac{1}{4}mr^2 + \frac{1}{2}mh^2$ $I_{x_1x_1} = I_{y_1y_1}$ $= \frac{1}{4}mr^2 + \frac{1}{6}mh^2$ $I_{zz} = \frac{1}{2}mr^2$ $\bar{I}_{zz} = \left(\frac{1}{2} - \frac{16}{9\pi^2}\right)mr^2$
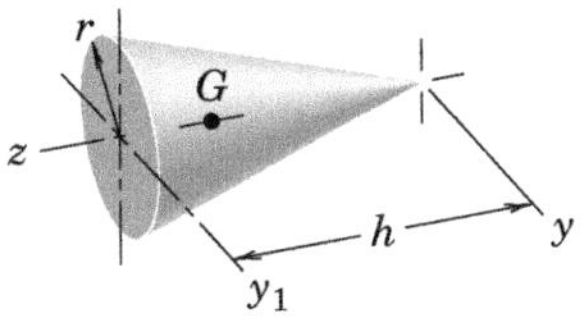 Right-Circular Cone	$\bar{z} = \frac{3h}{4}$	$I_{yy} = \frac{3}{20}mr^2 + \frac{3}{5}mh^2$ $I_{y_1y_1} = \frac{3}{20}mr^2 + \frac{1}{10}mh^2$ $I_{zz} = \frac{3}{10}mr^2$ $\bar{I}_{yy} = \frac{3}{20}mr^2 + \frac{3}{80}mh^2$

TABLE D/4 PROPERTIES OF HOMOGENEOUS SOLIDS ***Continued***
(m = mass of body shown)

BODY	MASS CENTER	MASS MOMENTS OF INERTIA
Half Cone	$\bar{x} = \frac{r}{\pi}$ $\bar{z} = \frac{3h}{4}$	$I_{xx} = I_{yy} = \frac{3}{20}mr^2 + \frac{3}{5}mh^2$ $I_{x_1x_1} = I_{y_1y_1} = \frac{3}{20}mr^2 + \frac{1}{10}mh^2$ $I_{zz} = \frac{3}{10}mr^2$ $\bar{I}_{zz} = \left(\frac{3}{10} - \frac{1}{\pi^2}\right)mr^2$
Semiellipsoid $\frac{x^2}{a^2} + \frac{y^2}{b^2} + \frac{z^2}{c^2} = 1$	$\bar{z} = \frac{3c}{8}$	$I_{xx} = \frac{1}{5}m(b^2 + c^2)$ $I_{yy} = \frac{1}{5}m(a^2 + c^2)$ $I_{zz} = \frac{1}{5}m(a^2 + b^2)$ $\bar{I}_{xx} = \frac{1}{5}m(b^2 + \frac{19}{64}c^2)$ $\bar{I}_{yy} = \frac{1}{5}m(a^2 + \frac{19}{64}c^2)$
Elliptic Paraboloid $\frac{x^2}{a^2} + \frac{y^2}{b^2} = \frac{z}{c}$	$\bar{z} = \frac{2c}{3}$	$I_{xx} = \frac{1}{6}mb^2 + \frac{1}{2}mc^2$ $I_{yy} = \frac{1}{6}ma^2 + \frac{1}{2}mc^2$ $I_{zz} = \frac{1}{6}m(a^2 + b^2)$ $\bar{I}_{xx} = \frac{1}{6}m(b^2 + \frac{1}{3}c^2)$ $\bar{I}_{yy} = \frac{1}{6}m(a^2 + \frac{1}{3}c^2)$
Rectangular Tetrahedron	$\bar{x} = \frac{a}{4}$ $\bar{y} = \frac{b}{4}$ $\bar{z} = \frac{c}{4}$	$I_{xx} = \frac{1}{10}m(b^2 + c^2)$ $I_{yy} = \frac{1}{10}m(a^2 + c^2)$ $I_{zz} = \frac{1}{10}m(a^2 + b^2)$ $\bar{I}_{xx} = \frac{3}{80}m(b^2 + c^2)$ $\bar{I}_{yy} = \frac{3}{80}m(a^2 + c^2)$ $\bar{I}_{zz} = \frac{3}{80}m(a^2 + b^2)$
Half Torus	$\bar{x} = \frac{a^2 + 4R^2}{2\pi R}$	$I_{xx} = I_{yy} = \frac{1}{2}mR^2 + \frac{5}{8}ma^2$ $I_{zz} = mR^2 + \frac{3}{4}ma^2$

Conversion Charts Between SI and U.S. Customary Units

mm | in.
500 | 20
400 | 15
300 | 10
200 | 5
100 |
0 | 0
Length

m | ft
30 | 100
 | 90
 | 80
 | 70
20 | 60
 | 50
 | 40
10 | 30
 | 20
 | 10
0 | 0
Length

km | mi
200 | 120
 | 110
 | 100
 | 90
 | 80
 | 70
100 | 60
 | 50
 | 40
 | 30
 | 20
 | 10
0 | 0
Length

kg | lbm
100 | 220
90 | 200
80 | 180
70 | 160
60 | 140
50 | 120
40 | 100
30 | 80
20 | 60
10 | 40
 | 20
0 | 0
Mass

N | lb
1000 |
900 | 200
800 |
700 | 150
600 |
500 | 100
400 |
300 | 50
200 |
100 |
0 | 0
Force

kPa | lb/in.^{2}
1000 | 140
900 | 130
800 | 120
 | 110
700 | 100
600 | 90
 | 80
500 | 70
400 | 60
 | 50
300 | 40
200 | 30
 | 20
100 | 10
0 | 0
Pressure or Stress

Conversion Charts Between SI and U.S. Customary Units (cont.)

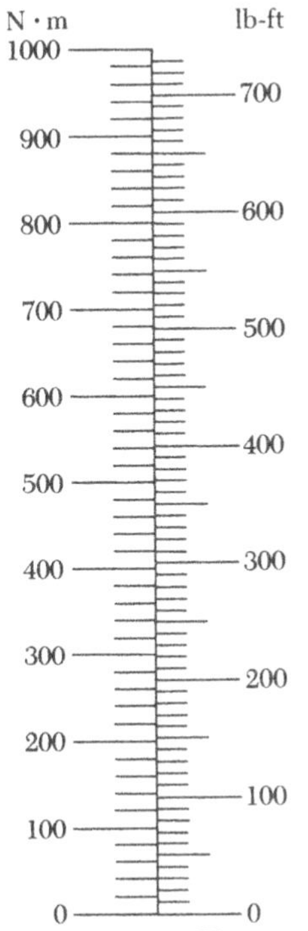

N·m | lb-ft
1000, 900, 800, 700, 600, 500, 400, 300, 200, 100, 0 | 700, 600, 500, 400, 300, 200, 100, 0
Moment or Torque

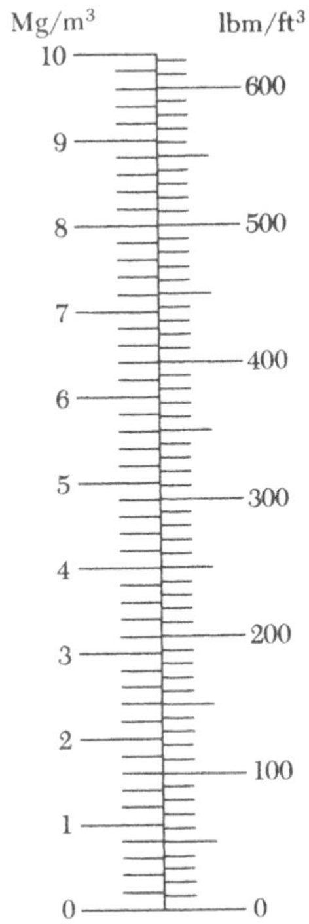

Mg/m³ | lbm/ft³
10, 9, 8, 7, 6, 5, 4, 3, 2, 1, 0 | 600, 500, 400, 300, 200, 100, 0
Density

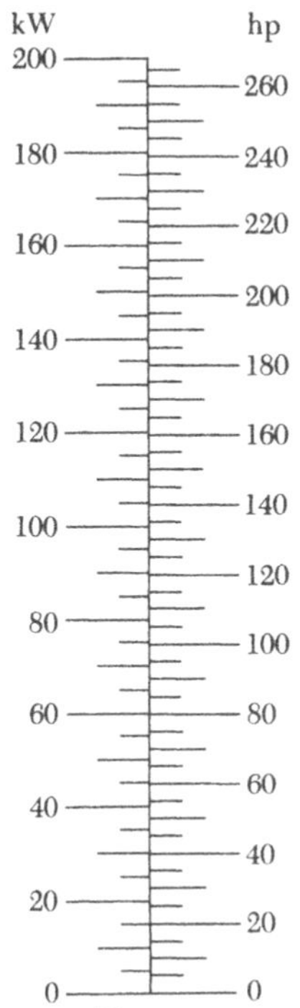

kW | hp
200, 180, 160, 140, 120, 100, 80, 60, 40, 20, 0 | 260, 240, 220, 200, 180, 160, 140, 120, 100, 80, 60, 40, 20, 0
Power

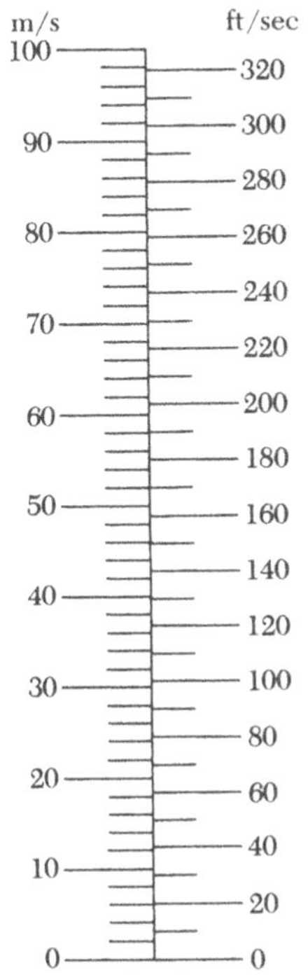

m/s | ft/sec
100, 90, 80, 70, 60, 50, 40, 30, 20, 10, 0 | 320, 300, 280, 260, 240, 220, 200, 180, 160, 140, 120, 100, 80, 60, 40, 20, 0
Velocity

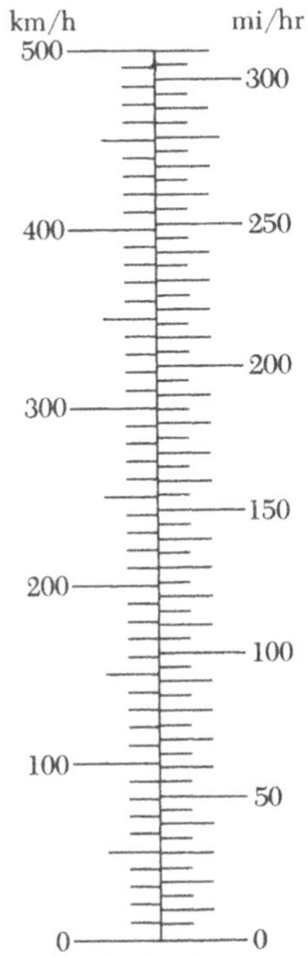

km/h | mi/hr
500, 400, 300, 200, 100, 0 | 300, 250, 200, 150, 100, 50, 0
Velocity

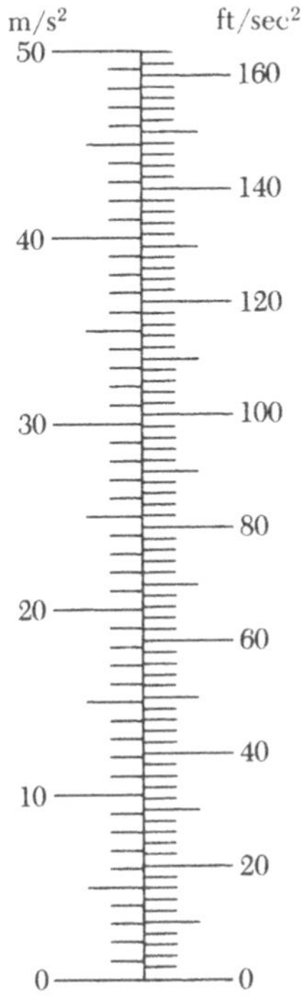

m/s² | ft/sec²
50, 40, 30, 20, 10, 0 | 160, 140, 120, 100, 80, 60, 40, 20, 0
Acceleration